Photofunctional Polymer Composites for Bioapplications

Photofunctional Polymer Composites for Bioapplications

Xiaoyu Wang and Lidong Li

WILEY-VCH

Authors

Xiaoyu Wang
University of Science and Technology Beijing
Beijing, 100083
People's Republic of China

Lidong Li
University of Science and Technology Beijing
Beijing, 100083
People's Republic of China

Cover Design: Wiley
Cover Image: © Rifqyhsn Design/Getty Images

Typesetting: Straive, Chennai, India

Library of Congress Card No.: applied for

British Library Cataloguing-in-Publication Data A catalogue record for this book is available from the British Library.

Bibliographic information published by the Deutsche Nationalbibliothek
The Deutsche Nationalbibliothek lists this publication in the Deutsche Nationalbibliografie; detailed bibliographic data are available on the Internet at <http://dnb.d-nb.de>.

Print ISBN: 978-3-527-35492-4
ePDF ISBN: 978-3-527-85114-0
ePub ISBN: 978-3-527-85113-3
oBook ISBN: 978-3-527-85115-7

Printed and bound by CPI Group (UK) Ltd, Croydon, CR0 4YY

C9783527354924_120326

Contents

Preface *xi*

1 Introduction *1*
1.1 Research Status of Photofunctional Polymer Composites *1*
1.1.1 Photofunctional Materials *1*
1.1.2 Photofunctional Composites *4*
1.1.2.1 Photoelectronic Devices Based on Photofunctional Polymer Pomposites *6*
1.1.2.2 Photocatalytic Application of Photofunctional Polymer Composites *7*
1.1.2.3 Biomedical Application of Photofunctional Polymer Composites *8*
1.1.3 Structure and Morphology of Photofunctional Polymer Composites *9*
1.2 Classification of Photofunctional Polymer Composites *11*
1.2.1 Nanocomposites *11*
1.2.1.1 Composite Nanoparticles *12*
1.2.1.2 Emulsion *18*
1.2.1.3 Liposome *20*
1.2.1.4 Micelle *22*
1.2.1.5 Capsule *24*
1.2.2 Fibers *26*
1.2.2.1 Natural Material-based Fibers *26*
1.2.2.2 Synthetic Polymer Fibers *31*
1.2.3 Films *35*
1.2.3.1 Fluorescent Film *36*
1.2.3.2 Photochromic Films *38*
1.2.3.3 Photoelectric Conversion Thin Film *40*
1.2.3.4 Near-infrared (NIR) Absorption Films *41*
1.2.4 Hydrogel *43*
1.2.4.1 Structure and Composition *45*
1.2.4.2 Physical Properties *46*
1.2.4.3 Fluorescent Hydrogel *46*
1.2.4.4 Photochromic Hydrogel *47*
1.2.4.5 Photothermal Hydrogel *48*

1.2.4.6 Photoelectric Conversion Hydrogel 50
1.2.4.7 Light-controlled Release Hydrogel 51
1.3 Prospects for Bioapplications of Photofunctional Polymer Composites 53
1.3.1 Biological Detection and Imaging 54
1.3.1.1 Fluorescence Imaging 54
1.3.1.2 Fluorescence Detection 56
1.3.2 Photodynamic Therapy (PDT) 60
1.3.2.1 Principle of Photodynamic Therapy 60
1.3.2.2 Application of Photofunctional Polymer Composites in Photodynamic Therapy 61
1.3.3 Light-controlled Release 64
1.3.3.1 Principle of Light-controlled Release 64
1.3.3.2 Application of Photofunctional Polymer Composites in Light-controlled Release 65
1.3.4 Photothermal Therapy (PTT) 67
1.3.4.1 Principle of Photothermal Therapy 67
1.3.4.2 Application of Photofunctional Polymer Composites in Photothermal Therapy 67
1.3.5 Photocatalytic Effect 70
1.3.5.1 Photocatalytic CO_2 Reduction 71
1.3.5.2 Photocatalytic Production of H_2 72
1.3.5.3 Light-driven Biotransformation 74
1.3.5.4 Photocatalytic Therapy 76
References 79

2 Structures and Properties of Photofunctional Materials 93
2.1 Small Organic Molecules 93
2.1.1 Photoluminescence 93
2.1.1.1 Fluorescence 93
2.1.1.2 Phosphorescence 97
2.1.2 Electroluminescence 100
2.1.3 Chemiluminescence 104
2.1.4 Mechanoluminescence 106
2.2 Conjugated Polymers 109
2.2.1 Main Chain Structure 109
2.2.2 Side Chain Structure 113
2.2.3 Conformation of Conjugated Polymers 116
2.2.4 Aggregation States of Conjugated Polymers 119
2.2.4.1 Static Aggregation 119
2.2.4.2 Dynamic Aggregation 120
2.2.4.3 Influencing Factors and Regulatory Strategies of Aggregation States 122
2.2.5 Luminescence Sensing Mechanism of Conjugated Polymers 123
2.2.5.1 Förster Resonance Energy Transfer Mechanism 124
2.2.5.2 Photoinduced Electron Transfer Mechanism 125

2.2.5.3 Exciplexes and Excimers 126
2.2.5.4 Metal-enhanced Fluorescence Effect 126
2.3 Noble Metal Nanomaterials 127
2.3.1 Noble Metal Nanoparticles 128
2.3.1.1 Gold Nanoparticles 129
2.3.1.2 Silver Nanoparticles 133
2.3.1.3 Copper Nanoparticles 135
2.3.2 Optical Properties of Noble Metal Nanoparticles 136
2.3.2.1 Metal-enhanced Fluorescence Effect (MEF) 136
2.3.2.2 Photothermal Effect 138
2.3.3 Noble Metal Nanoclusters 139
2.3.3.1 Gold Nanoclusters 139
2.3.3.2 Silver Nanoclusters 142
2.3.3.3 Copper Nanoclusters 143
2.3.4 Optical Properties of Noble Metal Nanoclusters 144
2.3.4.1 Photoluminescence 146
2.3.4.2 Two-photon Absorption 146
2.3.4.3 Aggregation-Induced Emission Effect 147
2.3.4.4 Photothermal Effect 148
2.3.4.5 Photosensitive Effect 149
2.4 Inorganic Nonmetallic Materials 150
2.4.1 Quantum Dots 151
2.4.2 Carbon Materials 154
2.4.2.1 Carbon Nanotube 154
2.4.2.2 Carbon Dots 155
2.4.2.3 Graphene 155
2.4.2.4 Fullerene 156
2.4.3 Carbon Nitride 156
2.4.3.1 Photocatalysis 159
2.4.3.2 Photoluminescence 160
2.4.3.3 Photothermal Effect 161
2.4.4 Transition Metal Disulfides 161
2.4.5 Perovskite 166
2.4.6 Titanium Dioxide 168
2.5 Other Materials 171
2.5.1 Metal-organic Framework (MOFs) 171
2.5.1.1 Structure of MOF 172
2.5.1.2 Optical Properties of MOF 172
2.5.2 Covalent Organic Framework (COFs) 176
2.5.2.1 Structure of COF 177
2.5.2.2 Optical Properties of COF 181
2.5.3 Photofunctional Protein 184
2.5.3.1 Light-harvesting Protein 184
2.5.3.2 Fluorescent Protein 187

2.5.4 Other Polymers *189*
2.5.4.1 Block Copolymers *189*
2.5.4.2 Dendrimers *191*
References *194*

3 Construction of Photofunctional Polymer Composites *211*
3.1 Construction of Photofunctional Organic Polymer Composite Systems *211*
3.1.1 Coprecipitation *212*
3.1.2 Self-assembly Method *218*
3.1.3 Microemulsion Method *225*
3.1.4 Direct Polymerization *227*
3.1.5 Covalent Modification *228*
3.1.6 Physical Mixing Method *229*
3.2 Construction of Photofunctional Organic/Inorganic Polymer Composite Systems *230*
3.2.1 Self-assembly Method *231*
3.2.2 Coprecipitation Method *235*
3.2.3 Microemulsion *236*
3.2.4 Covalent Modification *238*
3.2.5 In Situ Reaction Method *240*
3.2.6 Physical Mixing Method *243*
References *245*

4 Photofunctional Polymer Nanocomposites *253*
4.1 Structures of Photofunctional Polymer Nanocomposites *253*
4.1.1 Core-shell Structure *253*
4.1.2 Hollow Structure *259*
4.1.3 Hybrid Structure *261*
4.1.4 Emulsion *267*
4.1.5 Liposome *270*
4.1.6 Micelle *272*
4.1.7 Capsule *274*
4.2 Functions of Photofunctional Polymer Nanocomposites *276*
4.2.1 Metal Enhanced Fluorescence Effect *276*
4.2.2 Energy Transfer Mechanism *281*
4.2.3 Electron Transfer Mechanism *289*
4.2.4 Aggregation-induced Emission Effect *294*
4.2.5 Photosensitization Effect *298*
4.2.6 Photothermal Effect *303*
4.3 Bioapplications of Photofunctional Polymer Nanocomposites *309*
4.3.1 Cell Imaging *309*
4.3.2 Optical Therapies *317*
4.3.3 Optical Detections *325*
4.3.4 Drug Delivery *332*
4.3.5 Photosynthesis *334*
References *338*

5 Photofunctional Polymer Fiber Composites *345*
5.1 Preparation and Structures of Photofunctional Fiber Composites *346*
5.1.1 Electrospinning *346*
5.1.2 3D Printing *351*
5.1.3 Wet Spinning Process *352*
5.1.4 Stretching Process *354*
5.1.5 Capillary-templated Polymerization *355*
5.1.6 Microfluidic Method *357*
5.2 Functions of Photofunctional Fiber Composites *359*
5.2.1 Energy/Electron Transfer Effect *359*
5.2.2 Photochromic Effect *363*
5.2.3 Photosensitization Effect *364*
5.2.4 Photocatalytic Effect *365*
5.2.5 Photothermal Conversion Effect *370*
5.2.6 Light-guiding Properties *371*
5.3 Bioapplications of Photofunctional Fiber Composites *373*
5.3.1 Optical Detection *374*
5.3.2 Optical Therapy *377*
5.3.3 Optical Display *383*
5.3.4 Optogenetics *386*
References *390*

6 Photofunctional Polymer Composite Films *395*
6.1 Structures of Photofunctional Polymer Composite Films *395*
6.1.1 Self-Supported Film *396*
6.1.2 Substrate-Supported Film *406*
6.2 Functions of Photofunctional Polymer Composite Films *412*
6.2.1 Fluorescence Enhancement Effect *412*
6.2.2 Energy/Electron Transfer Effect *416*
6.2.3 Photoisomerization Effect *422*
6.2.4 Photothermal Effect *424*
6.2.5 Other Photofunctions *426*
6.3 Bioapplications of Photofunctional Polymer Composite Films *429*
6.3.1 Optical Detection *430*
6.3.2 Optical Therapy *437*
6.3.3 Energy Storage *443*
6.3.4 Food Packaging *446*
References *448*

7 Photofunctional Polymer Hydrogel Composites *453*
7.1 Structures of Photofunctional Polymer Hydrogel Composites *453*
7.1.1 The Network Structure of Chemical Cross-linking Hydrogels *454*
7.1.1.1 Permanent Covalent Bond *454*
7.1.1.2 Dynamic Covalent Bonds *457*
7.1.2 The Network Structure of Physical Cross-linking *464*
7.1.2.1 Hydrogen Bond Cross-Linking *465*
7.1.2.2 Metal Coordination Effect *468*

7.1.2.3 Hydrophobic Interaction *470*
7.1.2.4 Host–Guest Interaction *472*
7.1.2.5 Crystallization and Cross-Linking *476*
7.2 Functions of Photofunctional Polymer Hydrogel Composites *477*
7.2.1 Electron Transfer Mechanism *477*
7.2.2 Energy Transfer Mechanism *481*
7.2.3 Fluorescence Enhancement Effect *483*
7.2.4 Photothermal Conversion Effect *487*
7.2.5 Photosensitization Effect *490*
7.2.6 Photoisomerization Effect *493*
7.3 Bioapplications of Photofunctional Polymer Hydrogel Composites *495*
7.3.1 Optical Detection and Sensing *496*
7.3.2 Phototherapy *502*
7.3.3 Light-Controlled Drug Release *507*
7.3.4 Cell Culture *514*
References *518*

8 Future Development of Photofunctional Polymer Composites *525*
8.1 Structural Optimization of Photofunctional Polymer Composites *525*
8.1.1 Selection of Components for Intelligent Light Response *525*
8.1.2 In Situ Assembly Strategy *526*
8.1.3 In Situ Aggregation Strategy *531*
8.2 The Functional Improvement of Photofunctional Polymer Composites *533*
8.2.1 Multifunctional Integration *533*
8.2.2 Interdisciplinary Integration *540*
8.3 Expansion of Biological Applications *546*
8.3.1 Personalized Medicine *546*
8.3.2 Artificial Biological Photosynthesis *550*
References *556*

Index *563*

Preface

With the continuous advancement of materials science towards multifunctionality and intelligence, photofunctional materials, owing to their designable photoresponsiveness, have played a significant role in the biomedical field. As a crucial carrier for the intelligent development of materials science, photofunctional polymer composites are driving transformations in fields such as biomedicine through multidimensional innovation. The research on photofunctional polymer composite systems is highly interdisciplinary. By precisely modifying organic functional groups and regulating inorganic topological structures, comprehensive control over optoelectronic properties can be achieved. Utilizing multicomponent and polymer matrix interface engineering strategies, a multidimensional synergistic network for charge transport, energy conversion, and signal transduction can be constructed, effectively overcoming the limitations of stability and compatibility of single components in complex biological systems. Thanks to their photo-regulated intelligent characteristics, these materials exhibit unique advantages in biomedical fields such as optical imaging navigation, phototherapy, and photobioenergy conversion, marking a leap in material development from "performance optimization" to "functional programming" and providing a material foundation for the advancement of biomedicine.

This book focuses on research achievements in photofunctional polymer composites, starting from the fundamental physicochemical properties of photofunctional materials. It briefly introduces the construction methods of photofunctional polymer composites and categorically presents their optical functions and diverse biomedical applications. "Photofunctional Polymer Composites for Bioapplications" is an academic monograph that provides a detailed introduction to photofunctional polymer composites and their bioapplications. The book is divided into eight chapters. Chapter 1 gives an overview of the classification of photofunctional polymer composites and their various biomedical applications. Chapter 2 details the structures and photophysical and chemical properties of various organic and inorganic photofunctional materials. Chapter 3 discusses the construction strategies for photofunctional polymer composites. Chapters 4 to 7, following the classification of photofunctional polymer composites, introduce photofunctional polymer nanocomposites, photofunctional polymer fiber composites, photofunctional polymer composite films, and photofunctional polymer hydrogel composites, respectively, with detailed descriptions of their structures, photofunctions, and biomedical applications. Chapter 8 looks ahead to the future prospects of photofunctional polymer composites.

Although fundamental research on photofunctional polymer composites has made significant progress, their practical applications remain limited. Transforming these fundamental research achievements into practical products requires strengthened cooperation and communication between photofunctional materials researchers and product developers, further improving the biosafety of photofunctional polymer composites and developing more advanced materials. This is also the original intention of writing this book: to promote collaborative research across disciplines, realize the translational application of photofunctional polymer composites, and ultimately enable nanobiomaterials to benefit humanity.

The completion of this book would not have been possible without the efforts of all participants. We would like to thank Jiaren Liu, Jingkai Zhang, Donghao Qi, Ziwei Zhao, Jixuan Hu, Jiaxi Wang, Xuan Zhao, Jian Yang, Jinchi Xu, Wenchao Wang, Wenjia Li, Wenting Zhao, and Yonghui Zan, for their contributions to the collection of materials for each chapter.

Although every effort has been made to ensure accuracy during the writing of this book, given the novelty and breadth of the content, there may still be some imperfections or inappropriate discussions. We sincerely welcome criticism and corrections from fellow experts and readers.

Xiaoyu Wang and Lidong Li

1

Introduction

1.1 Research Status of Photofunctional Polymer Composites

1.1.1 Photofunctional Materials

Photofunctional materials refer to the optical materials that use the principle that the optical properties of the material itself (such as refractive index or induced electric polarization) change under the action of the external field (such as electricity, light, magnetic, thermal, acoustic, and force) to achieve the detection, modulation, and energy or frequency conversion of the incident light signal. With the rapid advancements in modern science and technology, photofunctional materials are driving profound changes in many high-tech fields. Their application not only broadens the channel of energy acquisition but also exerts far-reaching influence on information storage, display technology, medical diagnosis and treatment, and other fields [1–3].

At present, the world is facing problems such as energy crisis and environmental pollution, which seriously affect the sustainable development of society. Traditional materials have gradually revealed their limitations in dealing with these complex problems. First of all, the traditional materials used for energy conversion and storage are less efficient, and it is difficult to make full use of renewable energy, resulting in significant reliance on fossil fuels, exacerbating the energy crisis. Secondly, the problem of environmental pollution is becoming increasingly prominent, and the production and use of traditional materials are often accompanied by high energy consumption and a large number of pollutants, which is not in line with the eco-friendly development model. In addition, with the increasing aging of the global population, the demand for medical health has increased significantly, but traditional materials have shortcomings in biocompatibility, intelligent responsiveness, and versatility, and it is difficult to meet the requirements of precision medicine and personalized health management [4]. For example, traditional materials used for medical device implantation are less biocompatible and may trigger inflammatory responses. In the field of disease diagnosis and treatment, traditional materials have a single function, making it difficult to achieve multifunctional synergies and dynamic response effects. In addition, the development of modern information and

Photofunctional Polymer Composites for Bioapplications, First Edition. Xiaoyu Wang and Lidong Li.

intelligence requires materials with higher performance, while the shortcomings of traditional materials in terms of flexibility, lightweight, and intelligent response limit the further breakthrough of emerging technologies [5]. At the same time, with the acceleration of digitalization, traditional materials are also showing limitations in the face of high-speed data transmission, large-capacity storage, and digital medical needs. Therefore, new, efficient, and sustainable materials are urgently needed for social development.

As an important material, photofunctional materials stand out in many fields by virtue of their excellent photoelectric conversion efficiency, photosensitive response characteristics, and photocatalytic ability, and become one of the key materials to promote the sustainable development of society. In the field of green energy, solar cells based on light-functional materials are widely used, effectively reducing the dependence on fossil energy, promoting the use of renewable energy, thus easing the energy crisis. At the same time, the photocatalytic degradation technology of photofunctional materials has shown great potential in reducing air pollution and water pollution, significantly reducing the emission of industrial harmful substances, and promoting environmental protection and green development. In the field of information storage and communication technology, photofunctional materials play a central role because of their excellent optical properties and low-loss characteristics. Optical fiber communication uses photofunctional materials to transmit data, which not only greatly improves the transmission rate but also effectively avoids electromagnetic interference, and realizes high-speed, remote, and large-capacity real-time data transmission. In the area of display technology, materials used in organic light-emitting diode (OLED) have been gradually applied to mainstream display technology with their excellent self-luminous characteristics, high contrast, and low energy consumption advantages, and have achieved the popularity of ultrathin and flexible displays [6]. In addition, photofunctional materials have also made significant progress in biomedical fields such as disease diagnosis, precision treatment, and health monitoring. For example, photodynamic therapy (PDT) relies on photosensitizers to produce reactive oxygen species (ROS) upon illumination with light at specific wavelengths, thereby selectively killing cancer cells, which is a noninvasive treatment with low side effects [7]. With its high-resolution imaging capability, fluorescence imaging technology has achieved remarkable results in the accurate diagnosis of tumors and other diseases, greatly improving the efficiency and accuracy of diagnosis and treatment. With their excellent physical and chemical properties, photofunctional materials are gradually replacing traditional materials and have widespread use in areas like energy, environmental protection, information technology, and biomedicine. They are becoming an important force to cope with global challenges and promote sustainable development.

According to the composition and structural characteristics, photofunctional materials can be divided into inorganic materials and organic materials (Figure 1.1). Both have their own characteristics in molecular structure, optical properties, and application fields, and occupy an important position in modern science and technology. Inorganic photofunctional materials usually have a highly ordered crystal

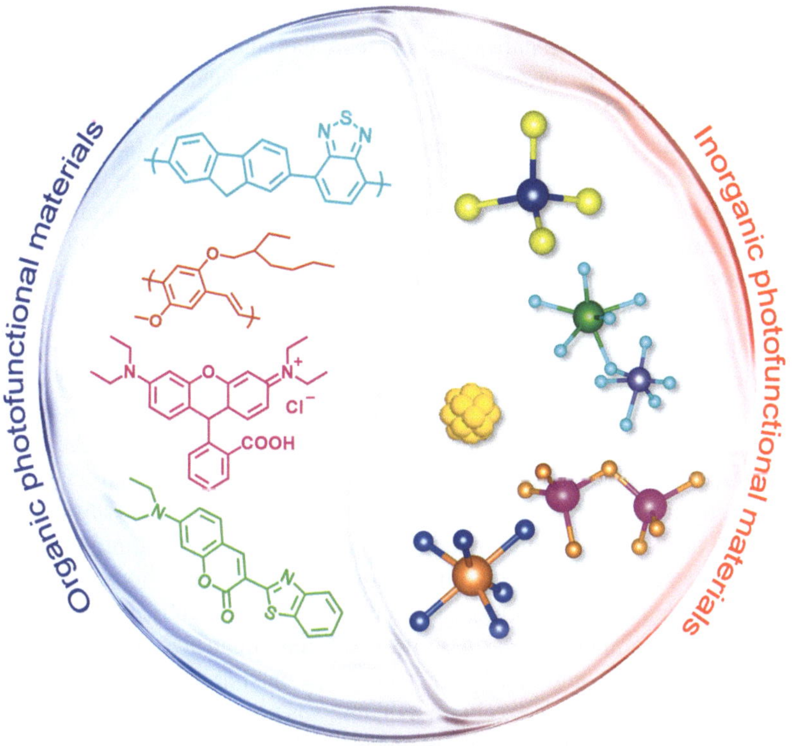

Figure 1.1 Inorganic and organic photofunctional materials.

structure, showing excellent stability, electrical conductivity, and optical properties, while organic photofunctional materials have unique advantages in some fields of application because of their good processability, flexibility, and adjustability.

Inorganic photofunctional materials are mainly composed of metals, semiconductors, or their compounds (such as oxides, nitrides, or halides). Their structure gives them excellent stability, electrical conductivity, and optical properties, making them play an important role in high-precision optoelectronic applications. For example, oxide materials (such as TiO_2 and ZnO) are widely used in the fields of photocatalysis, photodetectors, and sensors due to their excellent chemical stability and photocatalytic properties. Nitride materials, such as GaN and InN, play a key role in efficient light-emitting and high-power electronic devices, as the core material of modern LED technology. Perovskites are outstanding in the field of solar cells and photodetectors for their high photoelectric conversion efficiency and excellent machinability. In addition, gold nanoclusters (AuNCs), as a class of metallic nanomaterials with unique optical properties, have received considerable interest within the area of photofunctional materials lately. Due to their small particle size and quantum effect, AuNCs exhibit unique fluorescence characteristics and are widely used in biological imaging, sensors, and catalytic reactions.

Organic photofunctional materials are usually composed of carbon-containing molecules or polymers with conjugated π-electron systems in their molecular structures. This structure gives organic materials unique optical properties and good processing flexibility, making them show great potential in flexible electronics, wearable devices, display technology, and other emerging fields. For example, conjugated polymers (CPs) achieve efficient photoelectric conversion and luminous properties through the intramolecular π-electron delocalization effect. In addition, fluorescent dyes such as rhodamine and indole compounds are widely used in biological imaging, sensors, and laser labeling due to their excellent luminous efficiency and tunability. With the continuous development of organic materials, new organic semiconductors and fluorescent materials continue to emerge, further promoting the application of organic optoelectronic devices in the fields of intelligent display, wearable sensors, and optical communication technology.

Although inorganic and organic photofunctional materials show certain advantages in their respective application fields, there are also many shortcomings. These defects hinder their application in some fields. Inorganic materials are often brittle and difficult to adapt to flexible substrates, affecting their universality in flexible electronics, smart displays, and wearable technologies. In addition, the preparation of many inorganic materials is complex and costly, limiting the feasibility of large-scale, low-cost manufacturing. For example, although oxide and nitride materials perform well in photocatalytic and photoelectric conversion performance, the preparation process requires operating under high temperature and pressure, which makes manufacturing costly. Although the photoelectric conversion efficiency of perovskite is excellent, its stability is poor, and it is susceptible to deterioration under the influence of moisture, air, and ultraviolet light, and lacks long-term reliability. In contrast, organic materials have the advantages of flexibility and low cost, but poor stability and easy to be affected by the external environment (such as oxygen, moisture, and ultraviolet light). Organic photoelectric devices show high photoelectric conversion efficiency in the early use, but they are prone to performance degradation and short service life after long-term use. In addition, compared with inorganic materials, organic materials have lower carrier mobility, so their photoelectric conversion efficiency and luminous intensity are lower, limiting their application in high-power and high-performance optoelectronic devices.

1.1.2 Photofunctional Composites

In order to overcome the deficiency of single photofunctional materials, in recent years, researchers have developed photofunctional composites with better performance by combining inorganic materials with organic materials, giving full play to the advantages of both. These composites combine the high stability and excellent photoelectric performance of inorganic materials with the flexibility and adjustability of organic materials, showing significant advantages in many aspects [8]. Inorganic/organic composites not only have high photoelectric conversion efficiency but also improve the environmental stability of the material, and are widely used in flexible electronic devices and wearable devices. In addition, the preparation process of composites also has significant advantages. This feature greatly reduces the production cost and lays a foundation for the industrial application of composites. Therefore, the combination of inorganic and organic photofunctional materials not only makes up for their respective shortcomings but also has a wide scope for future applications [9], showing great potential in the fields of efficient optoelectronic devices, intelligent sensors, optical communication technology, and green energy.

The enhancement of performance by photofunctional polymer composites is chiefly manifested in the following ways.

Firstly, through interface optimization and electronic structure regulation, the carrier separation and migration efficiency of the composite are significantly improved. When inorganic materials with high conductivity and organic materials with excellent photoelectric properties are combined, favorable electron transport channels can be formed, which makes the transfer of charge carriers within the material more efficient. Optimizing the interface between inorganic materials and organic materials can improve the transmission path of electrons and holes, reduce interface defects and energy loss, and thus improve the photoelectric conversion efficiency.

Secondly, the combination of highly stable inorganic materials and flexible organic materials enables the composite to maintain excellent performance in extreme environments. Inorganic materials usually have high thermal and chemical stability, but lack flexibility, which is difficult to meet the application scenarios that require high flexibility, such as wearable devices. Although organic materials have good flexibility, they have poor stability and are easily degraded in high-humidity and high-temperature environments. By combining inorganic materials with organic materials, the stability advantages of inorganic materials and flexible characteristics of organic materials can be fully utilized to prepare photofunctional composites with both high mechanical flexibility and good environmental stability. This material can better adapt to complex external environments, extend service life, and maintain efficient photoelectric performance in flexible electronic devices, wearable devices, and other fields.

The multifunctional synergies of composites significantly enhance their application potential in various fields [10]. The composite of inorganic and organic photofunctional materials not only realizes complementary advantages in photoelectric conversion but also shows synergistic effects in many other functional fields. For example, by combining photocatalytic properties with photoelectric properties, composites can not only efficiently convert solar energy but also degrade environmental pollutants, thus serve as a key contributor to green energy and environmental governance.

In the field of biomedicine, composites of metal nanoclusters and organic polymers show unique versatility. For example, the excellent fluorescence properties of gold nanoclusters provide high resolution for bioimaging, while organic polymers give materials good biocompatibility and adjustability. These properties make composites show great potential in areas such as precision diagnosis, targeted therapy, and real-time imaging, which is expected to promote the development of smart medical technology.

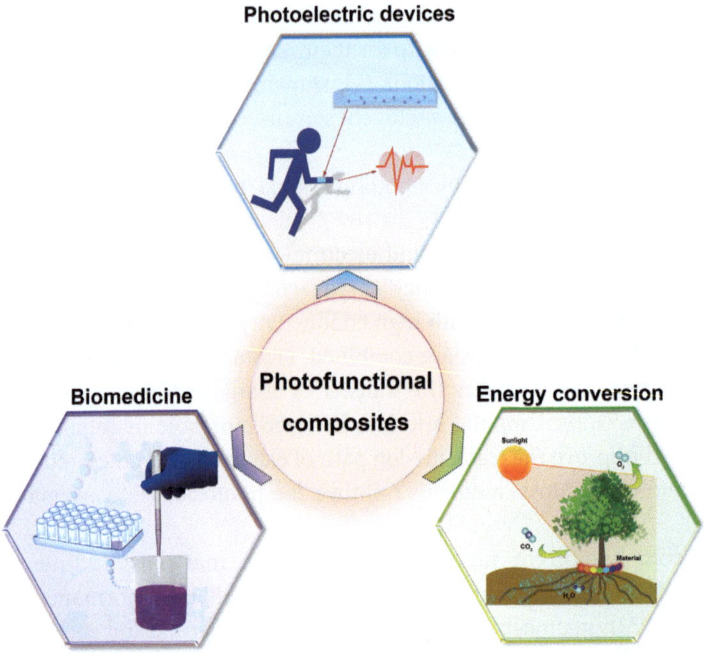

Figure 1.2 Application of photofunctional composites.

In addition, the synergistic effect of composites has also been extended to smart sensors, optical communication, and smart wearable devices, further enhancing the multifunctional integration and intelligent application performance of the material. With the steady evolution of material design and synthesis technologies, the multifunctional synergies of composites will accelerate their innovative applications in various technical fields (Figure 1.2).

1.1.2.1 Photoelectronic Devices Based on Photofunctional Polymer Pomposites

For organic photovoltaic cells, the photofunctional polymer composites can effectively increase the cell's efficiency in absorbing light and facilitate better charge transport. Photofunctional polymer materials usually have a highly conjugated structure and are able to act as donor materials and form efficient heterojunction structures with acceptor materials, thus facilitating the separation of excitons and the transport of electric charges. In addition, by introducing different side groups or adjusting the backbone structure, the energy level and electron mobility of the polymer can be adjusted, thereby improving the charge transport efficiency and the overall performance of the device [11]. However, photofunctional polymer materials still face some problems that limit their practical application, such as low photoelectric conversion efficiency, low carrier mobility, and susceptibility to photooxidation and thermal degradation under long-term light and environmental exposure conditions.

Photofunctional polymer composites usually broaden the light absorption range of the polymer matrix by introducing inorganic nanoparticles (such as perovskite [12] or quantum dots [13]) with wide spectral absorption properties, and improve the efficiency of photogenerated carrier generation. Certain metal complexes can effectively absorb specific wavelengths of light, thereby improving the light absorption efficiency of organic photovoltaic cells. At the same time, the inorganic nanoparticles form a conductive network in the polymer matrix, which helps to promote the separation and transmission of photogenerated charge, reduce the probability of charge recombination, and further improve the photoelectric conversion efficiency. In addition, photofunctional polymer composites can also be used in interface engineering to optimize the charge separation process by adjusting the energy level structure and the interface state density at the interface.

In OLEDs, photofunctional polymer materials are commonly used as the main material and charge transport layer of the light-emitting layer to provide efficient light emission. Inorganic nanoparticles (such as quantum dots) can be embedded in the light-emitting layer to widen the range of absorbed light and promote more efficient separation of photogenerated carriers [14]. However, the aging rate of photofunctional polymer materials is faster. Inorganic nanoparticles may also accumulate or migrate during long-term use, which affects the stability of the device. In addition, the interface stability between the inorganic nanoparticles and the organic light-emitting layer may reduce the charge injection and transport efficiency. From the perspective of material design, the introduction of inorganic nanoparticles into photofunctional polymer composites can broaden the emission spectrum, improve the photoluminescence quantum yield, and enhance the stability of materials [15].

1.1.2.2 Photocatalytic Application of Photofunctional Polymer Composites

In the field of photocatalytic decomposition of water, photofunctional polymer materials still have some problems. The light absorption range of these polymers is primarily confined to the visible region, resulting in poor absorption of ultraviolet and near-infrared light. To solve these problems, chemical modification of photofunctional polymer materials alone is not sufficient. By introducing inorganic nanoparticles, the light absorption range can be further broadened, and heterostructures can be constructed, using the interfacial electric field to promote the separation of photogenerated charge carriers, while enhancing the environmental stability of the material. For example, noble metal nanoparticles with local surface plasmon resonance (LSPR) effects can enhance the material's light absorption capacity in the ultraviolet to visible light range, thereby improving the efficiency of photocatalytic reactions [16]. For the inorganic nanoparticles themselves, they are prone to agglomeration or chemical corrosion in a long-term light and water environment, and there is also the problem of low efficiency of photogenic carrier separation. To this end, it is combined with a photofunctional polymer material, which will better help the inorganic nanoparticles disperse and effectively improve their stability under light [17].

1.1.2.3 Biomedical Application of Photofunctional Polymer Composites

In the field of biomedicine, photofunctional polymer composites are mainly used in diagnosis and imaging, drug delivery and therapy, tissue engineering, and regenerative medicine. Materials used in the biomedical field must first have good biocompatibility, that is, they do not cause inflammation or immune rejection after contact with biological tissues. In addition, it is best to design materials with self-adaptability and certain mechanical strength, so as to ensure their normal function in complex biological environments.

In terms of diagnosis and imaging, sensitive and highly specific biodetection and imaging require the combination of nanomaterials with excellent photoresponsive characteristics and biocompatible polymers to prepare biosafe photofunctional polymer composites [18]. Photofunctional polymer composites have been widely used in fluorescence imaging and photoacoustic imaging in biological imaging. For example, rare-earth–doped upconversion nanoparticles (UCNPs) can emit visible light under near-infrared excitation after being combined with polymers, which has the advantages of nonbiological tissue autofluorescence and no photobleaching [19]. In addition, metal nanoparticles such as gold nanoparticles can be used in photoacoustic imaging after being combined with polymers because of their good photothermal conversion ability and biocompatibility [20]. These materials can provide high-contrast and high-resolution imaging results in cell and small animal models and are suitable for a variety of biomedical research and clinical diagnosis.

In terms of biological detection, the ideal material has high sensitivity and specificity, while also enabling rapid detection. Photofunctional polymer composites are considered to be ideal materials because they combine the advantages of easy processing, flexibility, and biocompatibility of organic polymers with the advantages of inorganic materials, including high stability and special photoelectric properties, and are often used in the preparation and development of high-sensitivity biosensors. Compared with a single photofunctional polymer material, the composite prepared by introducing inorganic nanoparticles (for example, quantum dots and metal oxide nanoparticles) can significantly enhance the optical properties of the composite and improve the light absorption and luminescence efficiency [21, 22]. More importantly, by embedding specific inorganic nanoparticles in the polymer matrix, the prepared composite material can achieve specific recognition of biomolecules and improve the specificity of detection [23, 24].

In the field of therapy and drug delivery, high-load drug carriers and controllable and accurate drug release are important factors in designing materials [25]. The photofunctional polymer composite can achieve precise drug release and efficient therapeutic effect while adjusting the external light source. Compared with a single photofunctional polymer material, inorganic nanoparticles often have a high specific surface area and porous structure, which can significantly improve the drug loading capacity of the composite material and achieve intelligent drug release under the regulation of external light sources. A major use of photofunctional polymer composites is in the advancement of intelligent drug delivery systems. Through precise regulation of external light sources, controllable and accurate release of drugs can be achieved under light stimulation [26, 27]. In addition, 3D printing technology, combined with photopolymeric materials, can manufacture drug carriers and delivery devices with complex structures, such as microneedle patches and intracellular stents, which are able to precisely control the timing and location of drug release, enabling personalized treatment [28].

In terms of photothermal therapy (PTT), photofunctional polymer composites can convert light energy into heat energy for thermal ablation of tumors. In general, nanomaterials between 5 and 100 nm will have a longer residence time in tumor tissues, thus achieving better photothermal therapeutic effects [29]. Therefore, compared with inorganic nanoparticles alone, some photofunctional polymers themselves can regulate the size and structure of inorganic nanoparticles. This kind of photothermal treatment has the advantages of being noninvasive and highly selective, and can effectively reduce the damage to normal tissue. It is worth noting that by combining the photosensitizer with the polymer, ROS can be generated under specific wavelength light irradiation, so as to use PDT to kill tumor cells, which can improve the therapeutic effect and also have high selectivity [30].

Photofunctional polymer composites have a unique application in bone tissue regeneration. In particular, photofunctional polymer composites are often versatile and can integrate the photothermal effect, antibacterial property, and targeted therapy, so as to better promote the repair and growth of bone tissue [31]. For example, photoresponsive nanomaterials can promote the repair of infected bone defects through photothermal and photodynamic antibacterial action [32]. In addition, the mild photothermal stimulation generated by these materials under light can upregulate the expression of osteogenic genes and proteins, thus effectively enhancing the osteogenic effect [33]. This photoregulatory mechanism provides a new strategy for bone defect repair.

In soft tissue repair, photofunctional polymer composites are often used to prepare photoresponsive hydrogels and scaffolds. These materials are able to undergo structural changes when stimulated by specific light sources, thus promoting cell attachment and growth [34]. For example, hydrogels based on photoresponsive polymers can achieve dynamic regulation of the cellular microenvironment through photoregulation, providing ideal biocompatibility and mechanical properties for soft tissue regeneration [35]. Hydrogels that combine photofunctional polymers with inorganic nanoparticles (such as gold nanoparticles) can also be used for skin wound repair, speeding up the healing process and reducing the risk of infection [36].

1.1.3 Structure and Morphology of Photofunctional Polymer Composites

Photofunctional polymer composites have been widely used in energy, environment, biomedicine, and other fields because of their excellent performance in photoelectric properties, mechanical properties, and multifunctional synergies. Depending on their morphology and structure, these materials are divided into four principal categories: nanostructures, fibers, films, and hydrogels. Each class of materials presents unique performance advantages and application potential.

The size of photofunctional polymer nanocomposites is usually at the nanoscale and shows unique optical properties by virtue of a large specific surface area and quantum size effect. This kind of material is composed of a polymer matrix and inorganic nanoparticles (such as metal oxides, quantum dots, and precious metal nanoparticles), which combines the optical properties of inorganic materials with the flexibility and processability of polymers. For example, in solar cells, nanostructured composites can significantly improve the photoelectric conversion efficiency, while optimizing the light absorption and carrier transport performance of the material by regulating the size and distribution of the nanoparticles.

Photofunctional polymer fiber composite takes one-dimensional fibers as the main body, and realizes functionalization by embedding or coating photofunctional materials (such as inorganic nanoparticles or fluorescent dyes) in polymer fibers. Its high specific surface area, high strength, and good flexibility make it suitable for a variety of complex environments. In smart textiles, fiber composites can achieve the transmission and conversion of optical signals and realize multiparameter detection by integrating multiple functions.

Photofunctional polymer composite film is prepared by layer assembly or blending, and has uniform thickness and excellent optical properties. This kind of material not only has high transparency and good flexibility but also shows excellent light absorption and photoelectric conversion ability. In flexible displays and photoelectric sensors, thin-film composites enable high-resolution optical displays and sensitive environmental responses. Its performance and application range can be further enhanced by optimizing the interface design and material composition.

Photofunctional polymer hydrogel composites are three-dimensional networks, containing a large number of solvent molecules, by embedding photofunctional materials (such as fluorescent dyes, quantum dots, or metal nanoparticles) into the hydrogel network, giving it excellent optical properties and versatility. By adjusting

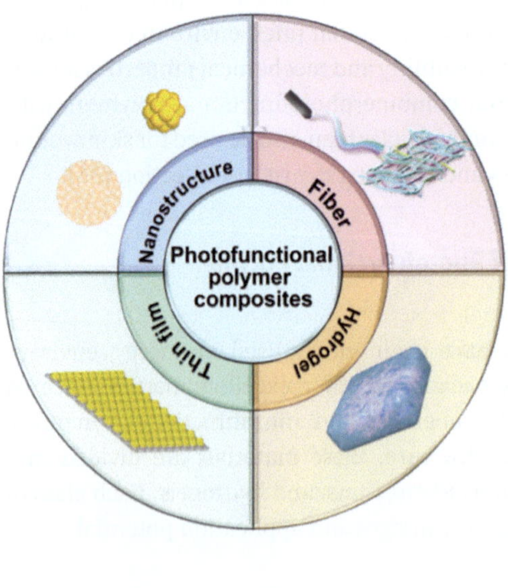

Figure 1.3 Four categories of photofunctional polymer composites.

the hydrogel network structure and the distribution of functional materials, the integration and optimization of multiple functions are realized.

The four categories of photofunctional polymer composites, namely nanostructure, fiber, film, and hydrogel, have their own characteristics (Figure 1.3). By optimizing the preparation process and interface design, the photoelectric properties and environmental adaptability of the materials can be further improved. In the future, with the continuous progress of materials science and nanotechnology, photofunctional polymer composites will achieve innovative applications in more fields, providing important support for solving the global energy crisis, environmental pollution, and medical and health problems.

1.2 Classification of Photofunctional Polymer Composites

1.2.1 Nanocomposites

Nanotechnology is instrumental in advancing the development of various key fields such as environmental energy, information science, and especially life science [37]. When the size of a substance approaches the nanoscale, its physical and chemical properties often undergo sudden changes, exhibiting unique properties that are distinct from both individual atoms/molecules and macroscopic materials, including size effects, surface effects, and quantum confinement [38]. These emerging characteristics enable nanoparticles to exhibit unprecedented functions at the microscopic scale, greatly broadening the horizons of scientific research. At present, researchers are dedicated to regulating atomic and molecular structures at the nanoscale to achieve precise control of material properties, thereby promoting the development of nanosensing, biomedicine, and other fields [39].

Compared with traditional materials, nanomaterials possess characteristics such as high catalysis, high photosensitivity, and high mechanical strength owing to their large surface area, interfacial dominant effect, and active surface state. At the nanoscale, surface tension, electrostatic force, and van der Waals interaction become dominant, correspondingly altering the photoelectrical and thermal response behaviors of the material. These characteristics not only endow nanomaterials with extensive application potential but also make them a research frontier in interdisciplinary fields such as materials science, chemistry, and physics.

Against this background, photofunctional polymer composites, as an important branch of nano-functional materials, have received increasing attention in recent years. Photofunctional polymers can be compounded with inorganic nanomaterials, bioactive molecules, etc., to prepare photofunctional composite systems. In addition, photofunctional nanocomponents such as fluorescent clusters, photosensitive nanoparticles, and photothermal conversion materials can be combined with polymer matrices to achieve multiple functional behaviors such as photoluminescence, photothermal response, light-controlled release, and photocatalysis, demonstrating broad application prospects in fields such as flexible electronics, biological imaging, and intelligent response systems.

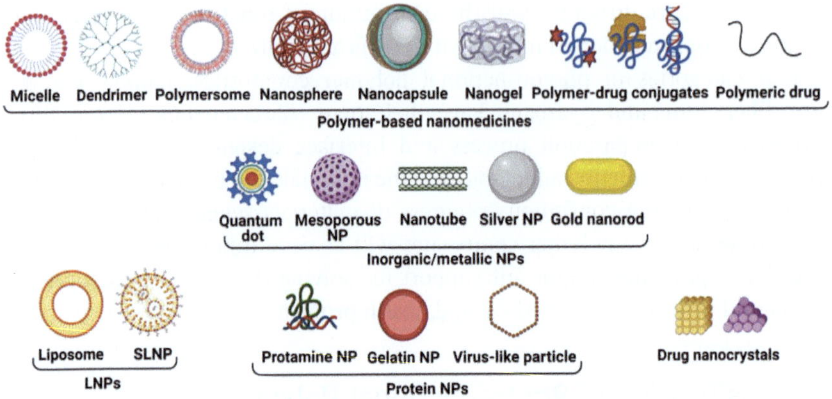

Figure 1.4 Graphical summary of common types of nanomedicines. *Source:* Younis et al. [40] / with permission of Elsevier.

It is worth noting that with the development of nano-self-assembly technology and molecular engineering, the construction methods of materials have shifted from the traditional "component-oriented" approach to the "structure-oriented" one. Morphological structural units represented by nanoparticles and clusters, capsules, micelles, liposomes, and emulsion systems are becoming the core platform for constructing photofunctional composites (Figure 1.4) [40]. For example, fluorescent composite materials composed of metal nanoclusters and polymer matrices can achieve the regulation of light response at the subnanometer scale. High-resolution vesicles and micelles have significant advantages in light-controlled drug release and imaging diagnosis due to their excellent self-assembly ability and environmental responsiveness. The emulsion system, on the other hand, provides a structural template for constructing photosensitive coatings, printable devices, and so on.

Therefore, the systematic classification and analysis of photofunctional polymer composites based on structural units not only helps to understand their performance sources and action mechanisms but also provides adjustable construction strategies for material design. Exploring its construction method, optical function realization mechanism and application prospects aim to provide theoretical support and idea reference for further research and development of this type of composite material.

1.2.1.1 Composite Nanoparticles

Conjugated polymers are a type of polymer with a main chain composed of alternating saturated and unsaturated bonds giving them excellent optical and electrical properties. They are widely used in the field of optical functions due to their unique π-conjugated structure. Through methods such as co-precipitation and microemulsion, these polymers can be further synergistically assembled with other functional components (such as inorganic nanomaterials, metal ions, and biomolecules) to construct composite nanoparticles with fluorescent properties, demonstrating multifunctional application potential in imaging, sensing, and photocatalysis. As one of the important construction units of composite nanoparticles, CP nanoparticles have attracted considerable interest because of their excellent optical properties and structural tunability. In addition, precious metal composite nanoparticles, as another important type of composite system, encompass two typical structural forms: precious metal nanoparticles and nanoclusters. The size of precious metal nanoparticles is usually 2–100 nm. Due to their strong LSPR effect, they show significant advantages in aspects such as light absorption, light scattering, and photothermal conversion. Precious metal nanoclusters with a size less than 2 nm exhibit excellent fluorescence properties and high catalytic activity by virtue of the quantum confinement effect and discrete energy level structure. By constructing composite structures with other nanomaterials or polymer components (such as oxides, carbon materials, and CPs), precious metal composite nanoparticles can not only achieve high controllability in morphology and interface regulation but also show potential advantages in optical, electrical, and catalytic performance.

Conjugated polymer nanoparticles (CPNs) integrate the characteristics of CPs with the properties of nanoparticles, demonstrating many excellent properties, including: good photostability, high brightness, low toxicity, outstanding biocompatibility, size effect, and the ability to sensitize and generate ROS. Due to its unique properties, CPNs are regarded as one of the biomaterials with great application prospects and value. By regulating the main chain structure, side chain substituents, and introducing specific functional groups of CPs, their optical, electrical, and interfacial properties can be precisely adjusted, thereby meeting the application requirements of different fields. Generally speaking, the properties of CPNs are profoundly affected by the type of chromophore, the configuration of the polymer chain, the aggregation state, and the size of the particles. Meanwhile, the charge state and the type of functional groups on the particle surface also regulate their biological interface behavior. With the enhancement of intrachain or interchain interactions in polymers, it often results in enhanced aggregation and more efficient energy transfer, thereby causing redshifts in absorption and emission wavelengths [41]. In contrast, CPNs constructed by means such as coprecipitation or microemulsion methods often show a blue shift in their absorption spectra due to factors such as polymer chain contraction and disordered stacking [42]. However, CPNs constructed through self-assembly exhibit redshift phenomena due to their more loose and orderly structure. These changes in optical characteristics are mainly attributed to the regulation of the aggregation state between the side chains of the polymer by the polarity changes in the solvent environment.

On the premise of having good optical basic properties, further modification of biomolecules (such as antibodies, DNA, and proteins) onto the surface of CPNs through covalent or noncovalent interactions can construct multifunctional composite CPNs with biological functions. Figure 1.5 presents the modification methods and functional molecules of biocomposite CPNs.

At present, the biological functionalization of CPNs mainly includes three functionalization strategies. The first functionalization approach is as follows: Firstly, the monomers are modified, such as by adding alkoxy chains, targeting groups, and charges to the monomers. Then, the monomers modified with biological elements are polymerized, and the resulting polymers possess certain biological functions. Subsequently, biocomposite CPNs are prepared. In this way, a variety

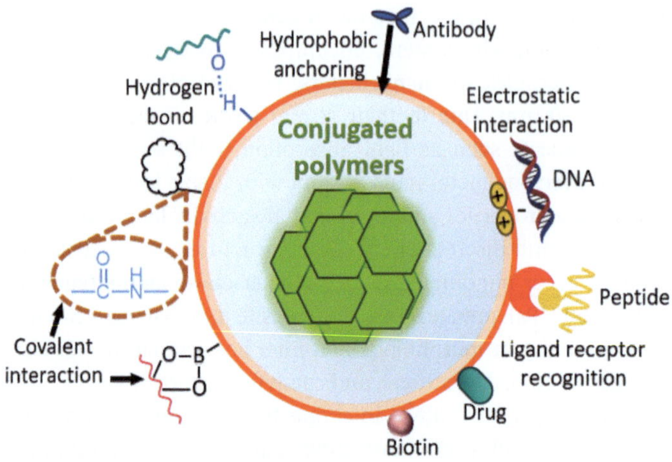

Figure 1.5 The functionalized groups of CPNs.

of CPNs that can be used for fluorescence imaging and clinical diagnosis can be prepared [43, 44].

Compared with the first functionalization method, the direct use of CPs that is easy to modify and commercialize is the most prominent feature of the second method. The specific functionalization mode is to form CPNs by encapsulating CPs with polymers containing biological functional groups through hydrophobic or electrostatic interaction, thereby achieving the biological functionalization of CPNs [45]. In addition, combining positively charged CPs with negatively charged biomolecules through electrostatic interaction is also an efficient way to obtain CPNs with the functions of biological elements. Multifunctional CPNs that can be used for cell imaging, drug delivery, and release were obtained [46, 47].

The third functional mode is an extension and deepening of the second one. Specifically, CPNs with surface carboxyl, amino, or azide modifications are first obtained through the second method. Then, through acid amine condensation catalyzed by EDC or Click coupling reaction catalyzed by Cu, molecules with specific biological functions are covalently linked to CPNs to obtain biocomposite CPNs. In this way, CPNs modified by antibodies, biotin, or affinity streptomycin can be obtained, and imaging of targeted tumor cells can be achieved [48]. CPNs have shown remarkable promise in various biological applications, including in vitro cellular and subcellular imaging, biosensing, as well as in vivo tissue imaging and disease therapy (Figure 1.6) [49].

Metal nanocomposites have shown broad application prospects in the biomedical field owing to their outstanding optical characteristics and interface regulation capabilities. By constructing multifunctional composite structures with polymers or fluorescent molecules, not only is the stability and responsiveness of the material enhanced but also precise control of optical properties is achieved, especially demonstrating unique advantages in imaging enhancement, PTT, and signal amplification. The metal-enhanced fluorescence (MEF) platform constructed based on such

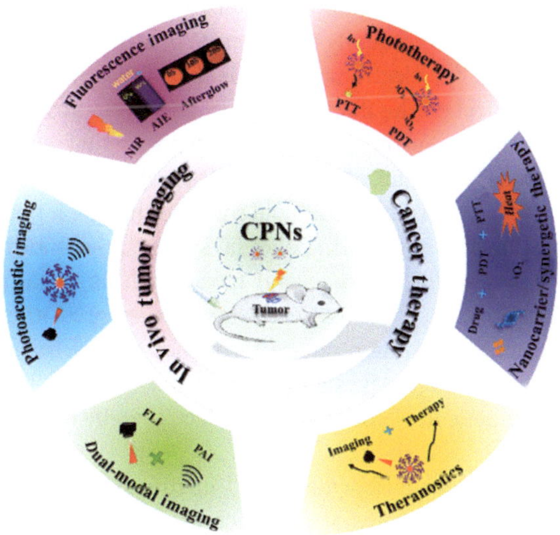

Figure 1.6 Schematic illustration of bioapplications of CPNs. *Source:* Reproduced with permission from Li et al. [49] / American Chemical Society.

a composite system has further expanded its application potential in high-sensitivity imaging and detection.

When a fluorescent substance approaches the surface of precious metal nanomaterials such as gold or silver, the phenomenon where its fluorescence emission intensity increases significantly compared to its free-state fluorescence emission intensity is called metal-enhanced fluorescence (MEF). The distance between the dyes and the nanometal, the morphology of the metal nanostructure, and the overlap degree of the emission peaks between the fluorescent substance and the metal nanomaterial are the primary factors influencing the enhanced fluorescence effect of the metal.

By exploring the electromagnetic field overlap and photoexcitation effects between gold nanomaterials and their surface dyes, it can be found that when the thickness of the silica coating between gold nanoparticles and dyes is thin, the intensity of fluorescent dyes undergoes quenching (Figure 1.7). When the thickness

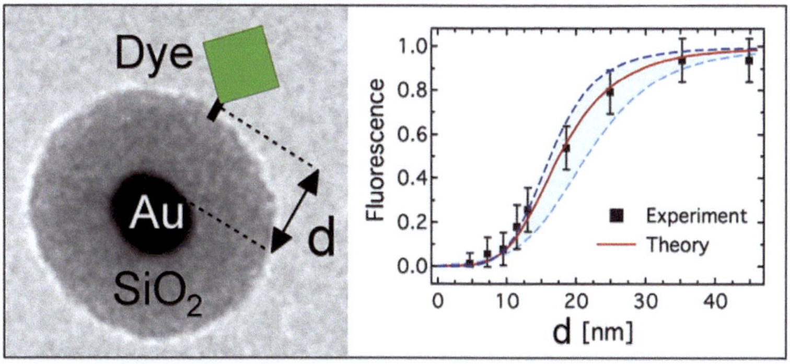

Figure 1.7 Transmission electron microscopy and fluorescence curves of Au@SiO_2-dye composite nanoparticles. *Source:* Reproduced with permission from Reineck et al. [50] / American Chemical Society.

of the silica coating gradually increases, the intensity of the fluorescent dye adsorbed on its surface gradually increases. When the thickness of the silica coating further increases, the electromagnetic field on the surface of the gold nanoparticles has a weak effect on the fluorescent dye, and its fluorescence intensity basically remains unchanged. Furthermore, there is no mutual electromagnetic field interaction between two adjacent dye molecules [50].

The MEF system in colloidal solutions has great potential in the field of biological imaging and sensing. In colloidal systems, the intermediate barrier layers of MEF can be classified into several types, such as silica, polymers, and biomolecules. These intermediate barrier layers can suppress the metal-induced fluorescence quenching and fluorescence resonance energy transfer (FRET) by regulating the distance between dyes and metal nanostructures, while optimizing the local electromagnetic field enhancement effect, thereby reducing energy loss and improving fluorescence efficiency.

Silicon dioxide is widely used in core-shell structured nanoparticles due to its multiple unique advantages, such as good light transparency, simple preparation, chemical inertness, low toxicity, thermal stability, and low cost [51]. In addition, the silica surface can be readily functionalized with various groups, endowing it with amino, thiol, and carboxyl groups, which is conducive to the subsequent modification of fluorescent molecules or biomolecules. As shown in Figure 1.8, the MEF colloid system using silicon dioxide as the intermediate barrier layer can be grouped into two types. One group is with metal nanoparticles as the core, silicon dioxide as the shell layer, and fluorescent molecules introduced in the outermost layer. Its preparation method is shown in Figure 1.8a. Another type is dye-coated silica or silica labeled with dye as the core, and metal nanoparticles [52]

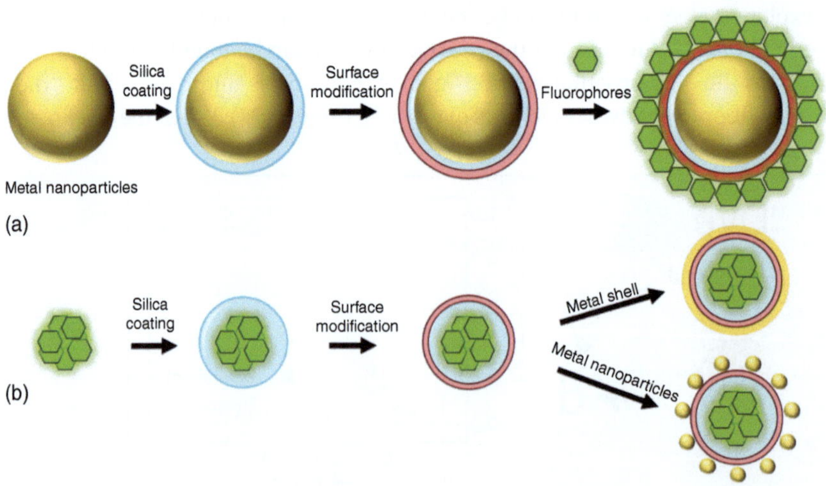

Figure 1.8 (a) With metal nanoparticles as the core, silicon dioxide as the shell layer, and fluorescent molecules located in the outermost layer; (b) with fluorescent materials as the core, silicon dioxide as the shell layer, and a metal shell or metal nanoparticles located in the outermost layer.

or continuous metal nanoshells [53] as the outermost layer. Its preparation method is shown in Figure 1.8b.

Compared with rigid materials, polymers are flexible chains in aqueous solutions, and their configuration and size can change with the variation of the surrounding environment. Therefore, when it serves as an intermediate barrier layer of MEF, the MEF effect can be regulated by changing its thickness and distance. There are several methods for introducing polymer shells on the surface of metal nanoparticles, such as in situ polymerization, the end-group functional connection method, and LbL self-assembly method. A thermally responsive poly(*N*-isopropylacrylamide) (PNIPAM) cross-linked shell was introduced on the surface of silver nanoparticles by in situ polymerization. The outermost layer introduced fluorescent molecules through electrostatic adsorption, and the distance between it and the metal surface could be controlled by changing the temperature. This not only achieved a temperature-adjustable MEF effect but also enhanced the stability of the fluorescent molecules at high temperatures [54]. Recently, the development of living radical polymerization has provided opportunities for designing and synthesizing polymers with diverse compositions and complex architectures. Polymers prepared via reversible addition–fragmentation chain transfer polymerization (RAFT) often feature end groups such as dithioesters, which can be anchored to metal surfaces through the affinity between metals and sulfur [55]. The MEF effect of this system shows reversible variation characteristics with the heating and cooling of temperature. In colloidal systems, LbL polyelectrolyte multilayer films can also be used to achieve the regulation of distance at the nanoscale [56]. Unlike on a flat surface, depositing multilayer films on a colloid requires multiple centrifugation purification steps.

Biomolecules are also often used to assemble and prepare composite nanoparticles with MEF effects, which improve their biocompatibility for application in organisms. Among them, DNA molecules or oligonucleotides are often used to regulate metal and fluorescent molecules at the nanoscale because their lengths can be precisely designed and controlled [57]. CdSe/ZnS quantum dots and gold nanoparticles were assembled using DNA molecules with thiol groups and biotin groups at both ends, respectively [58]. The surface of quantum dots is coated with streptavidin. Through biotin–streptavidin interaction and Au–S bond interaction, DNA molecules connect quantum dots and gold nanoparticles. Among them, when quantum dots are located between two gold nanoparticles, the maximum enhancement effect is achieved. This is because the electromagnetic fields of the two gold nanoparticles in this area interact with each other, generating a stronger magnetic field enhancement, namely the "hot spot" effect. DNA molecular chains remain relatively flexible in aqueous solutions [59]. This specific binding can be utilized to assemble fluorescent and metal nanomaterials. For example, assembling a layer of gold or silver nanoparticles through biotin–streptavidin interaction outside CdTe nanowires can show a significant fluorescence enhancement effect in aqueous solution [60].

Photofunctional composite nanoparticles have shown broad application potential in fields such as fluorescence imaging, PDT, and drug delivery owing to their excellent photostability, biocompatibility, and structural tunability. By introducing functional biomolecules, multiple targeting and integrated treatment can be achieved. Photofunctional composite nanoparticles are expected to achieve deeper integration in directions such as precision medicine, intelligent diagnosis and treatment, and multimodal imaging, promoting the development and transformation application of high-performance biological nanomaterials.

1.2.1.2 Emulsion

Emulsion is a relatively stable multiphase dispersion system, where one liquid is finely dispersed as small droplets in another liquid that is immiscible with it. Common emulsions always present an emulsified appearance. They are thermodynamically unstable multiphase dispersion systems. The dispersed phase, continuous phase, and emulsifier are the three main components that constitute an emulsion. Among them, the dispersed phase is the discontinuous phase that is dispersed in the emulsion in the form of droplets. The continuous phase is a continuous and integrated phase formed in the emulsion, also known as the dispersion medium.

According to the different dispersion situations, emulsions composed of oil and water can be divided into two major categories: the first is the type of emulsion where water encapsulates oil, that is, the situation where the oil phase is dispersed in water, which is expressed as oil/water or O/W. The second type is an emulsion where oil is coated with water, that is, the situation where the water phase is dispersed in the oil, which is expressed as water/oil or W/O. More complex emulsion types can also be formed, such as water containing oil and then water, that is, water/oil/water (W/O/W), or oil containing water and then oil, that is, oil/water/oil (O/W/O) (Figure 1.9).

Traditional emulsions are stabilized by surfactant molecules, which are often structured with hydrophilic head groups and hydrophobic chains. The stability of the emulsion is maintained by reducing the interfacial energy between the two phases. At the beginning of the last century, the mechanism by which solid particles stabilize emulsions was discovered and defined, respectively, by Ramsden and Pickering. The stabilization mechanisms of Pickering emulsions include the

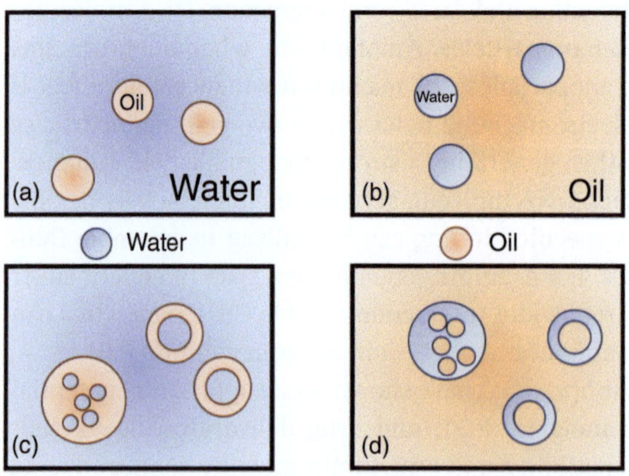

Figure 1.9 Schematic illustration of emulsion type: (a) O/W; (b) W/O; (c) W/O/W; (d) O/W/O.

interface film theory, the three-dimensional network structure theory, and the depletion interaction. Compared with surfactants, Pickering emulsions stabilized by solid particles have greater advantages, including better anticoalescence stability, minimizing the Ostwald ripening phenomenon, higher biocompatibility, and lower cytotoxicity [61]. Based on the above advantages, Pickering emulsion has broad application prospects in fields such as food, biomedicine, and cosmetics [62].

Light intensity (controlled at different wavelengths of near-infrared and ultraviolet) is another stimulating factor that has been studied for controlling Pickering emulsion systems. Although it has not attracted as much attention as more traditional stimuli (such as pH, temperature, and CO_2 concentration), it can be quite beneficial because it is usually nondestructive, precise, and easy to implement.

The use of interfacial active upconversion nano-phosphates as colloidal emulsifiers can achieve biocatalysis applications (Figure 1.10) [63]. The surface chemistry of the particles was regulated by coupling with photochromic isopropylpyran, and isopropylpyran underwent hydrophilic-lipophilic conformational changes under the action of ultraviolet light and visible light. After the introduction of near-infrared light into the system, it was observed that the upconversion nano-phosphor was excited and emitted ultraviolet photons. The emitted ultraviolet light triggered conformational changes in surface-grafted photochromic isopropylpyran. This leads to the particle surface becoming more hydrophilic, thereby stabilizing the Pickering

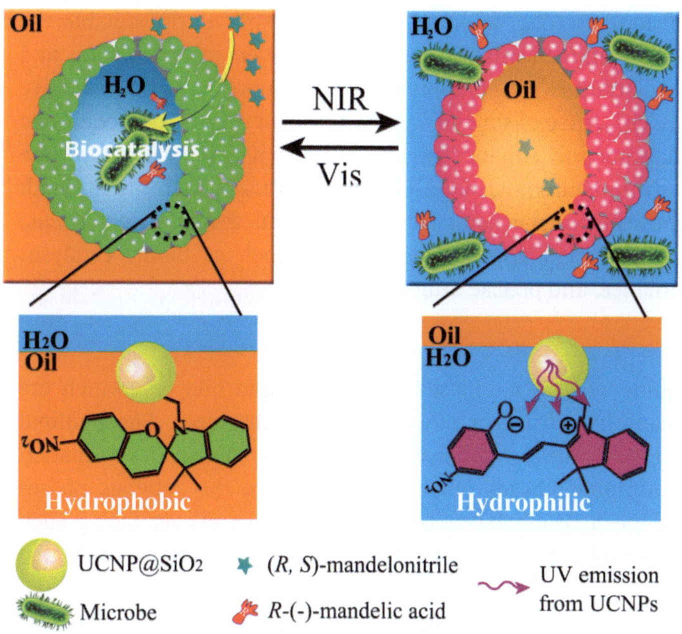

Figure 1.10 Schematic illustration of NIR/Visible light-induced phase inversion in emulsions stabilized by photochromic spiropyran-functionalized upconversion nanophosphors for optimized biocatalysis. *Source:* Chen et al. [63] / with permission of American Chemical Society.

of the o/w emulsion. Exposure to visible light will reverse the conformational changes on the particle surface, resulting in the transformation of the emulsion phase into w/o.

With the development of synthesis technology, the particles used to prepare Pickering emulsions have gradually evolved from hard particles to soft particles. Unlike rigid inorganic particles, polymer particles themselves are soft particles, and there are many types of monomers for synthesizing polymers. Through layer-by-layer (LbL) design during the synthesis process to adjust the properties of the particles and endow them with functionalization, emulsified materials with one or more specific functions or responsiveness can be developed, providing a new strategy for preparing functionalized particle emulsifiers. It has aroused extensive research by scholars.

An azobenzene monomer, (E)-6-(4-((4-(allyloxy)phenyl)diazenyl)phenoxy)hexyl methacrylate with asymmetric divinyl groups, was synthesized to create light-responsive particles [64]. When exposed to white light and ultraviolet light (365 nm), the particles undergo reversible, light-induced shape changes. Particles produced via precipitation polymerization and dispersion polymerization can respectively stabilize water-in-oil and oil-in-water Pickering emulsions, both of which exhibit reversible formation and deformation in response to light.

The application of photofunctional polymer composites in emulsion systems shows broad prospects. Relying on their response characteristics to light stimulation, reversible phase transformation, stability regulation, and multifunctional integration of emulsions can be achieved. They are particularly suitable for constructing intelligent emulsification systems, light-driven microreactors, and controllable release platforms. By compounding with inorganic photosensitive components (such as TiO_2, precious metal nanoparticles) or organic photochromic groups, these materials not only enhance the interfacial activity but also endow the emulsion system with tunability and reusability. In the future, with the in-depth development of light-response mechanisms and structural design, it is expected to achieve more breakthrough applications in fields such as bionic catalysis, environmental governance, and precise drug delivery.

1.2.1.3 Liposome

Liposomes are ultrafine vesicles composed of phospholipid bilayers, which are usually used to encapsulate drugs or other bioactive substances. Phospholipid molecules have hydrophilic heads and hydrophobic tails. During the formation of liposomes, the hydrophilic heads of phospholipids face the aqueous phases on both sides, while the hydrophobic tails form a plate-like bilayer facing each other. This structure enables liposomes to stably encapsulate water-soluble and liposoluble substances.

Liposomes can be categorized according to their structure, particle size, and the number of phospholipid bilayer membranes. Monolayer liposomes include small unilamellar vesicles (SUVs), which have particle sizes less than 100 nm, and large unilamellar vesicles (LUVs), with particle sizes ranging from 100 nm up to several micrometers. Multilamellar vesicles (MLVs) is composed of multiple layers of phospholipid bilayers, and the particle size is usually on the level of several hundred nanometers to micrometers. And multivesicular vesicle (MVV) is a large vesicle containing several discontinuous drug solution vesicles.

Based on the degree of modification, liposomes can be divided into conventional liposomes and targeted liposomes. The phospholipid bilayer of common liposomes is unmodified and prone to accumulate in the reticuloendothelial system (such as tissues and organs like the liver, spleen, lymphatic system, and lungs). The phospholipid bilayer of targeted liposomes has been modified to possess long-circulation characteristics (staying in the circulatory system for a relatively long time), active targeting characteristics (specifically binding to receptors), and sensitivity to changes in the physical and chemical environment.

Due to the environmental sensitivity of liposomes, they can be constructed as a multifunctional light-responsive nanoplatform, which not only plays an enhancing role in traditional PDT but also endows them with multiple photofunctional response characteristics through material engineering methods, providing an effective way for the integration of tumor diagnosis and treatment. The main optical functions can be seen in the following aspects.

Firstly, liposomes effectively improved the bioavailability and therapeutic performance of photosensitizers (PS). Most photosensitizers (such as PpIX, Ce6, mTHPC, and BPD) have strong hydrophobicity and are difficult to disperse directly in the aqueous phase. They tend to aggregate, leading to self-quenching and thereby affecting the generation of ROS and therapeutic effects. By encapsulating these photosensitizers into liposomes, their water solubility, stability, and cellular uptake efficiency can be significantly improved [65–68].

Secondly, liposomes have excellent light-responsive release performance, achieving precise controlled release in both space and time. By introducing lipids containing unsaturated double bonds (such as DOPC, DOTAP, and POPC), liposomes can respond to ROS, undergo oxidative disintegration of the membrane structure, and thereby release the drug-loaded components [69]. In addition, photothermal agents (such as ICG) are introduced into the thermosensitive liposome system. Under the irradiation of NIR laser (808 nm), local temperature rise occurs, triggering phase transformation of DPPC liposomes around 41 °C and achieving controlled release of the drug [70]. Such mechanisms have significantly enhanced the controllability and safety of treatment.

The liposome system supports the construction of an integrated imaging and treatment platform to achieve the function of "integrated diagnosis and treatment." The polymer liposomes constructed were co-loaded with Merocyanine 540 (MC540) and UCNPs, the latter of which can generate 540 nm visible light from 980 nm near-infrared excitation for activating MC540 (Figure 1.11). This is to achieve the dual functions of PDT and fluorescent imaging (FLI) of deep tissues [71].

In conclusion, liposomes not only solved the problems of water solubility and targeting of photosensitizers but also achieved light-controlled release, imaging guidance, oxygen environment regulation, and synergistic integration with other therapeutic methods through fine regulation of structure and function, fully demonstrating their multidimensional value in the field of photofunction. In the future,

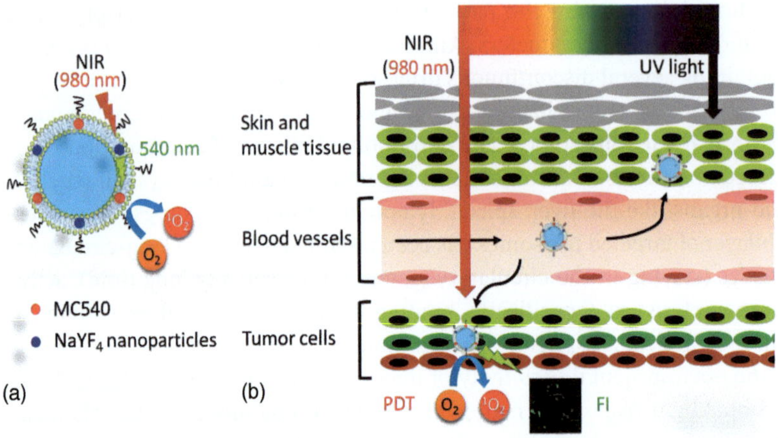

Figure 1.11 (a) NaYF4 UCN converted near-infrared irradiation to visible light at 540 nm, activating liposomal MC540 for (b) photodynamic therapy and fluorescence imaging. *Source:* Cheng et al. [71] / John Wiley & Sons / CC BY 4.0.

with the continuous optimization of liposome materials and the development of multicomponent collaborative systems, its application in personalized PDT, deep tissue treatment, and integrated diagnosis and treatment is expected to be further expanded, providing more precise, efficient, and safe solutions for clinical tumor treatment.

1.2.1.4 Micelle

Polymer micelles are dynamically stable colloids that result from the self-assembly of amphiphilic polymers in selective solvents [72]. The micellar process of polymer micelles is a thermodynamically kinetic equilibrium process. When the polymer concentration in the aqueous solution is low, the polymer exists in a free form. When the polymer concentration is higher than its critical micelle concentration (CMC), the free polymer spontaneously aggregates to form polymer micelles. Hydrophobic units cluster together to create a micelle core that minimizes exposure to the surrounding water, while hydrophilic blocks surround the core, forming a hydrophilic shell on the exterior of the micelle [73].

Polymer micelles can be classified in various ways based on their properties and structures. According to their self-assembly behaviors, they are categorized as spherical, worm-shaped, or rod-shaped micelles [74]. Based on the types of polymers used, classifications include block polymer micelles, polyelectrolyte copolymer micelles, dendritic polymer micelles, and graft copolymer micelles, among others [75]. When considering the structural characteristics of the polymers, polymer micelles can be divided into cross-linked, non–cross-linked, and noncovalently linked micelles [76]. Classification by topological structure yields linear and star-shaped polymer micelles [77]. In terms of polymer composition, they can be grouped into single-component micelles, single-molecule micelles, and mixed micelles [78]. Finally, based on functional characteristics, polymer micelles

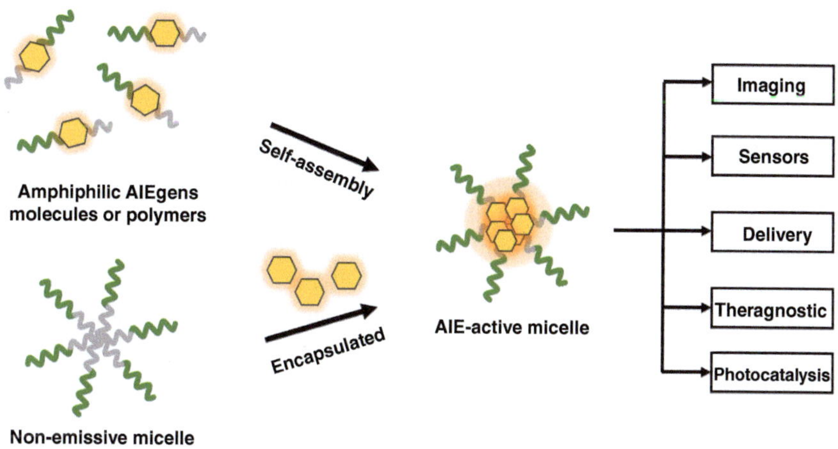

Figure 1.12 Schematic illustration of the preparation and bioapplications of AIE-active micelles.

are categorized as stimulus responsive, non-stimulus responsive, and targeted micelles, among others [79].

The optical performance of functional micelles based on aggregation-induced emission (AIE) molecules is different from the problem that traditional fluorescent materials are prone to aggregation-caused quenching (ACQ) in the aggregated state. After AIE molecules self-assemble and enter the micelle core, they instead show stronger fluorescence emission due to intramolecular rotational restriction (RIR mechanism). This enables the micelle system to have significant advantages in terms of optical functions. This type of AIE micelles not only solves the problem of low luminescence efficiency of traditional luminescent materials in biological and aqueous systems but also endows the micelles with new sensing, imaging, response, and therapeutic functions [80] (Figure 1.12).

Micelles not only possess optical properties but also have a responsive nature to external stimuli such as pH, temperature, CO_2, and light, making them an ideal platform for building "visual self-assembly systems." A CO_2-responsive block copolymer micelle, after the introduction of CO_2, transforms into vesicles, accompanied by obvious fluorescence enhancement and color change, which can reflect the transformation process of the self-assembled structure [80]. Similarly, the acid-sensitive micelles constructed by imine bonds undergo dissociation and reassembly under pH changes, and the corresponding fluorescence emission also changes accordingly [81]. Such systems can be used to monitor the drug release process, self-assembly behavior, and structural dynamic evolution in real time, and have application potential in the fields of smart materials and nanoreactors.

By selecting photofunctional molecules that have the potential for PDT or PTT, and co-loading the luminescent units with chemotherapy drugs in micelles, treatment and monitoring can be carried out simultaneously. For example, tetraphenylethylene (TPE)-modified polymers can stably load drugs after forming micelles and release drugs in response to environmental stimuli (such as pH or ROS), and fluorescence is used to visualize the release process [82]. In a specific design, dye molecules can also generate ROS under light, which are used for the photodynamic killing of tumor cells. Therefore, multifunctional micelles can achieve the "diagnosis and treatment integration" function, integrating imaging, drug loading, and treatment into one and have great prospects in the precise treatment of cancer.

The multiple roles of micelles in optical functions, ranging from enhancing basic optical properties to integrating stimulus response, imaging, biological detection, and treatment, demonstrate the multidimensional value of micelle materials. Thanks to the flexibility of molecular design and self-assembly regulation, micelles not only overcome the disadvantages of traditional luminescent materials in aqueous systems but also show broad application prospects in biological imaging, drug delivery, environmental sensing, smart materials, and light-controlled therapy. In the future, with the further development of precise structure control, multimodal signal output, and intelligent response mechanisms, micelles are expected to become an important platform for constructing high-performance photofunctional nanomaterials, providing new solutions for nanobiotechnology and intelligent diagnosis and treatment.

1.2.1.5 Capsule

A capsule is an empty micro-nano structure composed of polymers, lipids, or proteins, with a distinct shell and internal cavity. Capsules have good controllability, biocompatibility, and functional expansion ability. Its typical structure originates from the template-assisted LbL technology. Through successive deposition of polymer molecules with specific interactions (such as electrostatic interactions, hydrogen bonds, and covalent bonds) on the surface of the template, the core is finally removed to obtain hollow capsules with a stable geometric structure (Figure 1.13). This type of capsule can be precisely regulated in terms of size, shell thickness,

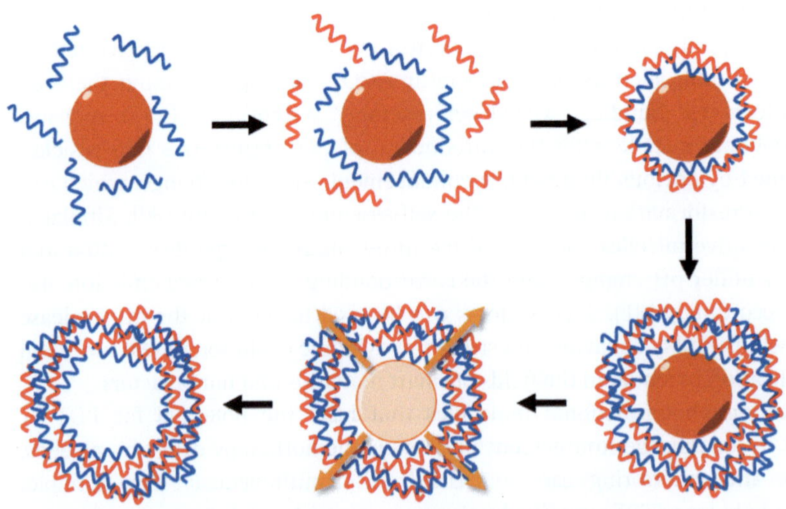

Figure 1.13 Schematic illustration of the preparation of capsules.

permeability, flexibility, and surface functional groups, and has broad application potential in nanomedicine, bionic materials, and intelligent release systems [83].

Capsules have significant physical and chemical responsiveness. Through material design, they can be endowed with the ability to respond to multiple external influences such as pH value, redox state, enzyme concentration, temperature, and light. Among them, light responsiveness is a direction that has attracted much attention in its functional expansion. By introducing photosensitive linking groups (such as azobenzenes, photosensitive amide bonds) into the vesicular shell or doping photoactive materials (such as gold nanoparticles, photosensitive dyes, semiconductor quantum dots), the capsules can undergo structural changes or chemical bond breaks under the action of light at specific wavelengths, thereby inducing changes in the permeability of the vesicular membrane, opening pores or overall deconstruction, and achieving the light-controlled release function [84].

Capsules can serve as important components of the controlled release system. Through the fine design of the shell molecular structure, a response system sensitive to visible light, near-infrared light, or ultraviolet light can be constructed, which releases encapsulated drugs, proteins, or signal molecules under local light stimulation. This light-response release process features noncontact, spatiotemporal controllability and high selectivity, and is particularly suitable for biological therapy scenarios that require precise positioning and reversible regulation [85].

In addition to drug delivery, the shell of the capsule can imitate the multilayer membrane structure of the natural thylakoid membrane, providing a structural template for artificial photosynthesis systems and photocatalytic platforms. By integrating photosensitive electron transfer molecules, organic dyes, or semiconductor materials into the capsule structure, the process of energy capture, transfer, and utilization after local photoexcitation can be achieved. Multichamber structure capsules can achieve photoactivated cascade reactions. Each chamber encapsulates different reactants or catalysts, which are successively released or activated under light stimulation, simulating the mechanism of multistep energy conversion and material synthesis in natural photosynthesis [86].

In multienzyme reactors or enzymatic catalytic platforms, light-responsive capsules can be used to activate specific reaction pathways. For example, the controllable initiation of multienzyme cascade reactions can be achieved by encapsulating a group of enzymes in capsules and only opening the capsules under specific light conditions to release the enzymes. This design can avoid mutual interference among enzymes and improve reaction efficiency and spatial resolution [87].

The diversified structure of the capsules further broadens their light-response ability. For example, spherical and ellipsoidal capsules exhibit different photoinduced release kinetics due to the difference in surface area; multilayer membrane structures can delay or control the release rate in segments. The combination of hydrophilic and hydrophobic film layers can achieve targeted protection and activation of different types of photosensitive molecules. This structure-function integration characteristic makes the capsule an ideal light-controlled delivery platform and optical functional bionic module [83].

With the development of nanomaterial science, photochemistry, and bioengineering, the application of capsules in optical functions still has broad prospects. How to achieve higher response efficiency, lower phototoxicity, stronger targeting ability, and more complex optical program control is an important direction for future research. Through molecular-level design, interdisciplinary material integration, and the introduction of advanced assembly technologies, capsules are expected to play a core role in multiple fields such as precision medicine, bionic light energy systems, optical detection, and environmental response materials.

1.2.2 Fibers

In the past few decades, remarkable advancements have been made in photofunctional polymer fibers. From the initial poly(methyl methacrylate) (PMMA)-based optical fibers to the current development of various biodegradable polymer optical fibers, these advancements have provided more possibilities for the application of optical fibers in the biomedical field. Especially, the degradable polymer optical fibers based on poly(lactic acid) (PLA), poly(ethylene glycol) (PEG), and poly(lactic-co-glycolic acid) (PLGA) can not only reduce the need for surgical intervention but also provide precise optical transmission and biological signal monitoring.

The advancement of medicine, especially the development of neuroscience, is urgently in need of multifunctional photofunctional fibers that possess higher biocompatibility, greater photoconductivity, and lower invasiveness. This is because the functions of cells or organs are coordinated and influenced by multiple signals, including electrical signals, chemical signals, and neurotransmitters, while traditional single-function probes are difficult to achieve synchronous monitoring and regulation. Photofunctional fibers enable the integration of waveguides, electrodes, and microfluidic channels, facilitating the simultaneous investigation of electrical, chemical, and mechanical signals with minimal disruption. More specifically, they can deliver drugs, nutrients, and viral vectors to deep tissues through hollow channels [88].

Two key parameters of optical fiber materials are their transparency in the spectral region required for waveguide operation and their processability in the selected manufacturing method. From an optical perspective, the materials used to manufacture optical waveguides need to have high light transmittance and an appropriate refractive index. The refractive index must be higher than that of the surrounding medium to support total internal reflection, thereby enabling the propagation of optical power within the waveguide. Meanwhile, the material should have a low absorbance and scattering to ensure low attenuation of optical power when propagating in the waveguide. In addition, the materials used to manufacture optical fibers should also have sufficient mechanical properties to meet the usage requirements [89].

1.2.2.1 Natural Material-based Fibers

Composition and Structure Natural polymer fibers are ideal candidates for biocompatible and biodegradable photofunctional fibers because they have distinctive optical and mechanical characteristics, nontoxicity, and inherent biodegradability, which are influenced by the diversity of their components. Biocompatible polymer fibers have been successfully obtained by using materials such as proteins, agar, different silks, and cellulose. Among them, the optical fibers based on bacteria and cells demonstrate superior biocompatibility and biodegradability, owing to their tissue-like characteristics. In addition, some natural fibers are applied in the biomedical field due to their excellent photofunctions, including good light conductivity and high transmittance. However, natural photofunctional polymer fibers inevitably need to evaluate the following issues [90]. For preparation, the molecular weight is difficult to control, resulting in poor stability among different batches. Furthermore, at temperatures below the melting point, this type of fiber is prone to decomposition or thermal modification, making thermoforming processes such as melt extrusion difficult. For biocompatibility, especially protein-based fibers, it may trigger an immune response. If not handled properly during collection, storage, or manufacturing, there is a risk of spreading infectious diseases.

At present, a variety of photofunctional polymer fibers based on natural materials have been developed for use in the biomedical field. Silk is a natural protein fiber material made from the silk spun by silkworms, with its main components being fibroin and sericin. Its fiber structure is unique. The external sericin protein wraps around the internal sericin protein, which is composed of β-folds, endowing the silk with high strength and luster [91]. As a biomaterial, silk, with its excellent biocompatibility and biodegradability, has become a commonly utilized fiber in clinical applications, including wound healing, suturing, tissue engineering, and textiles. In addition, by adjusting the proportion of various elements, the mechanical properties of the fiber based on the silk can be modified to suit particular requirements, and bioactive molecules can be linked to the fiber to make it more functional (Figure 1.14) [92].

Similarly, spider silk is also a kind of natural polymer fiber, which has higher tensile strength and toughness than silk, and has unique thermal, optical, and biocompatibility. Spider silk exhibits excellent light transmittance, making it potentially applicable in fields such as optical sensors. Its multilevel structure, including cross-linked structure, helical nanofiber structure, and core-shell structure, endows it with strength and toughness that surpass those of artificial fibers. The multilevel structure of spider silk, including the primary and secondary structures of the spider silk protein, the helical nanofiber structure, and the core-shell structure, endows it with strength and toughness that surpass those of artificial fibers (Figure 1.15) [93].

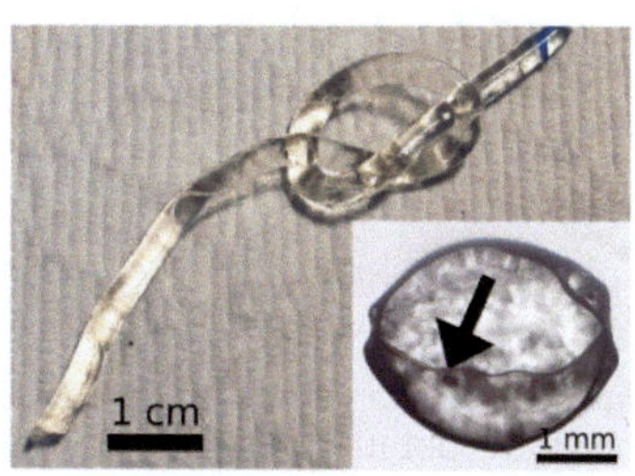

Figure 1.14 Silk waveguide with a knotted configuration and microstructured cross-section. *Source:* Applegate et al. [92] / Optica Publishing Group.

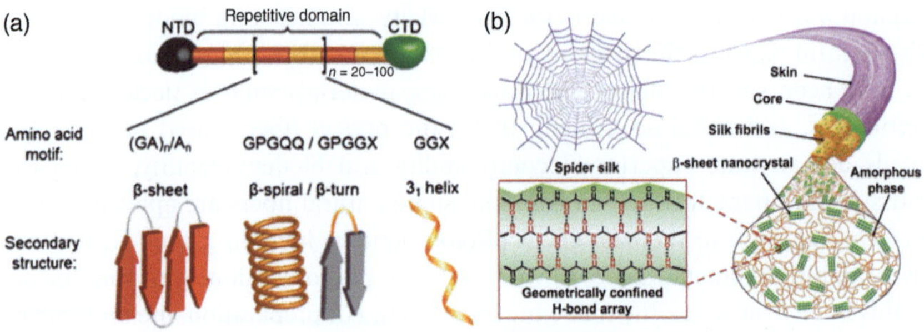

Figure 1.15 Spider silk multiscale organization: (a) molecular and secondary structural motifs, (b) hierarchical fibrous assembly. *Source:* Li et al. [93] / John Wiley & Sons / CC BY 4.0.

Meanwhile, the excellent biocompatibility of spider silk endows it with significant potential in the biomedical field, such as the treatment of cardiovascular diseases and the development of biomedical materials. Its self-repairing ability, cell affinity, and antibacterial performance further enhance its application prospects in these fields.

Cellulose, a natural material abundantly found in plants, shares many properties with silk and has been utilized in the fabrication of photofunctional fibers. Several previous studies have demonstrated that cellulose can be applied in chemical and electrochemical sensing, drug delivery, and so on. Photofunctional fibers made from cellulose exhibit high ion permeability and outstanding light transmittance across a broad spectral range. Notably, increasing the cellulose content in these materials enhances their modulus and tensile strength, allowing for the mechanical properties of the fibers to be tuned as needed. Similar to other natural fibers, cellulose-based fibers are both biocompatible and biodegradable. Furthermore, cellulose-based fibers with microstructures can serve many functions, offering significant potential for biomedical applications (Figure 1.16) [94].

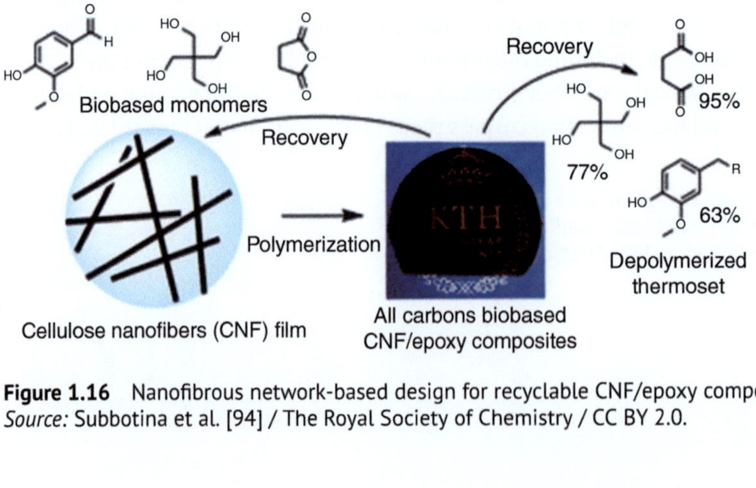

Figure 1.16 Nanofibrous network-based design for recyclable CNF/epoxy composites. *Source:* Subbotina et al. [94] / The Royal Society of Chemistry / CC BY 2.0.

Bioapplications The role of photofunctional polymer fibers in drug delivery is mainly reflected in their ability to achieve precise targeted drug release and enhance therapeutic effects [95]. Agarose hydrogel (derived from algal polysaccharides) coated with drugs combined with PTT can efficiently destroy tumor cells, with an effect similar to transcatheter arterial chemoembolization (TACE). As shown in Figure 1.17a, optical fibers not only deliver photons to deep tumor tissues but also carry photothermo-sensitizers and chemotherapeutic drugs for precision-targeted therapy. Figure 1.17b demonstrates that under laser irradiation, fiber probes doped with rare-earth ions generate localized high temperatures, activating a surface film to release the antitumor drug doxorubicin (Dox). Figure 1.17c illustrates a fiber-integrated sensor system incorporating Mach-Zehnder interferometer (MZI) and fiber Bragg grating (FBG), enabling real-time monitoring of drug release and lesion temperature. This delivery approach offers superior penetration depth and drug retention compared to conventional laser therapies, significantly enhancing

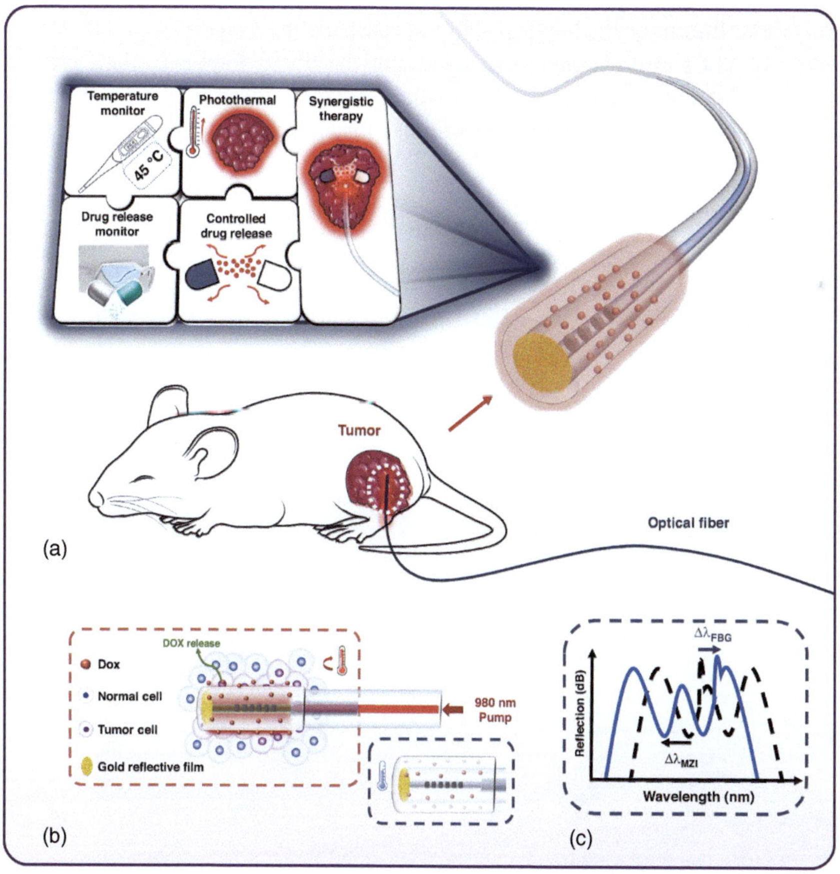

Figure 1.17 NIR-photothermal triggered fiber-optic theranostic probes with synergistic mechanism and closed-loop monitoring. *Source:* Zhang et al. [96] / Springer Nature / CC BY 4.0.

therapeutic efficacy for deep-seated tumors. These advantages enable photofunctional polymer fibers to go beyond simple therapeutic probes, providing efficient and compatible drug screening and evaluation strategies for the innovation of therapeutic drugs, and even reevaluating drugs that have been excluded due to systemic toxicity [96].

The application of natural materials in optogenetics is mainly reflected in the preparation of biocompatible fibers. Light-based stimulation is frequently contrasted with electrical stimulation for its ability to activate or suppress neuronal activity within specific brain regions. Supramolecular fibers generated via the interfacial polyelectrolyte complexation (IPC) of DNA and histones have been employed for the encapsulation and in situ differentiation of neural stem cells (NSCs) derived from mouse brain tissue (Figure 1.18) [97]. The encapsulated NSCs exhibit high viability, underscoring the good biocompatibility of these fibers as three-dimensional scaffolds. To further improve NSC encapsulation efficiency and restrict cell migration from the fibers, cell adhesion peptides (K6-PEG-RGD) are incorporated through electrostatic interactions, thereby enhancing spatial cellular organization. Immunocytochemical staining confirms the successful in situ differentiation of NSCs into oligodendrocytes within the fibers. Owing to their robust capacity for NSC encapsulation and in situ differentiation, these chromatin-inspired supramolecular fibers hold significant promise for applications in nervous system tissue engineering.

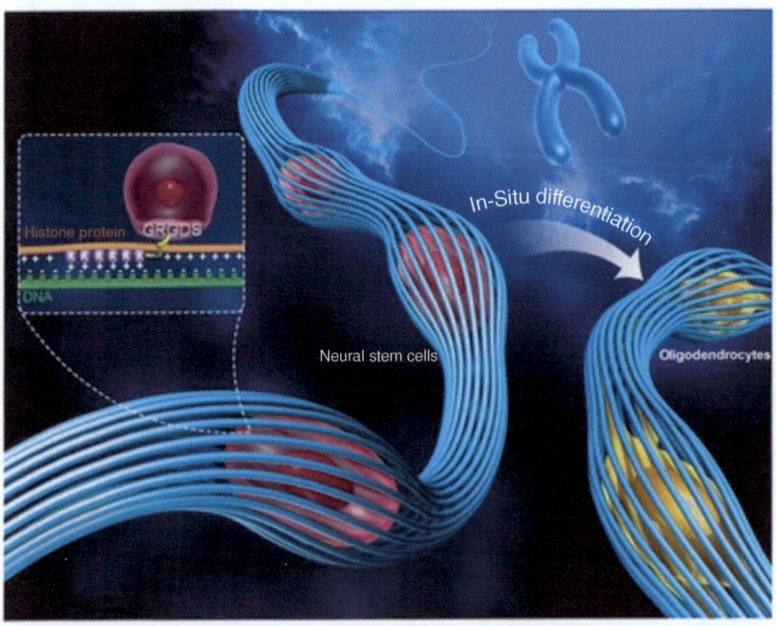

Figure 1.18 Schematic illustration of the supramolecular fibers for encapsulating NSCs and promoting their in situ differentiation. *Source:* Reproduced with permission from Zhao et al. [97] / John Wiley & Sons.

1.2.2.2 Synthetic Polymer Fibers

Synthetic polymer fibers are made from renewable resources and have better material designability because they are artificial products with adjustable physical, mechanical, and chemical properties. They can be made into degradable materials, which means they can be reabsorbed by the human body, and their degradation curves and optical properties can be adjusted and optimized to fulfill the requirements of various optical devices.

Generally speaking, chemical insertion and physical doping are two main methods for preparing photofunctional polymer fibers. In chemical methods, photoresponsive groups are inserted into polymer chains through various chemical means, including the copolymerization of photoresponsive monomers with other monomers and the postmodification of fibers. The former adopts diverse polymerization methods, including free radical, atomic transfer free radical polymerization, and reversible addition-break chain transfer. The photoresponsive part can be inserted into the polymer skeleton as the main chain or side group, and the photoresponsive dye can be inserted into the polymer chain. The latter requires functionalized photoresponsive dyes. Physical methods offer a simple and feasible preparation procedure. Photoreactive dyes are placed within or on the surface of nanofibers through physical interactions.

Poly(glycolic acid) (PGA), PEG, PLA, PLGA, poly(L-lactic acid) (PLLA), poly(dimethylsiloxane) (PDMS), and poly(vinyl alcohol) (PVA) are often used as the base materials of photofunctional polymer fibers due to their excellent biocompatibility and easy functionalization. These polymers have been widely studied and applied due to their controllable degradability, minimal propagation loss, and excellent mechanical properties. They can be degraded and absorbed in the body, leaving almost no toxic residue in a biophysical environment. Therefore, they are highly suitable for implantable medical devices, eliminating the need for additional surgery to extract the implant. These polymers exhibit excellent optical and mechanical properties in photofunctional fibers, making them have broad biomedical applications, especially in phototherapy, optogenetics, and biosensing.

Composition and Structure The application of PEG-based photofunctional polymer fibers in the biomedical field is increasing day by day, mainly owing to their specific physical and chemical properties. PEG photofunctional polymer fibers have become ideal materials in the biomedical field due to their high capacity for storing water, similarity to the extracellular matrix, and low toxicity. These fibers not only possess excellent light conductivity efficiency and optical transparency but also have mechanical flexibility and are easy to integrate, which makes them show great potential in biosensing and photoinduced therapy [98]. Glucose-responsive PEG hydrogel fiber optics can be used as a sensor to quantify the concentration of analytes, as shown in Figure 1.19.

Poly(lactic acid) (PLA) is a synthetic polyester formed by the polymerization reaction of lactic acid monomers and is one of the most widely used degradable polymers at present. Because PLA is derived from renewable resources like corn and sugarcane and has good biocompatibility and degradability, it is widely used in the fields

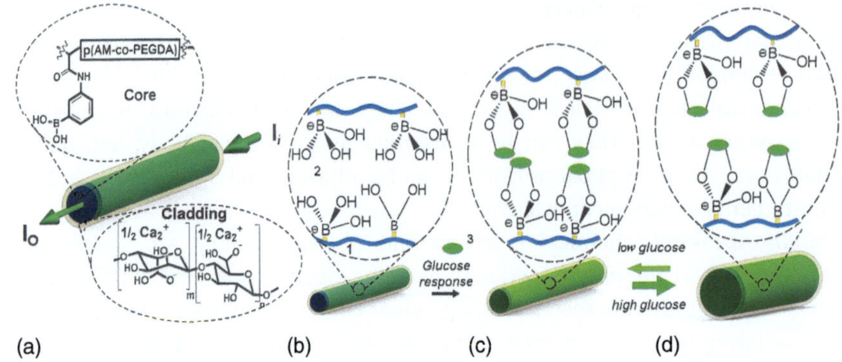

Figure 1.19 Glucose-responsive PEG hydrogel fibers for optical biosensing. *Source:* Yetisen et al. [98] / John Wiley & Sons / CC BY 4.0.

of medicine and environmental protection. The application of PLA in the biomedical field mainly benefits from its excellent mechanical properties, optical transparency, and controllable degradation characteristics. In the biological environment, PLA can gradually decompose into small lactic acid molecules through hydrolysis. These small molecules can be metabolized by the human body into carbon dioxide and water, with almost no toxic side effects [99].

Therefore, PLA has been approved for the manufacture of implantable medical devices, such as surgical sutures, fracture fixation plates, and drug sustained-release systems. In addition, PLA fibers have been developed into photofunctional polymer fibers for optical sensing and biological signal monitoring due to their low light propagation loss and high mechanical strength. PLA-based optical fibers can be used for light transmission in deep tissues to achieve photochemotherapy for tissue repair. The high transparency and moderate refractive index of PLA make it an ideal material for manufacturing waveguides. Research shows that PLA optical fibers can effectively transmit optical signals and achieve optical imaging and treatment in biological tissues. Its thermoplasticity also enables PLA to be easily processed through techniques such as melt extrusion and 3D printing, thereby meeting the demands of customized optical devices (Figure 1.20). In addition, by regulating the molecular weight and the proportion of copolymer monomers of PLA, its degradation rate and mechanical properties can be further optimized to adapt to specific biological application scenarios [100].

Elastomer photofunctional fibers such as poly(octamethylene carbonate-cooxymethylene carbonate) (POC-POMC) and PDMS have shown broad biomedical application potential due to their good mechanical flexibility and optical properties. POC-POMC photofunctional fiber is a kind of degradable elastomer material. Its transmittance of visible light (400–700 nm) is approximately 90%, and the elastic modulus is 0.1–1 MPa. It is suitable for flexible implant devices and shows excellent performance, especially in PDT. It can deliver photosensitizers and precisely act on the target tissues. Meanwhile, the degradation products are nontoxic and meet the safety requirements for in vivo implantation. PDMS photofunctional

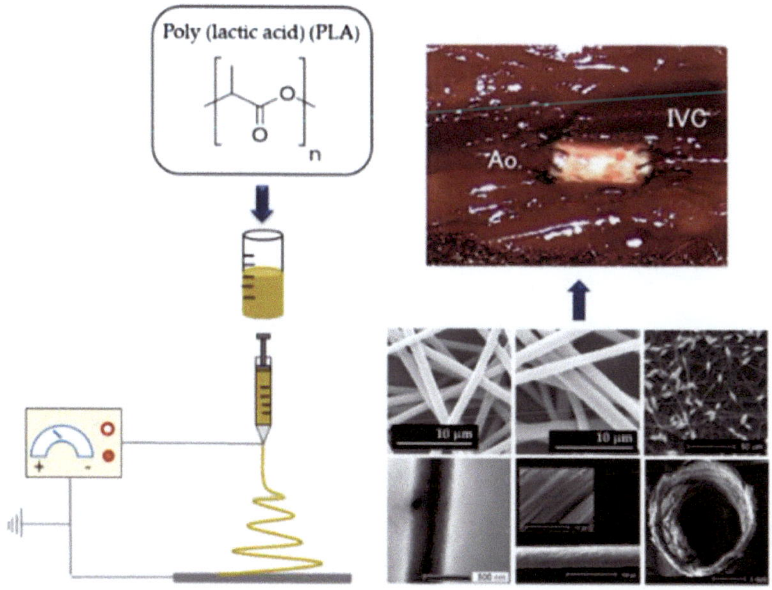

Figure 1.20 Bioapplications of hierarchically structured electrospun PLA materials. *Source:* Maleki et al. [100] / MDPI / CC BY 4.0.

fibers have garnered significant interest in the areas of optical imaging and biosensing due to their light transmittance of over 95% in the 400–800 nm band and low refractive index of approximately 1.4. Studies have shown that microstructured optical fibers based on PDMS have low optical loss (with an attenuation coefficient of approximately 0.2 dB/cm), can achieve efficient optical transmission in biological tissues, and are suitable for real-time monitoring and tissue repair. In conclusion, elastomer photofunctional fibers combine excellent optical performance and mechanical properties, providing efficient solutions for fields such as tissue engineering, phototherapy, and biosensing (Figure 1.21). Through further material modification and structural optimization, they are expected to expand their applications in complex medical scenarios [101].

Bioapplications Photofunctional polymer fibers based on synthetic materials have significant advantages in phototherapy [102]. Optical imaging offers high sensitivity, resolution, and speed, making it highly promising for investigating disease mechanisms, acquiring physiological information, and diagnosing various conditions. The role of photofunctional polymer fibers in biological imaging is mainly reflected in their optical properties of low-loss light conduction and biocompatibility, especially playing a crucial role in deep tissue biological imaging, which makes high-resolution imaging of deep structures in organisms possible. These fibers can penetrate the surface tissue and transfer light to the deep tissue, thereby imaging and analyzing the deep tissue (Figure 1.22) [103].

The citrate-based polymer fiber with photofunctions has been reported to be capable of in vivo imaging and deep tissue light propagation (Figure 1.23). As this kind of

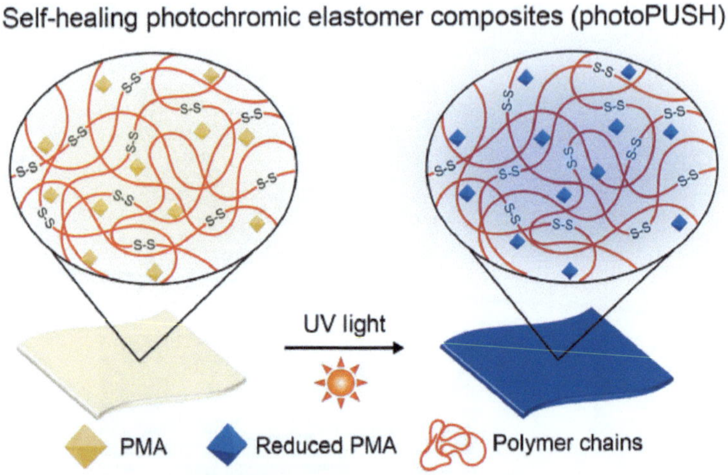

Figure 1.21 Photochromic performance of light-sensitive elastomer fiber. *Source:* Yimyai et al. [101] / with permission of John Wiley & Sons.

fiber is biodegradable and nontoxic, it can remain in the patient's body and naturally degrade after treatment without the need for surgical removal. This not only reduces the patient's pain and medical expenses but also improves the therapy effect. The degradation products of polymers at different concentrations were tested on 3T3 fibroblasts, and it was found that their cytotoxicity was low and there was a significant change compared with the FDA-approved PLGA. Subcutaneous implantation studies in rats revealed no significant difference in the in vivo foreign body response between POC or POMC and PLLA coating of optical fiber, which are widely used in biomedical implants [103].

In the future, research on photofunctional polymer fibers for biological applications will concentrate on substantially improving and optimizing various properties to overcome existing technological and market challenges. Firstly, for photothermal applications, flexible structures, and special load requirements, materials need to achieve breakthroughs in mechanical properties and thermal stability. Additionally, the potential of dye-doped fibers in sensing, wavelength conversion, and transmission applications is worth further exploration. The development of such fibers may become an important direction to fill the market gap in standard optical fiber applications. The development of specific polymer types remains highly uncertain and requires further research. For example, the batch stability of natural polymers, the controllability of the degradation process, and the potential immune response still need further study; meanwhile, the combination of natural and artificial fibers may become a new research trend.

From a long-term development perspective, the research on biocompatible optical fibers will advance in two directions in parallel: one is to develop optical fibers suitable for short-term implantation and that are fully degradable; the second is to design optical fibers that meet the long-term monitoring requirements and have sufficient stability. These advancements will rely on innovations in new materials and

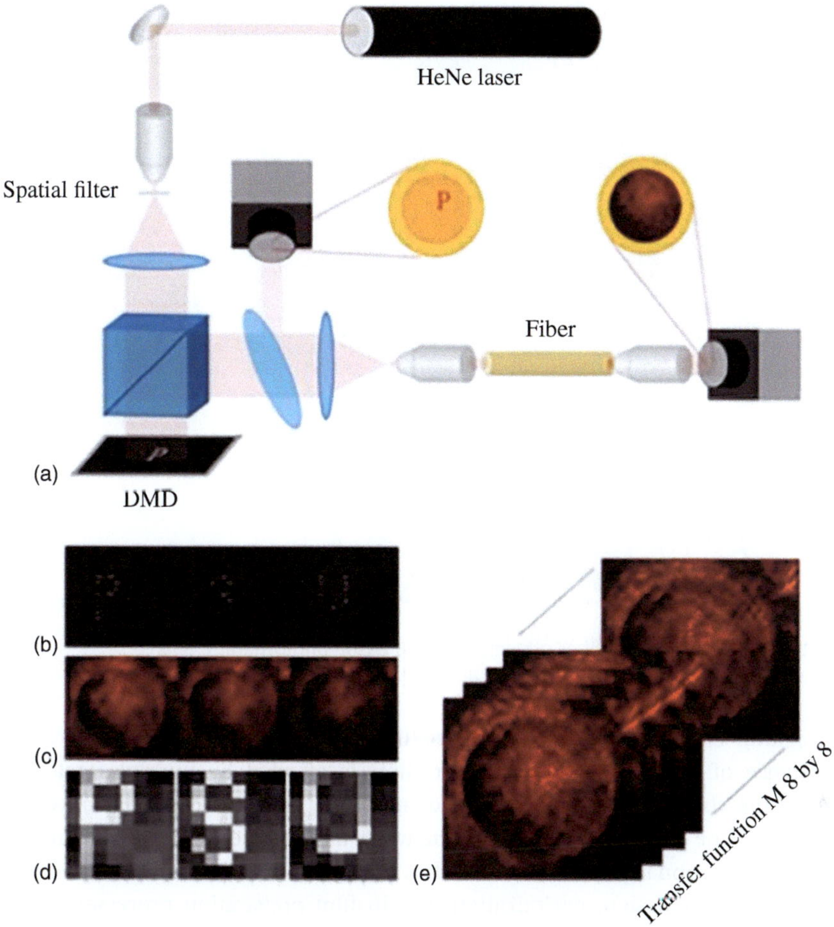

Figure 1.22 Bioimaging and image reconstruction utilizing degradable optical fibers. *Source:* Reproduced with permission from Shan et al. [103] / Elsevier.

manufacturing methods to meet the dual demands of the biomedical field for optical transmission performance and low immune response. Overall, the future of artificial polymer fibers will witness more breakthroughs in fields such as biosensing, optogenetics, and precision medicine through the deep integration of materials science, biomedical engineering, and advanced processing technologies.

1.2.3 Films

Photofunctional polymer films are a type of polymer material with special optical properties. These films achieve their functions by absorbing, emitting, reflecting, or scattering light. Due to their structural diversity and functional tunability, they are widely used in optical devices, biosensing, intelligent displays, solar cells, and biomedical fields. The focus of the research on photofunctional polymer films is to achieve an efficient light response and multifunctionalization of materials

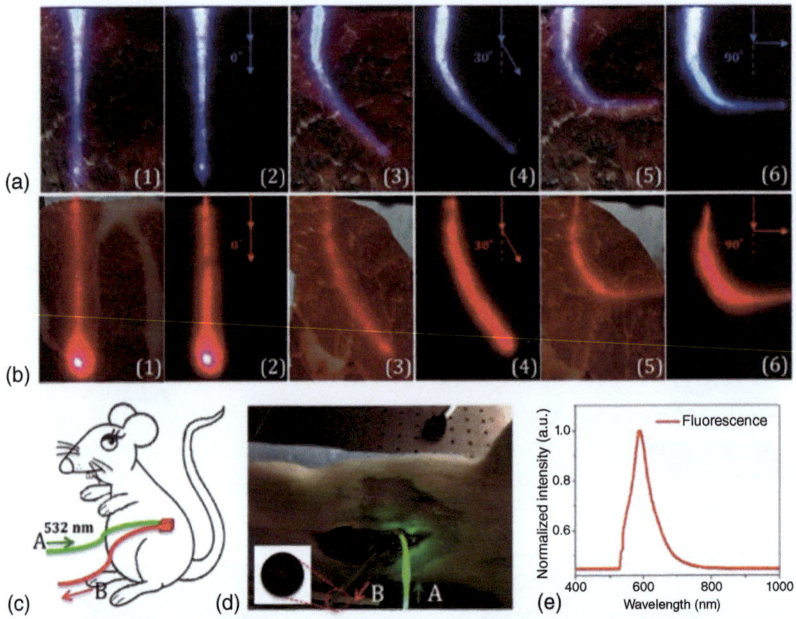

Figure 1.23 Flexible citrate fiber systems for deep tissue phototherapy. *Source:* Reproduced with permission from Shan et al. [103] / Elsevier.

by designing chemical structures and optimizing processing techniques. The performance of photofunctional polymer films is affected by multiple factors, including the chemical structure of the polymer, the thickness and uniformity of the film, and the interfacial interaction between the polymer and the substrate. The optical functional performance of films can be improved through methods such as molecular design, optimization of thin film preparation processes, and posttreatment techniques. For instance, by introducing polymers with specific functional groups, the light absorption efficiency and light stability of the film can be enhanced. By changing the thickness and uniformity of the film, the light scattering and light reflection characteristics of the film can be optimized. The adhesion and mechanical stability of the film can be improved by strengthening the interfacial interaction between the polymer and the substrate.

These films can be classified according to their functions and application fields, mainly including fluorescent films, photochromic films, photoelectric conversion films, and near-infrared (NIR) absorption films. These films have shown great application potential in fields such as biosensors, bioimaging, and drug delivery systems. In the following section, we will provide detailed information about the functional mechanisms, performance characteristics, and preparation methods of photofunctional polymer films and their potential in biological applications.

1.2.3.1 Fluorescent Film

Fluorescent films are materials that can absorb light and emit light of specific wavelengths, usually achieved by doping fluorescent dyes, quantum dots, or

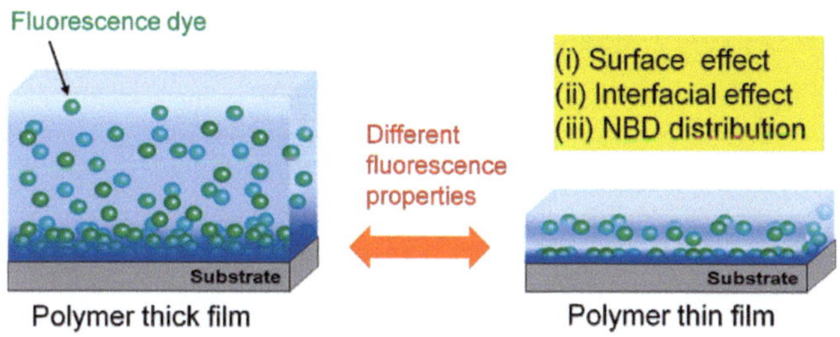

Figure 1.24 Thickness-dependent fluorescence anomalies in fluorescent films controlled by NBD interfacial aggregation. *Source:* Shundo et al. [104] / with permission of American Chemical Society.

luminescent polymers in a polymer matrix [104]. As shown in Figure 1.24, doping the fluorescent dye (6-(*N*-(7-nitrophenyl-2-oxa-1,3-diazol-4-yl)amino)hexanoic acid, NBD) into polymer films of polystyrene (PS) and hydrogenated PS produces unusual fluorescent behavior: the fluorescence lifetime and the maximum emission wavelength increase as the film thickness decreases. This is due to the fact that in thick films, NBD molecules form aggregates at the interface, but this behavior is not observed in thin films.

The mechanism of fluorescent films is mainly based on the emission properties of fluorescent segments, which enable the films to absorb light energy and reemit light of specific wavelengths. Fluorescent films usually contain photoactive components like organic dyes or inorganic quantum dots. When these photofunctional components are illuminated, they absorb light, causing electrons in the ground state to transition to an excited state. As these electrons relax back to the ground state, they release energy in the form of emission [105]. According to the various spin states in the radiation relaxation process, photoluminescence includes two emission types: fluorescence and phosphorescence. The term "fluorescence" refers to light emission that occurs between energy states with the same spin multiplicity, typically lasting no more than about 10 ns. In contrast, "phosphorescence" describes light emission between states with different spin multiplicities, and this process can persist from microseconds to several seconds [106]. The absorption and emission processes of photoluminescence can both be simplified in the Jablonski energy diagram shown in Figure 1.25.

Thin-film fluorescence sensing is an important form of fluorescence sensing. It has the advantages of high sensitivity, good designability, relatively simple instrument structure, low energy consumption, no involvement of radioactive sources, and easy portability, and has attracted special attention. The fluorescent sensing mechanism of the film mainly includes specific binding or reaction between the sensing unit and the detected molecules, as well as the disturbance of the microenvironment of the sensing unit by the detected molecules [107]. The structure of the active layer where the sensing unit is located greatly affects the sensing performance of the thin film, especially the sensing response dynamics. In addition, the fluorescence intensity and optical brightness of fluorescent films largely depend on preparation conditions, such as annealing temperature [108].

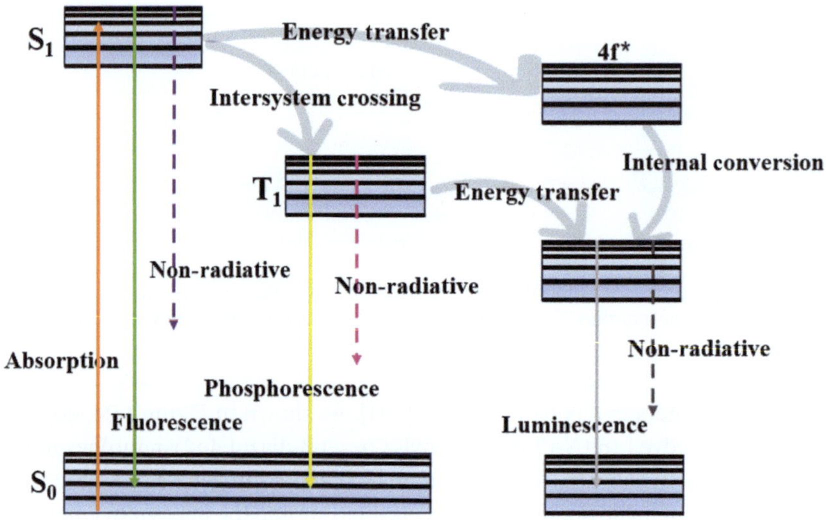

Figure 1.25 Jablonski diagram.

This type of film features high photoluminescence efficiency, spectral tunability, and excellent stability, and is widely applied in biological imaging, sensing, and display technologies. Fluorescent films are often used for real-time tracking of biomolecular dynamics, such as detecting specific proteins, DNA, or small-molecule metabolites. For example, thin films based on the FRET mechanism can be used for highly sensitive biosensing.

1.2.3.2 Photochromic Films

Photochromic functional films refer to thin film materials that can undergo specific chemical reactions under the irradiation of light of a certain wavelength, causing significant changes in their absorption spectra. However, under the irradiation of light of another wavelength or the effect of heat, they can restore their original functions (Figure 1.26) [109]. Due to its excellent optical sensitivity, it has broad application potential in fields such as sensors, erasable rewritable optical discs, camouflage, and anticounterfeiting.

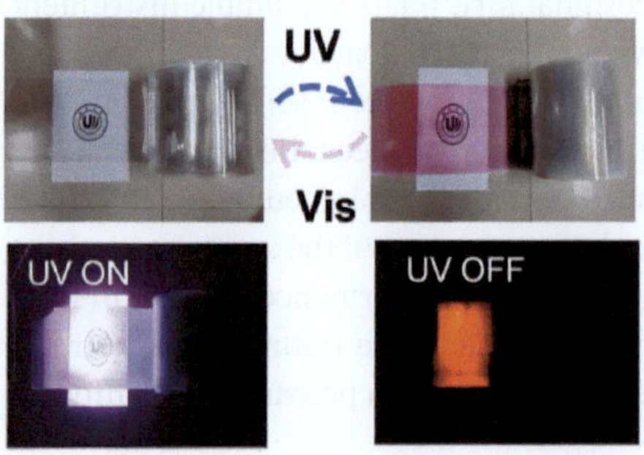

Figure 1.26 Photochromism in photofunctional polymer films. *Source:* Chen et al. [109] / John Wiley & Sons.

Photochromic functional film materials can be grouped into two major categories relying on their main components: inorganic photochromic functional films and organic photochromic functional films. Inorganic photochromic functional films, such as nano-WO_3 films [110], can be prepared by methods like electrodeposition, chemical vapor deposition, and sputtering. At present, the main components of the organic photochromic functional membrane materials under study are generally spiropyran and spirooxazine compounds, azo compounds, chloroacetic anhydride compounds, and diarylethylene compounds [111]. Compared with inorganic photochromic functional films, organic photochromic functional films can adjust their sensitive wavelengths by designing the molecular structure of photochromic compounds and are easy to be mechanically processed into films.

The color-changing mechanism of photochromic materials involves two processes: photoexcitation and structural change. When photochromic materials are exposed to light, the energy of photons is absorbed, causing excited-state transitions in the molecules within the material. This process is achieved through the excitation and transition of electrons within the molecule. When in the excited state, the molecule's electronic structure and energy levels are altered, leading to changes in both the absorption and reflection spectra of the material, thereby resulting in a color change. During the photoexcitation process, the molecular structure of photochromic materials changes, which is the basis of photochromism. The mechanism of photochromic films is mainly based on the photoinduced isomerization of molecules [112]. When these materials are exposed to light, their molecular structures change, thereby altering the light absorption properties of the materials. This structural change is reversible. When the light source is removed, the material will return to its original state. This phenomenon can be represented by the following chemical equation:

$$\text{A (Closed-ring)} + h\nu \rightleftarrows \text{B (Open-ring)}$$

$$\text{Or A (trans)} + h\nu \rightleftarrows \text{B (cis)}$$

Among them, A is the photochromic material, $h\nu$ is the photon energy. A and B can be reversibly converted under light irradiation.

Organic photochromic film materials have been widely studied and applied in some fields, but there are also some problems [113].

1) The synthesis of organic photochromic compounds is difficult, especially for some photochromic heterocyclic compounds, which leads to high costs of thin film materials and limits their application. The current research direction is to find simpler preparation methods to reduce costs.
2) The key problems in the organic light-induced chromic light storage system have not been completely solved and have not yet met the requirements for industrialization. The current research direction is to search for organic photochromic systems with reversible and detectable bistability at room temperature.
3) For most organic photochromic materials, including photochromic films prepared by various methods, the so-called reversible photochromic behavior is not unconditional and infinitely reversible. This is because during the long-term

photochromic cycle, these dye molecules will undergo side reactions such as oxidation and degradation, causing the loss of photochromic components, which in turn leads to the reduction or even loss of the photochromic performance of the material.

The fatigue resistance of photochromic materials is still not high enough. The current research directions include modifying the molecular structure of photochromism, seeking new photochromic varieties with better fatigue resistance, and enhancing their fatigue resistance through measures such as isolation and protection.

This type of film demonstrates significant value in fields such as optical switches, rewritable storage, and smart windows. This type of film features a fast light response speed, high repeatability, and excellent fatigue performance. By embedding photochromic molecules into a polymer matrix, sensors that respond to specific biomolecules can be fabricated for detecting pollutants or biomarkers in the environment. Photochromic films can also be applied for preparing intelligent systems of drug delivery, which control the release rate and time of drugs through light exposure. This system can be artificially controlled to achieve controllable and rapid release of therapeutic drugs at the lesion site (such as tumors), overcoming many shortcomings of traditional drug delivery systems (slow drug release, poor selectivity, low drug utilization rate, etc.), thereby significantly reducing damage to normal cells. For instance, films designed through azobenzene derivatives can achieve light-controlled release of drugs, effectively enhancing the targeting and safety of treatment [114].

1.2.3.3 Photoelectric Conversion Thin Film

Photoelectric conversion films are materials that harness the photovoltaic effect to directly transform solar energy into electrical energy. Their working principle mainly involves processes such as light absorption, electron excitation, and charge separation. In photoelectric conversion devices, when light illuminates the surface of a photoelectric conversion device, the semiconductor material within the device absorbs the energy of the incoming photons. In semiconductors, this photon energy can excite electrons, causing them to move from the valence band to the conduction band and thereby generate electron-hole pairs [115]. This process is the initial stage of photoelectric conversion and also the key step in converting light energy into electrical energy. Secondly, the excited electron-hole pairs will move in the semiconductor, forming electron and hole flows. Through appropriate structural design and material selection, electrons and holes can be collected on different electrodes, respectively, thereby forming voltage differences and currents. This is how light energy is converted to electrical energy.

Such films are usually prepared using π-conjugated polymers (such as polythiophene, polystyrene derivatives) or organic–inorganic composites (Figure 1.27) [116]. It features high photoelectric conversion efficiency, a wide spectral absorption range, and stable electrical performance.

Photoelectric conversion films can be used in biosensors and bioenergy collectors. By converting light energy into electrical energy, they achieve the detection of

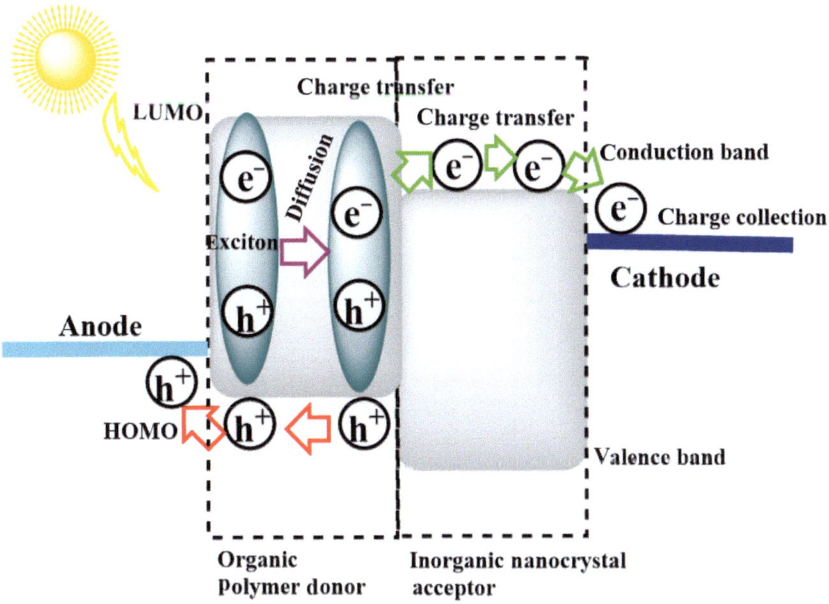

Figure 1.27 Schematic illustration of the working mechanism of organic-inorganic hybrid solar cells.

biological signals and the collection of energy. Flexible photonic devices can achieve implantable biosensing, imaging, and therapy by using biocompatible or biodegradable materials. These films bring more innovation and progress to medical health. In addition, photoconductive films have a broad spectrum of potential applications in the biomedical field. In the field of medical devices and equipment, they are one of the materials for the sensitive layer of biomedical sensors, and can also be used as a medical coating and electrode for electrophysiological monitoring equipment. In the field of diagnosis and imaging, they can be used in the key parts of photoacoustic imaging probes and also in optical coherence tomography components. In tissue engineering and regenerative medicine, they can serve as a cell culture platform and are also intelligent tissue engineering scaffold materials. Photoconductive thin films provide strong support for biomedical diagnosis, treatment, monitoring, and research.

1.2.3.4 Near-infrared (NIR) Absorption Films

The mechanism of near-infrared (NIR) absorption films is based on the material's absorption characteristics of near-infrared light. This type of film can absorb light in the near-infrared region (700–2500 nm) and convert it into other forms of energy, such as thermal energy. Near-infrared absorption films have extensive applications in fields such as solar energy utilization, thermal management, night vision equipment, and military stealth technology. Common NIR absorption film materials include organic materials and inorganic materials. Organic materials, such as some organic dyes and polymers, have good absorption performance and

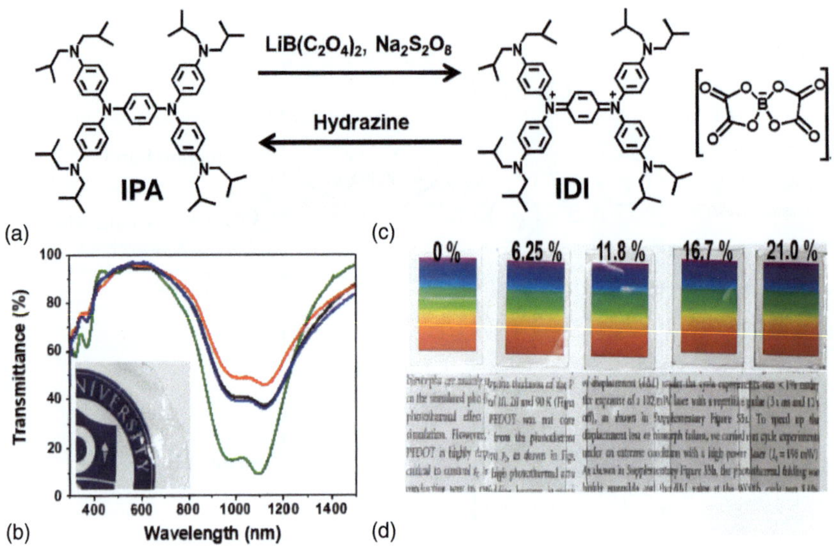

Figure 1.28 Optical properties of NIR-absorbing polymer film. *Source:* Reproduced with permission from Han et al. [117] / John Wiley & Sons.

processability. For instance, certain organic small molecule dyes can precisely absorb near-infrared light within a specific wavelength range (Figure 1.28) [117]. Inorganic materials: Semiconductor materials such as silicon, germanium can also be used as materials for NIR absorption films [118]. These materials can optimize their absorption performance by adjusting their chemical composition and crystal structure.

The mechanism of near-infrared absorption films can be explained in detail from the following aspects:

1) *Intramolecular charge transfer (ICT):* Some organic molecules have a strong absorption capacity in the near-infrared region, which is the result of ICT. This charge transfer can occur between different parts of a molecule, such as the donor–acceptor system, where electrons are transferred from the donor part to the acceptor part, leading to changes in the molecular polarizability and thereby enhancing the absorption of near-infrared light.
2) *Extension of the π-conjugated system:* By extending the π-conjugated system of a molecule, the absorption of near-infrared light by the molecule can be increased. In a π-conjugated system, electrons can move freely within the molecule. When the molecule absorbs near-infrared light, these electrons can be excited to higher energy levels, thereby achieving the absorption of light.
3) *Surface plasmon resonance (SPR) of metal nanostructures:* In metal nanostructures, SPR is a collective oscillating electron movement. When the resonant frequency of SPR of the metal matches the frequency of the incident light, enhanced light absorption occurs. This phenomenon can be used to design near-infrared absorption films. By adjusting the size, shape, and arrangement of the nanostructure, its absorption characteristics for near-infrared light can be controlled.

4) *Doping of organic materials:* Doping organic materials with specific chemical substances can modify their optical properties, increase their absorption of near-infrared light, and thereby effectively suppress infrared radiation.
5) *Thickness and structure of the film:* The thickness and microstructure of the film also affect its absorption of near-infrared light. By optimizing the thickness of the film structure, high-transparency infrared stealth films can be achieved, thereby effectively suppressing high-temperature infrared radiation.

NIR absorption films have numerous applications in the biomedical field: In PTT, NIR absorption films can convert the absorbed near-infrared light into thermal energy, which is used to kill tumor cells. For instance, polymer films doped with gold nanorods can be used in PTT, achieving efficient local heating and ablation of tumors through NIR laser irradiation [119]. NIR absorption films can also be used in drug-controlled-release systems, triggering drug release through NIR light exposure [120].

In conclusion, as a new type of polymer material, photofunctional polymer films have become a research hot spot in the intersection of materials science and biomedicine due to their unique optical properties and adjustability. From the highly sensitive biosensing of fluorescent films to the intelligent drug release of photochromic films, and then to the clean energy conversion of photoelectric conversion films and the precise PTT of near-infrared absorption films, these materials have achieved a leap from basic research to practical application through molecular design and process optimization. Their core advantage lies in achieving multifunctional integration through the precise regulation of chemical structures: Fluorescent films achieve real-time monitoring of biomolecules by mechanisms such as FRET. Photochromic films realize photocontrolled drug release through molecular isomerization. Photoelectric conversion films expand the spectral absorption range by means of π-conjugated systems. Near-infrared absorption films enhance the photothermal conversion efficiency by SPR. These characteristics not only break through the performance bottlenecks of traditional materials but also provide innovative solutions for biomedical engineering. However, the technological transformation of photofunctional polymer films still faces challenges. Problems such as the insufficient fatigue resistance of organic photochromic materials, the biocompatibility optimization of near-infrared absorption films, and the improvement of the stability of photoelectric conversion devices need to be solved urgently.

1.2.4 Hydrogel

Hydrogel, as a material with controllability, flexibility, and biocompatibility, is gradually becoming an important component of photofunctional composites [121, 122]. Traditional hydrogel is usually composed of natural polymers, for example, gelatin, chitosan, and sodium alginate. Hydrogel can be obtained from natural or synthetic polymers, such as polysaccharides and polyvinyl alcohol. These materials can form a three-dimensional network structure after absorbing water and have good swelling, elasticity, and mechanical properties [123, 124].

Hydrogels, with their unique physicochemical properties, are opening up new directions for the research of photofunctional composites. These materials are composed of a three-dimensional network structure, endowing them with a special swelling regulation ability [125]. Simply put, they can change their volume flexibly like a sponge according to the variations in the humidity, temperature, or pH of the surrounding environment. Meanwhile, hydrogels also have excellent flexibility. Even if they are repeatedly stretched and bent, they will not easily lose their original performance. This feature enables them to imitate the mechanical properties of biological tissues. In addition, hydrogels have excellent biocompatibility and have played a significant role in the biomedical field. Whether used as carriers for drug delivery or scaffolds for tissue engineering, they have shown broad application prospects [126]. From the perspective of practical application, gelatin-based hydrogels can provide a favorable growth environment for cells and promote cell adhesion and proliferation [127]. Chitosan hydrogels have both antibacterial and wound healing acceleration functions. Synthetic polymer hydrogels can achieve specific properties through precise molecular design, meeting various scientific research and application requirements.

Photofunctional hydrogels have become an emerging research trend in materials science, chemical engineering, and biomedicine over the past few years due to their unique optical properties and good processability [128]. Photofunctional hydrogels can achieve efficient processing and conversion of optical signals, and have the advantages of fast response speed, high sensitivity, and strong tunability. Furthermore, the three-dimensional network structure of hydrogels gives them unique advantages in fields such as optical sensing and optoelectronic devices [129].

Among them, the research on photofunctional hydrogel began with the exploration of light-responsive polymers, especially in applications such as light-induced deformation, swelling changes, and photocatalytic reactions. Early research focused on the molecular design of light-responsive polymers, such as the photoisomerization of azobenzene polymers, to achieve color switching and shape alteration of materials under light exposure [130]. Hydrogels have gradually demonstrated their unique intelligent responsiveness. They can respond to specific combinations of light intensities and wavelengths, and even achieve coordinated stimulus responses in multiple physical fields (light, electricity, magnetism, and force), ushering in the era of material intelligence [131].

The size effect opens up a new dimension for the performance regulation of photofunctional hydrogels. Nanoscale photofunctional hydrogels, by leveraging the quantum size effect, have a photoluminescence efficiency several times higher than that of traditional materials. Their surface plasmonic resonance peaks can be fully covered from ultraviolet to near-infrared through size adjustment, providing the possibility for optical imaging of deep tissues in organisms. However, the extremely high surface energy brought by the nanoscale size makes them prone to agglomeration during the preparation process and they have poor long-term stability. Photofunctional hydrogels with their moderate size have good processability and mechanical stability while ensuring certain optical properties and are suitable for the preparation of flexible optoelectronic device arrays [132]. The bulk photofunctional hydrogels achieve the improvement of the uniformity and stability of optical performance through the design of a three-dimensional network structure at the macroscopic scale. Multiple photofunctional units can be integrated inside to construct a complex optical signal system, meeting the strict requirements for high performance and high reliability of materials in high-end optical sensing and biomedical imaging equipment [133, 134].

1.2.4.1 Structure and Composition

Hydrogels are a kind of three-dimensional cross-linked polymer network system with macroscopic dimensions. Polymer chain segments form its internal network structure through either chemical cross-linking (such as covalent bonds) or physical cross-linking (including electrostatic interactions, hydrogen bonds, hydrophobic interactions, and more). This structure enables it to absorb water, expand, and form a stable structure similar to a colloid, while remaining insoluble. It can undergo significant and reversible volume changes between the dry and water-saturated states, providing a basis for the dynamic response of the material in different application scenarios. This process is not a simple physical adsorption but involves a complex thermodynamic equilibrium. The degree of swelling can be precisely regulated by parameters such as cross-linking density, the hydrophilicity of the polymer chain, and the osmotic pressure of the external environment [135, 136].

Hydrogel bulk materials can be grouped into natural and synthetic polymer hydrogels according to their composition [137]. Natural polymers include gelatin, chitosan, and alginate; synthetic polymers include polyvinyl alcohol and polymethacrylate.

From the perspective of material composition, natural and synthetic polymer hydrogels exhibit completely different performance characteristics. Gelatin is a protein extracted from animal tissues, featuring excellent biocompatibility and biodegradability. Its abundant functional groups, such as amino and carboxyl groups, can participate in cross-linking reactions. It can also undergo sol-gel transformation at body temperature. This characteristic enables temperature-triggered drug release in drug-controlled-release systems. Chitosan is the deacetylation product of chitin. As a natural cationic polysaccharide, it carries a positive charge in an acidic environment due to the protonation of its amino groups. It can not only interact electrostatically with negatively charged biomolecules but also exhibit broad-spectrum antibacterial activity. It is often used in wound dressings and tissue engineering scaffolds. Alginate is a polysaccharide extracted from brown algae that can form hydrogels in the presence of metal ions (such as Ca^{2+}) and is widely used in the field of biomaterials [138].

Synthetic polymer hydrogels include polyvinyl alcohol and polymethacrylate. The physical cross-linked network formed by PVA through the freeze-thaw cycle has high strength, high elasticity, and good chemical stability, and is often used to prepare gel sensors [139]. Polymethacrylate hydrogels, due to the modifiers of their side chains, can achieve light-controlled performance adjustment by grafting light-responsive groups [140].

1.2.4.2 Physical Properties

These hydrogels can exhibit adjustable physical properties, such as mechanical strength and swelling properties, under different environmental conditions. In terms of mechanical strength, the mechanical strength of the hydrogel can be significantly regulated by changing the cross-linking density, the molecular weight of the polymer chain, and the regularity of the network structure. For example, increasing the dosage of chemical cross-linking agents or enhancing the degree of physical cross-linking can make the hydrogel network tighter, thereby improving its mechanical properties, such as tensile and compressive resistance [141]. For swelling, the behavior of the hydrogel is influenced by the free volume of the network structure and the interaction between the polymer chains and the solvent molecules. In the solvent, increased free volume in the hydrogel network and strengthened affinity between the polymer chains and solvent molecules raised the swelling degree [142].

Hydrogel can respond to external environments like pH, temperature, and light. Take the temperature response as an example. Some hydrogels contain polymer segments that are sensitive to temperature, such as PNIPAM. When the environmental temperature changes, the conformation of the polymer chain will change, thereby causing the swelling or contraction of the hydrogel. In terms of pH response, the acidic or basic functional groups (such as carboxyl groups and amino groups) contained in the hydrogel material will undergo protonation or deprotonation reactions as the pH value of the solution changes [143]. This reaction will alter the charge distribution and intermolecular forces within the hydrogel, thereby causing changes in the hydrogel's volume. In addition, some hydrogels also have responsiveness to specific ions. When specific ions are present in the solution, the hydrogel will specifically bind to the ions, resulting in changes in the physical properties of the hydrogel [144].

In the field of photofunctional composites, hydrogel bulk materials can achieve light-responsiveness by embedding light-responsive molecules, such as photochromic molecules and photosensitive polymers. Photochromic molecules can undergo reversible structural changes after absorbing light of specific wavelengths, thereby altering the optical properties of the hydrogel, such as color and absorption spectra. Photosensitive polymers can undergo reactions such as cross-linking, de-cross-linking, or the movement of molecular chains under the irradiation of light, thereby altering the physical properties of the hydrogel, like mechanical strength and swelling degree. This light-responsiveness enables hydrogel bulk materials to impart them with remarkable flexibility and functional capabilities for biomedical uses [145–148].

1.2.4.3 Fluorescent Hydrogel

Fluorescent hydrogel is an advanced material that combines a three-dimensional network structure with fluorescent properties. It demonstrates unique optical response capabilities by embedding fluorophores into the hydrogel network. This material not only inherits the flexibility and environmental adaptability of traditional hydrogels but also possesses the functions of real-time monitoring and

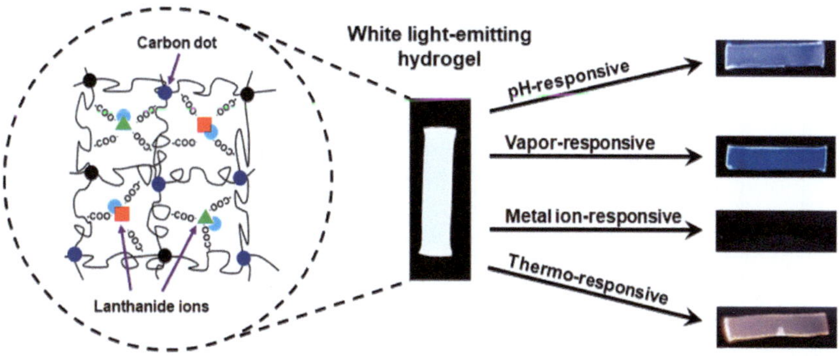

Figure 1.29 Multistimuli-responsive white-light hydrogel. *Source:* Reproduced with permission from Zhu et al. [149] / American Chemical Society.

signal feedback due to its fluorescent properties. The fluorescence behavior of fluorescent hydrogel can be regulated by external stimuli (Figure 1.29), enabling it to exhibit high adaptability and functionality in a variety of complex environments. The emergence of this material has brought new opportunities and challenges to fields like materials innovation, biomedicine, and information technology [149].

The emergence of fluorescent hydrogels stems from the demand for the expansion of the functions of traditional hydrogels. In the field of biomedicine, fluorescent hydrogels can be applied in cell labeling, drug delivery, and tissue engineering, helping researchers gain a deeper understanding of the physiological and pathological processes within organisms. In environmental monitoring, fluorescent hydrogels can detect pollutants in water bodies, providing technical support for environmental protection. In the field of information storage and encryption, the fluorescence signals of fluorescent hydrogels can be used to encode and decode information, achieving anticounterfeiting and information security. In addition, fluorescent hydrogels can also be used as smart materials to manufacture shape memory materials and self-healing materials, demonstrating great application potential [150, 151].

1.2.4.4 Photochromic Hydrogel

Photochromic hydrogel is a kind of smart material that can undergo reversible color changes under ultraviolet or visible light irradiation. This material combines the softness of hydrogel and the unique optical properties of photochromic molecules, making it show broad application prospects in a variety of fields. The color change mechanism of photochromic hydrogels is usually based on the conversion of photochromic molecules between different oxidation states or configurations, with Figure 1.30 illustrating the phase transition mechanism triggered by photothermal stimuli. For instance, inorganic materials such as tungstate and molybdate, as well as organic materials like spiropyrane and diarylethylene, undergo a transformation from colorless to colored under light exposure. This property makes them ideal optically responsive materials [152]. Significant advancements have been made in the investigation of photochromic hydrogels in recent years, especially in terms of mechanical properties, transparency, and response speed.

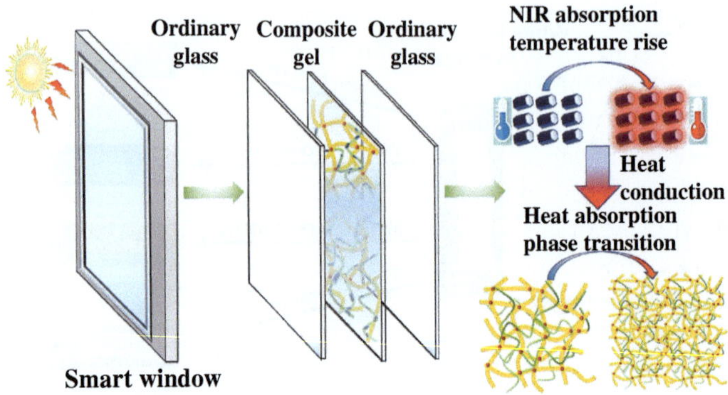

Figure 1.30 Phase transition mechanism of composite hydrogels triggered by photothermal stimuli. *Source:* Zhang et al. [152] / with permission of Elsevier.

The application scope of photochromic hydrogels is extensive, covering multiple fields from basic research to practical applications. In the field of smart windows, photochromic hydrogels can automatically adjust the light transmittance according to the light intensity, thereby achieving energy conservation and comfort [153]. In the biomedical field, photochromic hydrogels can be used for biological imaging and drug release, achieving precise drug delivery and cell labeling through light exposure [154]. In anticounterfeiting technology, photochromic hydrogels can be used to produce highly secure anticounterfeiting labels that display or hide information through light exposure. In addition, photochromic hydrogels can also be used in flexible electronic devices, such as wearable sensors and wireless communication devices, to achieve optical camouflage and information transmission [155]. These applications demonstrate the significance and potential of photochromic hydrogels in modern technology.

1.2.4.5 Photothermal Hydrogel

In today's society, the contradiction between energy demand and environmental protection is becoming increasingly prominent, and at the same time, the biomedical field is also facing many challenges. The emergence of photothermal hydrogel is timely. It can utilize renewable energy sources such as solar energy to achieve efficient photothermal conversion, providing a green and sustainable approach to solving energy problems [156]. In addition, in the field of biomedicine, photothermal hydrogels can be used for precise cancer treatment, antibacterial, wound healing, etc. Their noninvasive nature and high efficiency make them a highly promising therapeutic approach. For example, photothermal hydrogel can achieve local temperature rise through irradiation with near-infrared light, thereby killing cancer cells or bacteria while avoiding damage to the surrounding normal tissues [157]. In addition, photothermal hydrogels can also be used in the development of smart materials, such as photothermal-driven soft robots and smart sensors, providing a new material basis for the development of artificial intelligence and flexible electronics [158].

The realization of photothermal hydrogel primarily depends on the synergistic interaction between photothermal conversion units and hydrogel matrices. Photothermal conversion agents such as metal nanoparticles, semiconductor materials, and carbon-based materials can absorb light energy and convert it into thermal energy. These photothermal conversion agents are uniformly dispersed in the hydrogel matrix, such as polyvinyl alcohol and sodium alginate to form composite materials with photothermal conversion functions. When the photothermal hydrogel is exposed to light, the photothermal conversion agent absorbs light energy and generates heat. The heat is conducted through the hydrogel matrix, achieving a local temperature rise. For example, Figure 1.31 shows ultra-flexible

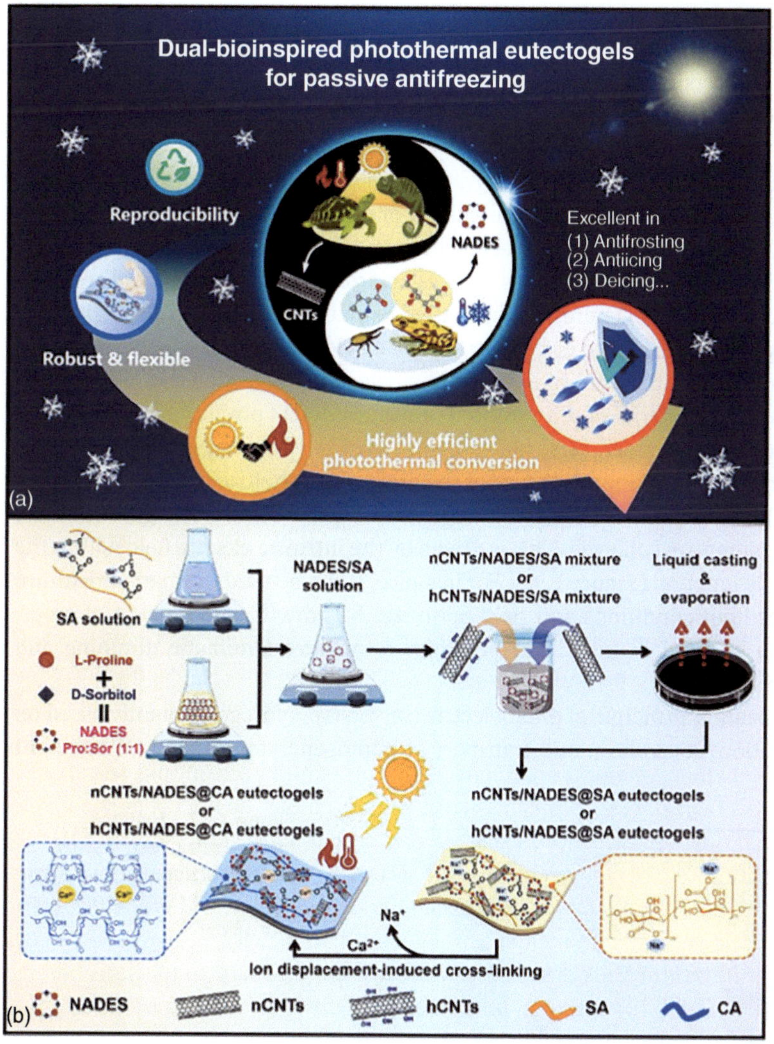

Figure 1.31 Ultra-flexible photothermal eutectogels with passive antifreezing. *Source:* Tian et al. [159] / Elsevier / CC BY 4.0.

photothermal eutectogels with passive antifreezing properties. These materials are being optimized through improved preparation processes to enhance performance and stability. Meanwhile, more in vivo experiments and clinical studies are also being carried out to verify the safety and efficacy of the photothermal hydrogel [159].

1.2.4.6 Photoelectric Conversion Hydrogel

Photoelectric conversion hydrogel is a special material capable of converting light energy into electrical energy. It is usually composed of a matrix with photosensitive properties and components capable of generating charge separation and transport. The core advantage of this material lies in its ability to directly convert light energy into electrical energy, enabling a self-power supply without the need for an external power source. It has broad application prospects in fields such as energy conversion, sensor technology, and intelligent devices, providing new possibilities for achieving efficient and sustainable energy utilization and intelligent systems.

With the development of technology, people's demand for energy is constantly increasing, and at the same time, the pursuit of environmentally friendly energy is becoming increasingly urgent. The acquisition and use of traditional energy sources are often accompanied by problems such as environmental pollution and resource depletion. In the field of energy, it can be used to manufacture solar cells, efficiently converting solar energy into electricity to provide clean energy for households and industries [160]. In the field of sensors, photoelectric conversion hydrogel can be used as the core material of self-powered sensors to detect physical quantities such as light intensity, temperature, and pressure [161]. By doping photoacid in the hydrogel diode and taking advantage of the property of photoacid releasing protons under light, the transmission of photogenerated carriers and the generation of electrical energy are achieved. This is driven by the intrinsic electric field within the hydrogel PN junction (Figure 1.32). For instance, it can be used in smart agriculture to monitor light conditions and help optimize the growth environment of crops. It can also be used in intelligent buildings to achieve automatic dimming and energy-saving control.

The realization principle of photoelectric conversion hydrogel is mainly based on the generation, separation, and transport of photogenerated carriers. When light

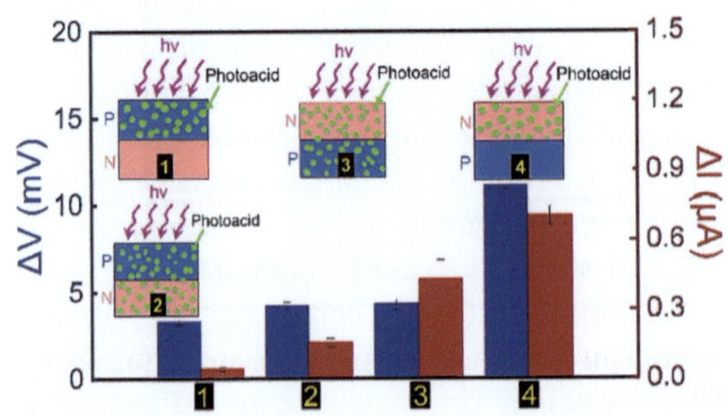

Figure 1.32 Hydrogel photoelectric response by photoacid doping. *Source:* Wu et al. [161] / with permission of Elsevier.

illuminates the hydrogel material, the photosensitive matrix absorbs photon energy, causing electrons to be excited from the valence band to the conduction band and forming electron-hole pairs. These photogenerated carriers are then separated and move in opposite directions under the influence of an internal electric field or concentration gradient within the material. By introducing appropriate electrode materials into the hydrogel, these migrating carriers can be collected, thereby generating current and voltage between the electrodes and achieving the conversion of light energy to electrical energy. For example, in ionic thermoelectric hydrogels, the absorption and conversion of light energy are achieved by combining photothermal materials and ionic thermoelectric effects. Photothermal materials convert light energy into thermal energy, generating a temperature gradient, which in turn drives the migration of ions in the hydrogel and generates a thermal voltage [162]. The photosensitive hydrogel utilizes the synergistic interaction between photothermal and thermoelectric effects. Through the photothermal layer, light energy is converted into heat, and then the thermoelectric layer converts heat into electrical energy, achieving the "photothermal—electrical" cascade conversion [163].

Furthermore, in the field of medical and healthcare, photoelectric conversion hydrogels can be used to develop wearable health monitoring devices, such as self-powered heart rate monitoring wristbands and blood glucose monitoring patches, providing convenience for people's health management [164].

1.2.4.7 Light-controlled Release Hydrogel

Light-controlled release hydrogel is an intelligent material that can achieve the on-demand release of drugs or other substances through light stimulation. The core advantage of this material lies in its precise control ability, which enables the release of drugs within a specific time and space, thereby enhancing therapeutic effects and reducing side effects. The emergence of light-controlled release hydrogels has brought brand-new solutions to fields such as drug delivery, biomedical engineering, and environmental science.

The application scope of light-controlled release hydrogels is extensive, covering multiple fields such as biomedicine, environmental science, and materials science. In the field of biomedicine, light-controlled release hydrogels can be applied in targeted drug delivery, cancer treatment, tissue engineering, and wound healing. In traditional drug delivery systems, achieving precise control over drug release is often challenging, which results in the accumulation of drugs at nontarget sites and increases the risk of side effects. Furthermore, for some diseases that require long-term or regular drug administration, traditional methods are difficult to meet the demands. The light-controlled release hydrogel achieves the precise spatiotemporal release of drugs through light exposure (Figure 1.33), solving these problems. For instance, in cancer treatment, light-controlled release hydrogels can precisely deliver drugs to the tumor site and release them on demand under light, thereby minimizing damage to normal tissues. By controlling drug release through light exposure, precise treatment can be achieved, reducing drug waste and side effects. In environmental science, light-controlled release hydrogels can

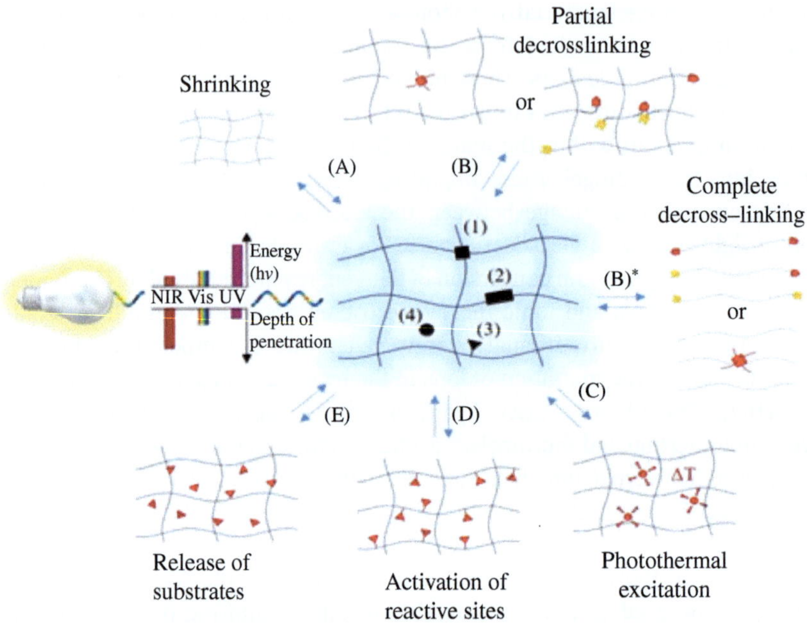

Figure 1.33 Light-activated release in hydrogels. *Source:* Li et al. [146] / with permission of John Wiley & Sons.

be used for the controllable release and degradation of pollutants, enhancing the efficiency of environmental remediation. In addition, light-controlled release hydrogels can also be used in the design of smart materials, such as light-controlled sensors and optical drivers, providing new ideas for the development of smart devices [145, 146].

Hydrogels have emerged as a prominent research focus in the biomedical field due to their excellent physicochemical properties and tunability in photo-functional composites. Hydrogel bulk materials have shown great potential in drug delivery, gene therapy, tissue engineering, PTT, and other aspects. With the continuous in-depth research on photofunctional materials, the biological applications of hydrogels will be further expanded, providing more advanced and efficient solutions for future intelligent medical care and precise treatment. There are numerous types of photofunctional bulk hydrogels, covering multiple fields such as photoresponse, photocatalysis, optical sensing, and photocontrolled drug release. These hydrogels have demonstrated great application potential in fields such as biomedicine, environmental protection, sensors, and smart materials through different photoresponsive mechanisms, combined with advanced optical materials like metal nanoparticles, quantum dots, and photoactive molecules. With the deepening of research, photofunctional bulk hydrogels will continue to play an important role in more precise light regulation and intelligent response materials. In intelligent manufacturing, it helps flexible manufacturing processes such as light printing and light molding [165].

1.3 Prospects for Bioapplications of Photofunctional Polymer Composites

Light, as an external source of stimulation, has many unique advantages. Its noninvasive nature allows it to penetrate biological tissues and reach the target area without disrupting the integrity of the surrounding tissues, providing a completely new approach to the diagnosis and treatment of diseases. At the same time, light also has high spatial resolution and precise time controllability, which makes it possible to achieve precise control of the behavior of light-responsive materials by adjusting the wavelength, intensity, and time of light irradiation [166]. Based on the light-responsive properties, photofunctional polymer composites, as an emerging research hot spot at the intersection of materials science and biomedicine, have received a great deal of attention recently. The ability of such materials to exhibit tunable physical, chemical, or biological responses to external light stimuli has brought new therapeutic and diagnostic tools to the biomedical field. By cleverly combining optical properties with other key functions such as drug release, tissue repair, and cell induction, photofunctional polymer composites have shown great potential for applications in a variety of cutting-edge fields such as precision medicine, smart drug delivery, noninvasive bio-imaging, and cancer therapy [167–169].

With the continuous progress of nanotechnology and the fine design of molecular structures, the microstructural design of photofunctional polymer composites has become increasingly delicate, providing a solid material foundation for their precise response to light stimulation [170]. The nanoscale material structure can significantly enhance the interaction between light and matter and improve the efficiency of light energy conversion [168]. The rapid development in the field of photoelectrochemistry further expands the application scope of photofunctional polymer composites, providing new ideas for their applications in biosensing [171]. In-depth research in biomaterials science has ensured the good compatibility, stability, and degradability of photofunctional polymer composites in living organisms, enabling them to safely and effectively serve biological applications.

The multifunctional integration of photofunctional polymer composites is also a core advantage in their biological applications. These materials are capable of combining various therapeutic means, such as photosensitive drug release, PDT, and PTT, to achieve synergistic therapeutic effects. The application in cancer treatment is especially remarkable. Cancer treatment has long faced challenges such as limited efficacy and significant side effects. Photofunctional polymer composites can integrate multiple therapeutic modes, among which PTT uses the material's absorption of light to convert light energy into heat energy, generating high temperatures locally to destroy tumor cells; PDT induces apoptosis of tumor cells by generating ROS through light excitation; and light-controlled release of chemotherapeutic drugs further enhances the killing effect on tumor cells. These three therapeutic modes synergize with each other, overcoming the limitations of a single therapeutic means and providing a brand new strategy for cancer treatment [172, 173].

Photofunctional polymer composites, with their unique light-responsive properties, intelligent responsiveness, and integrated multifunctionality, show a broad application prospect in the biological field. With the continuous development and improvement of related technologies, photofunctional polymer composites are expected to move from basic research to clinical applications, bringing revolutionary breakthroughs in precision medicine, disease diagnosis and treatment, and opening up a new chapter in the field of biomedicine.

In addition to the above applications in the biomedical field, by integrating photofunctional polymer composites with biological components (such as enzymes, cells, or tissues), efficient light energy conversion and intelligent environmental repair can be realized. In the field of light response, such systems not only mimic and enhance natural photosynthesis but also develop novel light-driven drug delivery systems and environmental sensing technologies, demonstrating the potential for interdisciplinary innovation. Although challenges remain in terms of stability and safety, their breakthroughs in the fields of energy, healthcare, and environmental protection provide a brand new technological path for sustainable development.

1.3.1 Biological Detection and Imaging

Biological detection and imaging technologies play a crucial role in modern medicine, drug development, and biological research. They help us to access the states and changes of molecules, cells, tissues, or organs in living organisms through different physical, chemical, and biological principles. These technologies usually combine physical principles such as optics, magnetism, and acoustics to realize real-time, noninvasive observations of structures and functions within organisms. With the continuous development of technology, these imaging techniques will play an increasingly important role in disease diagnosis, therapeutic monitoring, and drug development. For example, fluorescence imaging is based on the phenomenon that molecules emit light in the direction of long wavelengths after absorbing light energy. By exciting fluorescent molecules (e.g. dyes, quantum dots) in a sample and emitting fluorescence when they return to their ground state, the fluorescent signal is used to detect the presence and location of the target molecule. Different fluorescent dyes can emit different wavelengths of light, thus enabling multiple labeling and imaging [174]. By virtue of their special optical properties and fine design, photofunctional polymer composites can enhance the signal strength and contrast during biological detection and imaging through various mechanisms such as light response, surface enhancement effect, and quantum effect, thus significantly improving the sensitivity, resolution, and accuracy of the detection.

1.3.1.1 Fluorescence Imaging

Fluorescence bioimaging technology has found many applications in biomedical research due to its high sensitivity, noninvasiveness, and real-time imaging capability. However, traditional fluorescence imaging techniques have some limitations, such as monochromatic fluorescence being easily interfered with by biological background fluorescence, resulting in insufficient imaging accuracy.

In addition, fluorescent probes have poor intracellular stability, especially under ultraviolet irradiation, which is prone to photobleaching, limiting their application in long-term imaging.

Conjugated polymers, characterized by a rigid conjugated backbone and delocalized electronic structure, have side chains that can be modified and easily combined. These properties make them excellent fluorescent probes with outstanding light stability, strong fluorescence signals, and good biocompatibility [175].

Recently, two-photon fluorescence imaging (TPFI) has garnered a lot of attention from researchers as a noninvasive method for examining intricate biological issues. Deeper penetration, reduced autofluorescence, less scattering, and less phototoxicity are some of the major benefits of TPFI over traditional single-photon excitation methods because it is typically driven by a near-infrared excitation (700–1000 nm). The amphiphilic triblock copolymer F127 was used as a matrix to wrap the two-photon absorption (TPA) material PFBT to form micelles, and a protective shell layer of silicon dioxide was further grown on its surface (Figure 1.34a). The obtained PFBT-F127-SiO_2 NPs exhibited good dispersion, colloidal stability, good biocompatibility, a broad TPA cross-section, and a high quantum yield. The large TPA cross-section permits high imaging resolution to be achieved with less excitation power, further minimizing the photodamage of excitation light on fluorescent materials and biological targets. As shown in Figure 1.34b, high-resolution visualization of the deep cerebral vascular system can be achieved by intravenous administration in live mice, suggesting that PFBT-F127-SiO_2 NPs are useful TPA probes for vasculature imaging in vivo [176].

Furthermore, by co-encapsulating far-red/near-infrared (FR/NIR)-emitting conjugated polymer PFVBT, poly(lactic-*co*-glycolic-acid)-poly(ethylene glycol)-folate (PLGA-PEG-FOL), and PLGA, the resulting CP nanoparticles can serve as in vivo targeted fluorescent and magnetic resonance imaging dual-mode probes. As shown in Figure 1.35, the CPNs are smooth spherical structures with an average diameter of 180 nm. in vivo experiments, active endocytosis mediated by folate receptors and passive targeting brought on by the increased permeability and retention impact

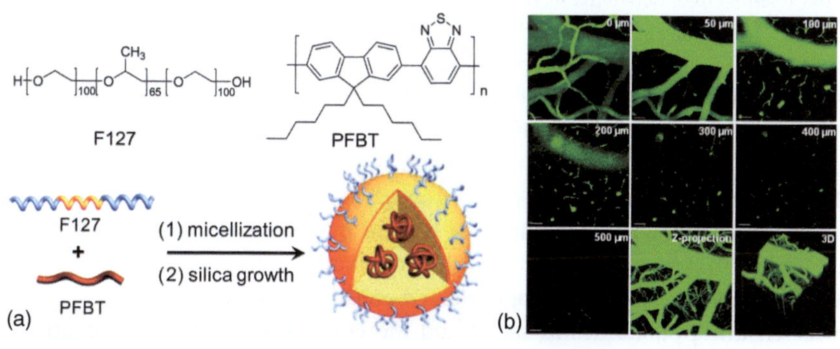

Figure 1.34 (a) Schematic diagram of the preparation of PFBT-F127-SiO_2 NPs; (b) fluorescence imaging of mice cerebrovascular system at different depths. *Source:* Reproduced with permission from Geng et al. [176] / American Chemical Society.

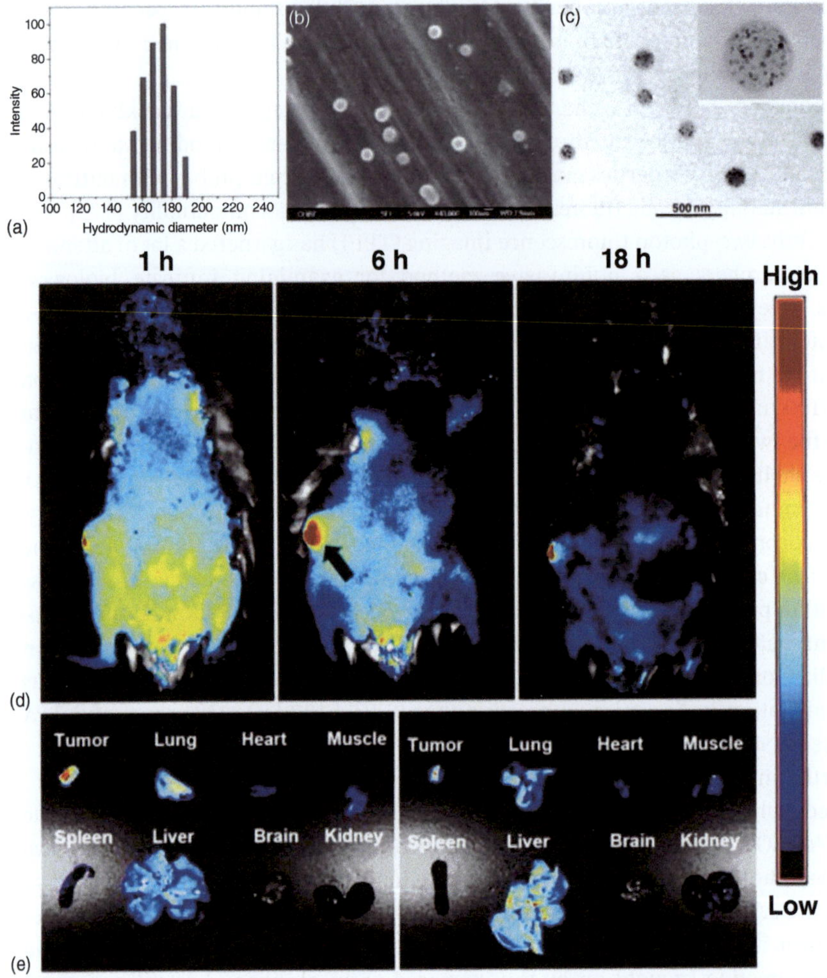

Figure 1.35 (a) Particle size distribution of CPNs; (b) and (c) SEM images of CPNs before and after modification; (d) and (e) in vivo imaging of CPNs in mice. *Source:* Reproduced with permission from Li et al. [177] / John Wiley & Sons.

were observed. The accumulation of CPNs within live tumor tissues can be observed in fluorescence imaging, enabling precise tumor monitoring [177].

1.3.1.2 Fluorescence Detection

Pathogen infections, which include those caused by bacteria, viruses, and fungi, pose a serious risk to public health worldwide. Rapid and accurate detection of pathogens is essential for early diagnosis and effective treatment of diseases, and even for ensuring public health safety and the effective prevention and control of infectious diseases. Traditional methods, such as bacterial culture and microscopic examination, have limitations in timeliness and sensitivity, making them inadequate for the immediate diagnostic and precise testing needs of clinical and public health settings. While PCR technology is highly sensitive, it is expensive and complex to use. Additionally, traditional methods can only detect the presence of pathogens individually, lacking the ability to quickly differentiate and process pathogens, which limits their efficiency and effectiveness in practical applications. In terms of visualization in pathogen detection, although fluorescence imaging technology has been widely adopted, it still faces challenges. For instance, the lack of specificity in fluorescent probes and poor stability of the fluorescence signal limit their effectiveness in pathogen detection. Therefore, by innovatively combining functional units with carrier materials, such as quantum dots, organic dyes, rare earth luminescent materials, and fluorescent polymers, which have excellent optical response properties (such as improved quantum yield, controllable emission wavelength, and good light stability), with carrier materials like nanoparticles, metal-organic frameworks, molecularly imprinted polymers, and biocompatible matrices, a novel light-functional polymer composite material fluorescence probe has been developed. This probe integrates signal amplification, target recognition, and specific response, offering a highly attractive solution for the development of next-generation microbial pathogen detection platforms that are highly sensitive, specific, rapid, convenient, and suitable for complex environments. This solution has significant practical implications.

By utilizing the differential fluorescence recovery of fluorescent quenching complexes after binding to bacteria, rapid detection and differentiation of various pathogenic bacteria can be achieved. For example, positively charged gold nanoparticles and negatively charged poly(*para*-phenyleneethynylene) (PPE) (Figure 1.36a) form electrostatic complexes in aqueous solutions, where the fluorescence of PPE can be effectively quenched by gold nanoparticles by a mechanism of energy or electron transfer. When a bacterial suspension is added to the formed fluorescent quenching complex, negatively charged bacteria can competitively bind to gold nanoparticles through electrostatic attraction and hydrophobic interactions, releasing the PPE fluorophore. Different types of bacteria have varying interactions with gold nanoparticles, leading to different abilities to displace PPE, which in turn induces varying degrees of fluorescence recovery (Figure 1.36). By adding 12 types of bacteria to the electrostatic complex system to detect characteristic fluorescence changes, and using linear discriminant analysis (LDA), the original fluorescence response signals can be converted into different standard scores. As shown in Figure 1.36c, different bacteria, including three types of *Escherichia coli*, are completely distinguished and fall into different specific regions on the discrimination map, demonstrating that minor differences in bacteria can cause significant variations in fluorescence response. Further, 64 unknown samples were selected, with 61 samples being correctly identified, achieving an accuracy of over 95%. This fluorescence-based strategy significantly reduces analysis time and allows for the simultaneous detection of multiple samples, providing a means for the differentiation and rapid diagnosis of pathogens [178].

DNA detection is of great significance in biomedical research and is widely used in genetic mutation analysis, diagnosis, and treatment of genetic diseases. Traditional methods of DNA detection, such as gel electrophoresis and PCR, have high

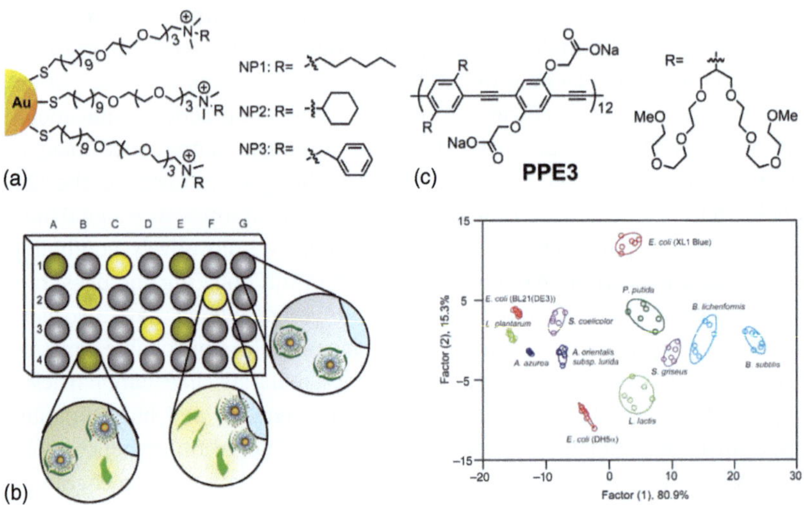

Figure 1.36 (a) Structure of gold nanoparticles and PPE3; (b) schematic diagram of fluorescence response signal of samples in microplate; (c) plot of canonical scores for the patterns of fluorescence response identified using LDA method. *Source:* Phillips et al. [178] / with permission of John Wiley & Sons.

sensitivity and specificity, but they also have the disadvantages of complex operation and long-time consumption.

In addition, fluorescence technology for biodetection has attracted much attention because of its high sensitivity and efficiency, but in practical applications, the probe's stability and biocompatibility still need to be further improved. Thus, a key area of current study is the creation of a quick, accurate, and easy DNA detection technique.

Due to the surface-enhanced Raman scattering effect, fluorescent dyes can undergo reversible adsorption or desorption on the surfaces of colloidal Au or Ag. Additionally, Au and Ag nanoparticles have a strong electron-donating ability, which allows them to undergo a PET process with dyes, quenching their fluorescence. By utilizing this principle, fluorescently labeled DNA can be covalently connected to the surface of colloidal Au nanoparticles, creating a highly sensitive DNA detection system (Figure 1.37). In this system, the fluorescent dyes are adsorbed onto the surface of colloidal Au, spontaneously forming a special ring structure, and their fluorescence is quenched by colloidal gold. After DNA hybridization, colloidal Au actively desorbs and releases fluorescein, restoring the fluorescence. This system enables highly sensitive and selective DNA detection [179].

Protein detection is of great significance in biomedical research, widely used in disease diagnosis, drug development, and cell biology studies. Conventional techniques for detecting proteins, include Western Blot and the enzyme-linked immunosorbent assay (ELISA), while highly specific, suffer from complex operations, long processing times, and limited sensitivity. Fluorescence detection technology, with its high sensitivity, rapid response, and noninvasive nature, has become a crucial method for protein detection. However, existing fluorescence probes still need improvement in terms of light stability, biocompatibility, and

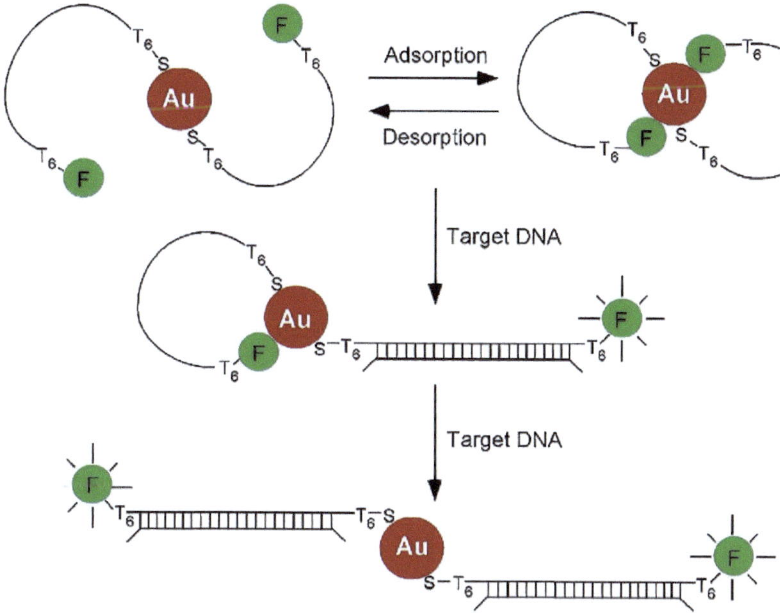

Figure 1.37 Schematic diagram of DNA detection using special "molecular beacon" induced by nano Au particles. *Source:* Maxwell et al. [179] / with permission of American Chemical Society.

detection limits. Therefore, developing organic molecular-based fluorescence platforms, which utilize their unique photophysical properties and controllable fluorescence signals, is essential for enhancing the efficiency and accuracy of protein detection.

The conformational changes of polymers can be utilized to detect proteins. By adding α-thrombin or nonspecific proteins (such as BSA) to single-stranded DNA, two composite systems are formed. Thrombin can induce ssDNA to form G-quadruplex structures, whereas nonspecific proteins do not have this ability. When cationic polyaniline derivatives are added to the solution, polyaniline forms complexes with G-quadruplex DNA through electrostatic interactions, enveloping the DNA and turning the solution orange. When polyaniline forms complexes with ssDNA, its conformation changes, enhancing coplanarity, and the solution turns purple (Figure 1.38). The color change in the solution induced by the conformational changes of CPs can be used to detect chymotrypsin directly, without the need for any substrate labeling [180].

In summary, the organic molecule-based fluorescence platform offers a new, efficient, sensitive, and stable method for protein detection. By optimizing the structure and detection mechanisms of fluorescent molecules, the efficiency and accuracy of protein detection can be significantly enhanced. In the future, with the advancement of nanotechnology and biotechnology, this fluorescence platform is expected to play a more significant role in the field of protein detection.

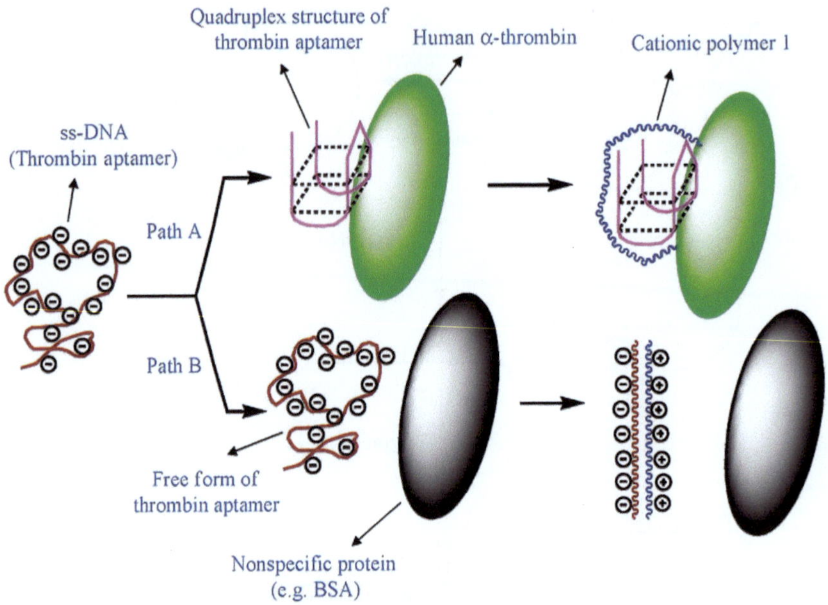

Figure 1.38 Schematic diagram of the specific detection principle of human α-thrombin using a cationic polymer and ss-DNA. *Source:* Ho et al. [180] / with permission of American Chemical Society.

In the future, as light-controlled fluorescence technology and nanomaterial science continue to advance, this new type of fluorescent probe is expected to play a more significant role in biomedical research and clinical diagnosis. Light-responsive materials have broad prospects in biological detection and imaging, particularly in precision medicine, real-time imaging, and early disease diagnosis. However, challenges such as biocompatibility, optical penetration, multifunctional integration, and standardization remain. As technology continues to evolve, especially with the integration of material science, nanotechnology, and biomedicine, light-responsive materials are poised to play an even more crucial role, potentially revolutionizing the healthcare industry.

1.3.2 Photodynamic Therapy (PDT)

1.3.2.1 Principle of Photodynamic Therapy

Photodynamic therapy (PDT) is a method that utilizes the interaction of photosensitizers and specific wavelengths of light to kill diseased cells or bacteria through the production of ROS (especially singlet oxygen). This treatment primarily relies on the reaction of photosensitizers with oxygen molecules under excitation light, generating destructive singlet oxygen, thus achieving therapeutic effects. The mechanism of photosensitizers can be divided into two types (Figure 1.39): Type I mechanism is also known as the free radical mechanism. ROS like superoxide anions ($O^{2-\cdot}$), hydrogen peroxide (H_2O_2), and hydroxyl radicals ($\cdot OH$) are produced by the direct electron exchange between photosensitizers and oxygen molecules

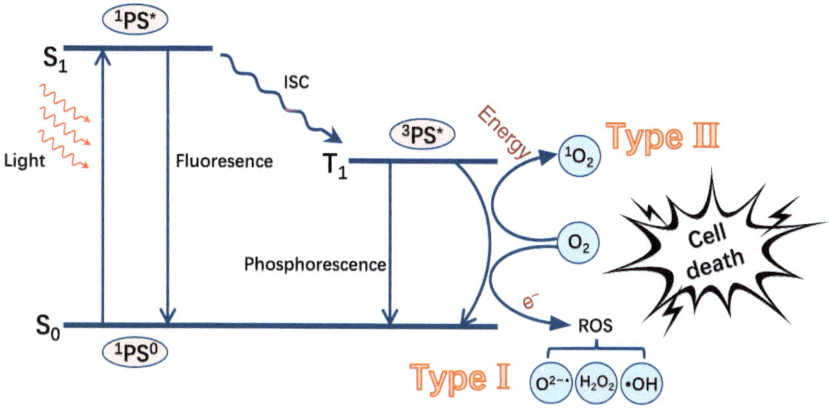

Figure 1.39 The ROS production mechanism of photosensitizers.

once they reach an excited state. These ROS are highly reactive and are capable of damaging proteins, nucleic acids, and cell membranes, resulting in cell death and damage. The Type II mechanism, also known as the singlet oxygen mechanism, is more common in PDT. Photosensitizers absorb photon energy, move from the ground state (S_0) to the excited state (S_1), and then intersystem cross (ISC) to the triplet excited state (T_1). These photosensitizers then interact with ground-state oxygen (O_2) in the triplet state, transferring energy to create singlet oxygen ($^1O_2$). $^1O_2$ is one extremely reactive ROS that can directly oxidize biomolecules, leading to the destruction of cell structure and function, ultimately causing cell apoptosis or necrosis.

1.3.2.2 Application of Photofunctional Polymer Composites in Photodynamic Therapy

Photodynamic therapy (PDT) has introduced new approaches to tumor diagnosis and treatment. Many fluorescent dyes are used not only for fluorescence imaging but also as photosensitizers, which produce ROS under light exposure, leading to cell damage. To address the issues of poor water solubility, easy aggregation, rapid metabolism by the body, and limited functionality in traditional photosensitizers, strategies such as modifying targeting molecules and encapsulating with polymer matrices can be employed to construct composite systems that better meet the clinical needs of PDT. For example, CS-2I was named by rhodamine-based photosensitizer with NIR emission, water-solubility, and an aggregate-induced emission (AIE) effect. By improving intersystem crossing, the heavy atom effect encourages the production of singlet oxygen. For bacterial imaging studies, co-culturing CS-2I with different bacteria demonstrated its unique staining capacity for gram-positive bacteria, making it useful for rapidly identifying and characterizing gram-positive bacteria in mixed bacterial cultures, with the possibility of long-term imaging with fluorescence tracking. When exposed to a 660 nm laser for 10 min, low concentrations of CS-2I can significantly kill drug-resistant bacteria, showing great potential as an alternative to antibiotic treatment. Furthermore, photosensitizer

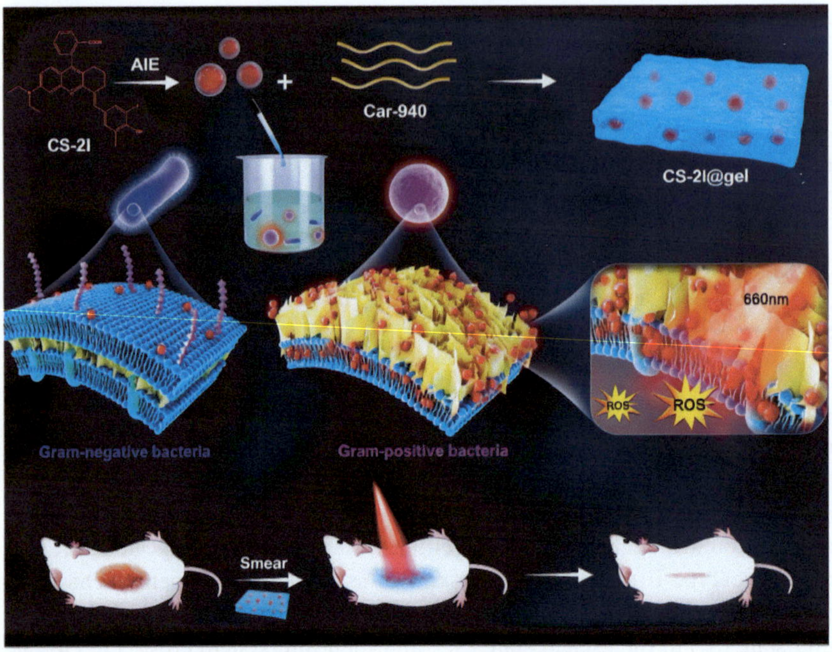

Figure 1.40 AIE photosensitizer hydrogel for selective elimination of gram-positive bacteria and promotion of wound healing. *Source:* Zeng et al. [181] / with permission of John Wiley & Sons.

CS-2I was mixed with carbomer 940 to create a hydrogel (CS-2I@gel) with PDT effect for antibacterial. The CS-2I@gel exhibits excellent adhesion, biocompatibility, and injectability. When applied to a mouse model of wound infection, it effectively inhibits bacterial proliferation through PDT, promoting wound healing (Figure 1.40). The composite hydrogel system is based on the PDT effect, and multiple biological functions work together to help antibacterial and wound treatment [181].

Porphyrin is a classic macromolecular heterocyclic compound with 26 π electrons on its ring, forming a highly conjugated system. As a natural photosensitive molecule, porphyrin exhibits excellent light capture capabilities and long excited state lifetimes. However, its practical applications are hindered by poor tumor selectivity and the tendency to self-aggregate, leading to quenching. To address these issues, a simple liquid-phase exfoliation method can be used to prepare COF nanodots. These COF nanodots form a highly ordered, periodically porous crystal material through strong covalent bonds, with porphyrin monomers precisely fixed within the framework. This arrangement effectively prevents self-aggregation and enhances PDT efficiency. Furthermore, by modifying the amphiphilic nanocarrier DSPE-PEG with hydrophobic interactions, the resulting CONDs-PEG exhibits excellent biocompatibility, physiological stability, the ability to produce ROS, and efficient tumor accumulation and renal metabolism in in vivo experiments [182].

To reduce the phototoxic effects of traditional photosensitizers on normal tissues, tumor-specific activatable dye molecules or composite systems can be designed according to the changes in the physiological microenvironment of tumor tissues (e.g. abnormal pH and GSH levels, overexpression of certain bioenzymes and nucleic acids) for use in bioimaging and PDT. For example, a case of pH-activatable photosensitive dye can be constructed by encapsulating diethylaminobenzene- and phenyl-substituted aza-BODIPYs in peptidomimetic micelles (cRGD-NEt_2Br_2BDP, NP). This photosensitive dye has an inhibited ROS production capacity under normal physiological conditions (pH = 7.4); targeting $\alpha v\beta_3$ overexpressing cancer cells through its cRGD nanomicellar shell, it was selectively activated in the weakly acidic environment of the tumors (pH < 7.0), restoring near-infrared fluorescence emission, while efficiently generating unilinear oxygen species in the presence of light ($\Phi_\Delta = 56\%$). In U87MG tumor-bearing mouse imaging experiments, the photosensitive dye showed bright NIR fluorescent signals in the tumor region, demonstrating the ability to achieve tumor-specific detection. In in vivo experiments, intravenous injection of photosensitizer micelles followed by light illumination resulted in a gradual reduction of tumor volume in mice, whereas the tumor volume of control mice remained unchanged, demonstrating the good PDT effect of the photosensitizing dye [183].

In atherosclerotic plaques, macrophages constitute about 20% of the lesion cells. According to the excessive ROS produced by activated macrophages at the lesion site (inducible nitric oxide synthase stimulates the production of peroxynitrite), ROS-activatable photosensitizer composite materials can be designed to target and kill these activated macrophages, facilitating PDT for atherosclerosis. By covalently attaching the photosensitizer Ce6 to hyaluronic acid (HA), amphiphilic HA-Ce6 conjugates can self-assemble into MacTNP nanoparticles in water. Due to the aggregation of Ce6 within the nanoparticles, conjugated stacking results in a self-quenching effect, which quenches the fluorescence of MacTNP and inhibits the generation of singlet oxygen. After MacTNP is taken up by activated macrophages, higher-than-normal levels of peroxynitrite degrade the backbone of HA, releasing Ce6, and strong NIR fluorescence can be observed inside the cell with a high label-to-target background ratio, which can be used for targeted imaging of macrophages (Figure 1.41). Additionally, under light exposure, MacTNP significantly reduces the survival rate of activated macrophages while showing no phototoxicity in other cells, indicating that the photosensitizer's photodynamic activity has been restored. This makes it highly valuable for selective NIR fluorescence imaging and PDT of atherosclerotic lesions in vivo [184].

Photodynamic therapy, as one possible therapeutic approach, has shown significant potential in treating tumors and infectious diseases. However, it still faces challenges such as the selectivity of photosensitizers, limitations in treatment depth, and unevenness in treatment. To overcome these challenges and achieve broader clinical application, further technological innovation and clinical research are needed. The development of photodynamic sterilization technology includes the creation of new photosensitizers and nanomaterials to enhance the efficiency and depth of photodynamic sterilization, as well as the clinical translation of this

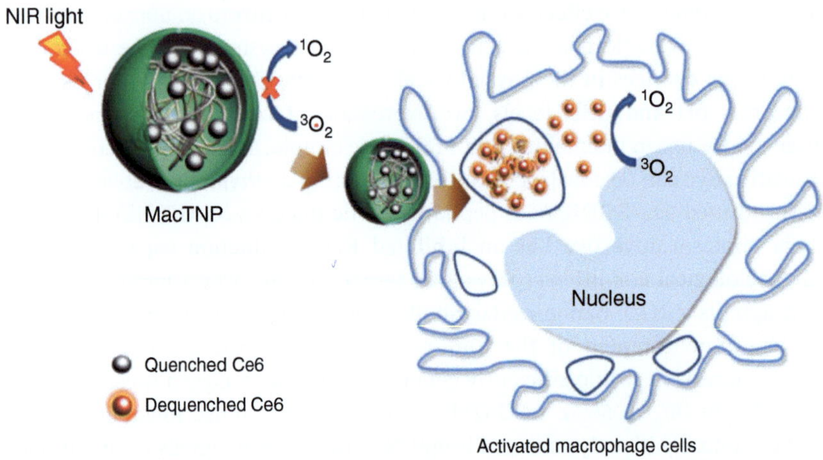

Figure 1.41 Schematic diagram of the process of activation of macrophages to activate the photosensitizer response. *Source:* Kim et al. [184] / Ivyspring International Publisher / CC BY 4.0.

technology. For example, integrating artificial intelligence and big data technologies can enable intelligent and personalized applications of photodynamic sterilization, providing more precise solutions for disease treatment.

1.3.3 Light-controlled Release

1.3.3.1 Principle of Light-controlled Release

The release mechanism triggered by light stimulation is a significant direction in the application of light-responsive materials, particularly in biomedical fields like drug delivery, gene therapy, and tissue engineering. By stimulating with external light, these materials can precisely control the release rate and timing of the substances, leading to more precise and controllable therapeutic outcomes. Common light-responsive release mechanisms include thermal response, photopolymerization, and photo-switching. The thermal response mechanism relies on the temperature changes in the material caused by light exposure. Light energy is converted into heat through photothermal conversion, which activates the material's temperature sensitivity. When the material is exposed to light, its temperature rises to a critical point, causing changes in the polymer chain conformation and triggering the drug release. This mechanism is commonly used in temperature-sensitive materials, such as thermoresponsive hydrogels or nanoparticles, which can produce localized heating effects under light, allowing drugs to be released from the carrier [185].

The photopolymerization release mechanism involves the depolymerization or cleavage of polymers triggered by light, leading to the release of drugs. For example, when a photosensitive group undergoes a depolymerization reaction under ultraviolet or visible light, it releases the drug or therapeutic molecule. This mechanism is commonly used in drug delivery systems made from photopolymerizable or photosensitive cross-linkable polymers. When an external light source illuminates the material, the polymer structure breaks down, allowing the drug to be released from the carrier [186].

The light switch release mechanism relies on the conformational changes or charge state alterations in materials triggered by light stimulation. These changes alter the material's hydrophilicity, solubility, or charge distribution, thereby controlling drug release. Such materials typically contain photosensitive switch groups, such as spiropyran [187]. This mechanism is widely applied in materials with reversible light regulation capabilities, such as certain nanoparticles or hydrogels containing photosensitive groups. These materials can undergo structural changes under different lighting conditions, thereby regulating the drug release rate.

1.3.3.2 Application of Photofunctional Polymer Composites in Light-controlled Release

Using photoisomeric molecules as the "switch" for drug loading and delivery systems, their structural and hydrophobic properties can be altered by light exposure, enabling noninvasive, precise, and on-demand drug administration. When exposed to ultraviolet light, spiropyran undergoes an opening reaction, transforming from a hydrophobic structure to a more hydrophilic flower-like structure. A mesoporous silica is loaded with the model drug curcumin (CUR). The closed-loop state of spiropyran that is anchored on the shell's surface can serve as a drug release switch. Given that high-energy UV light can damage tissues and has limited tissue penetration, directly using UV light to trigger the opening mechanism in clinical settings is impractical. Therefore, the drug-loaded silica shell incorporates UCNPs doped with lanthanides, which convert near-infrared incident light into UV light, inducing the opening of spiropyran and the release of the drug, effectively addressing this issue (Figure 1.42). Additionally, the UV light from UCNP can activate CUR to produce ROS, demonstrating significantly enhanced antitumor efficacy in in vivo experiments [188].

The light-controlled release system has significant potential in personalized treatment. By precisely controlling the timing and dosage of drug release, it can target specific lesion areas, reducing side effects in healthy tissues and thus enhancing treatment outcomes. This precise control gives the light-controlled release system a significant advantage in treating complex diseases such as cancer and inflammation, enabling the customization of treatment plans based on individual patient conditions.

Hydrogel scaffold, characterized by its abundant water content, adaptable mechanical characteristics, and availability of nutrients, closely matches the characteristics of natural extracellular matrices. They are frequently employed in tissue engineering and regenerative medicine as 3D cell carriers. By controlling the degradation of hydrogels, therapeutic cells can be delivered to injured or diseased tissues, promoting tissue repair and regeneration. In hydrogel degradation, the two most prevalent processes are hydrolysis and enzymatic degradation, but these methods cannot adjust or stop the degradation rate or control it spatially. In contrast, photodegradation allows for external control over the degradation

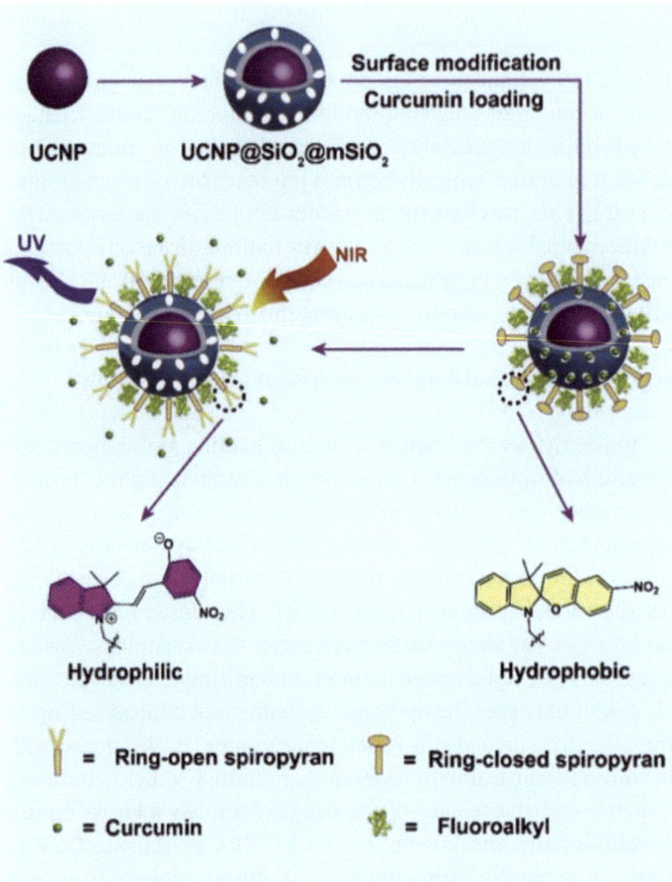

Figure 1.42 Schematic diagram of the infrared light-responsive drug release system based on spiropyran. *Source:* Liu et al. [188] / with permission of Elsevier.

process in terms of space and time, enabling controlled cell release. By introducing different photodegradable *o*-nitrobenzyl (*o*-NB) groups into the framework of PEG macromolecular monomers and initiating their copolymerization, photodegradable hydrogels can be formed, while encapsulating human mesenchymal stem cells (hMSC). After encapsulation, the cell viability is approximately 90%, indicating that this hydrogel has low cytotoxicity. Under 365 nm light irradiation, different *o*-NB linkers exhibit varying apparent degradation rate constants, allowing for different rates of hMSC release within various gels [189].

In addition to applications in tissue engineering, early cancer detection and diagnosis depend on the dynamic collection and release of cancer cells. Polymer molecular brushes refer to a type of special-structured polymer material formed by fixing one end of the macromolecular chains on the substrate surface (or polymer chains) through chemical or physical interactions. Due to the steric hindrance repulsion between high-density molecular chains, the polymer chains present a highly extended conformation, forming a structure similar to a "brush." As a powerful surface functionalization technique, it can be used to prepare artificial stimulus-responsive surfaces to mimic the dynamic functions of living systems. Poly(SP-*co*-NiPAAm) polymer brushes are grafted onto silicon nanowire arrays, which contain two response units: photoresponsive SP and thermoresponsive NiPAAm. Under visible light and above the phase transition temperature of PNiPAAm, the SP unit forms a hydrophobic closed loop configuration, while the PNiPAAm unit also assumes a hydrophobic spherical conformation. The hydrophobic surface can effectively adsorb biotin-BSA through hydrophobic interactions, and further bind to cancer cell capture agents biotin-anti-EpCAM by biotin–streptavidin interaction. It is capable of effectively and precisely identifying cancer cells (MCF-7) that express EpCAM. Under the illumination of 365 nm ultraviolet light and below the LCST temperature, the SP unit opens its loop, and PNiPAAm transforms into a coil conformation, thereby making the smart surface hydrophilic and leading to the desorption of biotin-BSA and biotin-anti-EpCAM, thus achieving the controllable release of cancer cells and demonstrating a dual response (Figure 1.43) [190].

1.3.4 Photothermal Therapy (PTT)

1.3.4.1 Principle of Photothermal Therapy

Photothermal therapy (PTT) is a novel cancer treatment that turns light energy into heat, thereby precisely targeting and destroying tumor cells. Compared to traditional treatments, PTT offers noninvasive, precise, and synergistic advantages. However, PTT still faces several challenges, such as low conversion efficiency, poor biocompatibility, and limited targeting ability [191]. These issues limit the clinical application of PTT, making the development of new PTT materials a key research focus. The core principle involves using specific light sources to irradiate photosensitive materials with photothermal properties, which absorb light and convert it into heat, leading to a localized temperature increase. This temperature rise can damage, coagulate, or kill tumor cells, achieving therapeutic effects. PTT relies on photosensitive materials, such as metal nanoparticles, carbon-based materials, and photosensitive molecules, which can efficiently absorb specific wavelengths of light (typically near-infrared light) and convert it into heat. Common PTT materials include gold nanoparticles (AuNPs), graphene, and carbon nanotubes [192]. Compared to traditional treatments like chemotherapy and radiotherapy, PTT offers noninvasive, controllable, and fewer side effects. By designing and optimizing photosensitive materials, PTT can effectively target and destroy tumor cells while minimizing damage to normal tissues. PTT is primarily used in cancer treatment, particularly for the local treatment of deep-seated tumors. In addition, it is effective in treating skin conditions, infectious diseases, and even as an adjuvant therapy in some wound healing processes.

1.3.4.2 Application of Photofunctional Polymer Composites in Photothermal Therapy

Compared to the more established PDT, PTT is hindered by the performance limitations of traditional photothermal agents, leading to slower clinical progress.

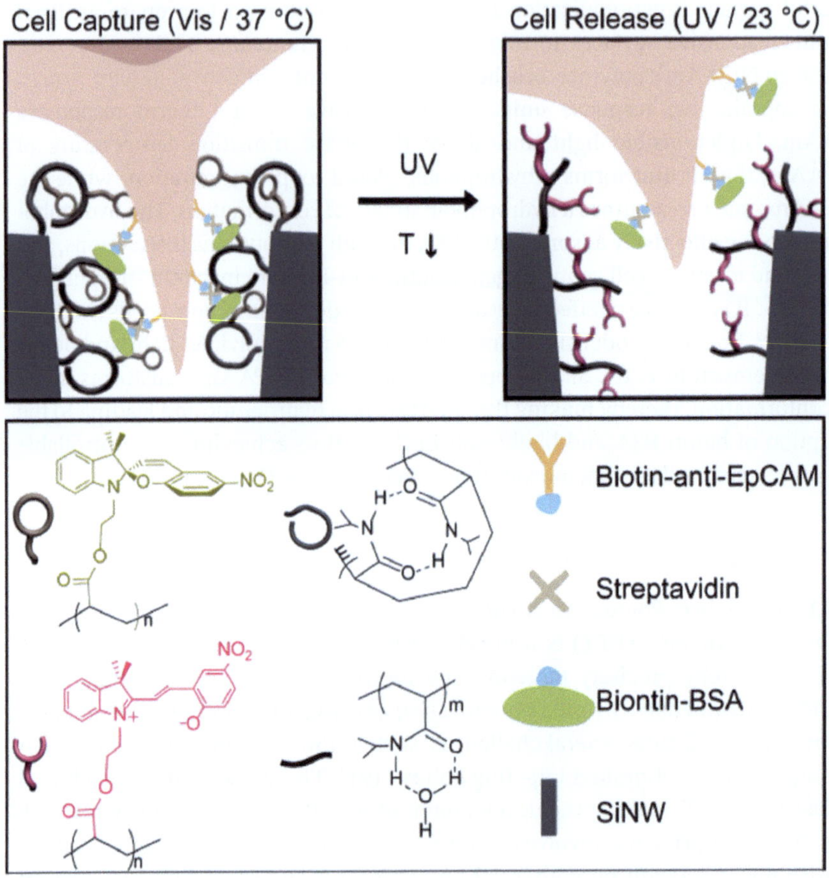

Figure 1.43 Schematic diagram of the principle of dual-responsive intelligent surface capturing and releasing cancer cells. *Source:* Hao et al. [190] / with permission of John Wiley & Sons.

For example, gold nanorods are known for their poor light stability, large size, and high cost; other inorganic photothermal materials, such as carbon nanomaterials, are nonbiodegradable and likely to stay in the body for a prolonged period of time, which raises questions regarding possible long-term toxicity. In contrast, organic polymers and small molecules offer excellent biocompatibility, designability, and ease of incorporation, making them highly promising for practical applications. Moreover, the excited photothermal agents under intense light may harm healthy tissues and skin. Therefore, it is crucial to prevent the decay of excitation energy through radiation transition pathways, enhance the photothermal conversion efficiency of photothermal agents, reduce the required intensity of excitation light, and ensure the safety of phototherapy in vivo.

For example, the -CF_3 group of tfm-BDP can rotate freely without any energy barrier, which promotes ultra-efficient nonradiative decay, enhancing the conversion of light energy into heat. DSPE-PEG is employed as the matrix for encapsulation

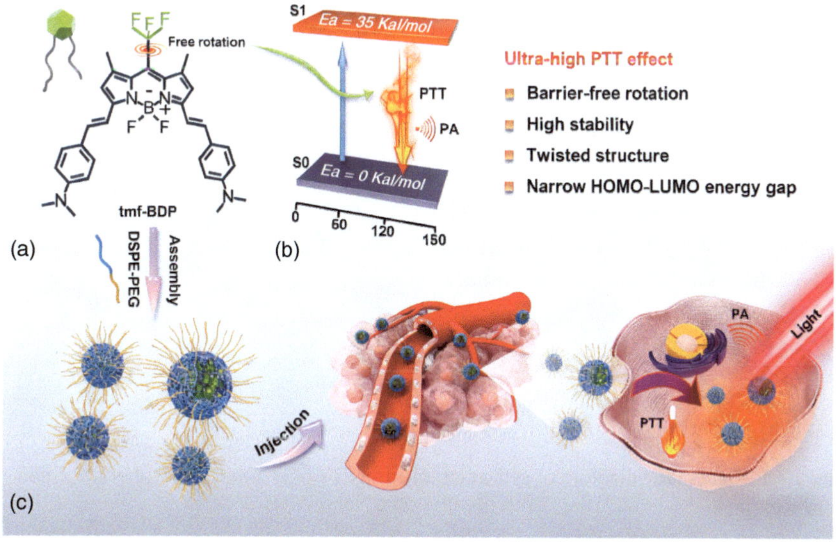

Figure 1.44 Schematic diagram of self-assembly and PTT of tmf-BDP nanoparticles. *Source:* Xi et al. [193] / with permission of John Wiley & Sons.

of tfm-BDP. The molecular twist in tfm-BDP impedes parallel π–π stacking (H aggregation), while still allowing the $-CF_3$ group to rotate freely. Because of their increased permeability and retention impact, PEG-coated nanoparticles can accumulate at tumor sites while maintaining greater stability in the circulation (Figure 1.44) [193].

In addition to chemical modification strategies, the aggregation of dye molecules can induce fluorescence quenching and inhibit photodynamic processes, thereby enhancing the material's photothermal conversion efficiency. For example, by introducing PEG as a hydrophilic segment into squaraine, an amphiphilic dye, PSQ, was synthesized. Due to its inherent planar structure, PSQ can form H-aggregates in aqueous solutions through π–π stacking and self-assemble into uniform nanospheres, PSQ-NSs, relying on hydrophilic–hydrophobic interactions with the PEG chain. The formation of these H aggregates inhibits fluorescence emission and intersystem crossing, resulting in PSQ-NSs exhibiting strong near-infrared (NIR) absorption, highly quenched fluorescence, good water solubility, physiological stability, and biocompatibility, while achieving an ultra-high photothermal conversion efficiency of 81.2%. Additionally, NIR light irradiation can generate strong photoacoustic (PA) signals from PSQ-NSs, which can be used as a PA contrast agent to guide the timing and location of irradiation for precise PTT. in vivo experiments showed that intravenous injection of PSQ-NSs followed by laser irradiation can ablate tumors in mice (Figure 1.45) [194].

Photothermal therapy (PTT) has made significant advancements in cancer treatment in recent years, particularly in the use of nanomaterials for targeted therapy. With the development of nanotechnology, a variety of photothermal

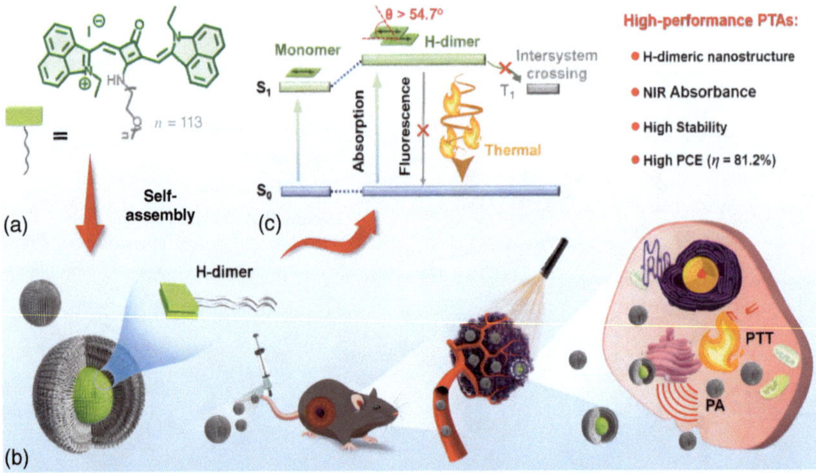

Figure 1.45 Schematic diagram of self-assembly and PTT of H dimer nanospheres formed by PSQ. *Source:* Wang et al. [194] / with permission of John Wiley & Sons.

materials, including gold nanoparticles, carbon-based nanomaterials, and sulfides, have been widely applied in PTT. However, several challenges remain. Firstly, the limited depth of light penetration restricts the effectiveness of treating deep-seated tumors. Secondly, precise temperature control is challenging, as both excessively high or low temperatures can affect treatment outcomes or harm healthy tissues. Additionally, the targeting ability of photosensitive materials is insufficient, potentially affecting normal tissues outside the treatment area and increasing side effects. The biocompatibility and long-term stability of these materials are also concerns, as some may trigger immune responses or toxicity issues. Lastly, the uniformity and reproducibility of treatments impact their effectiveness. Therefore, improving the precision and controllability of treatments remains a critical challenge for PTT.

1.3.5 Photocatalytic Effect

Photocatalysis is essentially a process where, under light exposure, a photocatalyst absorbs photon energy to accelerate specific chemical reactions, with the catalyst itself remaining unchanged before and after the reaction. The photocatalyst absorbs photon energy, causing electrons in the valence band (VB) to move to the conduction band (CB). This process also results in positively charged holes in the valence band, which form photogenerated electron-hole pairs. Photogenerated electrons and holes on the catalyst's surface can engage in reduction and oxidation processes, respectively. The core objective of photocatalysis is to mimic natural photosynthesis by using the most abundant and cleanest solar energy on Earth to drive valuable chemical reactions. By designing photofunctional polymer composites, it is possible to enhance energy conversion efficiency and reduce dependence on fossil fuels, providing a sustainable solution to address energy crises and environmental pollution.

The following are several key applications of photofunctional polymer composites in photocatalysis.

1.3.5.1 Photocatalytic CO_2 Reduction

The reduction and conversion of CO_2 is one of the effective methods to address climate change and reduce greenhouse gas emissions. Photofunctional polymer composites can convert CO_2 into fuel using solar energy. Under light conditions, the photocatalyst absorbs light energy, which excites electrons from the valence band to the conduction band, generating electron-hole pairs. The conduction band electrons are captured by CO_2 receptors on the catalyst surface and reduced to useful hydrocarbon fuels. Meanwhile, the valence band holes typically drive the oxidation of water to produce oxygen [195]. This process, known as photocatalytic CO_2 reduction, offers a new approach for biofuel production. By incorporating narrow-bandgap materials, plasma materials, upconversion materials, and dye-sensitized materials into photofunctional polymer composites, the spectral response range of the photocatalyst can be broadened. Combining light-absorbing materials with electron transport materials enhances the separation efficiency of photogenerated charges and accelerates the CO_2 reduction reaction rate. Protecting the sensitive components from corrosion through coating, core-shell structures, or by combining with stable carriers improves the stability of the photocatalyst, ensuring that the photocatalytic material maintains high catalytic performance over extended periods [196].

In the field of photocatalytic CO_2 reduction, CPs have emerged as a promising platform due to their tunable bandgap and ease of modification. Among these, conjugated microporous polymers excel in CO_2 adsorption and tunable photoelectric properties, making them an efficient organic semiconductor catalyst. To address issues such as spontaneous electron-hole recombination and low electron transfer efficiency within the catalyst, constructing conjugated microporous polymers with donor (D) and acceptor (A) components can significantly enhance charge separation and photocatalytic activity. For instance, ImI-CMP, a conjugated microporous polymer material with an imidate cation in its backbone, doped with Co(II) as a co-catalyst, exhibits a strong built-in electric field due to the presence of the imidate cation. This built-in electric field drives the separation of photogenerated charges and the directed transport of electrons. Additionally, the π-conjugated structure of the polymer provides intramolecular transport channels for electron transfer, enabling rapid charge transfer within the molecule during the photocatalytic process. Moreover, the strong electrostatic interaction between the imidate ring cation and carbon dioxide molecules facilitates the enrichment and activation of carbon dioxide. Using $[Ru(bpy)_3]Cl_2$ as a visible light photosensitizer and triethanolamine (TEOA) as the electron donor (Figure 1.46), ImI-CMP can effectively reduce CO_2 to CO under visible light irradiation [197].

The morphology of the three-dimensional porous CP network is highly advantageous for the efficient loading of photocatalysts. By supporting polyaniline (PANI) on a three-dimensional poly(*N*, *N*-dimethylacrylamide) (PDMA) framework, a novel flexible photoelectrode has been developed. The molecular channels of this electrode network provide a rich array of surface functional groups, which can

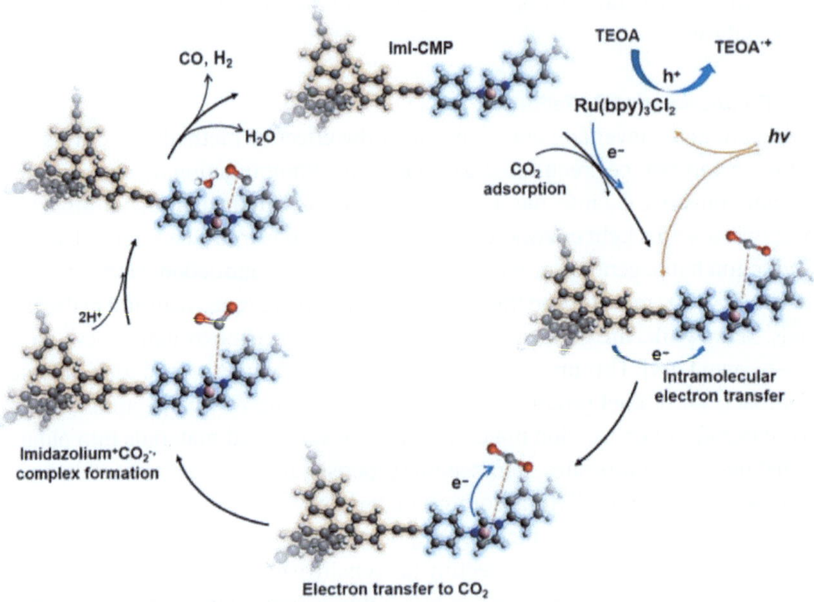

Figure 1.46 Mechanism of photocatalytic reduction of CO_2 by Iml-CMP. *Source:* Reproduced from Zhao et al. [197] / with permission of Elsevier.

assist in regulating the structure of photocatalytic self-assembly. This overcomes the structural and loading limitations of traditional inorganic photoelectrodes when loading functional molecules or materials, allowing for a stable and large-scale loading of Co(II) catalysts. Through strong chemical bonds throughout the network, a stable and high-density interface contact between the electrode substrate and the photocatalyst is achieved, enhancing the efficiency of photoelectron transfer. Under sunlight, this photoelectrode can reduce CO_2 to methanol with a conversion efficiency of 70%. When the area is expanded from 1 to 100 cm^2, the photoelectrode can continuously generate a stable photocurrent, demonstrating the excellent application prospects of this flexible photoelectrode material in large-scale solar energy conversion devices [198].

Photocatalytic CO_2 reduction technology offers new solutions to the challenges of global warming and the energy crisis. Through advancements in materials and technology, this technology promises to achieve efficient, stable, and sustainable CO_2 conversion. Future research will focus on developing new materials, gaining deeper insights into reaction mechanisms, and innovating technologies. It is expected that this technology will be broadly applied in energy, environmental protection, and chemical industries, providing crucial support for sustainable development.

1.3.5.2 Photocatalytic Production of H_2

Similar to the photocatalytic reduction of carbon dioxide, photocatalytic hydrogen production also requires a photocatalyst to absorb light energy and excite electrons in the conduction band, reducing H^+ to H_2. Hydrogen, a high-calorific value gas, is an environmentally friendly energy source that produces no greenhouse gases during combustion. Converting solar energy into chemical energy for hydrogen production through a photocatalytic system is a viable solution to environmental and energy challenges.

Enhancing the conjugation degree of photocatalyst molecules improves the photocatalytic hydrogen evolution process. Therefore, constructing a two-dimensional conjugated framework into a three-dimensional porous material significantly boosts photocatalytic activity. A novel two-dimensional fully π-conjugated donor–acceptor covalent organic framework (SP^2c-COF) has been developed using 1,3,6,8-tetrakis(4-formylphenyl)pyrene (TFPPy) and **poly(1,5-diaminonaphthalene)** (PDAN). Due to complete conjugation, the SP^2c-COF photocatalyst exhibits a narrow bandgap, enhancing its light absorption in the visible region. The introduction of electron-deficient units (–CN) endows the material with a D-A structure, which further strengthens conjugation through electronic effects, aiding in the separation of photogenerated excitons and the inhibition of charge recombination. To further optimize photocatalytic activity, 3-ethylrotenone (ERDN) groups, as electron-deficient units, were grafted onto the edges of the framework. The enhanced electronic effect more effectively splits excitons, thereby improving photocatalytic hydrogen evolution performance (2120 μmol/g/h). The ordered two-dimensional lattice features a rich one-dimensional channel network with high porosity, providing ample space for numerous reaction sites, promoting mass transport, and minimizing the electron transfer distance to the reaction center. These advantages are incorporated into the framework, enabling the material to effectively produce hydrogen gas even with low-energy photon input (Figure 1.47) [199].

The photocatalytic system based on organic conjugated molecules, due to the short excited state lifetime of these molecules, requires the involvement of a sacrificial agent to complete the catalytic process. In contrast, natural photosynthesis utilizes thylakoid membranes to spatially separate photogenerated charges, ensuring high light energy conversion efficiency. Inspired by this, the partitioning and transfer capabilities of liposomes can be utilized to overcome the rapid recombination of photogenerated charges, helping to construct a photocatalytic system that does not require a sacrificial agent, thus achieving the complete decomposition of water under light. In the half-reaction for hydrogen evolution reaction (HER), the catalytic center is [(TCPP)Pt], where TCPP = meso-tetra(4-carboxyphenyl)porphine, which has a monolayer thickness and is functionalized with hydrophobic pentafluoropropionic acid groups to insert into the hydrophobic region of the lipid bilayer. In the half-reaction for water oxidation reaction (WOR), Ir-bpy serves as the reaction center and is incorporated into the MOF based on $[Ru(bpy)_3]^{2+}$. HER- and WOR-MOF nanosheets are assembled in the hydrophobic and hydrophilic regions of liposomes composed of lecithin and cholesterol, respectively, using redox shuttle Fe^{3+} and tetrachlorobenzoquinone to link the two half-reactions in series. For the half-reaction of hydrogen evolution, the tetrachlorobenzoquinone system provides electrons and is oxidized itself. At the lipid/water interface, tetrachlorobenzoquinone oxidizes Fe^{2+} to Fe^{3+}. Fe^{3+} then oxidizes the photoactivated $[Ru(bpy)_3]^{2+}$ to form $[Ru(bpy)_3]^{3+}$, which continues the half-reaction of oxygen

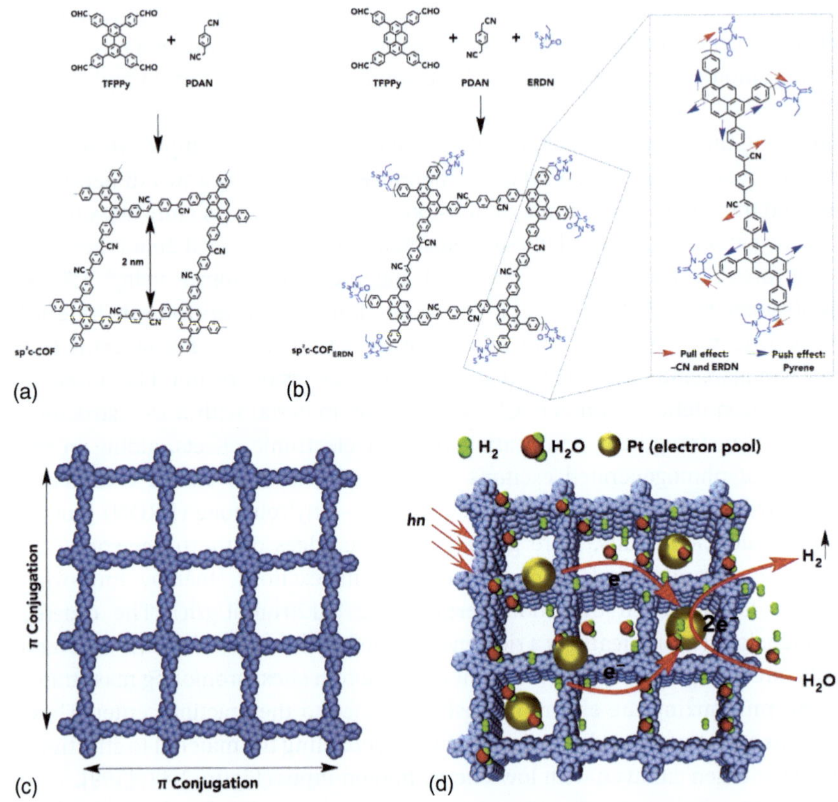

Figure 1.47 Schematic diagram of (a) sp^2c-COF structure without electron deficiency ERDN end group; (b) sp^2c-COF structure with electron deficiency ERDN end group; (c) π conjugation along x and y directions in the two-dimensional crystal layer of sp^2c-COF; (d) sp^2c-COF photocatalytic hydrogen evolution. *Source:* Jin et al. [199] / with permission of Elsevier.

evolution, completing the entire process. The full decomposition of photocatalytic water was achieved by ultrafast electron transport based on a lipid system and effective charge separation of the lipid bilayer, with an apparent quantum yield of (1.5±1)% [200].

1.3.5.3 Light-driven Biotransformation

The use of traditional fossil fuels has led to severe environmental pollution and greenhouse gas emissions, with limited reserves. Photobiological conversion technology harnesses solar energy to drive the biosynthesis process, converting greenhouse gases like carbon dioxide into valuable biofuels. This not only meets energy needs but also reduces environmental pollution, aligning with the principles of sustainable development [201]. The efficiency of light energy conversion in natural photosynthesis is relatively low. By constructing photobiological conversion systems, the capture and utilization of light energy can be optimized, enhancing overall light energy conversion efficiency and thus producing biofuels more efficiently. Microorganisms have limited metabolic pathways, making it challenging to directly and efficiently convert carbon dioxide into complex biofuels. Photobiological conversion systems, by incorporating photosensitive materials and optimizing microbial metabolic pathways, can achieve more efficient biosynthesis, expanding the range and yield of biofuels produced by microorganisms [202].

Light-driven biotransformation is a process where microorganisms or algae convert organic matter into biofuels using light energy. In this process, photofunctional polymer composites typically enhance light utilization efficiency, promote photosynthesis, and improve conversion efficiency. Improving the efficiency of photosynthesis with photofunctional polymer composites is a critical aspect of biofuel production. For example, algae such as spirulina and purple bacteria can use sunlight to perform photosynthesis, converting carbon dioxide into organic compounds, ultimately producing biofuels like biodiesel and acetic acid [203].

Perylene diimide derivatives (PDI) and poly(fluorene-co-phenylene) (PFP) were used as photosensitizers to coat the surface of the nonphotosynthetic bacterium *Moorella thermoacetica*, forming a P/N heterojunction (PFP/PDI) layer (Figure 1.48). This led to the creation of a hybrid photosynthetic system, integrating organic semiconductors with bacteria, that efficiently transforms CO_2 into acetic acid. The PFP/PDI heterojunction enhances the separation efficiency of photogenerated holes and electrons. The cationic side chains of PDI and PFP can penetrate the bacterial cell membrane, ensuring efficient electron transfer. *M. thermoacetica* can harvest photoexcited electrons from the PFP/PDI heterojunction, driving the Wood-Ljungdahl pathway to synthesize acetic acid from CO_2 under light, achieving an efficiency of 1.6%, comparable to previously reported inorganic–bacterial hybrid systems [204].

Organic optoelectronic composites, such as organic semiconductors combined with microorganisms or algae, can be used as a photosynthetic promoter. These materials can effectively capture light energy and transfer the energy to organisms through the charge separation process, promoting their photosynthesis and thus improving the production efficiency of biofuels.

In addition, the $WO_3/MoO_3/g\text{-}C_3N_4$ heterojunction, used as an electrode material for a photo-assisted biological cathode, was combined with the electrosynthesis bacterium *Serratia marcescens* Q1 to construct a photo-assisted microbial electrosynthesis system, achieving the efficient synthesis of acetic acid from inorganic carbon sources. The system demonstrated enhanced photocurrent and acetic acid production under light exposure, revealing the effective separation and transport of photogenerated charge carriers within the heterojunction. This discovery provides a new strategy for photo-driven microbial electrosynthesis [205].

As an emerging biotechnology, the photobiological conversion system has broad application prospects in biofuel production, carbon dioxide fixation, and the synthesis of high-value chemicals. By continuously optimizing the combination of photosensitive materials, microorganisms, and electron transfer media, and enhancing the overall performance and economic efficiency of the system, photobiological conversion technology is expected to offer more effective solutions for dealing with upcoming environmental and energy issues.

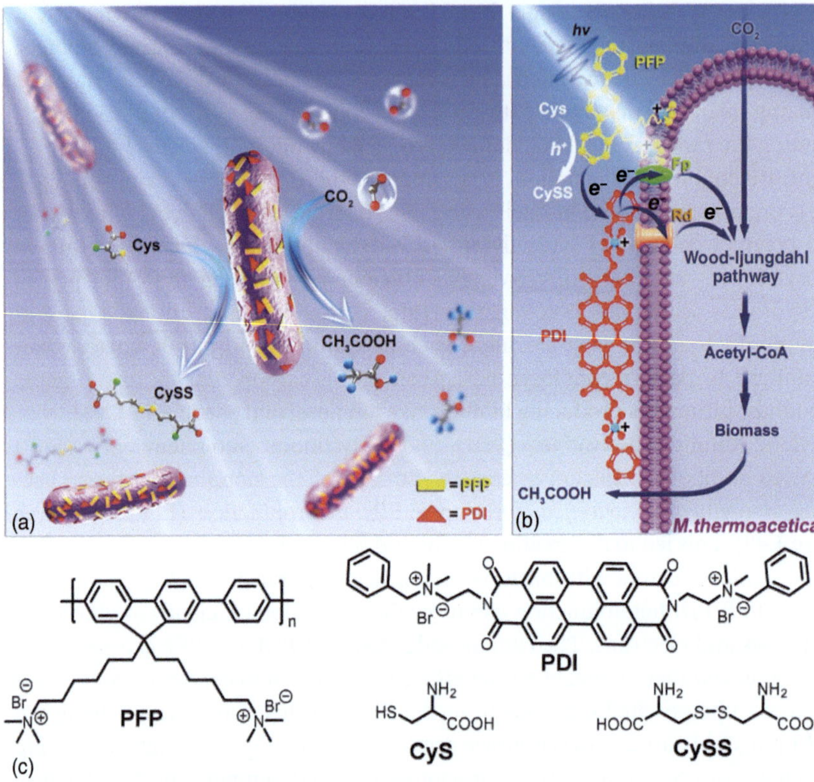

Figure 1.48 Schematic diagram of an organic semiconductor–nonphotosynthetic bacterial hybrid system producing acetic acid driven by solar energy. *Source:* Gai et al. [204] / with permission of John Wiley & Sons.

1.3.5.4 Photocatalytic Therapy

Photodynamic therapyhas the advantages of being spatiotemporally controllable and not limited by drug resistance, but its application scenarios have a high demand for oxygen. On the other hand, the eradication of pathogenic microorganisms by PTT requires a long period of high temperature, which inevitably has certain side effects on normal tissues. Therefore, it is very important and promising to develop a treatment modality that combines the advantages of both. The essence of life activities relies on various oxidation-reduction reactions and metabolic processes involving substance exchange. Thus, effective intervention in pathological processes can be achieved through photocatalytic modulation of electron transfer and ROS metabolism in organisms.

Conjugated polymers, due to their excellent photoelectric conversion efficiency, ease of functionalization, and good biocompatibility, are among the key contenders for in vivo catalysts for photo-mediated redox reactions. Unlike inorganic catalysts, which often require complex systems of multiple materials, CPs can achieve in vivo photocatalytic reactions more easily through simple doping or functionalization, showing potential for applications in the biomedical field. Three different CP-based

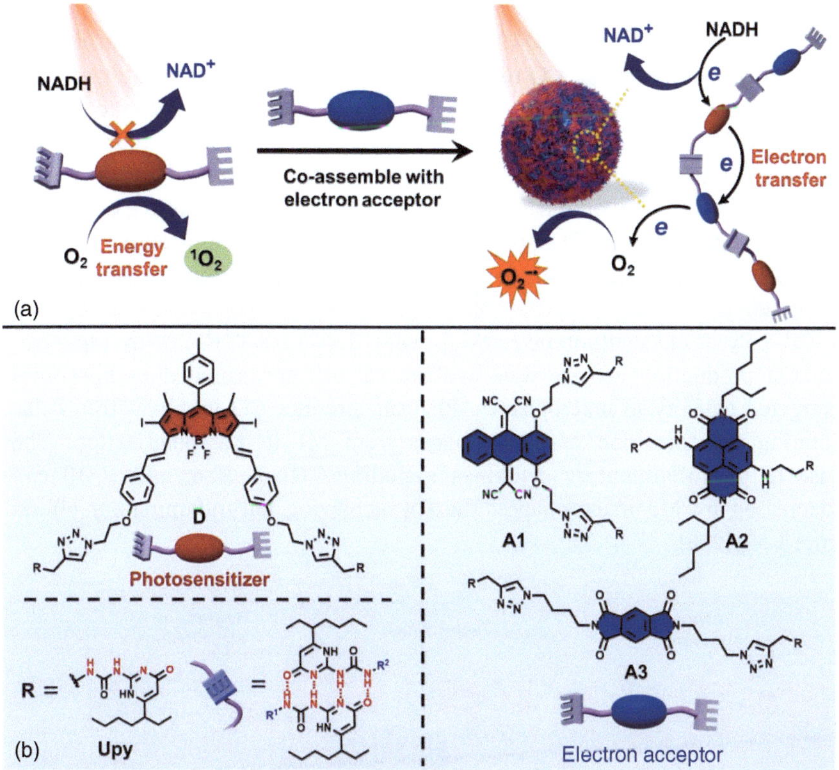

Figure 1.49 Schematic diagram of the structure and assembly of supramolecular composite photosensitizers and the photoinduced oxidation of NADH and the production of $O_2^{-\bullet}$. *Source:* Teng et al. [206] / ISpringer Nature / CC BY 4.0.

electron acceptors were designed and assembled with conventional BODIPY-based type II photosensitizers via quadruple hydrogen bonding to prepare supramolecular composite photosensitizers (Figure 1.49). Under light conditions, the shortened molecular distance and matched redox potentials between the photosensitizer molecules and electron acceptors facilitate transfer of the photoinduced electron of the photosensitizer to the acceptor, which further transfers electrons to oxygen to generate superoxide radicals. Additionally, this process generates strong oxidizing cationic radicals of the photosensitizer, which can effectively oxidize NADH in living cells, disrupting the coenzyme balance and breaking the mitochondrial electron transport chain, thereby achieving cell killing without relying on oxygen. This system exhibits the highest NADH oxidation turnover frequency, which surpasses that of conventional transition metal catalysts by one to three orders of magnitude. in vivo experiments using mouse models have shown significant tumor ablation effects. This method of utilizing photocatalytic reactions can inhibit the growth of pathogenic microorganisms by different principles from PDT and PTT, achieving disease treatment without being influenced by the tissue microenvironment [206].

Carbon monoxide (CO) is a crucial endogenous signaling molecule that significantly influences physiological processes such as immune regulation, inflammation, and oxidative stress. CO regulates the immune system by modulating the secretion of pro-inflammatory cytokines, thereby reducing the damage to normal cells caused by excessive pro-inflammatory cytokines secreted by macrophages in inflamed areas. Additionally, CO can selectively kill tumors by regulating mitochondrial function and triggering oxidative stress. However, gas therapy based on CO poses a risk of CO poisoning. Therefore, using intracellular endogenous CO_2 to catalyze the in situ generation of CO is a controllable and precise solution [207].

A CP poly(hexadecylfluorene-*co*-thiophene) derivative (PFT) with photocatalytic CO production activity was synthesized and encapsulated in liposomes for targeted delivery to macrophages. PFT can produce CO in situ within cells, facilitating the transition of macrophages from M1 to M2 polarization. The release of pro-inflammatory cytokines including TNF-α, IL-6, and IL-1β can be decreased by M2 macrophages, thereby achieving antiinflammatory effects (Figure 1.50) [208].

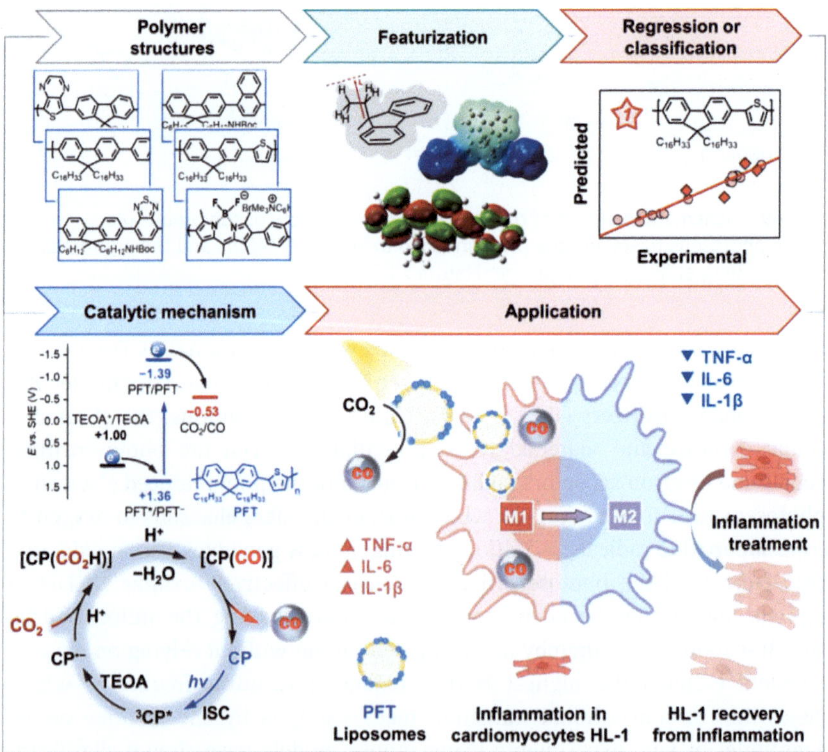

Figure 1.50 Schematic diagram of the principle of intracellular PFT photocatalytic CO production and immune regulation for inflammation treatment. *Source:* Zhu et al. [208] / with permission of American Chemical Society.

References

1 Wu, L.Z. (2023). Introduction to photofunctional materials and transformations. *Journal of Materials Chemistry A* 11 (25): 13063–13064.

2 De, R.Y., Tao, T.T., Tang, W.W., and Gong, J.B. (2024). Emerging biomineralization of organic photonic crystalline materials: ultrastructure, formation mechanism, and optical function. *Chemistry of Materials* 36 (13): 6321–6346.

3 Wu, H.L., Qi, M.Y., Tang, Z.R., and Xu, Y.J. (2023). Semiconductor quantum dots: a versatile platform for photoredox organic transformation. *Journal of Materials Chemistry A* 11 (7): 3262–3280.

4 Bu, H., He, C.L., Xu, Y.Q. et al. (2022). Emerging new-generation detecting and sensing of metal halide perovskites. *Advanced Electronic Materials* 8 (5): 2101204.

5 Zhang, B.X., Li, J.Z., Yang, X., and Yin, Z.F. (2025). Advances in multimodal electrohydrodynamic printing for high-resolution sensor fabrication: mechanisms, materials, and applications. *Nanoscale* 17 (25): 15132–15174.

6 Ren, X.X., Zhang, X., Xie, H.X. et al. (2022). Perovskite quantum dots for emerging displays: recent progress and perspectives. *Nanomaterials* 12 (13): 2243.

7 Zhang, M.L., Wang, S., Bai, Y.P. et al. (2024). A dual-function hemicyanine material with highly efficient photothermal and photodynamic effect used for tumor therapy. *Advanced Healthcare Materials* 13 (10): e2303432.

8 Ikake, H., Hara, S., and Shimizu, S. (2022). Skillful control of dispersion and 3D network structures: advances in functional organic-inorganic nano-hybrid materials prepared using the sol-gel method. *Polymers* 14 (16): 3247.

9 Zhang, X., Wang, Q., Liu, S. et al. (2021). Optical isomerization and photo-patternable properties of GeO_2/Ormosils organic-inorganic composite films doped with azobenzene. *Coatings* 11 (7): 818.

10 Gao, Q.F., Pan, P.J., Shan, G.R., and Du, M. (2021). Anisotropic bilayer hydrogels with synergistic photochromism behaviors for light-controlled actuators. *Journal of Materials Science* 56 (29): 16324–16338.

11 Abdollahi-Esfahlani, H., Pourmahdian, S., and Ameri, M. (2025). Boosting light harvesting of quaternary polymer solar cells through co-donor and co-acceptor with amine-functionalized graphene quantum dots and a ruthenium-based sensitizer. *Colloids and Surfaces A: Physicochemical and Engineering Aspects* 707: 135859.

12 Mondal, N., De, A., Das, S. et al. (2019). Ultrafast carrier dynamics of metal halide perovskite nanocrystals and perovskite-composites. *Nanoscale* 11 (20): 9796–9818.

13 Liu, M.X., Yazdani, N., Yarema, M. et al. (2021). Colloidal quantum dot electronics. *Nature Electronics* 4 (8): 548–558.

14 Wu, Q.Q., Cao, F., Yu, W.K. et al. (2025). Homogeneous ZnSeTeS quantum dots for efficient and stable pure-blue LEDs. *Nature* 639: 633–638.

15 Peng, B.X., Li, Q., Yu, B. et al. (2024). Dual nanofillers reinforced polymer-inorganic nanocomposite film with enhanced mechanical properties. *Small* 20 (48): e2406160.

16 Meng, X.G., Liu, L.Q., Ouyang, S.X. et al. (2016). Nanometals for solar-to-chemical energy conversion: from semiconductor-based photocatalysis to plasmon-mediated photocatalysis and photo-thermocatalysis. *Advanced Materials* 28 (32): 6781–6803.

17 Zhang, X. and Yang, P. (2024). Advances in noble metal-modified g-C_3N_4 heterostructures toward enhanced photocatalytic redox ability. *International Journal of Minerals, Metallurgy, and Materials* 31 (11): 2368–2389.

18 Chen, Q., Chang, Y., He, X. et al. (2025). Targeted delivery of mRNA with polymer-lipid nanoparticles for in vivo base editing. *ACS Nano* 19 (8): 7835–7850.

19 Zhang, J. and Chen, G. (2024). Progress in the application of rare-earth-doped upconversion nanoprobes in biological detection. *Chinese Journal of Inorganic Chemistry* 40 (12): 2335–2355.

20 García-Alvarez, R., Chen, L., Nedilko, A. et al. (2020). Optimizing the geometry of photoacoustically active gold nanoparticles for biomedical imaging. *ACS Photonics* 7 (3): 646–652.

21 Javaheri, S., Keshavarzi, F., and Karami, C. (2025). Development of co-doped ZnS-CdS quantum dots based composite sensor for the detection of cefixime (CXM) and tetracycline (TET), and application in real samples from local dairies. *Scientific Reports* 15: 10717.

22 Balakrishnan, T., Ang, W.L., Mahmoudi, E., and Sambudi, N.S. (2024). Recent development of quantum dots@metal-organic framework composites as potential chemical and biological luminescence nanosensors. *Physica B: Condensed Matter* 673: 415485.

23 Vanaraj, R., Arumugam, B., Mayakrishnan, G. et al. (2025). Chemical engineering innovations in nanoparticle-based biosensors for enhanced detection of biological molecules. *Chemical Engineering Journal* 507: 160081.

24 Tang, Y., Zhang, Q., Yuan, H. et al. (2025). Recent applications and challenges of inorganic nanomaterial-based biosensing devices for detecting nucleic acid biomarkers. *Advanced Sensor and Energy Materials* 4 (1): 100136.

25 Pakulska, M.M., Miersch, S., and Shoichet, M.S. (2016). Designer protein delivery: from natural to engineered affinity-controlled release systems. *Science* 351 (6279): aac4750.

26 Zan, Y., Liu, J., Zhao, Z. et al. (2025). A montmorillonite-based pickering nanoemulsion for the integration of photothermal therapy and NIR-responsive drug delivery. *ACS Applied Bio Materials* 8 (1): 652–660.

27 Adlsadabad, S.Y., Pourbadiei, B., Doroudian, M., and Pourjavadi, A. (2024). Photo-responsive doxorubicin delivery: nanogel systems based on azobenzene and host-guest interactions enhanced by squeezing action. *Polymer* 300: 126900.

28 Bernatoniene, J., Stabrauskiene, J., Kazlauskaite, J.A. et al. (2025). The future of medicine: how 3D printing is transforming pharmaceuticals. *Pharmaceutics* 17 (3): 390.

29 Maksymova, L., Pilger, Y.A., Nuhn, L., and Van Ginderachter, J.A. (2025). Nanobodies targeting the tumor microenvironment and their formulation as nanomedicines. *Molecular Cancer* 24 (1): 65.

30 Jin, W., Chen, Z., Wang, Y. et al. (2024). Nano metal-photosensitizer based on Aza-BODIPY-Cu complex for CDT-enhanced dual phototherapy. *Chinese Chemical Letters* 35 (7): 109328.

31 Tong, L., Liao, Q., Zhao, Y. et al. (2019). Near-infrared light control of bone regeneration with biodegradable photothermal osteoimplant. *Biomaterials* 193: 1–11.

32 Wu, Y., Xie, X., Luo, G. et al. (2025). Photothermal sensitive nanocomposite hydrogel for infectious bone defects. *Bone Research* 13: 22.

33 Hou, X., Zhang, L., Chen, Y. et al. (2025). Photothermal switch by gallic acid-calcium grafts synthesized by coordination chemistry for sequential treatment of bone tumor and regeneration. *Biomaterials* 312: 122724.

34 Kurian, A.G., Singh, R.K., Patel, K.D. et al. (2022). Multifunctional GelMA platforms with nanomaterials for advanced tissue therapeutics. *Bioactive Materials* 8: 267–295.

35 Zhang, R., Feng, J., Chen, H. et al. (2025). Hybrid hydrogel with photothermal stimulation elicits immunomodulation-mediated wound healing. *Advanced Functional Materials* 35 (7): 2419170.

36 Peng, J., Liu, C., Mo, M. et al. (2024). Construction of multifunctional hydrogel containing pH-responsive gold nanozyme for bacteria-infected wound healing. *International Journal of Biological Macromolecules* 283: 137746.

37 Daniel, M.C. and Astruc, D. (2004). Gold nanoparticles: assembly, supramolecular chemistry, quantum-size-related properties, and applications toward biology, catalysis, and nanotechnology. *Chemical Reviews* 104 (1): 293–346.

38 Stanisavljevic, M., Krizkova, S., Vaculovicova, M. et al. (2015). Quantum dots-fluorescence resonance energy transfer-based nanosensors and their application. *Biosensors & Bioelectronics* 74: 562–574.

39 Feng, L., Zhu, C., Yuan, H. et al. (2013). Conjugated polymer nanoparticles: preparation, properties, functionalization and biological applications. *Chemical Society Reviews* 42 (16): 6620–6633.

40 Younis, M.A., Tawfeek, H.M., Abdellatif, A.A.H. et al. (2022). Clinical translation of nanomedicines: challenges, opportunities, and keys. *Advanced Drug Delivery Reviews* 181: 114083–114083.

41 Nguyen, T.Q., Martini, I.B., Liu, J., and Schwartz, B.J. (2000). Controlling interchain interactions in conjugated polymers: the effects of chain morphology on exciton-exciton annihilation and aggregation in MEH-PPV films. *The Journal of Physical Chemistry. B* 104 (2): 237–255.

42 Pecher, J. and Mecking, S. (2007). Nanoparticles from step-growth coordination polymerization. *Macromolecules* 40 (22): 7733–7735.

43 Feng, X., Tang, Y., Duan, X. et al. (2010). Lipid-modified conjugated polymer nanoparticles for cell imaging and transfection. *Journal of Materials Chemistry* 20 (7): 1312–1316.

44 Xing, C., Liu, L., Tang, H. et al. (2011). Design guidelines for conjugated polymers with light-activated anticancer activity. *Advanced Functional Materials* 21 (21): 4058–4067.

45 Howes, P., Green, M., Levitt, J. et al. (2010). Phospholipid encapsulated semiconducting polymer nanoparticles: their use in cell imaging and protein attachment. *Journal of the American Chemical Society* 132 (11): 3989–3996.

46 Feng, X., Lv, F., Liu, L. et al. (2010). Conjugated polymer nanoparticles for drug delivery and imaging. *ACS Applied Materials & Interfaces* 2 (8): 2429–2435.

47 Chong, H., Nie, C., Zhu, C. et al. (2012). Conjugated polymer nanoparticles for light-activated anticancer and antibacterial activity with imaging capability. *Langmuir* 28 (4): 2091–2098.

48 Wu, C., Jin, Y., Schneider, T. et al. (2010). Ultrabright and bioorthogonal labeling of cellular targets using semiconducting polymer dots and click chemistry. *Angewandte Chemie, International Edition* 49 (49): 9436–9440.

49 Li, T., Wu, M., Wei, Q. et al. (2023). Conjugated polymer nanoparticles for tumor theranostics. *Biomacromolecules* 24 (5): 1943–1979.

50 Reineck, P., Gomez, D., Ng, S.H. et al. (2013). Distance and wavelength dependent quenching of molecular fluorescence by Au@SiO_2 core-shell nanoparticles. *ACS Nano* 7 (8): 6636–6648.

51 Chaudhuri, R.G. and Paria, S. (2012). Core/shell nanoparticles: classes, properties, synthesis mechanisms, characterization, and applications. *Chemical Reviews* 112 (4): 2373–2433.

52 Li, Z.Q., Chen, S., Li, J.J. et al. (2012). Plasmon-enhanced upconversion fluorescence in $NaYF_4$:Yb/Er/Gd nanorods coated with Au nanoparticles or nanoshells. *Journal of Applied Physics* 111 (1).

53 Fujii, M., Nakano, T., Imakita, K., and Hayashi, S. (2013). Upconversion luminescence of Er and Yb codoped $NaYF_4$ nanoparticles with metal shells. *The Journal of Physical Chemistry C* 117 (2): 1113–1120.

54 Tang, F., Ma, N., Wang, X. et al. (2011). Hybrid conjugated polymer-Ag@PNIPAM fluorescent nanoparticles with metal-enhanced fluorescence. *Journal of Materials Chemistry* 21 (42): 16943–16948.

55 Liu, J., Li, A., Tang, J. et al. (2012). Thermoresponsive silver/polymer nanohybrids with switchable metal enhanced fluorescence. *Chemical Communications* 48 (39): 4680–4682.

56 Schneider, G., Decher, G., Nerambourg, N. et al. (2006). Distance-dependent fluorescence quenching on gold nanoparticles ensheathed with layer-by-layer assembled polyelectrolytes. *Nano Letters* 6 (3): 530–536.

57 Fu, Y., Zhang, J., and Lakowicz, J.R. (2010). Plasmon-enhanced fluorescence from single fluorophores end-linked to gold nanorods. *Journal of the American Chemical Society* 132 (16): 5540–5541.

58 Cohen-Hoshen, E., Bryant, G.W., Pinkas, I. et al. (2012). Exciton-plasmon interactions in quantum dot-gold nanoparticle structures. *Nano Letters* 12 (8): 4260–4264.

59 Capehart, S.L., Coyle, M.P., Glasgow, J.E., and Francis, M.B. (2013). Controlled integration of gold nanoparticles and organic fluorophores using synthetically modified MS2 viral capsids. *Journal of the American Chemical Society* 135 (8): 3011–3016.

60 Lee, J., Govorov, A.O., Dulka, J., and Kotov, N.A. (2004). Bioconjugates of CdTe nanowires and Au nanoparticles: plasmon-exciton interactions, luminescence enhancement, and collective effects. *Nano Letters* 4 (12): 2323–2330.

61 Peito, S., Peixoto, D., Ferreira-Faria, I. et al. (2022). Nano- and microparticle-stabilized Pickering emulsions designed for topical therapeutics and cosmetic applications. *International Journal of Pharmaceutics* 615: 121455.

62 Brito de Carvalho-Guimaraes, F., Correa, K.L., de Souza, T.P. et al. (2022). A review of pickering emulsions: perspectives and applications. *Pharmaceuticals* 15.

63 Chen, Z., Zhou, L., Bing, W. et al. (2014). Light controlled reversible inversion of nanophosphor-stabilized Pickering emulsions for biphasic enantioselective biocatalysis. *Journal of the American Chemical Society* 136 (20): 7498–7504.

64 Itoh, T., Kimoto, M., Kuroda, N. et al. (2023). Light-responsive crosslinked polymer particles from heterogeneous polymerization of an asymmetric divinyl azobenzene monomer. *ACS Applied Polymer Materials* 5 (4): 2787–2797.

65 Przybylo, M., Glogocka, D., Dobrucki, J.W. et al. (2016). The cellular internalization of liposome encapsulated protoporphyrin IX by HeLa cells. *European Journal of Pharmaceutical Sciences* 85: 39–46.

66 Yang, Y.-T., Chien, H.-F., Chang, P.-H. et al. (2013). Photodynamic inactivation of chlorin e6-loaded CTAB-liposomes against Candida albicans. *Lasers in Surgery and Medicine* 45 (3): 175–185.

67 Lassalle, H.-P., Dumas, D., Graefe, S. et al. (2009). Correlation between *in vivo* pharmacokinetics, intratumoral distribution and photodynamic efficiency of liposomal mTHPC. *Journal of Controlled Release* 134 (2): 118–124.

68 Sobotta, L., Dlugaszewska, J., Kasprzycki, P. et al. (2018). *in vitro* photodynamic activity of lipid vesicles with zinc phthalocyanine derivative against *Enterococcus faecalis*. *Journal of Photochemistry and Photobiology, B: Biology* 183: 111–118.

69 Massiot, J., Makky, A., Di Meo, F. et al. (2017). Impact of lipid composition and photosensitizer hydrophobicity on the efficiency of light-triggered liposomal release. *Physical Chemistry Chemical Physics* 19 (18): 11460–11473.

70 Shemesh, C.S., Moshkelani, D., and Zhang, H. (2015). Thermosensitive liposome formulated indocyanine green for near-infrared triggered photodynamic therapy: *in vivo* evaluation for triple-negative breast cancer. *Pharmaceutical Research* 32 (5): 1604–1614.

71 Cheng, X., Gao, J., Ding, Y. et al. (2021). Multi-functional liposome: a powerful theranostic nano-platform enhancing photodynamic therapy. *Advanced Science* 8 (16): 2100876.

72 Cai, K., He, X., Song, Z. et al. (2015). Dimeric drug polymeric nanoparticles with exceptionally high drug loading and quantitative loading efficiency. *Journal of the American Chemical Society* 137 (10): 3458–3461.

73 Oerlemans, C., Bult, W., Bos, M. et al. (2010). Polymeric micelles in anticancer therapy: targeting, imaging and triggered release. *Pharmaceutical Research* 27 (12): 2569–2589.

74 Zheng, Y., Weng, C., Cheng, C. et al. (2020). Multiblock copolymers toward segmentation-driven morphological transition. *Macromolecules* 53 (14): 5992–6001.

75 Li, H., Liu, H., Nie, T. et al. (2018). Molecular bottlebrush as a unimolecular vehicle with tunable shape for photothermal cancer therapy. *Biomaterials* 178: 620–629.

76 Guo, M. and Jiang, M. (2009). Non-covalently connected micelles (NCCMs): the origins and development of a new concept. *Soft Matter* 5 (3): 495–500.

77 Wang, G. and Zhang, L. (2016). Synthesis, self-assembly and pH sensitivity of PDEAEMA-PEG-PDEAEMA triblock copolymer micelles for drug delivery. *Reactive & Functional Polymers* 107: 1–10.

78 Lin, W., Yao, N., Qian, L. et al. (2017). pH-responsive unimolecular micelle-gold nanoparticles-drug nanohybrid system for cancer theranostics. *Acta Biomaterialia* 58: 455–465.

79 Shang, Y., Zheng, N., and Wang, Z. (2019). Tetraphenylsilane-cored star-shaped polymer micelles with pH/redox dual response and active targeting function for drug-controlled release. *Biomacromolecules* 20 (12): 4602–4610.

80 Liu, Y., Chen, X., Liu, X. et al. (2023). Aggregation-induced emission-active micelles: synthesis, characterization, and applications. *Chemical Society Reviews* 52 (4): 1456–1490.

81 Wang, M., Xu, L., Lin, M. et al. (2021). Fabrication of reversible pH-responsive aggregation-induced emission luminogens assisted by a block copolymer *via* a dynamic covalent bond. *Polymer Chemistry* 12 (19): 2825–2831.

82 Zhang, C., Jin, S., Li, S. et al. (2014). Imaging intracellular anticancer drug delivery by self-assembly micelles with aggregation-induced emission (AIE micelles). *ACS Applied Materials & Interfaces* 6 (7): 5212–5220.

83 Quinn, J.F., Johnston, A.P.R., Such, G.K. et al. (2007). Next generation, sequentially assembled ultrathin films: beyond electrostatics. *Chemical Society Reviews* 36 (5): 707–718.

84 Wang, Y., Angelatos, A.S., and Caruso, F. (2008). Template synthesis of nanostructured materials via layer-by-layer assembly. *Chemistry of Materials* 20 (3): 848–858.

85 He, Q., Cui, Y., and Li, J. (2009). Molecular assembly and application of biomimetic microcapsules. *Chemical Society Reviews* 38 (8): 2292–2303.

86 Tong, W. and Gao, C. (2008). Multilayer microcapsules with tailored structures for bio-related applications. *Journal of Materials Chemistry* 18 (32): 3799–3812.

87 Shchukin, D.G. and Sukhorukov, G.B. (2004). Nanoparticle synthesis in engineered organic nanoscale reactors. *Advanced Materials* 16 (8): 671–682.

88 Guo, Y., Zheng, J., Wang, Z. et al. (2023). Biocompatible optical fiber for photomedical application. *Giant* 16: 100195.

89 Wang, P., Wang, Y., and Tong, L. (2013). Functionalized polymer nanofibers: a versatile platform for manipulating light at the nanoscale. *Light: Science & Applications* 2: e102.

90 Wang, Y., Huang, Y., Bai, H. et al. (2021). Biocompatible and biodegradable polymer optical fiber for biomedical application: a review. *Biosensors* 11 (12): 472.

91 Holland, C., Numata, K., Rnjak-Kovacina, J., and Seib, F.P. (2019). The biomedical use of silk: past, present, future. *Advanced Healthcare Materials* 8 (1): e1800465.

92 Applegate, M.B., Perotto, G., Kaplan, D.L., and Omenetto, F.G. (2015). Biocompatible silk step-index optical waveguides. *Biomedical Optics Express* 6 (11): 4221–4227.

93 Li, J., Li, S., Huang, J. et al. (2022). Spider silk-inspired artificial fibers. *Advanced Science* 9 (5): 2103965.

94 Subbotina, E., Montanari, C., Olsen, P., and Berglund, L.A. (2022). Fully bio-based cellulose nanofiber/epoxy composites with both sustainable production and selective matrix deconstruction towards infinite fiber recycling systems. *Journal of Materials Chemistry A* 10 (2): 570–576.

95 Ran, Y., Xu, Z., Chen, M. et al. (2022). Fiber-optic theranostics (FOT): interstitial fiber-optic needles for cancer sensing and therapy. *Advanced Science* 9 (15): e2200456.

96 Zhang, Y., Zheng, J., Jin, F. et al. (2024). Fiber-optic drug delivery strategy for synergistic cancer photothermal-chemotherapy. *Light: Science & Applications* 13 (1): 228.

97 Zhao, H., Xu, J., Peng, K. et al. (2020). Supramolecular nanofibers for encapsulation and in situ differentiation of neural stem cells. *Advanced Healthcare Materials* 9 (1): e1901295.

98 Yetisen, A.K., Jiang, N., Fallahi, A. et al. (2017). Glucose-sensitive hydrogel optical fibers functionalized with phenylboronic acid. *Advanced Materials* 29 (15): 1606380.

99 Fu, R., Luo, W., Nazempour, R. et al. (2018). Implantable and biodegradable poly(L-lactic acid) fibers for optical neural interfaces. *Advanced Optical Materials* 6 (3): 1700941.

100 Maleki, H., Azimi, B., Ismaeilimoghadam, S., and Danti, S. (2022). Poly(lactic acid)-based electrospun fibrous structures for biomedical applications. *Applied Sciences* 12 (6): 3192.

101 Yimyai, T., Crespy, D., and Pena-Francesch, A. (2023). Self-healing photochromic elastomer composites for wearable UV-sensors. *Advanced Functional Materials* 33 (20): 2213717.

102 Bansal, A., Yang, F., Xi, T. et al. (2018). In vivo wireless photonic photodynamic therapy. *Proceedings of the National Academy of Sciences of the United States of America* 115 (7): 1469–1474.

103 Shan, D.Y., Zhang, C.J., Kalaba, S. et al. (2017). Flexible biodegradable citrate-based polymeric step-index optical fiber. *Biomaterials* 143: 142–148.

104 Shundo, A., Okada, Y., Ito, F., and Tanaka, K. (2012). Fluorescence behavior of dyes in thin films of various polymers. *Macromolecules* 45 (1): 329–335.

105 Park, J.E., Kim, J., and Nam, J.M. (2017). Emerging plasmonic nanostructures for controlling and enhancing photoluminescence. *Chemical Science* 8 (7): 4696–4704.

106 Ma, H.L., Peng, Q., An, Z.F. et al. (2019). Efficient and long-lived room-temperature organic phosphorescence: theoretical descriptors for molecular designs. *Journal of the American Chemical Society* 141 (2): 1010–1015.
107 Shen, Z.Q., Li, W.N., Tang, W.Y. et al. (2024). Fluorophor embedded MOFs steering gas ultra-recognition. *Advanced Functional Materials* 34 (32): 2401631.
108 Qian, Y., Murayama, M., Guan, S., and Zhao, X. (2024). Annealing temperature-dependent luminescence color coordination in Eu-doped AlN thin films. *Journal of Materials Science and Chemical Engineering* 12: 20–28.
109 Chen, C.L., Zhang, W.J., Wang, Z. et al. (2025). Large-area, ultra-thin organic films with both photochromic and phosphorescence properties. *Angewandte Chemie, International Edition* 64 (18): e202501448.
110 Wei, J., Jiao, X.L., Wang, T., and Chen, D.R. (2016). Electrospun photochromic hybrid membranes for flexible rewritable media. *ACS Applied Materials & Interfaces* 8 (43): 29713–29720.
111 Balmond, E.I., Tautges, B.K., Faulkner, A.L. et al. (2016). Comparative evaluation of substituent effect on the photochromic properties of spiropyrans and spirooxazines. *The Journal of Organic Chemistry* 81 (19): 8744–8758.
112 Zhao, D., Li, Z.X., Li, Q.R. et al. (2025). Design, classification, and applications of organic, luminescent lanthanide-based photoresponsive materials: a review. *Coordination Chemistry Reviews* 540: 216770.
113 Zhang, J.J., Zou, Q., and Tian, H. (2013). Photochromic materials: more than meets the eye. *Advanced Materials* 25 (3): 378–399.
114 Akiba, U., Minaki, D., and Anzai, J. (2017). Photosensitive layer-by-layer assemblies containing azobenzene groups: synthesis and biomedical applications. *Polymers* 9 (11): 553.
115 Li, M., Igbari, F., Wang, Z.K., and Liao, L.S. (2020). Indoor thin-film photovoltaics: progress and challenges. *Advanced Energy Materials* 10 (28): 2000641.
116 He, M., Qiu, F., and Lin, Z.Q. (2013). Toward high-performance organic-inorganic hybrid solar cells: bringing conjugated polymers and inorganic nanocrystals in close contact. *Journal of Physical Chemistry Letters* 4 (11): 1788–1796.
117 Han, M., Kim, B., Lim, H. et al. (2020). Transparent photothermal heaters from a soluble NIR-absorbing diimmonium salt. *Advanced Materials* 32 (1): e1905096.
118 Sorianello, V., Perna, A., Colace, L. et al. (2008). Near-infrared absorption of germanium thin films on silicon. *Applied Physics Letters* 93 (11): 111115.
119 Wang, J., Zhu, C.H., Han, J. et al. (2018). Controllable synthesis of gold nanorod/conducting polymer core/shell hybrids toward in vitro and in vivo near-infrared photothermal therapy. *ACS Applied Materials & Interfaces* 10 (15): 12323–12330.
120 Singh, B., Chejara, M.R., and Park, M.H. (2023). Light-responsive layer-by-layer film containing gold nanorods for sequential drug release. *ACS Omega.* 8 (50): 48405–48412.
121 Maleki, A., He, J.H., Bochani, S. et al. (2021). Multifunctional photoactive hydrogels for wound healing acceleration. *ACS Nano* 15 (12): 18895–18930.

122 Russell, G.M., Kaneko, T., Ishino, S. et al. (2022). Transient photodegradability of photostable gel induced by simultaneous treatment with acid and UV-light for phototuning of optically functional materials. *Advanced Functional Materials* 32 (40): 2205855.

123 Huang, W., Chen, Y., Hu, J. et al. (2022). Algal sulfated polysaccharide-based hydrogels enhance gelling properties and in vitro wound healing compared to conventional hydrogels. *Algal Research* 65: 102740.

124 Wang, M.H., Bai, J.Z., Shao, K. et al. (2021). Poly(vinyl alcohol) hydrogels: the old and new functional materials. *International Journal of Polymeric Science* 2021: 1–16.

125 Wang, R., Cheng, C., Wang, H., and Wang, D. (2024). Swollen hydrogel nanotechnology: advanced applications of the rudimentary swelling properties of hydrogels. *ChemPhysMater* 3 (4): 357–375.

126 Hoare, T.R. and Kohane, D.S. (2008). Hydrogels in drug delivery: progress and challenges. *Polymer* 49 (8): 1993–2007.

127 Mohanto, S., Narayana, S., Merai, K.P. et al. (2023). Advancements in gelatin-based hydrogel systems for biomedical applications: a state-of-the-art review. *International Journal of Biological Macromolecules* 253: 127143.

128 Galindo, J.M., Tardío, C., Saikia, B. et al. (2023). Recent insights about the role of gels in organic photonics and electronics. *Gels* 9 (11): 875.

129 Zhang, J.X., Jin, J.H., Wan, J.Q. et al. (2021). Quantum dots-based hydrogels for sensing applications. *Chemical Engineering Journal* 408: 127351.

130 Nuñez, D.G., Fasce, D., Galante, M.J., and Oyanguren, P.A. (2021). Photo-induced changes in azobenzene-containing soft materials. *Optical Materials* 115: 111032.

131 Gang, F., Jiang, L., Xiao, Y. et al. (2021). Multi-functional magnetic hydrogel: design strategies and applications. *Nano Select* 2 (12): 2291–2307.

132 Bo, Y.J., Yang, M.H., Qian, Z.H. et al. (2025). Hydrogel flexible photodetector based on polarization of free water molecules and image sensor application. *Nano Energy* 138: 110889.

133 Cheng, Q.Q., Sheng, Z.Z., Ding, Y.F. et al. (2025). 3D printed colloidal aerogels: principle, process, performance, and perspective. *Progress in Materials Science* 152: 101456.

134 Moradian, A., Behzadnasab, M., Bagher, Z. et al. (2025). Photo- and thermal-crosslinked GelMA/chitosan hydrogels: a novel approach to enhanced mechanical and biological properties. *Carbohydrate Polymer Technologies and Applications* 10: 100834.

135 Yamamoto, T., Campbell, J.A., Panyukov, S., and Rubinstein, M. (2022). Scaling theory of swelling and deswelling of polymer networks. *Macromolecules* 55 (9): 3588–3601.

136 Dimitriyev, M.S., Chang, Y.W., Goldbart, P.M., and Fernández-Nieves, A. (2019). Swelling thermodynamics and phase transitions of polymer gels. *Nano Futures* 3 (4): 042001.

137 Satchanska, G., Davidova, S., and Petrov, P.D. (2024). Natural and synthetic polymers for biomedical and environmental applications. *Polymers* 16 (8): 1159.

138 Zhao, L.L., Zhou, Y.F., Zhang, J.Y. et al. (2023). Natural polymer-based hydrogels: from polymer to biomedical applications. *Pharmaceutics* 15 (10): 2514.

139 Xiong, X.K., Sun, J., Hu, D. et al. (2020). Fabrication of polyvinyl alcohol hydrogels with excellent shape memory and ultraviolet-shielding behavior via the introduction of tea polyphenols. *RSC Advances* 10 (58): 35226–35234.

140 Nicolescu, F.A., Jerca, V.V., Vuluga, D.M., and Vasilescu, D.S. (2010). Synthesis and characterization of side-chain poly(methacrylate)s bearing new azo-moieties. *Polymer Bulletin* 65 (9): 905–916.

141 Richbourg, N.R., Wancura, M., Gilchrist, A.E. et al. (2021). Precise control of synthetic hydrogel network structure via linear, independent synthesis-swelling relationships. *Science Advances* 7 (7): eabe3245.

142 Han, J., Fan, C.C., Li, D.S., and Jiang, L. (2025). A hydrogel electrolyte with dynamically Tunable mechanical properties for wide-temperature flexible supercapacitor. *Journal of Power Sources* 631: 236217.

143 Wang, J.Q., Yang, W.Y., Li, Y.T. et al. (2024). Dual-temperature/pH-sensitive hydrogels with excellent strength and toughness crosslinked using three crosslinking methods. *Gels* 10 (7): 480.

144 Leng, Y., Britten, C.N., Tarannum, F. et al. (2024). Stimuli-responsive phosphate hydrogel: a study on swelling behavior, mechanical properties, and application in expansion microscopy. *ACS Omega.* 9 (36): 37687–37701.

145 Yang, Y., Long, K.Q., Chu, Y.X. et al. (2024). Photoresponsive drug delivery systems: challenges and progress. *Advanced Functional Materials* 34 (38): 2402975.

146 Li, L., Scheiger, J.M., and Levkin, P.A. (2019). Design and applications of photoresponsive hydrogels. *Advanced Materials* 31 (26): 1807333.

147 Yang, Y.Q., Guan, L., Jiang, H.C. et al. (2018). A rapidly responsive photochromic hydrogel with high mechanical strength for ink-free printing. *Journal of Materials Chemistry C* 6 (28): 7619–7625.

148 Qi, J.J., Su, G.M.Y., and Li, Z. (2021). Gel-based luminescent conductive materials and their applications in biosensors and bioelectronics. *Materials* 14 (22): 6759.

149 Zhu, Q., Zhang, L., Van Vliet, K. et al. (2018). White light-emitting multistimuli-responsive hydrogels with lanthanides and carbon dots. *ACS Applied Materials & Interfaces* 10 (12): 10409–10418.

150 Zhang, Y.C., Le, X.X., Jian, Y.K. et al. (2019). 3D fluorescent hydrogel origami for multistage data security protection. *Advanced Functional Materials* 29 (46): 1905514.

151 Yang, S., Sarkar, S., Xie, X. et al. (2024). Application of optical hydrogels in environmental sensing. *Energy & Environmental Materials* 7 (3): e12646.

152 Zhang, H.Y., Liu, J.X., Shi, F. et al. (2022). A novel bidirectional fast self-responsive PVA-PNIPAM/LimCsnWO$_3$ composite hydrogel for smart window applications. *Chemical Engineering Journal* 431: 133353.

153 Chen, G.Q., Wang, K., Yang, J.H. et al. (2023). Printable thermochromic hydrogel-based smart window for all-weather building temperature regulation in diverse climates. *Advanced Materials* 35 (20): e2211716.

154 Cheng, H.B., Yoon, J., and Tian, H. (2018). Recent advances in the use of photochromic dyes for photocontrol in biomedicine. *Coordination Chemistry Reviews* 372: 66–84.

155 Shi, H.W., Wang, X., Guo, H.J. et al. (2024). Antiswelling photochromic hydrogels for underwater optically camouflageable flexible electronic devices. *ACS Applied Materials & Interfaces* 16 (35): 46810–46821.

156 He, H.J., Song, X.M., Huang, M.N. et al. (2023). A photothermal and conductive composite hydrogel membrane for solar-driven synchronous desalination and salinity power generation. *Green Chemistry* 25 (22): 9343–9350.

157 Croitoru, A.M., Ficai, D., and Ficai, A. (2024). Novel photothermal graphene-based hydrogels in biomedical applications. *Polymers* 16 (8): 1098.

158 Zhao, J., Zhai, X., Li, P. et al. (2025). From sea cucumbers to soft robots: a photothermal-responsive hydrogel actuator with shape memory. *ACS Applied Materials & Interfaces* 17 (4): 6979–6986.

159 Tian, Y., Sun, D.W., Xu, L. et al. (2024). Dual-bioinspired and ultra-flexible photothermal eutectogels for highly efficient passive anti-freezing. *Chemical Engineering Journal* 488: 151143.

160 Harankahawa, N., Perera, K., and Vidanapathirana, K. (2017). Use of gel polymer electrolytes to integrate photoelectric conversion and energy storage. *Journal of Energy Storage* 13: 96–102.

161 Wu, L.W., Qi, J.F., Zhang, L.L. et al. (2024). Self-powered photoelectric sensors based on hydrogel diodes doped with photoacid. *Chemical Engineering Journal* 489: 151215.

162 Li, C.C., Zhu, B., Liu, Z.X. et al. (2022). Polyelectrolyte-based photothermal hydrogel with low evaporation enthalpy for solar-driven salt-tolerant desalination. *Chemical Engineering Journal* 431: 134224.

163 Zang, P.Y., Du, Y.Q., Yu, C.H. et al. (2023). Photothermal-actuated thermoelectric therapy by harnessing janus-structured Ag-Ag_2S nanoparticles with enhanced antitumor efficacy. *Chemistry of Materials* 35 (18): 7770–7780.

164 Hu, Y.W., Wang, Y.H., Yang, F. et al. (2024). Flexible organic photovoltaic-powered hydrogel bioelectronic dressing with biomimetic electrical stimulation for healing infected diabetic wounds. *Advanced Science* 11 (10): e2307746.

165 Hussain, M.I., Xia, M., Ren, X.N. et al. (2025). Recent advances in photopolymerization 3D printing of alumina-ceramic. *Progress in Natural Science: Materials International* 35 (1): 1–30.

166 Timko, B.P. and Kohane, D.S. (2014). Prospects for near-infrared technology in remotely triggered drug delivery. *Expert Opinion on Drug Delivery* 11 (11): 1681–1685.

167 Du, B., Cao, X., Zhao, F. et al. (2016). Multimodal imaging-guided, dual-targeted photothermal therapy for cancer. *Journal of Materials Chemistry B* 4 (11): 2038–2050.

168 Lee, H.P. and Gaharwar, A.K. (2020). Light-responsive inorganic biomaterials for biomedical applications. *Advanced Science* 7 (17): 2000863.

169 Fooladi, S., Nematollahi, M.H., and Iravani, S. (2023). Nanophotocatalysts in biomedicine: cancer therapeutic, tissue engineering, biosensing, and drug delivery applications. *Environmental Research* 231: 116287.

170 Tsang, C.Y. and Zhang, Y. (2024). Nanomaterials for light-mediated therapeutics in deep tissue. *Chemical Society Reviews* 53 (6): 2898–2931.

171 Qiu, Z. and Tang, D. (2020). Nanostructure-based photoelectrochemical sensing platforms for biomedical applications. *Journal of Materials Chemistry B* 8 (13): 2541–2561.

172 Nasseri, B., Alizadeh, E., Bani, F. et al. (2022). Nanomaterials for photothermal and photodynamic cancer therapy. *Applied Physics Reviews* 9 (1): 011317.

173 Li, S., Zhang, G., Peng, Y. et al. (2023). Tyrosinase-activated nanocomposites for double-modals imaging guided photodynamic and photothermal synergistic therapy. *Advanced Healthcare Materials* 12 (23): e2300327.

174 Nuiachristos, V. (2006). Fluorescence molecular imaging. *Annual Review of Biomedical Engineering* 8: 1–33.

175 Li, K., Pan, J., Feng, S.-S. et al. (2009). Generic strategy of preparing fluorescent conjugated-polymer-loaded poly(DL-lactide-*co*-Glycolide) nanoparticles for targeted cell imaging. *Advanced Functional Materials* 19 (22): 3535–3542.

176 Geng, J., Goh, C.C., Tomczak, N. et al. (2014). Micelle/silica co-protected conjugated polymer nanoparticles for two-photon excited brain vascular imaging. *Chemistry of Materials* 26 (5): 1874–1880.

177 Li, K., Ding, D., Huo, D. et al. (2012). Conjugated polymer based nanoparticles as dual-modal probes for targeted in vivo fluorescence and magnetic resonance imaging. *Advanced Functional Materials* 22 (15): 3107–3115.

178 Phillips, R.L., Miranda, O.R., You, C.-C. et al. (2008). Rapid and efficient identification of bacteria using gold-nanoparticle–poly(para-phenyleneethynylene) constructs. *Angewandte Chemie, International Edition* 47 (14): 2590–2594.

179 Maxwell, D.J., Taylor, J.R., and Nie, S.M. (2002). Self-assembled nanoparticle probes for recognition and detection of biomolecules. *Journal of the American Chemical Society* 124 (32): 9606–9612.

180 Ho, H.A. and Leclerc, M. (2004). Optical sensors based on hybrid aptamer/conjugated polymer complexes. *Journal of the American Chemical Society* 126 (5): 1384–1387.

181 Zeng, S., Wang, Z., Chen, C. et al. (2022). Construction of rhodamine-based AIE photosensitizer hydrogel with clinical potential for selective ablation of drug-resistant gram-positive bacteria in vivo. *Advanced Healthcare Materials* 11 (17): e2200837.

182 Zhang, Y., Zhang, L., Wang, Z. et al. (2019). Renal-clearable ultrasmall covalent organic framework nanodots as photodynamic agents for effective cancer therapy. *Biomaterials* 223: 119462.

183 Tian, J., Zhou, J., Shen, Z. et al. (2015). A pH-activatable and aniline-substituted photosensitizer for near-infrared cancer theranostics. *Chemical Science* 6 (10): 5969–5977.

184 Kim, H., Kim, Y., Kim, I.-H. et al. (2014). ROS-responsive activatable photosensitizing agent for imaging and photodynamic therapy of activated macrophages. *Theranostics* 4 (1): 1–11.

185 Ruan, C., Liu, C., Hu, H. et al. (2019). NIR-II light-modulated thermosensitive hydrogel for light-triggered cisplatin release and repeatable chemo-photothermal therapy. *Chemical Science* 10 (17): 4699–4706.

186 Fonseca, R.G., De Bon, F., Pereira, P. et al. (2022). Photo-degradable, tough and highly stretchable hydrogels. *Materials Today Bio.* 15: 100325.

187 Barman, S., Das, J., Biswas, S. et al. (2017). A spiropyran-coumarin platform: an environment sensitive photoresponsive drug delivery system for efficient cancer therapy. *Journal of Materials Chemistry B* 5 (21): 3940–3944.

188 Liu, C., Zhang, Y., Liu, M. et al. (2017). A NIR-controlled cage mimicking system for hydrophobic drug mediated cancer therapy. *Biomaterials* 139: 151–162.

189 Griffin, D.R. and Kasko, A.M. (2012). Photodegradable macromers and hydrogels for live cell encapsulation and release. *Journal of the American Chemical Society* 134 (31): 13103–13107.

190 Hao, Y., Liu, H., Li, G. et al. (2018). Photo and thermo dual-responsive copolymer surfaces for efficient cell capture and release. *ChemPhysChem* 19 (16): 2107–2112.

191 Cai, Y., Chai, T., Nguyen, W. et al. (2025). Phototherapy in cancer treatment: strategies and challenges. *Signal Transduction and Targeted Therapy* 10 (1): 115.

192 Han, H.S. and Choi, K.Y. (2021). Advances in nanomaterial-mediated photothermal cancer therapies: toward clinical applications. *Biomedicine* 9 (3): 305.

193 Xi, D., Xiao, M., Cao, J. et al. (2020). NIR light-driving barrier-free group rotation in nanoparticles with an 88.3% photothermal conversion efficiency for photothermal therapy. *Advanced Materials* 32 (11): 1907855.

194 Wang, Y., Xia, G., Tan, M. et al. (2022). H-dimeric nanospheres of amphipathic squaraine dye with an 81.2% photothermal conversion efficiency for photothermal therapy. *Advanced Functional Materials* 32 (25): 2113098.

195 Ali, S., Razzaq, A., and In, S.-I. (2019). Development of graphene based photocatalysts for CO_2 reduction to C_1 chemicals: a brief overview. *Catalysis Today* 335: 39–54.

196 Sundar, D., Liu, C.-H., Anandan, S., and Wu, J.J. (2023). Photocatalytic CO_2 conversion into solar fuels using carbon-based materials-a review. *Molecules* 28 (14): 5383.

197 Zhao, W., Zhai, D., Liu, C. et al. (2022). Unblocked intramolecular charge transfer for enhanced CO_2 photoreduction enabled by an imidazolium-based ionic conjugated microporous polymer. *Applied Catalysis B: Environmental* 300: 120719.

198 Fang, Y., Gao, Y., Wen, Y. et al. (2024). Photoelectrocatalytic CO_2 reduction to methanol by molecular self-assemblies confined in covalent polymer networks. *Journal of the American Chemical Society* 146 (40): 27475–27485.

199 Jin, E., Lan, Z., Jiang, Q. et al. (2019). 2D sp^2 carbon-conjugated covalent organic frameworks for photocatalytic hydrogen production from water. *Chem* 5 (6): 1632–1647.

200 Hu, H., Wang, Z., Cao, L. et al. (2021). Metal-organic frameworks embedded in a liposome facilitate overall photocatalytic water splitting. *Nature Chemistry* 13 (4): 358–366.

201 Yunus, I.S., Wichmann, J., Woerdenweber, R. et al. (2018). Synthetic metabolic pathways for photobiological conversion of CO_2 into hydrocarbon fuel. *Metabolic Engineering* 49: 201–211.

202 Love, J. (2022). Microbial pathways for advanced biofuel production. *Biochemical Society Transactions* 50 (2): 987–1001.

203 Song, J., Lin, H., Zhao, G., and Huang, X. (2022). Photocatalytic material-microorganism hybrid system and its application-a review. *Micromachines* 13 (6): 861.

204 Gai, P., Yu, W., Zhao, H. et al. (2020). Solar-powered organic semiconductor-bacteria biohybrids for CO_2 reduction into acetic acid. *Angewandte Chemie, International Edition* 59 (18): 7224–7229.

205 Cai, Z., Huang, L., Quan, X. et al. (2020). Acetate production from inorganic carbon (HCO_3^-) in photo-assisted biocathode microbial electrosynthesis systems using $WO_3/MoO_3/g\text{-}C_3N_4$ heterojunctions and *Serratia marcescens* species. *Applied Catalysis B: Environmental* 267: 118611.

206 Teng, K.-X., Niu, L.-Y., Xie, N., and Yang, Q.-Z. (2022). Supramolecular photodynamic agents for simultaneous oxidation of NADH and generation of superoxide radical. *Nature Communications* 13 (1): 6179.

207 Zhang, E., Zuo, Z., Yu, W. et al. (2021). Photoactive conjugated polymer/graphdiyne nanocatalyst for CO_2 reduction to CO in living cells for hypoxia tumor treatment. *Materials Chemistry Frontiers* 5 (15): 5841–5845.

208 Zhu, C., Cheng, J., Lin, H. et al. (2024). Rational design of conjugated polymers for photocatalytic CO_2 reduction: towards localized CO production and macrophage polarization. *Journal of the American Chemical Society* 146 (36): 24832–24841.

2

Structures and Properties of Photofunctional Materials

2.1 Small Organic Molecules

Small organic molecules have become a research hot spot in the fields of materials science and molecular electronics due to their potential application value in optoelectronic devices like organic light-emitting diodes (OLEDs), photovoltaic devices, and biomedical devices such as fluorescence labeling and molecular probes. The photophysical process of this type of material is essentially that molecules in an excited state relax from high energy levels (such as singlet state S_1 or triplet state T_1) to the ground state (S_0) through radiative transitions, while releasing photons of characteristic wavelengths. According to the differences in the sources of excitation energy, its luminescence mechanisms can be classified into the following typical modes: photoluminescence (PL), electroluminescence (EL), chemiluminescence, and mechanoluminescence. In-depth research on such optical functional materials not only promotes theoretical breakthroughs in luminescence mechanisms but also gives rise to multiple technological innovations, such as high-efficiency OLED display devices, highly sensitive biosensors, intelligent responsive materials, and anticounterfeiting encryption technologies.

2.1.1 Photoluminescence

According to the Jablonski energy-level diagram (Figure 2.1), there are S_1 and T_1 electron spin states in the molecular excited state. Fluorescent radiation originates from the allowable transition of excitons in the excited state S_1 to the ground state S_0. Phosphorescence emission occurs when excitons are transferred to the T_1 state through intersystem crossing (ISC) and then return to the ground state. The difference between these two processes is essentially the result of the action of the spin selection rule.

2.1.1.1 Fluorescence

π-conjugated systems generally have fluorescence characteristics, but their luminescence behavior is significantly affected by the molecular aggregation state. Take perylene as an example (Figure 2.2). The planar configuration molecule shows blue fluorescence in the dispersed state, but in the aggregated state, because of the π–π

Photofunctional Polymer Composites for Bioapplications, First Edition. Xiaoyu Wang and Lidong Li.

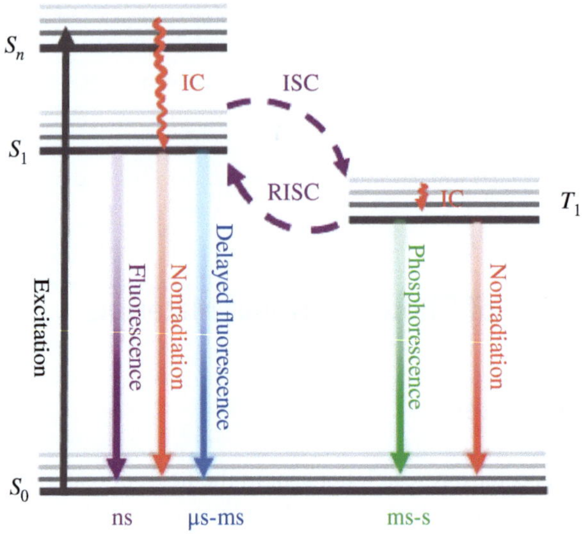

Figure 2.1 The schematic illustrating the fluorescence, ISC, phosphorescence, and reverse ISC processes.

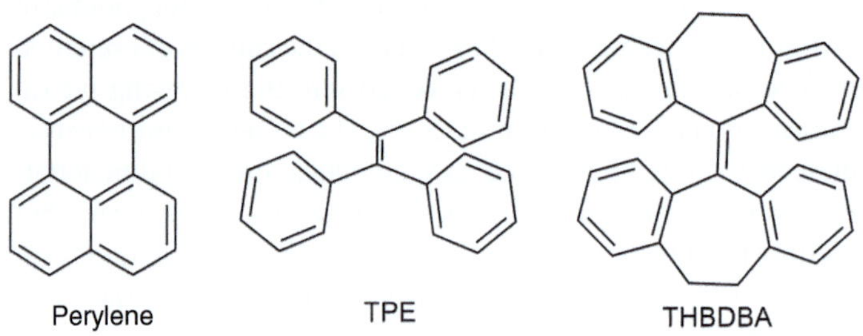

Figure 2.2 Molecular structures of perylene, TPE, and THBDBA.

interaction between molecules, nonradiative attenuation intensifies and the blue fluorescence almost disappears. This phenomenon is defined as aggregation-caused quenching (ACQ) [1]. Since the ACQ effect was first reported in 1954, the issue has posed a major obstacle to the utilization of luminescent molecules [2]. Although traditional solutions such as host-guest doping systems can partially alleviate the ACQ effect, they have inherent drawbacks such as complex preparation processes and poor phase separation stability [3].

Contrary to the ACQ effect, molecules with a propeller configuration exhibit a significant fluorescence enhancement effect in the aggregated state [1]. To explain this phenomenon, scientists have proposed many possible mechanisms. Currently, the mechanism generally recognized in the academic community is the theory of restriction of intramolecular motion (RIM). Its specific mechanism of action presents different characteristics due to the differences in molecular structure.

Restricted Intramolecular Rotation (RIR) Mechanism Tetraphenylethylene (TPE) serves as a representative example. Structurally, it features a central ethylene core attached to four benzene rings. In solution, the benzene rings can rotate freely, promoting nonradiative decay of the excited electrons and causing the loss of energy without fluorescence emission. Conversely, when TPE molecules aggregate, the close packing restricts the rotation of the benzene rings. This limitation reduces nonradiative decay and enables the excited-state energy to be released as fluorescence emission.

Restricted Intramolecular Vibration (RIV) Mechanism The molecular structure of 10,10′,11,11′-tetrahydro-5,5′-bidibenzo[a,d][7]annulenylidene (THBDBA) is illustrated in Figure 2.2, where the outer benzene ring is linked via an ethyl group. The energy dissipation in this molecule primarily arises from the vibrations of the molecular framework. The spatial constraints generated by molecular stacking in the aggregated state will suppress the vibrational degrees of freedom, thereby weakening the nonradiative attenuation and ultimately achieving enhanced fluorescence emission [4].

In addition to aggregation-induced emission (AIE), excimer/exciplex is another fluorescence phenomenon that is intimately associated with molecular stacking. Excimer/exciplex is a dimer of excited-state substances formed by the combination of excited-state molecules and ground-state molecules. As mentioned earlier, face-to-face π–π stacking often leads to ACQ. Because it suppresses radiative transitions and dissipates energy in a nonradiative form, π–π stacking has long been believed to quench fluorescence and reduce luminescence efficiency. However, excimers are a special exception.

A dimer is the simplest model among molecular aggregates, and the study of its association is conducive to a deeper understanding of the luminescence mechanism of aggregates. In very concentrated pyrene solutions, two emission bands are observed: a blue emission characterized by a clear fine structure, and a green emission lacking such features [5]. With the increase of concentration, the long-wavelength emission is enhanced but the absorption spectrum changes weakly, indicating that the interaction occurs in the excited state rather than the ground state. In 1960, such substances were officially named "excimers".

Figure 2.3 shows the emission properties and the generation process of excimers. The repulsive energy between two nearby molecules of the ground state increases dramatically, thereby causing the potential energy curve to rise sharply from flat to steep. Under excited conditions, when the excited-state molecules approach the ground-state molecules, a significant attractive force will be generated, thereby causing the potential energy curve of the association to decrease. When the excimer reaches the equilibrium state, the potential energy reaches the lowest value. A high redshift emission can be seen because the excimer's energy between the equilibrium position and the ground state is significantly lower than the monomer's. Meanwhile, due to the deep excitation potential energy and the steep ground state potential energy curve, thermal vibration will cause drastic fluctuations in the vibrational energy level, resulting in a wide emission range of excimers. Alternatively, owing to the dynamic interplay within the excited state, the lifetime of excimers is

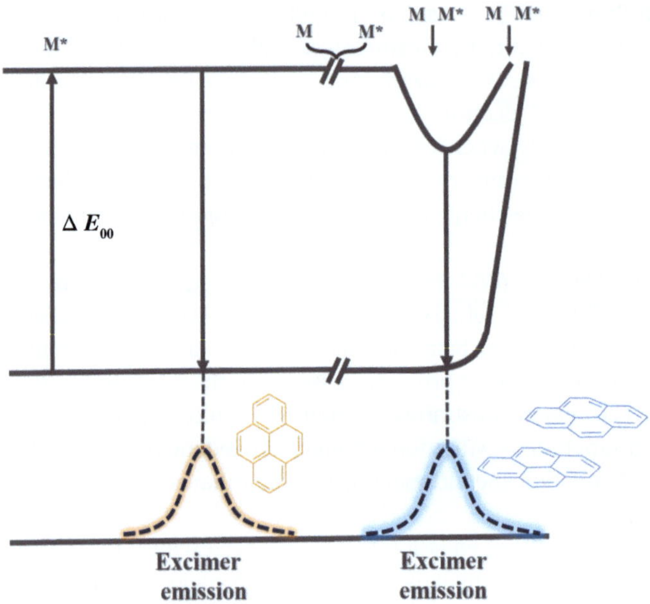

Figure 2.3 The excimer formation of pyrene.

often significantly longer than that of monomers. Therefore, by regulating the formation of excimers, it is possible to efficiently suppress nonradiative energy loss to improve luminescence efficiency as well as to control color and lifetime of the luminescence [6].

The vibration-induced emission (VIE) mechanism can dynamically regulate color, intensity, or lifetime by utilizing the vibration of molecules. *N*,*N*′-diaryl-dihydrodibenzo[a,c]phenazines (DPAC) is a typical saddle compound. Due to its VIE effect, it can exhibit an exceptionally large Stokes shift emission [7]. As illustrated in Figure 2.4, when the excited-state configuration of the VIE-active molecule in the solution changes from a bent structure to a planar shape, fluorescence change is observed. This planarization is impeded by the presence of K^+ ions, which results in the molecule's distinctive blue light being released instead. As a result, dynamic color tuning—which changes from orange-red to blue as well as white light—can be accomplished by adjusting the local environment of VIE-active molecules.

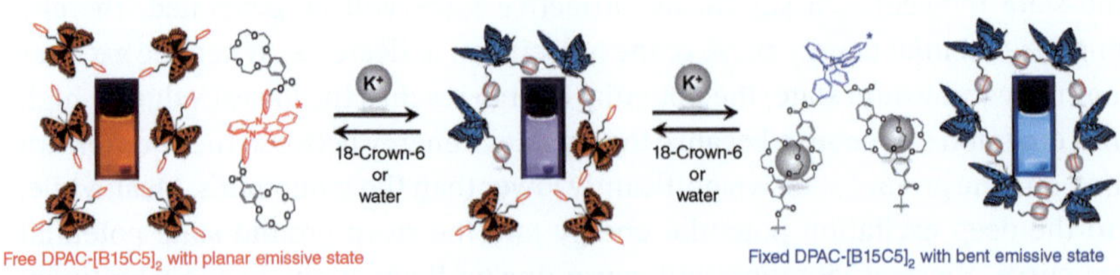

Figure 2.4 Schematic illustration of the photophysical behaviors of DPAC-[B15C5]$_2$. *Source:* Sun et al. [7] / with permission of American Chemical Society.

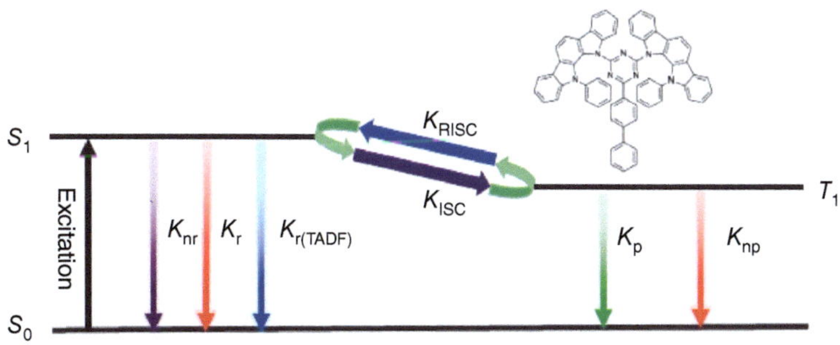

Figure 2.5 Schematic illustration of TADF of PIC-TRZ.

In contrast to regular fluorescence, thermal-activated delayed fluorescence (TADF) is a distinct form of fluorescence emission that has garnered a lot of interest because of its extended lifetime and improved efficiency in exciton utilization [8]. The Jablonski diagram reveals its luminescence mechanism. As shown in Figure 2.5, TADF is characterized by ISC and RISC transitions between the S_1 and T_1 states, before falling back to S_0 to emit fluorescence, while ordinary fluorescence directly falls back from S_1 to S_0. The different processes of the excited states result in significant differences in the behavior of TADF from the normal excited state [9]. 2-biphenyl-4,6-bis(12-phenylindolo[2,3-a]carbazole-11-yl)-1,3,5-triazine (PIC-TRZ) is reported to consist of an indolecarbazole donor and a triazine receptor [10]. With a small energy difference (ΔE_{ST} = 0.11 eV) between the S_1 and T_1 states, the ISC and RISC processes are effectively promoted.

From this, it can be known that for the emission of TADF, a small ΔE_{ST} is the most crucial element. Generally, it is considered that $\Delta E_{ST} < 0.3$ eV is necessary. Therefore, achieving a small ΔE_{ST} has become the focus of research. In luminescent molecules with strong spin-orbit coupling effects, the energy gap between S_1 and T_1 states is significantly large [11]. A spatially deformed donor–acceptor (D–A) structure proves advantageous. For example, the butterfly-shaped structure of Cz2BP shows a relatively large ΔE_{ST} (0.32 eV) due to the partial overlap between its HOMO and LUMO, while the ΔE_{ST} values of CC2BP and Px2BP are 0.10 and 0.01 eV, respectively [12]. Since the ΔE_{ST} value is small enough to achieve the upconversion from T_1 to S_1, all these compounds exhibit typical TADF effects, further demonstrating the effectiveness of the deformed D–A structure in achieving a small ΔE_{ST} and the TADF effect that follows.

2.1.1.2 Phosphorescence

Generally speaking, because of the poor spin-orbit coupling between S_1 and T_1 states, it is difficult to obtain phosphorescence from the T_1 state, whereas the excited electrons usually decay to S_0 by fluorescence. Heavy atoms (e.g. I, Br) or heteroatoms (e.g. N, O, S, P) are among the solutions to molecular design, which has been suggested to improve the ISC process between S_1 and T_1 states. Nevertheless, phosphorescence is often only observed for these molecules at low temperatures, like 77 K. At room temperature, nonradiative transitions of triplet excitons are dominant, so room-temperature phosphorescence is very rare. It was not until 2015 that scientists began to turn to the research of molecular stacking, and more room-temperature phosphorescent materials were developed. Efficient intersystem transitions and nonradiative transition processes that suppress triplet states are two key elements for achieving room-temperature phosphorescence.

Improve the ISC Efficiency The heavy-atom effect refers to the fact that heavy atoms can cause or enhance the spin-orbit coupling of solute molecules, which is conducive to the generation of phosphorescence and the increase of the quantum yield of phosphorescence. Based on the indole luminescent group, a series of novel room-temperature phosphorescent molecules (24CIM, 25CIM, 246CIM, and 26CIM) can be prepared by bridging indole with carbonyl groups and chlorine-substituted phenyl groups. As shown in Figure 2.6, by adjusting the substitution positions of chlorine atoms, the heavy atom enhances the spin-orbit coupling and ISC process, achieving high room-temperature phosphorescence efficiency. The Cl atoms at positions 2 and 6 efficiently support the hydrogen interaction between H and Cl atoms and facilitate the heavy-atom effect. The phosphorescence efficiency is improved from 2.9% (24CIM) to 48.9% (26CIM), which is 16 times higher [13]. This is an intramolecular space heavy-atom effect strategy for rationally constructing efficient room-temperature phosphorescent materials.

In addition to the heavy-atom effect, the El-Sayed rule is also one of the common strategies for constructing pure organic room-temperature phosphorescent materials [14]. As shown in Figure 2.7, this rule states that if the electron transition involves a change in orbital type, the occurrence rate of the ISC is relatively large. For example, in the ISC process of carbonyl compounds, the rate of $^1(\pi,\pi^*) \rightarrow {}^3(n,\pi^*)$

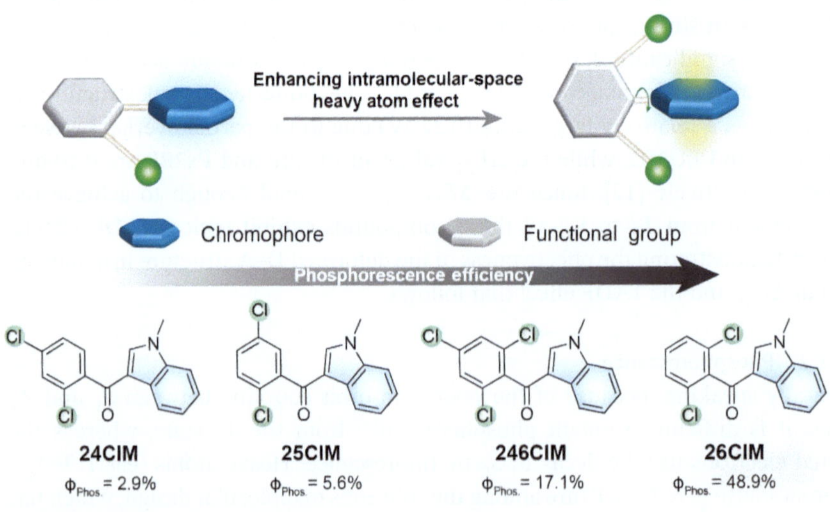

Figure 2.6 The structures of 24CIM, 25CIM, 246CIM, and 26CIM, along with the heavy-atom effect for promoting ISC efficiency. *Source:* He et al. [13] / with permission of John Wiley & Sons.

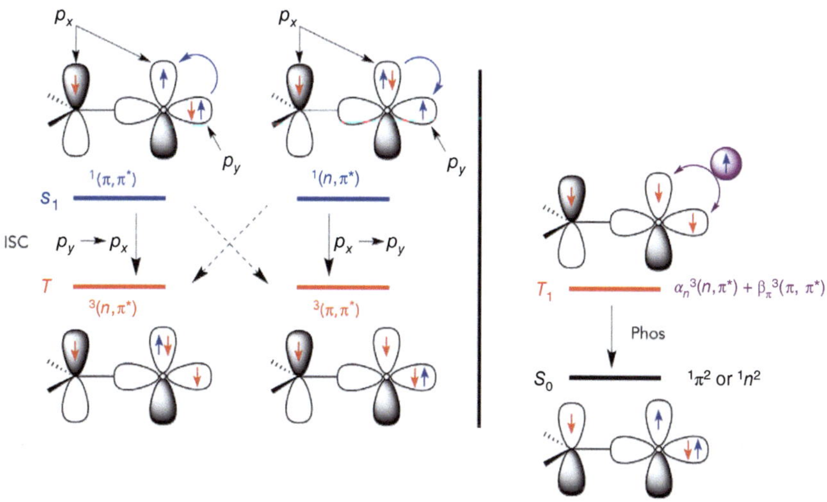

Figure 2.7 Schematic illustration of the El-Sayed rule. *Source:* Zhao et al. [15] / with permission of Elsevier.

is higher than that of $^1(\pi, \pi^*) \rightarrow {}^3(\pi, \pi^*)$, and the rate of $^1(n, \pi^*) \rightarrow {}^3(\pi, \pi^*)$ is higher than that of $^1(n, \pi^*) \rightarrow {}^3(n, \pi^*)$. This is based on the changes in oxygen atomic orbitals [15]. Therefore, introducing groups with lone pairs of electrons into pure organic molecules, such as aromatic carbonyl groups, heteroatoms (such as S, P), or introducing B atoms with empty *pz* orbitals, is a common approach to improving phosphorescent performance at room temperature [16].

Suppress the T_1 Nonradiation The crystal has a rigid structure and a clear packing mode, which can effectively avoid the nonradiation of excited electrons and enhance phosphorescence emission. The molecules in the crystal are arranged in an orderly manner, reducing the vibration and rotation between molecules. This helps stabilize the triple excited state and reduces the deactivation of energy in a nonradiative way. Secondly, the ordered arrangement in the lattice also limits the distance between molecules, reducing the nonradiation caused by intermolecular interactions. In addition, intramolecular or intermolecular interactions, such as hydrogen bonds, also have an impact in stabilizing the molecular structure and reducing nonradiative deactivation channels. These interactions enhance the rigidity of the molecule and further stabilize the triplet excited state. In addition, the dense crystal structure may stop outside elements like moisture and oxygen from penetrating, thereby reducing the collision and deactivation of triplet excited-state molecules with oxygen. This isolation effect is crucial for protecting excited-state molecules and prolonging the phosphorescence lifetime.

For example, the 1,4-diethoxybenzene (DEOPh) crystal structure, which illustrates the aggregation of H molecules [17]. Therefore, strong coupling between molecules can be achieved, providing an effective means to stabilize and protect the formed triplet excitons. In addition, the stable triplet excitons of DEOPh display an extended lifetime.

The effective concentration of luminescent molecules in the cocrystal system is significantly reduced, which helps to suppress the self-quenching phenomenon of triplet excitons. For example, 6-methyl-1,3,5-triazo-2,4-diamine (MTDA) can be assembled with aromatic dicarboxylic acids (IPA, TPA, and PA) based on noncovalent interactions to form a cocrystal system [18]. These ordered structures can effectively suppress the nonradiative emission of triplet excitons and enhance the ISC process. Consequently, these cocrystal systems exhibit notable afterglow, even at 150 °C.

Doping or embedding the guest emitter into a rigid matrix is a simple and effective strategy for obtaining phosphorescence. The appropriate energy levels of the host and guest can provide an effective transition path for energy transfer, which is conducive to the generation and stability of triplet excitons. Figure 2.8 illustrates that the strong electron-donating guest molecule TMB possesses a special structure endowing it with an extremely stable free radical cation. Meanwhile, the strong electron-accepting host molecule PPT has a high triplet state energy and provides a rigid amorphous environment for the system. This kind of environment is crucial for suppressing nonradiative deactivation. It helps to maintain the stability of the excited state, thereby prolonging the lifetime of phosphorescence. When TMB and PPT assemble, they form an exciplex [19]. Therefore, mixed exciplex emission with phosphorescence leads to a prolonged luminescence period.

2.1.2 Electroluminescence

Electroluminescence (EL) is a phenomenon of converting electrical energy into light, referring to the luminescence of functional molecules by exciton recombination excitation under the passage of current or the action of a strong electric field. Because organic luminescent bodies have advantages such as easy processing, high emission efficiency, and suitability for flexible display compared with inorganic luminescent bodies, organic EL has aroused great interest since its discovery. In the early stages of development, OLED devices, commonly known as organic EL devices, faced significant challenges in injecting charges into organic structures.

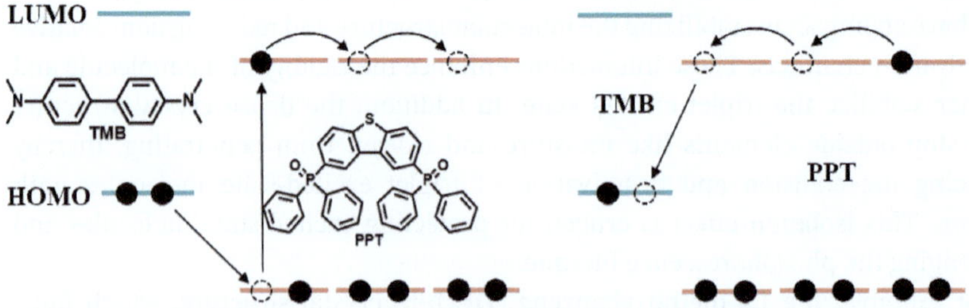

Figure 2.8 The molecular structures of TMB and PPT, along with the luminescence mechanism.

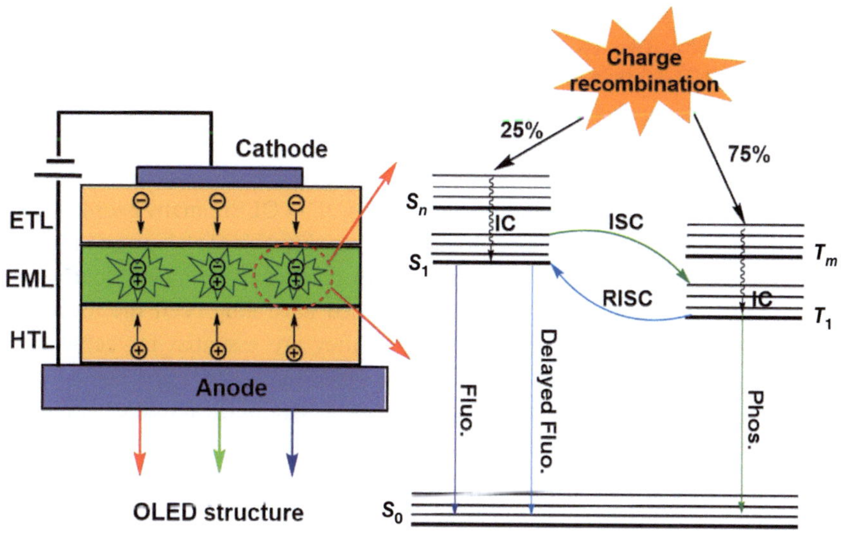

Figure 2.9 The fundamental structure of an OLED and its EL mechanism. *Source:* Fang et al. [21] / with permission of Elsevier.

This hindered their progress. However, in 1987, the use of a double-layer film containing tris-(8-hydroxyquinoline)-aluminum (Alq_3) as the emission layer achieved improved EL performance for the first time [20].

The sandwich configuration comprising two electrodes (Figure 2.9)—an electron transport layer, an emission layer, and a hole transport layer—is now the main structure used in OLED devices. Of these layers, the organic emitting layer is the most important as it has a direct impact on the overall performance of the OLED device. When voltage is applied, the electrodes inject electrons and holes into the electron and hole transport layers, respectively. Then, inside the emission layer, these charge carriers move and converge. According to spin statistics, at this point, electrons and holes recombine to yield 75% triplet excitons and 25% singlet excitons [21]. However, only singlet excitons are used to produce fluorescence in traditional fluorescent emitters, resulting in the loss of the triplet excitons as heat and less-than-ideal OLED performance.

To improve the utilization of excitons and enhance EL performance, advanced organic emitters with TADF, room-temperature phosphorescence, triplet–triplet annihilation (TTA), and hybridized local and charge transfer have been developed [22]. These mechanisms will be discussed in detail in the following sections.

As mentioned in the previous section on photoluminescence, TADF luminescent materials typically have a small energy difference (<0.3 eV) between S_1 and T_1 states. This characteristic has inspired the application of EL. TADF materials exhibit reverse ISC and display delayed fluorescence in a thermally activated state, effectively utilizing triplet excitons. This luminescence mechanism significantly improves the luminous efficiency of OLEDs. By employing carbazole as the donor and dicyanobenzene as the electron acceptor, a series of highly efficient TADF emitters was designed and synthesized [23]. Due to steric hindrance, the carbazolyl unit is significantly twisted relative to the dicyanobenzene plane, causing the HOMO and LUMO orbitals to be localized on the donor and acceptor, respectively, which results in a smaller ΔE_{ST}. Furthermore, the steric hindrance between the carbazole and dicyanobenzene groups within the molecule enhances its rigidity, thereby improving the luminescence efficiency of the resulting compounds.

Due to the promising application prospects of TADF in OLED, many people have devoted themselves to it, enabling it to achieve considerable development in the OLED field. Many color TADF excitation layers have been designed, achieving excellent OLED performance. Blue, green, and red have long been the focus of development in light-emitting devices [24], as they are essential for achieving full-color displays (Figure 2.10). For the blue TADF emission layer, the TADF emitter TDBA-DI, which was prepared using a rigid, symmetrical oxygen-bridged boron receptor and a diindolocarbazole (DI) donor, produces a blue color [25]. In 2018, TADF agents based on diboron, CzDBA [26], were reported for the green TADF luminescent layer. Carbazole derivatives were used as donors and 9,10-dihydro-9,10-diboranthracene (DBA) as the acceptor, achieving good quantum yields. Achieving high EL efficiency in the red TADF emission layer is typically challenging due to its small bandgap, which often results in significant nonradiative processes and a low quantum yield. To enhance the luminescence efficiency of red TADF emitters, a new receptor known as dibenzo[a,c]phenazine-3,6-dinitrile has been introduced [27]. This receptor has a large, planar structure that effectively suppresses nonradiation. Consequently, the TPA-PZCN red TADF emitter achieved an exceptional quantum yield. In conclusion, the TADF emission layer of the three primary colors has made considerable progress and has broad application prospects.

The room-temperature phosphorescent materials introduced earlier achieve their luminescence mechanism precisely through triplet exciton transitions, making them ideal for OLEDs. However, this approach also presents a challenge. While some room-temperature phosphorescent materials have been proven to have high quantum yields, these materials exist in a crystalline state, making it challenging to achieve high EL efficiency. It was not until 2019 that a new organic luminescent material, DPTZN, was developed, which can exhibit room-temperature phosphorescence in both the crystalline and amorphous thin-film states (Figure 2.11) [28]. In addition, the spin-orbit coupling can be promoted by incorporating selenium (Se) [29], and a series of PSe1, PSe2, and PSe3 phosphorescent agents have been designed

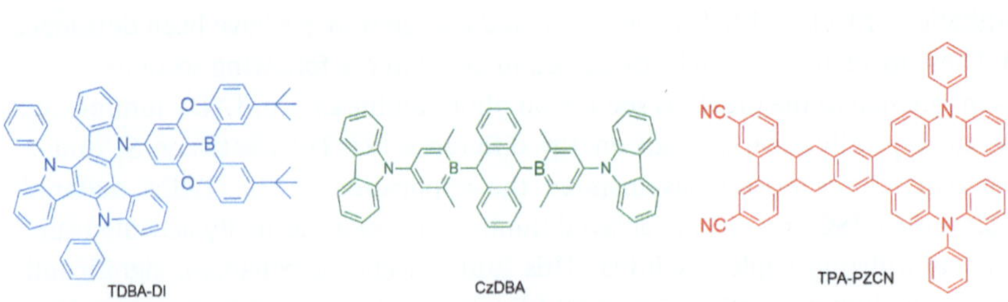

Figure 2.10 The molecular structures of TDBA-DI, CzDBA, and TPA-PZCN.

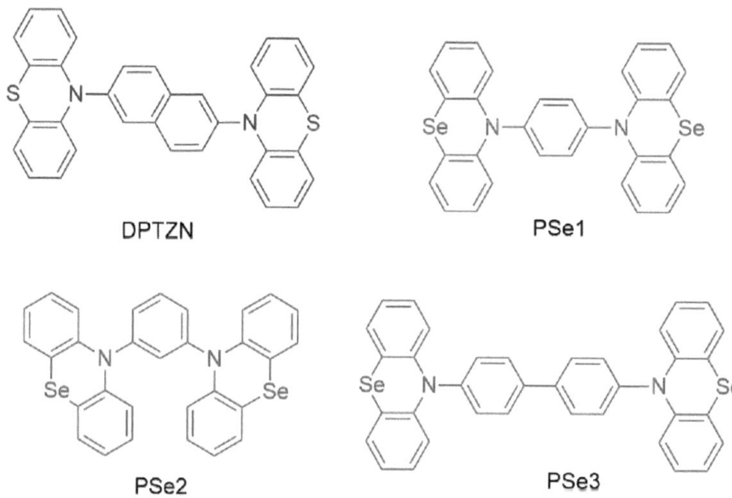

Figure 2.11 The molecular structures of DPTZN, PSe1, PSe2, and PSe3.

and synthesized. All of these compounds exhibited excellent room-temperature phosphorescence in the doped films.

In systems characterized by $2T_1 > S_1$ energy level properties, two triplet excitons can interact to form a singlet exciton and a ground-state molecule. This phenomenon is known as triplet–triplet annihilation (TTA). As the TTA effect usually occurs in polycyclic aromatic hydrocarbons, such as anthracene and phenanthrene, without significant intramolecular charge transfer (ICT), most TTA emitters are found in the blue light region [30, 31]. As shown in Figure 2.12, by introducing weak donor (—OCH_3) and receptor (—CN) units to 1,4-bis(10-phenylanthracene-9-yl)benzene, the EL efficiency based on these molecules can be increased from 8.9% of BD-1 to 9.5% of BD-2, and then to 12.0% of BD-3 [32]. In addition, a series of blue TTA emitters, including TSTA, TSMA, and TSNA, have been constructed using 2-(styrene-based) benzopyrene (TS) as the core component [33]. In addition, by replacing benzopyrene with pyrene, three other deep blue emitters, DPASP, DFASP, and PCzSP [34] were obtained. Their corresponding EL efficiencies could also exceed 10% to achieve an effective TTA effect.

The hybridized local and charge transfer excited states (HLCT) play a crucial role in enhancing the EL performance of emitting layers. HLCT combines two complementary characteristics: the local excitation feature exhibits a large transition probability, which significantly boosts the fluorescence radiation efficiency; charge transfer enhances reverse ISC efficiency, thereby improving exciton utilization. Moreover, the reverse ISC process in higher excited states occurs at a much faster rate, which helps mitigate exciton annihilation caused by high concentrations. As a result, HLCT emitters demonstrate excellent performance in nondoped OLED devices [35]. For example, an HLCT emitter consisting of TPA-PPI was created by linking the donor molecule triphenylamine (TPA) with the acceptor molecule 1,2-Diphenyl-1*H*-phenazol [9,10-d] imidazole (PPI) [36] (Figure 2.13). Deep blue

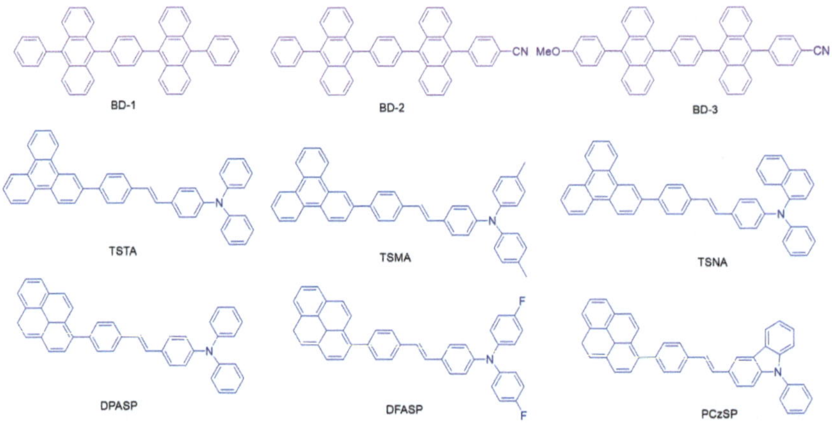

Figure 2.12 The molecular structures of BD-1, BD-2, BD-3, TSTA, TSMA, TSNA, DPASP, DFASP, and PCzSP.

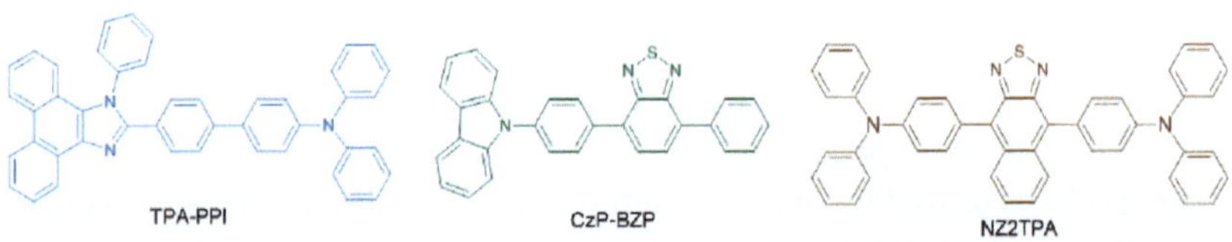

Figure 2.13 The molecular structures of TPA-PPI, CzP-BZP, and NZ2TPA.

emission with a high quantum yield could be produced by this emitter in both solution and solid states. Carbazole-phenyl (CzP) was used as the donor [37] and benzo [c][1,2,5] thiadiazole (BZ) as the acceptor to create the D–A molecule CzP-BZP. The undoped OLED based on CzP-BZP exhibits superior EL performance because of its more balanced HLCT properties. Furthermore, utilizing TPA as the donor and NZ as the receptor, near-infrared (NIR) luminous NZ2TPA with HLCT and AIE effects was created and synthesized [38]. NZ2TPA outperformed the others in terms of EL performance in NIR undoped OLEDs.

2.1.3 Chemiluminescence

Chemiluminescence is a process where the energy released during a chemical reaction excites the luminescent agents to generate an excited state, which then transitions back to the ground state and emits light, without the need for light excitation. Most chemiluminescence systems generate light through two consecutive chemical reactions. First, in the presence of an oxidant, the functional molecule is oxidized to form a high-energy intermediate, which is very unstable and immediately decomposes to produce another intermediate in an excited state. When it transitions to the ground state, it is accompanied by the release of light. Compared with photoluminescence, which requires the formation of an excited state through external light irradiation, chemiluminescence generates an excited state intermediate through a chemical reaction, and this process does not require the participation of an external light source. Therefore, eliminating the influence of spontaneous luminescence, light scattering, and photobleaching of biological structures can improve the detection sensitivity and imaging signal-to-noise ratio. Moreover, it has a fast response speed and a wide dynamic range, and is extensively employed in the domains of environmental pollutants and biomedical detection [39]. The common chemiluminescence systems mainly fall into three categories, namely luminol and its derivatives, oxalate peroxide esters, and dioxacyclobutane.

The chemiluminescence phenomenon of luminol was first proposed by Albrecht in 1928. The chemiluminescence principle of luminol is shown in Figure 2.14 [40]. In an alkaline environment, luminol is oxidized by ROS to form a cyclic peroxide intermediate. This intermediate has high energy and is unstable, and will immediately decompose to generate a 3-amino phthalate ion, which is in an excited state. Upon returning to its ground state, it emits blue light of around 425 nm. The luminol molecule is extensively applied in forensic medicine for detecting blood traces because its reaction can be catalyzed by iron in hemoglobin proteins [41, 42]. Recently, it has been the subject of much basic research in bioapplications [43].

The activation of luminol and its derivatives chemiluminescence probes and oxalate peroxide chemiluminescence probes all rely on reactive oxygen species (ROS), and thus are only suitable for the detection of ROS. However, dioxacyclobutane chemiluminescence probes are nonoxidization dependent. By connecting different recognition units, they can detect various bioactive molecules and have a very wide range of detection and applications. In recent years, with the continuous in-depth research of many research teams around the world on dioxacyclobutane chemiluminescence probes, their optical performance under physiological conditions has been greatly improved. The CL system is based on dioxacyclobutane (Figure 2.15), which could be activated by enzymes or selected analytes [40]. Based on this system, the application scope of CL's sensors has been greatly expanded.

However, Schaap's dioxetanes have an inherent problem: under aqueous conditions, they interact with water molecules, causing large nonradiative transitions and thus making their chemiluminescence very weak. To solve this problem, scientists have made many attempts. The simplest and most accessible method is to use surfactants to cover and encapsulate, providing a hydrophobic environment to reduce quenching. However, this physical encapsulation method can also cause some side effects, such as insufficient stability of micelles or toxicity of the active agent. To find a more appropriate solution, electron-withdrawing groups were introduced at

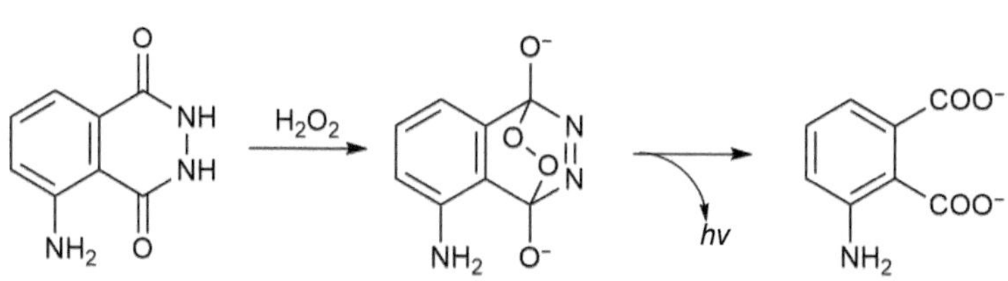

Figure 2.14 The schematic illustration for the chemiluminescence of luminols. *Source:* Zhang et al. [40] / with permission of John Wiley & Sons.

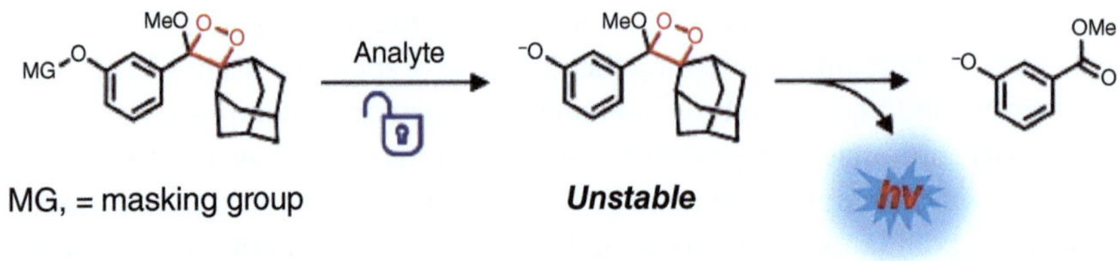

Figure 2.15 The schematic illustration for the chemiluminescence of Schaap's dioxetane. *Source:* Zhang et al. [40] / with permission of John Wiley & Sons.

the conjugated position of the phenolate donor of benzoate esters [44]. The emission of the released benzoate was greatly increased by this donor–acceptor arrangement.

Chemiluminescence is capable of effectively avoiding the background signal generated by the excitation light due to the fact that it does not necessitate light excitation for luminescence. Due to this attribute, it has extensive potential in the field of biological imaging. The most effective method for achieving deep tissue imaging is to choose emitted light with extended wavelengths, particularly those with NIR luminescence. In order to accomplish this, chemical luminescent substances can be modified by increasing intramolecular energy transfer and decreasing the energy gap. The CL color of Schaap's dioxetanes can be altered by altering the substituents at the ortho-position of the phenol donor or by increasing energy acceptors.

2.1.4 Mechanoluminescence

Mechanoluminescence (ML) refers to the phenomenon where the crystal structure of optical functional materials changes under stress (such as grinding, friction, or shaking), resulting in luminescence. There is a significant correlation between the ML effect and the molecular packing mode. The analysis of single crystal structures reveals that a well-organized intramolecular and intermolecular hydrogen bond network plays a crucial role in minimizing energy loss during nonradiation. By effectively suppressing intermolecular slippage induced by mechanical stimulation, this network significantly enhances the performance of ML.

For instance, CDpP-B crystals of CDpP molecules were obtained from a mixture of dichloromethane and methanol. Blue ML emissions were observed in the locally excited state of the crystals, which had a greater number of C—H…π interactions. This is because the twisted process of the excited state is constrained by the C—H…π interaction. By contrast, CDpP-G, which has fewer C—H…π interactions, is more likely to exhibit a twisted ICT excited state in the freely rotating diphenylamine part. This leads to green ML emission (Figure 2.16) [45].

In the ML process, it is also essential to have a noncentral symmetric arrangement. For example, Ptz-Cho is produced by combining the chiral cholesterol group with the phenothiazine derivative, a typical ML structural unit. In its crystalline state, Ptz-Cho demonstrates a vigorous ML emission; however, this emission is lost when the compound is extensively ground. Thus, the corresponding ML effect has been significantly influenced by the specific molecular packaging with the chiral space group [46].

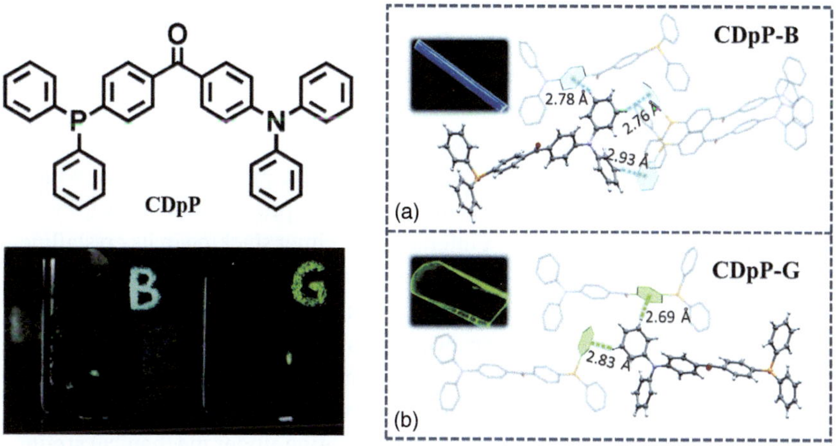

Figure 2.16 The molecular structure of CDpP and ML behavior of different crystal structures. *Source:* Xie et al. [45] / The Royal Society of Chemistry / CC BY 3.0.

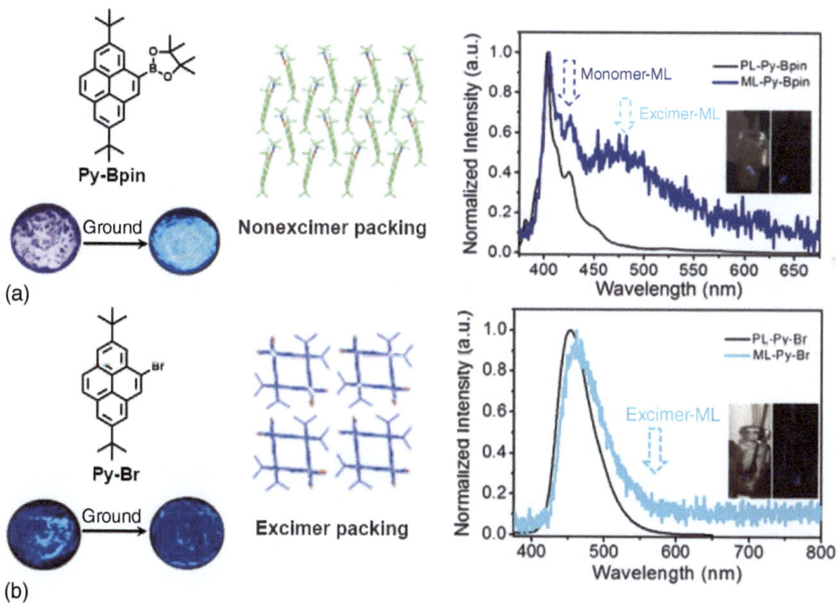

Figure 2.17 The different molecular structures, packings, and ML behaviors of (a) Py-BPin and (b) Py-Br. *Source:* Gong et al. [47] / with permission of John Wiley & Sons.

As introduced in the previous text about photoluminescence, the formation of excimers is also related to molecular packing. Using pyrene as the main construction unit, two organic luminescent agents with excimer potential could be constructed (Figure 2.17) [47]. The Py-Bpin crystal presented dual ML emission of monomer–excimer, while the Py-Br crystal presented single ML emission of excimer. The primary factors driving differences in ML behaviors are variations in molecular packing. These are induced by mechanical stimulation. This has been revealed by a detailed analysis of their single crystal structures and the observations of photoluminescence behavior changes during grinding. Py-Bpin's crystalline state is characterized by intermolecular hydrogen interactions and a nonexcimer-type clustering. Nevertheless, the transition to excimer-type packaging is demonstrated by variations in emission following the grinding process, which explains the presence of dual emissions during the ML process. In contrast, Py-Br exhibits the typical intermolecular π–π interaction excimer stacking in its crystalline state, a configuration that remains stable even after grinding. Consequently, only excimer emission is observed in both the photoluminescence and ML processes. These findings emphasize the significance of molecular packing changes under mechanical stimulation in influencing ML emissions.

In addition to molecular packing, the energy levels of luminescent compounds are closely related to changes in molecular conformation under mechanical strain, and these changes can significantly affect ML emission. Phenothiazine derivatives of FCO-CzS, which exhibit a bimolecular conformation, and the transformation between them can be achieved by grinding solid samples (Figure 2.18) [48]. In the crystalline state, the FCO-CzS compound with a quasi-axial conformation exhibits a blue emission. However, after grinding in the PL process, it turns white and then yellow, with quasi-equatorial conformation. Similarly, under the mechanical stimulation of changing the molecular conformation, dynamic ML emission from blue to white and then to yellow can be achieved. This should be the first example of dynamic mechanical luminescence, which will greatly promote the in-depth understanding of the properties of ML, especially the close relationship between molecular conformation and the ML process. Therefore, the influence of molecular packing changes under mechanical stimulation on ML emission is significant.

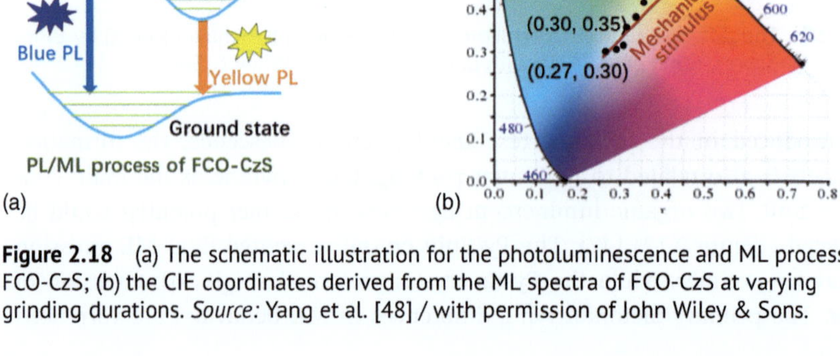

Figure 2.18 (a) The schematic illustration for the photoluminescence and ML processes of FCO-CzS; (b) the CIE coordinates derived from the ML spectra of FCO-CzS at varying grinding durations. *Source:* Yang et al. [48] / with permission of John Wiley & Sons.

2.2 Conjugated Polymers

Conjugated polymers are organic, macromolecular materials with a distinctive electronic structure. Their molecular backbone consists of alternating single and double or triple carbon–carbon bonds, as well as unsaturated bonds, such as aromatic hydrocarbons. This structure allows electrons to delocalize along the entire polymer chain, forming large π bonds, which endow conjugated polymers with a series of unique optoelectronic properties. Due to the chemical reactivity and modifiability of organic compounds, the side chains of conjugated polymers can be designed into various structures. This diversity of side chains not only enriches the chemical composition of conjugated polymers but also has a substantial effect on their physical characteristics. By altering the type, length, and functional groups of the side chains, the solubility, stability, electron mobility, and optoelectronic properties of conjugated polymers can be adjusted, thereby demonstrating broad application potential in fields like OLEDs, solar cells, sensors, and biomedicine. Therefore, the study and development of conjugated polymers hold an important position in the realms of materials science and chemistry, and the regulation of their structure and properties is key to achieving their application in high-tech fields.

2.2.1 Main Chain Structure

The main chain of conjugated polymers has an extended π-electron conjugated system, which determines their optical properties, including intramolecular energy and electron transfer efficiency, the wavelength of absorption and emission, the quantum yield and lifetime of fluorescence and phosphorescence, and the ability to generate ROS. Under normal circumstances, the main chain of conjugated polymers is composed of alternating carbon–carbon double and triple bonds, and the precise design of donor and acceptor units is the core strategy for regulating the optoelectronic properties of the material. Figure 2.19 illustrates how this strategy is implemented through the use of six conjugated units within a donor–acceptor–donor framework. The precise engineering of such systems largely determines key optical properties, including the absorption and emission bands and the fluorescence quantum yield. Alongside carbon–carbon double and triple bonds, the backbone of conjugated polymers commonly contains fused aromatic rings and unsaturated heterocycles comprising atoms such as N and S. The various combinations and arrangements of these units endow conjugated polymers with significant diversity in their optoelectronic properties. This enables their absorption and emission spectra to span almost the entire visible light range, generating various electronic band structures. Adjusting the structure of the main and side chains enables the design of conjugated polymers with unique photophysical and photochemical properties, providing new materials for use in biosensing and disease diagnosis applications. Next, we will introduce the common structural units found in the main chains of conjugated polymers [49].

The core function of donor structures is to provide hole transport channels through π-electron delocalization, and they usually have electron-rich characteristics. Common donor elements include the following:

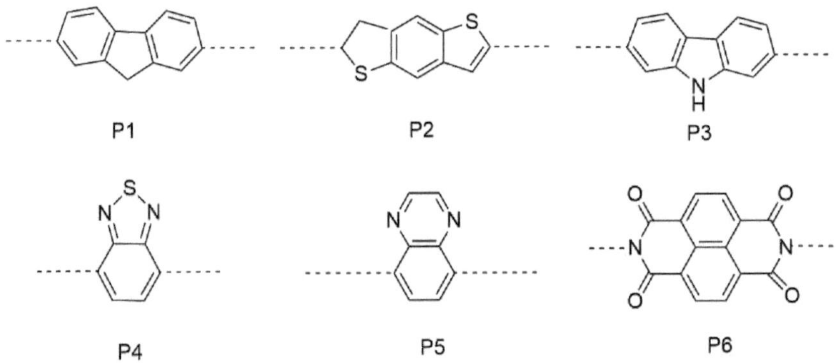

Figure 2.19 The molecular structures of six conjugated units for investigating the donor–acceptor–donor model.

Fluorene (P1), which is renowned for its rigid planar biphenyl structure and large conjugated system, is one of the most common backbone structures of conjugated polymer main chains. This structure endows fluorene compounds with unique photoelectric properties and biological activities, and gives them extensive application potential in fields such as photoelectric materials and medicine [50]. The C9 site of fluorene is sp^3 hybridized, enabling the convenient modification of side chains with different functions to alter the polymer's physical properties. The C2 and C7 sites of fluorene are also highly active and can be easily homopolymerized or copolymerized through methods such as Suzuki coupling, which further expands the range of applications of fluorene-based polymers. Polyfluorene itself exhibits blue fluorescence and an extremely high fluorescence quantum yield, rendering it an ideal organic fluorescent material. However, polyfluorene's absorption mainly occurs in the ultraviolet (UV) region, limiting its application in certain fields. To overcome this, researchers have copolymerized fluorene with other aromatic electron-deficient groups to shift the absorption spectrum to the visible or NIR regions, using pathways such as energy and electron transfer. This strategy expands the application of polyfluorene in fields such as biomedicine and enhances its performance in optoelectronic devices [51].

The fused polycyclic conjugated skeleton of benzodithiophene (BDT, P2) significantly enhances the molecular orbital delocalization effect and achieves wide-spectral domain light capture capability through dimensional expansion of the π-electron system. Compared with a single thiophene structural unit, this polycyclic system extends the optical response range from the visible light region to the NIR region via band regulation, creating a continuous photon energy gradient. The regioselective reaction characteristics exhibited at the C2, C7, and C9 sites in the molecule provide a platform for precisely tailoring the electronic structure of the material [52]. The site-specific anchoring of electron-withdrawing and electron-donating groups can therefore be used to directionally reconstruct the intramolecular charge distribution pattern. The wavelength of the photoelectric response can also be tuned by regulating the energy level difference between the highest occupied molecular orbital (HOMO) and the lowest unoccupied molecular orbital (LUMO). This molecular engineering strategy not only optimizes intrinsic photophysical properties but also modulates the intermolecular force network by designing side-chain steric hindrance and polarity. This achieves synergistic optimization of crystal growth orientation and exciton transport paths. Notably, the rearrangement of the electron cloud density between the benzene ring and the dithiophene unit in the polycyclic system induces a strong ICT effect. This results in a significant redshift of the absorption spectrum and an increase in the concentration of photogenerated carriers. The superiority of molecular planarity also promotes the ordered assembly of π–π stacks, forming anisotropic carrier transport channels in the solid state. The fluorescence emission originates from the relaxation process of the excited state configuration. The synergistic effect of long-wavelength radiation and a wide Stokes shift range effectively reduces energy dissipation during optical cycling, providing a molecular design basis for constructing an efficient photoelectric conversion system.

The N-heterocyclic aromatic hydrocarbon skeleton of carbazole (P3) forms a distinctive electron transport channel via its electron-deficient nitrogen center. Within the molecular topological structure, the N atom and its neighboring positions (e.g. the C3 and C6 sites) demonstrate significant regional chemical reactivity. This site-selective characteristic provides a precise platform for the modification of molecular functionality. The degree of delocalization of nitrogen lone pair electrons can be regulated through *N*-alkylation or arylation, and introducing substituents at the C3 and C6 sites can effectively adjust the electron cloud distribution of the polycyclic system. Its intramolecular charge distribution characteristics give the material significant hole-dominated transport properties. Meanwhile, the coupling effect of π–π and n–π transitions forms a characteristic absorption band in the UV-blue-violet light region. The rigid polycyclic skeleton inhibits energy dissipation caused by molecular vibrational relaxation and reduces the splitting of the energy levels between the singlet and triplet states through spin-orbit coupling. This promotes the reverse ISC process and generates microsecond-level delayed fluorescence emission. This blue-green light emission exhibits a significant Stokes shift relative to the excitation wavelength, which reduces the photodegradation rate while maintaining a high fluorescence quantum yield. Notably, chemically modifying the active sites can directionally regulate the formation and dissociation pathways of triplet excitons, providing a molecular design basis for constructing long-lived excited state systems.

Receptor units typically exhibit defective carrier characteristics within their structure. Through ICT effects and energy level matching, they can have a significant impact on the light-capture ability, exciton-separation efficiency, and carrier mobility of conjugated polymers. Common receptor units include the following:

The electron-deficient property of benzothiadiazole (BT, P4) stems from the synergistic effect of its fused benzene and dithiazole rings. The strong electronegativity of the heteroatoms, combined with the polarization effect of the π-conjugated system, forms a highly electron-deficient molecular framework. This significantly reduces the LUMO of the material, constructs a gradient electron distribution in the conjugated main chain, drives the ICT effect to a saturated state, and promotes the extension of the light absorption boundary toward the NIR region. It also exhibits broad-spectrum absorption characteristics. The rigid planar skeleton creates a highly organized molecular structure via π–π stacking. This close intermolecular contact provides a continuous transport channel for electron migration. Introducing steric hindrance groups into the molecular design strategy can balance the competitive relationship between packing density and exciton diffusion, thereby suppressing the nonradiative recombination process. During the relaxation process of the excited state, significant reconstruction of the molecular geometric configuration decouples the energy level matching between the emission and absorption spectra, generating a wide Stokes shift. This weakens the spectral overlap between the ground and excited states, significantly reducing energy self-loss in the optical cycle.

The molecular configuration of quinoxoline (P5) is characterized by the insertion of a dinitrogen atom into the ortho position of the benzene ring. Its asymmetric electron distribution significantly enhances receptor strength by directing the delocalization of nitrogen's lone pair electrons. The electron-withdrawing properties of the nitrogen atoms mean that the C5, C8, C2, and C3 sites of quinoxaline exhibit unique chemical reactivity, providing key sites for molecular functionalization. The electron cloud density distribution of the conjugated system can be precisely regulated through site-selective substitution. For instance, introducing electron-withdrawing groups at the C5 and C8 positions polarizes the molecular orbitals (MOs) further, whereas grafting steric hindrance groups at the C2 and C3 positions adjusts the molecular solubility and stability. In comparison to the classical BT receptor, the slight decrease in this unit's conjugation degree stabilizes its light absorption edge within the mid-visible range, leading to a distinct narrowband, high-intensity absorption peak. This selective light-capture property stems from the dynamic balance between the local excited state and the charge-transfer state. The molecule's inherent dipole moment promotes exciton dissociation by creating an internal electric field, while the rigid framework suppresses nonradiative transition channels. This enables the system to achieve high carrier mobility while maintaining excellent fluorescence quantum efficiency. Moderate molecular planarity forms a dynamic equilibrium π–π stacking mode. Through the steric hindrance effect and polarity regulation of side-chain engineering, a synergistic optimization mechanism can be established between solution-processing performance and solid-state photoactivity. This provides an ideal receptor platform for constructing high-efficiency, high-stability, flexible optoelectronic devices [53].

Naphthalenediimide (NDI, P6) uses a fused naphthalene ring to form a rigid framework for electron transport. The imide groups at both ends form a highly delocalized planar conjugated system through the synergistic electron-withdrawing effect of the nitrogen and oxygen heteroatoms. This structural feature causes the molecular orbital energy levels to shift significantly downward, forming deep electron traps and endowing the material with intrinsic N-type semiconductor properties and excellent electron transport capabilities. The material's photophysical behavior is characterized by two absorptions in the UV-visible region, resulting from the synergistic effect of π–π^* transitions and charge transfer excitons. Meanwhile, long-wavelength fluorescence emission is closely related to the relaxation and reorganization of the molecular configuration in the excited state. While the planar conjugated skeleton facilitates intermolecular π-orbital overlap to form an electron transport network, overly strong π–π interactions can induce exciton quenching in aggregates. To this end, introducing branched alkyl side chains at imide sites precisely regulates molecular packing density via steric hindrance, optimizing processing performance in solutions and suppressing nonradiative recombination. This molecular design strikes a balance between carrier migration and exciton diffusion, enabling the NDI system to efficiently regulate photophysical processes in solid-state films while maintaining a long excited state lifetime [54].

The molecular topological structure of thiophene takes the sulfur atom in the five-membered aromatic heterocyclic ring in the first position as the core. Its α (2,5) and β (3,4) positions show significant differences in regional chemical reactivity. Due to its direct conjugation with sulfur atoms, the α site has a higher electron density and is often the preferred site for electrophilic substitution. In contrast, the β site is more suitable for nucleophilic modification or steric hindrance design. When used as an electron-rich unit, the alternating D–A structure formed with electron-deficient receptors can induce strong ICT, achieving a red shift in the light absorption boundary and improving exciton dissociation efficiency. However, electron-withdrawing substitution strategies, such as fluorination or cyanoylation, can transform it into an electron-acceptor module, constructing an asymmetric charge distribution system. This structural flexibility enables thiophene derivatives to form a donor–acceptor–donor (D–A–D) molecular topology. By regulating the conjugation length and degree of energy level matching, the photoelectric conversion threshold and carrier transport path of the material can be optimized precisely. In the field of organic photovoltaics in particular, conjugated systems based on thiophene can simultaneously achieve a wide spectral response and efficient charge collection through energy level gradient design and phase separation regulation, thus becoming the core structural unit for constructing high-performance optoelectronic devices [55].

2.2.2 Side Chain Structure

Side chains are typically used to alter the solubility, processability, dispersibility, and intermolecular interactions of conjugated polymers without affecting the optical properties of their backbones directly. Having clarified the effect of the main chain on the performance of conjugated polymers, it is evident that the structure of the side chain plays a key role in determining their functionality. This directly influences the stacking and aggregation of polymers, which significantly alter their photophysical properties and play a significant role in fields such as optoelectronic devices and biomedicine. Side chain modification aims to give the conjugated polymer richer functional properties. Common side-chain modification strategies are based on the principles of covalent bonding and noncovalent interactions. Covalent modification can directly connect specific functional groups to the side chains of polymers. Through reactions such as esterification and amidation, functional groups with specific reactivity, such as carboxyl and amino groups, can be introduced. This enables conjugated polymers to specifically bind to biomolecules or other materials. In contrast, noncovalent modification utilizes weak interactions, such as π–π packing and hydrogen bonds, to functionalize side chains without altering the polymer's main-chain chemical structure. This approach is more flexible and allows the side chain structure to be rapidly adjusted according to need [56].

By precisely regulating the length and density of side chains, the intermolecular interactions can be effectively regulated. Appropriately shortening the length of side chains or reducing their density can help enhance the crystallinity of polymers and the tight packing between molecules, thereby increasing the charge mobility, a principle structurally represented by the alkyl chain variations in semiconducting polymers P1–P9 (Figure 2.20) [57]. For example, reducing the density or shortening the length of the side chains can make the polymer chains more regularly arranged, forming an ordered molecular packing structure. This has been verified in research on polythiophene polymers. Additionally, adjusting the position at which the side chain is replaced is an effective strategy. Moving the side chains from the middle to the ends of the polymer chain, the main chain's distortion is reduced, and the face-to-face stacking between molecules is improved, thereby increasing charge mobility. This effect has been observed in polystyrene-based polymers. Adjusting the position of the side-chain branch points is equally crucial. Moving the branch points away from the main chain reduces steric hindrance and keeps the polymer chain straight and regular, enhancing π–π stacking between molecules and promoting efficient charge transfer. This has been demonstrated in research into polyfluorene-conjugated polymers. The partial substitution strategy significantly improves the crystallinity and molecular packing of the polymer, enhancing charge mobility by replacing some branch side chains with linear ones while maintaining solubility. These strategies offer efficient methods for enhancing the efficacy of conjugated polymers.

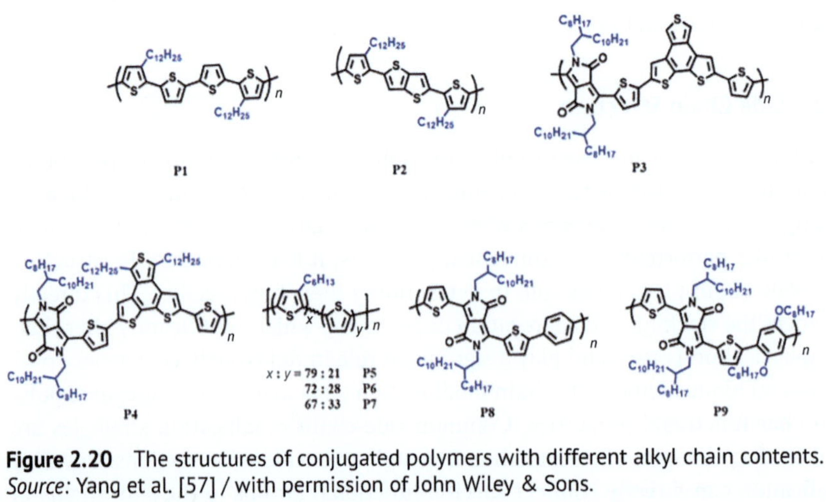

Figure 2.20 The structures of conjugated polymers with different alkyl chain contents. *Source:* Yang et al. [57] / with permission of John Wiley & Sons.

The rigid main-chain structure of conjugated polymers makes them difficult to dissolve in water, limiting their application in homogeneous aqueous solutions. However, Heeger's research combined the water-soluble conjugated polymer polythiophene, thereby facilitating the study of water-soluble conjugated polymers [58]. It has been discovered that modifying the side chains of conjugated polymers with hydrophilic ionic groups can improve their solubility. The prepared water-soluble conjugated polymers can be classified into three categories: [59] cationic, anionic, and nonionic (Figure 2.21). Cationic polymers are obtained by modifying the positively charged functional groups on the side chains, such as quaternary ammonium salts or amino groups. This provides excellent solubility and biological activity. These polymers are often used to interact with negatively charged biomolecules, such as DNA and RNA, and thus have broad application prospects in gene delivery, nucleic acid sensing, and cell imaging. Anionic water-soluble conjugated polymers achieve solubility in water by incorporating negatively charged groups like sulfonic acid, carboxylic acid, or phosphate groups into the side chains. These negatively charged groups stabilize the polymer in an aqueous solution and form coordination bonds with positively charged biomolecules or metal ions. This makes them widely used in chemical sensors, metal ion detection, and the field of biomarkers. Nonionic water-soluble conjugated polymers achieve solubility in water by incorporating nonionic hydrophilic groups (e.g. polyethylene glycol chains or sugar groups) into the side chains. This type of polymer avoids the charge shielding or aggregation problems that may occur in ionic polymers. Nonionic water-soluble conjugated polymers (WSCPs) have demonstrated outstanding performance in biological imaging, drug delivery, and biosensor applications, particularly in complex biological environments.

In addition, molecular engineering of side chains enables precise manipulation of conjugated polymers in biological systems via multidimensional regulation. Terminal functionalization modification utilizes active groups, such as carboxyl and amino groups, to construct biocoupling anchor points and form directional covalent connections with target ligands via click chemistry. This effectively enhances the ability of molecular probes to recognize specific targets. The introduction of pH-, enzyme-, or redox-sensitive units endows the material with intelligent characteristics that respond dynamically to microenvironmental changes, triggering conformational rearrangement to achieve functions such as controlled drug release, diagnosis, and treatment synergy. In terms of optical performance regulation, the push-pull electron effect of the side chain works with the main chain's conjugate

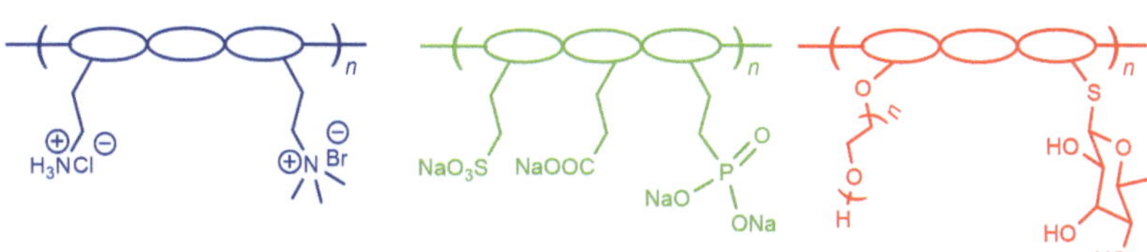

Figure 2.21 The side chains of water-soluble conjugated polymers.

system to optimize the exciton transition path. This promotes the shift of the absorption/emission peak to the NIR region, thereby enhancing tissue penetration depth. Subcellular interface behavior is controlled by the balance between the hydrophobic and hydrophilic properties of the side chains. Hydrophobic alkyl side chains enhance endocytosis efficiency through membrane insertion mechanisms, while biomimetic phospholipid side chains can simulate natural membrane structures to reduce immune rejection. Charged side chains regulate intracellular transport pathways through electrostatic mediation and temperature-responsive units achieve spatiotemporal programming of organelle localization. This multi-level functional design integrates molecular recognition, dynamic response, and interface regulation mechanisms. It enables conjugated polymer systems to cross application boundaries, from in vitro detection to in vivo diagnosis and treatment. This provides a molecular design paradigm for constructing intelligent biomedical platforms [60, 61].

2.2.3 Conformation of Conjugated Polymers

The conformation of a conjugated polymer refers to the arrangement and morphology of its molecular chains in space. Due to the presence of conjugated units, these polymers have a rigid main-chain structure that makes it easy for them to adopt specific conformations in solution. Types of chain conformation include wire clusters, rings, and rods (Figure 2.22). These conformations significantly influence the physical and chemical properties of polymers. The rigidity of a polymer's chain conformation is usually characterized by the length of a molecular chain before it loses its directional association. For conjugated polymers, this value is typically large, indicating high rigidity of the main chain [62, 63].

The conformation of conjugated polymer chains is influenced by multiple factors, including the torsional potential energy between the backbone units, the degree of polymerization, and the solvent properties. These factors jointly determine the rigidity of the polymer chain, which in turn affects its optical, electrical, and mechanical properties [64]. From the perspective of the intrinsic properties of molecular structure, the topological design of the backbone units is the core factor determining the rigidity of the chain:

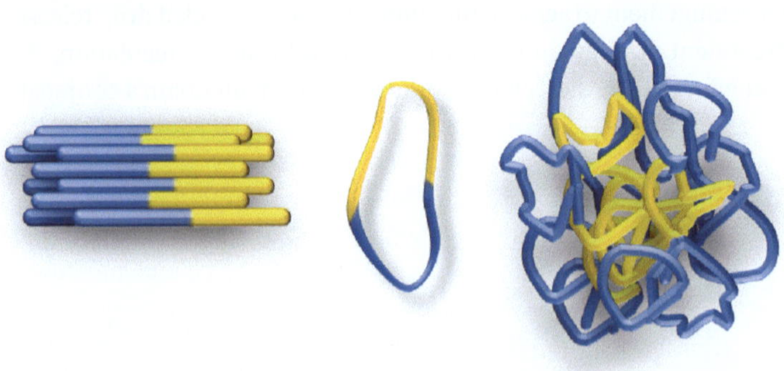

Figure 2.22 The conformation of conjugated polymer chains.

Aromatic heterocycles, such as thiophene, exhibit nonzero torsional potential energy due to angular strain within intra-ring bonds. This strain induces periodic bending of chain segments, thereby reducing the persistent length and promoting flexible conformations. In contrast, the planar bond angle arrangement characteristic of hexagonal ring systems, such as benzene, is primarily governed by bond-bending potential energy. This arrangement restricts conformational degrees of freedom, facilitating the maintenance of a linear, rigid molecular structure.

The degree of polymerization, as a critical molecular-scale parameter, significantly influences the conformational distribution of polymer chains. Through the dynamic interplay between chain length and flexibility, an increase in polymerization degree leads to a nonlinear rise in the total free energy barrier of the molecular chain. This suppresses the thermal motion of chain segments and drives conformational ordering. Additionally, the interchain interactions associated with high degrees of polymerization stabilize extended electron delocalization within the molecular conformation, enhancing electronic properties.

External environmental factors reshape the conformational equilibrium of polymer chains via energy exchange mechanisms. In good solvents, strong solvation effects reduce the cohesion between chain segments, promoting chain extension to maximize conformational entropy. Conversely, poor solvents enhance cohesive energy, driving chain folding to minimize surface energy. Temperature gradients modulate conformational diversity by influencing the system's kinetic energy, while solvent polarity reconstructs intramolecular dipole interactions through dielectric shielding effects, indirectly altering charge distribution. The coupling of these multi-scale factors ultimately dictates the material's band structure, exciton diffusion efficiency, and carrier transport pathways, establishing a comprehensive structure-property relationship that bridges molecular design and macroscopic performance.

The conformation of conjugated polymer chains plays a critical role in dictating the photophysical processes and charge transport properties of materials by modulating the dimensional and spatial distribution of the π-electron system [65]. In the extended chain conformation, the polymer backbone adopts a maximally planar arrangement, enabling π orbitals to achieve full conjugated overlap along the molecular axis, thereby forming a continuous electron delocalization channel. This quasi-one-dimensional electronic structure reduces exciton binding energy significantly and induces a redshift in the absorption and emission spectra. This phenomenon arises from the band contraction effect, which is attributed to the uniform distribution of the electron cloud within the extended conjugated system [66]. Furthermore, the rigid planar conformation is stabilized by intramolecular noncovalent interactions, such as S···O or C—H···π bonds, effectively minimizing energy disorder caused by torsional motion of chain segments and facilitating the formation of highly delocalized exciton states. While tight π–π stacking in this conformation can construct a 3D charge transport network, excessive exciton-phonon coupling may lead to nonradiative energy transfer, reducing luminescence efficiency [67].

In contrast, the folded chain conformation disrupts the long-range conjugation order through bending or helical distortions of the polymer backbone, resulting in a localized electronic structure. The deformation reduces π orbital overlap and generates discrete exciton-localized states, which manifest as bandgap broadening and blue shifts in the absorption spectrum. Under this conformation, steric hindrance between chain segments weakens intermolecular π–π interactions. Although this conformation suppresses exciton quenching and enhances fluorescence quantum yield, it compromises carrier mobility. Importantly, the dynamic conformational changes of conjugated polymers can be reversibly controlled by external stimuli such as solvent polarity or temperature, enabling a "conformational switching" effect. For instance, conjugated chains extended in good solvents spontaneously fold in poor solvents (Figure 2.23) [65]. This conformation-dependent photoelectric response provides a promising strategy for the development of intelligent optical functional materials [68].

The evolution of polymer chain conformation plays a pivotal role in reconstructing the molecular orbital distribution. Specifically, an extended chain conformation enhances π-electron delocalization, thereby raising the energy level of the HOMO and lowering the energy level of the LUMO, resulting in a reduced bandgap

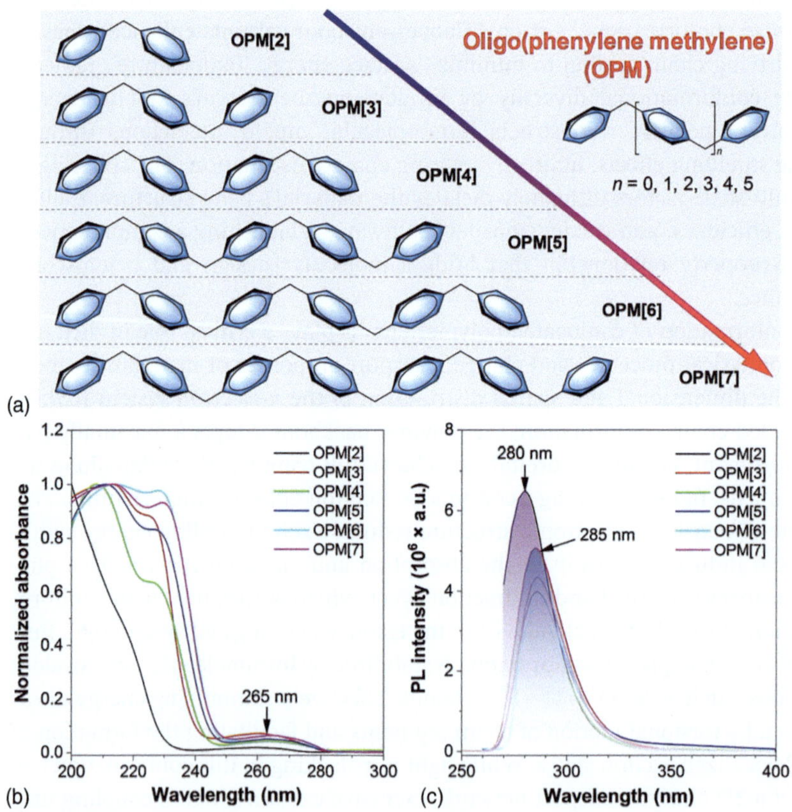

Figure 2.23 (a) Molecular structures of OPM[2]–OPM[7]. (b) Normalized absorption and (c) photoluminescence spectra of OPM[2]-OPM[7] in acetonitrile solution. *Source:* Wang et al. [65] / with permission of John Wiley & Sons.

characteristic. Conversely, a folded chain conformation disrupts conjugation, leading to an opposite regulation of the orbital energy level difference. This dynamic tunability of the electronic band structure highlights the importance of precisely controlling chain conformation through strategies such as side-chain engineering or main-chain topology design to optimize the photoelectric properties of conjugated polymers [69]. For instance, the incorporation of steric hindrance groups into the polymer backbone can induce controlled torsion within molecular segments, effectively suppressing exciton recombination while maintaining a balanced carrier mobility. Similarly, the use of rigid, rod-like conjugated units can stabilize the extended conformation, facilitating the construction of anisotropic charge transport channels. These approaches underscore the critical role of molecular design in achieving advanced functional materials.

2.2.4 Aggregation States of Conjugated Polymers

The aggregated state structure of conjugated polymers is crucial in determining their properties, making it an important area of research and development for new functional materials. These structures exhibit great diversity through intermolecular interactions and changes in external conditions, and can be classified into two main types: static and dynamic. These structural types are closely related to optical, electrical, and mechanical properties [70].

2.2.4.1 Static Aggregation

Static aggregation occurs from specific molecular packing arrangements and mainly has two forms: J-aggregates and H-aggregates (Figure 2.24).

J-aggregation is typically formed when molecules are arranged in a head-to-tail pattern. This configuration improves exciton coupling and the delocalization of electronic states, resulting in an increase in the redshift of the absorption spectrum and the fluorescence quantum yield. J-aggregation has significant applications in the field of optoelectronics. For example, in organic photovoltaics, its ability to transfer energy efficiently is used for light capture. In OLEDs, it uses its narrow-band emission spectrum to achieve high-color-purity emission. Furthermore, some J-aggregates exhibit AIE, whereby molecules significantly enhance fluorescence upon aggregation. This characteristic opens up new possibilities for designing highly sensitive sensors and imaging probes.

Unlike J-aggregation, H-aggregation is formed through face-to-face packing of molecules. Its key characteristics are the blue shift of the absorption spectrum and the inhibition of fluorescence. The tight intermolecular packing that limits electron delocalization results in relatively low luminescence efficiency in H-aggregation. However, this structure has unique advantages in other respects. For instance,

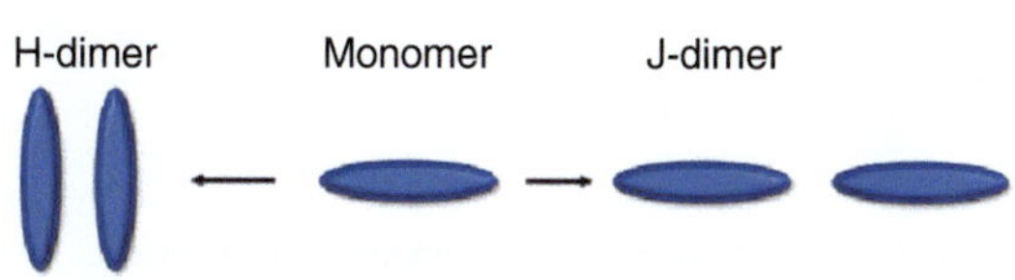

Figure 2.24 A diagram of the static aggregation.

H-aggregation is widely used in electronic devices such as organic field-effect transistors thanks to its ability to transfer charges efficiently. Additionally, its optical properties make it suitable for use as an optical filter and modulator, and its excellent thermal and structural stability meet usage requirements in high-temperature or harsh environments.

In addition to J-aggregation and H-aggregation, conjugated polymers can form a variety of different aggregation states under different external conditions. This polymorphism is affected by multiple factors such as solvent polarity, temperature, ionic strength, and processing methods. For instance, polymers can transform between highly ordered crystalline states and disordered amorphous forms, where the crystalline state typically exhibits a higher charge mobility, while the amorphous form offers superior mechanical flexibility. Furthermore, the selection of solvents and changes in polarity can regulate whether J-aggregation or H-aggregation is formed, thereby endowing the material with adjustable optical and electronic properties. Under temperature induction, the aggregation state of polymers will also change. For instance, high temperatures can disrupt π–π packing, leading to a transformation from an ordered structure to a disordered one. These characteristics provide the possibility for the development of temperature-responsive materials.

The static aggregation state of conjugated polymers presents extensive opportunities for advancing optoelectronic applications, arising from the intricate interplay among molecular packing configurations, optical characteristics, and environmental parameters. By systematically manipulating factors such as ionic strength, solvent composition, temperature, and processing techniques, it is possible to finely tune the static aggregation state. This approach not only facilitates the enhancement of material performance but also contributes critical theoretical insights and practical guidelines for the rational design and development of next-generation photofunctional materials.

2.2.4.2 Dynamic Aggregation

The dynamic aggregation state of conjugated polymers refers to the reversible transformation of the molecular aggregation structure under the action of external stimuli such as temperature, pressure, solvent polarity, and ionic strength (Figure 2.25). These aggregated states exhibit controllable optical, electrical, and mechanical properties through the response capabilities of intermolecular interactions and packing structures to environmental changes. The characteristics of dynamic aggregation provide significant advantages for the development of materials with adaptability

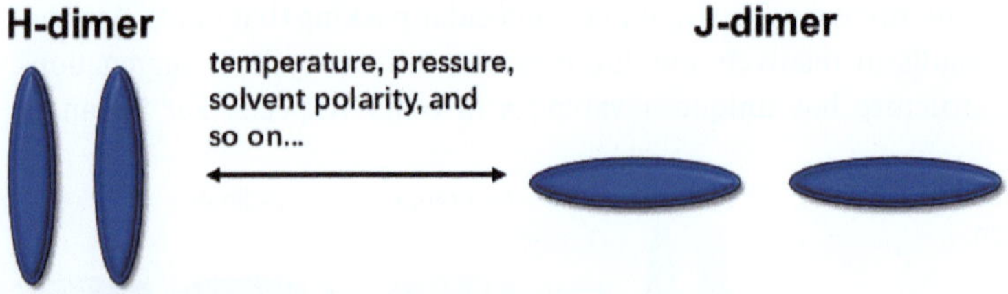

Figure 2.25 A diagram of the dynamic aggregation.

and intelligent response and are an effective supplement to functions that static aggregation cannot achieve.

A key characteristic of dynamic aggregation is temperature-induced transformation. As the temperature rises, the intermolecular forces weaken and the ordered molecular arrangement is disrupted. This transformation usually leads to the polymer changing from a crystalline state with good π–π packing to an amorphous state with a more disordered molecular arrangement. The material's optical and electrical characteristics are greatly impacted by this alteration. For instance, the transition from a crystalline state to an amorphous state typically broadens the absorption spectrum, reduces the fluorescence intensity, and alters the charge transport characteristics, making the material suitable for applications such as flexible electronic devices and sensors that require temperature response [71]. Another remarkable feature is the reversible transformation between J-aggregation and H-aggregation. By altering the polarity of the solvent, the ionic strength, or introducing specific additives, the polymer can switch between these two aggregation states, each of which exhibits unique optical and electronic properties. J-aggregation usually presents as fluorescence with redshift absorption and enhancement, while H-aggregation shows fluorescence with blue-shift absorption and inhibition. This reversible transformation can precisely regulate the material properties, making it possible for the application of optical switches, adaptive sensors, and reconfigurable devices.

The ion-specific effect also has an important influence on the dynamic aggregation state. For example, ions with high charge density or specific hydration energy can promote the closer packing of molecules, which is conducive to the formation of H-aggregation. Larger ions or ions with lower hydration energy may lead to more delocalized molecular arrangements, thereby promoting J-aggregation. These effects indicate that precisely regulating the ionic environment is crucial for controlling aggregation behavior, especially in dye-sensitized solar cells or electrolyte-based devices [72]. Moreover, dynamic aggregation states hold immense promise for the creation of materials that respond to stimuli. Under the action of external stimuli (such as light intensity, pH value, or solvent composition), the aggregation state and its properties undergo reversible changes [73]. This characteristic is crucial for advanced applications, such as self-healing materials, reconfigurable optoelectronic devices, and adaptive sensors. For instance, materials that adapt to changes in the external environment, such as the polarity of the solvent, can be employed in drug delivery systems, chemical sensors, and dynamic displays.

In conclusion, the dynamic aggregation states of conjugated polymers provide a new direction for exploring the adaptability and responsiveness of materials. By taking advantage of temperature-induced transitions, reversible switching of the J-H aggregation state, and ion-specific effects, researchers can design materials with highly tunable performance. In-depth research on the mechanism of dynamic aggregation will drive innovation in the field of smart materials and provide solutions for advanced applications in sensing, energy, and optoelectronics.

2.2.4.3 Influencing Factors and Regulatory Strategies of Aggregation States

The aggregation behavior of conjugated polymers is influenced by a variety of key factors, among which the molecular structure is one of the core factors determining the aggregation state. Adjusting the conjugation length has a significant influence on the carrier transport performance. For instance, oligomers based on benzodithiophene dione can effectively regulate their charge transport performance by altering the conjugation length. Similarly, the geometric configuration of molecules and the side-chain properties also have significant influence on aggregation behavior. A smaller steric hindrance helps the conjugated unit maintain coplanarity, thereby enhancing the delocalization degree of π electrons and improving the light absorption capacity and electronic performance. The properties of solvents are equally crucial for the aggregation behavior of conjugated polymers [74]. In good solvents, polymer chains usually present an extended conformation, with a high degree of π-electron delocalization, a small bandgap, and redshifts in absorption and emission spectra. Conversely, in poor solvents, polymer chains tend to fold and form aggregates, thereby leading to fluorescence quenching. The molecular structure of OPPV is shown in Figure 2.26. Poly(*p*-phenylene vinylene) (PPV) presents an extended conformation in good solvents, while in poor solvents, the chain structure folds and forms a core-shell structure.

Temperature is another important factor affecting aggregation. Lower temperatures usually reduce thermal motion, thereby enhancing intermolecular interactions and promoting the aggregation of polymer chains. Studies have shown that as the proportion of ethanol in the solution increases, poly(9,9-dioctylfluorene) (PFO) gradually collapses from the rigid extended chain and forms a curled state, eventually developing into an aggregated structure at the micrometer level. Figure 2.27 demonstrates the corresponding α and β conformations of PFO [75]. Aggregation kinetics is also an important factor influencing aggregation behavior, including the dynamic processes of polymer chains folding, aggregating, and depolymerizing in solution. By regulating the aggregation dynamics, precise control of the aggregation state of polymers can be achieved. For example, changing the composition of the solvent or the temperature can induce the polymer chain to transform from an extended conformation to a folded conformation, or the

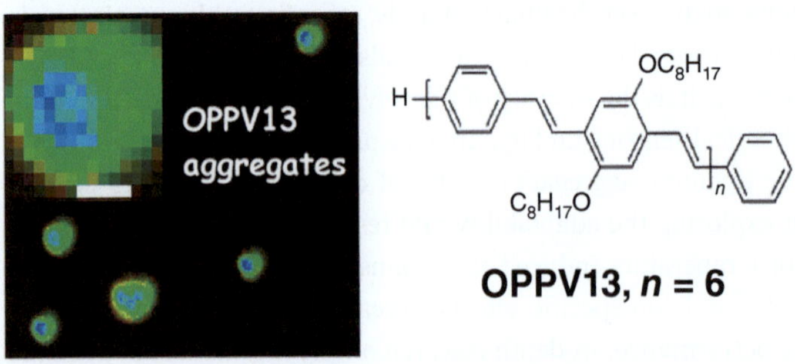

Figure 2.26 The molecular structure and aggregation structure of OPPV. *Source:* Reproduced with permission from Peteanu et al. [74] / American Chemical Society.

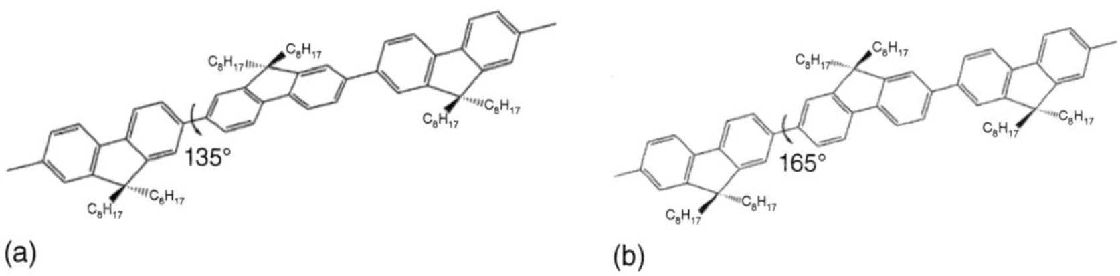

Figure 2.27 (a) α conformation and (b) β conformation of PFO. *Source:* Reprinted with permission from Li et al. [75].

reverse transformation, thereby achieving customized regulation of the aggregation behavior.

To achieve the regulation of the aggregation state of conjugated polymers, researchers have proposed a series of strategies. Molecular design is an effective method. By introducing functional groups or adjusting the geometric configuration of polymers, the rigidity and flexibility of chains can be changed, thereby affecting the aggregation behavior [76]. For example, introducing oxygen atoms with high electronegativity into the polymer main chain can lower the HOMO energy level and adjust its optical properties. Solvent engineering regulates the solubility and aggregation behavior of polymers by selecting appropriate solvents or mixed solvent systems. For example, polymer chains can aggregate when a small quantity of a poor solvent is added to a good solvent. Temperature control is also a commonly used method. By adjusting the solution temperature, the thermal motion and aggregation behavior of polymer chains can be influenced. For instance, the tendency of polymer chains to aggregate can be increased by reducing the solution temperature.

The regulation of the aggregation state of conjugated polymers has brought new opportunities for biosensing, cell imaging, and disease treatment. By regulating the aggregation state of polymers, highly sensitive detection of biomolecules can be achieved. In cell imaging, the regulation of aggregation states can increase the fluorescence quantum yield of polymers, thereby enhancing the imaging effect. In disease treatment, by regulating the aggregation state of polymers, precise control of drug release can be achieved, thereby improving the therapeutic effect. For instance, conjugated polymers based on aggregation-induced emission can undergo changes in aggregation state in response to a specific biological environments such as pH value or enzyme concentration, which gives them significant advantages in targeted diagnosis and treatment. Further research into the quantitative relationship between aggregation states and properties is needed to inform the design and application of new high-performance materials.

2.2.5 Luminescence Sensing Mechanism of Conjugated Polymers

The fluorescence signal amplification properties of conjugated polymers enable their interaction with analytes through both nonspecific mechanisms, such as electrostatic and hydrophobic interactions, as well as specific binding facilitated by intermolecular recognition. These interactions result in pronounced alterations in the photofunctions of conjugated polymers, thereby providing a robust platform for the detection and analysis of analytes. Changes in fluorescence signals, including variations in fluorescence intensity, spectral shape, and emission wavelength, are typically attributed to three primary mechanisms: Förster resonance energy transfer (FRET), photoinduced electron transfer, and excimer/exciplex-based fluorescence mechanisms. Additionally, metal-enhanced fluorescence (MEF) also contributes to these signal changes [77].

2.2.5.1 Förster Resonance Energy Transfer Mechanism

Förster resonance energy transfer (FRET) is an intermolecular energy transfer mechanism governed by dipole–dipole interactions between a donor and an acceptor molecule (Figure 2.28). Upon absorption of light at a specific wavelength, the donor molecule transitions to an excited electronic state. From this state, it can transfer energy to the acceptor molecule, which remains in its ground state, through a nonradiative process. This energy transfer occurs efficiently only within a specific distance range, leading to the excitation of the acceptor molecule. The process's efficiency depends on several critical factors, the most important of which is a substantial overlap between the donor's emission spectrum and the acceptor's absorption spectrum. Then, the distance between the donor and the acceptor needs to be maintained within the range of 1–10 nm. Finally, the dipole orientations of the donor and the acceptor need to be appropriate. These conditions jointly determine the efficiency of energy transfer.

A notable feature of the FRET phenomenon is its extremely short action distance and high spatial sensitivity, which enables it to precisely reflect the interactions, conformational changes, or dynamic binding events among biomolecules.

The energy transfer rate is given by k_{ET}:

$$k_{\mathrm{ET}} \propto \frac{1}{r^6} \cdot \kappa^2 \cdot J(\lambda)$$

the spectral overlap integral ($J(\lambda)$) requires a match between the donor's emission spectrum and the acceptor's absorption spectrum. The distance between the donor and the acceptor is represented by r. The ideal value of the dipole orientation factor κ^2 is 2/3. In a rigid conjugated polymer system, the value of κ^2 can be stabilized through molecular design to reduce the efficiency fluctuation caused by random orientation. The precise regulation of these parameters makes FRET a powerful tool, which is widely used in the fields of chemical and biological detection. For example, by designing donor and acceptor molecules with specific spectral characteristics, specific detection of the target analyte can be achieved. Upon binding of the target

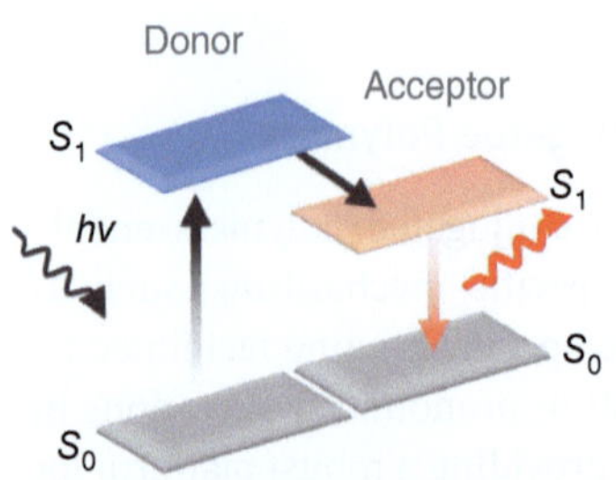

Figure 2.28 Schematic representation of FRET mechanism.

analyte to the receptor, modifications occur in the receptor's electronic properties or spatial conformation. These alterations influence the efficiency of energy transfer between the donor and receptor, thereby leading to measurable changes in the fluorescence signal. The changes in this fluorescence signal can be captured by highly sensitive detection equipment, thereby enabling quantitative analysis of the target analyte. In addition, the FRET mechanism can also be used to study the dynamic processes of biomolecules, such as conformational changes of proteins, intermolecular interactions, and intracellular signal transduction. By labeling donor and acceptor molecules on target biomolecules, their dynamic changes within cells can be monitored in real time, providing important information for biomedical research [78].

2.2.5.2 Photoinduced Electron Transfer Mechanism

In the sensing mechanisms of conjugated polymers, photoinduced electron transfer (PET) plays a pivotal role in modulating intramolecular electron transition pathways, thereby enabling dynamic responses in fluorescence signals. This phenomenon has emerged as a fundamental mechanism for the development of highly sensitive detection systems. This process involves the directional transfer of electrons within or between excited-state molecules (Figure 2.29). When the electron donor is excited by light, the electrons will transfer to LUMO of the acceptor, thereby forming a charge separation state. This electron transfer causes the energy of the excited state to dissipate via nonradiative pathways, thereby leading to the fluorescence quenching phenomenon. In conjugated polymer systems, by precisely designing the donor–acceptor (D–A) structure, the electron transfer barrier can be optimized. By regulating the energy level matching degree of MOs, the switching response of fluorescence signals can be achieved. For instance, when the analyte is bound to the receptor, it can alter the electron affinity of the receptor, thereby inhibiting or restoring the PET process, which is manifested as a reversible change in fluorescence intensity. It is worth noting that the rigid framework and π-electron delocalization characteristics of conjugated polymers can effectively stabilize the charge separation state, extend the lifetime of the excited state, and at the same time, regulate the intermolecular forces through side chain engineering to further optimize the electron transfer efficiency. This PET-based sensing mechanism establishes the theoretical foundation for developing fluorescent probes with enhanced spatiotemporal resolution and reduced background noise. Especially in complex biological environments, it can achieve precise identification and dynamic tracking of trace analytes.

Figure 2.29 Schematic representation of PET mechanism.

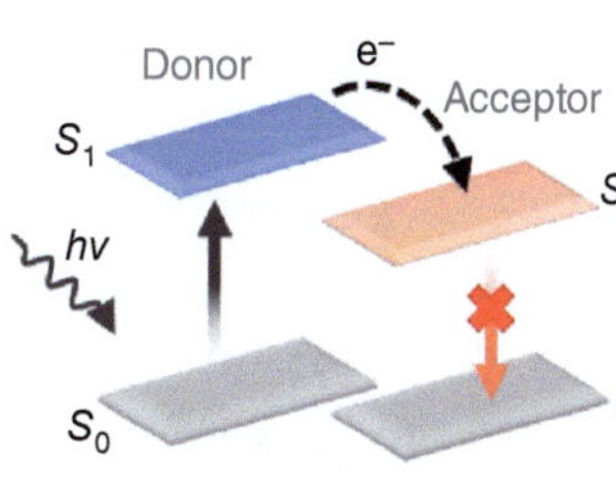

2.2.5.3 Exciplexes and Excimers

The emission of exciplexes and excimers is a special fluorescence phenomenon based on the interaction between organic molecules, and it has unique advantages in the design of fluorescence platforms. Excimers are usually formed by two identical molecules in an excited state, while exciplexes are formed when an electron donor and acceptor interact in an excited state (Figure 2.30). These two structures are formed through noncovalent interactions, such as π–π stacking, electrostatic attraction, and hydrogen bonding, which enable effective charge transfer in the molecules' excited state. This generates significantly different redshift fluorescence from that emitted by monomers. Because the formation of exciplexes and excimers requires close contact between molecules and specific arrangement patterns, their fluorescence intensity and emission wavelength are highly sensitive to the distance and relative orientation between molecules. This characteristic makes the fluorescence platform based on exciplexes and excimers have extremely high sensitivity and selectivity in detecting environmental changes, intermolecular interactions, and conformational changes of biomolecules. In addition, the fluorescence lifetime of exciplexes and excimers is relatively long, which makes them have potential application value in time-resolved fluorescence detection. By rationally designing the molecular structure and regulating the intermolecular interactions, the emission characteristics of exciplexes and excimers can be precisely regulated, thereby developing high-performance fluorescent probes suitable for various chemical and biological detection requirements [79]. Exciplex/excimer emission competes with monomer emission, and its formation is based on the light-induced electron transfer process. This luminescence mechanism is different from the luminescence characteristics of individual molecules. Exciplex emission has significant research value and application prospects in the field of organic luminescent materials and devices. By rationally designing the structures of donor and acceptor molecules and optimizing the complex system, efficient and stable luminescence performance can be achieved, laying important theoretical support and practical guidance for the development of new organic luminescent devices.

2.2.5.4 Metal-enhanced Fluorescence Effect

The optical phenomenon known as metal-enhanced fluorescence (MEF) is caused by the interaction of fluorescent molecules with metal nanostructures. Its core lies in significantly enhancing the luminescence efficiency of fluorescent molecules

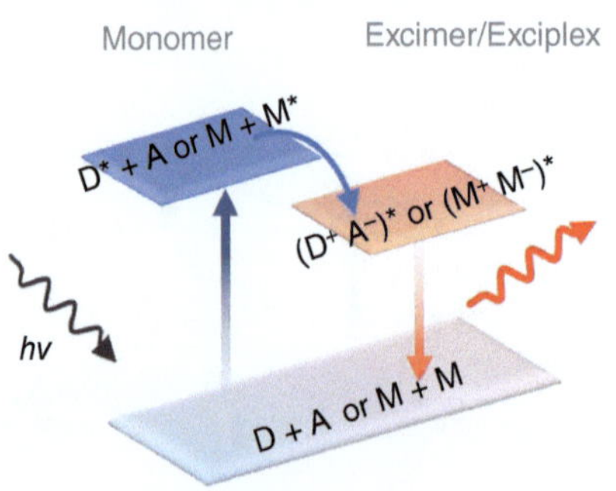

Figure 2.30 Schematic representation of excimer or exciplex formation.

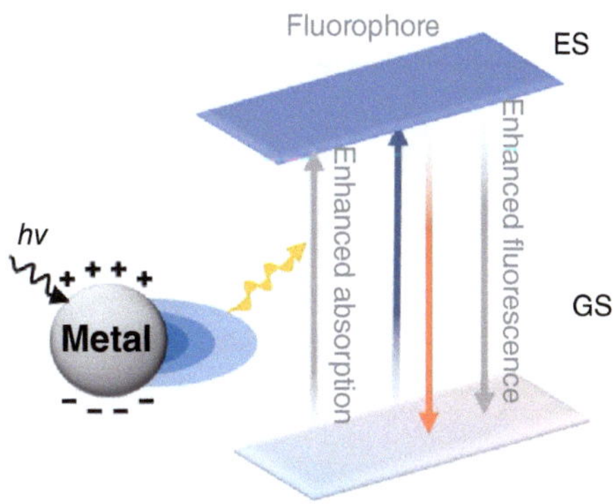

Figure 2.31 Schematic representation of MEF mechanism.

by utilizing the surface plasmon resonance effect (SPR) of metal nanoparticles (Figure 2.31). In the MEF mechanism, when metal nanoparticles, including gold, silver, or copper nanoparticles, are excited by light of specific wavelengths, their free electrons will oscillate collectively to form surface plasmons, thereby generating a strong local electric field near the metal surface. When fluorescent molecules approach these metal nanoparticles, their fluorescence emissions can be significantly enhanced, with the enhancement amplitude reaching several times or even thousands of times. The mechanism underlying the MEF phenomenon is fundamentally influenced by several critical factors, including the spatial separation between metal nanostructures and fluorescent molecules, as well as the size, shape, and composition of the nanoparticles. It has been reported that the increase in brightness is most obvious when the separation between fluorescent molecules and metal nanoparticles is between 5 and 90 nm [77]. Furthermore, the size and morphology of the metal nanoparticles play a pivotal role in determining their SPR effect, which directly modulates the fluorescence enhancement effect.

The MEF mechanism is widely used in the development of highly sensitive fluorescence sensors. Fixing fluorescent molecules to the surface of metal nanoparticles allows for the specific detection of the target analyte. When the target analyte interacts with fluorescent molecules, it will change the luminescence characteristics of the fluorescent molecules. This change can be captured by highly sensitive detection equipment, thereby enabling quantitative analysis of the target analyte. Biological imaging is another area in which the MEF mechanism may be used. By combining metal nanoparticles with fluorescent molecules, high-resolution imaging of intracellular biomolecules can be achieved. By rationally designing the size, shape of metal nanoparticles, and the interaction distance with fluorescent molecules, precise regulation of fluorescence signals can be achieved, thereby developing high-performance fluorescent probes suitable for a variety of chemical and biological detection requirements.

2.3 Noble Metal Nanomaterials

In recent years, nanotechnology has developed rapidly. This has promoted the exploration and research of new optical functional materials. Noble metal nanoparticles and nanoclusters have been the focus of considerable attention in domains

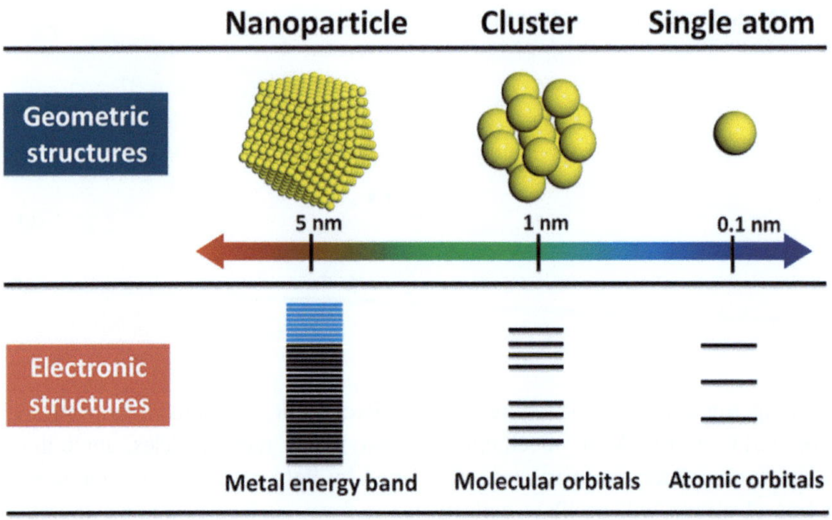

Figure 2.32 Schematic diagram and electronic structures of single atom, nanoclusters, and nanoparticles. *Source:* Liu and Corma [80] / American Chemical Society / CC BY 4.0.

such as materials science, chemistry, and biotechnology, owing to their distinctive optoelectronic and catalytic characteristics. Gold, silver, and copper nanoparticles, for example, exhibit excellent optical properties due to their local surface plasmon resonance (LSPR) effect, and can be used in sensor design and photothermal therapy (PTT) (Figure 2.32). By contrast, noble metal nanoclusters typically consist of several to several hundred metal atoms with linked organic ligands on their surfaces. Their size ranges generally below 2 nm and they have discrete electronic energy levels. Ground-state electrons exchange energy with photons to produce transitions. Inter-energy level transitions can generate various molecular optical properties, such as strong emission, two-photon absorption (TPA), photothermal conversion, and photosensitization effects [80].

This section will systematically introduce the structure and optical properties of noble metal nanoparticles and nanoclusters, with a focus on analyzing their different structural characteristics and the significance of their corresponding optical properties in materials science research. By studying these materials, we can gain a deeper understanding of how to influence their physical and chemical properties by adjusting their size, shape, and surface chemistry.

2.3.1 Noble Metal Nanoparticles

Noble metal nanoparticles are particles composed of noble metals such as gold, silver, and copper. They are typically nanoscale in size. At this scale, surface and quantum effects become pronounced, leading to significant alterations in such nanoparticles' physical and chemical properties. Their unique optoelectronic and catalytic characteristics have made noble metal nanoparticles the subject of extensive research in biology, chemistry, and materials science. The principle that "structure determines function" is particularly relevant for metal nanocrystals, where shape and size critically influence behavior. Variations in the structural parameters of these nanocrystals result in distinct physicochemical properties. Specifically, gold, silver, and copper nanoparticles exhibit different functionalities due to their unique structural features.

2.3.1.1 Gold Nanoparticles

Gold nanoparticles (AuNPs) are highly favored due to their excellent stability and ease of surface functionalization. Their LSPR effect grants them unique optical properties, making them widely used in biological imaging, sensors, and drug delivery [81].

Gold nanoparticles can be prepared into various shapes and sizes, such as spherical, rod-shaped, cage-shaped, and star-shaped. Gold nanoparticles of different shapes exhibit different optical behaviors. For example, the LSPR absorption peak of 10 nm spherical gold nanoparticles acts at 520 nm and migrates with the change of particle size, but the migration range is limited. In the nonspherical gold system, due to its anisotropic structure, it exhibits bimodal LSPR characteristics. The resonant absorption peak can continuously redshift as the ratio of its long and short axes increases. Because their size and shape may be accurately regulated by modifying the synthesis conditions, gold nanoparticles are frequently used in the domains of biomedicine, sensing, and catalysis.

Gold Nanorods Gold nanorods are the most studied anisotropic AuNPs. Due to the different aspect ratios, gold nanorods have two SPR absorption bands. Of these, shorter wavelength absorption relates to the SPR in the transverse axis direction, occurring in the visible light region of 500–550 nm. The longer wavelength absorption peak relates to the SPR in the longitudinal axis direction. This band is relatively sensitive to changes in the aspect ratio, with the wavelength gradually shifting toward the red end of the spectrum as the aspect ratio increases [82]. For example, when the aspect ratios of gold nanorods are 2.2, 3.5, 4.1, 5.1, and 6.3, their longitudinal absorption peaks are located at 650, 760, 840, 920, and 1000 nm, respectively [83]. Gold nanorods with different aspect ratios exhibit photothermal effects. Gold nanorods with sizes of 28×7 and 77×16 nm show absorption peaks at 808 and 1000 nm, respectively. The photothermal conversion efficiency of the 28×7 nm gold nanorods is approximately 40% higher than that of the 77×16 nm gold nanorods [84]. The photothermal conversion efficiency of 10×38 nm gold nanorods is 95%, whereas that of 77×173 nm nano-bipyramidal structures is only 51% [85].

Therefore, when the aspect ratios are inconsistent, gold nanorods with smaller particle sizes/volumes usually have better photothermal effects, mainly related to the absorption/extinction ratio. Of course, merely understanding the influence of the absorption/extinction ratio on the photothermal effect is usually insufficient. The absorption spectrum and photothermal conversion efficiency of gold nanorods are also affected by their actual size, as well as their aspect ratio. When the aspect ratio was consistent, the size of the gold nanorods was changed to 38×11, 28×8,

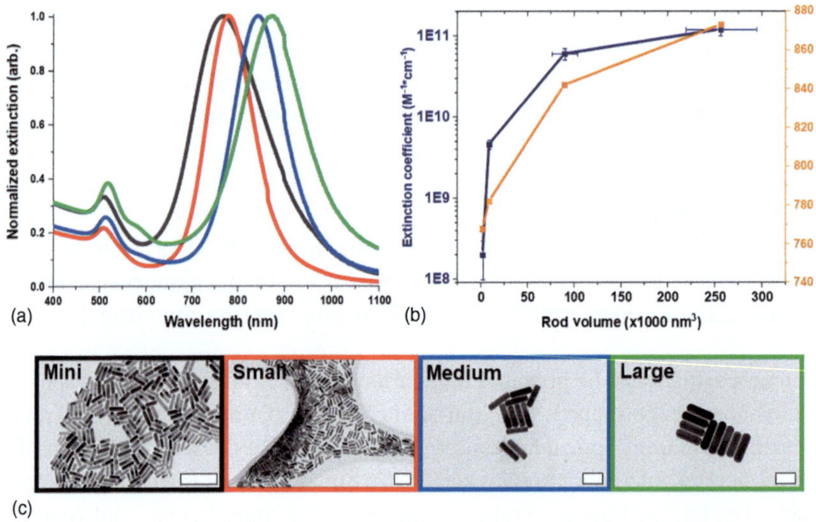

Figure 2.33 (a) Normalized UV–vis extinction spectra, (b) extinction coefficients along with plasmonic maxima, and (c) TEM images of different nanorods (scale bars: 100 nm). *Source:* Meyer et al. [87] / with permission of American Chemical Society.

and 17 × 5 nm, and the longitudinal absorption peaks were located at 740, 770, and 755 nm, respectively. When irradiated with the same NIR light, the photothermal effect of 28 × 8 nm gold nanorods is best [86].

Among a series of gold nanorods with varying volumes but the same aspect ratio, the smallest gold nanorods theoretically exhibit the highest absorption/extinction ratio, but their light absorption efficiency is very poor. In contrast, slightly larger gold nanorods are advantageous because they have higher absorption efficiency and the ability to generate similar heat at low concentrations (Figure 2.33) [87].

Hollow Gold Nanoparticles The two most representative forms of hollow gold nanoparticles are gold nanoshells and gold nanocages, respectively [88, 89]. Gold nanoshells typically consist of a core with a diameter of around 100 nm surrounded by a gold shell with a thickness of 5–20 nm. The SPR absorption range of the gold nanoshell is determined by the thickness of the shell. As shown in Figure 2.34, when the shell thickness decreases from 20 to 5 nm, the maximum absorption wavelength of the gold nanoshell redshifts from 700 nm to approximately 1050 nm [88], which is suitable for PTT excited by NIR. In addition to the thickness of the gold shell, the size of the nanoshell also affects the photothermal effect of the gold nanoshell. Among the gold nanoshells with nanocore sizes of 63, 112, and 264 nm, respectively, the gold nanoshell with a nanocore of 63 nm has the highest photothermal conversion efficiency, and the photothermal temperature can increase by about 35 °C under the same conditions [90]. In addition, the more complete the gold nanoshell is, the higher the photothermal conversion efficiency will be. Gold nanoshells can be prepared by two methods. One is to directly reduce $HAuCl_4$ on the lamellar structure, and the resulting gold nanoshells have an incomplete

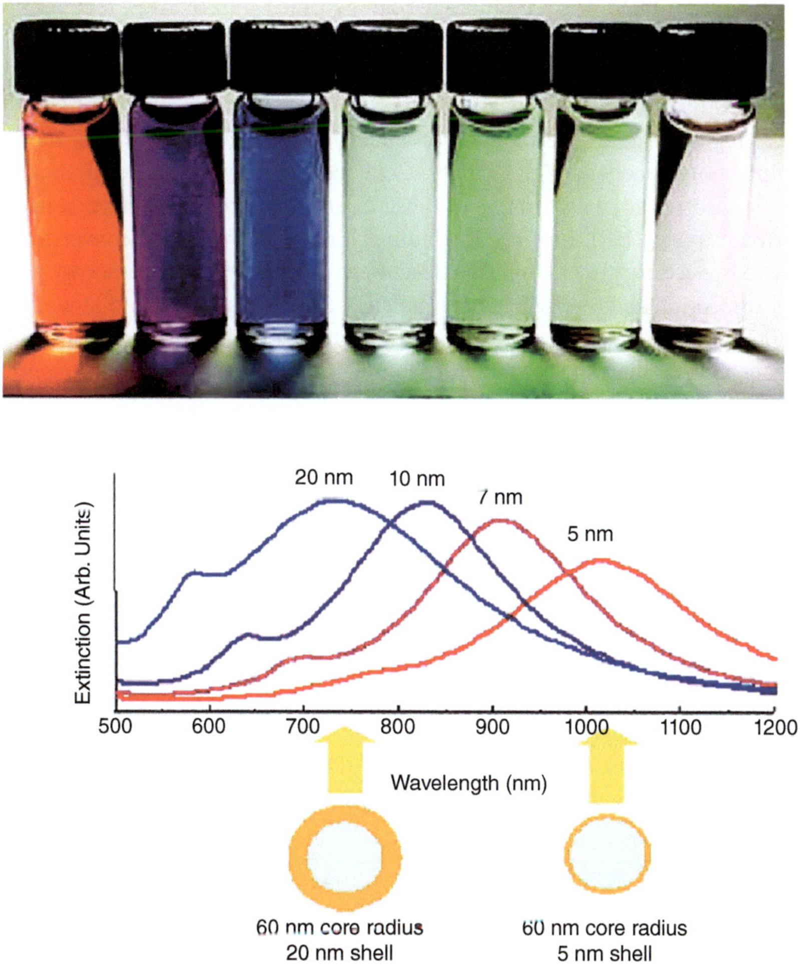

Figure 2.34 Photographs and extinction spectra of different gold nanoshell solutions. *Source:* Stewart et al. [88] / with permission of American Chemical Society.

structure. Another method is to prepare relatively complete gold nanoshells on the surface of the lamellar body by the seed growth method. The temperature rise of the latter is about 15 °C higher than that of the former, and it has a better thermal effect [91].

Gold nanocages are a unique type of nanoparticle characterized by a hollow interior and porous walls, which are generally prepared through a galvanic replacement reaction between gold salt solutions and Ag nanostructures [92]. Ag nanocubes can be used as reaction templates to undergo redox reactions with $AuCl_4^-$, and the resulting gold is confined to the surface of the nanocubes. During the gold deposition process, the internal Ag is oxidized and removed, and the alloying and de-alloying processes occur simultaneously, generating a hollow structure and eventually forming a gold nanocage with a porous structure [89, 93].

The SPR absorption band of gold nanocages can be tuned into the NIR region by optimizing the thickness of the gold layer—specifically by adjusting the amount of the Ag/Au ratio. Increasing the amount of $HAuCl_4$ added to the silver nanocubes shifts the maximum absorption peak of the gold nanocages gradually into the NIR optical region. Due to the unique porous structure of gold nanocages, they often exhibit more excellent photoactivity. Gold nanocages with a size of about 25 nm show absorption peaks similar to those of gold nanorods, but have better photothermal capacity [94]. In order to obtain a higher photothermal conversion efficiency, it is necessary to reasonably design the size of the gold nanocage to control its absorption and scattering characteristics. Studies have shown that when the absorption cross-section of gold nanocages is smaller (side length <45 nm), a greater proportion of incident light is absorbed by the nanopracticles. Conversely, when the absorption cross-section is wider, light scattering dominates [95], which can be used as the theoretical basis for the design of gold nanocages.

Gold Nanostars Gold nanostars usually contain a gold core and a gold tip on the surface. The SPR absorption peak of gold nanostars shifts to longer wavelengths (red-shifts) as the core size, number of tips, tip length, and tip sharpness increase [96]. For example, by employing polymers as templates and varying the surfactant and ion types in the growth solution, various forms of gold nanoparticles can be synthesized [97]. As the density and sharpness of surface tips increase, the SPR absorption can be tuned from around 700 nm up to over 1100 nm. Furthermore, by adjusting the buffer type and the ratio of buffer to gold salt concentration, gold nanoparticles with different core sizes and tip lengths can be produced, allowing the SPR absorption to be modulated from the visible to the NIR region (Figure 2.35) [98].

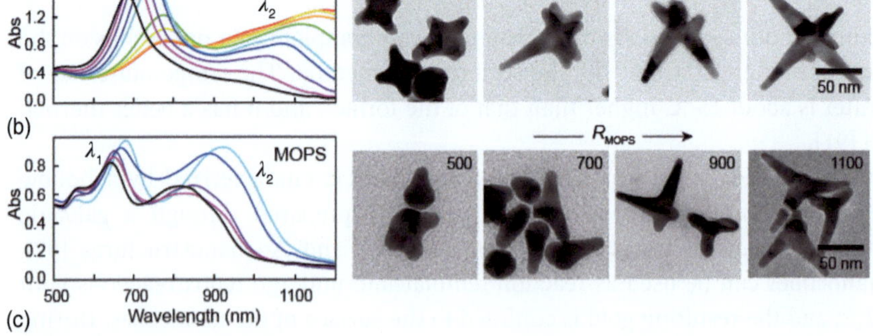

Figure 2.35 Absorbance spectra and TEM images of various gold nanostars synthesized using (a) RHEPES, (b) REPPS, and (c) RMOPS buffer. *Source:* Chandra et al. [98] / with permission of American Chemical Society.

The special surface tips of gold nanostars not only affect the absorption of SPR, but also contribute significantly to the photothermal effect. The tip points on the surface of gold nanostars can enhance the photothermal conversion efficiency, which can reach 75.5%, much higher than that of gold nanoshells (16%) and gold nanorods (22%) prepared by the same method [99]. With the increase in tip sharpness, the absorption peak of SPR moves toward the long-wave direction, and the photothermal conversion efficiency gradually increases, which are 18.7%, 26.2%, and 28.2%, respectively [100]. It is speculated that this is related to the redshift of the spectrum and the electromagnetic field enhancement caused by the tipped branch.

2.3.1.2 Silver Nanoparticles

Silver is the metal with the best electrical conductivity. It has excellent chemical and catalytic properties, as well as outstanding antibacterial and biocompatibility properties. Therefore, it is widely used in industries such as electronics, chemical engineering, biomedicine, medicine, and daily necessities. Due to their small volume, large surface area, and unique physical and chemical properties, silver nanomaterials can currently be used as wires and switches in nanoelectronic devices, or to develop new conductive or biomedical composites and highly efficient catalysts. However, in addition to this, silver nanomaterials possess optical properties that bulk silver does not have, which makes their applications more extensive and they have received much attention among various nanomaterials. The performance of silver nanomaterials largely depends on their shape, size, composition, crystallinity, and structure [101].

The reduction of silver ions by citric acid was the earliest method for preparing silver nanoparticles. However, this method generally only exists in extremely dilute solutions, and stabilizers need to be added when the concentration increases. The morphology control of the products can be achieved by using the preparation method developed by Murphy et al. [102, 103]. Controlling the ratio of $AgNO_3$ to the coating agent cetyltriethylammonium bromide (CTAB), adjusting the pH value, and adding a reducing agent enables the formation of a silver nanorod/wire structure with a controllable length-to-diameter ratio. The preparation of seed crystals uses low-concentration silver salts, trisodium citrate as the stabilizer, and $NaBH_4$ solution as the reducing agent. The reaction is typically carried out at room temperature for between 2 and 5 h, and silver nanoparticles of different diameters can be obtained.

The polyol method can be used to prepare silver nanomaterials of different shapes and structures. By using ethylene glycol as the solvent and reducing agent, polyvinylpyrrolidone (PVP) as the stabilizer and inducer, and metal salts as the precursors, metal nanomaterials of different shapes can be prepared by adjusting the reaction process. These include spherical nanoparticles, nanocubes, nanowires/rods, and nanotriangles, among others [104]. Figure 2.36 shows a typical growth process of silver nanomaterials, which includes several stages such as reduction, nucleation, growth, and product. The seed crystal provides the active points for crystal growth and the original crystal planes, while PVP plays a role in inducing the anisotropic growth of crystals during the growth process. Single crystal seeds

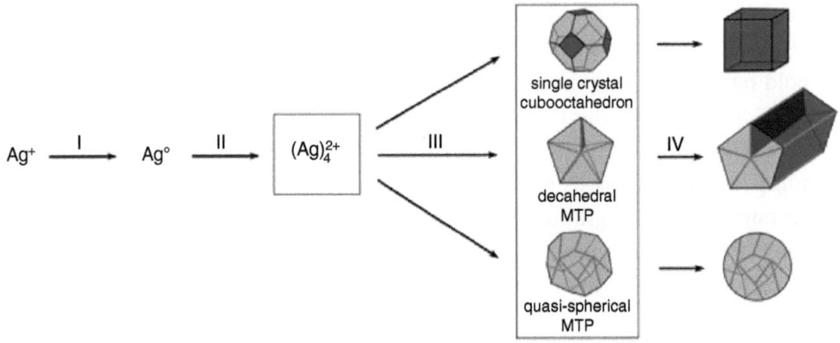

Figure 2.36 Schematic illustrating the preparation of silver nanocubes, nanorods or nanowires, and nanospheres. *Source:* Wiley et al. [101] / with permission of John Wiley & Sons.

easily yield single crystal products. If it is a single-crystal cuboctahedron seed, then the final growth obtained is mostly nanocubes. The twinned seeds with decahedral shape are prone to yield products of nanowire/rod structures. While polycrystalline seeds (similar to spherical shapes) will yield polyhedral nanoparticles without regular shapes [101].

The absorption peak of the spherical silver structure with a diameter of around 10 nm occurs at approximately 391 nm, whereas the absorption peak of the sphere with a diameter of 80 nm occurs at around 430 nm. The silver triangular nanoparticle with a side length of 30 nm has three absorption peaks at 688, 485, and 334 nm. Similarly, the silver nanocube with a side length of 80 nm has three absorption peaks at 476, 400, and 350 nm, respectively.

Maxwell's equations can provide a rigorous explanation of the color of metallic colloids, but this equation is only applicable to the theoretical derivation of the shapes of spheres, spheroids, cylinders, and spherical shells. Based on this, a theory that can calculate the absorption, scattering, and extinction properties of various other morphologies of metal particles was developed—the discrete dipole approximation (DDA) [105]. Combined with the spectra measured experimentally, it has become an important means for understanding the optical properties of nanoparticles at present.

It can be seen from Figure 2.37a that the silver nanospheres of a size of 40 nm mainly exhibit the absorption of blue light, thus making the silver nanosphere colloid present a superimposed color of red and green light, i.e. yellow [106]. At 410 nm in the spectrum, there is a dipole oscillation characteristic absorption peak, and at 370 nm, there is a very weak shoulder peak, which is a quadrupole oscillation absorption peak. Figures 2.37b–d show, respectively, the spectra of a silver cube, a tetrahedron, and a regular octahedron, all with a particle size of 40 nm, as calculated by DDA theory. The position of the strongest absorption peak in the spectra undergoes redshift [106, 107]. This phenomenon is commonly seen in nanoparticles with sharp angular structures. This is because the surface charges of metals tend to accumulate at the tip structures, thereby enhancing charge separation and reducing the restoring force required for electron oscillation [108–110].

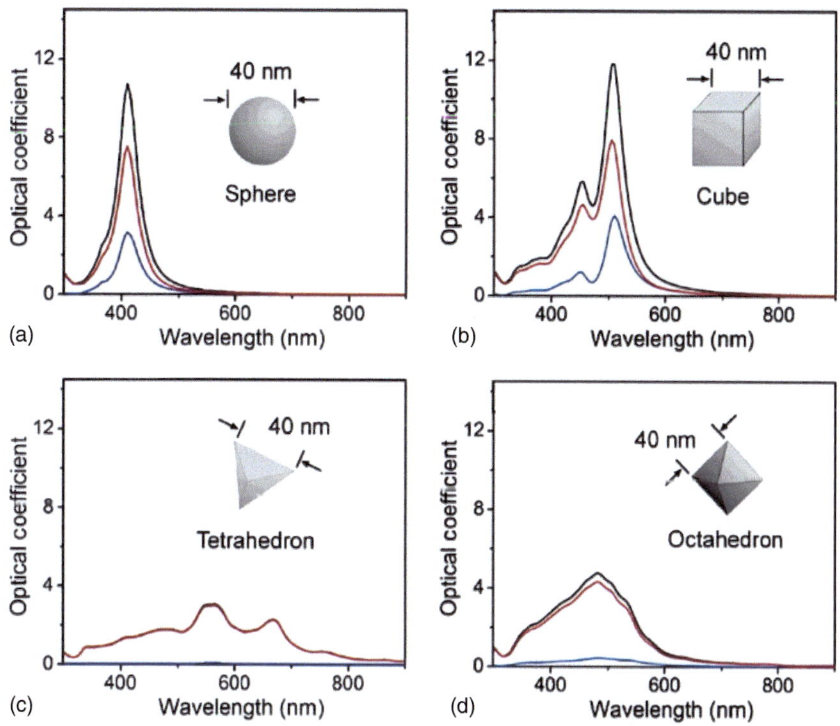

Figure 2.37 Theoretical extinction spectra (black), absorption spectra (red), and experimental scattering spectra (blue) corresponding to (a) silver sphere, (b) silver cube, (c) silver tetrahedron, and (d)silver octahedron. *Source:* Wiley et al. [106] / with permission of American Chemical Society.

2.3.1.3 Copper Nanoparticles

Copper is located in the transition group and has excellent physical properties, such as electrical conductivity and ductility. In particular, its electrical conductivity is second only to that of silver. Copper nanocrystals also have similar optical and catalytic properties to gold and silver, and are low in price. They are expected to replace noble metal nanocrystals in most situations. However, due to the relatively lower redox potential of copper compared to noble metals, on the one hand, copper nanocrystals are prone to oxidation, and on the other hand, there is a lack of suitable reducing agents. Therefore, controlling the structural shape of copper nanocrystals synthesized in the liquid phase is more difficult and complex than for other noble metals [111].

The preparation of metallic copper nanomaterials by "wet chemistry" technology is a classic preparation method. Wet chemical methods typically use reducing agents to provide electrons and reduce copper precursors to prepare copper nanomaterials. During the preparation process, end-capping agents and stabilizers are generally added to regulate the growth process of the nanostructure and stabilize the product nanoparticles. For example, copper nanowires were synthesized using glucose as the reducing agent and cetylamine as the end-capping agent. The final product yield of

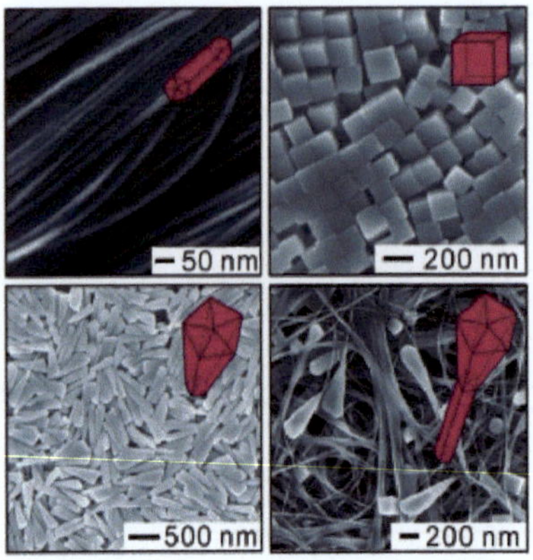

Figure 2.38 SEM images of different copper nanostructures. *Source:* Reproduced with permission from Jin et al. [112] / John Wiley & Sons.

this method can reach 93%, and the copper nanowires prepared have the characteristics of high purity, uniform size, and high flexibility. Moreover, the size distribution and morphology of the products can be controlled by adjusting the concentration ratio of glucose and cetylamine. This results in various copper nanostructures, such as band-like nanowires, cubes, nanocrystals and, tadpole-shaped nanowires (Figure 2.38) [112].

The polyol approach can also be used to produce copper nanoparticles. In this method, diethylene glycol functions as the solvent, while PVP acts as a stabilizing agent. Sodium hypophosphite (NaH_2PO_2) serves as the chemical reducing agent. During synthesis, an aqueous copper sulfate solution is gradually introduced into the preheated reaction mixture, triggering the reduction of copper ions and leading to the formation of spherical nanoparticles. By adjusting variables such as the amount of reducing agent, the reaction temperature, and the rate at which the copper precursor is added, it is possible to fine-tune the size of the resulting copper nanoparticles [113].

Copper nanocrystals can continuously transform among single crystals, polycrystals, and twins under a certain number of atoms. Multi-twin metal nanocrystals are often stable at extremely small sizes, but as the size increases, they often evolve into single crystals. The SPR peak of copper nanocrystals varies with the particle size. When the dimensions of a metal are reduced to below the wavelength of visible light, its SPR effect becomes more pronounced. Compared with bulk metals, metal nanoparticles have a very low reflectivity and exhibit deeper and more distinctive colors. Therefore, unlike bulk copper, copper nanoparticles can present colors such as deep red and black. Based on the optical properties of copper nanoparticles, they are applied to materials for high-efficiency photothermal and photoelectric conversion, etc.

2.3.2 Optical Properties of Noble Metal Nanoparticles

2.3.2.1 Metal-enhanced Fluorescence Effect (MEF)

Because the metal atomic nucleus has a relatively weak attraction to its outermost electrons, there are many free electrons inside the metal. Free electrons carry a

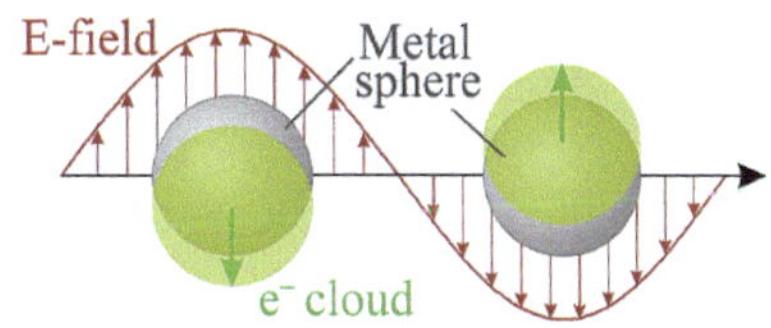

Figure 2.39 Schematic illustration of plasmon oscillation for metal sphere. *Source:* Kelly et al. [114] / with permission of American Chemical Society.

negative charge and are delocalized among the positively charged atomic nuclei, shared by the entire system. In a system where the positive and negative charges inside a metal are equal in quantity and do not show electrical properties externally, remaining in a balanced state, this system is usually referred to as a plasma in metals. Under the influence of some external interference, the freely moving electrons inside the metal structure will shift relative to their original equilibrium positions, resulting in the separation of positive and negative charges. At this time, the electrons will be subjected to the Coulomb force at the center of the positive charge, causing them to return to their original equilibrium positions. The resulting restoring force causes the free electrons to oscillate back and forth at the centers of the positive and negative charges. This phenomenon is called plasmon oscillation (Figure 2.39) [114]. The electric field of light creates a rhythmic movement of free electrons in metal nanoparticles. A resonant interaction occurs when the frequency of the incoming light coincides with the natural frequency at which these electrons vibrate together. Localized surface plasmon resonance is the term used to describe this phenomenon.

Metal-enhanced fluorescence refers to the remarkable increase in fluorescence emission observed when fluorescent molecules are positioned on or near metallic surfaces—such as those of gold or silver—or in proximity to their nanoparticles. This enhancement is highly sensitive to the spatial relationship between the fluorophores and the metal surface. If the fluorescent molecules are situated too close to the metal, nonradiative energy transfer dominates, leading to quenching of fluorescence rather than enhancement. Conversely, if the distance is too great, the metal exerts little to no influence, and the fluorescence remains at its baseline level. The MEF effect is thus most pronounced within a specific optimal distance from the metal surface, where the intensified local electromagnetic fields generated by the metal can interact efficiently with the fluorophores. This interaction boosts the excitation rate of the fluorescent molecules, resulting in a substantial amplification of their emission intensity [115].

To achieve a more comprehensive understanding of the phenomenon, a large amount of electrodynamic theoretical research is still needed. Based on the existing experimental phenomena, it is mainly manifested in three aspects: (1) fluorescent species will undergo fluorescence quenching (k_m) at a very close distance from the metal surface; (2) the excitation efficiency of fluorescent species has been greatly improved (E_m); (3) the increase in the radiation attenuation rate of fluorescent species (Γ_m) (shown in Figure 2.40). As the distance between the fluorescent species and the metal surface increases, these three phenomena may be manifested successively [116].

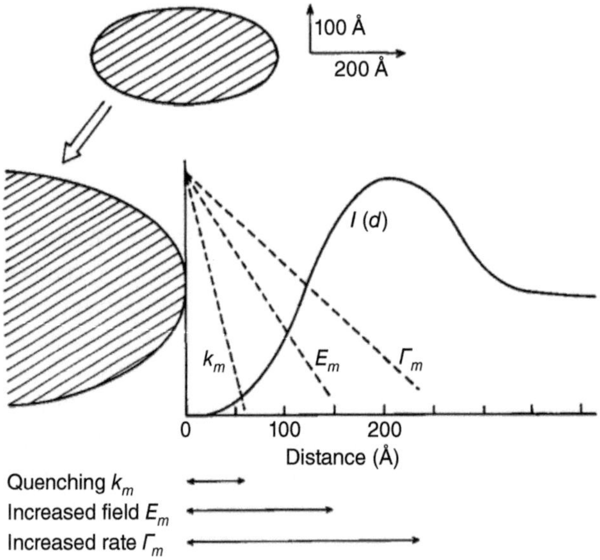

Figure 2.40 Distance effects of metallic particles on fluorescent species: quenching of fluorescence k_m, enhancement of the local electric field E_m, and increase in the rate of radiative decay Γ_m. *Source:* Lakowicz [116] / with permission of Elsevier.

2.3.2.2 Photothermal Effect

Photothermal therapy (PTT) has advantages such as minimally invasive characteristics, high selectivity and precision, and has been applied in biomedical domains, particularly for cancer therapies and fighting bacterial infections. Gold nanoparticles, copper nanoparticles, etc. have unique SPR phenomena, which can conveniently adjust their absorption and scattering characteristics, and have high photothermal conversion efficiency. They are usually used as photothermal agents for PTT [117]. After absorbing light, the activated electrons of gold nanoparticles reach the excited state and undergo nonradiative transitions. The temperature of the AuNPs' surface increases through electron-electron scattering. Then, heat is transferred through electron-phonon relaxation to cool the particles. The heat is transferred to the surrounding environment through phonon–phonon relaxation, increasing the temperature of the nearby medium [118]. This produces a photothermal effect (Figure 2.41).

The photothermal conversion efficiency of noble metal nanoparticles as photothermal agents is highly dependent on the ability of the nanoparticles themselves to generate heat and the laser irradiation parameters. In biomedical applications, people usually choose NIR light as the external light source because NIR light can safely and deeply penetrate healthy tissues and stimulate AuNPs at the lesion site to function. The first NIR window (NIR-I, 650–850 nm) is beneficial to superficial tissues. NIR light of these wavelengths can treat tumors and bacterial infection sites 2–3 cm on the surface or under the skin [119]. The second NIR window (NIR-II, 950–1350 nm) is beneficial for deeply embedded tumors, and the maximum penetration distance can reach 10 cm [120]. Laser parameters, including

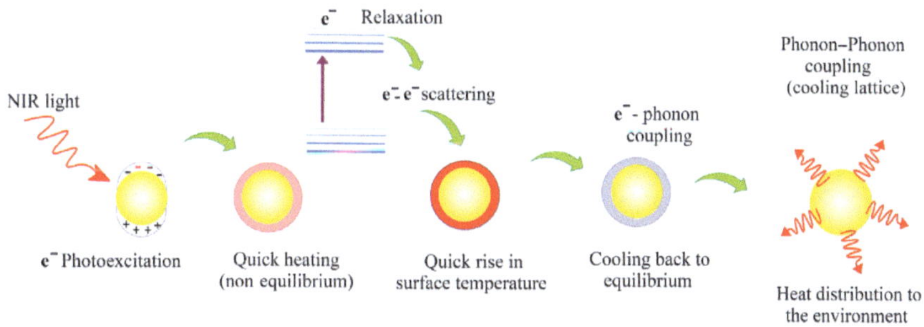

Figure 2.41 Scheme of photothermal effect of AuNPs under NIR laser irradiation. *Source:* Kadkhoda et al. [118] / with permission of Elsevier.

wavelength, power density, irradiation duration, and laser distance, also have an impact on the photothermal effect of nanoparticles [121]. Irradiation with lasers of higher power density can increase the excitation of nanoparticles and improve the efficiency of photothermal conversion. However, excessively high power density can produce side effects such as skin burns and tissue damage. Additionally, optimizing the duration of laser irradiation and the number of irradiations can enhance the photothermal effect.

2.3.3 Noble Metal Nanoclusters

Noble metal nanoclusters are tiny structures made up of a few noble metal atoms, and their size is usually less than 2 nm, located in the transitional region between molecules and nanoparticles. Unlike larger nanoparticles, noble metal nanoclusters exhibit obvious quantum confinement effects, and their electronic energy level structures show discretization, endowing them with unique optoelectronic properties.

Noble metal nanoclusters usually exist in specific geometric structures including icosahedra, octahedra, cubes, and triangles. However, these geometric structures are not fixed but are significantly influenced by ligands. The types, concentrations, and distribution patterns of ligands directly affect the lattice structure of nanoclusters and the arrangement of surface atoms. They can precisely regulate the geometric structure of nanoclusters, thereby endowing them with different functional characteristics. This characteristic is of great significance in fields such as catalytic reactions, optical performance regulation, and sensor design, providing an important theoretical basis and technical path for the development of noble metal nanoclusters with specific functions.

2.3.3.1 Gold Nanoclusters

Gold nanoclusters are the most widely studied type of noble metal nanoclusters, featuring excellent stability and controllable optical properties. The number of atoms ranges from several to dozens, such as Au_{13}, Au_{25}, and Au_{44}. Each nanocluster has a unique geometric structure and optical characteristics. Gold nanoclusters show good application potential in the fields of biomarkers, sensors, and catalysis.

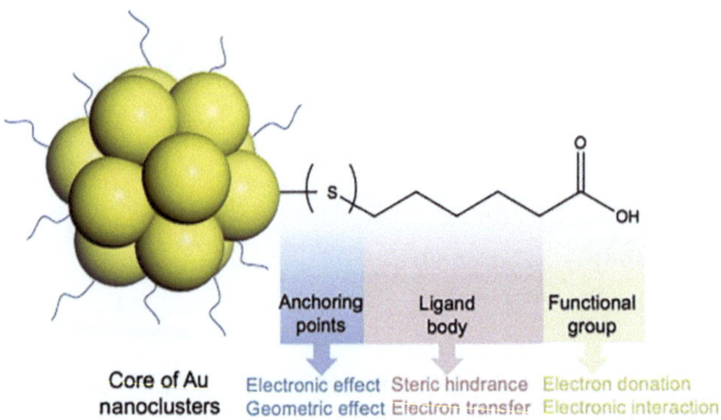

Figure 2.42 The scheme of the three segments of organic ligands of Au NCs (anchoring points, ligand body, and functional group), each of which can affect the structural and physicochemical properties of the Au NCs. *Source:* Zhang et al. [122] / with permission of John Wiley & Sons.

AuNCs have a clear molecular formula and structure, such as $[M_nL_m]^q$, where M represents a gold atom, L represents the ligand protecting AuNCs, and q represents the net charge of AuNCs. The organic ligands that are located on the surface of AuNCs are made up of three constituent parts. The innermost part acts as an anchorage point, linked directly to the gold core and protecting it. The middle part is the main body of the organic ligand, which can provide conjugation and affect the properties of AuNCs through the steric hindrance effect. The outermost part is the functional groups, which affect the hydrophilicity and hydrophobicity of AuNCs (Figure 2.42) [122].

Different protective ligands can be used to synthesize AuNCs with different structures and properties. Gold nanoclusters, in the absence of ligands or with a small amount of ligands, usually tend to form relatively compact structures, such as icosahedral or octahedral shapes. However, when specific ligand molecules, such as thiol compounds or amino acid molecules, coat the surface of metal nanoclusters, these ligand molecules change the geometry of the nanoclusters by forming coordination bonds with metal atoms, which may lead to the formation of more complex triangular or tetrahedral structures of the nanoclusters.

Taking thiol ligands as an example, the synthesized Au_{102} nanocluster used *p*-mercaptobenzoic acid as the ligand [123], the synthesized Au_{25} nanocluster using phenylethanethiol as the ligand [124], the synthesized Au_{23} nanocluster using cyclohexanethiolate as the ligand [125], the synthesized Au_{30} nanocluster using 1-adamantanethiolate as the ligand [126], and the Au_{42} nanocluster using 4-*tert*-butylbenzenelthiolate as the ligand [127]. As shown in Figure 2.43, although they are all ligands belonging to the thiol family, the prepared AuNCs exhibit different size characteristics. In addition, their structures also vary greatly. At the center of the Au_{25} nanocluster, there is an icosahedral Au_{13} core, with the remaining 12 gold atoms forming an incomplete shell around it. The thiol ligands on the surface are bonded to the Au_{25} core in a bridging manner. The structure of the Au_{23} nanocluster includes a bipyramidal Au_{15} core based on a cubic prism,

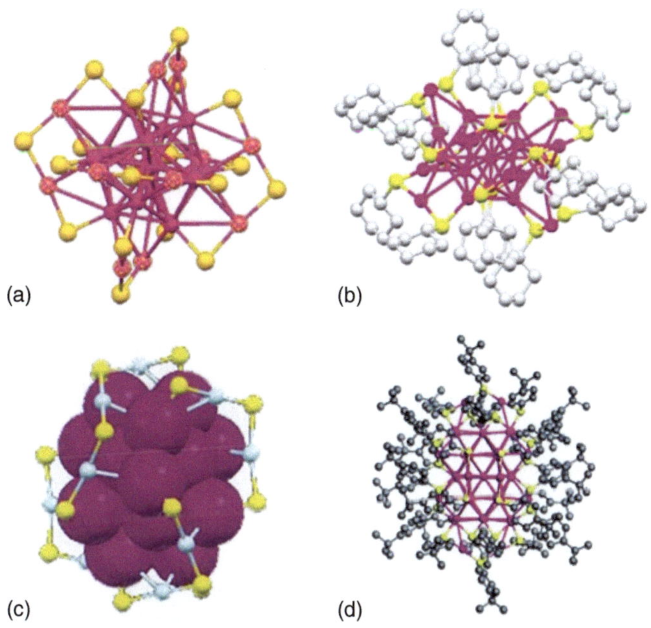

Figure 2.43 The structure of (a) Au_{25} nanocluster. *Source:* Zhu et al. [124] / with permission of American Chemical Society. (b) Au_{23} nanocluster. *Source:* Das et al. [125] / with permission of American Chemical Society. (c) Au_{30} nanocluster. *Source:* Higaki et al. [126] / with permission of John Wiley & Sons. (d) Au_{42} nanocluster. *Source:* Zhuang et al. [127] / The Royal Society of Chemistry / CC BY 4.0.

with its surface protected by two spike trimers $Au_3(SR)_4$, two monomer $Au(SR)_2$, and four planar bridged SR ligands. The Au_{30} nanocluster consists of a hexagonal, closely arranged Au_{18} core protected by the six $Au_2(SR)_3$. The core of the Au_{42} nanocluster is composed of four cubic bodies and contains 34 Au atoms. The structural differences of these AuNCs naturally lead to differences in their properties, thus making them applicable to different fields.

The properties of AuNCs can be finely tuned by methods like ligand exchange. For example, $Au_{38}(PET)_{24}$ undergoes ligand exchange with 4-*tert*-butylbenzenelthiolate (TBBT), transforming into $Au_{36}(TBBT)_{24}$ (Figure 2.44) [128]. During the exchange, larger TBBT ligands replace PET on Au_{38}, which eventually induces a structural rearrangement in the cluster. This rearrangement leads to the loss of two gold atoms, resulting in the formation of $Au_{36}(TBBT)_{24}$ featuring a tetrahedral geometry.

In recent years, AuNCs protected by N-donor ligands have attracted people's attention because N-donors are a new type of protecting ligand compared with thiol, *N*-heterocyclic carbenes, and alkyne groups, which may enhance the structural and functional diversity of AuNCs. Pyridyl, amido group, amidinate group, and amino acid are typical N-donor ligands [129]. The synthesis of a phosphine-protected $[Au_{20}(PPhpy_2)_{10}Cl_4]Cl_2$ nanocluster (Au_{20}) was achieved through the direct reduction of $Au(PPhpy_2)Cl$ using $NaBH_4$ [130]. The Au_{20} skeleton can be considered as two incomplete icosahedra (Au_{11}), fused together at two vertices. Crucially,

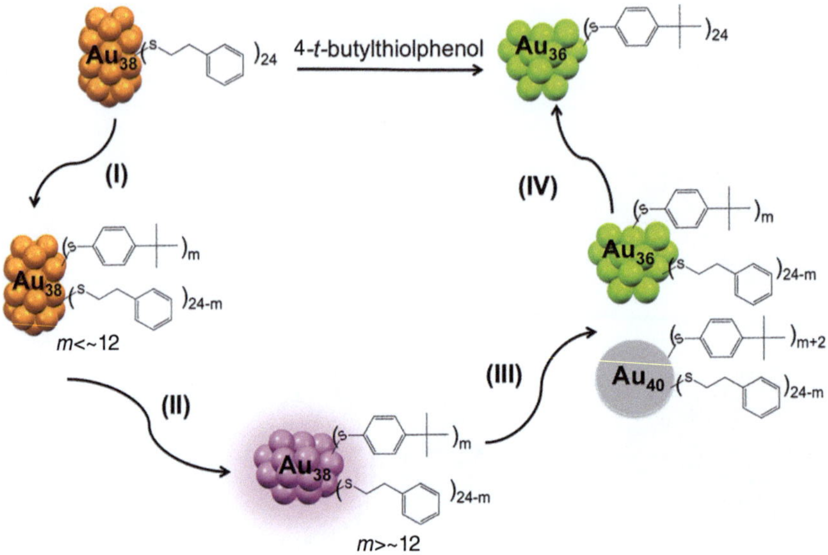

Figure 2.44 Mechanistic pathway for the transformation from $Au_{38}(PET)_{24}$ to $Au_{36}(TBBT)_{24}$. *Source:* Zeng et al. [128] / with permission of American Chemical Society.

the core's waist is connected by four P–N bridges, which significantly improve the stability of the nanocluster. Through structural analysis of the synthesis of a $[Au_{10}Ag(2\text{-pyC}{\equiv}C)_3(dppy)_6](BF_4)_5$ nanocluster [131], $Au_{10}Ag_2$ core has a triangular $Au_4(2\text{-py-C}{\equiv}C)_3$ unit. In this unit, an Au atom is attached to an alkyne group and an N atom on the pyridine, demonstrating the formation of the Au—N bond.

2.3.3.2 Silver Nanoclusters

Silver nanoclusters have attracted attention due to their excellent optical properties and strong fluorescence. Similar to gold nanoclusters, the optical behavior of silver nanoclusters is strongly influenced by their size and structure. Ag_7, Ag_{13}, etc. are common silver nanoclusters in research. Silver nanoclusters have unique advantages in biological imaging and antibacterial applications.

The geometric structure of silver nanoclusters has certain similarities with that of gold nanoclusters, but their stability and structural characteristics depend to a certain extent on the size and synthesis conditions of the nanoclusters [132]. $Ag_{14}(SC_6H_3F_2)_{12}(PPh_3)_8$ was synthesized by the mixed protection of thiol ligands and phosphine ligands [133]. The most significant feature is that the thiol ligands form bonds with three silver atoms simultaneously, without forming a structural pattern similar to that of motifs found in gold nanoclusters. The core is made up of a perfect octahedral Ag_6^{4+} unit.

In addition, by stabilizing and assisting nanocluster crystallization with Ph_3P, the single crystal nanocluster of $Ag_{29}(BDT)_{12}(TPP)_4^{3-}$ could be obtained [134]. Single crystal diffraction analysis reveals that the core of the nanocluster is an icosahedral Ag_{13}. Another 12 silver atoms are arranged in four chair configurations directly above the icosahedral core. The remaining four silver atoms cap the core. Thus,

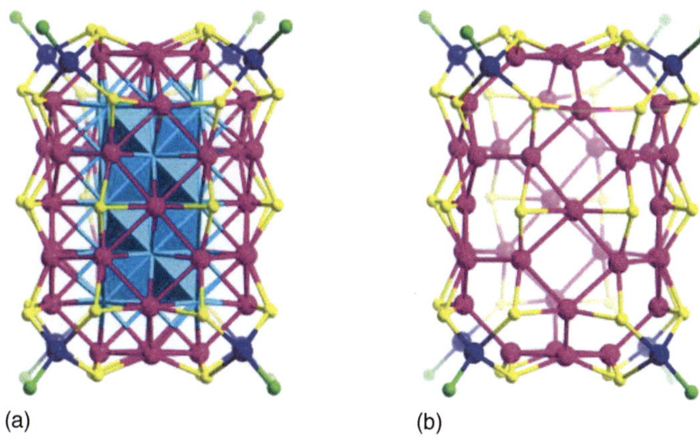

Figure 2.45 (a) Total structure of $[Ag_{67}(SPhMe_2)_{32}(PPh_3)_8]^{3+}$ nanocluster and (b) the Ag_{44} framework. *Source:* Alhilaly et al. [135] / with permission of American Chemical Society.

the core is passivated. In addition, by co-protecting with triphenylphosphine and thiolate ligands, the shape and size of the nanoclusters were well controlled, resulting in a box-shaped $[Ag_{67}(SPhMe_2)_{32}(PPh_3)_8]^{3+}$ structure (Figure 2.45) [135]. Ag_{23} core is composed of two cubic octahedral Ag_{13} units arranged in a box shape. The formation of a protective framework around the core was achieved by the outside $Ag_{44}(SPhMe_2)_{32}(PPh_3)_8$.

By regulating with phosphine ligands and sulfur ligands, the size and shape of the silver nanoclusters were well controlled, and cubic silver nanoclusters with different numbers of nuclei were obtained [136]. The phosphine ligand occupies 8 vertex positions, while the thiol ligand occupies its 12 edges and 6 faces. The size of the nanoclusters can be well regulated by controlling the proportion of phosphine ligands. To enhance the stability of silver nanoclusters, $[Ag_{21}\{S_2P(OiPr)_2\}_{12}]^+$ nanoclusters were synthesized by protecting them with their characteristic disulfur ligands. An icosahedral Ag_{13} core is covered with eight silver atoms to form an Ag_{21} cage [137]. The unique electronic structure of this core-shell configuration gives the silver nanocluster good stability.

2.3.3.3 Copper Nanoclusters

Gold and silver nanoclusters have been extensively researched and used due to their advantages, such as material stability, excellent optical properties, and good biocompatibility. However, their costs are relatively high and they are difficult to achieve large-scale industrialization. Therefore, in recent years, copper nanoclusters have gradually begun to attract the interest of researchers. This is not only because copper nanoclusters can have a high yield under mild conditions, but also because copper has a larger reserve in the Earth's crust and a relatively lower refining cost. However, due to the late start of the research on copper nanoclusters, there are inevitably some problems: (1) A lack of profound understanding of the structure and optical properties of copper nanoclusters results in poor control over the fluorescence intensity and color of copper nanoclusters. (2) The relatively low stability

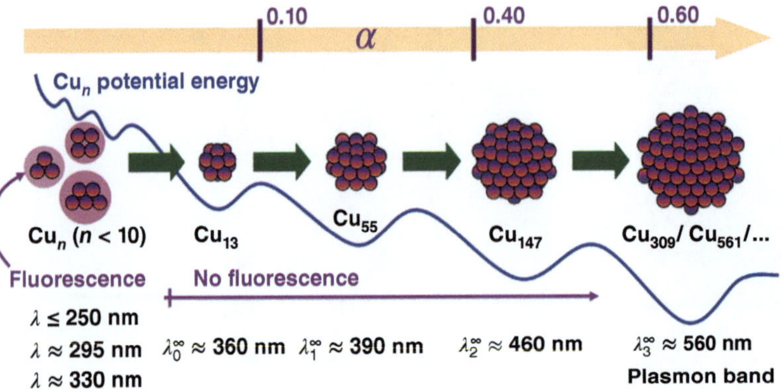

Figure 2.46 Schematic illustration of the progression of copper nanocluster with rising α values. *Source:* Vázquez-Vázquez et al. [138] / with permission of American Chemical Society.

of the material, particularly with regard to placement, severely restricts its use in both fundamental research and applications.

Based on the microemulsion method, $Cu(AOT)_2$ and AOT were used as surfactants, and $NaBH_4$ or hydrazine was used as reducing agents to synthesize ultrafine copper nanoclusters. These nanoclusters can easily be made larger or smaller by adjusting the percentage of reducing agent used (taking $NaBH_4$ as an example). As shown in Figure 2.46, the ratio of supplied $NaBH_4$ to $NaBH_4$ required for the stoichiometric reduction of Cu^{2+} is represented by α. When the proportion of the reducing agent used is relatively high, the size of the synthesized copper nanoclusters is relatively large (about 3 nm), while when the proportion of the reducing agent used is low, the size of the obtained copper nanoclusters is less than 2 nm and has the characteristics of UV photoluminescence [138].

Under the condition of constant current, Cu^{2+} generated during the anode electrolysis process undergoes a reduction reaction in the solution through the electrons provided by the electrodes, forming stable copper nanoclusters. In the presence of 0.1 mol/L tetrabutylammonium nitrate, the current density was adjusted to 25 mA/cm^2, and copper nanoclusters with the structure of Cu_{13} were obtained through sacrificial anode copper sheets [139]. When the current density is reduced to a minimum value, copper nanoclusters with the structure of Cu_5 can be obtained through electrochemical synthesis [140].

2.3.4 Optical Properties of Noble Metal Nanoclusters

When metallic particles are reduced to ultrasmall dimensions, quantum confinement effects begin to dominate their behavior. Unlike their larger nanoparticle counterparts, noble metal nanoclusters—comprising anywhere from a handful to about a hundred gold atoms—do not display the characteristic LSPR. Instead, they possess distinctive optical absorption and photoluminescence traits. These nanoclusters, which bridge the gap between individual atoms and bulk materials, typically measure less than 2 nm—comparable to the Fermi wavelength of conduction electrons. This minute scale leads to the formation of discrete energy states rather than the continuous bands seen in larger particles, causing the plasmonic features to vanish and giving rise to fluorescence that varies with nanocluster size. When illuminated with light matching the energy gap of the nanocluster, electrons in the ground state absorb this energy and transition to higher, excited states, resulting in unique photophysical responses.

Noble metal nanoclusters are known for exhibiting several absorption peaks in the UV-visible spectrum, a phenomenon stemming from their quantized electronic structures. Taking the $Au_{25}(SH)_{18}^-$ cluster as a representative example, Figure 2.47a illustrates how its Kohn-Sham MOs, energy levels, and atomic orbital (AO) contributions are distributed. The HOMO and the three LUMOs predominantly arise from the 6sp AOs of gold atoms (indicated in green), collectively forming what is referred to as the sp band. In contrast, the orbitals ranging from HOMO-1 to HOMO-5 are mainly derived from the $5d^{10}$ AOs of gold (shown in blue), which together constitute the d band. Notably, both the HOMO and LUMO groups also exhibit significant sulfur 3p (yellow) character, highlighting the influence of the ligand environment. The nanocluster's first optically active excited state appears at 1.52 eV, corresponding to a transition from the HOMO to the LUMO (sp ← sp), which is an intra-band excitation (Figure 2.47a). Among the occupied orbitals, only the threefold degenerate HOMO orbitals exhibit a greater s-character than d, so transitions from other occupied orbitals (HOMO-*n*) tend to involve inter-band processes (sp ← d). Further up in energy, the absorption peak at 2.63 eV (labeled "b" in Figure 2.47b) results from both intra-band (sp ← sp) and inter-band (sp ← d) transitions, while the peak at 2.91 eV (labeled "c") is primarily attributed to inter-band transitions (sp ← d).

Importantly, the HOMO, LUMO, HOMO-1, and LUMO+1 are almost exclusively composed of AOs from the Au_{13} core. As such, the initial absorption peak can be directly linked to electronic transitions within the Au_{13} core, underscoring the dominant role of the core's electronic and geometric structure. This behavior stands in stark contrast to that of clusters with a bi-icosahedral motif. These observations collectively underscore how the precise atomic arrangement and core structure of nanoclusters critically dictate their optical properties [124].

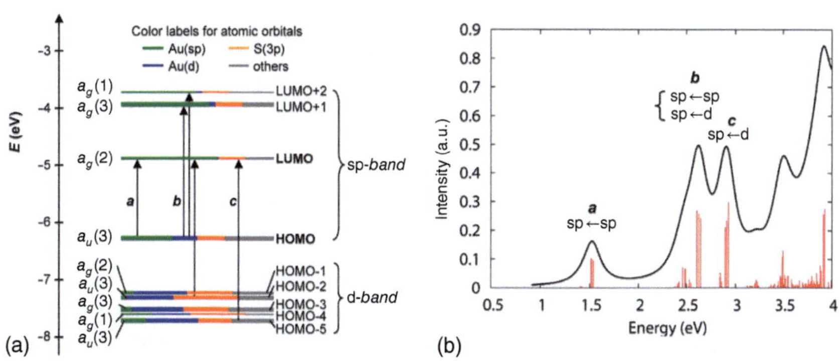

Figure 2.47 (a) Kohn–Sham orbital energy-level diagram and (b) the theoretical absorption spectrum of $Au_{25}(SH)_{18}^-$. *Source:* Zhu et al. [124] / with permission of American Chemical Society.

2.3.4.1 Photoluminescence

Photoluminescence is an important characteristic of noble metal nanoclusters. Excited electrons are unstable and return to the ground state by emitting light with an energy equivalent to the width of the bandgap. Regarding the luminescence mechanism of gold nanoclusters, the current consensus is that fluorescence is produced by a complex of Au(I) and thiol ligands. This is due to charge transfer between the S atom in the thiol ligand and the metal center (ligand-to-metal charge transfer, LMCT), or ligand-to-metal-metal charge transfer (LMMCT) from the ligand to the metal-metal interaction center (Figure 2.48) [141]. The latter explanation mainly considers the Au–Au affinity bond at the metal center. The transfer process of electrons in LMCT or LMMCT originates from S–Au or Au–Au. Therefore, the ability of the surface ligands of nanoclusters to donate electrons will affect their quantum yield.

Generally speaking, enhancing the capacity of ligands to donate electrons will increase the fluorescence emission intensity of gold nanoclusters. Studies have shown that partially replacing the ligands on the surface of nanoclusters with long-chain polypeptides containing a large amount of N and O can enhance fluorescence. This indicates that the surface ligands' electron-rich atoms or groups may increase the fluorescence emission intensity of the nanoclusters. Ligand perturbation can also be regulated to enhance the luminescence of AuNCs. For instance, when tetraethylammonium cations are added to the substrate of AuNCs, a significant improvement in their luminescence quantum efficiency can be observed [142]. Similarly, if arginine is used to fix 6-aza-2-thiothymine (ATT)-modified AuNCs, a significant increase in luminescence intensity and quantum yield will also be observed. This is because arginine limits the vibration and rotation of ATT-protected ligands, reduces the nonradiation, and thereby improves the luminescence efficiency of ATT-AuNCs [143].

2.3.4.2 Two-photon Absorption

Two-photon absorption (TPA) is a light absorption mechanism first proposed by Goppert-Mayer in 1931. Its basic principle is that two photons are absorbed by a

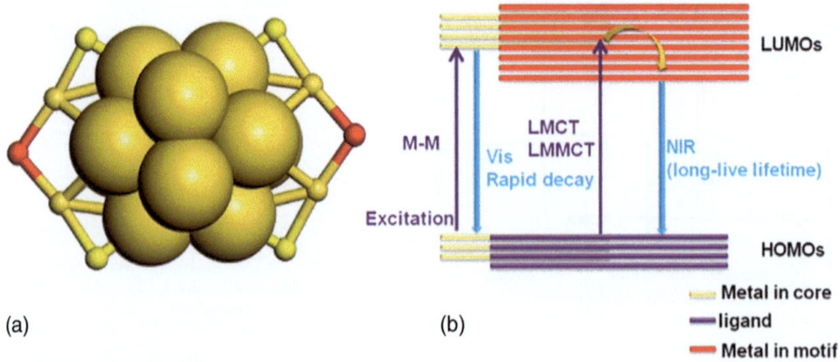

Figure 2.48 Schematic illustration of the (a) structure and (b) photoluminescence process of $Au_{25}(SR)_{18}$. *Source:* Liu et al. [141] / with permission of Elsevier.

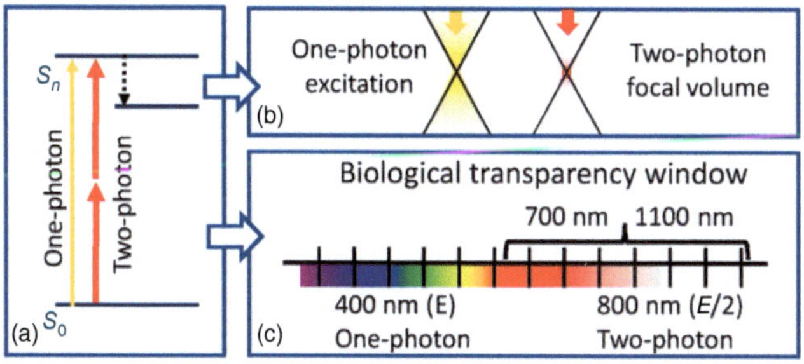

Figure 2.49 (a) Illustration of two-photon absorption; (b) excitation volumes for one-photon and two-photon absorption. (b) Comparison of excitation volumes for one-photon versus two-photon absorption. (c) The excitation of fluorophores using near-infrared light by two-photon absorption. *Source:* Galas et al. [145] / MDPI / CC BY 4.0.

substance, exciting electrons to the excited state [144]. In the TPA process, a sufficiently strong incident light is required to make the electrons in the material pass through the intermediate virtual energy state to the excited state, and at the same time, high-frequency upconversion fluorescence may occur. When the intensity of the incident light increases and there are sufficient electrons, TPA will reach saturation and will no longer increase in response to an increase in light intensity (Figure 2.49) [145].

Two-photon excitation (TPE) based on NIR light can increase the penetration depth and spatial resolution of biological tissues due to its lower scattering and minimized autofluorescence [146]. Au_{25} nanocluster has TPA characteristics in hexane. When the excitation wavelength is 800 nm, TPE fluorescence can be observed in the visible light region, and the cross-sectional measurement value of TPA is 42,700 GM (Goppert-Mayer units, $10^{-50} \cdot cm^4 \cdot s$) [147]. The TPA cross-section of water-soluble 11-MUA-protected gold nanodots in water is approximately 8761 GM [148]. Recently, strong two-photon emissions from AuNCs protected by water-soluble glutathione have also been observed. The TPA cross-section of these AuNCs in water was found to be 189,740 GM, a figure much higher than that of typical organic dyes and quantum dots [149]. Gold nanoclusters have become effective absorbents for multi-photon biological imaging and other nonlinear optical applications due to their large TPA cross-sections [150].

2.3.4.3 Aggregation-Induced Emission Effect

The aggregation-induced emission (AIE) effect refers to the phenomenon where substances that emit almost no light in solution significantly enhance their luminescence in an aggregated state or under a solid film. For metal nanoclusters, the surface ligands form stable structures through strong metal-ligand interactions. The enhancement of aggregated luminescence of AuNCs is closely related to the RIM.

As shown in Figure 2.50, the assembly of oligomeric Au(I)–thiolate complexes can be orchestrated through two distinct strategies: solvent-mediated aggregation and cation-mediated aggregation. In the solvent-mediated approach, aggregation is

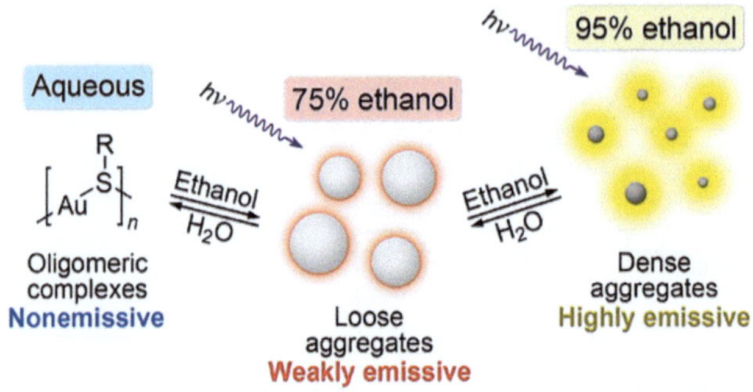

Figure 2.50 Schematic illustration of the AIE effect of oligomeric Au(I)–thiolate complexes. *Source:* Luo et al. [151] / with permission of American Chemical Society.

triggered by introducing a weakly polar solvent, such as ethanol. This is in stark contrast to the highly polar environment provided by water. When a substantial amount of ethanol is added to an aqueous solution containing the Au(I)–thiolate complexes, the polarity of the medium decreases significantly. This change disrupts the hydration shell that normally stabilizes the Au(I)–thiolate species in water, leading to a reduction in solvation and a subsequent neutralization of charges on the complex. As a result, the electrostatic repulsion between individual complexes diminishes, thereby promoting their aggregation into larger assemblies.

Alternatively, aggregation can be induced through the addition of specific cations, a process referred to as cation-mediated aggregation. This method exploits the strong electrostatic and coordination interactions between the introduced cations and the negatively charged or electron-rich sites on the Au(I)–thiolate complexes. In addition to neutralizing the negative charge on the complex, the crosslinking agent also brings the Au(I)–thiol complex closer and promotes the formation of lipophilic bonds and dense aggregates [151].

2.3.4.4 Photothermal Effect

The photothermal effect of AuNCs is due to their ability to absorb light energy efficiently and convert it into heat when exposed to light of specific wavelengths. Typically, NC with less than 50 gold atoms has continuous absorption peaks in the visible light region. When the size increases (>100 Au atoms), as the energy gap decreases, the long absorption tail extends into the NIR. The absorption intensity of NC drops sharply at longer wavelengths. Therefore, effective photothermal conversion cannot be carried out through NIR excitation.

$Au_{42}(SCH_2Ph)_{32}$ has a hexagonal close packing (hcp)Au_{20} core with an aspect ratio of 6.2. It has a strong absorption at 815 nm and a molar absorptivity of 1.4×10^5/M/cm. The Au_{20} core is protected at the terminal by two pairs of interlocked $Au_4(SR)_5$ motifs and around its main body by six $Au(SR)_2$ motifs. Under the excitation of 808 nm, it shows effective photothermal conversion. The NIR peak of Au_{42} corresponds to the electron transition along the long axis of the Au_{20} core. The intense photothermal conversion is attributed to the restriction of rod-shaped intracellular excitons. Therefore, the longitudinal excitation analogous

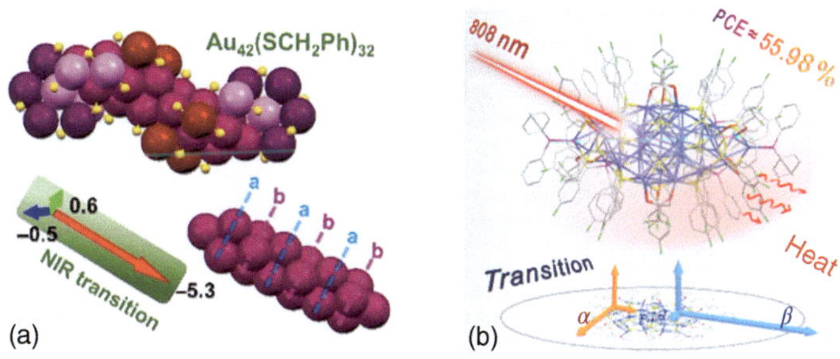

Figure 2.51 Au_{42} and Cu_{62} photothermal conversion. (a) *Source:* Li et al. [152] / with permission of American Chemical Society. (b) *Source:* Yan et al. [153] / American Chemical Society / CC BY 4.0.

to that in plasmonic gold nanorods can be achieved in atomically precise ultrafine nanorods with exciton states smaller than 2 nm. Although there is no SPR, the NIR exciton absorption may be very strong and lead to intense photothermal conversion (Figure 2.51a) [152].

The $Cu_{62}(4\text{-F-PhS})_{30}(CF_3COO)_8(PPh_3)_6H_{10}$ nanocluster features a metal core with an aspect ratio of 2.6. Distinguished by its windmill-like core, this copper-based nanocluster is further stabilized by an orderly arrangement of organic ligands. Its distinctive sheet-like architecture imparts notable optical properties, particularly strong absorption in the NIR region at 760 nm. When exposed to an 808 nm laser, $Cu6_2$ demonstrates an impressive photothermal conversion efficiency, reaching up to 55.98% (Figure 2.51b) [153].

2.3.4.5 Photosensitive Effect

Through surface functionalization, the unique photophysical properties of AuNCs enable them to be used as photosensitizers in photodynamic therapy. PDT is achieved by irradiating photosensitizers or photoactive materials with an appropriate light source. Among them, the photoactivated photosensitizers damage cancer cells by reacting with O_2 to generate singlet oxygen (type II light reaction pathway) or by interacting with substrate molecules to generate free radicals (type I light reaction pathway) [118]. AuNCs can effectively absorb light of specific wavelengths and excite electrons to the excited state. Singlet electrons can cross between systems to reach the triplet state. When electrons are in the triplet state, they can transfer energy or electrons to surrounding oxygen molecules, generating various ROS including $^1O_2$, superoxide anions (O^{2-}), and hydroxyl radicals (OH). These ROS have strong oxidizing properties and can kill cancer cells in photodynamic therapy.

The effects of Au(I) species and surface charges on singlet oxygen were investigated using thiol–polyethylene glycol terminally modified with methoxy (—OCH_3) or carboxyl (—COOH) groups as surface ligands to synthesize AuNCs. It was found that AuNCs with more surface positive charges and more Au(I) species contribute to the generation of a higher content of singlet oxygen under light irradiation. This is an

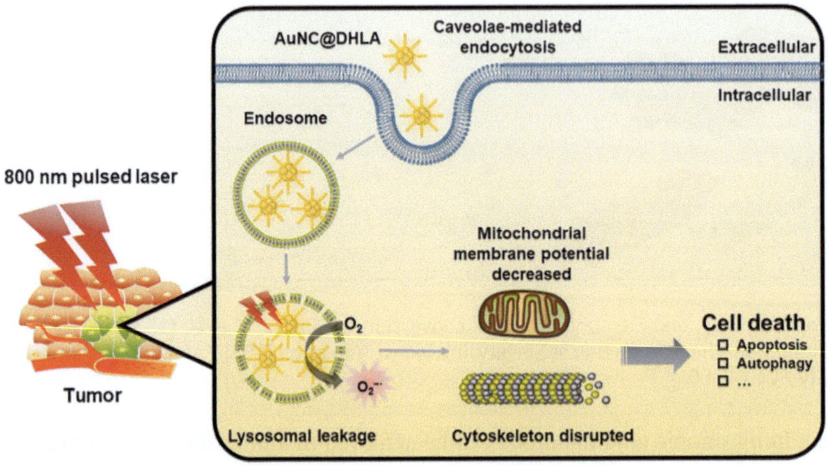

Figure 2.52 Schematic illustration of photosensitization effect of AuNC@DHLA for cancer therapy. *Source:* Han et al. [156] / with permission of American Chemical Society.

effective strategy for enhancing the PDT effect of AuNCs through simple regulation of ligands [154].

Due to the fact that in an oxygen-deficient environment, the type II light reaction pathway dependent on O_2 is often restricted, and solutions to this problem have emerged accordingly. One approach is to combine AuNCs with catalase, which can catalyze hydrogen peroxide in the physiological environment, enabling O_2 to be self-sufficient. For example, AuNCs (AuNCs@HSA/CAT) co-modified with human serum albumin and catalase can convert endogenous hydrogen peroxide to O_2. Under irradiation with the NIR-II laser, the O_2 produced in situ transfers energy with AuNCs to produce singlet oxygen [155]. The second method is to design AuNCs that can achieve the PDT effect through the photoreaction pathway of type I electron transfer. AuNCs coated with dihydrolipoic acid (AuNC@DHLA) exhibit excellent two-photon optical properties and photodynamic performance, and it was confirmed that these AuNCs generate superoxide free radicals under laser irradiation, which cause apoptosis and necrosis through the type I photoreactive pathway (Figure 2.52) [156].

2.4 Inorganic Nonmetallic Materials

The optical properties of inorganic nonmetallic materials are primarily determined by their unique crystal and electronic structures, as well as the interaction between light and matter. These materials' optical properties are primarily reflected in how they absorb, emit, reflect, refract, and propagate light. Firstly, inorganic nonmetallic materials usually have regular crystal structures, such as titanium dioxide (TiO_2), whose periodic structure can affect the propagation and refraction of light, making these materials have special optical properties. Second, the electronic energy level structure of inorganic nonmetallic materials is usually simpler, especially wide bandgap semiconductors (e.g. zinc oxide ZnO), which can effectively control the absorption and emission process of light. When a photon interacts with these materials, the electron can jump to a high-energy state, producing the photoemission phenomenon. In addition, the quantum effect is particularly significant in certain inorganic nonmetallic materials (e.g. quantum dots), where the quantum size effect allows the materials to exhibit optical properties that are different from those of macroscopic materials, such as enhanced fluorescence emission and surface plasmon resonance. These properties endow the materials with important applications in bioimaging, sensors, and other fields.

Overall, the optical properties of inorganic nonmetallic materials originate from their unique crystalline and electronic structures. By modulating these factors, new materials with specific optical functions can be designed and developed for use in optoelectronics, environmental protection, and medical treatment. This subsection will focus on inorganic nonmetallic photofunctional materials.

2.4.1 Quantum Dots

Quantum dots (QDs) are tiny semiconductor nanoparticles, typically ranging in size from a few nanometers to tens of nanometers. The electrons and holes of these nanoparticles are quantum confined in three dimensions, resulting in quantum dots with unique optoelectronic properties. These properties are closely related to their size, shape, material constituent elements, and surface modifications, and often exhibit significant quantum effects. Quantum dots have different luminescence properties (e.g. spectral range, brightness) at different wavelengths, and can be categorized as fluorescent, luminescent, and light-absorbing quantum dots.

The luminescence mechanism of quantum dots is usually closely related to their size, shape, and surface treatment. The luminescence mechanism primarily involves electron–hole recombination process within the quantum dots:

1) Electronic excitation: Upon absorption of a photon, a quantum dot undergoes electronic excitation wherein the photon's energy is transferred to an electron. This results in the promotion of the electron from the valence band (VB) to the conduction band (CB), simultaneously generating a corresponding hole in the VB.
2) Electron–hole recombination: The photoexcited electron and the generated hole may subsequently recombine via a radiative recombination process. During this process, the energy released is emitted as a photon, manifesting as photoluminescence. The energy (or equivalently, the wavelength) of the emitted photon is strongly dependent on the quantum dot's size. Specifically, smaller quantum dots typically emit photons of higher energy (shorter wavelength, e.g. blue light), whereas larger quantum dots emit lower-energy photons (longer wavelength, e.g. red light).

Semiconductor nanocrystals known as quantum dots have strong quantum confinement effects. This phenomenon, which was initially noticed in colloidal

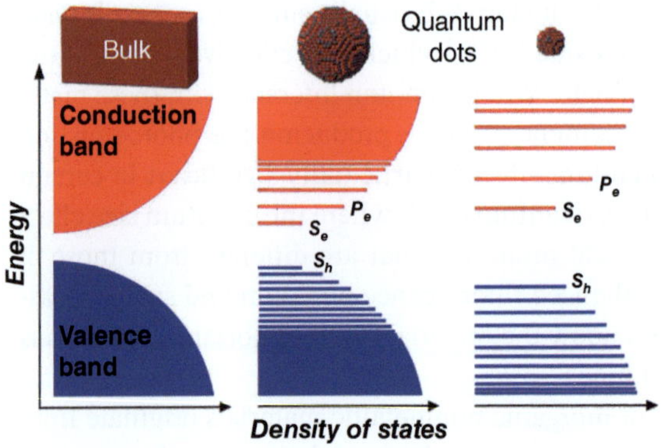

Figure 2.53 The electronic structure of bulk semiconductors and quantum dots. *Source:* Veeramani et al. [157] / with permission of Elsevier.

quantum dots created using solution-phase techniques, occurs when the quantum dot's spatial dimensions go close to or drop below the equivalent bulk semiconductor's exciton Bohr radius. Under these conditions, the energy levels within the quantum dot become quantized, fundamentally altering their optical and electronic properties compared to the bulk material, as shown in Figure 2.53. Due to the discretization of the energy levels, quantum dots can narrow full-width at half-maximum of the emission. The emission wavelength of quantum dots is profoundly influenced by their physical dimensions [157]. As the size of the quantum dot increases, the emission wavelength shifts to longer values. This tunability of emission wavelength by controlling particle size is a key advantage of quantum dots for applications in display technology, optoelectronic devices, and related fields.

The motion of electrons and holes in quantum dots is strongly restricted, making the energy level structure appear discrete. This discrete energy level structure allows quantum dots to absorb a wider range of photon wavelengths, thus achieving a wider range of absorption compared to dyes [158]. Quantum dots typically exhibit maximum optical absorption in the UV region (230–320 nm), accompanied by a tail that extends into the visible light [159]. One notable optical characteristic of quantum dots is their ability to emit light with a very narrow spectral linewidth, typically quantified by the full width at half maximum (FWHM). While it might be tempting to attribute the fluorescence linewidth solely to variations in quantum dot size, research indicates that intrinsic factors within the quantum dots themselves also play a significant role [160]. For instance, a single indium phosphide/zinc selenide/zinc sulfide quantum dot can exhibit a linewidth of 38 meV at ambient temperature. However, phenomena such as elastic phonon scattering can further influence the FWHM, especially at elevated temperatures, by narrowing the zero-phonon line [161]. In applications like optoelectronic devices and advanced display systems, achieving a narrower emission linewidth in quantum dots leads to reduced spectral broadening, which in turn enhances color purity and saturation. This property not only boosts device efficiency but also delivers superior visual quality and a broader, more vivid color palette in display technologies.

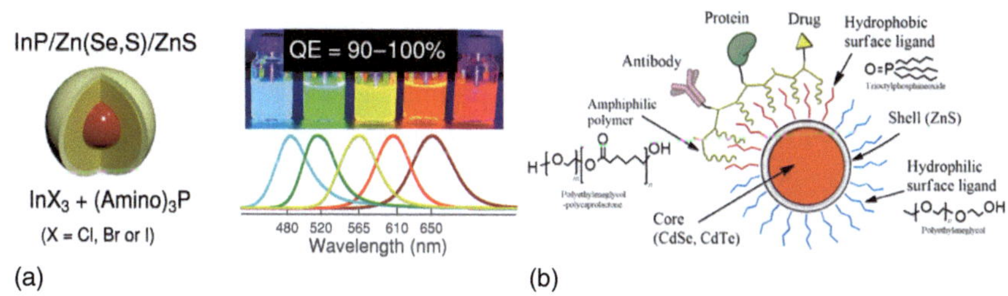

Figure 2.54 (a) High photoluminescence quantum yields (>90%) of quantum dots. *Source:* Van Avermaet et al. [163] / with permission of American Chemical Society. (b) Surface-dependent luminescence properties of quantum dots. *Source:* Maysinger et al. [166] / Frontiers / CC BY 4.0.

Another unique advantage of quantum dots is the ability to precisely control their emission wavelength by adjusting parameters such as size and composition, allowing their optical output to be fine-tuned from the UV through the visible and into the infrared spectrum [162]. Furthermore, quantum dots can achieve high luminescence efficiency through strategies like core-shell engineering or surface ligand passivation. For instance, coating an indium phosphide core with a zinc-containing shell provides effective surface passivation, leading to consistently high photoluminescence quantum yields across a broad spectral range (Figure 2.54a). This makes quantum dots highly promising for applications in advanced lighting technologies, such as high color-rendering white LEDs and next-generation micro-LED displays [163].

Longer fluorescence lifetime of semiconductor quantum dots compared to fluorescent dyes [164]. Unlike conventional organic dyes, which generally exhibit extremely brief fluorescence lifetimes on the scale of just a few nanoseconds, quantum dots are capable of maintaining their fluorescence for much longer periods—often ranging from 20 to 100 ns. This pronounced difference in excited state duration is particularly advantageous in optoelectronic applications. The prolonged lifetime of quantum dots enhances charge transfer efficiency, which is instrumental in developing advanced photovoltaic cells and robust light-emitting diodes that deliver superior performance and reliability [165]. Moreover, in biomedical imaging and sensing, the extended emission duration of quantum dots allows for highly multiplexed imaging and persistent monitoring of biological events, offering remarkable improvements in both sensitivity and temporal resolution.

The last advantage of quantum dots lies in the extraordinary versatility of their interfaces, which can be meticulously engineered to fine-tune their optoelectronic and chemical behaviors (Figure 2.54b) [166, 167]. By manipulating interfacial chemistry, it is possible to reduce surface imperfections and suppress nonradiative losses, resulting in enhanced quantum efficiency and greater photostability. Additionally, tailored surface modifications allow for the attachment of biomolecules, opening the door to precise targeting and advanced imaging techniques in biomedical contexts.

Quantum dots exhibit high fluorescence efficiency, narrow emission spectra, good photostability, and a high quantum yield. They also have tunable absorption spectra and the ability to adjust the fluorescence color by altering the size and surface modification, achieving coverage from UV to infrared. They also exhibit good photoluminescence and EL properties, making them promising for a variety of applications in display technology, bioimaging, optoelectronic devices, and more.

2.4.2 Carbon Materials

For millennia, carbon's primary application was as a reducing agent in metallurgy, while only three forms of carbon were recognized: graphite, diamond, and amorphous carbon. This understanding remained unchanged until the 1980s. The landscape shifted dramatically in 1985 with the identification of fullerenes, sparking intense scientific interest in low-dimensional carbon-based materials. Over the past three decades, this enthusiasm has led to the creation of a diverse array of carbon nanostructures, including carbon nanotubes (CNTs), carbon dots, graphene, and fullerenes, all illustrated in Figure 2.55.

2.4.2.1 Carbon Nanotube

Carbon nanotubes (CNTs) are nanoscale tubular structures consisting of single or multiple layers of graphene convoluted with unique optical properties. The optical properties of CNTs can exhibit significant differences depending on their structural differences (e.g. single-walled versus multi-walled), diameters, how they are curled, and the number of outer layers.

When it comes to optical absorption, CNTs exhibit significant optical absorption in a specific wavelength range in the visible and infrared regions due to their special electronic structure (one-dimensional quantum effect). Single-walled CNTs are particularly characterized by selective absorption, with absorption spectra dependent on their chirality (curl direction) and diameter, which determines their mode of interaction with light.

In photoluminescence, CNTs demonstrate distinctive luminescent characteristics. The photoluminescence observed in CNTs primarily arises from the recombination of electron–hole pairs (excitons) within the nanotube structure. The emission wavelength is highly dependent on the nanotube's structural parameters, such as chirality and diameter. Consequently, CNTs with different structural configurations exhibit variable luminescent behaviors, with emission wavelengths spanning the visible to NIR regions of the electromagnetic spectrum.

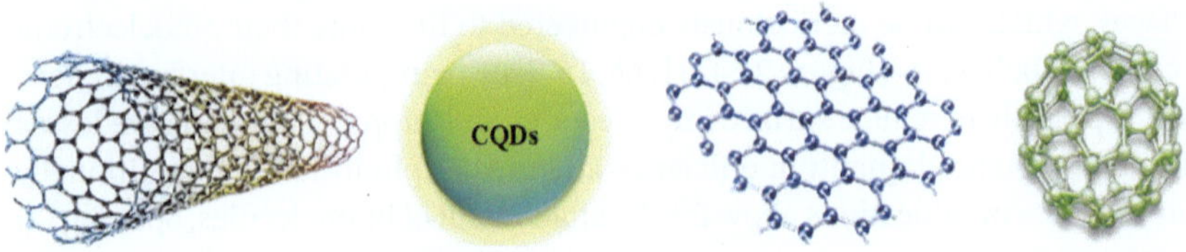

(a) (b) (c) (d)

Figure 2.55 Schematic illustration of (a) carbon nanotubes, (b) carbon dots, (c) graphene, and (d) fullerenes. *Source:* Asil et al. [168] / John Wiley & Sons / CC BY 4.0.

In addition, CNTs exhibit unique behavior in terms of optical nonlinear effects, such as strong optical second harmonic generation and four-wave mixing. This makes them potentially useful for optical communications, sensors, and optoelectronic devices.

2.4.2.2 Carbon Dots

Carbon dots are carbon-based materials with nanoscale dimensions, typically between a few nanometers and tens of nanometers. They exhibit unique optical properties, which are closely related to their size, shape, and surface functional groups.

The optical properties of carbon dots are primarily influenced by quantum confinement effects, resulting in absorption and luminescence properties that differ from those of other carbon materials. They are capable of absorbing energy under UV or visible light irradiation and emitting light in a specific wavelength range. Their photoluminescence properties usually have broad emission peaks, and the emission wavelength can be modulated by adjusting the size or surface modification of the quantum dots. In general, the luminescence of carbon dots can extend from the UV to visible wavelengths and even into the NIR region, with strong fluorescence quantum efficiency. The surface functional groups of carbon dots (e.g. carboxyl groups and amino groups) also have a major effect on their optical properties. These surface groups not only modulate the luminescence properties but also provide additional electron transport channels and improve the stability of the quantum dots. In addition, carbon dots have strong photostability and biocompatibility, making them suitable for applications in bioimaging, sensors, and optoelectronic devices.

2.4.2.3 Graphene

The optical structure of graphene is determined by its unique two-dimensional (2D) monolayer arrangement of carbon atoms that are hybridized by sp^2 to form a hexagonal honeycomb lattice structure. Each carbon atom bonds with its neighboring carbon atoms via σ-bonds, while the remaining π-electrons create a delocalized electron cloud spread across the graphene plane. This unique arrangement endows graphene with exceptional optical properties, particularly in the UV, visible, and infrared spectral regions.

In terms of optical absorption, graphene has a good ability to absorb electromagnetic radiation across a broad spectrum. Graphene exhibits strong light absorption, especially in the UV and visible ranges. Its absorption spectrum exhibits a unique "Planck absorption edge," where the monolayer structure of graphene allows it to exhibit an absorption coefficient close to a constant in these regions. This is because light energy may be absorbed by graphene's π-electrons, which can be stimulated into the CB.

In addition, graphene has strong nonlinear optical properties. For example, under high-intensity light irradiation, graphene exhibits nonlinear effects such as second harmonic generation and four-wave mixing. Graphene also has excellent photoconductivity and high electron mobility, giving it significant potential for applications in optoelectronic devices, optical sensors, and ultrafast optics.

Overall, the 2D structure and optical properties of graphene allow it to excel in modern optical and electronic applications.

2.4.2.4 Fullerene

Fullerenes, such as C_{60}, possess a highly symmetrical spherical structure made up of 60 carbon atoms organized into 12 pentagons and 20 hexagons, resembling the pattern of a soccer ball. The optical structure of fullerenes is closely related to their unique molecular geometry and electronic structure. The carbon atoms in its molecule form a strong π-electron conjugation system through covalent bonding, which gives fullerenes significant light absorption properties.

The optical characteristics of fullerenes are predominantly defined by their pronounced ability to absorb electromagnetic radiation across both UV and visible spectral regions. In the UV domain, fullerenes exhibit intense and well-defined absorption bands, attributable to π–π^* electronic transitions within the conjugated carbon framework. Conversely, in the visible region, the absorption is markedly weaker, a result of the symmetry-forbidden nature of the corresponding electronic transitions. The spectral position, bandwidth, and intensity of these absorption features are highly sensitive to the molecular size, geometric structure, and degree of conjugation within the fullerene species, as well as to the presence of functional groups or heteroatom doping.

Moreover, fullerenes are capable of exhibiting photoluminescence, albeit with relatively low fluorescence quantum yields under ambient conditions. This limited emission efficiency is generally attributed to rapid nonradiative decay processes facilitated by the high density of electronic states and efficient ISC to triplet states. Nevertheless, through molecular engineering or interaction with specific host matrices, it is possible to enhance both the absorption and emission efficiencies, thereby expanding their potential utility in light-harvesting and emissive applications.

Beyond linear optical absorption, fullerenes possess notable nonlinear optical properties. Under high-intensity light irradiation, they can manifest phenomena such as second harmonic generation, TPA, and optical limiting. These effects arise from the unique electronic delocalization and high third-order nonlinear susceptibility intrinsic to the fullerene cage.

The diverse suite of optical properties endowed by the unique structural, chemical, and electronic configuration of fullerenes has catalyzed their integration into a broad spectrum of advanced technologies. These include, but are not limited to, optoelectronic devices, photovoltaic cells, photodetectors, optical sensors, and nonlinear optical components [168].

2.4.3 Carbon Nitride

Carbon nitride appeared for the first time in 1834. In the research progress since then, different structures of carbon nitride have been obtained according to different synthesizing methods and different hydrogen contents. In 2004, graphite-phase carbon nitride came into people's view again, and researchers began to focus on its catalytic properties as a 2D material. In 2006, graphite-phase carbon nitride was first used for the photolysis of water to obtain hydrogen gas.

As a result of the increasing attention, carbon nitride has been prepared in a more complementary way. As research progressed, more carbon nitride structures were confirmed and discovered. C_xN_y has largely expanded the concept of carbon nitride. With continuous development, CN(C_2N_2, C_3N_3, C_4N_4), C_2N, C_3N, C_3N_2, C_3N_4, C_3N_5, C_3N_6, C_3N_7, C_4N, C_4N_3, C_5N, C_5N_8, C_6N_6, C_6N_7, C_6N_8, C_7N_6, C_9N_4, C_9N_7, $C_{10}N_3$, $C_{10}N_9$, and so on, have emerged. Different carbon and nitrogen ratios have brought about different carbon nitride structures, as shown in Figure 2.56, and their properties and applications are also different [168, 169].

Carbon nitride has a highly conjugated carbon-nitrogen network structure, which is usually layered. Its optical properties are affected by the nitrogen content, doping sites, and surface defects. Carbon nitride has a bandgap width of about 2.7 eV, giving it good light absorption in the visible range. Carbon nitride has a high luminescence quantum efficiency, and its luminescence wavelength can be structurally tuned from the UV to the visible region. Carbon nitride exhibits excellent performance in photocatalytic degradation of organic matter, especially in visible light-driven photocatalytic reactions. In addition, carbon nitride exhibits good photoluminescence properties and photothermal conversion characteristics, which could make it beneficial for applications in bioimaging, photocatalysis, and PTT.

Among the many carbon-to-nitrogen ratios of carbon nitride, the graphite-phase configuration of carbon nitride (g-C_3N_4) has garnered significant notice because of its excellent nonmetallic semiconductor properties. With the smallest bandgap energy, this is the most stable form of carbon nitride at room temperature. In g-C_3N_4 materials, the C and N atoms are combined by sp^2 hybridization to form a heptazine ring (C_6N_7) structure, which belongs to the P6m2 space group in the crystal structure. The N atoms of the end groups are interconnected between the rings to form a π–π conjugated mesh structure, and the layers interact with each other through van der Waals forces, which usually include two graphitic phase structures, triazine (C_3N_3) and heptazine (C_6N_7) [170].

A critical aspect of g-C_3N_4's electronic configuration arises from the presence of nitrogen atoms bearing nonbonding lone pairs. These lone pairs contribute to the formation of a unique VB, predominantly derived from nitrogen's p_z orbitals. Meanwhile, the CB is mainly constituted by the p_z orbitals of carbon atoms, which play a pivotal role in defining the material's semiconducting properties. This orbital arrangement underpins the n-type semiconducting behavior of g-C_3N_4, facilitating electron transport and making it suitable for various photoelectronic and catalytic applications. Figure 2.57 shows the charge transfer in g-C_3N_4. As a photocatalyst, g-C_3N_4 has the ability to separate holes and electrons. The bandgap of 2.7 eV allows it to absorb sunlight and hence it can be used for water purification, hydrogen production, and solar cell applications [171].

The photofunctional properties of carbon nitride mainly include photoluminescence, photocatalysis, and photothermal conversion.

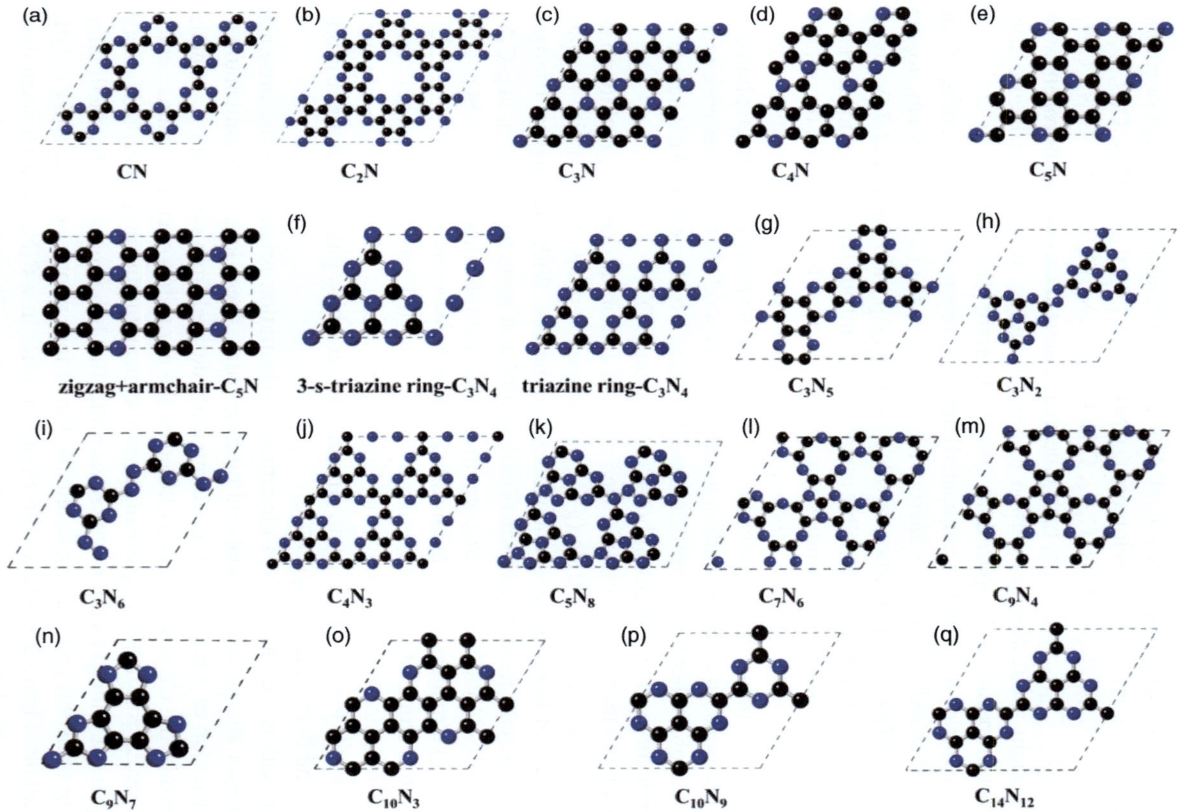

Figure 2.56 Schematic illustration of the C_xN_y structure. (a) CN, (b) C_2N, (c) C_3N, (d) C_4N, (e) C_5N, (f) triazine ring-C_3N_4, (g) C_3N_5, (h) C_3N_2, (i) C_3N_6, (j) C_4N_3, (k) C_5N_8, (l) C_7N_6, (m) C_9N_4, (n) C_9N_7, (o) $C_{10}N_3$, (p) $C_{10}N_9$, (q) $C_{14}N_{12}$. *Source:* Chen et al. [169] / Elsevier / CC BY 4.0.

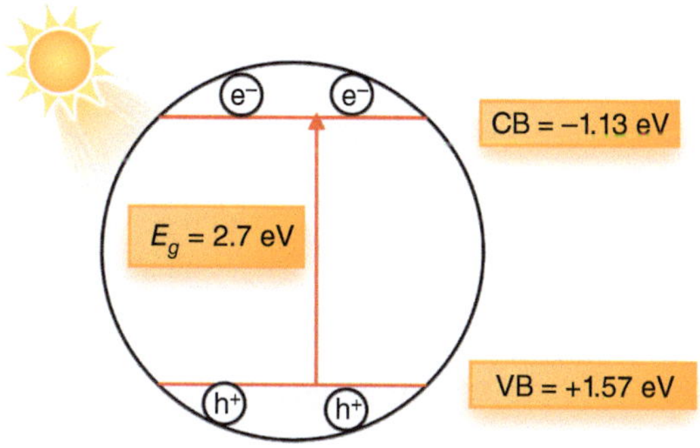

Figure 2.57 Typical charge transfer scheme for g-C_3N_4.

2.4.3.1 Photocatalysis

g-C_3N_4 is used as a promising photocatalyst owing to its moderate bandgap and efficient separation of photoexcited charge carriers. The ability to harness visible light enables the generation of ROS, such as superoxide and hydroxyl radicals, which play a crucial role in the breakdown of organic contaminants. This property has led to remarkable results in the degradation of pollutants like rhodamine B and tetracycline using g-C_3N_4-based materials. Structurally, g-C_3N_4 is characterized by a layered architecture with tertiary amine linkages, resulting in a graphite-like stacking and a unique conjugated system. The coexistence of carbon and nitrogen atoms in various oxidation states contributes to the material's complex electronic band structure and enhances its electrical conductivity.

Despite these advantages, g-C_3N_4 faces several significant challenges that hinder its practical application. Its relatively large bandgap restricts the absorption of a broad range of solar wavelengths, limiting its photocatalytic efficiency. Moreover, the rapid recombination of photogenerated electron–hole pairs further diminishes its catalytic performance, while low carrier mobility impedes effective charge transport. Other drawbacks include a modest specific surface area and insufficient stability under operational conditions, as extended exposure to light and reactive species can degrade its activity over time.

To address these issues, researchers have explored diverse modification strategies aimed at boosting the photocatalytic capabilities of g-C_3N_4. These approaches encompass the construction of heterojunctions, elemental doping, incorporation of co-catalysts, morphological engineering, metal deposition, and the intentional introduction of defects [172], as shown in Figure 2.58. Such modifications are designed to broaden light absorption, promote charge separation, enhance surface reactivity, and improve the overall stability of g-C_3N_4-based photocatalysts.

The development of heterostructures has become the most effective strategy for improving the photocatalytic performance of g-C_3N_4. Through the construction of heterostructures, the bandgap of g-C_3N_4 can be efficiently modified, leading to enhanced light absorption and more effective separation of electron–hole pairs. Incorporating metallic, nonmetallic, or other nanomaterials into the g-C_3N_4 structure offers a promising route to further boost its photocatalytic activity.

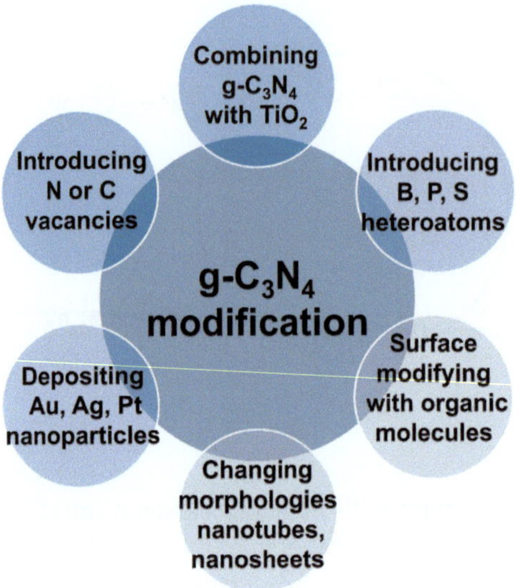

Figure 2.58 Regulation strategies of the photocatalytic properties of g-C_3N_4.

2.4.3.2 Photoluminescence

C_3N_4 is renowned for its remarkable photoluminescent characteristics, which stem primarily from its distinctive electronic configurations and the presence of surface defects. By manipulating factors such as chemical composition, crystal structure, and defect concentration, it is possible to finely tune the photoluminescent behavior of carbon nitride. For instance, introducing elements like nitrogen or sulfur into the carbon nitride lattice can significantly enhance and modulate its luminescent response. This material boasts high quantum efficiency in luminescence, and its emission wavelength can be precisely shifted from the UV to the visible range through targeted structural adjustments. Owing to these versatile emission properties, carbon nitride is increasingly explored for applications in fields such as biological imaging, optoelectronics, and advanced display technologies.

g-C_3N_4, in particular, exhibits both downconversion and upconversion photoluminescence (PL). Downconversion involves the absorption of a high-energy photon followed by the emission of two photons with lower energy, while upconversion occurs when the material absorbs two NIR photons and emits a photon in the visible spectrum. At ambient temperatures, g-C_3N_4 typically emits a strong blue luminescence, with emission wavelengths spanning from 430 to 550 nm. Its relatively wide bandgap (about 2.7 eV) usually results in violet-to-blue emission with modest quantum yields. Nonetheless, through synthetic modifications, researchers have achieved g-C_3N_4 variants that emit in the UV, violet, and green regions. Since longer-wavelength emissions are often more suitable for imaging and sensing, engineering g-C_3N_4 to emit green or even red luminescence is of significant interest, especially for optoelectronic device fabrication.

Moreover, g-C_3N_4 is capable of upconversion luminescence via TPA. When excited simultaneously by two NIR photons—such as with a 780 nm red laser—it can emit a vivid green light (Figure 2.59). This upconversion process leads to a redshift in the

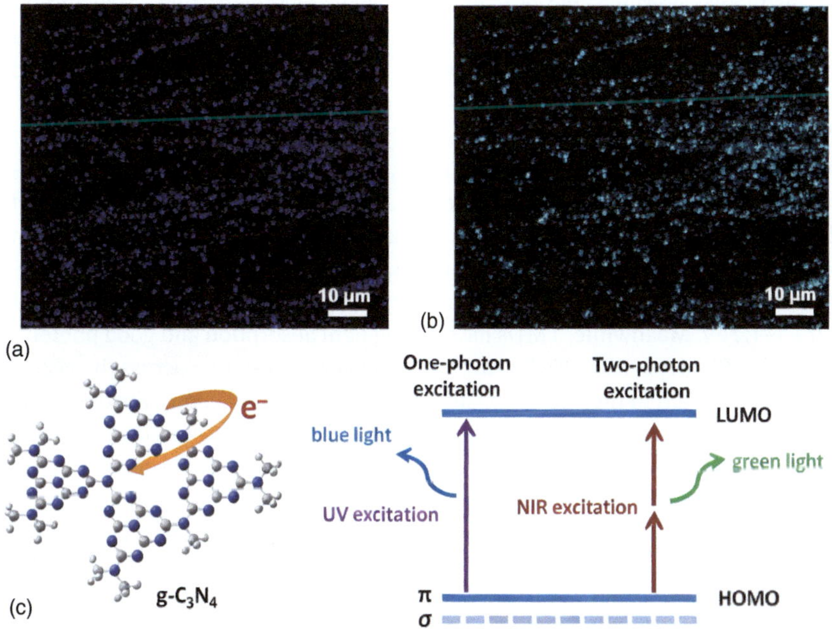

Figure 2.59 (a) Single-photon and (b) two-photon luminescence images of g-C_3N_4 quantum dots. (c) Photoluminescence mechanism of g-C_3N_4 quantum dots. *Source:* Roy et al. [173] / with permission of John Wiley & Sons.

emission spectrum compared to single-photon excitation. Energy calculations have shown that bilayer g-C_3N_4 is energetically more stable than its monolayer counterpart; however, monolayer g-C_3N_4 exhibits higher reactivity under laser excitation, making it more favorable for TPA processes than multilayer structures [173].

2.4.3.3 Photothermal Effect

Carbon nitride also shows some potential for application in photothermal conversion. Its photothermal conversion mechanism is mainly attributed to its absorption of light and nonradiative relaxation process of electrons. Structural modulation and surface modification can improve the photothermal conversion efficiency of carbon nitride. For example, by introducing defects or compounding other materials, its ability to absorb light can be enhanced, thus improving the photothermal effect. The photothermal conversion effect of carbon nitride makes it potentially valuable for applications in fields such as PTT and photothermal-driven chemical reactions.

2.4.4 Transition Metal Disulfides

Transition metal dichalcogenides (TMDs) represent a class of 2D materials consisting of transition metal atoms (such as Mo, W) coordinated with chalcogen elements (such as S, Se). Prototypical examples include molybdenum disulfide (MoS_2) and tungsten disulfide (WS_2). The remarkable optical behavior of TMDs arises from their distinctive crystal symmetry and electronic band structure. Structurally, TMDs exhibit a layered architecture, where each monolayer comprises a plane of transition metal atoms sandwiched between two planes of chalcogen atoms, and adjacent layers are held together by weak van der Waals interactions.

The optical characteristics of TMDs are highly sensitive to parameters such as layer thickness, crystallographic phase (e.g. 1T, 2H, or 3R polytypes), and surface topology. Notably, the bandgap of TMDs undergoes a pronounced transition from indirect to direct as the material is thinned down to a monolayer. Many TMDs, especially in the single-layer state, behave as direct bandgap semiconductors with strong photovoltaic conversion capability, while multilayer TMDs behave as indirect bandgap of about 1.1–1.2 eV. Meanwhile, TMDs have strong light absorption and good photoluminescence (PL) properties. TMDs exhibit strong light absorption across the visible spectrum, attributed to their high density of states at the band edges. Furthermore, these materials possess exceptionally large exciton binding energies due to reduced dielectric screening in two dimensions, rendering excitonic effects central to their optical responses and photophysical processes. Moreover, the optical properties vary significantly with the number of layers, with single-layer TMDs typically exhibiting stronger photoluminescence and larger bandgaps, while multilayer materials exhibit weaker photoemission and smaller bandgaps.

In addition, TMD materials also exhibit good photoluminescence properties and photocatalytic properties, which make them have a wide range of application prospects in the fields of optoelectronic devices, photocatalysis, and bioimaging. Figure 2.60 illustrates the structure of Janus 2D TMDs [174, 175].

Transition metal dichalcogenides stand out for their intense interaction with light, a feature that underpins their remarkable ability to generate and manipulate excitons with high efficiency. Thanks to this inherent characteristic, TMDs are increasingly recognized as versatile materials for next-generation optoelectronic components, advanced photodetectors, and light-emitting devices. In such systems, light not only triggers but also emerges from the device, enabling a host of unconventional applications.

1) Illumination acts as a powerful external trigger in TMDs, injecting additional charge carriers and thereby enhancing electrical conductivity (stimulus: light; response: current). For example, micromechanically exfoliated MoS_2 has demonstrated outstanding photodetection performance, combining minimal dark currents with strong photoluminescence quantum yields. However, the interplay between incident photons, the intrinsic properties of TMDs, and applied gate voltages in phototransistor architectures leads to complex, device-specific behaviors that can complicate reproducibility and performance benchmarking.
   Beyond charge generation, light can also be harnessed for efficient photothermal conversion (stimulus: light; response: heat production). Chemically exfoliated MoS_2, in particular, excels as a photothermal material in the NIR spectrum, highlighting the multifunctional nature of TMDs under optical excitation.
2) A number of TMDs, including monolayer 2H-MoS_2, 2H-WS_2, and WSe_2, display strong visible-wavelength photoluminescence (PL) at ambient conditions—a direct manifestation of stimulus-response behavior where light serves both as

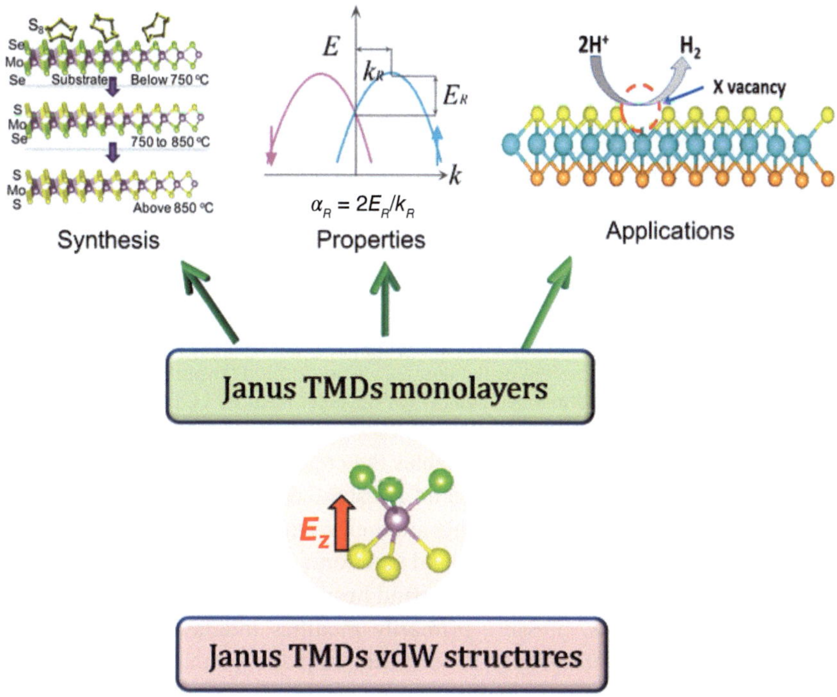

Figure 2.60 The structure of Janus TMDs. *Source:* Tang and Kou [175] / with permission of John Wiley & Sons.

the input and output. To further broaden their utility as responsive materials, researchers have explored various strategies for reversibly tuning these optical responses. One particularly effective method involves electrical gating: in a typical field-effect transistor setup, applying a gate voltage modulates the population and emission of excitons. For instance, in $MoSe_2$ monolayers, the PL intensity of both neutral and charged excitons (around 1.65 eV) can be precisely controlled by adjusting the gate bias, enabling dynamic, electrically driven modulation of the emitted light (stimulus: gate voltage; response: photoluminescence).

3) The application of mechanical strain offers a robust route for tuning the electronic and optical properties of semiconductors, a technique referred to as strain engineering. In TMDs, in-plane stretching or compression alters the metal-chalcogen bond lengths and the overall lattice symmetry, leading to significant changes in the band structure and optoelectronic behavior. For example, stretching MoS_2 increases bond distances, reduces orbital overlap, and narrows the bandgap. Theoretically, even modest tensile strains (1–2%) can shift monolayer 2H-MoS_2 from a direct to an indirect bandgap, with further strain potentially inducing metallicity. Experimental evidence confirms that the conductivity of MoS_2 can be systematically tuned by applying uniaxial strain (Figure 2.61). Although compressive strain effects are less well characterized, biaxial compression—achieved, for example, by cooling TMD

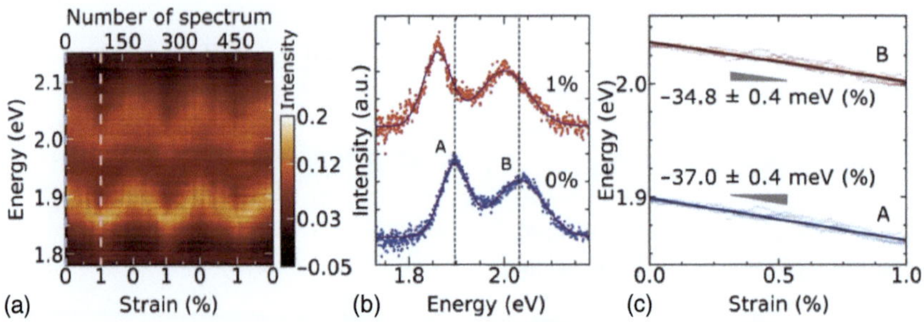

Figure 2.61 Color map displaying reflectivity intensity (color scale) as a function of uniaxial strain (%) (a). Spectral comparison at 0% (blue dashed line) and 1% (red dashed line) strain (b). Plot of A and B exciton peak energies versus applied tensile strain (c). *Source:* Torres-Cavanillas and Forment-Aliaga [176] / Springer Nature / CC BY 4.0.

monolayers on flexible substrates—enables the creation of photodetectors with strain-dependent spectral responses and substantial shifts in neutral exciton energies. Hydrostatic pressure presents another avenue for mechanical manipulation, affecting both the crystalline and electronic frameworks of TMDs. The consequences of pressure application depend heavily on the sample's thickness: in multilayer structures, pressure primarily modifies interlayer interactions, while in monolayers, it directly distorts the in-plane atomic arrangement, resulting in marked alterations to the electronic band structure [176].

Molybdenum disulfide (MoS_2), a typical material for TMDs, consists of hexagonal crystalline layers, each consisting of triatomic S–Mo–S stacks, similar to a sandwich structure as shown in Figure 2.62a, with a thickness of about 6.5 Å. The layers of molybdenum atoms are connected to the layers of S atoms by covalent bonding [177]. Each S–Mo–S layer is connected to the others by van der Waals forces. Each molybdenum atom in the MoS_2 cell is surrounded by six S atoms, forming a honeycomb structure. This structure makes electron transport in the horizontal direction much easier than in the vertical direction, while bringing about a higher planar conductivity. Inspired by this, the electrical properties have been enhanced by reducing the number of MoS_2 layers.

A distinctive characteristic of MoS_2, as compared to graphite, lies in its polymorphism, which imparts diverse physicochemical properties. Polymorphism in MoS_2 arises from variations in the coordination geometry of Mo and S atoms and the stacking sequence of the layers. Four principal polymorphs have been identified: the distorted tetragonal phase (1T), two hexagonal phases (1H and 2H), and the rhombohedral phase (3R). The nomenclature reflects both the stacking sequence (the digit, indicating the number of layers in the repeating unit) and the symmetry (H for hexagonal, T for tetragonal, R for rhombohedral). For instance, the 1T and 1H phases consist of a single layer per unit cell, the 2H phase comprises two layers, and the 3R phase consists of three layers per repeating unit (Figure 2.62b). While the S–Mo–S layer arrangement (triangular prismatic coordination) is consistent across the 1H, 2H, and 3R phases, the stacking order and symmetry of the repeating units differ. Typically, 1H-MoS_2 refers to a monolayer structure with octahedral

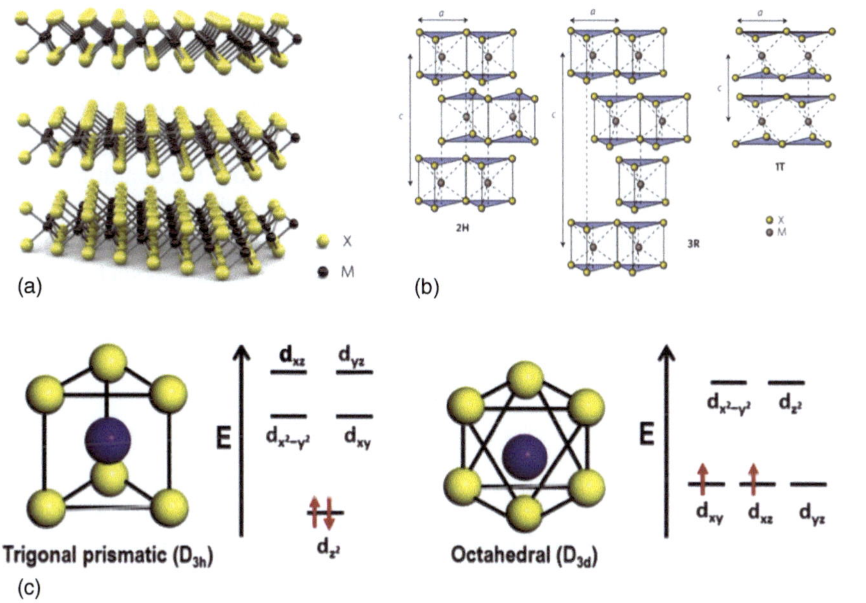

Figure 2.62 (a) Three-dimensional structure of MoS_2. (b) Structural polymorphism of 2H, 3R, and 1T phases and (c) the corresponding electron occupation in the Mo 4d orbitals. *Source:* Wang et al. [177] / with permission of Elsevier.

coordination of Mo by S. The 2H phase is characterized by hexagonal symmetry, the 3R phase by rhombohedral symmetry, and the 1T phase by a distorted octahedral (tetragonal) symmetry. Notably, 2H- and 3R-MoS_2 exhibit trigonal prismatic coordination, whereas 1T-MoS_2 is defined by octahedral coordination. In practice, MoS_2 nanosheets predominantly crystallize in the 2H or 1T phases.

The different phases of MoS_2 have unique properties due to differences in energy band structure. Due to the similarity in energy band structure, the 2H and 3R phases of MoS_2 have similar properties, such as antimagnetism and semiconducting properties; on the contrary, 1T-MoS_2 is paramagnetic and metallic. The 2H and 3R phases can be found in nature, where the 2H phase is the most abundant polymorph and is abundant in the earth's crust, whereas the 3R phase is only found in bulk or multilayered MoS_2. The 2H phase is a stabilized phase, while the 1T phase is a metastable phase 3R-MoS_2 and 1T-MoS_2 can be easily transformed into 2H-MoS_2, e.g. the 3R phase can be converted into the 2H phase by mild heating, while 1T-MoS_2 can be transformed into the 2H-MoS_2 form by annealing, strain application, and electron doping.

The fundamental distinction between the 2H and 1T phases stems from the electronic structure of molybdenum, specifically the occupancy and crystal field splitting of Mo 4d orbitals. As depicted in Figure 2.62c, for the 2H phase, the two Mo d electrons preferentially occupy the lower-energy dz^2 orbital, stabilizing the semiconducting phase at ambient conditions. The incomplete filling of the d_{xy} and d_{xz}

orbitals in 1T-MoS_2 results in metallic conductivity but also renders this phase less stable.

Variations in the energy band structure and bandgap of MoS_2 give it different optical properties. Bulk MoS_2 exhibits weak photoluminescence due to its low quantum yield. In contrast, monolayer MoS_2 with a direct bandgap, with its large optical gap (1.8~1.9 eV), can be observed as photoluminescence. And the quantum yield of monolayer MoS_2 is much higher in comparison, with a 10^4-fold increase in the photoluminescence quantum yield of MoS_2 from bulk to monolayer, which makes monolayer MoS_2 an important candidate for photovoltaic applications such as transistors. The luminescence of MoS_2 can be controlled by the intercalation of interlayer positive ions. The fewer the layers, the higher the luminescence efficiency of the nanosheets, which is useful in biomedical applications.

2.4.5 Perovskite

Perovskites were initially identified as a class of minerals with the prototypical composition of calcium titanate ($CaTiO_3$). However, the scientific understanding and definition of perovskites have undergone a profound transformation. Today, "perovskite" broadly encompasses a vast family of materials characterized by the general formula ABX_3, where the A-site is typically occupied by a monovalent cation (which may be inorganic, such as Cs^+, or organic, such as methylammonium $CH_3NH_3{}^+$), the B-site is a divalent metal cation (commonly Pb^{2+}, but also Sn^{2+}, Er^{2+}, etc.), and the X-site is a halide anion (Cl^-, Br^-, or I^-) [178].

Perovskite materials are distinguished by their exceptional absorption coefficients spanning the visible to NIR spectrum. By judiciously varying the composition of the A, B, and X constituents, the bandgap can be continuously tuned—typically from 1.1 to 1.5 eV for photovoltaic applications—thereby maximizing solar spectral utilization and enhancing device efficiencies. Structurally, perovskites adopt the archetypal ABX_3 configuration, wherein the A and B cations and the X anion form a highly symmetric and versatile lattice. This architecture endows perovskites with remarkable crystallographic tunability, enabling the realization of diverse phases (cubic, tetragonal, orthorhombic) and the engineering of electronic band structures. At the nanoscale, perovskite nanocrystals (typically 10–100 nm in diameter) offer a high specific surface area and enhanced quantum confinement effects, which are instrumental in optimizing light absorption and emission characteristics. The deliberate control over nanocrystal morphology and dimensionality allows for precise tuning of optical properties, such as absorption edge and photoluminescence (PL) emission.

As shown in Figure 2.63a, the physicochemical properties of perovskites can be extensively tailored through compositional doping and ion substitution at the A, B, or X sites. Notably, both homovalent and heterovalent doping at the B-site can yield novel perovskite derivatives, such as double perovskites, thereby expanding the material design space and offering new avenues for property modulation. Furthermore, the incorporation of bulky organic cations can induce the formation of quasi-2D layered perovskites, which exhibit unique optoelectronic and stability characteristics.

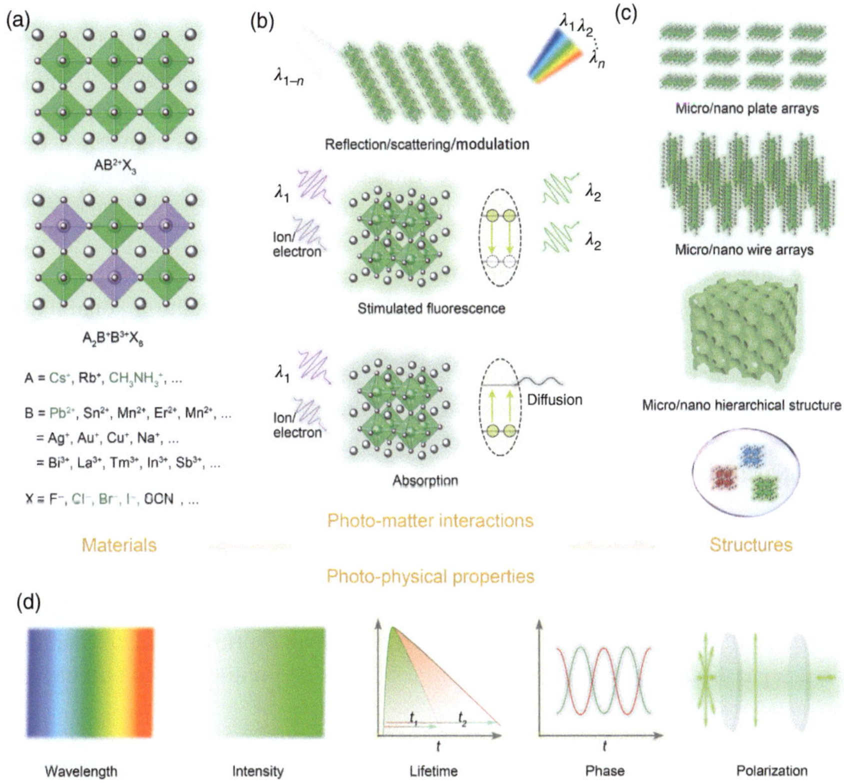

Figure 2.63 (a) Perovskite structures: single-halide perovskites (ABX_3) and various lead-free double-halide perovskites ($A_2B_2X_6$). (b) Reflection, scattering, modulation, emission, and absorption of perovskite. (c) Patterned perovskite structures like micro- and nanoplates, wire arrays, and hierarchical structures. (d) Wavelength, intensity, lifetime, phase, and polarization of perovskite. *Source:* Liu et al. [178] / with permission of Elsevier.

Advanced control over light-matter interactions in perovskite systems enables the selective encoding, modulation, sensing, and decoding of photonic information, thus facilitating multifunctional device architectures (Figure 2.63b). The photonic/optical performance of perovskites can be dynamically tuned via modifications at multiple structural levels—from the atomic arrangement and crystal symmetry to nano/microstructuring and hierarchical patterning (Figure 2.63c). These structural interventions can profoundly influence photon management processes, including reflection, scattering, trapping, extraction, and modulation, as well as light-induced phenomena such as fluorescence, absorption, and photoelectric conversion. Doping during presynthesis, in situ growth, or postsynthetic processing, this enables control over the crystal structure, electrical characteristics, excitonic dynamics, and ultimately the photonic and photovoltaic functionalities.

The above effects can be synergistically combined with the intrinsic emission and absorption properties of perovskites to expand the accessible photonic dimensions, including wavelength, intensity, lifetime, phase, and polarization (Figure 2.63d).

Perovskites offer exceptional wavelength tunability, spanning the visible to NIR regions, which is advantageous for applications ranging from human vision to biophotonics. Through compositional engineering—via controlled incorporation of heteroatoms and postsynthetic treatments—the emission intensity, spectral position, and PL lifetime can be finely regulated. The PL lifetime, governed by mechanisms such as thermally assisted detrapping of self-trapped excitons, can be exploited for lifetime-encoded imaging, minimizing spectral crosstalk and background noise.

Furthermore, the phase and polarization of emitted or transmitted light can be manipulated by engineering the perovskite micro/nanostructure, enabling advanced functionalities such as phase-modulated signal processing and polarization-sensitive detection. The integration of perovskite micro/nanostructures thus paves the way for next-generation photonic applications that transcend traditional paradigms based solely on color and intensity.

The intense and spectrally tunable photoluminescence of perovskite materials is pivotal for the development of high-performance light-emitting diodes (LEDs), where perovskite-based devices have achieved remarkable brightness and EL efficiencies. In the realm of photocatalysis, perovskites show promise for applications such as the photodegradation of organic pollutants and photocatalytic hydrogen evolution. Their superior photocatalytic activity is primarily attributed to efficient separation of photogenerated charge carriers and reduced recombination rates.

2.4.6 Titanium Dioxide

Titanium dioxide (TiO_2) consists of the elements titanium and oxygen. It was first discovered by William Gregor in 1791, and Tadao Araki first proposed that it has photocatalytic properties in the 1970s. He proved through a series of experimental studies that TiO_2 can catalyze the decomposition of water to generate hydrogen and oxygen when irradiated with UV light, which became the basis for further research into TiO_2 photocatalysis.

The optical structure of TiO_2 is mainly determined by its crystal structure. There are two main crystal forms of TiO_2: anatase and rutile. Both forms of TiO_2 have unique optical properties, but anatase-type TiO_2 has attracted more attention because of its higher photocatalytic activity. The anatase structure is composed of TiO_6 octahedra, each sharing four edges and four corners. The conventional unit cell consists of 12 atoms, and each octahedron is connected to eight surrounding octahedra, which gives TiO_2 a high refractive index and good optical stability. The size of TiO_2 nanoparticles is usually in the range of 10–100 nm, and this nanoscale structure allows for a higher specific surface area and better light absorption in optical applications.

TiO_2, particularly in its anatase phase, is characterized by a relatively large bandgap, typically ranging from 3.0 to 3.2 eV. Upon exposure to UV radiation, TiO_2 absorbs photons with sufficient energy to promote electrons from the VB up to the CB, simultaneously generating positive charge carriers, or "holes," in the VB

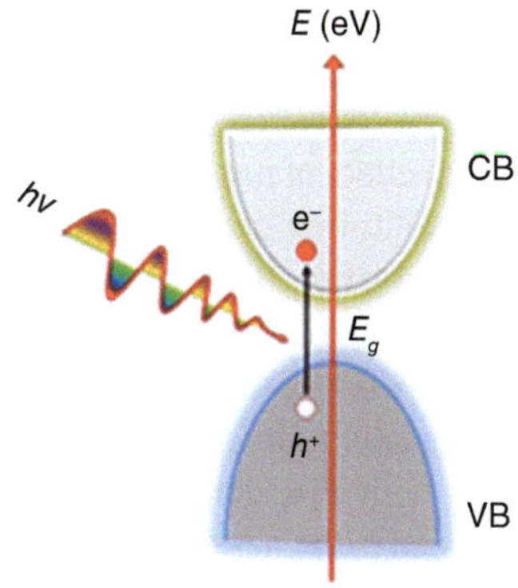

Figure 2.64 Schematic diagram of photoinduced excitation of electrons in TiO_2.

(Figure 2.64). This band structure not only governs the material's ability to absorb photons but also plays a pivotal role in determining its photocatalytic activity, as it influences the efficiency of redox reactions under illumination.

The photocatalytic mechanism of TiO_2 operates such that when the energy of incident photons exceeds the bandgap between the VB and CB, electrons in the VB absorb this energy and are excited to the CB, generating photogenerated electrons (e^-) and leaving behind holes (h^+) in the VB. Since the TiO_2 energy band is discontinuous, e^- and h^+ can move under the action of an electric field, migrate to various positions on the TiO_2 surface, and participate in redox reactions as adsorbed species on the TiO_2 material. The photogenerated electron e^- is easily captured by dissolved oxygen and other oxidizing substances in water, generating O^{2-}, and the hole h^+ reacts with the water molecule H_2O adsorbed on the surface of TiO_2 to generate hydroxyl radical ·OH. ·OH and $\cdot O^{2-}$ have strong oxidizing properties, which will oxidize most of the organic and inorganic pollutants in the water and convert them into inorganic small molecules, CO_2 and H_2O, and other harmless substances. During the migration process of e^--h^+ pairs, some of them inevitably undergo in vivo complexation and surface complexation, which affects the efficiency of photocatalytic degradation of organic matter [179]. The reactions occurring on the TiO_2 surface are shown in Figure 2.65.

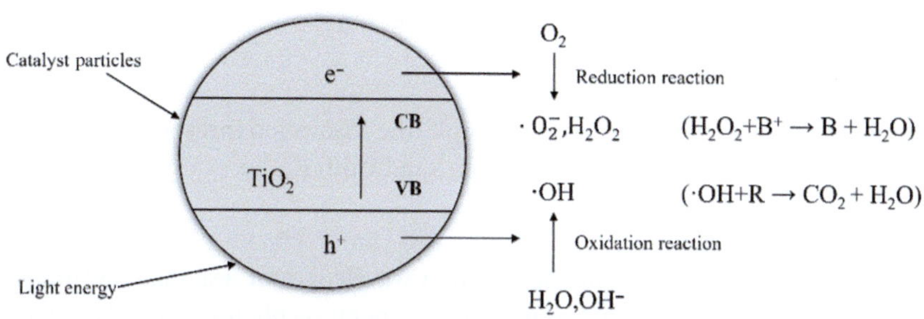

Figure 2.65 Schematic diagram of TiO_2 photocatalysis process.

The influencing factors of photocatalysis are as follows:

1) Nanoparticle size

The effect of nanoparticle size on the photocatalytic performance of TiO_2 is complex. Within a certain range, the smaller the TiO_2 particle size is, the higher the number of particles per unit mass is, and the larger the specific surface area is. Increasing the surface area strengthens the adsorption performance of the TiO_2 catalyst, increases the number of active spots per unit area, and increases the likelihood of a reaction, thus improving its photocatalytic activity. The photocatalytic effect of TiO_2 with a particle size of less than 20 nm is optimal, but when the particle size is reduced to the nanoscale, the likelihood of electron–hole recombination increases as the specific surface area increases. Once the recombination plays a major role, it will also reduce the catalytic efficiency. The larger the quantum size effect, the wider the forbidden band width and the greater the blue shift of the absorption spectral line. This reduces light utilization and lowers photocatalytic efficiency.

2) Crystal structure and morphology

The different crystal structures of TiO_2 are formed by connecting TiO_6 octahedra in different ways, and the octahedra are distorted to different degrees, making the three crystals have different energy band structures. Compared with anatase and rutile types, the former has a slightly larger bandgap and higher oxidizing ability; in addition, anatase TiO_2 has smaller grains with a larger specific surface area, stronger adsorption ability, and exhibits higher photocatalytic activity. Because TiO_2 brookite contains a large number of C—H hydrophobic bonds and is itself structurally unstable, it is hardly photocatalytic.

With the deepening of the research on TiO_2, a range of TiO_2 catalysts with different morphologies have been synthesized, mainly including spherical, porous, linear, and core-shell. The synthesized core-shell microspheres have improved photocatalytic activity due to their ability to reflect UV light at multiple levels inside their shells. The linear pore structure has higher photocatalytic activity because it is favorable for the direct propagation of light, and the pore structure becomes an ideal channel for UV light, and the parallel pore channel is more favorable for light absorption, which in turn improves its photocatalytic activity. Nowadays, it is possible to prepare TiO_2 materials with rose-like micro-nanostructures, which have strong absorption ability for light in the wavelength range of 300–800 nm, covering the whole visible light region, and have good photocatalytic performance.

3) Light response range

Taking the anatase crystal type as an example, the absorption threshold λ_g of TiO_2 for light is utilized in relation to its forbidden bandwidth E_g, Eq:

$$\lambda_g\,(\text{nm}) = 1240/E_g\,(\text{eV})$$

It has been calculated that only light with a wavelength of less than 387 nm is absorbed when the surface of TiO_2 is irradiated, generating photogenerated carriers and exhibiting photocatalytic activity. Therefore, only UV irradiation, accounting for 5% of sunlight, can stimulate the photocatalytic ability of TiO_2 semiconductor materials. If TiO_2 expands its light response range, i.e. reduces its forbidden band width, this favors the enhancement of photocatalytic ability.

Depositing a controlled quantity of noble metal onto the surface of TiO_2 brings about two key benefits [180]. Firstly, it facilitates the efficient separation of photogenerated electrons and holes. Secondly, it decreases the overpotential required for reduction reactions, such as proton reduction and dissolved oxygen reduction, thereby significantly enhancing the catalyst's overall activity. As the amount of gold introduced to TiO_2 increases, there is a corresponding improvement in its photocatalytic performance. This enhancement is primarily attributed to the ability of well-dispersed gold nanoparticles to efficiently trap photogenerated electrons, thus promoting the separation of electron–hole pairs. Additionally, noble metal nanoparticles extend the light absorption of TiO_2 into the visible spectrum, further boosting its photocatalytic efficiency.

2.5 Other Materials

In the previous sections, small-molecule optical materials, conjugated polymeric materials, noble metal nanomaterials, and inorganic nanomaterials with specific structural and optical characteristics have been emphasized in the domains of photocatalysis, biosensing, and energy conversion. More recently, advanced materials such as metal-organic frameworks (MOFs) and covalent organic frameworks (COFs), biomolecular materials, and polymers with nonconjugated structures have been endowed with unique optical properties through structural customization design. MOFs are highly efficient photocatalysts owing to their large specific surface area and customizable pore structures, whereas COFs have emerged as promising materials in optoelectronic devices and sensing applications, thanks to their well-ordered crystalline architecture and outstanding optoelectronic characteristics. Biomolecular materials (e.g. proteins) possess natural biofunctionality and environmental friendliness, and are suitable for application scenarios such as light-controlled drug release and bioimaging. Block copolymers and hyperbranched polymers, on the other hand, enable precise modulation of optoelectronic properties through self-assembly or molecular design. These materials also provide building blocks for the construction of photofunctional polymer composites.

This subsection will focus on the following photofunctional materials.

2.5.1 Metal-organic Framework (MOFs)

Metal-organic frameworks (MOFs) represent a class of crystalline porous materials constructed through the coordination of metal ions or clusters with organic ligands. These materials have garnered significant attention in recent years within the field of photofunctional material research, owing to their distinctive architectures and exceptional physicochemical properties. The inherent flexibility in selecting both metal and organic components, coupled with the vast array of combinatorial possibilities, endows MOFs with unparalleled structural diversity compared to other porous materials.

2.5.1.1 Structure of MOF

MOFs are modular materials consisting of metal nodes, as shown in Figure 2.66. These nodes are connected by organic ligands, forming an infinite network in principle [181]. Thanks to their inherent chemical tunability, combining various metals, ligands, and topologies has resulted in the creation of hundreds of thousands of MOFs [182]. The structural diversity of MOFs originates from the modular combination of metal nodes and organic ligands, which enables precise control of their porosity, topology, and functionality. The choice of organic ligands directly affects the light absorption capacity of MOF structures. The metal nodes are usually transition metals (e.g. Zn^{2+}, Zr^{4+}, Fe^{3+}) or lanthanides (e.g. Tb^{3+}, Eu^{3+}), which determine the stability of the framework and the electronic properties.

2.5.1.2 Optical Properties of MOF

In photocatalytic processes, the bandgap energy of a material should be less than 3.2 eV to meet the thermodynamic requirements for the reaction. Furthermore, the alignment of the VB and CB plays a critical role in determining the feasibility and specificity of chemical reactions. MOFs typically exhibit large bandgaps, which restrict their ability to absorb visible light, thereby limiting their photocatalytic efficiency. To address this limitation, strategies to modulate the bandgap structure are essential for enhancing light absorption. In MOF-based photocatalysis, organic ligands often function as "light-harvesting antennas," capturing visible light and facilitating the transfer of photogenerated electrons to the metal clusters within the framework [183]. Partially metal clusters can also directly absorb light to generate photogenerated electrons, which together constitute the core elements of the MOF photocatalytic mechanism. In addition, the separation rate of photogenerated carriers is closely related to the survival lifetime of electrons and holes. The faster the separation rate, the longer the survival time of electrons and holes, and the more carriers will migrate to the surface of the MOF. This promotes the occurrence of more redox reactions, thus enhancing the photocatalytic performance of the MOF. Therefore, improving the carrier separation efficiency in MOFs is also an important

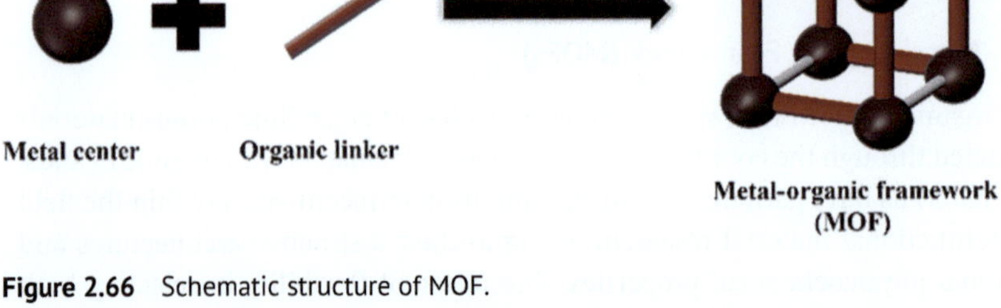

Figure 2.66 Schematic structure of MOF.

way to optimize photocatalytic performance. In summary, strategies for modulating the photocatalytic performance of MOF materials mainly focus on two aspects: tuning the bandgap and improving the separation efficiency of photogenerated carriers. The following section will introduce four strategies:

1) Selection of organic ligands with high conjugation or strong visible light absorption

Porphyrins and their derivatives are widely used as organic ligands in studies of the photocatalytic performance of MOFs and their applications. In this context, the photocatalytic process of MOF typically involves the porphyrin absorbing photons and transitioning between the ground and excited states, thereby inducing the generation of singlet oxygen ($^1O_2$). Meanwhile, the porous structure and large surface area of MOF increase its contact with oxygen, thus enhancing the production of $^1O_2$. The construction of MOFs with porphyrin as an organic ligand not only overcomes the defects of porphyrin, such as its poor stability and susceptibility to self-extinction, but also significantly improves its photocatalytic efficiency and expands its application scope [184]. In addition, pyrene and its derivatives with tetracyclic densely packed planar electron-rich frameworks are favorable for the generation of ROS, especially $^1O_2$, via energy transfer due to their low triplet state energy and enhanced ISC. For example, using the pyrene-based organic ligand 4,4′,4′,4″-((2,7-di-tert-butylpyrene-4,5,9,10-tetrayl)tetrakis(ethyne-2,1-diyl))-tetrabenzoate ($BPETB^{4-}$) and Hf ions constructed PCN-822 (Hf) MOF, as shown in Figure 2.67 [185]. PCN-822(Hf) is highly responsive to visible light with a broadband absorption of 225–650 nm due to the $BPETB^{4-}$ with π-conjugated pyrene moiety, which is highly visible light absorbing. The heavy metal effect of Hf improves the ISC process of the material, which promotes the production of $^1O_2$ in PCN-822(Hf). Due to the possible photocatalytic energy transfer and photoinduced charge separation to drive $^1O_2$ production, PCN-822 (Hf) has excellent photocatalytic properties for amine oxidation.

Due to the low price and simple structure of terephthalic acid and its derivatives, they are widely used in the construction of MOFs and photocatalytic studies. Typical MOFs such as MIL-101 (Fe)/NH_2-MIL-101 (Fe), UiO-66-NH_2, and MIL-125/NH_2-MIL-125 have been widely used in photocatalytic studies, such as photocatalytic H_2O decomposition, CO_2 reduction, or compounded with various materials (e.g. precious metals and semiconductors) for photocatalytic studies [186–188]. Terephthalic acid can be modified with various functional groups (e.g. OH, Br, CH_3, and SH) or grafted with diverse side chains to produce a wide range of derivative ligands. These derivatives have been extensively utilized in the synthesis of MOFs [189, 190]. In summary, terephthalic acid and its derivatives represent the most widely used organic ligands in MOF construction.

2) Selection of metal clusters

(1) Doping metal ions around the metal oxide clusters constituting the MOF [191, 192]. The metal clusters usually act as photoactive sites, and their electronic structures are susceptible to external ion effects. The doping of metal ions can

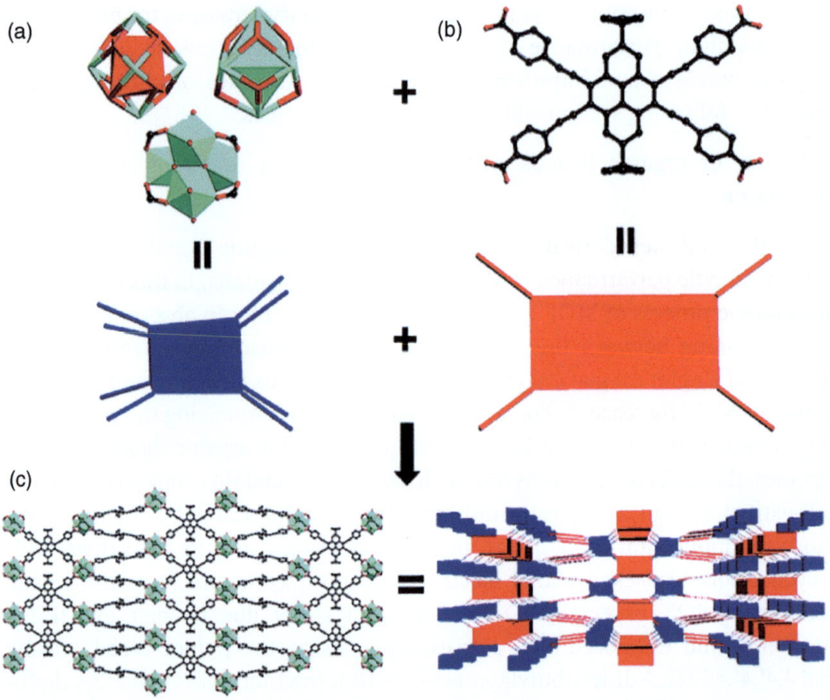

Figure 2.67 (a) The connectivity of the 8-connected Zr_6/Hf_6 clusters; (b) the topology of tetratopic $BPETB^{4-}$ ligand; (c) the resulting (4,8)-connected PCN-822. *Source:* Zhang et al. [185] / The Royal Society of Chemistry / CC BY 4.0.

change the electron density around the metal clusters, which in turn affects the bandgap of the MOF. (2) Doping metal ions into metal clusters to form partially substituted heterometallic clusters [193]. The introduction of the second metal ion significantly alters the electron density around the metal clusters and enhances the absorption of visible light by the MOF, thereby reducing the bandgap. (3) The use of metal ions partially replaces the metal nodes that form the MOF [194]. Partial introduction of metal ions helps in better mixing with organic ligands, thereby reducing the bandgap of the MOF while improving the responsiveness to visible light. (4) Metal clusters with different valence states (e.g. Cu^{2+}/Cu^{+}, Fe^{3+}/Fe^{2+}, and Ce^{4+}/Ce^{3+}) were utilized to change the bandgap of the MOF [195–197]. (5) Tuning the bandgap of semiconductor MOFs by changing the size of metal clusters [198]. For example, the addition of Ni^{2+} dopant can alter the electronic structure of the clusters in the NH_2-MIL-125-Ti, enhancing electron transport and charge separation and thereby improving the photocatalytic activity of the compound (Figure 2.68) [191].

3) Creating defects

Defects are irregular arrangements of atoms in a body of products, the presence of which can significantly affect the energy band and carrier concentration of semiconductors. The periodicity and tunability of the MOF structure make it an ideal

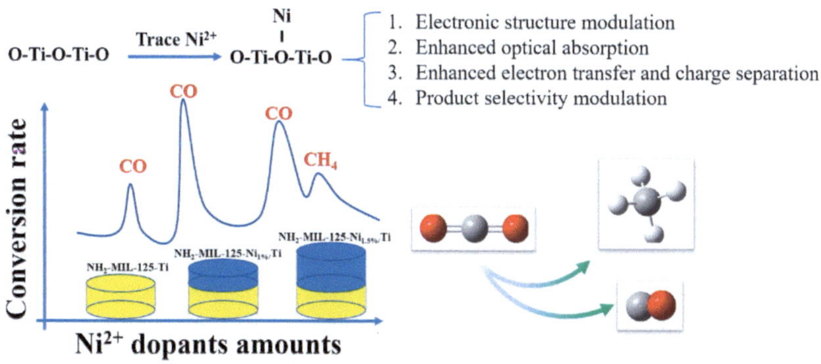

Figure 2.68 The enhanced photocatalytic performance of NH_2-MIL-125-Ti by Ni^{2+} doping. *Source:* Chen et al. [191] / with permission of Elsevier.

model for defect studies. Currently, the introduction of defects in MOFs is mainly carried out by two types of methods: One approach is to induce ligand deletion at the synthesis stage using methods such as acid modulation, the mixed ligand method and molecular imprinting method to form uncoordinated and highly active metal sites; The other approach is to construct defective structures using postprocessing techniques such as plasma treatment, acid and alkali etching, ligand exchange and temperature treatment [199–204]. The acid modulation method involves introducing a modulator into the precursor solution, which interferes with the equilibrium reaction by leading to ligand deletion and the formation of unsaturated metal sites. Such defects can significantly enhance the rate of photogenerated carrier separation and, consequently, the photocatalytic activity. For example, Pt@UiO-66-NH_2-100 exhibits excellent photocatalytic activity when it has structural defects (Figure 2.69). Creating structural defects using a modulator is essential for promoting efficient electron–hole separation [199].

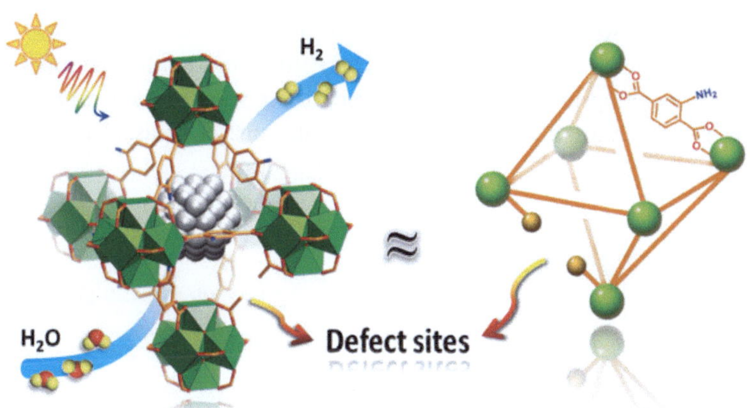

Figure 2.69 The structural defects within Pt@UiO-66-NH_2-X. *Source:* Ma et al. [199] / with permission of John Wiley & Sons.

4) Morphology change

The size and morphology of the MOF determine the diffusion distance of photogenerated electron–hole pairs. This in turn affects the photocatalytic efficiency. In crystalline materials, the surface energy and photocatalytic efficiency vary significantly across different crystallographic planes. Consequently, understanding and optimizing the exposure of crystallographic surfaces has become a critical area of investigation in the field of MOF-based photocatalysis [205–207]. In addition to the effect of the crystal surface, the photocatalytic efficiency of MOF is also significantly affected by its size. Usually, larger-volume MOFs expose fewer active catalytic sites, which limits the further improvement of their catalytic performance. Therefore, it is of significant research value to downsize multidimensional large-volume MOFs to low-dimensional small-volume, especially ultra-thin 2D dimensions. This is because the smaller the dimension, the more active catalytic sites are exposed by the MOF and the higher the photocatalytic efficiency [208, 209]. Meanwhile, the distance between active sites also affects the photocatalytic efficiency. For example, ultrathin zirconium porphyrin MOF nanosheets with an atomic thickness were synthesized in a one-pot solvothermal reaction (Figure 2.70) [208]. These prepared nanosheets showed $^1O_2$ generation and excellent photocatalytic activity due to their intrinsic 2D nature and efficient light harvesting by the porphyrin units. In summary, how to controllably change the morphology of MOFs, reduce the distance between active sites, and construct crystalline surfaces with the highest photocatalytic efficiency has attracted extensive attention from scientists. In this context, synthesizing MOFs with specific sizes or shapes has become increasingly popular.

2.5.2 Covalent Organic Framework (COFs)

Covalent organic frameworks (COFs) are crystalline materials that can adopt either 2D or 3D architectures. They are constructed from light elements such as carbon, hydrogen, oxygen, nitrogen, and boron, which are linked together via strong covalent bonds to form extended organic building units [210]. These frameworks are synthesized by assembling the building blocks according to well-defined topological principles, resulting in highly ordered structures. Owing to their tunable and predictable architectures, high specific porosities and surface areas, as well as

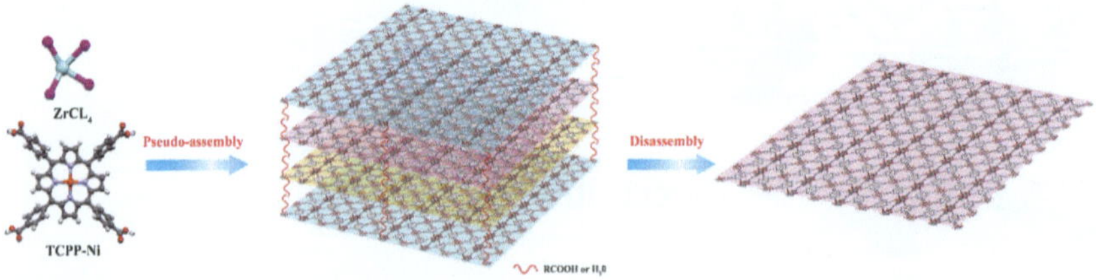

Figure 2.70 Schematic illustration of the prepared nanosheet of MOF via a pseudo-assembly–disassembly strategy. *Source:* He et al. [208] / with permission of John Wiley & Sons.

remarkable thermal and chemical stability, COFs have found broad applications in gas adsorption and separation, heterogeneous catalysis, optics, electrochemistry, and biomedicine.

2.5.2.1 Structure of COF

In terms of spatial dimensions, COFs can be categorized into one-dimensional (1D COFs), two-dimensional (2D COFs), and three-dimensional (3D COFs). Among them, 2D COFs are usually assembled from building units with rigid planar structures by stacking modes such as AA, AB, or ABC to form a 2D laminar framework structure. This structural feature makes 2D COFs usually have a highly conjugated large π-system. In 3D COF structures, an orthogonal geometry is used to guide the polymer backbone's extension into a 3D, covalently connected spatial framework. Compared to 2D COFs, 3D COFs typically exhibit a number of unique properties, such as higher specific surface area, lower density, and active sites exposed by interconnected pores. However, the construction of 3D COFs faces more challenges due to the limited diversity of monomers, lack of rigid planar structures, and the interlayer π–π interaction.

In a COF network, the way in which the building units are connected depends on the type of bonding reaction. Different connection modes can lead to significant differences in the physicochemical properties and functional performance of COFs. COFs are a class of long-range ordered skeletal materials based on dynamic covalent chemical construction. Reversible reactions are the intrinsic mechanism by which the framework structure is able to build highly crystalline COFs through its own error-correcting behavior. Several reversible condensation reactions have been successfully applied to the construction of crystalline COFs, such as borate condensation, Schiff base condensation, spiro-borane condensation, Knoevenagel condensation, Suzuki coupling, Aldol condensation, transesterification, Zincke reaction, Michael addition, cyano self-polymerization, and so on. Based on these reactions, the reported connection modes mainly include boronate ester, triazine, imine, and hydrazone [211].

1) Boron-based linkage

Boron-based COFs are usually classified into two categories: those constructed based on the self-condensation reaction of a single unit, and those constructed through the co-condensation reaction of two or more units. The first 3D COFs were successfully constructed through the self-condensation of tetrahedral molecules of boron groups or their analogs (Figure 2.71) [212]. The boronic acids TBPM and TBPS are used as tetrahedral building units. They undergo self- or co-condensation with a planar triangular HHTP unit based on B_3O_3 or C_2O_2B rings to form COFs. Boron link-containing COFs have been widely used in several applications due to their low density and high specific surface area. However, the hydrolysis and oxidation sensitivity of boroxanes and borate esters limit their practical applications. Nevertheless, boron-containing COFs are of great importance in studying the reaction mechanism.

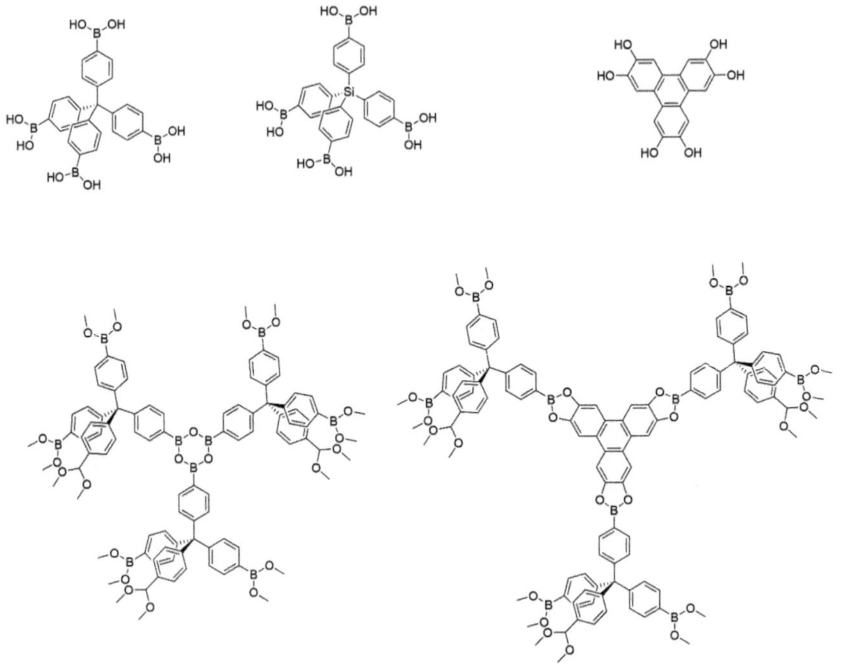

Figure 2.71 The molecular structures of TBPM, TBPS, HHTP, B_3O_3 ring, and C_2O_2B ring.

2) Triazine linkage

COFs with triazine linkages are usually synthesized at harsh reaction conditions, which limit the choice of building units and reduce the diversity of triazine-based COFs (CTFs). In order to broaden the choice of building units, researchers have developed other, milder synthesis strategies. For example, aldehyde monomers obtained via in situ oxidation of alcohols were polycondensed with terephthalamidine to prepare CTFs with high crystallinity (Figure 2.72). Among them, the abundant nitrogen atoms in the main chain can serve as anchor sites for the active centers, while the infinite number of organic building blocks endow them with tunable energy band structures and light absorption ranges, thus enabling the optimization of photocatalytic performance [213].

3) Imine linkage

Imine linkages are constructed through aldehyde–amine condensation reactions, and this type of connection is widely used in COFs. Among them, layered imine-bonded COFs share some structural similarities with borate ester COFs (Figure 2.73) [214], but imine-bonded COFs are significantly more stable to water than borate ester COFs. The relatively mild process of imine bond formation, which can usually be accomplished under milder conditions, allows a wider variety of building units to participate in the reaction, thus providing a wider choice of options for the design and synthesis of COFs. In addition, the dynamic nature of imine bonding endows COFs with a certain self-healing ability, which helps to improve

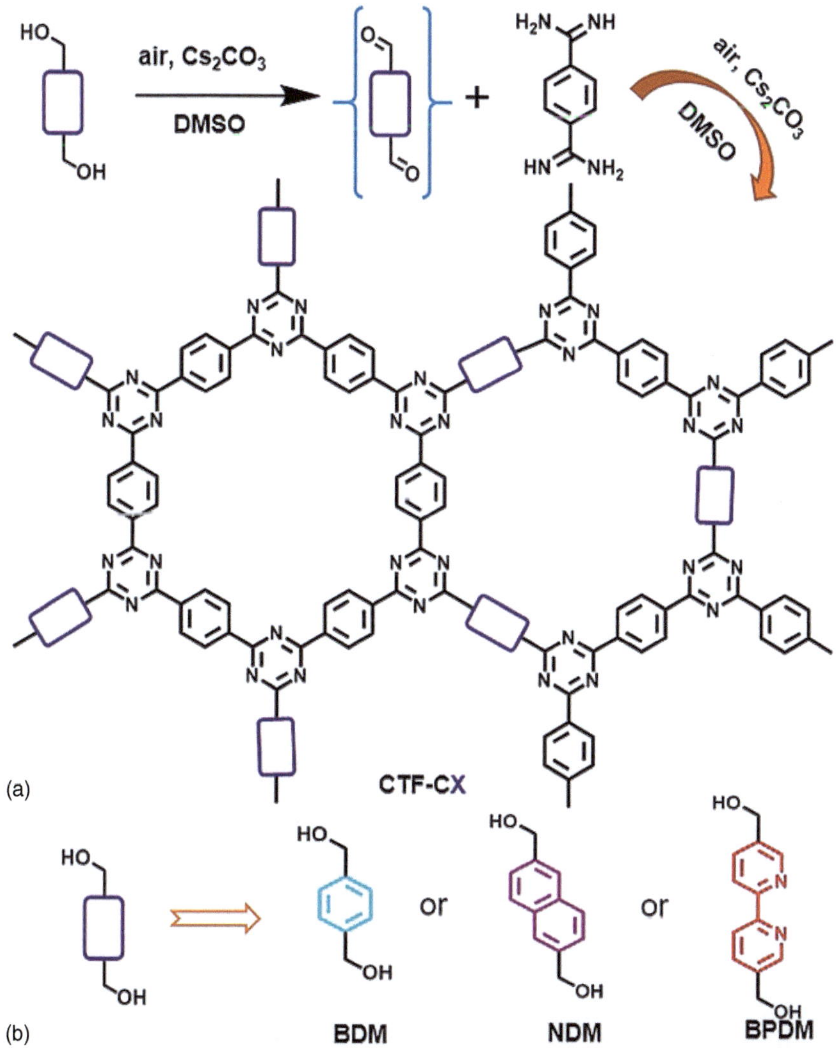

Figure 2.72 (a) The schematic illustration of aldehyde monomers polycondensed with terephthalamidine. (b) The aldehyde monomers obtained from BDM, NDM, and BPDM. *Source:* Liu et al. [213] / with permission of John Wiley & Sons.

the crystallinity and stability of the materials. Therefore, imine-bonded COFs not only excel in structural stability, but also show great potential in functional diversity, making them an important direction in COFs research.

4) Hydrazone linkage

The hydrazone-linked COFs are formed by the reversible condensation reaction of hydrazides with aldehydes. As shown in Figure 2.74, the hydrazone-linked COFs (COF-42) were reported in 2011 [215]. The distinctive feature of this linkage is that

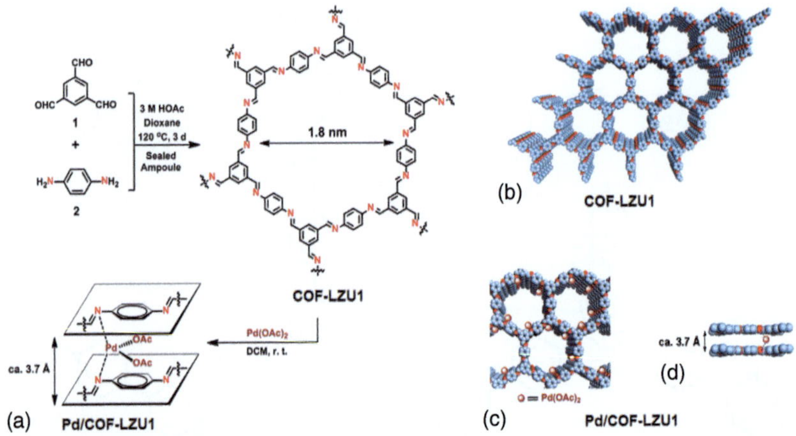

Figure 2.73 (a) Schematic illustration of the preparation of COF-LZU1 and Pd/COF-LZU1. Structural models of (b) COF-LZU1 and (c, d) Pd/COF-LZU1. *Source:* Ding et al. [214] / with permission of John Wiley & Sons.

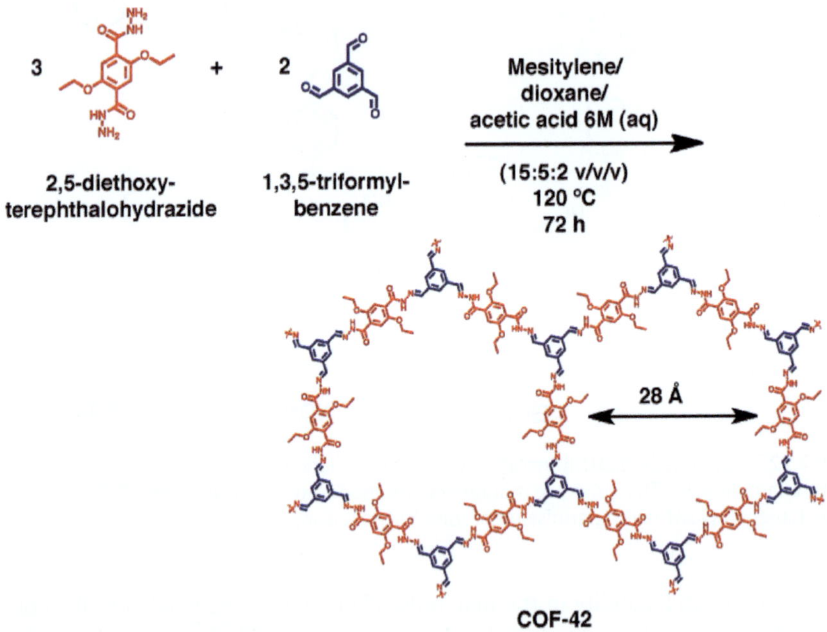

Figure 2.74 Synthesis of COF-42 by 2,5-diethoxyterephthalohydrazide with trigonal-planar 1,3,5-triformylbenzene. *Source:* Uribe-Romo et al. [215] / with permission of American Chemical Society.

the hydrazone bond structure exhibits good stability in water (H_2O) and organic solvents. In addition, there are abundant hydrogen-bonding interactions between the hydrazone units, which not only contribute to the formation of a stable stacking structure but also promote carrier transport, thus enhancing the functional properties of the materials. This hydrazone bonding-based connection provides new possibilities for the design and synthesis of COFs, especially in application scenarios that require a balance between stability and functionality.

2.5.2.2 Optical Properties of COF

As an emerging organic semiconductor functional material, COFs have garnered a lot of interest in catalysis due to their many excellent properties. Firstly, COFs have the advantage of atomic-level precision tuning, which can precisely regulate the framework composition at the atomic level, thus realizing the high tunability of the energy gap and active sites. Second, their high level of porosity and large surface area significantly enhance substrate accessibility and enrich accessible catalytic sites, ensuring effective adsorption and rapid mass transfer to reaction substrates. In addition, the high crystallinity of COFs can form a defect-free framework structure, which effectively inhibits the recombination of photogenerated carriers, while their extended π-conjugated structure facilitates the separation and transport of carriers. COFs can also be used for the adsorption and activation of reaction substrates. Furthermore, COFs are able to integrate metal single atoms, metal clusters, and inorganic catalysts into their backbones to meet different catalytic needs. Meanwhile, their high chemical and thermal stability ensures the robustness in the catalytic process and avoids the photodegradation of the photoactive units immobilized in the framework, thus prolonging the lifetime of the excited state.

The poor separation and transfer of photogenerated charge carriers significantly limit the photocatalytic performance of COFs. In recent years, considerable research efforts have focused on devising strategies to control the behavior of photogenerated charges in COFs. These include extending the absorption range into the visible spectrum, optimizing the electronic band structure, and minimizing the recombination of photogenerated electron–hole pairs.

1) Coupling of functional building units

The diversity of the functional building units of COFs and their connection modes provides a means of improving the stability of materials. More importantly, optimizing bandgap energies and valence/conducting band potentials can be achieved by introducing building units with unique chemical properties. For example, the number of N atoms in the central aromatic ring of COFs increased from 0 to 3 (Figure 2.75) [216]. An increase in nitrogen atoms decreases the dihedral angle between the benzene ring and the aromatic ring, significantly increasing the degree of planarity and conjugation. This partially counteracts the effect of the central aromatic ring's lack of electrons on the bandgap, resulting in a narrower bandgap. In addition, with the increase of nitrogen content, the crystallinity and structural clarity of the material are significantly enhanced, and the layers are more tightly

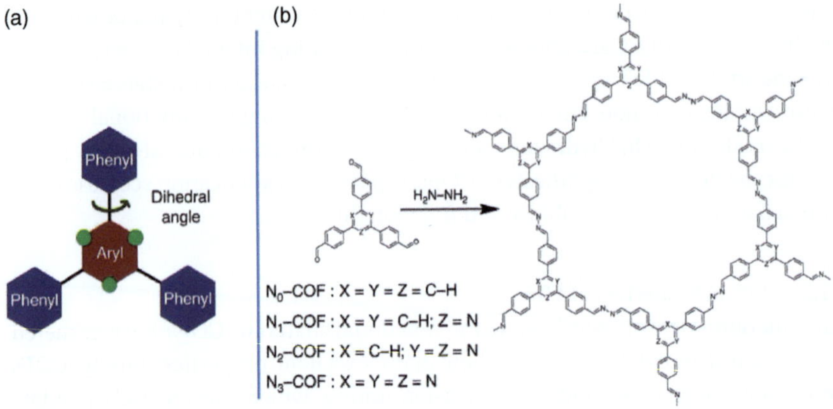

Figure 2.75 (a) The diagram of building units and (b) structures of N_x–COFs. *Source:* Vyas et al. [216] / Springer Nature/ CC BY 4.0.

stacked, which endows the N_3-COF with stronger exciton mobility in the in-plane and along the stacking direction.

2) Element doping

The doping of nonmetallic elements (e.g. N, P, S), metal single-atom sites, or precious metal co-catalysts is an effective strategy to control the photoelectric properties of COFs. For example, adjusting the ratio of M-substituted terephthalimidamide (where M = S or Se) enables the precise control of the amount of sulfur and selenium doping in a series of COF materials (Figure 2.76) [217]. It was shown that the introduction of S or Se could form a heterogeneous electronic configuration inside the COFs skeleton. Therefore, the photocatalytic performance

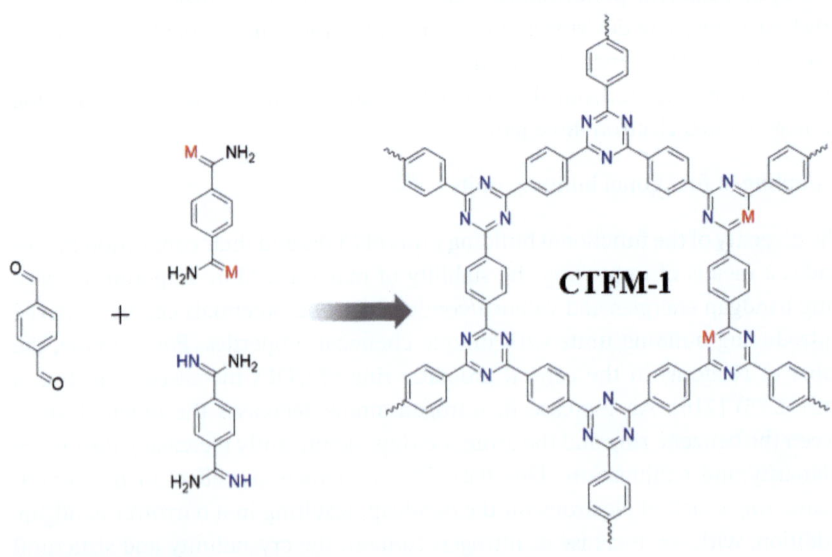

Figure 2.76 Synthesis of CTFM-1.

was significantly enhanced by enhancing the strength of the built-in electric field, promoting the transfer of photogenerated carriers, and effectively inhibiting their recombination.

3) Introduction of photosensitive groups

The introduction of photosensitive groups into the framework of COFs can significantly improve light harvesting capacity and promote efficient separation of photogenerated charges. For example, a novel photocatalyst (Re-COF) was prepared by introducing Re(bpy)$(CO)_3$Cl into a 2D triazine COF (Figure 2.77) [218]. Introducing Re(bpy)$(CO)_3$Cl was shown to extend the UV-visible absorption range of the COF and prolong the lifetime of the charge transfer state within the excited molecule, which effectively inhibits charge recombination. Catalytic experiments show that the material is capable of efficiently reducing CO_2 to CO under visible light with a selectivity of up to 98%.

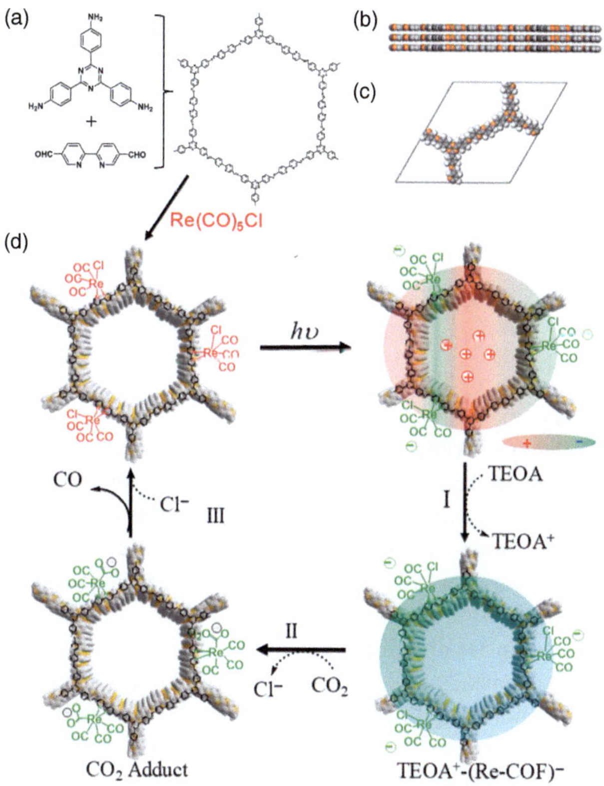

Figure 2.77 (a) Synthesis, (b) side view, and (c) unit cell structure of Re-COF; (d) catalytic mechanism of Re-COF for CO_2 reduction. *Source:* Yang et al. [218] / with permission of American Chemical Society.

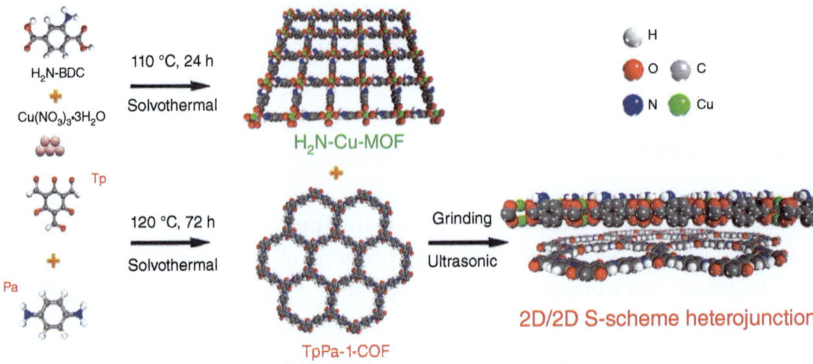

Figure 2.78 Scheme of the preparation of H_2N-Cu-MOF/TpPa-1-COF heterojunction. *Source:* Hu et al. [219] / with permission of John Wiley & Sons.

4) Heterojunction construction

The construction of heterojunctions is a traditional and effective method to modulate the optoelectronic behavior of COFs. Heterojunctions are usually composed of two or more different materials and can retain the advantages of a single semiconductor while solving the problem of its lack of activity. The most common types of heterojunction are type II, Z-type, and S-type. For example, a 2D/2D S-scheme heterojunction was successfully constructed by combining with TpPa-1-COF (Figure 2.78) [219]. This heterojunction promotes the separation of electron–hole pairs and maintains a high redox capacity, significantly improving photocatalytic performance. Experiments have shown that the photocatalytic conversion efficiency of the complex is greater than that of either bare H_2N-Cu-MOF or TpPa-1-COF.

2.5.3 Photofunctional Protein

2.5.3.1 Light-harvesting Protein

Light-harvesting proteins are widely distributed in plants, algae, and photosynthetic bacteria, and consist of a protein backbone complexed with pigment molecules. The complex exhibits significant structural heterogeneity in different biological systems, ranging from highly ordered and symmetric to relatively disordered configurations. The pigment also varies widely in type and number, ranging from a few to hundreds, and the complexity of the embedded protein environments also varies. The absorption and transfer of light energy are mainly carried out by pigment molecules, which are chemically divided into four main groups: chlorophyll, bacteriochlorophyll, carotenoids, and phycobilins. Chlorophyll is based on a cyclic tetrapyrrole ring, in which four nitrogen atoms coordinate with a central magnesium atom. The structural diversity of the chlorophyll family is achieved by linking different side-chain groups, and this side-chain variation also leads directly to changes in the absorption spectral properties of the pigment molecule. Bacteriochlorophyll is structurally very similar to chlorophyll and also exhibits

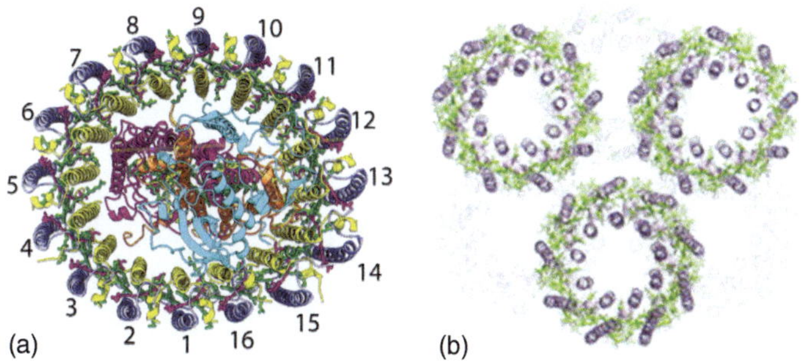

Figure 2.79 Structure of (a) LH1. *Source:* Swainsbury et al. [220] / American Association for the Advancement of Science / CC BY 4.0; (b) LH2. *Source:* Cherezov et al. [221] / with permission of Elsevier.

structural diversity, with the main differences being in its side chains. Phycobilins are linear, open-chain tetrapyrroles, while carotenoids exist in various isomeric forms.

The light-harvesting system of purple bacteria consists of two complexes, LH1 and LH2 (Figure 2.79) [220, 221]. The LH2 (also known as B800-850) is localized on the membrane and presents two significant absorption peaks at 800 and 850 nm. In contrast, the LH1 has a single absorption peak only at 875 nm and is tightly coupled to the reaction center. Both the LH2 and LH1 exhibit a symmetrical ring conformation with a structural basis of a repeating α-β monomer unit. Each monomer contains two transmembrane α-helices, α-aproprotein, β-aproprotein, three bacteriochlorophylls, and one carotenoid molecule. The ring diameter of the LH2 complex is about 65 Å, and the crystal structure exhibits a centrosymmetric octamer (with 24 bacteriochlorophylls) or a nonamer (with 27 bacteriochlorophylls). In contrast, the LH1 ring is larger, with a diameter of about 120 Å, and can accommodate 16–18 monomers, each containing two bacteriochlorophylls.

The light-harvesting system in plants is mainly composed of the light-harvesting complex (LHC) proteins. The members of these proteins are structurally similar and are all based on three transmembrane α-helices. They are named light-harvesting complex I (LHC I) and light-harvesting complex II (LHC II), respectively, according to the type of photosystem to which they are connected (photosystem I, PSI or photosystem II, PSII) (Figure 2.80) [222, 223]. LHC I contains four monomers that form two heterodimers, one of which has a half-moon structure toward PSI. LHC I includes 45 chlorophyll a, 12 chlorophyll b, and 13 carotenoids. In contrast, LHC II forms a trimeric structure with 8 chlorophyll a, 6 chlorophyll b, and 3–4 carotenoids in each monomer. The spectral properties of LHC complexes are mainly regulated by three factors: the electronic coupling strength between pigment molecules, pigment-protein electrostatic interactions, and the intrinsic properties of each pigment molecule itself. Therefore, accurate prediction of their spectral behavior relies on the precise calculation of the local excitation energies and electronic coupling strengths of the pigment molecules [224, 225].

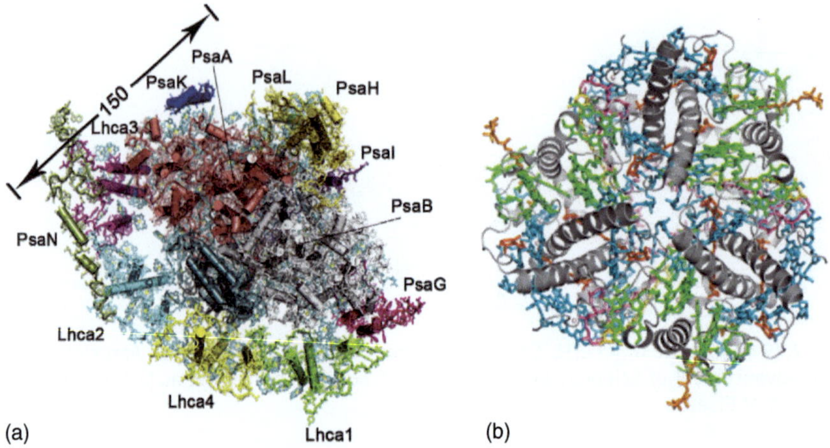

Figure 2.80 (a) Structure of Plant PSI-LHCI. *Source:* Amunts and Nelson [222] / with permission of Elsevier. (b) Structure of LHC-II. *Source:* Barros and Kühlbrandt [223] / with permission of Elsevier.

In photosynthetic green sulfur bacteria, absorbed light energy is delivered to the reaction center via the FMO protein. The spectral characteristics of FMO proteins derive from the fine-tuning of the local excitation energies of the pigment molecules they contain by the surrounding protein environment. The FMO proteins are homotrimers with a high degree of symmetry and each monomer contains eight pigment molecules (Figure 2.81) [226]. Current studies have focused on two FMO proteins of different origins, *Prosthecochloris aestuarii* and *Chlorobaculum tepidum*. The residue sequences of the two FMOs are largely identical, as are the relative positions of their pigment molecules. However, their absorption spectra differ significantly. The intensity of the absorption peak at 815 nm is higher than that at 806 nm for the *P. aestuarii* FMO protein, whereas the *C. tepidum* FMO protein exhibits the opposite feature.

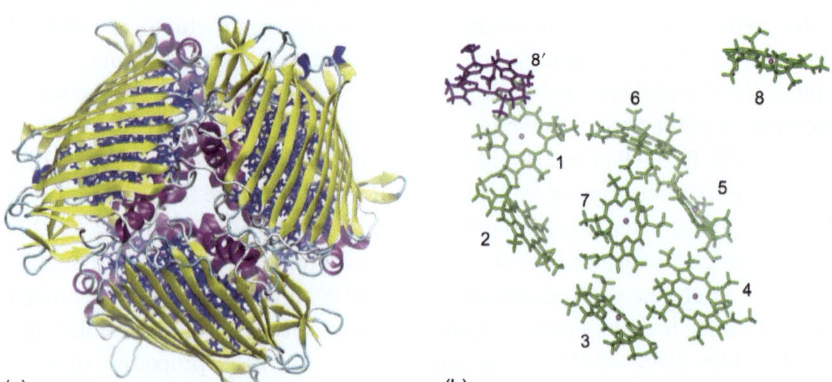

Figure 2.81 (a) The FMO trimer displaying the protein backbone. (b) Spatial organization of the BChl molecules. *Source:* Jia et al. [226] / Springer Nature / CC BY 4.0.

2.5.3.2 Fluorescent Protein

The green fluorescent protein (GFP), originally isolated from jellyfish and characterized at the protein level, has become an important tool in biological and medical research. The GFP contains 238 amino acids and emits green light when excited by blue light, with maximum excitation and emission wavelengths of 395 and 508 nm, respectively [227]. The spatial structure is a β-barrel with a diameter of about 24 Å and a height of about 42 Å (Figure 2.82) [228]. GFP and its fusion proteins have had a profound impact on the basic research of life sciences, and as fluorescent markers, they are of irreplaceable value in the localization of proteins and in the study of their dynamic processes and functions. The modification of the GFP chromophore has greatly expanded the photophysical properties of fluorescent proteins (FPs), including spectral range, quantum yield, photostability, and photoswitching ability [229, 230]. Currently, the emission spectrum of the FP covers the entire visible region and extends into the far-infrared region. Their color tuning can be achieved by a variety of mechanisms, such as π-conjugation systems that regulate chromophores, protonation states, π–π stacking, electrostatic interactions, and specific interactions with neighboring amino acid residues [228, 231].

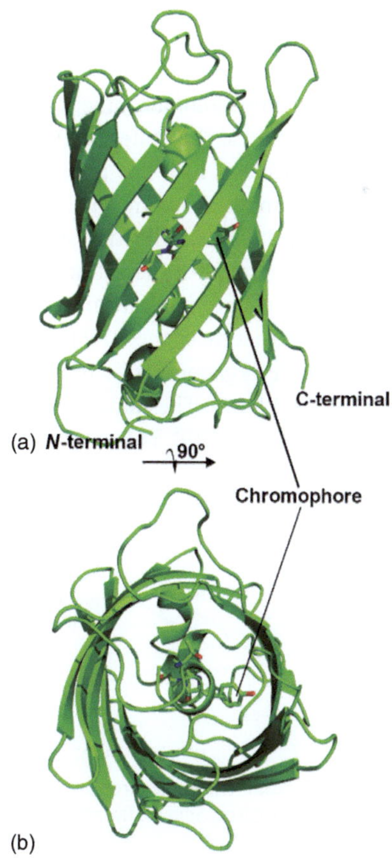

Figure 2.82 (a) Front view and (b) top view of GFP structure. *Source:* Kong et al. [228] / with permission of Elsevier.

When detached from their natural protein cavities, FP chromophores often exhibit fluorescence bursts due to rapid nonradiative relaxation caused by twisted intramolecular charge transfer (TICT) [232, 233]. In order to restore their fluorescence, several strategies have been developed to inhibit the fluorescence of the chromophore molecules. Strategies have been developed to inhibit the TICT process within the chromophore molecules, which are mainly categorized into two types of methods: covalent and noncovalent modifications.

(1) Covalent modification strategies

The free rotation of the aryl–alkene bond and the cis–trans isomerization in the GFP chromophore led to its weak fluorescence. Taking advantage of the ease of chemical modification of FP chromophores, covalent modification strategies have been widely used to enhance the fluorescence intensity of free FP chromophores. Two high-brightness planar-configuration GFP chromophore analogs have been reported: reversible *p*-HOPyDI:Zn and irreversible *p*-HOBDI-BF_2 molecules (Figure 2.83) [232]. This strategy effectively inhibits the nonradiative relaxation process by locking the benzene ring in the same plane as the imidazolidinone via BF_2 group or Zn^{2+} ions, thus enhancing the fluorescence intensity. In addition, a series of derivatives can be obtained by chemically modifying the core structure of the BF_2 group-locked GFP chromophore [234]. By locking the twisted double bond of the chromophore with a BF_2 group, the quantum yield of the chromophore is increased. In addition, a "double donor–single acceptor" strategy is employed, where electron-donating and electron-withdrawing groups are introduced to the phenol and imidazolinone moieties, respectively, to synergistically tune the energies of ground and excited states, thereby achieving red-shifted emission. This design concept provides a new way to develop GFP chromophore derivatives with a large Stokes shift and high quantum yield.

(2) Noncovalent modification strategies

Noncovalent modification strategies utilize noncovalent interactions to restore the fluorescence of chromophore analogs by encapsulating them in molecular cages that mimic the structure of the GFP β-barrel. Based on the HBI analogue, H_2BDC-HBI, artificial GFP organometallic framework materials (GF MOF) can be constructed (Figure 2.84) [235]. The maximum emission wavelength of such MOFs is similar to that of natural GFP. This study presents a new approach to designing artificial

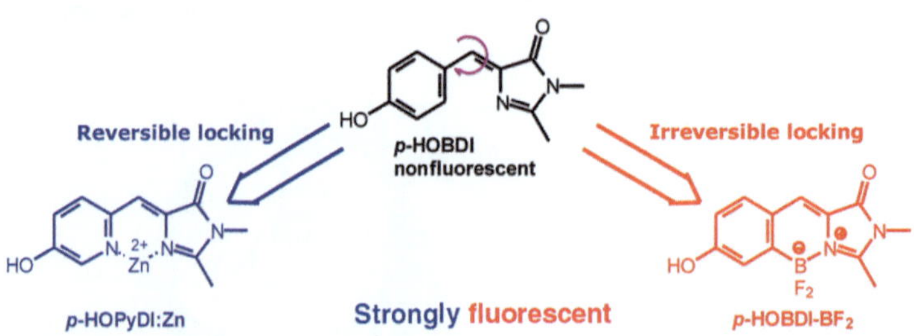

Figure 2.83 GFP chromophores and the conformational change of GFP chromophore derivative structures. *Source:* Baranov et al. [232] / with permission of American Chemical Society.

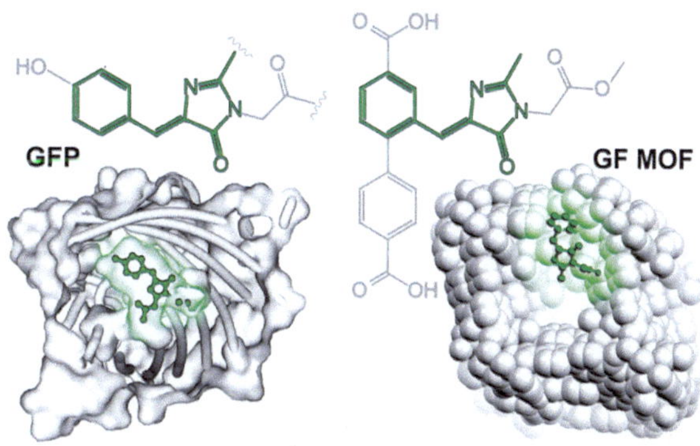

Figure 2.84 (a) Structures of GFP and the HBI chromophore. (b) GF MOF constructed from the H_2BDC–HBI. *Source:* Williams et al. [235] / with permission of American Chemical Society.

systems that mimic the environment of natural chromophores. In addition, the FP chromophore derivative TBI [236] was developed using a structure-directed strategy. The Broccoli aptamer exhibits higher thermostability and lower magnesium ion dependence. The enhanced binding rate of TBI to the aptamer enhances the cycling efficiency of the fluorophore, which significantly improves the brightness and photostability of the TBI-Broccoli complex.

2.5.4 Other Polymers

Researchers have created a new class of materials that achieve excellent photofunctional properties by designing molecular structures and modulating the assembly behavior at the nanoscale. With the recent development of nanotechnology and polymer chemistry, the range of applications for these materials in photofunctional processes has gradually increased. Novel materials such as block copolymers and dendritic polymers have unique molecular structures and self-assembly behaviors, and are capable of forming ordered nanostructures. Their potential in bioapplications has been demonstrated by significantly improving photoelectric conversion efficiency and enhancing light trapping ability through nanoscale self-assembly [237].

2.5.4.1 Block Copolymers

Block copolymers are composed of two or more types of polymer chain segments that are linked by covalent bonds. They are capable of self-assembling to form a variety of nanostructures, such as spherical, columnar, or layered structures. This diversity endows them with the capability of photofunctional modulation. Its structural advantage lies in the ability of the layered and columnar structures to optimize light absorption pathways and improve photoelectric conversion efficiency. By adjusting the block ratio and molecular weight, the spectral response range of the material can be precisely controlled. Amphiphilic block copolymers are typically made up of both a hydrophilic and a hydrophobic block, and are also referred to as polymeric surfactants. Block copolymers can form various forms of aggregates by self-assembly in selective solvents to reduce the free energy of the system. Compared to small-molecule surfactant micelles, amphiphilic block copolymer micelles have

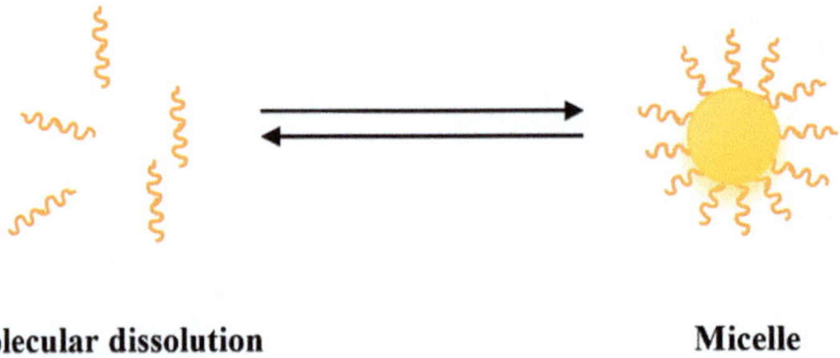

Figure 2.85 Schematic illustration of the self-assembly process of amphiphilic block copolymers.

lower critical micellization concentration (CMC) and slower micelle–monomer conversion rates (Figure 2.85) [238].

By introducing photofunctional groups, this polymer can be grouped into two categories based on its light-responsive properties: reversible and nonreversible. Reversible photochromic polymers are typically represented by polymers containing spiropyran (SP) groups. Under 365 nm UV light irradiation, the colorless and hydrophobic SP undergoes a structural transformation to form a hydrophilic amphiphilic molecule, merocyanine. On the contrary, under 620 nm visible light irradiation, the hydrophilic merocyanine can be reversed to hydrophobic SP, and this process is completely reversible. For example, the block copolymer PEO-*b*-SP containing SP moieties exhibits a unique self-assembly and de-assembly behavior under light stimulation, as shown in Figure 2.86 [239]. Under visible light irradiation, PEO-*b*-PSP was able to self-assemble into micelles due to the hydrophobicity of SP. When irradiated with UV light, the SP moiety in the polymer was transformed into a hydrophilic merocyanine, leading to the disintegration of the micellar structure. When switched to visible light irradiation again, the polymer is reconverted to hydrophobic SP, and the micelle structure is re-formed. Such SP-based photoresponsive polymers have significant potential for drug delivery applications. By precisely controlling the wavelength of light, the loading, release, and re-encapsulation of drug molecules can be realized, thus providing an effective strategy for precise spatiotemporal delivery of drugs.

In addition to the polymers mentioned above, there are a number of other types of polymers that exhibit light-induced self-assembly and disassembly behavior. For example, polyethylene glycol with a terminal functional group of malachite green (PEG-MG) can form vesicle structures in water [240]. However, under UV irradiation, the hydrophobic malachite green is transformed into positively charged hydrophilic molecules, leading to the disassembly of the vesicular structure. This light-induced disassembly process provides an effective means to modulate the aggregated state structure of polymers. In addition, supramolecular amphiphilic polymers can be formed between poly(*N*-isopropylacrylamide) modified with β-cyclodextrin (CD-PNIPAAm) and phenyl ether dendrimer molecules containing an azobenzene moiety (Gx-Azo) due to the host-guest interaction [241]. At 25 °C, this supramolecular amphiphilic polymer is able to self-assemble into vesicles.

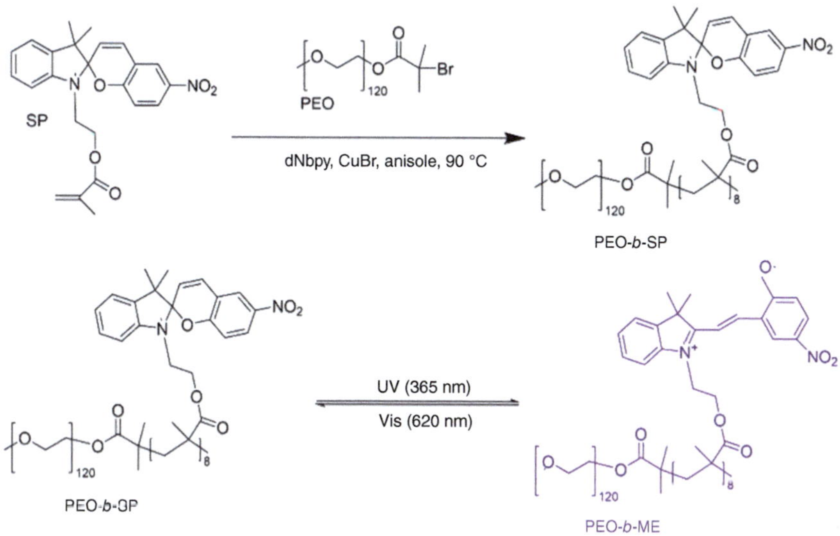

Figure 2.86 Reversible light-responsive PEO-*b*-SP block copolymer. *Source:* Lee et al. [239] / with permission of John Wiley & Sons.

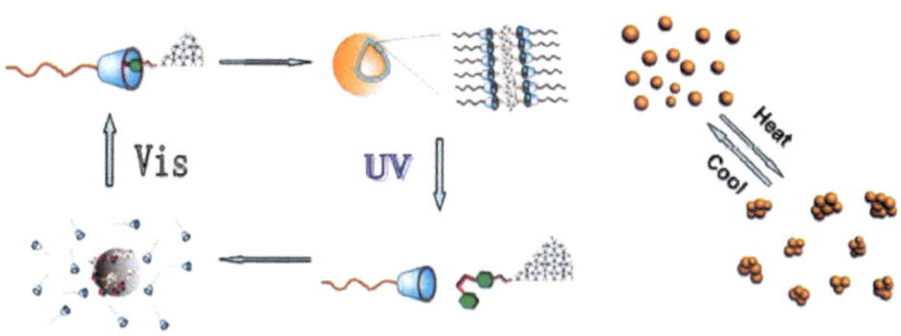

Figure 2.87 Schematic illustration of self-assembly and disassembly of amphiphilic supramolecular polymer. *Source:* Zou et al. [241] / with permission of American Chemical Society.

When irradiated with UV light, the azobenzene moiety shifted from trans to cis configuration, leading to the release of its encapsulation with cyclodextrins, which led to the decomposition of the supramolecular amphiphilic polymer and vesicle disassembly. When irradiated with visible light, the azobenzene group changes from the cis to the trans configuration. This results in the release of the cyclodextrin encapsulation and the formation of an amphiphilic supramolecular polymer, followed by the reassembly of the vesicles (Figure 2.87).

2.5.4.2 Dendrimers

Dendrimers comprise a class of highly branched, spherical, nanoscale polymer compounds, named after their unique, 3D, and dendritic structure. Dendrimers have been the focus of much attention in a variety of fields, including organic chemistry, inorganic chemistry, life science, and materials science, since their successful synthesis for the first time.

Dendrimers consist of a central nucleus, a terminal group, and a repeating unit connecting the two. By adjusting the number of repeating units, their generation (G), and molecular weight can be precisely controlled, resulting in deterministic molecular mass and size. Taking the common polyamide-amine (PAMAM) dendrimer as an example, up to G10 generations of PAMAM can be synthesized at present, covering a wide range of molecular weights and sizes, providing a wealth of options for research and applications. The monomers of dendritic macromolecules are usually of the AB_n type (in which the B-terminal end contains a group that can continue the reaction, and n is generally 2 or 3), and this structure gives them hyperbranched properties. The main methods of their synthesis include divergent (synthesized from the core outward), convergent (synthesized from the shell inward), and a combination of both (Figure 2.88) [242]. Dendrimers in the early stages of development usually have a planar structure, but as they progress through the generations, their molecular weight, degree of branching, and density of surface functional groups increase, causing them to gradually form a 3D spherical conformation. Due to the role of intramolecular resistance, the interior of the sphere is not completely solid; rather, there are many holes, or cavities, and the number of these increases with each generation. Dendrimers exhibit a sparse interior and a dense exterior; the closer to the outer layer, the greater the number of chemical bonds and the smaller the cavity size.

Light-responsive dendrimer undergoes transformation of chemical or physical properties after absorbing light at specific wavelengths. As a noninvasive and easily spatio-temporally controlled means of stimulation, light-stimulated response modification has broad application prospects in material design. Depending on the wavelength, light used to trigger a material response can be categorized as UV, visible, or NIR. For example, a dendrimer modified with a porphyrin core (Figure 2.89) produces ROS when irradiated with UV light. These species are then

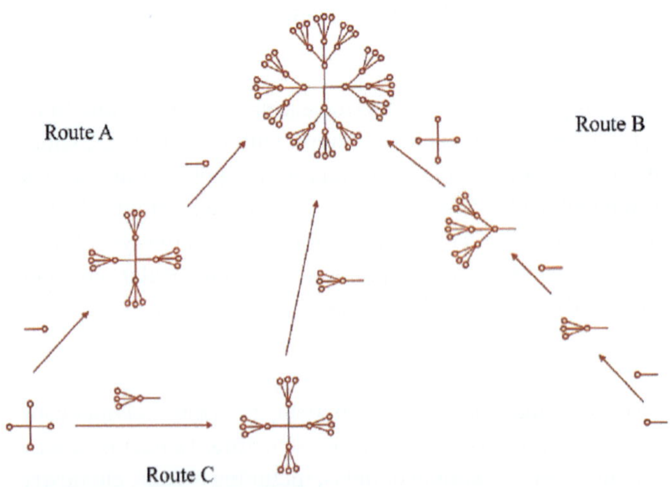

Figure 2.88 The synthetic strategies for dendrimer construction. *Source:* Newkome and Shreiner [242] / with permission of American Chemical Society.

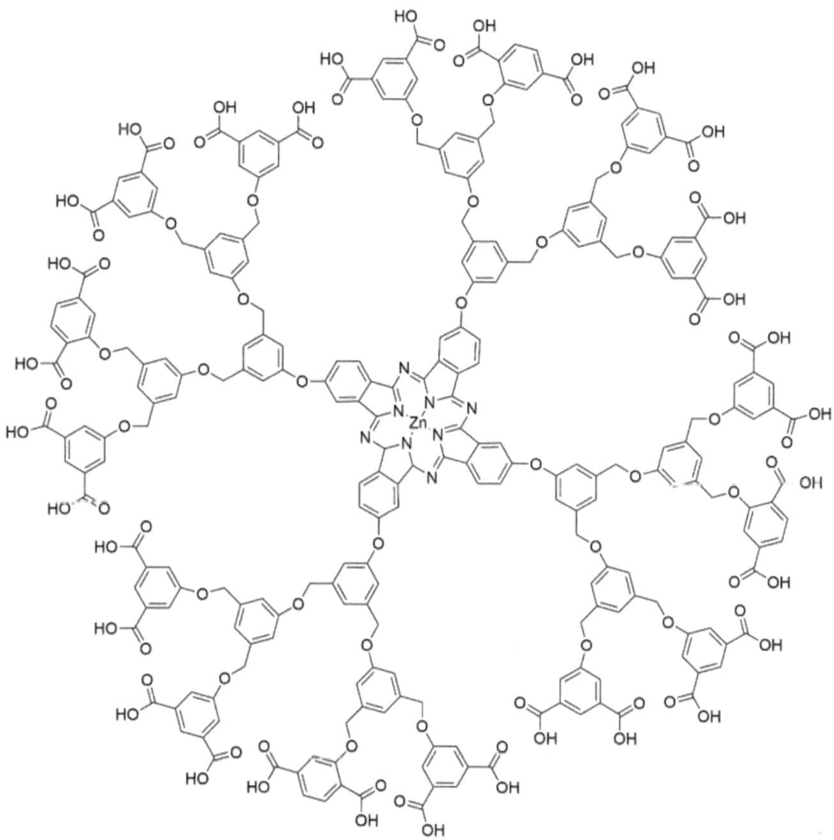

Figure 2.89 The chemical structure of dendrimer with a porphyrin core.

used to disrupt the endosomal membrane, facilitating the cytoplasmic delivery of drugs and genes [243]. In in vitro experiments, the complex increased transgene expression efficiency by more than 100-fold while reducing photocytotoxicity. The system specifically induces gene expression under laser irradiation and is suitable for specific regions such as conjunctival tissues or tumors. In addition, the modified azobenzene moiety on the G1–G5 generation dendrimer undergoes *cis-trans* conformational transitions when exposed to UV or visible light [244]. Compared with unmodified dendrimers, dendrimers with azobenzene terminal modifications had a significant retardation effect on the guest molecule of salicylic acid. The drug release rate was significantly higher under visible light irradiation than under UV light irradiation.

In summary, a variety of new material systems are emerging in the field of photofunctional materials, including MOFs, COFs, biomolecular materials, and nanostructured polymers. These materials show great potential in photocatalysis, biosensing, energy conversion, and bioimaging thanks to their high surface area, tunable pore structure, excellent optoelectronic properties, and biocompatibility. MOFs achieve precise tuning of pore and optical properties through modular design of metal nodes and organic ligands. COFs form ordered porous structures through covalent bonding to optimize light absorption and electron transport. Biomolecular materials utilize natural photosensitive units and chromophores to achieve environmentally friendly functionalization. Nanostructured polymers enhance light trapping and photovoltaic conversion efficiency through self-assembly behavior. However, environmental stability, cost control, and long-term biocompatibility remain challenges for these materials.

Future development trends will focus on constructing multi-material composite systems to enhance comprehensive performance through the synergistic effects of MOFs and polymers, COFs and photosensitizers. At the same time, breakthroughs in precise synthesis technologies will enable the large-scale preparation of complex nanostructures, and integrating quantum dots, perovskites, and other emerging photofunctional units will further expand application boundaries. As intelligent demand grows, the integration of light-responsive materials and micro-nano processing technology is expected to lead to the development of adaptive optical devices and implantable optical diagnostic systems. This will promote the optimization of light-functional materials, moving from single-performance to multi-functional integration and ultimately realizing the leap from basic research to industrial applications.

References

1 Mei, J., Leung, N.L.C., Kwok, R.T.K. et al. (2015). Aggregation-induced emission: together we shine, united we soar! *Chemical Reviews* 115 (21): 11718–11940.

2 Jablonski, A. (1933). Efficiency of anti-stokes fluorescence in dyes. *Nature* 131 (3319): 839–840.

3 Chatterjee, T. and Wong, K.-T. (2019). Perspective on host materials for thermally activated delayed fluorescence organic light emitting diodes. *Advanced Optical Materials* 7 (1): 1800565.

4 Leung, N.L.C., Xie, N., Yuan, W. et al. (2014). Restriction of intramolecular motions: the general mechanism behind aggregation-induced emission. *Chemistry – A European Journal* 20 (47): 15349–15353.

5 Förster, T. (1969). Excimers. *Angewandte Chemie International Edition in English* 8 (5): 333–343.

6 Tang, X., Cui, L.-S., Li, H.-C. et al. (2020). Highly efficient luminescence from space-confined charge-transfer emitters. *Nature Materials* 19 (12): 1332–1338.

7 Sun, G., Pan, J., Wu, Y. et al. (2020). Supramolecular assembly-driven color-tuning and white-light emission based on crown-ether-functionalized dihydrophenazine. *ACS Applied Materials & Interfaces* 12 (9): 10875–10882.

8 Zhang, Y., Zhang, D., Tsuboi, T. et al. (2019). Simultaneous enhancement of efficiency and stability of OLEDs with thermally activated delayed fluorescence materials by modifying carbazoles with peripheral groups. *Science China Chemistry* 62 (3): 393–402.

9 Li, X., Shen, S., Zhang, C. et al. (2021). Small-molecule based thermally activated delayed fluorescence materials with dual-emission characteristics. *Science China Chemistry* 64 (4): 534–546.

10 Endo, A., Sato, K., Yoshimura, K. et al. (2011). Efficient up-conversion of triplet excitons into a singlet state and its application for organic light emitting diodes. *Applied Physics Letters* 98 (8): 083302.

11 Xu, S., Duan, Y., and Liu, B. (2020). Precise molecular design for high-performance luminogens with aggregation-induced emission. *Advanced Materials* 32 (1): e1903530.

12 Lee, S.Y., Yasuda, T., Yang, Y.S. et al. (2014). Luminous butterflies: efficient exciton harvesting by benzophenone derivatives for full-color delayed fluorescence OLEDs. *Angewandte Chemie International Edition* 53 (25): 6402–6406.

13 He, Y., Wang, J., Li, Q. et al. (2023). Highly efficient room-temperature phosphorescence promoted via intramolecular-space heavy-atom effect. *Advanced Optical Materials* 11 (14): 2201641.

14 El-Sayed, M.A. and Triplet state. (1968). Its radiative and nonradiative properties. *Accounts of Chemical Research* 1 (1): 8–16.

15 Zhao, W., He, Z., Lam Jacky, W.Y. et al. (2016). Rational molecular design for achieving persistent and efficient pure organic room-temperature phosphorescence. *Chem* 1 (4): 592–602.

16 Wang, Z., Cheng, X., Xie, Y. et al. (2023). Recent advances in organic room-temperature phosphorescence of heteroatom (B/S/P)-containing chromophores. *CCS Chemistry* 5 (2): 292–309.

17 An, Z., Zheng, C., Tao, Y. et al. (2015). Stabilizing triplet excited states for ultra-long organic phosphorescence. *Nature Materials* 14 (7): 685–690.

18 Singh, M., Shen, K., Ye, W. et al. (2024). Achieving high-temperature phosphorescence by organic cocrystal engineering. *Angewandte Chemie International Edition* 63 (14): e202319694.

19 Kabe, R. and Adachi, C. (2017). Organic long persistent luminescence. *Nature* 550 (7676): 384–387.

20 Tang, C.W. and VanSlyke, S.A. (1987). Organic electroluminescent diodes. *Applied Physics Letters* 51 (12): 913–915.

21 Fang, M., Yang, J., and Li, Z. (2022). Light emission of organic luminogens: generation, mechanism and application. *Progress in Materials Science* 125: 100914.

22 Wei, Q., Fei, N., Islam, A. et al. (2018). Small-molecule emitters with high quantum efficiency: mechanisms, structures, and applications in OLED devices. *Advanced Optical Materials* 6 (20): 1800512.

23 Uoyama, H., Goushi, K., Shizu, K. et al. (2012). Highly efficient organic light-emitting diodes from delayed fluorescence. *Nature* 492 (7428): 234–238.

24 Hatakeyama, T., Shiren, K., Nakajima, K. et al. (2016). Ultrapure blue thermally activated delayed fluorescence molecules: efficient HOMO–LUMO separation by the multiple resonance effect. *Advanced Materials* 28 (14): 2777–2781.

25 Maeng, J.H., Ahn, D.H., Lee, H. et al. (2020). Rigid indolocarbazole donor moiety for highly efficient thermally activated delayed fluorescent device. *Dyes and Pigments* 180: 108485.

26 Wu, T.-L., Lei, J., Hsieh, C.-M. et al. (2022). Substituent engineering of the diboron molecular architecture for a nondoped and ultrathin emitting layer. *Chemical Science* 13 (44): 12996–13005.

27 Zhang, Y.-L., Ran, Q., Wang, Q. et al. (2019). High-efficiency red organic light-emitting diodes with external quantum efficiency close to 30% based on a novel thermally activated delayed fluorescence emitter. *Advanced Materials* 31 (42): e1902368.

28 Wang, J., Liang, J., Xu, Y. et al. (2019). Purely organic phosphorescence emitter-based efficient electroluminescence devices. *The Journal of Physical Chemistry Letters* 10 (19): 5983–5988.

29 Lee, D.R., Lee, K.H., Shao, W. et al. (2020). Heavy atom effect of selenium for metal-free phosphorescent light-emitting diodes. *Chemistry of Materials* 32 (6): 2583–2592.

30 Xing, L., Zhu, Z.-L., He, J. et al. (2021). Anthracene-based fluorescent emitters toward superior-efficiency nondoped TTA-OLEDs with deep blue emission and low efficiency roll-off. *Chemical Engineering Journal* 421: 127748.

31 Pu, Y.-J., Kamiya, A., Nakayama, K.-i., and Kido, J. (2010). Arylamino-9,10-diphenylanthracenes for organic light emitting devices. *Organic Electronics* 11 (3): 479–485.

32 Hu, J.-Y., Pu, Y.-J., Satoh, F. et al. (2014). Bisanthracene-based donor–acceptor-type light-emitting dopants: highly efficient deep-blue emission in organic light-emitting devices. *Advanced Functional Materials* 24 (14): 2064–2071.

33 Chou, P.Y., Chou, H.H., Chen, Y.H. et al. (2014). Efficient delayed fluorescence via triplet–triplet annihilation for deep-blue electroluminescence. *Chemical Communications* 50 (52): 6869–6871.

34 Chen, Y.-H., Lin, C.-C., Huang, M.-J. et al. (2016). Superior upconversion fluorescence dopants for highly efficient deep-blue electroluminescent devices. *Chemical Science* 7 (7): 4044–4051.

35 Qiu, X., Xu, Y., Wang, C. et al. (2019). Synergistic effects of hydrogen bonds and the hybridized excited state observed for high-efficiency, deep-blue fluorescent emitters with narrow emission in OLED applications. *Journal of Materials Chemistry C* 7 (18): 5461–5467.

36 Li, W., Liu, D., Shen, F. et al. (2012). A twisting donor-acceptor molecule with an intercrossed excited state for highly efficient, deep-blue electroluminescence. *Advanced Functional Materials* 22 (13): 2797–2803.

37 Wang, C., Li, X., Pan, Y. et al. (2016). Highly efficient nondoped green organic light-emitting diodes with combination of high photoluminescence and high exciton utilization. *ACS Applied Materials & Interfaces* 8 (5): 3041–3049.

38 Liu, T., Zhu, L., Zhong, C. et al. (2017). Naphthothiadiazole-based near-infrared emitter with a photoluminescence quantum yield of 60% in neat film and external quantum efficiencies of up to 3.9% in nondoped OLEDs. *Advanced Functional Materials* 27: 1606384.

39 Marquette, C.A. and Blum, L.J. (2006). Applications of the luminol chemiluminescent reaction in analytical chemistry. *Analytical and Bioanalytical Chemistry* 385 (3): 546–554.

40 Zhang, Y., Yan, C., Wang, C. et al. (2020). A sequential dual-lock strategy for photoactivatable chemiluminescent probes enabling bright duplex optical imaging. *Angewandte Chemie International Edition* 59 (23): 9059–9066.

41 Giussani, A., Farahani, P., Martínez-Muñoz, D. et al. (2019). Molecular basis of the chemiluminescence mechanism of luminol. *Chemistry – A European Journal* 25 (20): 5202–5213.

42 Song, G., Yang, Z., Cheng, J. et al. (2025). Chemiluminescence resonance energy transfer for targeted photoactivation of ion channels in living cells. *Angewandte Chemie International Edition* 64 (15): e202423792.

43 Liu, S., Yuan, H., Bai, H. et al. (2018). Electrochemiluminescence for electric-driven antibacterial therapeutics. *Journal of the American Chemical Society* 140 (6): 2284–2291.

44 Vacher, M., Fdez. Galván, I., Ding, B.-W. et al. (2018). Chemi- and bioluminescence of cyclic peroxides. *Chemical Reviews* 118 (15): 6927–6974.

45 Xie, Z., Yu, T., Chen, J. et al. (2018). Weak interactions but potent effect: tunable mechanoluminescence by adjusting intermolecular C–H···π interactions. *Chemical Science* 9 (26): 5787–5794.

46 Tian, Y., Yang, X., Gong, Y. et al. (2021). The initial attempt to reveal the emission processes of both mechanoluminescence and room temperature phosphorescence with the aid of circular dichroism in solid state. *Science China Chemistry* 64 (3): 445–451.

47 Gong, Y.-B., Zhang, P., Gu, Y.-r. et al. (2018). The influence of molecular packing on the emissive behavior of pyrene derivatives: mechanoluminescence and mechanochromism. *Advanced Optical Materials* 6 (16): 1800198.

48 Yang, J., Qin, J., Geng, P. et al. (2018). Molecular conformation-dependent mechanoluminescence: same mechanical stimulus but different emissive color over time. *Angewandte Chemie International Edition* 57 (43): 14174–14178.

49 Zhou, B., Hu, Z., Jiang, Y. et al. (2018). Theoretical exploitation of acceptors based on benzobis(thiadiazole) and derivatives for organic NIR-II fluorophores. *Physical Chemistry Chemical Physics* 20 (30): 19759–19767.

50 Cui, Q., Wang, X., Yang, Y. et al. (2016). Binding-directed energy transfer of conjugated polymer materials for dual-color imaging of cell membrane. *Chemistry of Materials* 28 (13): 4661–4669.

51 Ding, L., Yu, Z.-D., Wang, X.-Y. et al. (2023). Polymer semiconductors: synthesis, processing, and applications. *Chemical Reviews* 123 (12): 7421–7497.

52 He, Z., Shrives, H.J., Fernández-Salas, J.A. et al. (2018). Synthesis of C2 substituted benzothiophenes via an interrupted Pummerer/[3,3]-sigmatropic/1,2-migration cascade of benzothiophene *S*-oxides. *Angewandte Chemie International Edition* 57 (20): 5759–5764.

53 Gedefaw, D., Prosa, M., Bolognesi, M. et al. (2017). Recent development of quinoxaline based polymers/small molecules for organic photovoltaics. *Advanced Energy Materials* 7 (21): 1700575.

54 Al Kobaisi, M., Bhosale, S.V., Latham, K. et al. (2016). Functional naphthalene diimides: synthesis, properties, and applications. *Chemical Reviews* 116 (19): 11685–11796.

55 Kaloni, T.P., Giesbrecht, P.K., Schreckenbach, G., and Freund, M.S. (2017). Polythiophene: from fundamental perspectives to applications. *Chemistry of Materials* 29 (24): 10248–10283.

56 Liu, Z., Zhang, G., and Zhang, D. (2018). Modification of side chains of conjugated molecules and polymers for charge mobility enhancement and sensing functionality. *Accounts of Chemical Research* 51 (6): 1422–1432.

57 Yang, Y., Liu, Z., Zhang, G. et al. (2019). The effects of side chains on the charge mobilities and functionalities of semiconducting conjugated polymers beyond solubilities. *Advanced Materials* 31 (46): e1903104.

58 Patil, A.O., Ikenoue, Y., Wudl, F., and Heeger, A.J. (1987). Water soluble conducting polymers. *Journal of the American Chemical Society* 109 (6): 1858–1859.

59 Wang, J., Lv, F., Liu, L. et al. (2018). Strategies to design conjugated polymer based materials for biological sensing and imaging. *Coordination Chemistry Reviews* 354: 135–154.

60 Jiao, G.-S., Zhang, Y., Cheng, H. et al. (2023). Role of side chains in the packing structure and dynamics of conjugated polymers. *Macromolecules* 56 (14): 5248–5258.

61 Wang, Z., Lin, H., Zhang, M. et al. (2023). Water-soluble conjugated polymers for bioelectronic systems. *Materials Horizons* 10 (4): 1210–1233.

62 Schwartz, B.J. (2003). Conjugated polymers as molecular materials: how chain conformation and film morphology influence energy transfer and interchain interactions. *Annual Review of Physical Chemistry* 54: 141–172.

63 Meng, L., Shang, Y., Li, Q. et al. (2010). Dynamic Monte Carlo simulation for highly efficient polymer blend photovoltaics. *The Journal of Physical Chemistry B* 114 (1): 36–41.

64 Zhang, Y., Cheng, D., Li, T. et al. (2024). Concentration effect on the chain structure and photoelectric properties of conjugated polymer precursor solutions and thin films: a mini review. *Journal of Polymer Science* 62 (6): 1156–1174.

65 Wang, L., Xiong, Z., Zhi Sun, J. et al. (2024). How the length of through-space conjugation influences the clusteroluminescence of oligo(phenylene methylene)s. *Angewandte Chemie International Edition* 63 (8): e202318245.

66 Sun, X., Bai, J.-K., Yang, Y.-D. et al. (2024). Controlled interconversion of macrocyclic atropisomers via defined intermediates. *Nature Communications* 15 (1): 6559.

67 Liu, C., Yang, W., Jiang, H. et al. (2022). Chain conformation and liquid-crystalline structures of a poly(thieno)thiophene. *Macromolecules* 55 (7): 2892–2903.

68 Sabater i Serra, R., Torregrosa Cabanilles, C., Meseguer Dueñas, J.M. et al. (2018). Conformational changes and dynamics during adsorption of macromolecules with different degree of polymerization studied by Monte Carlo simulations. *Macromolecular Theory and Simulations* 27: 1800012.

69 Sulatha, M.S. and Natarajan, U. (2012). Molecular dynamics simulations of PAA–PMA polyelectrolyte copolymers in dilute aqueous solution: chain conformations and hydration properties. *Industrial & Engineering Chemistry Research* 51 (33): 10833–10839.

70 Zhang, H., Zhao, Z., Turley, A.T. et al. (2020). Aggregate science: from structures to properties. *Advanced Materials* 32 (36): 2001457.

71 Zhang, J., Li, A., Zou, H. et al. (2020). A "simple" donor–acceptor AIEgen with multi-stimuli responsive behavior. *Materials Horizons* 7 (1): 135–142.

72 Xu, Z., Park, K.S., Kwok, J.J. et al. (2022). Not all aggregates are made the same: distinct structures of solution aggregates drastically modulate assembly pathways, morphology, and electronic properties of conjugated polymers. *Advanced Materials* 34 (32): 2203055.

73 Hu, L., Duan, Y., Xu, Z. et al. (2016). Stimuli-responsive fluorophores with aggregation-induced emission: implication for dual-channel optical data storage. *Journal of Materials Chemistry C* 4 (23): 5334–5341.

74 Peteanu, L.A., Sherwood, G.A., Werner, J.H. et al. (2011). Visualizing core–shell structure in substituted PPV oligomer aggregates using fluorescence lifetime imaging microscopy (FLIM). *The Journal of Physical Chemistry C* 115 (31): 15607–15616.

75 Li, T., Cheng, D., Liu, L. et al. (2025). Facile control of PFO β conformation in blended polymers. *Polymer* 319: 128057.

76 Li, Q.-Y., Yao, Z.-F., Wang, J.-Y., and Pei, J. (2021). Multi-level aggregation of conjugated small molecules and polymers: from morphology control to physical insights. *Reports on Progress in Physics* 84 (7): 076601.

77 Wang, X., Liu, L., Zhu, S., and Li, L. (2019). Fluorescent platforms based on organic molecules for chemical and biological detection. *Physica Status Solidi RRL: Rapid Research Letters* 13: 1800521.

78 Zhu, S., Wang, X., Yang, Y. et al. (2018). Conjugated polymer with aggregation-directed intramolecular Förster resonance energy transfer enabling efficient discrimination and killing of microbial pathogens. *Chemistry of Materials* 30 (10): 3244–3253.

79 Liu, L., Liu, R., Wang, X. et al. (2018). Facile preparation of fluorescent nanoparticles with tunable exciplex emission and their application to targeted cellular imaging. *ACS Applied Bio Materials* 1 (1): 185–192.

80 Liu, L. and Corma, A. (2018). Metal catalysts for heterogeneous catalysis: from single atoms to nanoclusters and nanoparticles. *Chemical Reviews* 118 (10): 4981–5079.

81 Zhang, Q. and Wang, H. (2014). Facet-dependent catalytic activities of Au nanoparticles enclosed by high-index facets. *ACS Catalysis* 4 (11): 4027–4033.

82 Lohse, S.E. and Murphy, C.J. (2013). The quest for shape control: a history of gold Nanorod synthesis. *Chemistry of Materials* 25 (8): 1250–1261.

83 Ogarev, V.A., Rudoi, V.M., and Dement'eva, O.V. (2018). Gold nanoparticles: synthesis, optical properties, and application. *Inorganic Materials: Applied Research* 9: 134–140.

84 Maestro, L.M., Camarillo, E., Sánchez-Gil, J.A. et al. (2014). Gold nanorods for optimized photothermal therapy: the influence of irradiating in the first and second biological windows. *RSC Advances* 4 (96): 54122–54129.

85 Chen, H., Shao, L., Ming, T. et al. (2010). Understanding the photothermal conversion efficiency of gold nanocrystals. *Small* 6 (20): 2272–2280.

86 Mackey, M.A., Ali, M.R.K., Austin, L.A. et al. (2014). The most effective gold nanorod size for plasmonic photothermal therapy: theory and in vitro experiments. *The Journal of Physical Chemistry B* 118 (5): 1319–1326.

87 Meyer, S.M., Pettine, J., Nesbitt, D.J., and Murphy, C.J. (2021). Size effects in gold nanorod light-to-heat conversion under femtosecond illumination. *The Journal of Physical Chemistry C* 125 (29): 16268–16278.

88 Stewart, M.E., Anderton, C.R., Thompson, L.B. et al. (2008). Nanostructured plasmonic sensors. *Chemical Reviews* 108 (2): 494–521.

89 Skrabalak, S.E., Chen, J., Sun, Y. et al. (2008). Gold nanocages: synthesis, properties, and applications. *Accounts of Chemical Research* 41 (12): 1587–1595.

90 Liu, H., Liu, T., Li, L. et al. (2012). Size dependent cellular uptake, in vivo fate and light-heat conversion efficiency of gold nanoshells on silica nanorattles. *Nanoscale* 4 (11): 3523–3529.

91 Abbasi, A., Park, K., Bose, A., and Bothun, G.D. (2017). Near-infrared responsive gold–layersome nanoshells. *Langmuir* 33 (21): 5321–5327.

92 Qiu, J., Xie, M., Wu, T. et al. (2020). Gold nanocages for effective photothermal conversion and related applications. *Chemical Science* 11 (48): 12955–12973.

93 Skrabalak, S.E., Au, L., Li, X., and Xia, Y. (2007). Facile synthesis of ag nanocubes and Au nanocages. *Nature Protocols* 2 (9): 2182–2190.

94 Durán-Meza, A.L., Moreno-Gutiérrez, D.S., Ruiz-Robles, J.F. et al. (2016). Synthesis and characterization of extremely small gold nanoshells, and comparison of their photothermal conversion capacity with gold nanorods. *Nanoscale* 8 (21): 11091–11098.

95 Yang, W., Liang, H., Ma, S. et al. (2019). Gold nanoparticle based photothermal therapy: development and application for effective cancer treatment. *Sustainable Materials and Technologies* 22: e00109.

96 Khoury, C.G. and Vo-Dinh, T. (2008). Gold nanostars for surface-enhanced Raman scattering: synthesis, characterization and optimization. *The Journal of Physical Chemistry C* 112 (48): 18849–18859.

97 Sanchez-Gaytan, B.L., Qian, Z., Hastings, S.P. et al. (2013). Controlling the topography and surface plasmon resonance of gold nanoshells by a templated surfactant-assisted seed growth method. *The Journal of Physical Chemistry C* 117 (17): 8916–8923.

98 Chandra, K., Culver, K.S.B., Werner, S.E. et al. (2016). Manipulating the anisotropic structure of gold nanostars using good's buffers. *Chemistry of Materials* 28 (18): 6763–6769.

99 Song, J., Yang, X., Yang, Z. et al. (2017). Rational design of branched nanoporous gold nanoshells with enhanced physico-optical properties for optical imaging and cancer therapy. *ACS Nano* 11 (6): 6102–6113.

100 Ma, H., Liu, Z., Wei, Y., and Jiang, L. (2019). Controlled morphology evolution of branched Au nanostructures and their shape-dependent catalytic and photo-thermal properties. *Colloids and Surfaces A: Physicochemical and Engineering Aspects* 582: 123889.

101 Wiley, B., Sun, Y.G., Mayers, B., and Xia, Y.N. (2005). Shape-controlled synthesis of metal nanostructures: the case of silver. *Chemistry – A European Journal* 11 (2): 454–463.

102 Jana, N.R., Gearheart, L., and Murphy, C.J. (2001). Wet chemical synthesis of silver nanorods and nanowires of controllable aspect ratio. *Chemical Communications* (7): 617–618.

103 Murphy, C.J. and Jana, N.R. (2002). Controlling the aspect ratio of inorganic nanorods and nanowires. *Advanced Materials* 14 (1): 80–82.

104 Sun, Y., Mayers, B., and Xia, Y. (2003). Transformation of silver nanospheres into nanobelts and triangular nanoplates through a thermal process. *Nano Letters* 3 (5): 675–679.

105 Draine, B.T. and Flatau, P.J. (1994). Discrete-dipole approximation for scattering calculations. *Journal of the Optical Society of America. A, Optics, Image Science, and Vision* 11 (4): 1491–1499.

106 Wiley, B.J., Im, S.H., Li, Z.-Y. et al. (2006). Maneuvering the surface Plasmon resonance of silver nanostructures through shape-controlled synthesis. *The Journal of Physical Chemistry B* 110 (32): 15666–15675.

107 Haes, A.J., Haynes, C.L., McFarland, A.D. et al. (2005). Plasmonic materials for surface-enhanced sensing and spectroscopy. *MRS Bulletin* 30 (5): 368–375.

108 Fuchs, R. (1974). Theory of the optical properties of small cubes. *Physics Letters A* 48 (5): 353–354.

109 Kottmann, J.P., Martin, O.J.F., Smith, D.R., and Schultz, S. (2001). Plasmon resonances of silver nanowires with a nonregular cross section. *Physical Review B* 64 (23): 235402.

110 Aizpurua, J., Bryant, G.W., Richter, L.J. et al. (2005). Optical properties of coupled metallic nanorods for field-enhanced spectroscopy. *Physical Review B* 71 (23): 235420.

111 Lee, S., Wang, S., Wern, C., and Yi, S. (2021). The green synthesis of 2D copper nanosheets and their light absorption. *Materials* 14 (8): 1926.

112 Jin, M., He, G., Zhang, H. et al. (2011). Shape-controlled synthesis of copper nanocrystals in an aqueous solution with glucose as a reducing agent and hexadecylamine as a capping agent. *Angewandte Chemie International Edition* 50 (45): 10560–10564.

113 Park, B.K., Jeong, S., Kim, D. et al. (2007). Synthesis and size control of monodisperse copper nanoparticles by polyol method. *Journal of Colloid and Interface Science* 311 (2): 417–424.

114 Kelly, K.L., Coronado, E., Zhao, L.L., and Schatz, G.C. (2003). The optical properties of metal nanoparticles: the influence of size, shape, and dielectric environment. *The Journal of Physical Chemistry B* 107 (3): 668–677.

115 Lakowicz, J.R., Ray, K., Chowdhury, M. et al. (2008). Plasmon-controlled fluorescence: a new paradigm in fluorescence spectroscopy. *The Analyst* 133 (10): 1308–1346.

116 Lakowicz, J.R. (2001). Radiative decay engineering: biophysical and biomedical applications. *Analytical Biochemistry* 298 (1): 1–24.

117 Ding, X., Liow, C.H., Zhang, M. et al. (2014). Surface plasmon resonance enhanced light absorption and photothermal therapy in the second near-infrared window. *Journal of the American Chemical Society* 136 (44): 15684–15693.

118 Kadkhoda, J., Tarighatnia, A., Barar, J. et al. (2022). Recent advances and trends in nanoparticles based photothermal and photodynamic therapy. *Photodiagnosis and Photodynamic Therapy* 37: 102697.

119 Riley, R.S. and Day, E.S. (2017). Gold nanoparticle-mediated photothermal therapy: applications and opportunities for multimodal cancer treatment. *WIREs Nanomedicine and Nanobiotechnology* 9 (4): e1449.

120 Tchounwou, C., Sinha, S.S., Viraka Nellore, B.P. et al. (2015). Hybrid theranostic platform for second near-IR window light triggered selective two-photon imaging and photothermal killing of targeted melanoma cells. *ACS Applied Materials & Interfaces* 7 (37): 20649–20656.

121 Deng, H.-y., Wang, L., Tang, D. et al. (2021). Review on the laser-induced performance of photothermal materials for ignition application. *Energetic Materials Frontiers* 2 (3): 201–217.

122 Zhang, B., Chen, J., Cao, Y. et al. (2021). Ligand design in ligand-protected gold nanoclusters. *Small* 17 (27): e2004381.

123 Jadzinsky, P.D., Calero, G., Ackerson, C.J. et al. (2007). Structure of a thiol monolayer-protected gold nanoparticle at 1.1 Å resolution. *Science* 318 (5849): 430–433.

124 Zhu, M., Aikens, C.M., Hollander, F.J. et al. (2008). Correlating the crystal structure of a thiol-protected Au_{25} cluster and optical properties. *Journal of the American Chemical Society* 130 (18): 5883–5885.

125 Das, A., Li, T., Nobusada, K. et al. (2013). Nonsuperatomic $[Au_{23}(SC_6H_{11})_{16}]^-$ nanocluster featuring bipyramidal Au_{15} kernel and trimeric $Au_3(SR)_4$ motif. *Journal of the American Chemical Society* 135 (49): 18264–18267.

126 Higaki, T., Liu, C., Zeng, C. et al. (2016). Controlling the atomic structure of Au_{30} nanoclusters by a ligand-based strategy. *Angewandte Chemie International Edition* 55 (23): 6694–6697.

127 Zhuang, S., Liao, L., Zhao, Y. et al. (2018). Is the kernel–staples match a key–lock match? *Chemical Science* 9 (9): 2437–2442.

128 Zeng, C., Liu, C., Pei, Y., and Jin, R. (2013). Thiol ligand-induced transformation of $Au_{38}(SC_2H_4Ph)_{24}$ to $Au_{36}(SPh\text{-}t\text{-}Bu)_{24}$. *ACS Nano* 7 (7): 6138–6145.

129 Yuan, S.-F., Liu, W.-D., Liu, C.-Y. et al. (2022). Nitrogen donor protection for atomically precise metal nanoclusters. *Chemistry – A European Journal* 28 (24): e202104445.

130 Wan, X.-K., Lin, Z.-W., and Wang, Q.-M. (2012). Au_{20} nanocluster protected by hemilabile phosphines. *Journal of the American Chemical Society* 134 (36): 14750–14752.

131 Lei, Z., Guan, Z.-J., Pei, X.-L. et al. (2016). An atomically precise $Au_{10}Ag_2$ nanocluster with red–near-IR dual emission. *Chemistry – A European Journal* 22 (32): 11156–11160.

132 Yang, H., Wang, Y., and Zheng, N. (2013). Stabilizing subnanometer ag(0) nanoclusters by thiolate and diphosphine ligands and their crystal structures. *Nanoscale* 5 (7): 2674–2677.

133 Yang, H., Lei, J., Wu, B. et al. (2013). Crystal structure of a luminescent thiolated ag nanocluster with an octahedral Ag_6^{4+} core. *Chemical Communications* 49 (3): 300–302.

134 AbdulHalim, L.G., Bootharaju, M.S., Tang, Q. et al. (2015). $Ag_{29}(BDT)_{12}(TPP)_4$: a tetravalent nanocluster. *Journal of the American Chemical Society* 137 (37): 11970–11975.

135 Alhilaly, M.J., Bootharaju, M.S., Joshi, C.P. et al. (2016). $[Ag_{67}(SPhMe_2)_{32}(PPh_3)_8]^{3+}$: synthesis, total structure, and optical properties of a large box-shaped silver nanocluster. *Journal of the American Chemical Society* 138 (44): 14727–14732.

136 Yang, H., Yan, J., Wang, Y. et al. (2017). Embryonic growth of face-center-cubic silver nanoclusters shaped in nearly perfect half-cubes and cubes. *Journal of the American Chemical Society* 139 (1): 31–34.

137 Dhayal, R.S., Liao, J.-H., Liu, Y.-C. et al. (2015). $[Ag_{21}\{S_2P(OiPr)_2\}_{12}]^+$: an eight-electron superatom. *Angewandte Chemie International Edition* 54 (12): 3702–3706.

138 Vázquez-Vázquez, C., Bañobre-López, M., Mitra, A. et al. (2009). Synthesis of small atomic copper clusters in microemulsions. *Langmuir* 25 (14): 8208–8216.

139 Vilar-Vidal, N., Blanco, M.C., López-Quintela, M.A. et al. (2010). Electrochemical synthesis of very stable photoluminescent copper clusters. *The Journal of Physical Chemistry C* 114 (38): 15924–15930.

140 Huseyinova, S., Blanco, J., Requejo, F.G. et al. (2016). Synthesis of highly stable surfactant-free Cu_5 clusters in water. *The Journal of Physical Chemistry C* 120 (29): 15902–15908.

141 Liu, Z., Wu, Z., Yao, Q. et al. (2021). Correlations between the fundamentals and applications of ultrasmall metal nanoclusters: recent advances in catalysis and biomedical applications. *Nano Today* 36: 101053.

142 Pyo, K., Thanthirige, V.D., Kwak, K. et al. (2015). Ultrabright luminescence from gold nanoclusters: rigidifying the Au(I)–thiolate shell. *Journal of the American Chemical Society* 137 (25): 8244–8250.

143 Yang, L., Zhang, B., Fu, L. et al. (2019). Efficient and monochromatic electrochemiluminescence of aqueous-soluble Au nanoclusters via host–guest recognition. *Angewandte Chemie International Edition* 58 (21): 6901–6905.

144 Masters, B.R. and So, P.T.C. (2004). Antecedents of two-photon excitation laser scanning microscopy. *Microscopy Research and Technique* 63 (1): 3–11.

145 Galas, L., Gallavardin, T., Bénard, M. et al. (2018). "Probe, sample, and instrument (PSI)": the hat-trick for fluorescence live cell imaging. *Chemosensors* 6 (3): 40.

146 So, P.T.C., Dong, C.Y., Masters, B.R., and Berland, K.M. (2000). Two-photon excitation fluorescence microscopy. *Annual Review of Biomedical Engineering* 2: 399–429.

147 Ramakrishna, G., Varnavski, O., Kim, J. et al. (2008). Quantum-sized gold clusters as efficient two-photon absorbers. *Journal of the American Chemical Society* 130 (15): 5032–5033.

148 Liu, C.-L., Ho, M.-L., Chen, Y.-C. et al. (2009). Thiol-functionalized gold nanodots: two-photon absorption property and imaging in vitro. *The Journal of Physical Chemistry C* 113 (50): 21082–21089.

149 Polavarapu, L., Manna, M., and Xu, Q.-H. (2011). Biocompatible glutathione capped gold clusters as one- and two-photon excitation fluorescence contrast agents for live cells imaging. *Nanoscale* 3 (2): 429–434.

150 Larson, D.R., Zipfel, W.R., Williams, R.M. et al. (2003). Water-soluble quantum dots for multiphoton fluorescence imaging in vivo. *Science* 300 (5624): 1434–1436.

151 Luo, Z., Yuan, X., Yu, Y. et al. (2012). From aggregation-induced emission of Au(I)–thiolate complexes to ultrabright Au(0)@Au(I)–thiolate core–shell nanoclusters. *Journal of the American Chemical Society* 134 (40): 16662–16670.

152 Li, Y., Song, Y., Zhang, X. et al. (2022). Atomically precise Au_{42} nanorods with longitudinal excitons for an intense photothermal effect. *Journal of the American Chemical Society* 144 (27): 12381–12389.

153 Yan, B., Sun, J., Liu, J. et al. (2025). Flat-shaped copper nanoclusters with near-infrared absorption for enhanced photothermal conversion. *JACS Au* 5 (4): 1884–1893.

154 Liu, L., Luo, X., and Liu, J. (2020). Bidirectional regulation of singlet oxygen generation from luminescent gold nanoparticles through surface manipulation. *Small* 16 (14): 2000011.

155 Chen, Q., Chen, J., Yang, Z. et al. (2018). NIR-II light activated photodynamic therapy with protein-capped gold nanoclusters. *Nano Research* 11 (10): 5657–5669.

156 Han, R., Zhao, M., Wang, Z. et al. (2020). Super-efficient in vivo two-photon photodynamic therapy with a gold nanocluster as a type I photosensitizer. *ACS Nano* 14 (8): 9532–9544.

157 Veeramani, V., Bao, Z., Chan, M.H. et al. (2019). Quantum dots for light conversion, therapeutic and energy storage applications. *Journal of Solid State Chemistry* 270: 71–84.

158 Etgar, L., Park, J., Barolo, C. et al. (2012). Enhancing the efficiency of a dye sensitized solar cell due to the energy transfer between CdSe quantum dots and a designed squaraine dye. *RSC Advances* 2 (7): 2748–2752.

159 Namdari, P., Negahdari, B., and Eatemadi, A. (2017). Synthesis, properties and biomedical applications of carbon-based quantum dots: an updated review. *Biomedicine & Pharmacotherapy* 87: 209–222.

160 Yuan, F.L., Yuan, T., Sui, L.Z. et al. (2018). Engineering triangular carbon quantum dots with unprecedented narrow bandwidth emission for multicolored LEDs. *Nature Communications* 9: 2249.

161 Berkinsky, D.B., Proppe, A.H., Utzat, H. et al. (2023). Narrow intrinsic line widths and electron-phonon coupling of InP colloidal quantum dots. *ACS Nano* 17 (4): 3598–3609.

162 Medintz, I.L., Uyeda, H.T., Goldman, E.R., and Mattoussi, H. (2005). Quantum dot bioconjugates for imaging, labelling and sensing. *Nature Materials* 4 (6): 435–446.

163 Van Avermaet, H., Schiettecatte, P., Hinz, S. et al. (2022). Full-spectrum InP-based quantum dots with near-unity photoluminescence quantum efficiency. *ACS Nano* 16 (6): 9701–9712.

164 Resch-Genger, U., Grabolle, M., Cavaliere-Jaricot, S. et al. (2008). Quantum dots versus organic dyes as fluorescent labels. *Nature Methods* 5 (9): 763–775.

165 Khan, F. and Kim, J.H. (2019). Emission-wavelength-dependent photoluminescence decay lifetime of *N*-functionalized graphene quantum dot downconverters: impact on conversion efficiency of Cu(In, Ga)Se_2 solar cells. *Scientific Reports* 9 (1): 10803.

166 Maysinger, D., Ji, J., Hutter, E., and Cooper, E. (2015). Nanoparticle-based and bioengineered probes and sensors to detect physiological and pathological biomarkers in neural cells. *Frontiers in Neuroscience* 9: 480.

167 Song, J.Z., Li, J.H., Xu, L.M. et al. (2018). Room-temperature triple-ligand surface engineering synergistically boosts ink stability, recombination dynamics, and charge injection toward EQE-11.6% perovskite QLEDs. *Advanced Materials* 30 (30): 1800764.

168 Asil, S.M., Guerrero, E.D., Bugarini, G. et al. (2023). Theranostic applications of multifunctional carbon nanomaterials. *View* 4 (2): 20220056.

169 Chen, M.S., Zhang, H.R., Li, H. et al. (2024). C_xN_y-based materials as gas sensors: structure, performance, mechanism and perspective. *Coordination Chemistry Reviews* 503: 215653.

170 Kroke, E. and Schwarz, M. (2004). Novel group 14 nitrides. *Coordination Chemistry Reviews* 248 (5–6): 493–532.

171 Rono, N., Kibet, J.K., Martincigh, B.S., and Nyamori, V.O. (2021). A review of the current status of graphitic carbon nitride. *Critical Reviews in Solid State and Materials Sciences* 46 (3): 189–217.

172 Ahmed, M.A., Mahmoud, S.A., and Mohamed, A.A. (2024). Unveiling the photocatalytic potential of graphitic carbon nitride (g-C_3N_4): a state-of-the-art review. *RSC Advances* 14 (35): 25629–25662.

173 Roy, R., Chacko, A.R., Abraham, T. et al. (2022). Recent advances in graphitic carbon nitrides (g-C_3N_4) as photoluminescence sensing probe: a review. *ChemistrySelect* 7 (36): e202200876.

174 Zhang, L., Xia, Y., Li, X.D. et al. (2022). Janus two-dimensional transition metal dichalcogenides. *Journal of Applied Physics* 131 (23): 230902.

175 Tang, X. and Kou, L.Z. (2022). 2D Janus transition metal dichalcogenides: properties and applications. *Physica Status Solidi B* 259 (4): 2100562.

176 Torres-Cavanillas, R. and Forment-Aliaga, A. (2024). Design of stimuli-responsive transition metal dichalcogenides. *Communications Chemistry* 7 (1): 241.

177 Wang, H., Feng, J., Xing, H. et al. (2024). Progress in the use of MoS_2-based composites for microwave absorption. *Materials Science and Engineering: R: Reports* 161: 100838.

178 Liu, Y., Li, F.S., and Huang, W. (2023). Perovskite micro–/nanoarchitecture for photonic applications. *Matter* 6 (10): 3165–3219.

179 Kamat, P.V. (2012). TiO_2 nanostructures: recent physical chemistry advances. *The Journal of Physical Chemistry C* 116 (22): 11849–11851.

180 Zhang, G., Miao, H., Hu, X. et al. (2017). A facile strategy to fabricate Au/TiO_2 nanotubes photoelectrode with excellent photoelectrocatalytic properties. *Applied Surface Science* 391: 345–352.

181 Zhang, J., Li, Z., Qi, X.L., and Wang, D.Y. (2020). Recent progress on metal-organic framework and its derivatives as novel fire retardants to polymeric materials. *Nano-Micro Letters* 12 (1): 173.

182 Moghadam, P.Z., Li, A., Wiggin, S.B. et al. (2017). Development of a Cambridge structural database subset: a collection of metal-organic frameworks for past, present, and future. *Chemistry of Materials* 29 (7): 2618–2625.

183 Wu, X.P., Gagliardi, L., and Truhlar, D.G. (2018). Cerium metal-organic framework for photocatalysis. *Journal of the American Chemical Society* 140 (25): 7904–7912.

184 Nie, X.L., Wu, S.L., Mensah, A. et al. (2020). Insight into light-driven antibacterial cotton fabrics decorated by in situ growth strategy. *Journal of Colloid and Interface Science* 579: 233–242.

185 Zhang, Y.M., Pang, J.D., Li, J.L. et al. (2019). Visible-light harvesting pyrene-based MOFs as efficient ROS generators. *Chemical Science* 10 (36): 8455–8460.

186 Wang, D.K., Huang, R.K., Liu, W.J. et al. (2014). Fe-based MOFs for photocatalytic CO_2 reduction: role of coordination unsaturated sites and dual excitation pathways. *ACS Catalysis* 4 (12): 4254–4260.

187 Silva, C.G., Luz, I., Xamena, F. et al. (2010). Water stable Zr-benzenedicarboxylate metal-organic frameworks as photocatalysts for hydrogen generation. *Chemistry – A European Journal* 16 (36): 11133–11138.

188 Fu, Y.H., Sun, D.R., Chen, Y.J. et al. (2012). An amine-functionalized titanium metal-organic framework photocatalyst with visible-light-induced activity for CO_2 reduction. *Angewandte Chemie International Edition* 51 (14): 3364–3367.

189 Xu, H.Q., Yang, S.Z., Ma, X. et al. (2018). Unveiling charge-separation dynamics in CdS/metal-organic framework composites for enhanced photocatalysis. *ACS Catalysis* 8 (12): 11615–11621.

190 Xu, X.Q., Liu, R.X., Cui, Y.H. et al. (2017). PANI/FeUiO-66 nanohybrids with enhanced visible-light promoted photocatalytic activity for the selectively aerobic oxidation of aromatic alcohols. *Applied Catalysis B: Environmental* 210: 484–494.

191 Chen, S.Y., Hai, G.T., Gao, H.Y. et al. (2021). Modulation of the charge transfer behavior of Ni(II)-doped NH_2-MIL-125(Ti): regulation of Ni ions content and enhanced photocatalytic CO_2 reduction performance. *Chemical Engineering Journal* 406: 126886.

192 Xu, C.Y., Pan, Y.T., Wan, G. et al. (2019). Turning on visible-light photocatalytic C-H oxidation over metal-organic frameworks by introducing metal-to-cluster charge transfer. *Journal of the American Chemical Society* 141 (48): 19110–19117.

193 Dong, L.Z., Zhang, L., Liu, J. et al. (2020). Stable heterometallic cluster-based organic framework catalysts for artificial photosynthesis. *Angewandte Chemie International Edition* 59 (7): 2659–2663.

194 Chen, M., Long, Z., Dong, R. et al. (2020). Titanium incorporation into Zr-porphyrinic metal–organic frameworks with enhanced antibacterial activity against multidrug-resistant pathogens. *Small* 16 (7): 1906240.

195 Gao, Y.J., Zhang, L., Gu, Y.M. et al. (2020). Formation of a mixed-valence Cu(I)/Cu(II) metal-organic framework with the full light spectrum and high selectivity of CO_2 photoreduction into CH_4. *Chemical Science* 11 (37): 10143–10148.

196 Guo, J.D., Jia, H.S., Zhang, A.Q. et al. (2021). MIL-100 (Fe) with mix-valence coordinatively unsaturated metal site as Fenton-like catalyst for efficiently removing tetracycline hydrochloride: boosting Fe(III)/Fe(II) cycle by photoreduction. *Separation and Purification Technology* 262: 118334.

197 Gong, Y.N., Mei, J.H., Liu, J.W. et al. (2021). Manipulating metal oxidation state over ultrastable metal-organic frameworks for boosting photocatalysis. *Applied Catalysis B: Environmental* 292: 120156.

198 Lin, C.K., Zhao, D., Gao, W.Y. et al. (2012). Tunability of band gaps in metal-organic frameworks. *Inorganic Chemistry* 51 (16): 9039–9044.

199 Ma, X., Wang, L., Zhang, Q., and Jiang, H.L. (2019). Switching on the photocatalysis of metal-organic frameworks by engineering structural defects. *Angewandte Chemie International Edition* 58 (35): 12175–12179.

200 Yuan, S., Qin, J.S., Zou, L.F. et al. (2016). Thermodynamically guided synthesis of mixed-linker Zr-MOFs with enhanced tunability. *Journal of the American Chemical Society* 138 (20): 6636–6642.

201 Wan, L.B., Liu, H., Huang, C.X., and Shen, X.T. (2020). Enzyme-like MOFs: synthetic molecular receptors with high binding capacity and their application in selective photocatalysis. *Journal of Materials Chemistry A* 8 (48): 25931–25940.

202 Xiang, W.L., Ren, J., Chen, S. et al. (2020). The metal-organic framework UiO-66 with missing-linker defects: a highly active catalyst for carbon dioxide cycloaddition. *Applied Energy* 277: 115560.

203 Taddei, M., Schukraft, G.M., Warwick, M.E.A. et al. (2019). Band gap modulation in zirconium-based metal-organic frameworks by defect engineering. *Journal of Materials Chemistry A* 7 (41): 23781–23786.

204 Zhang, B.X., Zhang, J.L., Tan, X.N. et al. (2018). MIL-125-NH_2@TiO_2 core-shell particles produced by a post-solvothermal route for high-performance photocatalytic H_2 production. *ACS Applied Materials & Interfaces* 10 (19): 16418–16423.

205 Guo, F., Guo, J.H., Wang, P. et al. (2019). Facet-dependent photocatalytic hydrogen production of metal-organic framework NH_2-MIL-125(Ti). *Chemical Science* 10 (18): 4834–4838.

206 Cheng, X.M., Dao, X.Y., Wang, S.Q. et al. (2021). Enhanced photocatalytic CO_2 reduction activity over NH_2-MIL-125(Ti) by facet regulation. *ACS Catalysis* 11 (2): 650–658.

207 Yang, W., Wang, H.J., Liu, R.R. et al. (2021). Tailoring crystal facets of metal-organic layers to enhance photocatalytic activity for CO_2 reduction. *Angewandte Chemie International Edition* 60 (1): 409–414.

208 He, T., Ni, B., Zhang, S.M. et al. (2018). Ultrathin 2D zirconium metal-organic framework nanosheets: preparation and application in photocatalysis. *Small* 14 (16): e1703929.

209 Zheng, H.L., Huang, S.L., Luo, M.B. et al. (2020). Photochemical in situ exfoliation of metal-organic frameworks for enhanced visible-light-driven CO_2 reduction. *Angewandte Chemie International Edition* 59 (52): 23588–23592.

210 Geng, K.Y., He, T., Liu, R.Y. et al. (2020). Covalent organic frameworks: design, synthesis, and functions. *Chemical Reviews* 120 (16): 8814–8933.

211 Zhang, Z.W., Jia, J., Zhi, Y.F. et al. (2022). Porous organic polymers for light-driven organic transformations. *Chemical Society Reviews* 51 (7): 2444–2490.

212 El-Kaderi, H.M., Hunt, J.R., Mendoza-Cortés, J.L. et al. (2007). Designed synthesis of 3D covalent organic frameworks. *Science* 316 (5822): 268–272.

213 Liu, M.Y., Huang, Q., Wang, S.L. et al. (2018). Crystalline covalent triazine frameworks by in situ oxidation of alcohols to aldehyde monomers. *Angewandte Chemie International Edition* 57 (37): 11968–11972.

214 Ding, S.Y., Gao, J., Wang, Q. et al. (2011). Construction of covalent organic framework for catalysis: Pd/COF-LZU1 in Suzuki-Miyaura coupling reaction. *Journal of the American Chemical Society* 133 (49): 19816–19822.

215 Uribe-Romo, F.J., Doonan, C.J., Furukawa, H. et al. (2011). Crystalline covalent organic frameworks with hydrazone linkages. *Journal of the American Chemical Society* 133 (30): 11478–11481.

216 Vyas, V.S., Haase, F., Stegbauer, L. et al. (2015). A tunable azine covalent organic framework platform for visible light-induced hydrogen generation. *Nature Communications* 6: 8508.

217 Chen, M.H., Xiong, J., Li, X.Y. et al. (2023). In-situ doping strategy for improving the photocatalytic hydrogen evolution performance of covalent triazine frameworks. *Science China: Chemistry* 66 (8): 2363–2370.

218 Yang, S.Z., Hu, W.H., Zhang, X. et al. (2018). 2D covalent organic frameworks as intrinsic photocatalysts for visible light-driven CO_2 reduction. *Journal of the American Chemical Society* 140 (44): 14614–14618.

219 Hu, H.J., Zhang, X.Y., Zhang, K.L. et al. (2024). Construction of a 2D/2D crystalline porous materials based S-scheme heterojunction for efficient photocatalytic H_2 production. *Advanced Energy Materials* 14 (11): 2303638.

220 Swainsbury, D.J.K., Qian, P., Jackson, P.J. et al. (2021). Structures of *Rhodopseudomonas palustris* RC-LH1 complexes with open or closed quinone channels. *Science Advances* 7 (3): eabe2631.

221 Cherezov, V., Clogston, J., Papiz, M.Z., and Caffrey, M. (2006). Room to move: crystallizing membrane proteins in swollen lipidic mesophases. *Journal of Molecular Biology* 357 (5): 1605–1618.

222 Amunts, A. and Nelson, N. (2009). Plant photosystem I design in the light of evolution. *Structure* 17 (5): 637–650.

223 Barros, T. and Kühlbrandt, W. (2009). Crystallisation, structure and function of plant light harvesting complex II. *Biochimica et Biophysica Acta-Bioenergetics* 1787 (6): 753–772.

224 Liu, Z.F., Yan, H.C., Wang, K.B. et al. (2004). Crystal structure of spinach major light-harvesting complex at 2.72 Å resolution. *Nature* 428 (6980): 287–292.

225 Qin, X.C., Suga, M., Kuang, T.Y., and Shen, J.R. (2015). Structural basis for energy transfer pathways in the plant PSI-LHCI supercomplex. *Science* 348 (6238): 989–995.

226 Jia, X.Y., Mei, Y., JZH, Z., and Mo, Y. (2015). Hybrid QM/MM study of FMO complex with polarized protein-specific charge. *Scientific Reports* 5: 17096.

227 Tsien, R.Y. (1998). The green fluorescent protein. *Annual Review of Biochemistry* 67: 509–544.

228 Kong, J., Wang, Y., Qi, W. et al. (2020). Green fluorescent protein inspired fluorophores. *Advances in Colloid and Interface Science* 285: 102286.

229 Walker, C.L., Lukyanov, K.A., Yampolsky, I.V. et al. (2015). Fluorescence imaging using synthetic GFP chromophores. *Current Opinion in Chemical Biology* 27: 64–74.

230 Chen, H., Liu, L.J., Qian, K. et al. (2022). Bioinspired large stokes shift small molecular dyes for biomedical fluorescence imaging. *Science Advances* 8 (32): eabo3289.

231 Day, R.N. and Davidson, M.W. (2009). The fluorescent protein palette: tools for cellular imaging. *Chemical Society Reviews* 38 (10): 2887–2921.

232 Baranov, M.S., Lukyanov, K.A., Borissova, A.O. et al. (2012). Conformationally locked chromophores as models of excited-state proton transfer in fluorescent proteins. *Journal of the American Chemical Society* 134 (13): 6025–6032.

233 Baldridge, A., Samanta, S.R., Jayaraj, N. et al. (2011). Steric and electronic effects in capsule-confined green fluorescent protein chromophores. *Journal of the American Chemical Society* 133 (4): 712–715.

234 Chen, C., Baranov, M.S., Zhu, L.D. et al. (2019). Designing redder and brighter fluorophores by synergistic tuning of ground and excited states. *Chemical Communications* 55 (17): 2537–2540.

235 Williams, D.E., Dolgopolova, E.A., Pellechia, P.J. et al. (2015). Mimic of the green fluorescent protein β-barrel: photophysics and dynamics of confined chromophores defined by a rigid porous scaffold. *Journal of the American Chemical Society* 137 (6): 2223–2226.

236 Li, X., Wu, J.H., and Jaffrey, S.R. (2021). Engineering fluorophore recycling in a fluorogenic RNA aptamer. *Angewandte Chemie International Edition* 60 (45): 24153–24161.

237 Chung, Y.H. and Oh, J.K. (2024). Research trends in the development of block copolymer-based biosensing platforms. *Biosensors* 14 (11): 542.

238 Kumar, L., Horechyy, A., Bittrich, E. et al. (2019). Amphiphilic block copolymer micelles in selective solvents: the effect of solvent selectivity on micelle formation. *Polymers* 11 (11): 1882.

239 Lee, H.I., Wu, W., Oh, J.K. et al. (2007). Light-induced reversible formation of polymeric micelles. *Angewandte Chemie International Edition* 46 (14): 2453–2457.

240 Jiang, Y.G., Wang, Y.P., Ma, N. et al. (2007). Reversible self-organization of a UV-responsive PEG-terminated malachite green derivative: vesicle formation and photoinduced disassembly. *Langmuir* 23 (7): 4029–4034.

241 Zou, J., Guan, B., Liao, X.J. et al. (2009). Dual reversible self-assembly of PNIPAM-based amphiphiles formed by inclusion complexation. *Macromolecules* 42 (19): 7465–7473.

242 Newkome, G.R. and Shreiner, C. (2010). Dendrimers derived from 1 → 3 branching motifs. *Chemical Reviews* 110 (10): 6338–6442.

243 Nishiyama, N., Iriyama, A., Jang, W.-D. et al. (2005). Light-induced gene transfer from packaged DNA enveloped in a dendrimeric photosensitizer. *Nature Materials* 4 (12): 934–941.

244 Gao, M., Wang, B., Jia, X. et al. (2008). Photo and H^+ dual-responsive behaviors of azobenzene-shelled poly(aminoamine) dendrimers. *Acta Polymerica Sinica* 1: 32–40.

3

Construction of Photofunctional Polymer Composites

Photofunctional materials possess the properties of luminescence, photocatalysis, photovoltaic conversion, photothermal conversion, and photosensitization by absorbing or converting light energy. The optical properties of these materials can be optimized through customized modulation of their composition and structure, and thus they are widely used in biomedical fields such as biochemical sensing, bioimaging, disease diagnosis, and optical therapy. However, conventional materials are susceptible to aggregation or decomposition in complex physiological environments, resulting in the degradation or even loss of optical properties and possible biotoxicity. In addition, the optical functions provided by a single material can hardly meet the increasingly diversified needs of biological applications, which limits its further development in the biological field.

The systems based on photofunctional polymer composites offer a new strategy to address the limitations of traditional materials by integrating the complementary properties of polymer materials and photofunctional components, combining excellent biocompatibility and unique optical functions. As one of the most widely used materials in the biomedical field, the spatial protection of polymers with long molecular chains can significantly reduce the loss of active groups in organisms. The abundant organic functional groups (e.g. charged groups, hydrophobic regions, hydrogen-bonding sites) of polymers can establish flexible and strong connections with biological interfaces, thereby maximizing the application value of photofunctional materials in disease diagnosis and treatment. The structural diversity and easy modification of the polymers support efficient complexation with various organic/inorganic photomaterials through physical/chemical interactions.

3.1 Construction of Photofunctional Organic Polymer Composite Systems

Photofunctional organic polymer composite systems consist of polymer molecules and organic photofunctional materials. The components of the composite system are organic molecules with good compatibility with each other, which can form a stable and strong bond through a variety of physical or chemical interactions.

Photofunctional Polymer Composites for Bioapplications, First Edition. Xiaoyu Wang and Lidong Li.

Organic optical functional materials are characterized by a clear structure and easy modification, and their optical properties can be regulated by designing the molecular structure, which is widely used in chemical detection and bioimaging. Organic photofunctional materials used in the construction of polymer composite systems for biological applications usually contain specific color-emitting groups and exhibit certain conjugated structures, which mainly include organic fluorescent small molecules [1], conjugated polymers [2], and covalent organic frameworks (COFs).

The construction methods of photofunctional organic polymer composite systems mainly include coprecipitation, self-assembly, microemulsion, direct polymerization, covalent modification, and physical mixing.

3.1.1 Coprecipitation

Coprecipitation is a method of self-aggregation to form nanoparticles by taking advantage of the different solubility of the same substance in different solvents. In this process, photoactive monomers or photofunctional materials and polymers are usually dissolved in a solvent to form photofunctional complexes in solution by coprecipitation.

Conjugated polymers (CPs) are characterized by their π-conjugated, hydrophobic backbones. By precisely controlling the experimental conditions, CP nanoparticles of different sizes, which show high and stable fluorescence and limited scintillation, can be prepared by coprecipitation. Further encapsulation can provide these CP nanoparticles with functional surfaces for biocoupling and specific target detection.

The strategy of using polymer molecules with good water solubility and biocompatibility to encapsulate conjugated polymers in composite nanoparticles can effectively improve the optical properties of conjugated polymers in aqueous solution. For example, the amphiphilic polymers polystyrene polyethylene glycol (PS-PEG) and poly(styrene-*co*-maleic anhydride) (PSMA) wrapped around the conjugated polymer poly[(9,9-dioctylfluorenyl-2,7-diyl)-co-(1,4-benzo-{2,1′,3}-thiadiazole)] (PFBT) using the coprecipitation method, and composite nanoparticles that can be uniformly and stably dispersed in aqueous solution can be prepared.

Polystyrene-poly(ethylene glycol)-carboxyl (PS-PEG-COOH), for example, is composed of a hydrophobic polystyrene backbone and multiple hydrophilic ethylene oxide side chains terminated with carboxyl groups. During the formation of nanoparticles, the hydrophobic segments tend to be embedded within the nanoparticles, while the hydrophilic segments extend outward into the surrounding aqueous environment. Unlike physical adsorption, this method ensures a permanent anchoring of the PEG and carboxyl groups to the surface of conjugated polymer dots. The PEG chains create a biocompatible layer that reduces nonspecific interactions and serve as spatial barriers to prevent nanoparticle aggregation. Additionally, the functional carboxyl groups can be readily covalently linked to biomolecules using the EDC/NHS reaction (Figure 3.1) [3].

PSMA polymers can also be used to produce carboxyl-functionalized polymer dots [4]. Polymer dot formation is primarily driven by hydrophobic interactions,

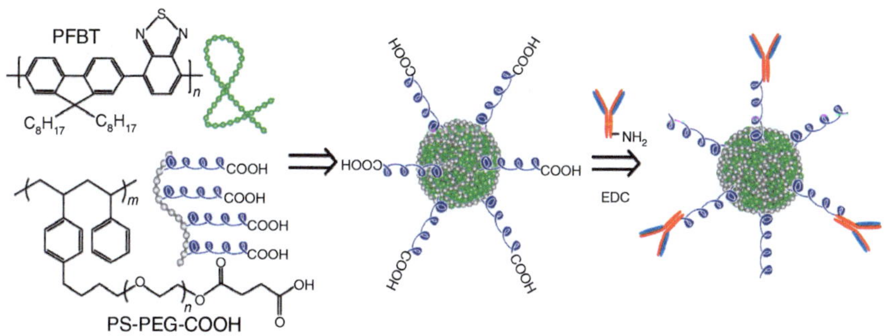

Figure 3.1 Surface functionalization of conjugated polymer dots. *Source:* Wu et al. [3] / with permission of American Chemical Society.

allowing amphiphilic polymers with hydrophilic groups to co-condense into individual polymer dots during nanoparticle synthesis. The hydrophilic groups on these amphiphilic polymers serve as useful handles for functionalizing polymer dots, enabling their coupling with biomolecules. The general-purpose copolymer PSMA was utilized to functionalize the conjugated polymer PFBT, resulting in the formation of polymer dots. During the polymer dot formation process, the hydrophobic polystyrene component of PSMA is likely embedded within the polymer dots, while the maleic anhydride groups are positioned on the surface. In an aqueous environment, these maleic anhydride groups undergo hydrolysis, producing carboxyl groups on the polymer dot surface. These carboxyl groups enable further surface coupling reactions.

Coprecipitation of composite nanoparticles prepared by coprecipitation of two conjugated polymers, poly(diketopyrrolopyrrole-thienothiophene) (PDPPTT), which is photothermal, and poly[2-methoxy-5-((2-ethylhexyl)oxy)-p-phenylenevinylene] (MEH-PPV), which is a photosensitizer, simultaneously encapsulated in PSMA (Figure 3.2) [5]. The nanoparticles are dispersed in water through hydrophobic interactions, avoiding the aggregation of photothermal and

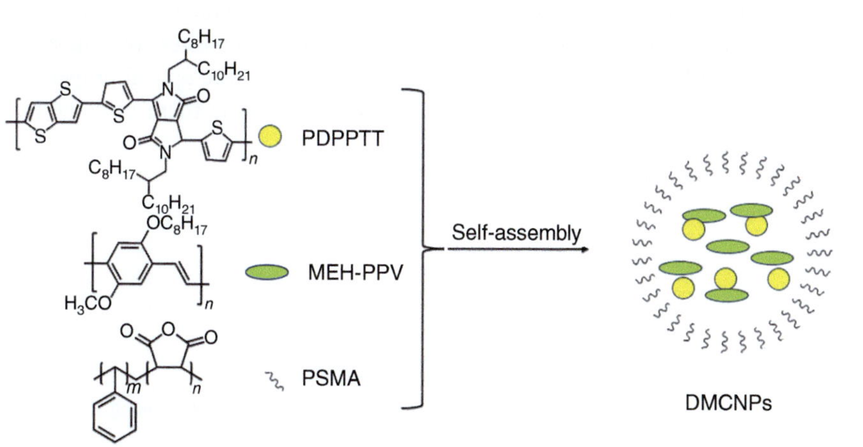

Figure 3.2 Schematic representation of the DMCNPs preparation.

photosensitizer, which is consistent with the nanoparticles' simultaneous photothermal effect and ability to sensitize the surrounding oxygen to produce reactive oxygen species (ROS) upon irradiation, and the inhibition rate of *E. coli* reaches 93%. Compared with the complicated process of water-soluble side-chain modification, the preparation of water-soluble nanoparticles by wrapping conjugated polymers in polymers with hydrophilic groups is simpler and easier, which greatly expands the scope of conjugated polymers in the field of biological applications.

The extent of hydrophilic functionalization is critical in determining the stability and fluorescent properties of nanoparticles, necessitating a low density of hydrophilic groups. To form nanoparticles, the PFBT-NH_2 polymer, containing side-chain amine groups, was cross-linked with the polymer PIMA through coprecipitation [6]. PIMA provides covalent cross-links with semiconductor polymers while introducing carboxyl groups for biocoupling. This approach yielded stable, carboxyl-functionalized PFBT nanoparticles with a smaller size, as depicted in Figure 3.3.

Photofunctional small molecules are organic small molecule materials that can respond to light and exhibit specific optical properties, and are commonly used in the fields of light absorption, luminescence, and photovoltaic conversion. They usually have small molecular weights and can exhibit properties such as fluorescence, photocatalysis, or photosensitive reaction under different light conditions. In the coprecipitation method, photofunctional small molecules are usually involved in the composite process of materials as functional ligands or additives. By forming coordination complexes with metal ions, photofunctional small molecules coprecipitate with metal ions or other components during the precipitation process to form composite materials. These composites usually have a high degree of structural homogeneity, and the photofunctional small molecules can significantly affect the optical properties of the final material, such as enhancing fluorescence intensity and optimizing photocatalytic performance. The coprecipitation method is not only simple and low-cost, but also can effectively control the morphology, size, and dispersion of the materials, which is an important method to prepare the photofunctional small molecule composites. By adjusting the reaction conditions, such as pH value and solvent type, the optical properties of the resulting materials can be precisely controlled, making them valuable for various photofunctional applications.

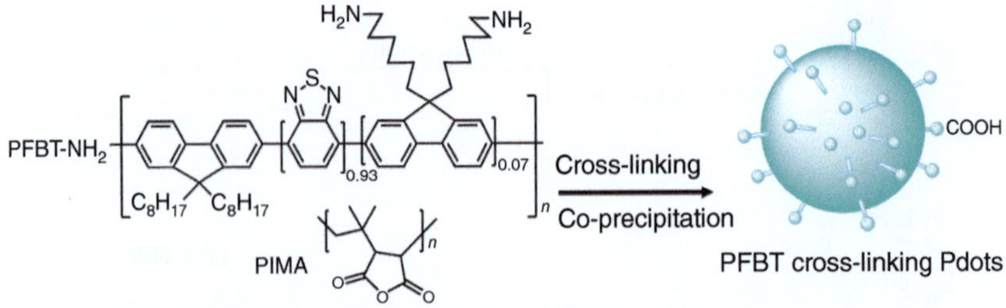

Figure 3.3 Schematic representation of the preparation of PFBT-NH-PIMA dots.

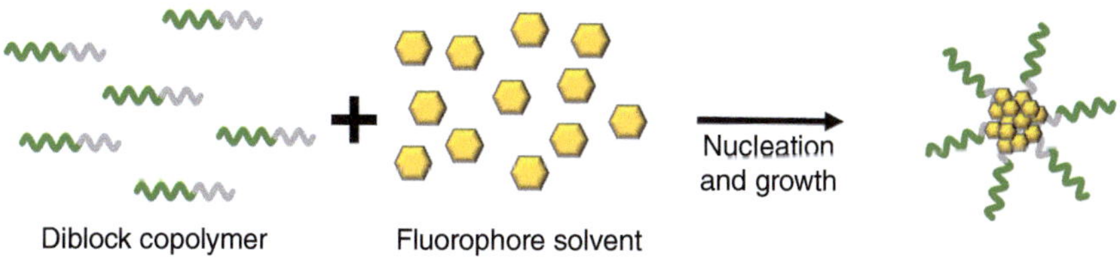

Figure 3.4 Schematic representation of the flash nanoprecipitation method to form nanoparticles.

Hydrophobically or hydrophobically modified fluorescent dyes (Nile Red, coumarin, pyrene, etc.) are encapsulated in amphiphilic block copolymer nanoparticles, with fluorescence emission peaks covering 370–720 nm, and the composite fluorescent nanoparticles are more stable and do not photo-bleach as compared to free dyes [7]. Here flash nanoprecipitation (FNP), a coprecipitation method, was employed to synthesize nanoparticles with high quantum yields, tailored excitation wavelengths, surface functionalization, and controlled particle sizes. In this method, known as FNP, hydrophobic fluorophores and amphiphilic polymers are dissolved in a water-miscible organic phase. These components are rapidly mixed with aqueous counter-solvent streams, inducing precipitation of reactive substances (Figure 3.4). The resulting nanoparticles exhibit a tunable, narrow size distribution ranging from 30 to 800 nm. Successful nanoparticle formation requires two key conditions. First, high supersaturation mixing must occur within milliseconds to initiate precipitation. Second, the adsorption time scale of the block copolymer must align with the solute precipitation rate. The polymer assembles quickly to form a micelle, depleting the stabilizing polymer concentration and compromising particle stability. Conversely, if polymer assembly is too slow, nanoparticles may grow excessively before sufficient polymer adsorption halts further growth, resulting in oversized particles. The hydrophilic portion of the block copolymer protects the nanoparticles from aggregation and spatially stabilizes them.

Ultrasmall nanoparticles prepared by a charge-controlled nanoprecipitation method can be efficiently loaded with dyes and further stabilized by surfactants. In detail, by introducing negatively charged carboxyl groups, sulfonic acid groups, or positively charged trimethylammonium groups into the polymer molecular chain, ultra-small polymer nanoparticles (20 nm) can be prepared and effectively encapsulate ionic dyes, with a fluorescence intensity up to 10 times brighter than that of quantum dots of the same size. The method is universally applicable to polylactic acid-glycolic acid (PLGA), PCL, and PMMA, as shown in Figure 3.5, and reducing the concentration of the polymers can further lead to particles of 15 nm [8].

Encapsulation of hydrophobic aggregation-induced luminescent dye EDP into poly(ethylene glycol)-*b*-poly(caprolactone) block copolymer nanoparticles resulted in nanoparticles with high fluorescence brightness. Here, the FNP method was also employed to prepare nanoparticles in 1 s using a multi-inlet vortex mixer (MIVM) system [9].

In the FNP process, the organic solute and amphiphilic block copolymer are mixed in a water-miscible organic solvent. This solution is then rapidly mixed

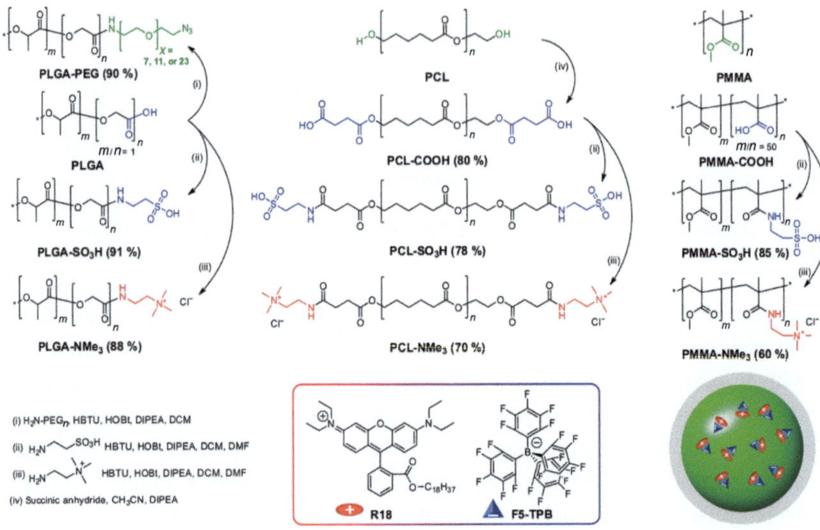

Figure 3.5 The synthesis of chemically modified polymers—PLGA-SO_3H, PLGA-NMe_3, PLGA-PEG, PCL-SO_3H, PCL-NMe_3, PMMA-SO_3H, and PMMA-NMe_3—along with the preparation of dye-loaded nanoparticles (R18, F5-TPB). *Source:* Reisch et al. [8] / with permission of American Chemical Society.

with a counter-solvent, typically water, in the device, resulting in the formation of nanoparticles encapsulated by the organic solute (Figure 3.6). FNP produces oriented or engineered assemblies that are driven by external force. The rapid change in solvent polarity during the process governs the kinetics of nanoparticle assembly. By adjusting the solvent ratio and flow rate, the size of nanoparticles can be precisely controlled while maintaining their stability. Moreover, this method is easily scalable, offering significant advantages for the controlled fabrication of nanoparticles with exceptional optical properties.

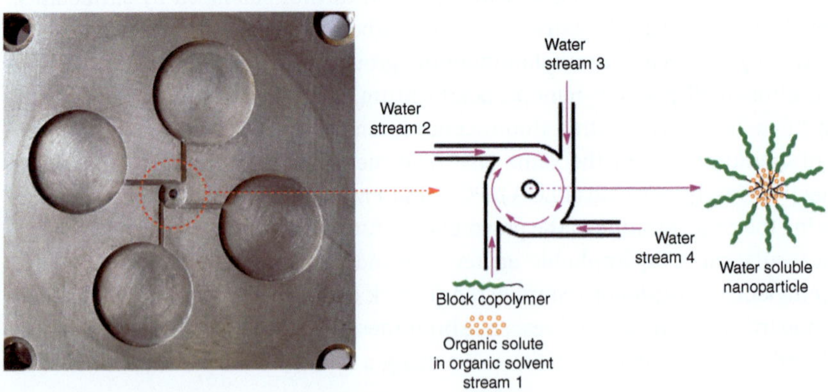

Figure 3.6 Schematic illustration of MIVM system and FNP method. *Source:* Wang et al. [9] / with permission of American Chemical Society.

In the coprecipitation method, photofunctional conjugated polymers and photofunctional small molecules are mixed and precipitated together with metal ions or other components in solution to form composites. This process can effectively enhance the light emission efficiency, absorption range, and stability of the composites. This is achieved by adjusting the reaction conditions (e.g. solvent, pH, and temperature) and leveraging the electron transfer or synergistic luminescence effect between the photofunctional conjugated polymer and the small molecules. The advantage of this method is that it can closely combine the conjugated polymer with the photofunctional small molecules, give full play to the synergistic effect of the two, and improve the photoelectric properties of the material.

Compared with designing new molecular structures, it is easier and more efficient to construct composite systems to modulate the optical properties of materials through intermolecular energy transfer or electron transfer. For example, using PEG-coated conjugated polymer poly[2,6-(4,4-bis(2-ethylhexyl)-4H-cyclopenta[2,1-b;3,4-b′]dithiophene)-alt-4,7-(2,1,3-benzothiadiazole)] (PCPDTBT) and fullerene material PC70BM with strong electron-absorbing ability, composite nanoparticles can be prepared by coprecipitation to modulate the optical properties of the conjugated polymer, as shown in Figure 3.7 [10]. By selecting energy level-matched organic molecules, the highest occupied orbital energy level (HOMO) and the lowest unoccupied orbital energy level (LUMO) of the electrons of the host material PCPDTBT are higher than those of the dopant unit PC70BM, and an intermolecular photoluminescent electron transfer system is constructed in the closed nanostructure formed by PEG encapsulation. Under the excitation light, the fluorescence quenching but nonradiative transition of the conjugated polymer is significantly enhanced, which ultimately results in enhanced photothermal conversion efficiency and photoacoustic imaging signals.

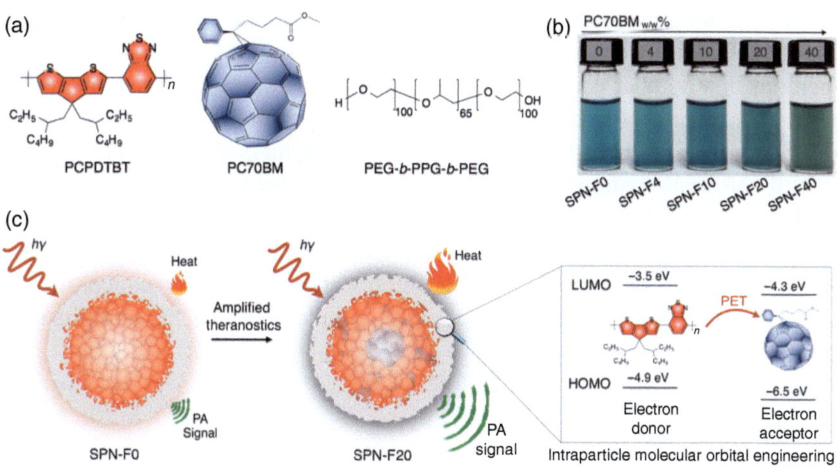

Figure 3.7 (a) The structures of PCPDTBT, PC70BM, and PEG-*b*-PPG-*b*-PEG. (b) The solution of different SPNs. (c) Diagram illustrating amplification theranostics of SPN with PET process. *Source:* Lyu et al. [10] / with permission of American Chemical Society.

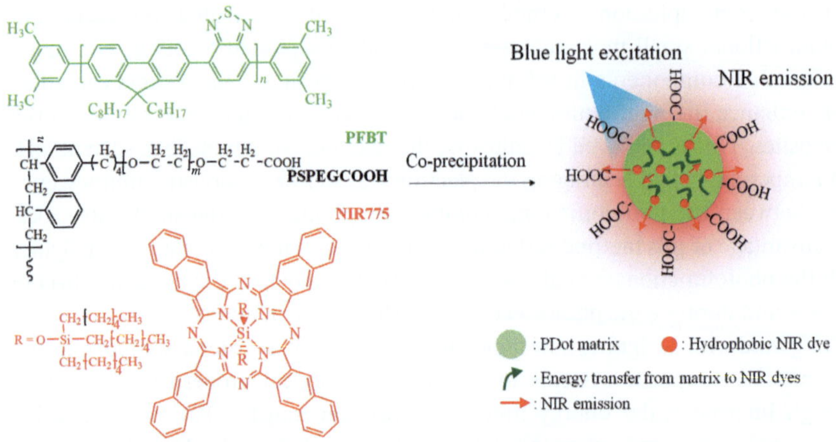

Figure 3.8 Preparation of NIR dye-doped polymer dots using a precipitation method. *Source:* Jin et al. [11] / with permission of American Chemical Society.

Doping of near-infrared dye NIR775 in 9,9-dioctylfluorene-*co*-benzothiadiazole (PFBT) nanoparticles resulted in a significant enhancement of the fluorescence brightness of NIR775 due to the strong light-absorbing capacity of PFBT and fluorescence resonance energy transfer (FRET), which was fourfold brighter compared with the same-sized near-infrared quantum dot ITK Qdot800. NIR775 was incorporated into PFBT dots through a nanoprecipitation technique. PFBT serves as the hydrophobic core of the polymer dots and functions as a host for the NIR dye. Simultaneously, during the synthesis process, the surface of the polymer dots was carboxyl-functionalized by blending the PFBT matrix with PS-PEG-COOH with functional carboxyl groups, as illustrated in Figure 3.8. This modification not only provides a hydrophilic surface but also introduces a wealth of functional groups, enabling further bioconjugation of the polymer dots [11].

There are many other examples, such as the temperature-responsive dye rhodamine B covalently attached to amine-capped amphiphilic polystyrene ends, mixed and stirred with hydrophobic conjugated polymers PFBT or PFPV, and Pdot-RhB composite nanoparticles can be simply obtained by nanoprecipitation [12]. PFBT/PFPV with RhB has up to 60% of the FRET efficiency. In addition, since the emission of Pdot is insensitive to temperature, whereas the emission of RhB is highly dependent on temperature, the fluorescence intensity of nanoparticles under single-wavelength excitation, $I_{573\,\text{nm}}/I_{510\,\text{nm}}$ ratio, varies highly and linearly with the temperature over 10–70 °C ($R^2 = 0.999$).

3.1.2 Self-assembly Method

Self-assembly method is a technique of spontaneous ordered aggregation of molecules driven by noncovalent interactions (e.g. electrostatic interactions, host–guest recognition, and π–π stacking). The core of the method is to utilize autonomous interactions between components to achieve controlled assembly of materials without external intervention, and ultimately form multilevel structures with specific sizes (nanoparticles, nanowires, etc.) and morphologies (thin films, three-dimensional networks, etc.). The method allows for an aqueous-based reaction system that confers excellent hydrophilicity and stability to the products. Moreover, the fluorophores can be directly embedded into the assemblies through co-assembly or premodification strategies to meet the needs of biological applications. The intermolecular forces can be precisely regulated by adjusting the solution concentration, temperature, pH, etc., to tailor the morphology and properties of the structures. The ordered structures constructed by this method at the nanoscale can not only enhance the optoelectronic properties of the materials but also provide a molecular-level regulatory space for their functionalization (e.g. biosensing and drug-delivering systems), which is an ideal strategy for constructing photofunctional organic polymer composite systems.

Utilizing electrostatic interactions between oppositely charged molecules is one of the strategies to construct composite systems. Modification of bisimidazolium salt groups in the polymer side chain yields positively charged PF-DBT-BIMEG fluorescent polymers [13]. Due to electrostatic interactions, the polymer can self-assemble with negatively charged ATP to form nanoparticles, and the higher the content of ATP, the stronger the electrostatic interaction and the stronger the polymer aggregation. The polymer backbone structure in this nano-aggregate can undergo effective energy transfer and fluorescence color change, which can efficiently detect the content of ATP molecules.

The surface of biological systems is rich in functional groups like carboxyl and hydroxyl groups, which usually show stable negative electrical properties. This property provides an important basis for the assembly of photofunctional materials. Through intermolecular interaction forces such as hydrogen bonding and electrostatic interactions, biomolecules or their structural units can spontaneously form ordered structures. These ordered structures can be integrated into composite functional materials with light-responsive properties. Self-assembly to build composites can be performed using biotin and avidin systems, the steps of which are shown in Figure 3.9 [14]. This results in a biofunctionalized surface. Among other things, PFPE provides excellent resistance to contamination, prevents nonspecific biomolecule adsorption, and enhances the binding selectivity of the final biomolecule array. The strong specific binding of the biotin-avidin enables precise and spatial assembly of biomolecules on the surface. The composites constructed by this method possess excellent selectivity and optical sensing capabilities.

The important principle of the host–guest self-assembly method is the selective binding of a host molecule (e.g. cucurbituril) with a specific cavity structure to a guest molecule with a complementary structure. The method relies on three main features: (1) binding of the host and guest molecules driven by hydrophobic interactions, (2) structure/size matching as a molecular recognition signal, and (3) reversibility of noncovalent interactions. This dynamic binding can spontaneously form functionally controllable composite systems [15].

As an example, cucurbit[7]uril (CB[7]) has a barrel cavity consisting of glycouril units, and the hydrophobic inner cavity can specifically encapsulate the quaternary

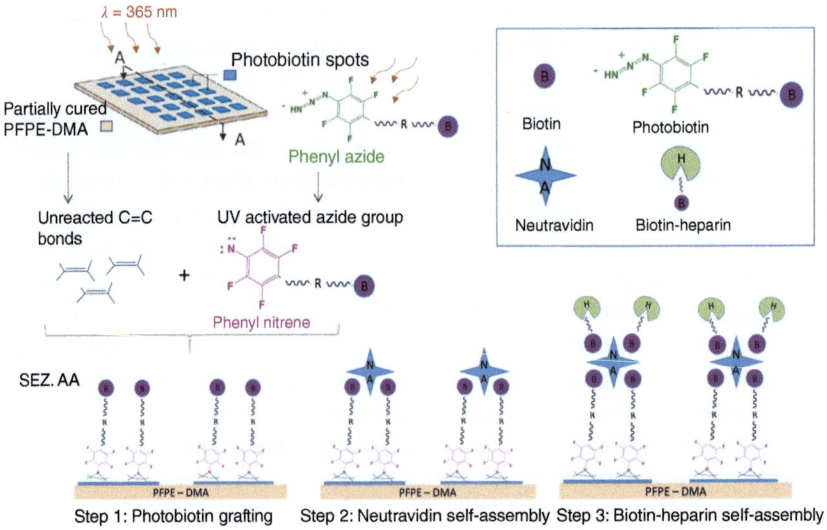

Figure 3.9 Schematic illustration of the three-step strategy for self-assembly of biomolecules on PFPE-DMAs substrates. *Source:* Credi et al. [14] / with permission of Elsevier.

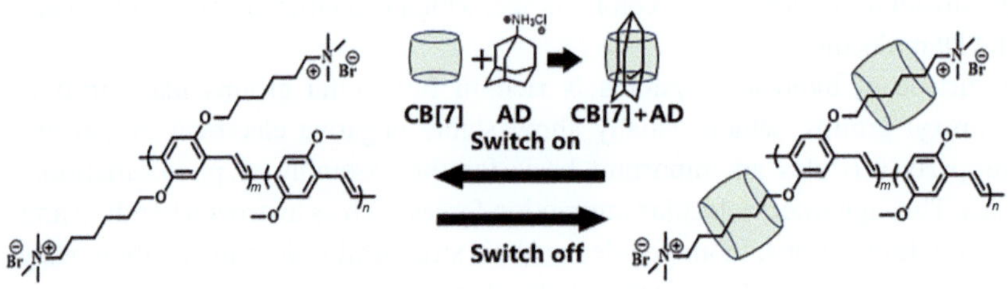

Figure 3.10 Supramolecular assembly and disassembly process of conjugated polymer PPV with CB[7].

ammonium side chain of the cationic conjugated polymer PPV (Figure 3.10). This binding has a dual inhibitory effect: (1) spatial blockage prevents PPV from binding to bacteria; and (2) restricts the ability of the PPV backbone to produce ROS under white light irradiation, which puts the system in an "antimicrobial shutdown" state. When a competing guest molecule, adamantane (AD), is introduced, the higher binding constant of AD and CB[7] triggers a host–guest reorganization: CB[7] dissociates from the PPV side chain and binds preferentially to AD, which restores the bacterial binding capacity and ROS generation of PPV, and achieves the "antimicrobial switch." This molecular competition mechanism, based on binding energy difference, constructs a responsive antimicrobial switch system [16].

Based on the unique molecular recognition ability of cyclodextrin cavities, diverse poly(pseudo)rotaxane structures can be constructed through host–guest interactions. The core lies in the size-matching effect between the hydrophobic cavities of cyclodextrins and the guest molecules: when hydrophobic fragments of linear polymer chains (e.g. polyethylene glycol) enter into the cyclodextrin cavities, they form quasi-rotaxanes similar to the "molecular beads"; if they are further capped by sterically bulky groups, they are transformed into stabilized poly(rotaxanes) [17]. Such assemblies can also be expanded into more complex topological architectures: (1) poly(pseudo)rotaxane molecular pipelines formed on the basis of intermolecularly cross-linked cyclodextrins; (2) chain-like structures constructed on the basis of alternating α- and β-cyclodextrin derivatives, where the α- and β-cyclodextrins are connected to guest molecules via covalent bonds, respectively, and recognize each other's guest molecules in an alternating manner; (3) polycatenane molecular pipelines which can reversibly switch between linear and cyclic structures of polysoxanes; (4) drug carriers that can be released by endocytosis, in which the drug functional group is modified on the hydroxyl group of the cyclodextrin through a linking group; (5) cross-linked star polymers possessing polyrotaxanes with multiple arms, which can be capped by click chemistry to immobilize the cyclodextrin structure in place; and (6) biodegradable cyclodextrin-poly(pseudo) rotaxanes capsules, etc., prepared by a template approach.

As another important class of macrocyclic hosts, calix[*n*]arenes exhibit remarkable structural designability. By tuning the number of phenol units (from calix[4]arene to calix[8]arene), the cavity size can be precisely tuned [18]. For sulfonated calix[*n*]arene, the sulfonation modification of the upper edge benzene ring not only confers water solubility, but the electrostatic interaction of the sulfonic acid group with the guest molecule enhances the binding constant. Typically, for example, *p*-sulfonated calix[4]arene, their electron-rich cavities can efficiently capture organic cationic drug molecules, and through the synergistic effect of π–π stacking and hydrogen bonding, highly efficient drug-carrying nanocapsules can be constructed [19]. The dual advantages of adjustable cavity size and controllable interfacial properties make calix[*n*]arene irreplaceable in the fields of molecular sensing and drug release.

Amphiphilic molecules are the core materials for the construction of self-assembled systems. These molecules consist of two parts: a hydrophobic segment that contains the photofunctional ligand or is used to link or encapsulate the photofunctional ligand, and a hydrophilic segment that is responsible for enhancing its solubility and biocompatibility. When amphiphilic polymers with a small number of hydrophilic groups are utilized, the resulting photofunctional polymer composites typically exhibit a compact structure, characterized by a hydrophobic core and a hydrophilic surface. This approach provides significant design flexibility, allowing precise control over the morphology, size, and hierarchy by adjusting the polymer's molecular structure, functional group types, and solvent environment. For instance, as illustrated in Figure 3.11, the cationic conjugated polymer PFO can be self-assembled into solid nanoparticles through electrostatic interactions with anionic poly(L-glutamic acid). [20] Amphiphilic oligomers, where hydrophobic functional groups are covalently bonded to hydrophilic ones, often self-assemble into dense particles. In such cases, hydrophobic interactions serve as the primary driving force behind the self-assembly process. For example, amphiphilic oligomers with π-conjugated structures were synthesized by modifying gallic acid derivatives

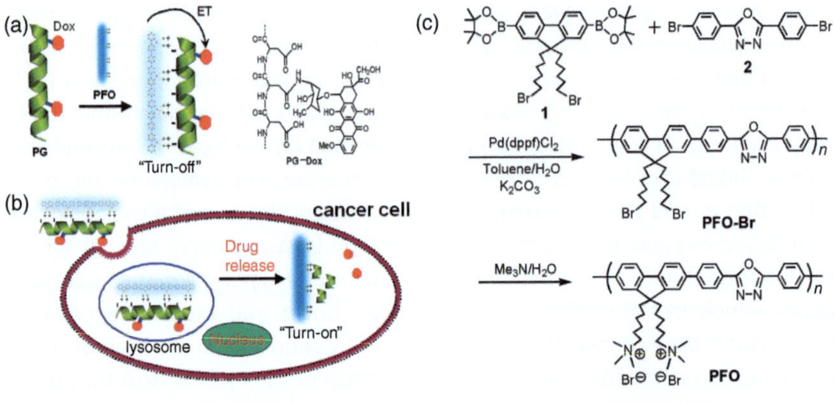

Figure 3.11 (a) Schematic illustration of PFO fluorescence quenching by Dox, resulting in a signal "turn-off" state. (b) Schematic depiction of lysosomal hydrolysis of poly-L-glutamic acid, releasing Dox to activate PFO fluorescence and achieve the "turn-on" state. (c) Synthesis route of the PFO. *Source:* Feng et al. [20] / with permission of American Chemical Society.

with alkyl tailing groups at one end of the hydrophobic pigment and the other end of the ethylene glycol chain; functional groups were subsequently introduced at the periphery of the ethylene glycol side chain. Fluorescent nanoparticles with tunable compositions and densities were prepared by co-assembling different functionalized oligomers [21]. In addition, amine groups can be incorporated into the glycol side chain, and the density of amine groups on the surface can be regulated by adjusting the ratio of amine group-modified oligomers to unmodified oligomers. Thus, it is possible to prepare amine group-modified nanoparticles of tunable size (40–232 nm) [22]. In another example of research, fluorescent organic nanoparticles can be prepared by synthesizing star-shaped glycosylated conjugated oligomers, and the nonionic oligomers spontaneously form nanoparticles with an average diameter of 61 nm, driven by the high density of hydrophilic sugar side chains [23].

In addition, when amphiphilic molecules (molecules with hydrophilic and hydrophobic regions) reach a certain concentration in water, they spontaneously aggregate through hydrophobic forces and other interactions to form vesicle structures. The structure of a vesicle consists of an outer hydrophilic portion and an inner hydrophobic portion forming a membrane package, with the aqueous phase or solvent filling its interior. For example, tetrastyrene modification of aggregation induced emission (AIE) molecules on amphiphilic polymer molecules allows the preparation of nanomicelles with fluorescent tracer and drug delivery functions [24]. As shown in Figure 3.12, in aqueous solution, amphiphilic molecules form vesicles to give the composite system good biostability and compatibility to enter cells. The tetraphenylene molecules in the hydrophobic interior of the vesicles aggregate to produce a fluorescent signal that can be used for imaging tracing and other purposes. When the composite system enters an acidic environment, such as a cellular lysosome, the covalent bond between the tetraphenylene and the polymer

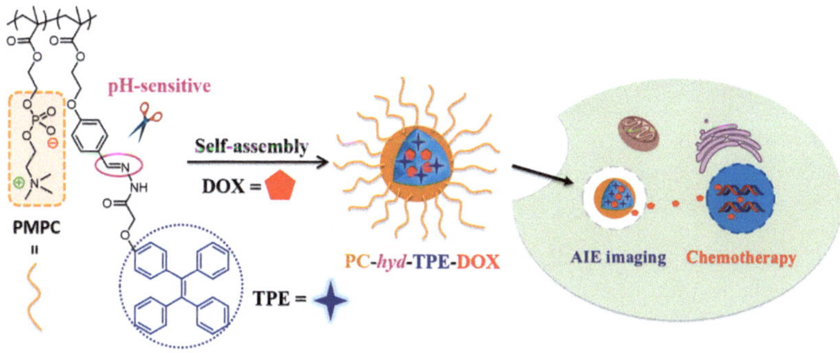

Figure 3.12 Schematic illustration of self-assembled vesicles formed by phosphatidylcholine-TPE conjugates. *Source:* Chen et al. [24] / with permission of American Chemical Society.

molecules is broken, resulting in the disintegration of the vesicles and the release of the loaded anticancer drug to achieve therapeutic effects.

Enzyme-responsive self-assembly is especially appealing due to its excellent biocompatibility and sensitivity, making it a promising approach for applications in biomaterials and drug delivery systems. An enzyme-responsive supramolecular vesicle can be constructed using sulfonated cucurbit[8]uril as a host and naturally enzymatically cleaved myristoylcholine as a guest molecule [25]. Enzyme-responsive supramolecular vesicles are used as carriers to encapsulate a standard water-soluble cholinesterase inhibitor, such as tacrine. Enzyme-induced cleavage of myristoylcholine disrupts the affinity equilibrium between the two parts of the binary supramolecular, leading to vesicle disassembly and release of the encapsulated drug. Notably, tacrine-loaded vesicles underwent only partial disassembly in the presence of butyrylcholinesterase, in contrast to free vesicles, which fully disassembled under the same conditions. This controlled release mechanism enhances the drug's efficacy while reducing side effects caused by the excessive release of tacrine.

Liposomes are spherical, enclosed structures composed of a phospholipid bilayer that separates one aqueous medium from another. Incorporating drugs into liposomes enhances therapeutic efficacy by minimizing drug concentrations in normal tissues while maximizing their accumulation in tumor sites. Figure 3.13 shows the evolution of liposomes [26]. An early conventional phospholipid "normal" liposome in which the water-soluble drug is encapsulated inside the aqueous phase of the liposome, while the water-insoluble drug is incorporated into the liposome membrane (Figure 3.13a). Figure 3.13b shows antibody-targeted immunoliposomes in which the antibody binds to reactive phospholipids in the membrane via covalent bonding or is embedded in the liposome membrane via hydrophobic anchoring, having been previously modified by hydrophobic moieties. Long-circulating liposomes are grafted with a protective polymer (e.g. polyethylene glycol), which shields the surface of the liposome from interacting with phagocytic proteins (Figure 3.13c). Figure 3.13d illustrates long-circulating immunoliposomes equipped with both a protective polymer and an antibody, which can either bind to the liposome surface

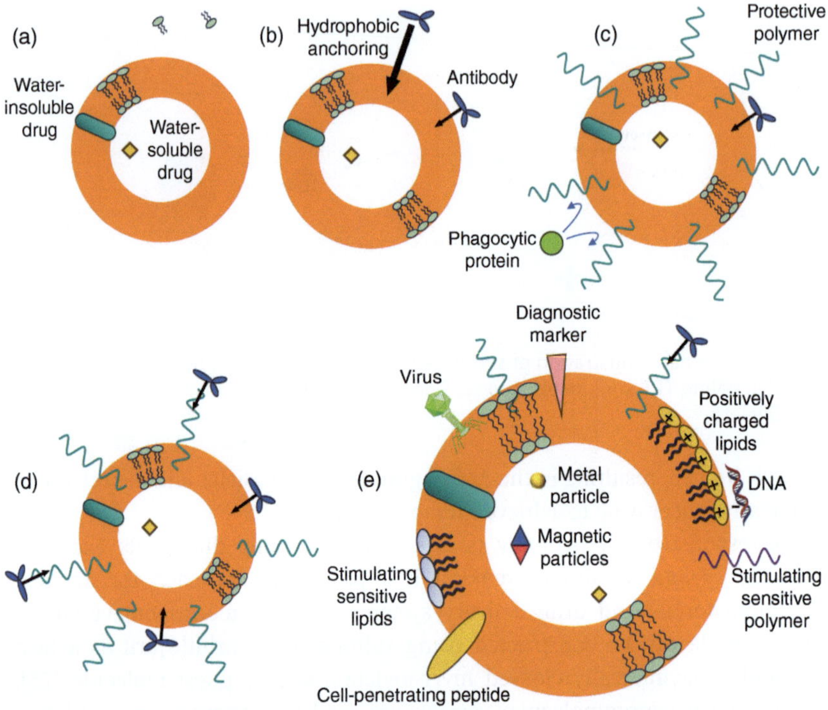

Figure 3.13 Evolution of liposomes. (a) Liposomes loaded with drugs. (b) Targeted liposomes. (c) Long-circulating liposomes. (d) Liposomes with both targeting and long-circulating properties. (e) Stimuli-responsive liposomes co-loaded with multiple components.

or, preferably, to the distal end of the grafted polymer chain. Figure 3.13e shows a new generation of liposomes whose surfaces can be modified in different ways (individually or simultaneously). These modifications include: attaching protective polymers or linking protective polymers to targeting ligands (e.g. antibodies); attaching or embedding diagnostic markers; incorporating positively charged lipids to facilitate complex formation with DNA; embedding stimulus-sensitive lipids; integrating stimulus-sensitive polymers; incorporating cell-penetrating peptides; and embedding viral components. In addition to drugs, liposomes can be infused with magnetic particles for magnetic targeting or with colloidal gold or silver nanoparticles for use in electron microscopy. Enhanced permeability and retention (EPR) effect can be achieved by the liposome method.

For example, Doxil is a commercial anticancer drug designed to extend the circulation time of Doxil and minimize its cardiotoxicity. However, loading insoluble anticancer drugs into the cavity of liposomes remains challenging. While the hydrophobic membranes of liposomes can encapsulate insoluble drugs, their stability is often poor, with structures prone to aggregation or decomposition. To address this, water-soluble complexes can be efficiently encapsulated by liposomes through the formation of host–guest complexes based on cyclodextrin (CD) and the drug. Ionizable β-CDs, containing weakly basic or acidic groups on the solvent-exposed surfaces, can be engineered to encapsulate insoluble chemotherapeutic drugs [27]. Cyclodextrin-drug complexes are remotely loaded into liposomes using a pH gradient, significantly improving the water solubility of the drugs while producing liposomes that are less toxic and exhibit enhanced therapeutic activity.

The impossibility made possible by the supramolecular strategy which provides a novel approach to prepare nanomedicines using anticancer drugs that have failed in the clinic due to their troublesome formulation, and involves bringing the drugs back into use. Alternatively, liposomes can form complexes with DNA by self-assembly for applications in nonviral gene delivery systems, where the self-assembly process relies on electrostatic interactions between the negative charge of the DNA molecule and cationic liposomes (e.g. DOTAP), resulting in the formation of a stable structural body [28]. Among the common self-assembled structures are lamellar multilayer membrane structures and reverse hexagonal structures. In multilayer membrane structures, DNA monolayers are sandwiched between lipid bilayers to form a multilayer sandwich-like arrangement, which usually consists of DOTAP and the neutral lipid DOPC, which is used to regulate the flexibility and charge density of the membrane.

3.1.3 Microemulsion Method

The microemulsion is a dynamically stable, transparent or semi-transparent colloidal dispersion system. It consists of at least three components, such as water, oil, and surfactant, and is isotropic in nature. The microemulsion method operates on the principle of utilizing the combined effects of surfactants and cosurfactants to lower the interfacial tension between oil and water. This can even result in an instantaneous negative surface tension, generating stable microemulsions. Amphiphilic polymers function as surfactants in this process, facilitating the formation of stable droplets and contributing to particle generation. However, this limits the loading concentration of photoactive substances, such as fluorophores. Furthermore, the proportion of hydrophobic chain within the amphiphilic polymer significantly influences the densification of the particle core through hydrophobic interactions between the encapsulated matrix and the integrated fluorophore. The quantum yield of fluorescent nanoparticles is highly dependent on the properties of amphiphilic polymers, as the density of the particle core affects both the efficiency of energy transfer within the particle and the isolation from external quenching substances [29].

Phospholipids are commonly utilized in microemulsion techniques due to their low critical micelle concentration and high kinetic stability. These molecules arrange themselves such that their hydrophilic ends interact with water, while their hydrophobic acyl chains remain within dichloromethane droplets, forming micelles containing the dichloromethane solution. As the dichloromethane begins to evaporate, the micelles shrink as the solvent exits. During this shrinking process, hydrophobic polymers are trapped and compressed within the micelles, leading to the formation of nanoparticle cores [30]. This mechanism ultimately enables the stable distribution of conjugated polymer nanoparticles encapsulated in phospholipids within water, as shown in Figure 3.14.

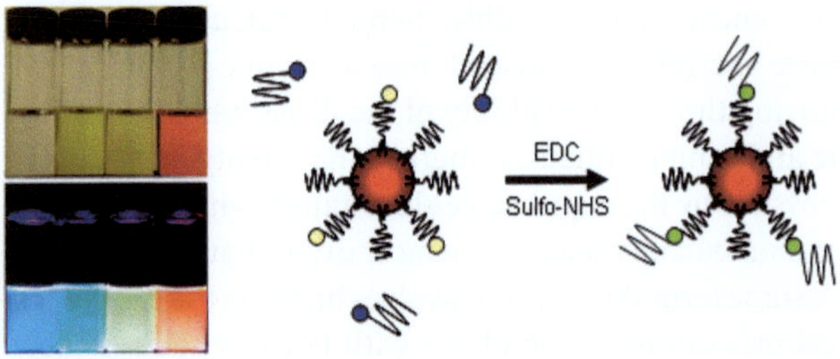

Figure 3.14 Schematic illustration of a phospholipid-encapsulated emulsion. *Source:* Howes et al. [30] / with permission of American Chemical Society.

To ensure the stability of the emulsion droplets during the preparation process, the organic solvents used in the microemulsion method are usually immiscible with water. When the polymer-containing raw material is introduced into the aqueous solution, the hydrophilic segments of the amphiphilic polymer molecules extend into the water, while the hydrophobic segments aggregate to form a microcavity. This microcavity encapsulates the organic solvent along with the dissolved photofunctional material. Finally, the organic solvent is removed by evaporation, which allows the surfactant to be firmly attached to the surface of the nanoparticles through hydrophobic interaction. Using a modified microemulsion method, PLGA molecules encapsulate a variety of conjugated polymers to obtain organic nanoparticles with stable emission in aqueous solution [31]. Functionalization of PFV-loaded nanoparticles (NPs) was achieved through an *N*-(3-dimethylaminopropyl)-*N′*-ethylcarbodiimide hydrochloride (EDAC)-mediated coupling reaction, which occurs at the PLGA matrix between the carboxyl group (COOH) and the amino group of the aminated folic acid.

Water-soluble polymer nanoparticles can be prepared using the microemulsion method [32]. The basic principle of this method, shown in Figure 3.15, is to dissolve the polymer in a good solvent (e.g. carbon tetrachloride) and then inject it into

Figure 3.15 Schematic representation of conjugated polymer nanoparticles prepared by coprecipitation and microemulsion methods. *Source:* Koralli et al. [32] / The Royal Society of Chemistry / CC BY-NC 3.0.

an aqueous phase, where it forms a microemulsion. By this method, the resulting polymer particles can be self-assembled into nanoparticles to form composites with high light stability and a smaller size. In this regard, conjugated polymers with different degrees of fluorination (Probe37-42) can be used as the main materials. The optical functions of the nanoparticles can be enhanced by changing the number and position of the fluorine atoms through the design of the polymers to adjust their optoelectronic properties. In addition, the use of amphiphilic block copolymers, mPEG-*b*-PLGA, as surfactants improved the solubility of the polymer nanoparticles in the aqueous phase and enhanced their biocompatibility through encapsulation. Through the method, the prepared nanoparticles with smaller particle sizes are more stable in aqueous solution. The size and shape of the nanoparticles can be regulated by controlling the components of the microemulsion, which provides a more precise particle control capability.

3.1.4 Direct Polymerization

Direct polymerization is a method of constructing organic composites with specific functions and structures through direct polymerization reactions between monomers or small molecule precursors. During the polymerization process, functionalized monomers or photofunctional groups (e.g. luminescent groups, photosensitive molecules) can be introduced directly into the polymer chain, thereby imparting specific optical properties to the final material. Advantages of using direct polymerization to construct photofunctional composites include (1) covalent cross-linking between functionalized monomers or photofunctional groups and particle matrices can effectively prevent them from leaking out of the composites, especially in complex biological environments, and (2) by adding functionalized monomers to the reaction, the optical properties and surface chemistry of the composites can be precisely tuned, which facilitates the construction of photofunctional reagents suitable for application. Based on this approach, it is possible to construct a wide range of photofunctional composites with excellent optical functionality as well as excellent biocompatibility.

Copolymerization of poly(ethylene glycol) vinyl monomer containing diphosphate groups with *N*-(3-aminopropyl) methacrylamide hydrochloride and tetra(ethylene glycol) diacrylate by direct polymerization yields an NIR-fluorescent bisphosphonate nanoparticle with a narrow size distribution (Figure 3.16) [33]. In this case, the first monomer provides the BP moiety for bone targeting; APMA provides the primary amine moiety for subsequent NIR dye binding; and TTEGDA is used to cross-link the resulting polymer matrix. The resulting fluorescent nanoparticles have a relatively high concentration of polyethylene glycol groups, which can be injected intravenously to extend the blood half-life of the device for bone-targeting applications, and have good biocompatibility with skeletal cells.

Fluorescent and biocompatible nanoparticles based on methyl methacrylate (MMA) can be synthesized through free-radical polymerization, utilizing a fluorescent polymer created by conjugating rhodamine B with hydroxyethyl methacrylate [34]. During polymerization, the free radicals generated by the radical

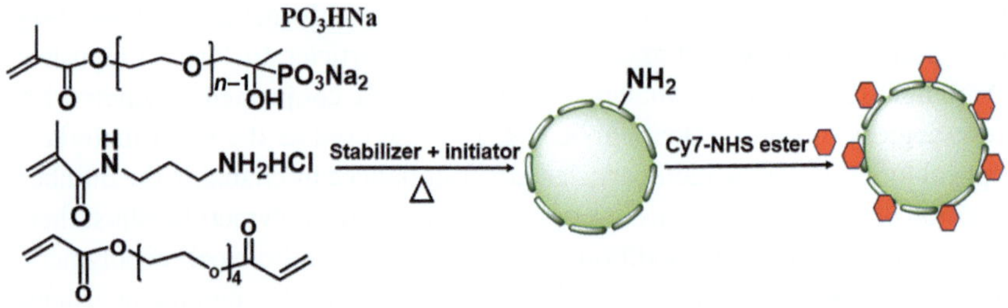

Figure 3.16 Preparation of NIR fluorescent bisphosphonate nanoparticles.

initiator will first react with the double bonds in MMA and HEMA-RhB to form an active center. The reactive center can continue to react with other monomers to form long-chain polymers. Through free radical reactions, MMA and HEMA-RhB are alternately linked in the polymer chain, resulting in the formation of fluoNPs, a photofunctional complex material based on poly(methyl methacrylate) (PMMA) that contains fluorescent rhodamine B. MMA, as a base monomer, mainly provides the polymer framework. While HEMA-RhB, as a fluorescent monomer, not only imparts the fluorescent properties of the nanoparticles but also ensures the stability of the fluorescent molecules in the nanoparticles through its co-combination with the MMA copolymerization ensured the stability of the fluorescent molecules in the nanoparticles. This free radical copolymerization reaction ensures that the final product has both good fluorescence tracking ability and biocompatibility and stability.

3.1.5 Covalent Modification

Covalent modification is one of the most commonly used methods for preparing optically functional systems. It involves forming covalent bonds between groups located on the side chains of molecules or the surface of materials without altering the material's main structure. This approach is typically used to link components with complete functions and enables precise control over the distribution and density of functional groups, thereby imparting the material with superior and tunable optical properties.

Covalent attachment of hemoglobin to conjugated polymer nanoparticles enables the preparation of photodynamic polymer composite systems with self-luminescence and self-feeding capabilities [35]. Nanoparticles with surface-bound carboxyl groups were first prepared by coprecipitation of the commercially available conjugated polymer MEH-PPV with PSMA. Then hemoglobin was covalently linked to this composite nanoparticle by carbodiimide reaction. The hemoglobin in this system was able to catalyze the chemiluminescence of luminol, and the conjugated polymer nanoparticles could absorb the chemiluminescence of luminal through chemiluminescence resonance energy transfer. Oxygen provided by hemoglobin is then sensitized again to produce ROS and kill cancer cells. The system does not require an external light source and solves the problem of insufficient ROS production capacity under anoxic conditions.

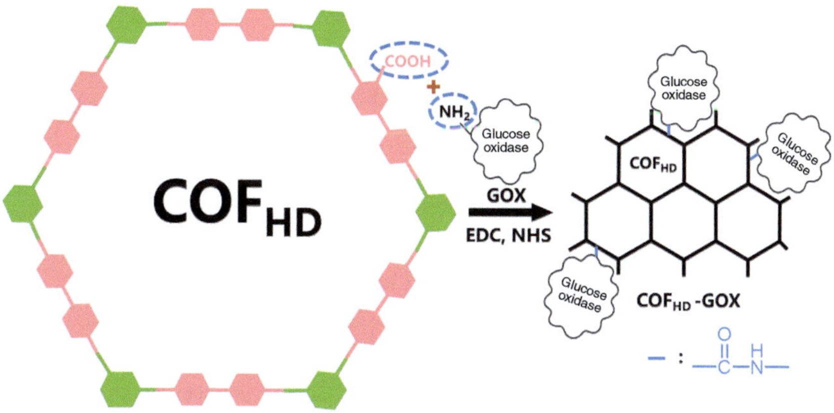

Figure 3.17 Schematic of preparation of COF_{HD} and COF_{HD}-GOX.

COF materials with carboxyl-functionalized groups are covalently bonded to glucose oxidase (GOX) to prepare COF-enzyme composites (Figure 3.17) [36]. Among them, the COF forms a stable covalent linkage through the reaction of functionalized groups with carboxyl groups and the amino groups in GOX. COF, as a porous organic framework material with high chemical stability and mechanical strength, can effectively immobilize GOX as a support and prevent the loss of activity of free enzymes due to environmental factors. The composite can be applied to color-responsive sensors for easy and rapid detection of glucose concentration with a detection limit of 0.54 mM and a linear range from 0.005 to 2 mM.

3.1.6 Physical Mixing Method

Physical blending of photofunctional organic polymers is a method of mixing polymers, fillers, additives, and other materials by means of a nonchemical reaction that utilizes the physical properties of the different components (e.g. fluidity, solubility, and diffusivity). The method relies on physical principles such as mechanical force, heat, solvent action, electrostatic force, and surface treatment, which can improve the performance of the polymer or impart new properties. In contrast to chemical reaction methods, physical blending does not involve breaking or reorganizing chemical bonds, thereby eliminating the risk of by-products and the need for complex reaction conditions. This makes physical mixing methods simpler and more controllable for preparing photofunctional organic polymer materials, while also minimizing environmental and safety concerns during processing. As a result, these methods are widely applied in fields such as optoelectronics, sensors, and photocatalysis.

Introduction of COFs into a polymer matrix leads to the formation of mixed matrix membranes (MMMs) [37]. The incorporation of COFs into a polymer matrix can greatly improve the compatibility between the continuous polymer matrix and the dispersed phase. COF-polymer composite-based MMMs are commonly fabricated using the nonsolvent-induced phase separation (NIPS) technique. In this process,

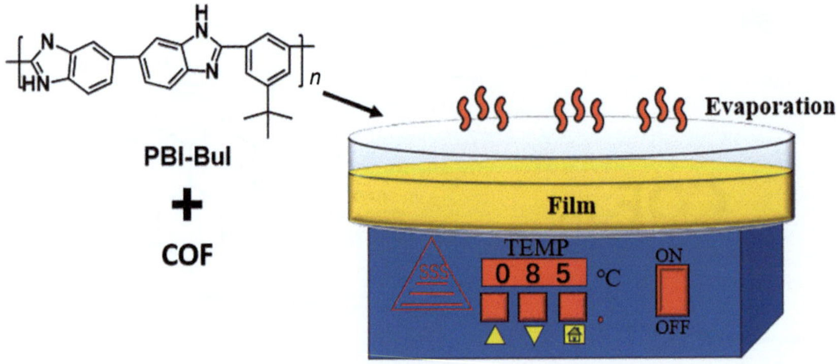

Figure 3.18 Schematic diagram of the preparation of COF@PBI-BuI hybrid film.

COFs and polymers are blended in an organic solvent to create a uniform solution. This solution is then spread onto a glass or nonwoven fabric substrate to form a film of desired thickness, which undergoes phase separation upon immersion in water, resulting in the formation of a membrane. Finally, MMMs based on COF-polymer composites were obtained by removing the remaining solvent by immersing them in deionized water. As shown in Figure 3.18, two COFs, TpPa-1 and TpBD, were synthesized by the solvent-thermal method and physically mixed with PBI-BuI polymer, respectively, and the composite thin-film MMMs could be prepared by the above methods [38]. With this approach, the COFs can achieve higher loadings and the MMMs remain highly flexible.

3.2 Construction of Photofunctional Organic/Inorganic Polymer Composite Systems

Photofunctional organic/inorganic polymer composite systems use polymers as carriers and inorganic nanomaterials as photofunctional units, integrating the excellent properties of inorganic, organic, and nanomaterials [39]. Inorganic photofunctional materials are often able to strongly interact with light, but are less compatible with living organisms. Polymers possessing biocompatible, flexible, and processable characteristics can significantly enhance the physical and chemical surface properties of inorganic photofunctional materials. These improvements promote better compatibility between inorganic materials and living organisms, enabling their application in various biomedical fields, including biosensing, disease detection, antibacterial activity, and cancer diagnosis and treatment [40]. Inorganic photofunctional materials are very diverse, but the physical and chemical properties depend strongly on their size [41]. Among the structures often used for biological applications are those dominated by nanomaterials, including conventional metal nanoparticles with surface plasmon resonance (SPR) [42], metal nanoclusters with optical properties similar to those of organic molecules, semiconductor quantum dots with high fluorescence intensities and quantum yields [43], and carbon nitride with photocatalytic activity.

The construction methods of photofunctional organic/inorganic polymer composite systems mainly include self-assembly, coprecipitation, microemulsion, covalent modification, in situ reaction, and spin coating.

3.2.1 Self-assembly Method

In the development of photofunctional organic/inorganic composites, the self-assembly method serves as a crucial strategy. It not only facilitates the effective integration of organic and inorganic components but also enables the regulation of the structure and functionality of the composites by controlling self-assembly interactions. In the composite, inorganic materials such as metal nanoparticles, metal nanoclusters, and nanotubes are often combined with organic molecules through noncovalent or covalent bonds to form composites with unique optical, photovoltaic, and mechanical properties. Compared with self-assembly methods in purely organic composite systems, self-assembly methods in organic/inorganic composite systems not only need to consider the self-assembly behavior of the organic components, but also need to deal with the compatibility and interfacial bonding between inorganic and organic materials.

In a report on cellular imaging based on the MEF effect of silver nanomaterials, silver nanoparticles can be capped with silica, thus controlling the distance between the fluorescent PFV and the metal nanoparticles, and PFV/Ag@SiO_2 nanocomposites can be prepared by self-assembly through electrostatic interactions [44]. Zeta potential measurements show that the surface of Ag@SiO_2-2 nanoparticles is negatively charged. Therefore, the positively charged PFV can be adsorbed onto the anionic Ag@SiO_2 surface by electrostatic attraction. This assembly approach preserves the superior monodispersity of the silver nanoparticles and the intense fluorescence of PFV. Moreover, the cationic surface and optimal size of these nanocomposites enable their rapid internalization into cancer cells.

In addition, electrostatic interactions can also be used to prepare composites based on gold nanoclusters (AuNCs). Gold nanoclusters modified with glutathione (GSH-AuNCs) are widely utilized for electrostatic self-assembly because of their strong negative charge, ease of preparation, and excellent stability. The sandwich structure based on GSH-AuNC can be formed through electrostatic interaction [45]. Specifically, negatively charged GSH-AuNCs were capped by cationic polymer PEI, and subsequently, negatively charged 8Q5S was adsorbed on the surface of GSH-AuNCs/PEI, resulting in the preparation of GSH-AuNCs/PEI/8Q5S composites. Furthermore, the optical properties of AuNCs were improved through electrostatic interactions. Specifically, when GSH-AuNCs were bound to positively charged nitrogen-doped quantum dots (N-GQDs) via electrostatic forces (Figure 3.19), the FRET effect was introduced to enhance the fluorescence emission of AuNCs. As a result, the fluorescence quantum yield of the composite was higher compared to that of the bare GSH-AuNCs [46]. Apart from negatively charged GSH-AuNCs, other positively charged ligand-modified gold nanoclusters, such as lysozyme-protected AuNCs [47] and bovine serum albumin (BSA)-protected gold nanoclusters (BSA-AuNCs) [48], are also suitable for electrostatic assembly.

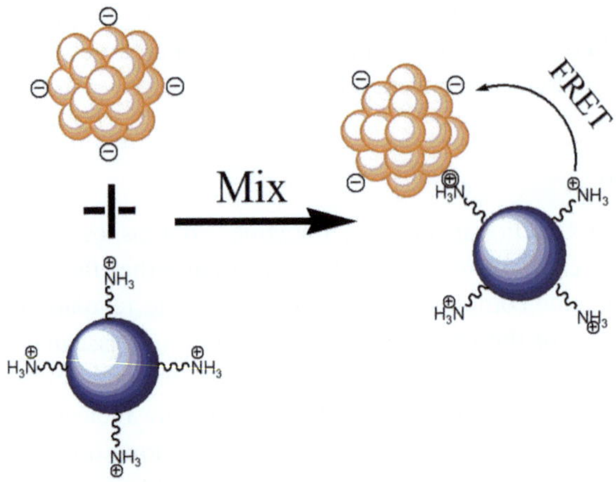

Figure 3.19 Electrostatic interaction of GSH-AuNCs with N-GQDs. *Source:* Gao et al. [46] / with permission of Elsevier.

Given the unique "gold nucleus-ligand shell" structure, the ligands on the surface of AuNCs can induce a variety of intermolecular forces, enabling the effective assembly of AuNCs with a variety of small molecules to form nanoscale composites. Common ligands, such as BSA, glutathione (GSH), and thiol-containing compounds, can provide AuNCs with a negative surface charge and promote the electrostatic assembly of AuNCs. For example, fluorescence emission can be improved by electrostatic interactions of the cationic polyelectrolyte PDDA with single-stranded DNA (ssDNA) templated AuNCs, demonstrating enhanced microenvironmental stability induced by conformational changes in ssDNA [49].

Relying on electrostatic interaction, a composite structure of gold nanoparticle-metal thin film can be constructed via self-assembly [50]. The method utilized is the deposition technique of layer-by-layer self-assembly, which involves the alternating deposition of positively and negatively charged polyelectrolytes (PE) to construct a multilayer spacer layer. By controlling the number of deposition cycles, the nanometer-scale spacing between the gold nanoparticles and the metal film can be precisely regulated, as shown in Figure 3.20. The process starts with the deposition

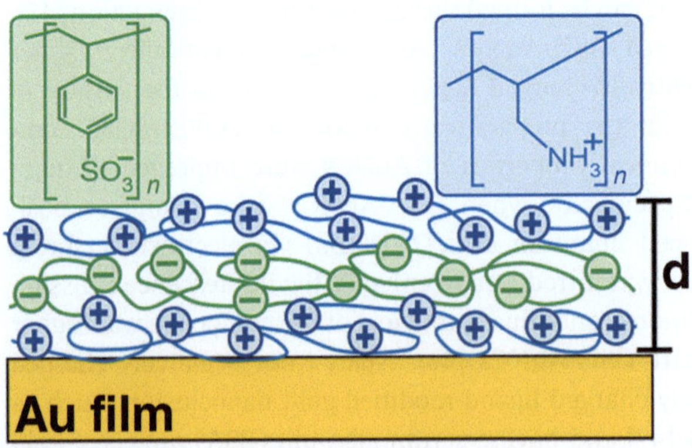

Figure 3.20 Alternating PE layers assembly on a positively charged surface of gold films. *Source:* Mock et al. [50] / with permission of American Chemical Society.

of a metal film about 45 nm thick on the substrate, followed by the layer-by-layer deposition of PE layers to form a thickness interval ranging from 0.64 to 22.3 nm, and finally the gold nanoparticles with a diameter of 60 nm are immobilized on the topmost layer of PE by electrostatic adsorption. The metal film assumes the function of supporting and exciting the surface plasmon polarization wave (SPP) in the structure. The polyelectrolyte layer is used to regulate the distance between the particles and the film, thus modulating the plasmon coupling strength, and the gold nanoparticles provide the localized surface plasmon resonance (LSP) response as a highly sensitive optical detector element. The resulting composite exhibits a sandwich structure of "film-mesolayer-nanoparticle," whose plasmonic resonance behavior is extremely sensitive to nanoscale distance variations, with the ability to induce a resonance wavelength drift of about 5.6–7.8 nm per 1 nm of distance variation. This structure not only exhibits considerable optical coupling properties but also has potential for applications in high-resolution biosensing, nanoscale distance measurement, and detection of molecular conformational changes.

Weak electrostatic interactions give rise to the dissociation of AuNCs from the composites when the solution charge changes. Based on this strategy, AuNC composites are used for detection applications [51]. For example, negatively charged BSA-AuNCs and positively charged hollow silica nanoparticles (MSNs) were assembled. Based on electronic interactions at pH 7, two drugs were loaded, respectively. Upon lowering the pH to 5, these electrostatic interactions were disrupted, resulting in the dissociation of the AuNC composites and subsequent drug release [52]. Additionally, some strongly charged substances also affect the charge, including negatively charged heparin and positively charged amino-functionalized MSNs. MSN/AuNC composites can be prepared by electrostatic interaction [53], leading to greater fluorescence emission. When negatively charged heparin is assembled on the MSN surface by electrostatic interaction, the fluorescence enhancement is suppressed. This allows for the sensitive detection of heparin concentration by observing changes in fluorescence.

AuNCs, with ligands composed of benzene ring-containing compounds or biomolecules, can undergo self-assembly with other benzene ring-containing substances through π–π interactions. For example, supramolecular chiral hydrogels with enhanced luminescent properties can be prepared. Through π–π stacking and hydrogen bonding, nonchiral TBA-AuNCs are co-assembled with chiral histidine molecules [54]. Notably, dopamine polymers (polydopamine) have been used as functional materials with π–π stacking due to their phenolic moieties. Additionally, RNA-modified AuNCs can be assembled onto the surface of dopamine nanoparticles through π–π stacking interactions, resulting in fluorescence quenching. For instance, as illustrated in Figure 3.21, AuNCs may be functionalized by attaching two distinct RNA strands that specifically target miR-21 and let-7a, respectively [55]. When AuNCs attach to polydopamine nanoparticles through π–π stacking interactions, their fluorescence is quenched as a result of the FRET effect. However, when the target RNA strand is present, the nanoclusters detach from the polydopamine surface, leading to the recovery of fluorescence. This "on-off" fluorescence approach, based on π–π stacking, was thus employed to detect two different microRNAs.

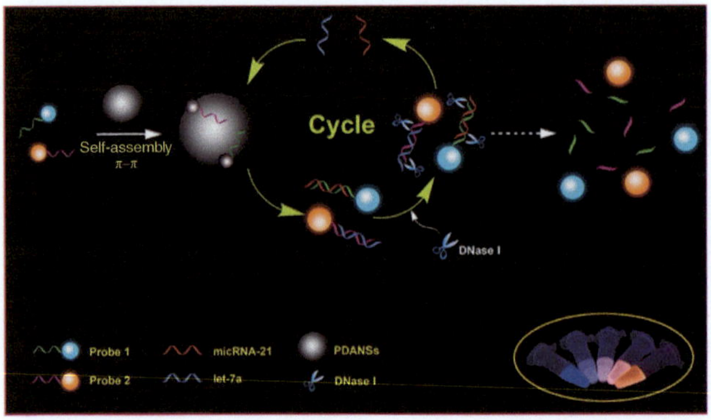

Figure 3.21 Self-assembly of AuNCs with two RNA strands targeting miR-21 and let-7a by π–π stacking. *Source:* Xu et al. [55] / with permission of American Chemical Society.

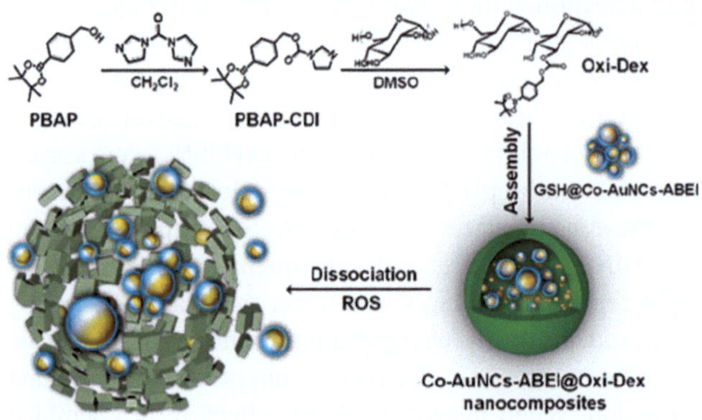

Figure 3.22 Schematic illustration of ROS-responsive polymer nanocomposites assembled with GSH@Co-AuNCs-ABEI. *Source:* Gao et al. [56] / with permission of American Chemical Society.

Hydrophobic AuNCs can be assembled into AuNC composites by surface adsorption of hydrophobic materials. As shown in Figure 3.22, gold nanoclusters modified by hydrophobic ligands (Co-AuNCs-ABEI) were encapsulated in borate-modified dextran (Oxi-Dex) to form fluorescence-quenched Co-AuNCs-ABEI@Oxi-Dex composites [56]. Owing to its pronounced sensitivity to ROS, the shell layer Oxi-Dex rapidly disintegrates upon exposure to hydrogen peroxide, triggering the release of the encapsulated AuNCs and restoring the previously quenched fluorescence. In a similar manner, hydrophobic GSH-AuNCs [57], which exhibit both colloidal stability and photoluminescent properties in a range of organic solvents, can be synthesized and subsequently encapsulated within PLGA nanocarriers through the hydrophobic effect.

Macrocyclic molecules are capable of promoting the self-assembly of nanomaterials. By means of host–guest interactions, the nanoclusters can bind to the macrocyclic molecules through the surface ligands to realize the self-assembly of the nanoclusters. There exists a self-assembly of gold nanoclusters (FGGC-AuNCs)

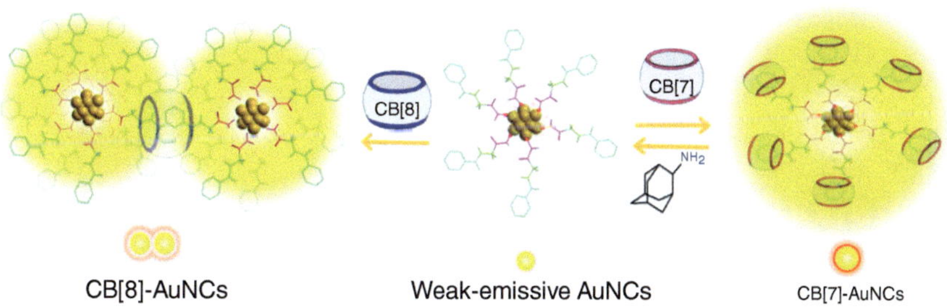

Figure 3.23 Schematic illustration of the host–guest assembly of CB[7] or CB[8] with FGGC-AuNCs. *Source:* Jiang et al. [58] / The Royal Society of Chemistry / CC BY-NC 3.0.

modified by Phe-Gly-Gly-Cys peptides, a process triggered by CB[8] [58]. The size of the nanorods formed ranged from 30 to 80 nm (Figure 3.23). Similarly, CB[8] has been considered as a "glue" for joining gold nanorods with a gap of 3 nm. In addition, succinylcholine can induce the assembly of BSC4A-AuNPs into SuCh-BSC4A-AuNPs polymers, owing to the strong affinity between bis(p-calixarene) (BSC4A) and succinylcholine [59].

Self-assembly using discrete DNA strands relies on the highly specific binding between complementary DNA to create rationally designed, programmable DNA nanostructures. In these systems, the distance between nanoparticles can be precisely tuned by adjusting the length of the DNA linker. DNA origami techniques employ long DNA scaffold strands that are folded into desired shapes with the help of hundreds of specifically designed short oligonucleotides. For example, using double-sided DNA origami as a template, gold nanorods (AuNRs) can be guided to self-assemble to form 3D nano-architectures [60]. Fine-tuning of the spatial configuration by adjusting the position of AuNRs on the origami template enables the preparation of individual nano-architectures of a 3D nanorod dimer with tunable optical chirality.

The layer-by-layer self-assembly technique enables the fabrication of COF@SiO_2 composites characterized by a large surface area, uniform size, excellent stability, facile modification, and high adsorption ability [61]. In this process, SiO_2-TPDA is first synthesized by reacting SiO_2-NH_2 with terephthalaldehyde. Next, SiO_2-TAM is produced by reacting SiO_2-TPDA with tetra(4-anilyl)methane. By repeating these steps for several cycles, a polymer@SiO_2 composite is built up. Finally, a solvothermal method is applied to convert the amorphous polymer coating on SiO_2 into crystalline COF-300, resulting in the formation of COF-300@SiO_2.

3.2.2 Coprecipitation Method

The principle of coprecipitation of photofunctional organic polymer systems has been introduced, and the principle of coprecipitation of photofunctional organic/inorganic polymer composite systems is similar to that of photofunctional organic polymer systems. Among the two co-precipitation methods, the coprecipitation of photofunctional organic/inorganic polymer composite systems focuses on synergistic interaction between organic and inorganic materials to optimize

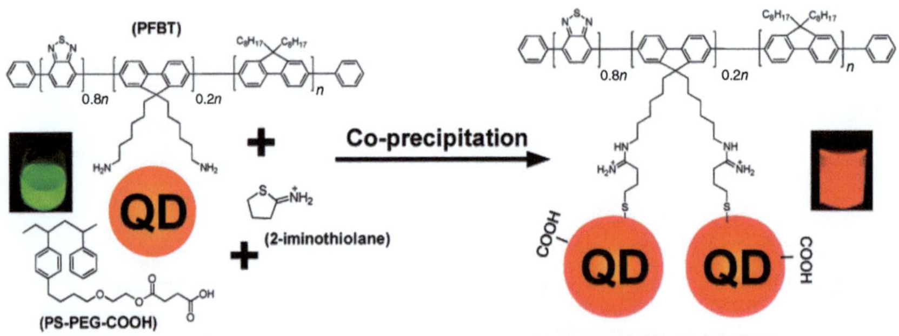

Figure 3.24 Schematic illustration of the preparation of Pdot-Qdot nanocomposites. *Source:* Chan et al. [62] / with permission of American Chemical Society.

the properties of the materials through co-precipitation, fabricating many types of high-performance materials.

For example, coprecipitation of the conjugated polymer PFBT with the inorganic quantum dot CdSe allows the preparation of composite nanoparticles with strong light absorption of the conjugated polymer and strong fluorescence emission and good monochromaticity of the quantum dot (Figure 3.24) [62]. PFBT bearing amino functional groups was initially synthesized, after which these amino groups were converted to thiol groups by using 2-iminothiolane. This modification enabled an efficient ligand exchange on the amino-capped quantum dot. Following ligand exchange, PFBT was mixed with a PS-PEG-COOH copolymer, which supplies the carboxyl groups necessary for subsequent bioconjugation of the quantum dots. The resulting polymer-quantum dot mixture was then coprecipitated into water under ultrasonic treatment, yielding the final Pdot-Qdot nanocomposites.

Three-dimensional carbon nanosheets (3D-CNS) were synthesized using water-soluble biomass materials—such as proteins, starch, and glucose—as carbon sources, and inorganic salts like K_2SO_4, NaCl, and KCl as soft templates [63]. Initially, the water-soluble biomass is dissolved and mixed with a saturated solution of the selected salt. This homogeneous mixture is then added to an organic solvent (e.g. ethanol or isopropanol) that is miscible with water. During this step, both the salts and the biomass coprecipitate layer by layer, since they are insoluble in the organic solvent. The resulting coprecipitates are subsequently subjected to carbonization at various temperatures. Following the removal of the soft templates, a 3D-CNS structure is obtained. The produced 3D-CNS exhibits a seamlessly interconnected network, characterized by a high surface area, remarkable photothermal conversion efficiency for infrared radiation, and excellent solar evaporation.

3.2.3 Microemulsion

The introduction of inorganic fillers can make the stability and dispersion behavior of emulsions more complex. In addition to the emulsification of organic polymers by surfactants, surface modification, coupling agents, or grafting of polymers are required to enable the inorganic particles to be uniformly dispersed in the emulsion.

The stability of emulsions and the properties of composites are greatly influenced by the interactions between inorganic components and organic polymers, such as electrostatic interactions, hydrogen bonding, van der Waals forces, and the coordination of metal ions to polymers.

The microemulsion is dynamically stable. First, the tension at the oil-water interface is controlled by selecting suitable solvents and surfactants to form stable microemulsions. Then, light-functional inorganics (e.g. titanium dioxide nanoparticles and zinc oxide), which have photocatalytic or fluorescent properties, are introduced into the system. Next, organic polymers are added to the microemulsion, and the polymers combine with the inorganics in the system through chemical reactions or physical adsorption to form composites. The microemulsion method provides a homogeneous dispersion environment, which contributes to the fine structure modulation of the inorganic and organic polymers, resulting in excellent optical properties, such as photocatalytic properties and fluorescence emission. The method is widely used in the fields of photocatalysis, optoelectronics, and sensors.

Multifunctional nanocapsules based on double emulsions are widely used in the microemulsion method. Hydrophilic AuNCs were encapsulated within the inner aqueous phase, while the hydrophobic indocyanine green (ICG) molecule was primarily incorporated into the oil phase. The polymer shell of the nanocapsules was formed by polymer mPEG-PLGA [64]. During the preparation of these hybrid nanocapsules, a combination of poly(acrylic acid) (PAA) and poly(vinyl alcohol) (PVA) in the outer aqueous phase served as a surfactant to improve the stability of the nanocapsules in aqueous solution (Figure 3.25). These nanocapsules can co-deliver both probes and drug molecules for tumor applications.

Pickering emulsion is an emulsion system with solid particles (e.g. nanoparticles and inorganics) as stabilizers. Unlike conventional emulsions, Pickering emulsions do not rely on surfactants (e.g. phospholipids and sodium dodecylbenzene sulfonate) to stabilize the oil-water interface, but rather stabilize the emulsion through the adsorption and alignment of solid particles at the oil-water interface. These solid particles typically possess high surface energy, enabling them to spontaneously adsorb at the oil-water interface. This results in the formation of a stable interface that inhibits the aggregation or separation of oil droplets. Due to the solid particles, Pickering emulsions have better thermal stability, anticoalescence, and strong drug

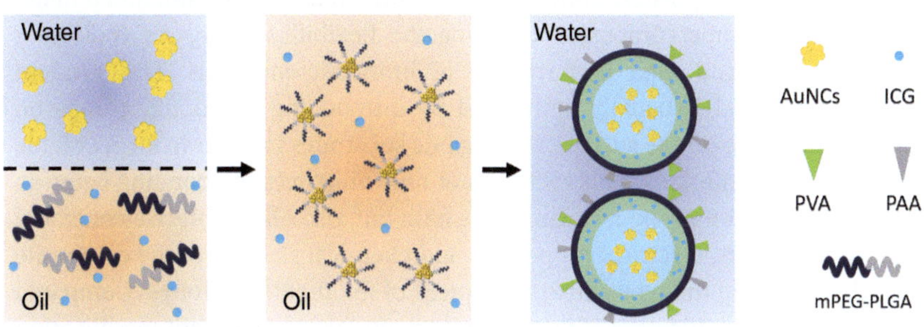

Figure 3.25 Schematic illustration of nanocapsule preparation with AuNC-ICG-mPEG-PLGA.

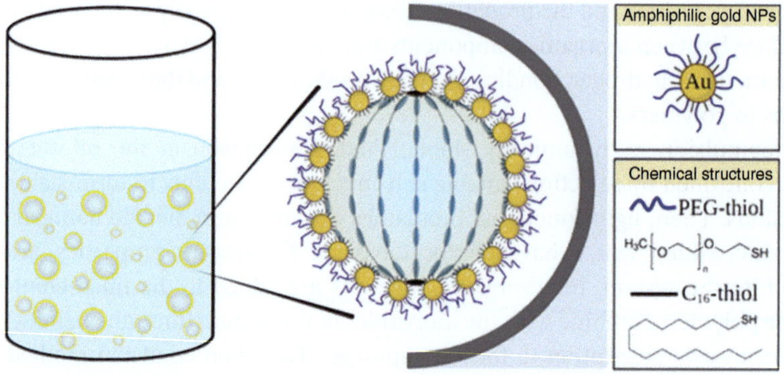

Figure 3.26 Schematic illustration of the structure of Pickering emulsion using PEG/C_{16}-thiol-coated AuNPs (left panel). The structures of the PEG and C_{16} thiol molecules (right panel). *Source:* Piñeres-Quiñones et al. [65] / with permission of American Chemical Society.

carrier function, and are widely used in drug delivery, food industry, and cosmetics. The preparation of Pickering emulsions belongs to a common application of the microemulsion method.

To increase the stability of the Pickering emulsion, the droplets of thermotropic liquid crystals (LC) could be decorated with gold nanoparticles (AuNP), which were chemically functionalized using 1-Hexadecanethiol (C_{16}-thiol) and sulfhydryl-capped polyethylene glycol methyl ether (PEG-thiol) (Figure 3.26) [65]. It has been shown that the accumulation of AuNP functionalized with PEG/C_{16}-thiol at the hexadecane/water interface stabilizes hexadecane in aqueous emulsions, and that the combination of PEG-thiols and C_{16}-thiols renders AuNP amphiphilic and surface active. Thus, water-in-water emulsions can be prepared using a similar PEG/C_{16}-thiol-AuNP system.

3.2.4 Covalent Modification

The covalent modification method is not only widely used in pure organic photofunctional composites, but also an important method in the construction of organic/inorganic composites. Compared with pure organic photofunctional materials, organic/inorganic composites combine the flexibility of organic molecules with the stability and high performance of inorganic materials. The covalent modification method is usually performed by introducing functional groups on organic molecules or on the surface of inorganic substrates in order to form a stable organic–inorganic interface. This method not only enhances the bonding between the organic material and the inorganic substrate but also allows precise control of the optical properties, electrical properties, and chemical stability of the composite material by adjusting the type and arrangement of the modified groups. Compared with the covalent modification of pure organic materials, the covalent modification method in composites requires higher selectivity and control to ensure a tight and stable bonding between organic molecules and inorganic materials.

Covalent modification strategies are extensively utilized to integrate gold nanoclusters (AuNCs) with functional materials, offering advantages such as enhanced stability and strong binding. Among these strategies, amide bond formation is one of the most widely used covalent interactions for preparing gold nanocluster composites. This process typically involves EDC and NHS, which facilitate the dehydration condensation reaction between the carboxyl group of the AuNCs and the amino groups of the functional material. This reaction results in the formation of an amide bond, effectively linking AuNCs with functional materials to produce AuNC composites. Various functional materials have been covalently bound to AuNCs, including quantum dots, fluorescent dyes, and hydrogels. For instance, colorimetric fluorescent probes (CQDs-AuNCs) were synthesized by coupling amide-modified carbon quantum dots (CQDs) with BSA-AuNCs through amide bonding. The AuNC composites can leverage the FRET effect for dopamine detection [66].

By treating aniline via the co-condensation method, PANI/g-C_3N_4 composites can be obtained using APS as an activator [67]. In this process, the 3D hierarchical structure was constructed by covalently modifying the polyaniline nanofibers to form between the carbon nitride nanosheets (CNNS) layers. As the main semiconductor, the CNNS were able to provide electronic structure and high chemical stability. The introduction of CNNS not only enhanced the light absorption performance of the photocatalysts but also enabled a substantial increase in the efficiency of photogenerated charge separation and transport through the 3D structure. Due to the presence of aniline, the electron/hole separation was enhanced, and the photodegradation ability was consequently enhanced. The construction of the 3D hierarchical structure enabled the composite hydrogel to have a higher surface area and stronger adsorption ability, and at the same time, it was able to effectively avoid the aggregation of the nanomaterials, which improved its efficiency in the photocatalytic reaction. The 3D structure also increased the multiple reflections of light, which helped to increase the efficiency of the utilization of visible light.

Preparation of gel-like polymer films using carbon nitride mixed with *N,N*-dimethylacrylamide in a mixed system of ethylene glycol and water [68]. DMA monomer was added to the g-CN dispersion and the photoinitiated polymerization reaction was initiated by visible light irradiation to form the g-CN prepolymer to obtain the aerogel, as shown in Figure 3.27. PDMA, as a hydrophilic monomer, was able to effectively increase the mechanical strength and elasticity of the aerogel after covalent binding with carbon nitride. 3-Sulfopropyl methacrylate potassium salt (SPMA) was further chosen for light-mediated surface modification of carbon nitride and for the covalent cross-linking of the prepolymer. As a charged monomer, SPMA imparts a lubrication property to the gels and can be used to adjust the solubility of the carbon nitride and change its hydrophilicity. By the above methods, the mechanical properties of hydrogels can be significantly improved and the lubrication property can be imparted to the hydrogels, which provides them with the potential to be used in alternative applications to joints.

Covalently cross-linked nanosilver gel/polydimethylsiloxane bilayer composites with a bilayer structure can be simply prepared by UV photopolymerization [69].

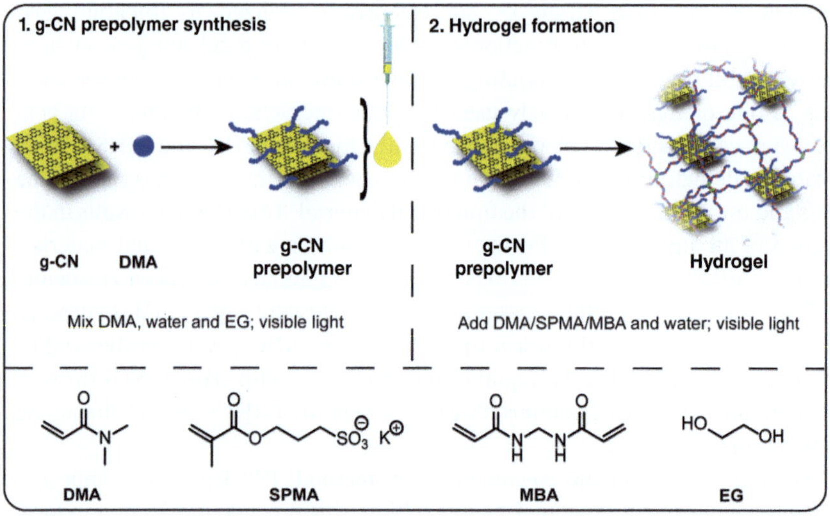

Figure 3.27 Schematic illustration of the preparation of g-CN polymer hydrogel. *Source:* Kumru et al. [68] / The Royal Society of Chemistry / CC BY 4.0.

The composite exhibits excellent mechanical properties, with a large tensile stress and tensile strain. Additionally, the bilayer composites demonstrate good biocompatibility: the nanosilver gel layer provides effective bactericidal activity, while the polydimethylsiloxane layer effectively prevents bacterial adhesion. The presence of a bilayer structure effectively enhances the mechanical strength of the two materials. Compared to the nanosilver gel layer with a separate polydimethylsiloxane layer, the bilayer nanosilver gel/polydimethylsiloxane composite is in a topologically entangled state due to the formation of covalent interconnections between the interfaces and the two networks are in a topologically entangled state. The strong interactions enable the polydimethylsiloxane network to transfer high applied stresses into the hydrogel matrix. Energy dissipation occurs through the breaking of hydrogen bonds between the gelatin and the solvent in the nanosilver gel, thereby enhancing the mechanical properties of the nanosilver gel/polydimethylsiloxane bilayer composites.

3.2.5 In Situ Reaction Method

The in situ reaction method is a synthetic method that generates target products by directly initiating reactions during the synthesis process, and is widely used in the preparation of organic/inorganic photofunctional composites. In this method, the reaction usually occurs in a reaction system without additional pretreatment steps, and the reaction conditions (e.g. temperature, time, and solvent) are controlled to promote the reaction between the organic molecules and the inorganic precursors to generate the desired inorganic components directly at the material interface. The in situ reaction method can achieve an effective combination of organic and inorganic components with high controllability and efficiency, which is particularly suitable for optimizing the optical, electronic, and catalytic properties of materials. This method not only simplifies the material preparation process but also significantly improves the stability and functionality of the composite materials and has become an important technical tool in the synthesis of organic/inorganic composites.

Firstly, gold nanoparticles have excellent optical properties and catalytic ability, which can significantly enhance the SPR effect and enhance light absorption and scattering, thus accelerating the polymerization reaction in response to light. Meanwhile, the high activity of the gold nanoparticle surface provides a good binding with polymer monomers, which promotes the smooth polymerization process. In addition, the tunable surface functionalization of gold nanoparticles can meet the needs of different photofunctional polymer materials, which are widely used in the fields of optoelectronics, sensing, etc.

Gold nanoparticles do not require ligand exchange, and the stability of monovalent and multivalent functionalized nanoparticles—both colloidal and chemical—relies entirely on the connectivity of the thiol groups. To improve their stability against aggregation and disaggregation, three types of thiolated PEGs were introduced into the reaction mixture immediately after nanoparticle formation, enabling in situ functionalization of the "naked" gold nanoparticles. This process utilized mono-, di-, and tri-thiolated PEG ligands to investigate how ligand multivalency influences the stability of the colloidal solution (Figure 3.28) [70]. Gold nanoparticles functionalized in situ exhibited significantly higher resistance to small nucleophilic reagents, such as sodium cyanide, compared to nanoparticles coated with citrate ligands and functionalized via ligand exchange. The synthesis of ligand- and chloride-free gold nanoparticles ("naked" NPs), combined with in situ stabilization using PEG thiol ligands, greatly enhanced their chemical and colloidal stability.

Hydrogels are widely employed as functional materials for the in situ synthesis of gold nanoclusters (AuNCs). The synthesis of AuNCs within protein hydrogel leverages the triple functionality of proteins, which act as a gelling agent, reducing agent, and structural template. Similarly, hyaluronic acid hydrogels can be utilized for the in situ synthesis of AuNCs, resulting in multifunctional hybrid nanogels known as MHA-gold clusters (MHA-GC) [71].

Carbon nitride has a wide range of light absorption, especially in the ultraviolet and visible regions, efficiently harnessing light energy to generate electron-hole pairs, which in turn facilitate polymerization reactions. Its high chemical stability and strong photocatalytic activity make it a great potential for application in the synthesis of photofunctional polymers. In addition, the conductivity and surface

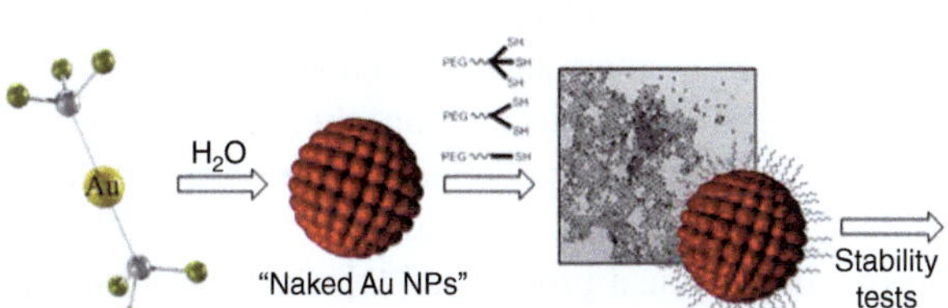

Figure 3.28 In situ encapsulation of "naked Au NPs" with univalent and multivalent thiol PEG ligands. *Source:* Zopes et al. [70] / with permission of American Chemical Society.

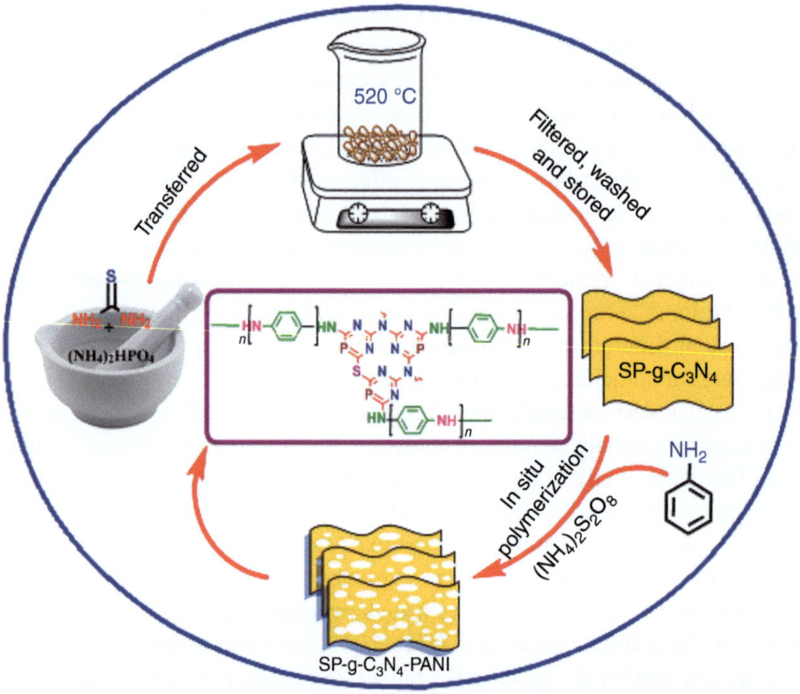

Figure 3.29 Schematic illustration of the preparation of SP-g-C_3N_4-PANI nanocomposites. *Source:* Vellaichamy and Periakaruppan [72] / with permission of American Chemical Society.

tunability of carbon nitride help to improve the efficiency of the polymerization reaction, which is particularly suitable for photocatalysis and photovoltaics.

Novel metal-free nanocomposites, sulfur (S)-phosphorus (P) co-doped graphitic carbon nitride (SP-g-C_3N_4) covalently grafted onto polyaniline (PANI), can be synthesized by in situ oxidative polymerization (Figure 3.29) [72]. By loading sulfur and phosphorus atoms onto carbon nitride and performing in situ oxidation of aniline, PANI/carbon nitride catalysts were successfully synthesized. Compared to their individual components (i.e. PANI, S-g-C_3N_4, P-g-C_3N_4, and SP-g-C_3N_4), the nanocomposite catalysts significantly enhanced the generation of photogenerated electrons and improved charge separation efficiency between electrons and holes, thereby boosting their catalytic degradation performance.

UiO-66-NH_2 is suitable as a photofunctional polymer material for in situ polymerization due to its unique chemical structure and functionality. Its amino group (—NH_2) can react with photosensitive monomers such as methyl methacrylate to form polymerizable active sites. Under UV light irradiation, these sites can trigger in situ polymerization reactions, allowing the polymer to grow on the surface or within the pores of UiO-66-NH_2, resulting in the formation of photofunctional polymer materials that confer excellent optical properties [73]. Covalently linked MOF and PIM-1 (PIM = intrinsic microporous polymer) were synthesized through a one-pot reaction, where chemical bonds were formed between the NH_2 groups in the MOF and the monomer (DCTB) [74].

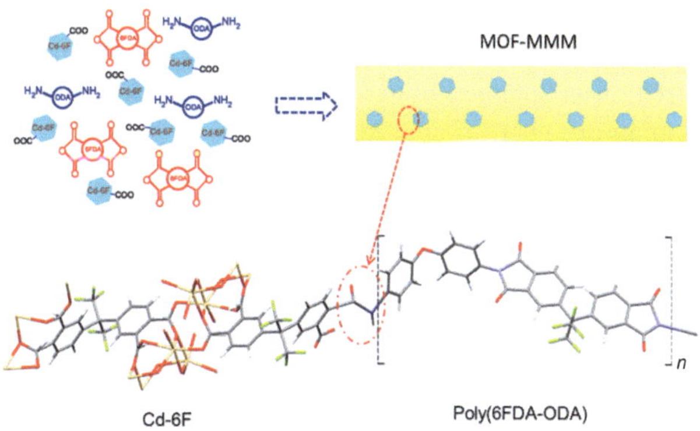

Figure 3.30 Schematic illustration of covalent link of MOF in MMM-A. *Source:* Lin et al. [75] / with permission of American Chemical Society.

Although the prospects are promising, the poor compatibility between MOFs and polymers remains a significant challenge to address. A notable strategy is the direct attachment of MOFs to polymers. As illustrated in Figure 3.30, defect-free, cross-linked MOF-polymer composites can be achieved through in situ copolymerization using CD-6F (MOFs derived from 6FDA ligands) and 6FDA-ODA polymer [75]. To evaluate the properties of composites based on their interfacial interactions with MOFs, four types of MMMs were prepared: MMM-A, derived from covalently linked MOF-polymer composites; MMM-B, created by introducing MOFs after polymerization; MMM-C, formed from a physical mixing of MOFs and polymers; and the original 6FDA-ODA polymer. All the composites demonstrated improved CO_2 permeability due to the incorporation of porous MOFs compared to pure polymer membranes. Among these, only MMM-A exhibited selective enhancements.

3.2.6 Physical Mixing Method

The physical mixing method, a widely used material preparation technique, involves the straightforward physical blending of organic materials with inorganic components. This method offers several advantages, including simplicity of operation, lower cost, and the absence of complex chemical reactions. Unlike its application in pure organic photofunctional composites, where the focus is solely on organic components, the physical mixing method in organic/inorganic composites primarily relies on physical interactions between the organic and inorganic materials, such as van der Waals forces and electrostatic interactions. By carefully optimizing physical mixing techniques—such as spin-coating, adsorption, or simple blending—the optical, mechanical, and electrical properties of the composites can be significantly enhanced. Furthermore, the physical mixing method in organic/inorganic composites preserves the intrinsic properties of the inorganic components, ensuring their superior performance in photofunctional applications. As a result, this approach is commonly employed in the fabrication of high-performance composites.

A photothermal nanoplatform can be constructed by combining carbon materials, especially reduced graphene oxide (rGO), with mesoporous silica (MS) through physical mixing [76]. First, rGO was obtained by reducing graphene oxide with dopamine, which enables it to absorb near-infrared light and convert it into heat. Next, mesoporous silica was coated on the surface of rGO by physical means to construct the rGO-MS composite, which has a large surface area and adjustable pore size to effectively encapsulate photodynamic therapeutic drugs (e.g. chlorin e6) in the pore space by physical adsorption. Finally, hyaluronic acid was further modified on the surface of the composite to prevent the premature release of the drug, and it was also able to enhance the targeting of the nanoplatform by binding to the CD-44 receptor of cancer cells.

The coating method can be utilized for applying PEG copolymers to metal nanoparticles. Photofunctional nanocomposites can be created using acetal-PEG-*b*-PAMA block copolymers [77]. This approach relies on the self-reduction of the tertiary amino groups in the PAMA segment, which also facilitates the anchoring of PEG onto the surface of gold nanoparticles.

In addition to conventional coating methods, the physical drop-coating technique is employed to prepare films. The drop-coating process involves adding solutions dropwise onto a substrate without rotation, followed by heating or annealing for some time to remove water. Using ligands of varying lengths—such as poly(ethylene glycol), TGL, or DTG—colloidal PbS quantum dots can be combined with graphene to construct photofunctional composites via the drop-coating method [78]. In this process, the PbS quantum dot solution is first added dropwise onto the graphene surface and subsequently dried under vacuum, forming a continuous overlayer approximately 100 nm thick, as illustrated in Figure 3.31. In these composites, graphene serves as the substrate material, playing a pivotal role in efficient photovoltaic conversion by providing high electron mobility and excellent electrical conductivity, while quantum dots contribute photosensitivity. The photoresponsivity of the composites can be influenced by the size, composition, and type of surface ligands of the quantum dots. The drop-coating method allows for the formation of a high-quality quantum dot layer on the graphene surface, optimizing interfacial properties. Furthermore, the photovoltaic performance of the composites can be tailored by adjusting the functionalization, morphology of the quantum dots, and other parameters.

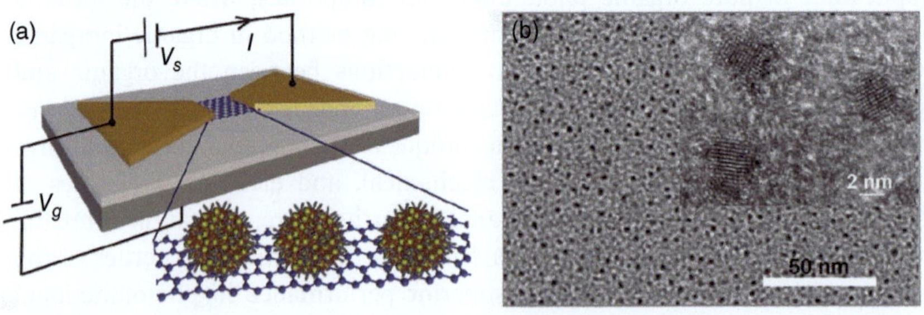

Figure 3.31 (a) The structure of the device. (b) TEM and HRTEM (inset) images of QD combined with graphene oxide. *Source:* Turyanska et al. [78] / John Wiley & Sons / CC BY 4.0.

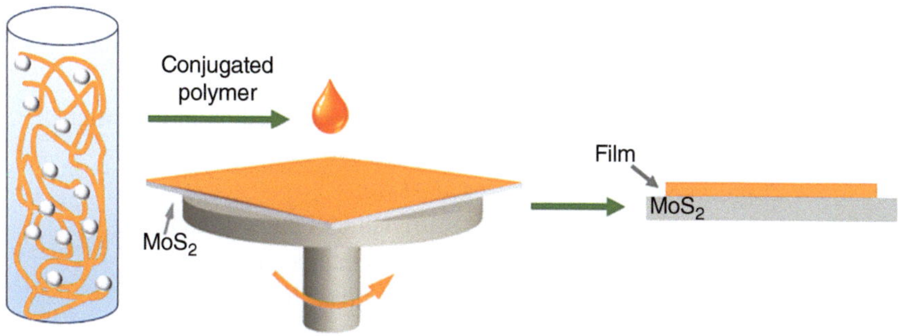

Figure 3.32 Schematic diagram of composite film prepared by spin-coating method.

Spin coating is an equally effective alternative to drop coating. Unlike drop coating, spin coating involves the use of a rotating substrate for applying the coating. The film thickness can be precisely controlled by adjusting the rotation speed and duration of the process. Reduced graphene oxide composites modified with germanium quantum dots can also be constructed by spin-coating germanium quantum dot solutions onto graphene substrates [79]. In addition, the spin-coating method can also be used to prepare organic/two-dimensional (2D) semiconductor heterojunction composites [80]. As shown in Figure 3.32, conjugated polymer solutions (e.g. P3HT, PCDTBT, and PTB7) were dissolved in appropriate solvents and coated on the surface of transferred molybdenum disulfide film by spin-coating to constitute a heterojunction. The spin-coating method can precisely control the coating thickness and uniformity of the polymers and can effectively promote the charge transfer process, thus realizing higher photovoltaic performance.

The physical mixing method can be easily achieved through simple sonication. For instance, heterojunction composites of conjugated polymers (such as PFBT, PFFBT, and PF2FBT) with TiO_2 were prepared using physical mixing techniques [81]. Initially, TiO_2 powder and polymers (e.g. PFBT, PFFBT, and PF2FBT) were combined with anhydrous chloroform solvent in specific proportions to form a mixture, which was then transformed into a stable suspension via ultrasonication. Following the evaporation of the suspension, the resulting polymer/TiO_2 powder was further dried at 50 °C to produce the final composite material. The integration of TiO_2 with the conjugated polymer significantly improves photocatalytic performance under visible light, while the incorporation of fluorine enhances the efficiency of the heterojunction materials in photocatalytic reactions, particularly by boosting H_2O_2 production.

References

1 Liu, Y.N., Wang, H., Li, S.L. et al. (2020). In situ supramolecular polymerization-enhanced self-assembly of polymer vesicles for highly efficient photothermal therapy. *Nature Communications* 11 (1): 1724.

2 Wang, X.Y., Li, S.L., Zhang, P.B. et al. (2015). An optical nanoruler based on a conjugated polymer-silver nanoprism pair for label-free protein detection. *Advanced Materials* 27 (39): 6040–6045.
3 Wu, C.F., Schneider, T., Zeigler, M. et al. (2010). Bioconjugation of ultrabright semiconducting polymer dots for specific cellular targeting. *Journal of the American Chemical Society* 132 (43): 15410–15417.
4 Wu, C.F., Jin, Y.H., Schneider, T. et al. (2010). Ultrabright and bioorthogonal labeling of cellular targets using semiconducting polymer dots and click chemistry. *Angewandte Chemie International Edition* 49 (49): 9436–9440.
5 Zhang, H.J., Liang, Y.C., Zhao, H. et al. (2020). Dual-mode antibacterial conjugated polymer nanoparticles for photothermal and photodynamic therapy. *Macromolecular Bioscience* 20 (2): 1900301.
6 Yu, J.B., Wu, C.F., Zhang, X.J. et al. (2012). Stable functionalization of small semiconducting polymer dots via covalent cross-linking and their application for specific cellular imaging. *Advanced Materials* 24 (26): 3498–3504.
7 Akbulut, M., Ginart, P., Gindy, M.E. et al. (2009). Generic method of preparing multifunctional fluorescent nanoparticles using flash nanoprecipitation. *Advanced Functional Materials* 19 (5): 718–725.
8 Reisch, A., Runser, A., Arntz, Y. et al. (2015). Charge-controlled nanoprecipitation as a modular approach to ultrasmall polymer nanocarriers: making bright and stable nanoparticles. *ACS Nano* 9 (5): 5104–5116.
9 Wang, M.W., Yang, N., Guo, Z.Q. et al. (2015). Facile preparation of AIE-active fluorescent nanoparticles through flash nanoprecipitation. *Industrial & Engineering Chemistry Research* 54 (17): 4683–4688.
10 Lyu, Y., Fang, Y., Miao, Q.Q. et al. (2016). Intraparticle molecular orbital engineering of semiconducting polymer nanoparticles as amplified theranostics for *in vivo* photoacoustic imaging and photothermal therapy. *ACS Nano* 10 (4): 4472–4481.
11 Jin, Y.H., Ye, F.M., Zeigler, M. et al. (2011). Near-infrared fluorescent dye-doped semiconducting polymer dots. *ACS Nano* 5 (2): 1468–1475.
12 Ye, F.M., Wu, C.F., Jin, Y.H. et al. (2011). Ratiometric temperature sensing with semiconducting polymer dots. *Journal of the American Chemical Society* 133 (21): 8146–8149.
13 Cui, Q.L., Yang, Y., Yao, C. et al. (2016). Aggregation-induced energy transfer of conjugated polymer materials for ATP sensing. *ACS Applied Materials & Interfaces* 8 (51): 35578–35586.
14 Credi, C., De Marco, C., Molena, E. et al. (2016). Heparin micropatterning onto fouling-release perfluoropolyether-based polymers via photobiotin activation. *Colloids and Surfaces B: Biointerfaces* 146: 250–259.
15 Kim, K., Selvapalam, N., Ko, Y.H. et al. (2007). Functionalized cucurbiturils and their applications. *Chemical Society Reviews* 36 (2): 267–279.
16 Bai, H.T., Yuan, H.X., Nie, C.Y. et al. (2015). A supramolecular antibiotic switch for antibacterial regulation. *Angewandte Chemie International Edition* 54 (45): 13208–13213.

17 Al Kobaisi, M., Bhosale, S.V., Latham, K. et al. (2016). Functional naphthalene diimides: synthesis, properties, and applications. *Chemical Reviews* 116 (19): 11685–11796.

18 Ikeda, A. and Shinkai, S. (1997). Novel cavity design using calix[*n*]arene skeletons: toward molecular recognition and metal binding. *Chemical Reviews* 97 (5): 1713–1734.

19 Wang, L.H., Guo, D.S., Chen, Y., and Liu, Y. (2006). Thermodynamics of interactions between organic ammonium ions and sulfonatocalixarenes. *Thermochimica Acta* 443 (1): 132–135.

20 Feng, X.L., Lv, F.T., Liu, L.B. et al. (2010). Conjugated polymer nanoparticles for drug delivery and imaging. *ACS Applied Materials & Interfaces* 2 (8): 2429–2435.

21 Petkau, K., Kaeser, A., Fischer, I. et al. (2011). Pre- and postfunctionalized self-assembled π-conjugated fluorescent organic nanoparticles for dual targeting. *Journal of the American Chemical Society* 133 (42): 17063–17071.

22 Fischer, I., Petkau-Milroy, K., Dorland, Y.L. et al. (2013). Self-assembled fluorescent organic nanoparticles for live-cell imaging. *Chemistry: A European Journal* 19 (49): 16646–16650.

23 Wang, G., Pu, K.Y., Zhang, X.H. et al. (2011). Star-shaped glycosylated conjugated oligomer for two-photon fluorescence imaging of live cells. *Chemistry of Materials* 23 (20): 4428–4434.

24 Chen, Y.J., Han, H.J., Tong, H.X. et al. (2016). Zwitterionic phosphorylcholine-TPE conjugate for pH-responsive drug delivery and AIE active imaging. *ACS Applied Materials & Interfaces* 8 (33): 21185–21192.

25 Guo, D.S., Wang, K., Wang, Y.X., and Liu, Y. (2012). Cholinesterase-responsive supramolecular vesicle. *Journal of the American Chemical Society* 134 (24): 10244–10250.

26 Torchilin, V.P. (2005). Recent advances with liposomes as pharmaceutical carriers. *Nature Reviews Drug Discovery* 4 (2): 145–160.

27 Loftsson, T., Vogensen, S.B., Brewster, M.E. et al. (2007). Effects of cyclodextrins on drug delivery through biological membranes. *Journal of Pharmaceutical Sciences* 96 (10): 2532–2546.

28 Safinya, C.R. (2001). Structures of lipid-DNA complexes: supramolecular assembly and gene delivery. *Current Opinion in Structural Biology* 11 (4): 440–448.

29 Peng, H.S. and Chiu, D.T. (2015). Soft fluorescent nanomaterials for biological and biomedical imaging. *Chemical Society Reviews* 44 (14): 4699–4722.

30 Howes, P., Green, M., Levitt, J. et al. (2010). Phospholipid encapsulated semiconducting polymer nanoparticles: their use in cell imaging and protein attachment. *Journal of the American Chemical Society* 132 (11): 3989–3996.

31 Li, K., Pan, J., Feng, S.S. et al. (2009). Generic strategy of preparing fluorescent conjugated-polymer-loaded poly(DL-lactide-*co*-glycolide) nanoparticles for targeted cell imaging. *Advanced Functional Materials* 19 (22): 3535–3542.

32 Koralli, P., Nega, A.D., Vagiaki, L.E. et al. (2020). New conjugated polymer nanoparticles with high photoluminescence quantum yields for far-red and near infrared fluorescence bioimaging. *Materials Chemistry Frontiers* 4 (8): 2357–2369.

33 Gluz, E., Grinberg, I., Corem-Salkmon, E. et al. (2013). Engineering of new crosslinked near-infrared fluorescent polyethylene glycol bisphosphonate nanoparticles for bone targeting. *Journal of Polymer Science Part A: Polymer Chemistry* 51 (20): 4282–4291.

34 Cova, L., Bigini, P., Diana, V. et al. (2013). Biocompatible fluorescent nanoparticles for *in vivo* stem cell tracking. *Nanotechnology* 24 (24): 245603.

35 Jiang, L.Y., Bai, H.T., Liu, L.B. et al. (2019). Luminescent, oxygen-supplying, hemoglobin-linked conjugated polymer nanoparticles for photodynamic therapy. *Angewandte Chemie International Edition* 58 (31): 10660–10665.

36 Yue, J.Y., Ding, X.L., Wang, L. et al. (2021). Novel enzyme-functionalized covalent organic frameworks for the colorimetric sensing of glucose in body fluids and drinks. *Materials Chemistry Frontiers* 5 (10): 3859–3866.

37 Yuan, S.S., Li, X., Zhu, J.Y. et al. (2019). Covalent organic frameworks for membrane separation. *Chemical Society Reviews* 48 (10): 2665–2681.

38 Biswal, B.P., Chaudhari, H.D., Banerjee, R. et al. (2016). Chemically stable covalent organic framework (COF)-polybenzimidazole hybrid membranes: enhanced gas separation through pore modulation. *Chemistry: A European Journal* 22 (14): 4695–4699.

39 Ananikov, V.P. (2019). Organic-inorganic hybrid nanomaterials. *Nanomaterials* 9: 9, 1197.

40 Wu, W. and Zhou, S. (2010). Hybrid micro-/nanogels for optical sensing and intracellular imaging. *Nano Reviews* 1: 5730.

41 Zhang, L.B. and Wang, E.K. (2014). Metal nanoclusters: new fluorescent probes for sensors and bioimaging. *Nano Today* 9 (1): 132–157.

42 Homola, J. (2008). Surface plasmon resonance sensors for detection of chemical and biological species. *Chemical Reviews* 108 (2): 462–493.

43 Zhang, L.B., Jean, S.R., Ahmed, S. et al. (2017). Multifunctional quantum dot DNA hydrogels. *Nature Communications* 8: 381.

44 Tang, F., He, F., Cheng, H.C. et al. (2010). Self-assembly of conjugated polymer-Ag@SiO_2 hybrid fluorescent nanoparticles for application to cellular imaging. *Langmuir* 26 (14): 11774–11778.

45 Zhang, Z.P., Feng, J.Y., Huang, P.C. et al. (2019). Ratiometric fluorescent detection of phosphate in human serum with functionalized gold nanoclusters based on chelation-enhanced fluorescence. *Sensors and Actuators B-Chemical* 298: 126891.

46 Gao, X., Ma, Z.Y., Sun, M.J. et al. (2022). A highly sensitive ratiometric fluorescent sensor for copper ions and cadmium ions in scallops based on nitrogen doped graphene quantum dots cooperating with gold nanoclusters. *Food Chemistry* 369: 130964.

47 Li, M.Q., Lao, Y.H., Mintz, R.L. et al. (2019). A multifunctional mesoporous silica-gold nanocluster hybrid platform for selective breast cancer cell detection using a catalytic amplification-based colorimetric assay. *Nanoscale* 11 (6): 2631–2636.

48 Kong, X.J., Tian, J.X., Fang, Y.Z. et al. (2022). Terbium metal-organic framework/bovine serum albumin capped gold nanoclusters-based dual-emission

reverse change ratio fluorescence nanoplatform for fluorimetric and colorimetric sensing of heparin and chondroitin sulfate. *Sensors and Actuators B-Chemical* 356: 131331.
49 Wu, N.N., Chen, L.G., Xiao, M.Z. et al. (2023). Determination of trypsin using protamine mediated fluorescent enhancement of DNA templated Au nanoclusters. *Microchimica Acta* 190 (4): 158.
50 Mock, J.J., Hill, R.T., Degiron, A. et al. (2008). Distance-dependent plasmon resonant coupling between a gold nanoparticle and gold film. *Nano Letters* 8 (8): 2245–2252.
51 Barman, M.K., Paramanik, B., Bain, D., and Patra, A. (2016). Light harvesting and white-light generation in a composite of carbon Dots and dye-encapsulated BSA-protein-capped gold nanoclusters. *Chemistry: A European Journal* 22 (33): 11699–11705.
52 Croissant, J.G., Zhang, D.Y., Alsaiari, S. et al. (2016). Protein-gold clusters-capped mesoporous silica nanoparticles for high drug loading, autonomous gemcitabine/doxorubicin co-delivery, and in-vivo tumor imaging. *Journal of Controlled Release* 229: 183–191.
53 Ma, L., Zhang, M.Y., Yang, A.J. et al. (2018). Sensitive fluorescence detection of heparin based on self-assembly of mesoporous silica nanoparticle-gold nanoclusters with emission enhancement characteristics. *Analyst* 143 (22): 5388–5394.
54 Park, H.K., Park, J.H., Lee, H. et al. (2020). Material-selective polydopamine coating in dimethyl sulfoxide. *ACS Applied Materials & Interfaces* 12 (43): 49146–49154.
55 Xu, S.H., Nie, Y.Y., Jiang, L.P. et al. (2018). Polydopamine nanosphere/gold nanocluster (Au NC)-based nanoplatform for dual color simultaneous detection of multiple tumor-related microRNAs with DNase-I-assisted target recycling amplification. *Analytical Chemistry* 90 (6): 4039–4045.
56 Gao, B.J., Haghighatbin, M.A., and Cui, H. (2020). Polymer-encapsulated cobalt/gold bimetallic nanoclusters as stimuli-responsive chemiluminescent nanoprobes for reactive oxygen species. *Analytical Chemistry* 92 (15): 10677–10685.
57 Alkilany, A.M., Alsotari, S., Alkawareek, M.Y. et al. (2019). Facile hydrophobication of glutathione-protected gold nanoclusters and encapsulation into poly(lactide-*co*-glycolide) nanocarriers. *Scientific Reports* 9: 11098.
58 Jiang, T., Qu, G.J., Wang, J. et al. (2020). Cucurbiturils brighten Au nanoclusters in water. *Chemical Science* 11 (13): 3531–3537.
59 Li, P.Y., Chen, Y., Chen, C.H. et al. (2019). Multi-charged bis(*p*-calixarene)/pillararene functionalized gold nanoparticles for ultra-sensitive sensing of butyrylcholinesterase. *Soft Matter* 15 (41): 8197–8200.
60 Lan, X., Chen, Z., Dai, G.L. et al. (2013). Bifacial DNA origami-directed discrete, three-dimensional, anisotropic plasmonic nanoarchitectures with tailored optical chirality. *Journal of the American Chemical Society* 135 (31): 11441–11444.
61 Qian, H.L., Yang, C., and Yan, X.P. (2018). Layer-by-layer preparation of 3D covalent organic framework/silica composites for chromatographic separation of position isomers. *Chemical Communications* 54 (83): 11765–11768.

62 Chan, Y.H., Ye, F.M., Gallina, M.E. et al. (2012). Hybrid semiconducting polymer dot-quantum dot with narrow-band emission, near-infrared fluorescence, and high brightness. *Journal of the American Chemical Society* 134 (17): 7309–7312.

63 Xie, J.Y., Fan, F.F., Chen, Y.J. et al. (2021). Gram-scale synthesis of porous three-dimensional carbon nanosheets for high efficiency clean water production. *Materials Research Letters* 9 (4): 176–182.

64 Gu, W., Zhang, Q., Zhang, T. et al. (2016). Hybrid polymeric nano-capsules loaded with gold nanoclusters and indocyanine green for dual-modal imaging and photothermal therapy. *Journal of Materials Chemistry B* 4 (5): 910–919.

65 Piñeres-Quiñones, O.H., Oñate-Socarras, M.K., Wang, F.R. et al. (2024). Pickering emulsions of thermotropic liquid crystals stabilized by amphiphilic gold nanoparticles. *Langmuir* 40 (7): 3923–3934.

66 Zhang, Y.Y., Xu, H.D., Yang, Y.L. et al. (2021). Efficient fluorescence resonance energy transfer-based ratiometric fluorescent probe for detection of dopamine using a dual-emission carbon dot-gold nanocluster nanohybrid. *Journal of Photochemistry and Photobiology A: Chemistry* 411: 113195.

67 Jiang, W., Luo, W., Zong, R. et al. (2016). Polyaniline/carbon nitride nanosheets composite hydrogel: a separation-free and high-efficient photocatalyst with 3D hierarchical structure. *Small* 12 (32): 4370–4378.

68 Kumru, B., Molinari, V., Hilgart, M. et al. (2019). Polymer grafted graphitic carbon nitrides as precursors for reinforced lubricant hydrogels. *Polymer Chemistry* 10 (26): 3647–3656.

69 Kang, N., Zhang, S., Tang, F. et al. (2022). Silver-hydrogel/PDMS film with high mechanical strength for anti-interference strain sensor. *Colloids and Surfaces A: Physicochemical and Engineering Aspects* 654: 130071.

70 Zopes, D., Stein, B., Mathur, S. et al. (2013). Improved stability of "naked" gold nanoparticles enabled by in situ coating with mono and multivalent thiol PEG ligands. *Langmuir* 29 (36): 11217–11226.

71 Lin, Y., Li, C., Liu, A. et al. (2021). Responsive hyaluronic acid-gold cluster hybrid nanogel theranostic systems. *Biomaterials Science* 9 (4): 1363–1373.

72 Vellaichamy, B. and Periakaruppan, P. (2018). Synergistic combination of a novel metal-free mesoporous band-gap-modified carbon nitride grafted polyaniline nanocomposite for decontamination of refractory pollutant. *Industrial & Engineering Chemistry Research* 57 (19): 6684–6695.

73 Molavi, H., Shojaei, A., and Mousavi, S.A. (2018). Improving mixed-matrix membrane performance *via* PMMA grafting from functionalized NH_2-UiO-66. *Journal of Materials Chemistry A* 6 (6): 2775–2791.

74 Nguyen, T.B., Rodrigue, D., and Kaliaguine, S. (2018). In-situ cross interface linking of PIM-1 polymer and UiO-66-NH_2 for outstanding gas separation and physical aging control. *Journal of Membrane Science* 548: 429–438.

75 Lin, R.J., Ge, L., Hou, L. et al. (2014). Mixed matrix membranes with strengthened MOFs/polymer interfacial interaction and improved membrane performance. *ACS Applied Materials & Interfaces* 6 (8): 5609–5618.

76 Jiang, W., Mo, F., Jin, X. et al. (2017). Tumor-targeting photothermal heating-responsive nanoplatform based on reduced graphene oxide/mesoporous silica/hyaluronic acid nanocomposite for enhanced photodynamic therapy. *Advanced Materials Interfaces* 4 (20): 1700425.

77 Nagasaki, Y. (2010). Engineering of poly(ethylene glycol) chain-tethered surfaces to obtain high-performance bionanoparticles. *Science and Technology of Advanced Materials* 11 (5): 054505.

78 Turyanska, L., Makarovsky, O., Svatek, S.A. et al. (2015). Ligand-induced control of photoconductive gain and doping in a hybrid graphene-quantum dot transistor. *Advanced Electronic Materials* 1 (7): 1500062.

79 Liu, X., Ji, X.B., Liu, M.J. et al. (2015). High-performance Ge quantum dot decorated graphene/zinc-oxide heterostructure infrared photodetector. *ACS Applied Materials & Interfaces* 7 (4): 2452–2458.

80 Petoukhoff, C.E., Kosar, S., Goto, M. et al. (2019). Charge transfer dynamics in conjugated polymer/MoS_2 organic/2D heterojunctions. *Molecular Systems Design & Engineering* 4 (4): 929–938.

81 Hong, Y., Cho, Y., Go, E.M. et al. (2021). Unassisted photocatalytic H_2O_2 production under visible light by fluorinated polymer-TiO_2 heterojunction. *Chemical Engineering Journal* 418: 129346.

4

Photofunctional Polymer Nanocomposites

Nanoscale structures, as a special state between macroscopic matter and atoms/molecules, break through the performance boundaries of traditional materials. Unlike bulk matrix materials with constant physical properties, the size effect of nanoparticles significantly regulates their physicochemical behavior. The quantum effects induced by electron hybridization and charge transfer often endow nanomaterials with unconventional properties. Thanks to the innovation of chemical/physical synthesis methods, it is now possible to precisely construct complex-structured photofunctional nanoparticles from polymeric multiphase systems. Nanoparticles with core-shell structures can achieve functional modularization through the partition integration of heterogeneous materials. The cavity load characteristics of hollow structures expand applications in drug delivery and molecular encapsulation. The ordered assembly of multiple components in a hybrid structure generates a synergistic effect of $1+1>2$. Biomimetic structures such as emulsions, liposomes, micelles, and capsules not only enhance biocompatibility but also strengthen system stability through intermolecular forces, providing a new paradigm for photofunction regulation. The diverse structural types have greatly enriched the photofunctions of polymer nanocomposites, making them an important research direction in the current field of nanotechnology and functional materials. In this chapter, we will systematically analyze the relationship between structure design, optical properties, and application scenarios of photofunctional polymer nanocomposites with detailed data to demonstrate the great potential of photofunctional polymer nanocomposites in biomedicine, environmental treatment, and opto-electronics due to their distinctive structural design and varied optical characteristics.

4.1 Structures of Photofunctional Polymer Nanocomposites

4.1.1 Core-shell Structure

The selection of materials for the core and the shell in core-shell nanostructures has an impact on different aspects of functionality. Integrating multiple compo-

Photofunctional Polymer Composites for Bioapplications, First Edition. Xiaoyu Wang and Lidong Li.

nents into a single functional system through modification, encapsulation, or functionalization can protect active substances from harsh environments and bring improved performance, including stability, biocompatibility, multifunctionality, and dispersibility, all of which are not easily obtained from a single component alone. More importantly, the heterogeneous interfaces between the various components in the core-shell structure may generate enhanced or new physicochemical properties [1].

For example, nanoparticles with different SiO_2 shell thicknesses can be prepared by the improved sol-gel method, labeled as Ag@SiO_2-n (n = 1 – 6) (Figure 4.1). The thickness of the SiO_2 shell layer in the experiment can be controlled within the range of 2–39 nm. In the TEM images, the darker nanoparticles are Ag cores, and the lighter material around them is the SiO_2 shell layer. The as-prepared Ag@SiO_2 nanoparticles exhibit excellent monodispersity, with nearly all nanocomposites containing a single Ag core. The Ag@SiO_2 nanoparticles, which are negatively charged on their surface, were assembled with positively charged water-soluble polyfluorene-based polymer (PFV) via electrostatic adsorption, producing fluorescent nanoparticles based on conjugated polymers. The adjustable SiO_2 shell layer serves as an isolation layer between the fluorescent species and the metal surface, keeping the fluorescent groups at a certain distance from the metal surface [2].

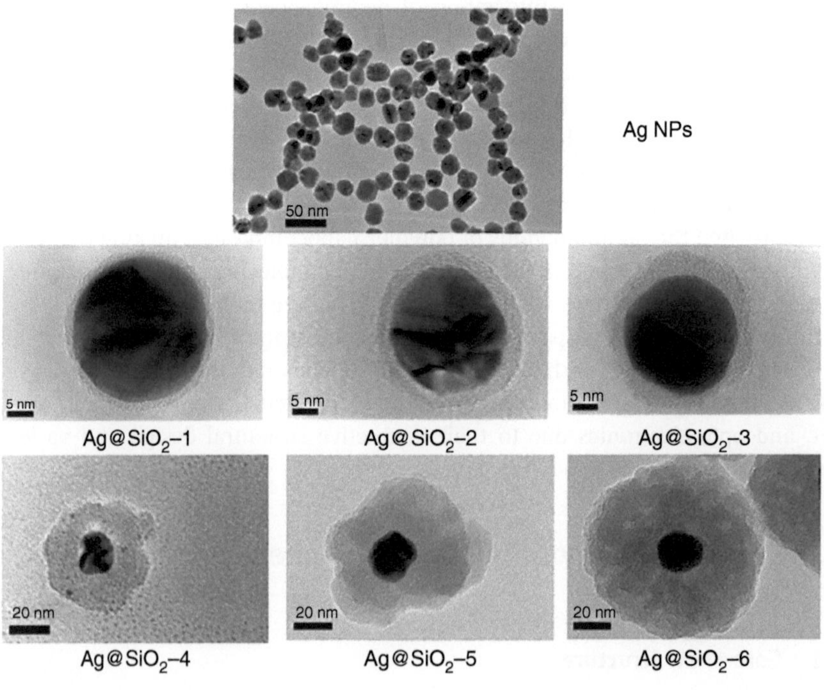

Figure 4.1 TEM images of Ag nanoparticles and Ag@SiO_2–1, Ag@SiO_2–2, Ag@SiO_2–3, Ag@SiO_2–4, Ag@SiO_2–5, Ag@SiO_2–6 nanoparticles. *Source:* Reproduced with permission from Tang et al. [2] / American Chemical Society.

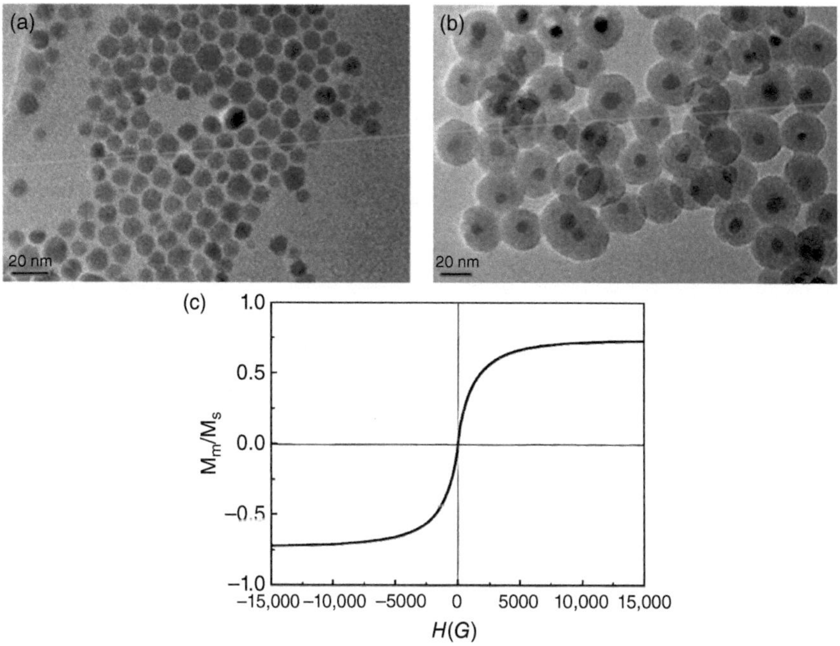

Figure 4.2 TEM images of (a) $CoFe_2O_4$ nanoparticles and (b) $CoFe_2O_4@SiO_2$ nanoparticles; (c) hysteresis loop of $CoFe_2O_4$ nanoparticles. *Source:* Reproduced with permission from Wang et al. [3] / Elsevier.

Using the identical sol-gel method, $CoFe_2O_4@SiO_2$ core-shell nanoparticles were fabricated as substrates. Water-soluble conjugated polymers (CPs) were then coated onto these nanoparticles via the layer-by-layer (LbL) assembly technique, thereby constructing a novel water-soluble fluorescent-magnetic bifunctional nanocomposite. The typical size of the synthesized $CoFe_2O_4$ nanoparticles is 12 ± 1 nm (Figure 4.2a). Almost all the prepared $CoFe_2O_4@SiO_2$ core-shell nanoparticles were uniformly coated, and the average thickness of the silica shell layer was 8.5 ± 1.5 nm (Figure 4.2b). By keeping their size below 16 nm, the nanoparticles can exhibit stable superparamagnetization at room temperature (Figure 4.2c). The SiO_2 shell not only effectively prevents the fluorescence quenching effect of magnetic nanoparticles but also endows the core-shell nanoparticles with dual functions of magnetism and fluorescence [3].

A cross-linked PNIPAM layer of a certain thickness can be synthesized on the surface of Ag nanoparticles through free radical polymerization reaction. Subsequently, the fluorescent conjugated polymer PFV is adsorbed on its surface (Figure 4.3a) to prepare temperature-sensitive core-shell fluorescent nanocomposites (PFV/Ag@PNIPAM) [4]. Ag nanoparticles were produced by reducing $AgNO_3$ using tannic acid, with an average particle size of 33 nm (Figure 4.3b). By taking advantage of the strong interaction between silver and disulfide bonds, *N*,*N*′-bis-acryloyl cystamine (BAC) is selected to interact with silver nanoparticles, thereby introducing groups with C=C double bonds on the surface

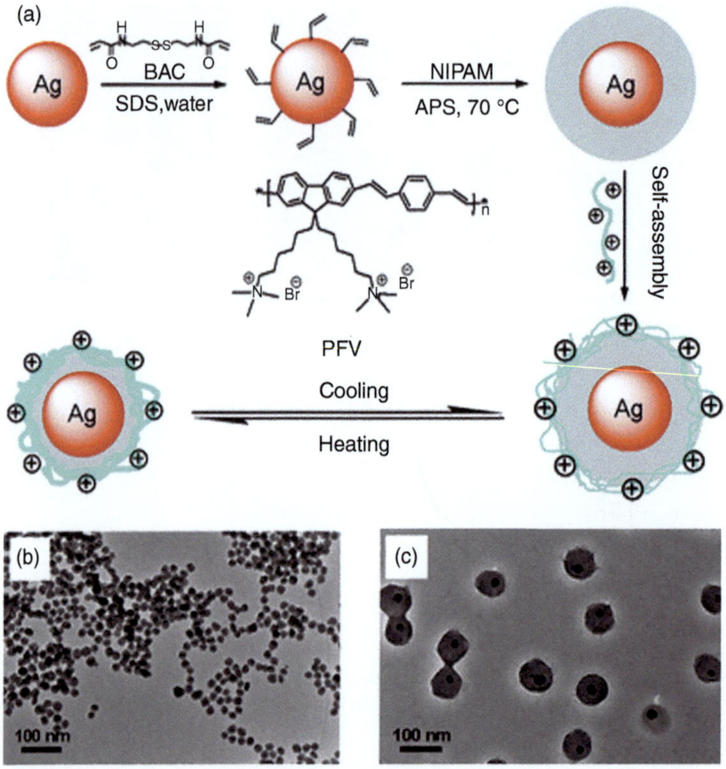

Figure 4.3 (a) Preparation of PFV/Ag@PNIPAM nanocomposites. TEM images of (b) Ag nanoparticles and (c) Ag@PNIPAM nanocomposites. *Source:* Reproduced with permission from Tang et al. [4] / The Royal Society of Chemistry.

of silver nanoparticles. Meanwhile, BAC also serves as a cross-linking agent in the NIPAM polymerization process. During the polymerization process, PNIPAM polymer chains grow on the surface of silver nanoparticles and form a uniform cross-linked polymer shell layer under the action of the cross-linking agent BAC. Moreover, the thickness of the PNIPAM shell can be adjusted by changing the amount of NIPAM monomer used and the time of the polymerization reaction [5]. Figure 4.3c reveals that the external surfaces of the silver nanoparticles are all uniformly coated with a PNIPAM shell of about 24 nm, and the core-shell structure is very obvious in the TEM image.

Similarly, amino groups are introduced via disulfide-linked cystamine dihydrochloride bound to the Ag nanoparticles' surface. The cross-linking agent 3,3′-dithiobis(sulfosuccinimidylpropionate) can be directly used to cross-link the side chains of PLL, the Ag nanoparticles, and DOX molecules, thereby preparing the Ag@PLL-DOX core-shell structure. Characterization reveals that the diameter of Ag nanoparticles is approximately 30 nm, and the size distribution is relatively uniform. After being coated with a shell layer, the composite system exhibits a clear core-shell structure, with a shell thickness of about 7 nm [6, 7].

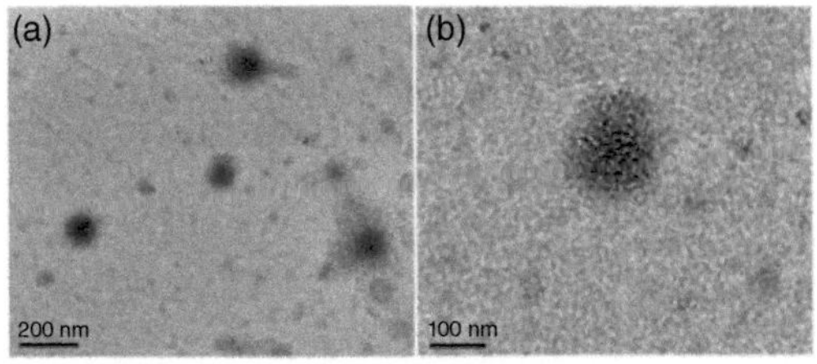

Figure 4.4 TEM images of (CTS–Au)@CMSD nanocomposites. *Source:* Reproduced with permission from Xia et al. [8] / American Chemical Society.

Biomacromolecules such as chitosan have demonstrated unique application value in the field of nanotechnology, especially in the in situ preparation of gold nanoparticles. Chitosan not only serves as a reducing agent but also acts as a stabilizer. During the formation process of chitosan-gold nanoparticles, gold nanoparticles are embedded in the coiled chitosan molecular chains. The TEM image clearly shows this structure: the edges of the nanoparticles are distributed with light-colored polymers, while the middle is uniformly distributed with dark-colored gold nanospheres (Figure 4.4). Negatively charged carboxymethyl dialdehyde starch is adsorbed onto the surface of CTS–Au nanoparticles through electrostatic interaction and further undergoes a cross-linking reaction to form a stable nanoparticle (CTS–Au)@CMSD. In addition, the negative surface endows these nanoparticles with the ability to adsorb positively charged fluorescent conjugated polymers, thereby successfully preparing fluorescent nanocomposites with core-shell structures [8].

Core materials for such structures are not limited to inorganic nanoparticles; organic polymers also constitute viable alternatives. Polymer nanoparticles with carboxyl-rich surfaces were fabricated by co-precipitating a blend of conjugated polymer PDPP-DBT and poly(styrene-co-maleic anhydride) (PSMA). Surface functionalization was subsequently carried out via amide bond formation between the amine groups of the antimicrobial peptide tachyplesin-I (AMP) and the carboxyl groups of PSMA (Figure 4.5a,b) [9]. This structure utilizes reducing groups such as sulfhydryl group (—SH), amino group (—NH_2), and phenolic hydroxyl group (—PhOH) in the AMP molecule, allowing it to function both as a reducing agent, facilitating the reduction of Au(III) to Au(0) and Au(I), and as a ligand to stabilize and protect AuNCs. Under NIR light irradiation, the highly efficient photothermal conversion performance of PDPP-DBT converts light energy into local high temperature, which not only significantly accelerates the reduction rate of gold ions by AMP but also promotes molecular motion, thereby greatly enhancing the generation efficiency of core-shell structured Au(0) NC@Au(I)-AMP nanoclusters, and simultaneously increasing the fluorescence quantum yield of the composite system.

SEM images reveal that the incorporation of AuNCs significantly enhances the surface conductivity of the nanocomposites, resulting in higher imaging contrast (Figure 4.5c). Dynamic light scattering analysis further demonstrates that the CNP@AMP-AuNCs have an average size of approximately 66 nm (Figure 4.5d). TEM images distinctly illustrate the core-shell architecture of CNP@AMP-AuNCs, characterized by a low-contrast gray spherical organic core (PDPP-DBT@AMP) surrounded by high-contrast, black dot–like AuNCs. These AuNCs, with an average diameter of 2.1 ± 0.2 nm, are uniformly distributed throughout the outer shell

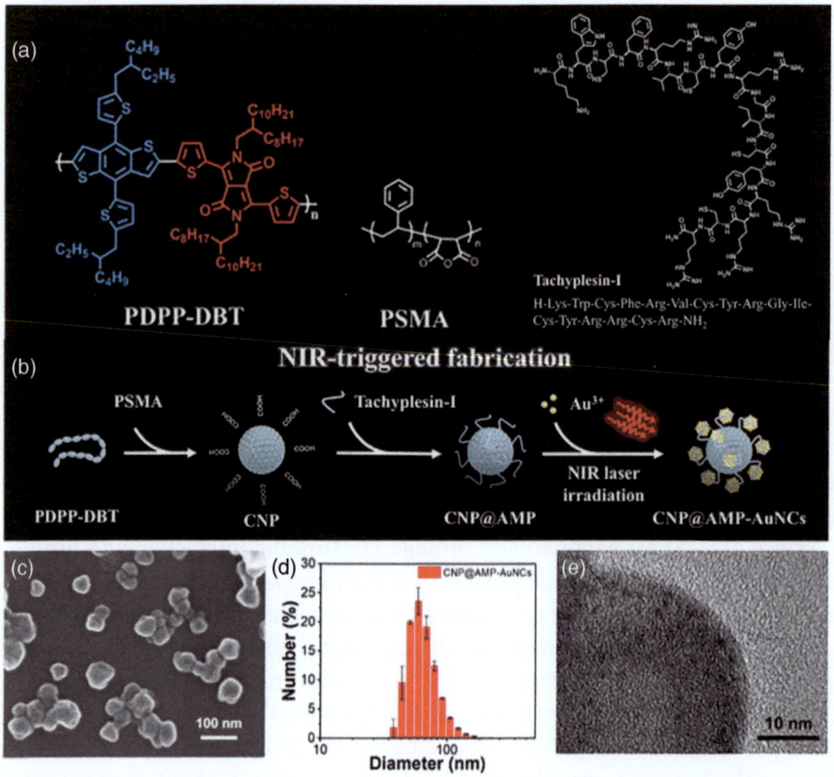

Figure 4.5 (a) The molecular structures of PDPP-DBT, PSMA and tachyplesin-I; (b) preparation process, (c) SEM images, (d) size distributions, and (e) TEM image of CNP@AMP-AuNCs. *Source:* Reproduced with permission from Zhu et al. [9] / American Chemical Society.

of the nanoparticles, while their presence in the core is minimal (Figure 4.5e). This core-shell configuration not only streamlines the synthesis of AuNCs but also enables the creation of an efficient and controllable nanocomposite system regulated by near-infrared light.

In addition to the in situ reduction method, the fluorescent conjugated polymer MEH-PPV and photothermal AuNCs can also be combined through covalent grafting to construct bifunctional nanocomposites (Figure 4.6a) [10]. Fluorescent nanoparticles were prepared by co-precipitation of the commercially produced conjugated polymer MEH-PPV. Polyethyleneimine (PEI) was then adsorbed on its surface by electrostatic interactions to modify the nanoparticle surface with abundant amino groups (Figure 4.6b). Photothermal AuNCs containing carboxylate ligands were covalently modified on the nanoparticle surfaces via the amidation reaction that forms amide bonds. As shown in Figure 4.6c, the as-prepared AuNCs have a uniform size distribution, with an average diameter of approximately 2 nm [11]. Distinct contrast differences are observed: the light gray spherical areas with low contrast correspond to the relatively large organic matrix composed

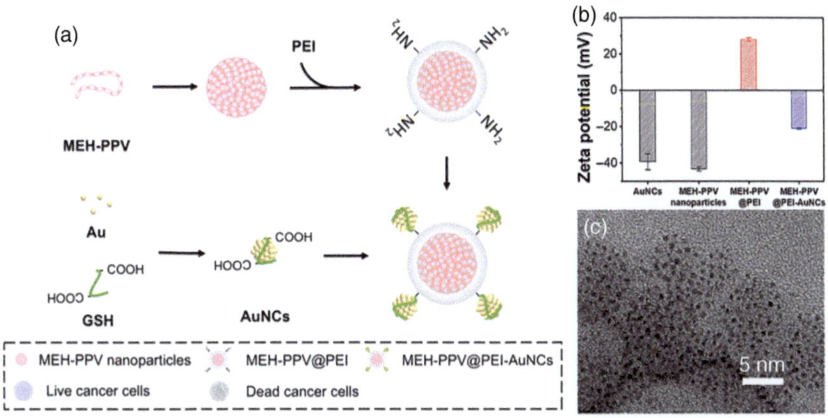

Figure 4.6 (a) The preparation process, (b) corresponding change in zeta potential, and (c) TEM image of the MEH-PPV@PEI-AuNCs. *Source:* Reproduced with permission from Zhu et al. [10] / Elsevier.

of MEH-PPV and PEI polymers, while the high-contrast black punctate areas represent the AuNCs. These metallic nanoclusters are uniformly distributed within the organic matrix, forming a characteristic core-shell structure.

4.1.2 Hollow Structure

Hollow structured nanocomposites, due to their unique topological structure and complete morphology, possess characteristics such as large specific surface area, low density, wide internal space, excellent stability, and good permeability. Inorganic and organic hollow spheres possess distinct structural and performance characteristics, thus exhibiting diverse properties.

Based on the co-precipitation of organic conjugated oligomers (CB) and PSMAB, unique hollow-structured nanoparticles (CB/PSMAB-SA NP) were successfully prepared (Figure 4.7). Before the addition of CB, PSMAB formed solid spheres with a diameter of 20–30 nm and a uniform particle size. As the CB content increased from 0.1:1 to 0.5:1, more and larger-sized hollow nanoparticles appeared in the system. When the mass ratio of CB to PSMAB reaches 1:1, most nanoparticles exhibit a hollow structure with a size of approximately 100 nm. This phenomenon is due to the fact that CB, as a hydrophobic agent, helps to ensure the uniform distribution of nanoparticles in the system, which conforms to Ostwald's maturation theory [12].

Based on simulation calculation results, the distance between the two CDT "wings" within a single CB molecule (d_1) ranges from 2.8 to 4.0 Å, which is conducive to intramolecular π–π interactions to reduce the adverse effects of steric hindrance and form a stable "U"-shaped molecular configuration. On the other hand, CB dimers tend to adopt a face-to-face configuration of stacks, with a vertical distance (d_2) of 3.5–3.8 Å between the two BODIPY cores, which is conducive to the occurrence of intermolecular π–π interactions. SEM and TEM analyses demonstrated that the CB/PSMAB-SA nanoparticles were nearly spherical in

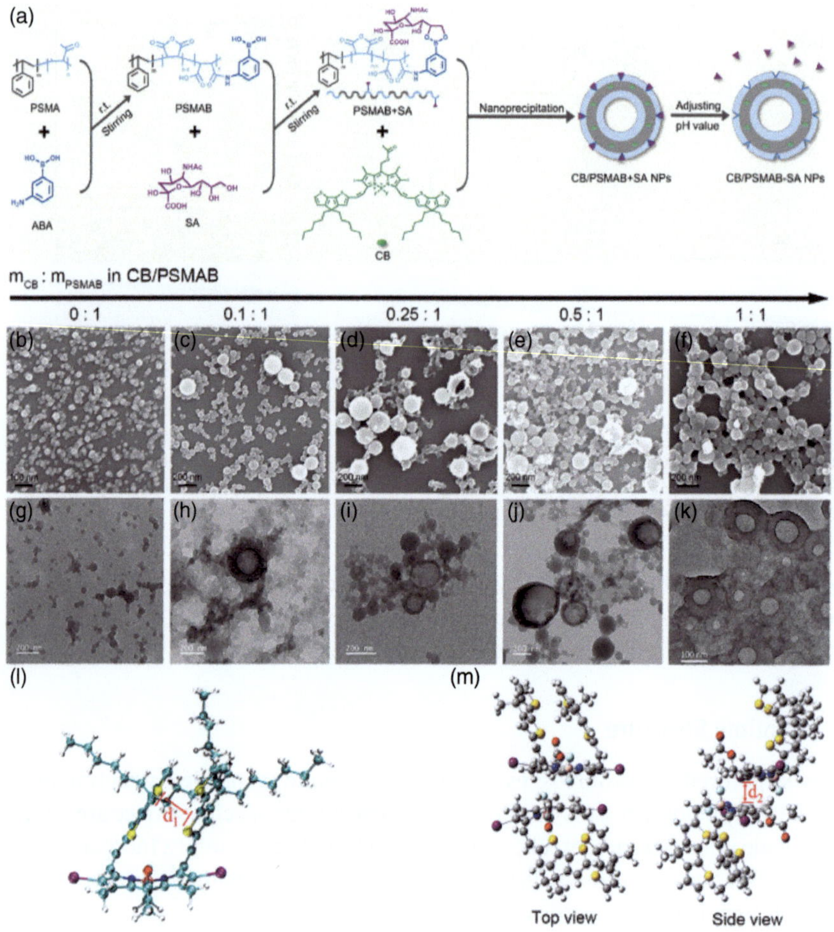

Figure 4.7 (a) The preparation process for CB/PSMAB-SA nanoparticles. (b–f) SEM and (g–k) TEM images of CB/PSMAB-SA NPs prepared with varying CB/PSMAB ratios. (l) Illustration of the stacking structure of CB. (m) Top and side views of a CB dimer. *Source:* Reproduced with permission from Peng et al. [12] / John Wiley & Sons.

shape, with an average size of approximately 100 nm. It is worth noting that some of the nanoparticles in the SEM images show a collapsed state, while the TEM clearly reveals their hollow structure, indicating that nanoparticles with cavities have been successfully prepared. Based on these data, it can be concluded that strong intramolecular and intermolecular π–π stacking interactions promote the self-assembly of CB molecules, which further guides the co-assembly of PSMAB and the subsequent formation of hollow nanoparticles.

Inorganic hollow nanoparticles are often prepared via templating methods. Monodisperse silica nanoparticles, with controllable size distribution and ease of large-scale synthesis, are often used as solid templates. Porous silica shells were prepared with SiO_2 as the core and *N*-cetyltrimethylammonium bromide (CTAB) as the soft template. After etching the core SiO_2, hollow mesoporous silica

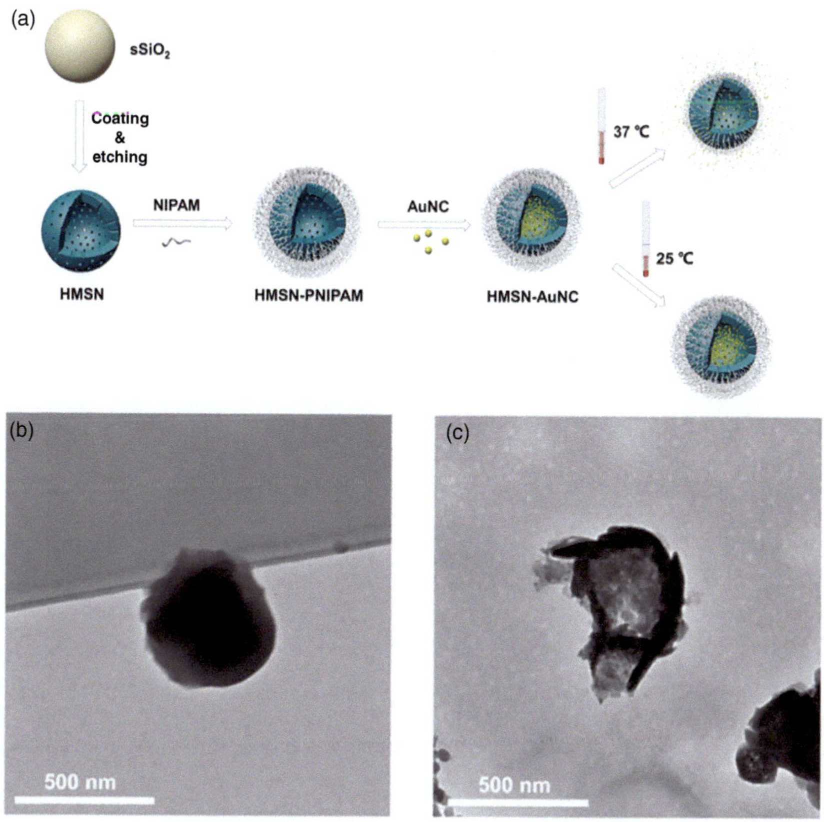

Figure 4.8 (a) The preparation of HMSN-AuNC. TEM images of (b) $sSiO_2$@CTAB-SiO_2 and (c) HMSN. *Source:* Reproduced with permission from Kang et al. [13] / Elsevier.

nanoparticles (HMSN) were obtained. Subsequently, thermally responsive PNIPAM was grafted onto its surface to prepare thermosensitive nanocomposites. The size of the ultra-small AuNCs matches well with the aperture of the HMSN. PNIPAM undergoes reversible phase transformation at different temperatures, accompanied by changes in apparent volume. This enables the "open/close" switching of the HMSN pores, allowing for the effective loading and encapsulation of AuNCs. Consequently, photofunctional nanocomposites loaded with AuNCs are constructed successfully (Figure 4.8) [13].

4.1.3 Hybrid Structure

By establishing physical interactions among multiple components and various substances (such as polymers and nanoparticles), uniform composites can be achieved through simple blending, thereby enabling the preparation of multifunctional nanomaterials. Furthermore, by adjusting the proportions or combinations of these components, precise modulation of the nanoparticle properties can be realized.

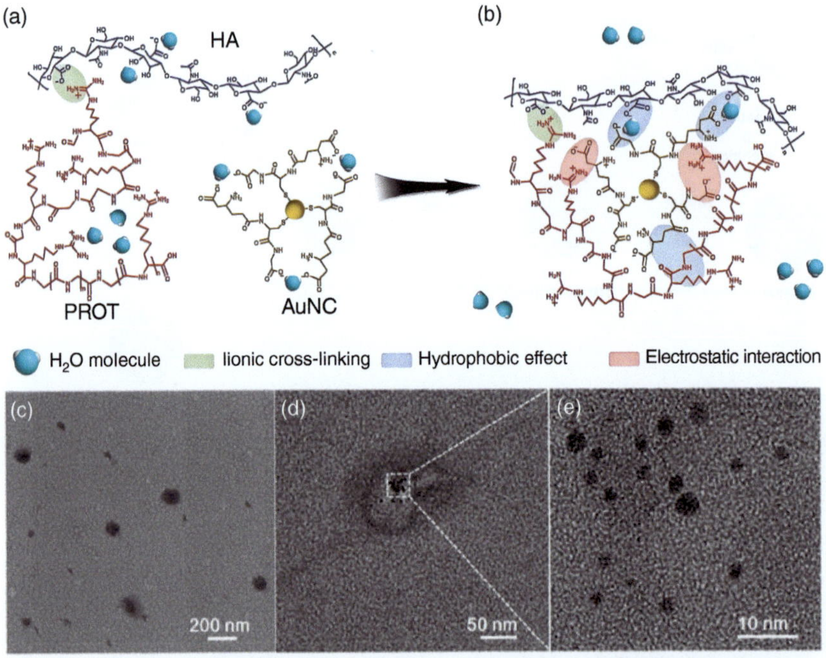

Figure 4.9 (a, b) The reparation of AuNC-HA-PROT nanocomposites via different interactions. TEM images of (c) AuNC-HA-PROT, (d) individual AuNC-HA-PROT, and (e) the loaded AuNC. *Source:* Reproduced with permission from Xia et al. [14] / American Chemical Society.

As shown in Figure 4.9a,b, through ionic cross-linking, the highly cationic protein protamine (PROT) and the negatively charged polymer hyaluronic acid (HA) form large supramolecular aggregates. Upon the addition of glutathione (GSH)-capped AuNCs, hydrophobic associations occur between the AuNC surface and the hydration layer of HA. Hydrophobic interactions simultaneously form between the hydrophobic amino acid side chains of PROT and GSH on AuNCs, while electrostatic attraction occurs between the positively charged PROT and the negatively charged GSH on AuNCs. Together, the hydrophobic effects and electrostatic interactions drive the formation of the AuNC-HA-PROT nanocomposite, resulting in a dense and stable structure [14]. TEM images confirmed that AuNCs were successfully encapsulated in the complex, forming uniform nanoparticles with a particle size of approximately 80 nm (Figure 4.9c–e). Moreover, AuNC-SP, capped with 11-mercaptoundecanoic acid (MUA), exhibits a negatively charged surface. Chitosan, as an amino polysaccharide, has abundant amino groups on its chain that carry positive charges. By simply mixing AuNC-SP and chitosan, AuNC-SP@CS nanocomposites can be obtained through electrostatic interactions [15].

Polymers can act as soft templates and bond with precursors through non-covalent bonds, thereby controlling the morphology and structure of inorganic materials. The microgel spheres prepared by cross-linked PNIPAM-co-PAA can efficiently adsorb a large number of positively charged silver ions due to the rich negative charges on their surfaces. The silver ions adsorbed on the surface of the microgel are reduced to silver nanoparticles under the action of a reducing agent and stably dispersed in the microgel matrix, thereby forming structurally stable nanocomposites [16].

Moreover, positively charged protamine can aggregate with molybdate ions through electrostatic attraction to form complexes. During the hydrothermal reaction process, biomolecules have an impact on MoS_2 nucleation, preventing the creation of large MoS_2 aggregates and producing smaller, evenly distributed MoS_2 nanoparticles [17]. The abundant amino and guanidine groups in protamine not only enable physical adsorption but also facilitate chemical bonding through the formation of Mo—N bonds. This interaction regulates the crystal structure of MoS_2 nanoparticles (Figure 4.10a).

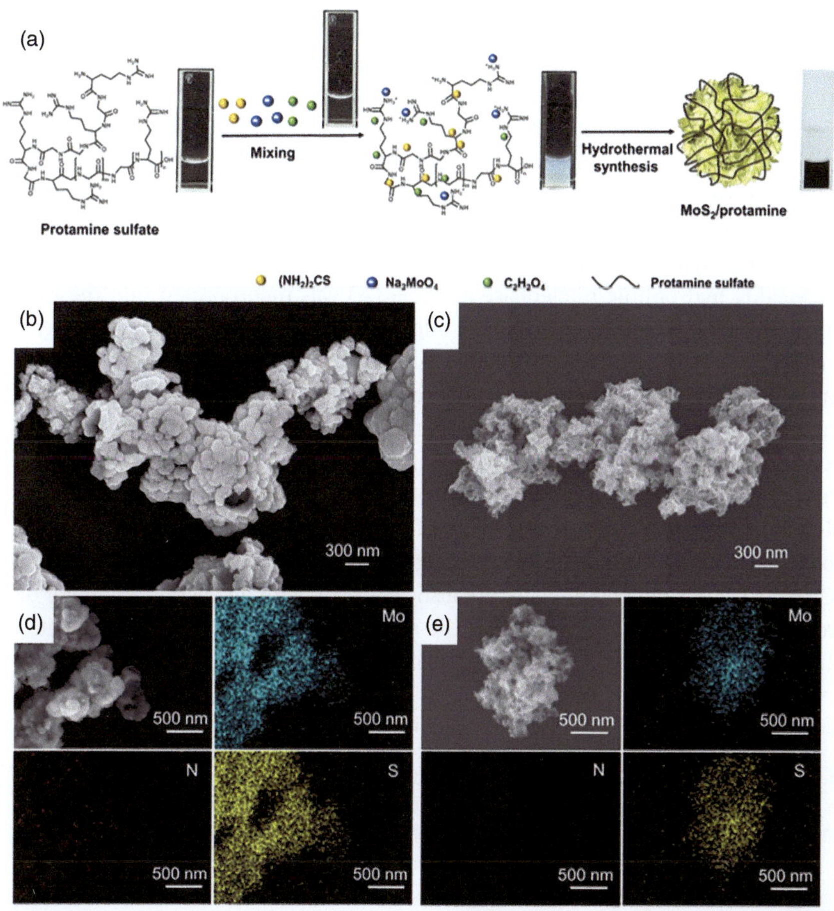

Figure 4.10 (a) The preparation process for MoS_2/protamine nanocomposites. (b) SEM images of MoS_2/protamine nanocomposites and (c) pure MoS_2 nanoparticles; (d) elemental maps of Mo, S, and N in MoS_2/protamine nanocomposites and (e) pure MoS_2 nanoparticles. *Source:* Reproduced with permission from Kang et al. [17] / American Chemical Society.

The MoS_2/protamine nanocomposites exhibit a spherical morphology with an average size of approximately 139 nm, and no distinct layered structure is observed (Figure 4.10b). In contrast, large aggregates are visible in the MoS_2 nanoparticle image (Figure 4.10c), where layered structures can be seen stacked on the surface of the aggregates. Additionally, the elemental mapping of MoS_2/protamine nanocomposites reveals consistent distributions of Mo, S, and N elements (Figure 4.10d), with the N signal confirming the presence of protamine. For MoS_2 nanoparticles, Mo and S elements are evenly distributed, but no N element is detected (Figure 4.10e). These findings confirm the incorporation of protamine into the MoS_2/protamine nanocomposites. The smaller size of these nanocomposites is attributed to the inhibitory effects of protamine.

Photofunctional molecules with conjugated structures have a certain degree of hydrophobicity, which also makes it easy for them to form stable intermolecular interactions through π–π stacking. Polymer PVK, with carbazole as its structural unit, is a typical electron donor material. Due to the hydrophobic interaction of molecules, PVK can co-precipitate with electron acceptor molecules (in a 1:1 mass ratio) to form nanoparticles. The SEM images in Figure 4.11 demonstrate that PVK, PVK/TPOTZ, PVK/BPOTZ, and PVK/BPOTZ-OR nanoparticles all exhibit a spherical morphology, with average diameters of 45 ± 3, 52 ± 5, 56 ± 7, and 53 ± 4 nm, respectively. Incorporating electron-acceptor small molecules into the fluorescent polymer PVK aggregates facilitates electron transfer, leading to the formation of exciplex nanoparticles [18].

The weakly hydrophilic nucleobase adenine (A) was incorporated into the side chain of the fluorescent conjugated oligomer 4,7-16(9,9′-bis(6-adeninehexyl) fluorenyl)-2,1,3-benzothiadiazole (OFBT-A). As shown in "Method A" in Figure 4.12a, OFBT-A nanoparticles obtained via the co-precipitation method are approximately spherical, with an average diameter of 88.6 nm (Figure 4.12b) and a ζ-potential of 26.6 mV in aqueous solution. It can simultaneously carry negatively charged DNA or RNA. Using the same co-precipitation method, short-chain DNA labeled with Texas Red (TR-DNA) can be efficiently encapsulated within OFBT-A nanoparticles. The SEM image shown in Figure 4.12c demonstrates that the resulting OFBT-A/TR-T5 nanocomposites are also spherical, with an average diameter of approximately 94.1 nm, indicating that the size and shape of the nanocomposites are little affected by the presence of TR-T5. After loading TR-T5, the ζ potential

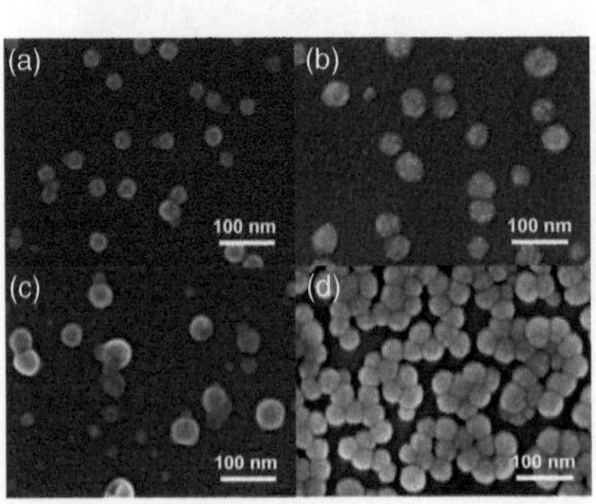

Figure 4.11 SEM images of (a) PVK, (b) PVK/TPOTZ, (c) PVK/BPOTZ, and (d) PVK/BPOTZ-OR nanoparticles. *Source:* Reproduced with permission from Liu et al. [18] / American Chemical Society.

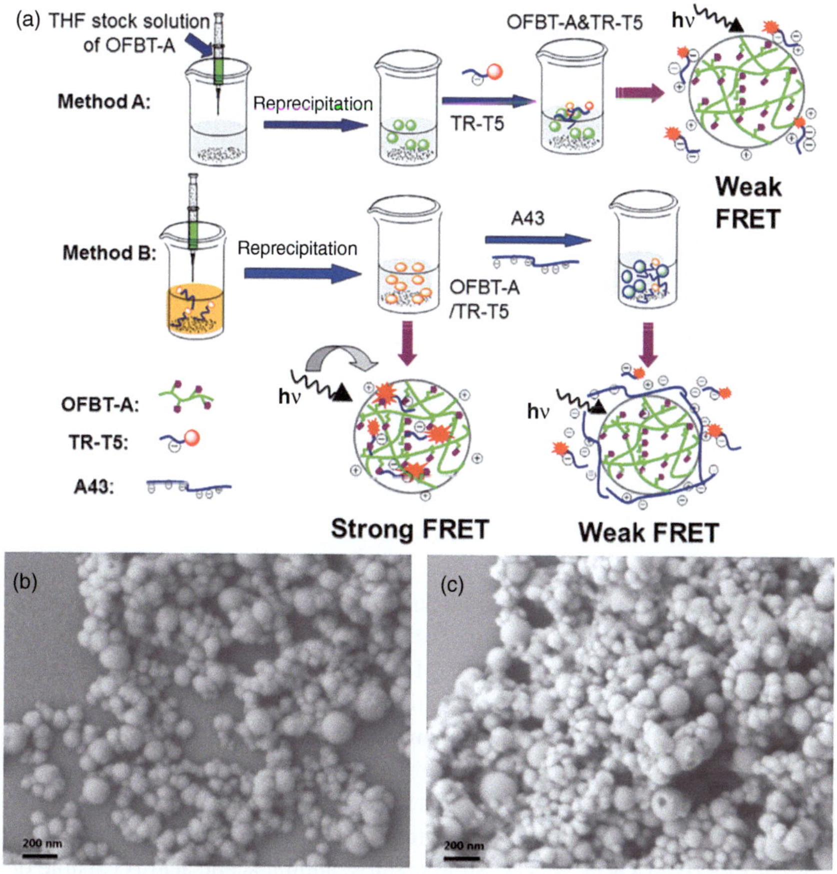

Figure 4.12 (a) Schematic diagram of the preparation of OFBT-A nanoparticles and OFBT-A/TR-T5 nanocomposites. SEM images of (b) OFBT-A nanoparticles and (c) OFBT-A/TR-T5 nanocomposites. *Source:* Reproduced with permission from Wang et al. [19] / American Chemical Society.

of the nanocomposites dropped from 26.6 to −7.1 mV. This nanoparticle is not dependent on base matching and can nonselectively load oligonucleotides of different sequences. However, the loading capacity of this nanoparticle is related to the charge density of DNA. A small amount of TR-T20 chains carries more negative charges, so the amount loaded onto nanoparticles is less than that of TR-T5 [19].

Combining the co-precipitation method with a molecular imprinting strategy enables the preparation of photofunctional nanocomposites with specific recognition capabilities. As illustrated in Figure 4.13a, poly(9-vinylcarbazole) (PVK) was selected as both the template and fluorescence signal molecule. Due to its rich carbazole group, it can combine with paracetamol (PCM) molecules through hydrogen bonds. PVK was dispersed using PSMA, which serves to reduce aggregation and mitigate subsequent self-quenching effects. In addition, PSMA partially hydrolyzes into carboxyl groups in water, introducing carboxyl groups into the system. This not

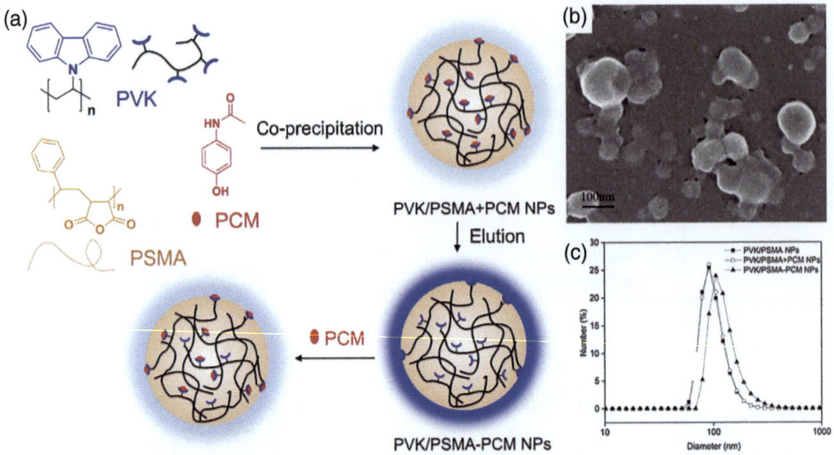

Figure 4.13 (a) The preparation process and (b) SEM image of PVK/PSMA-PCM NPs. (c) The size of PVK/PSMA NPs, PVK/PSMA + PCM NPs, and PVK/PSMA-PCM NPs. *Source:* Reproduced with permission from Wang et al. [20] / Elsevier.

only enhances the hydrophilicity of nanoparticles but also increases their binding affinity with template molecules. The formation of nanoparticles through the co-precipitation of PVK, PSMA, and PCM is facilitated by the combined influence of π–π stacking, hydrophobic interactions, and hydrogen bonding. The rich presence of aromatic groups in each component enhances both hydrophobicity and π–π stacking. Additionally, the carbazole groups in PVK and the acidic groups in PSMA establish numerous hydrogen bonds with PCM molecules. Therefore, when PCM molecules are embedded in the polymer matrix, the resulting nanoparticles are named PVK/PSMA + PCM NPs. Subsequently, adding acetic acid is capable of disrupting the binding between PVK and PCM, removing the template molecules, and finally obtaining PCM-imprinted polymer nanoparticles, which are named PVK/PSMA-PCM NPs [20].

SEM images reveal that all three nanoparticle types possess roughly spherical morphologies (Figure 4.13b). The measured particle sizes are as follows: PVK/PSMA nanoparticles average 73 ± 10 nm, PVK/PSMA + PCM nanoparticles average 78 ± 15 nm, and PVK/PSMA-PCM nanoparticles average 85 ± 25 nm (Figure 4.13c). It is worth noting that the size of PVK/PSMA-PCM nanoparticles is significantly larger than that of the other two types. This is because during the elution of the template molecules, the addition of acetic acid further hydrolyzes the anhydride groups, enhancing the hydrophilicity of the nanoparticles and causing the polymer nanoparticles to expand. The obtained nanoparticle dispersion showed no obvious aggregation when stored for at least 1 month, demonstrating good stability.

Similarly, the conjugated polymer PFBT-PBA can be obtained by employing PFBT as the backbone of the fluorescent polymer and introducing phenylboronic acid groups onto its side chains, which exhibit affinity for sialic acid molecules. The preparation of sialic acid-imprinted nanoparticles mainly consists of two steps: the co-precipitation method and template removal process. Specifically, first dissolve the polymer PFBT-PBA and sialic acid molecules in a good solvent, allowing the two to bond through a stable boronic ester; Subsequently, it is injected into its poor solvent, water. Due to the hydrophobic property of the conjugated polymer backbone, it will rapidly aggregate to form polymer nanoparticles. In this step, PFBT-PBA nanoparticles with embedded sialic acid molecules are obtained. The pH of the polymer nanoparticle dispersion was set to an acidic level. Most of the sialic acid molecules left the nanoparticle surface through the dissociation of the boronic ester. By dialysis, these free sialic acid molecules were removed to obtain nanoparticles with sialic acid imprinting. These sialic acid–imprinted nanoparticles have pores. The size, shape, and abundant binding sites of the pores make them have recognition specificity, thus making it easier for them to bind to sialic acid molecules or cancer cells with overexpression of sialic acid [21].

4.1.4 Emulsion

Pickering emulsions are emulsion systems stabilized by solid particles, where stability originates from the selective adsorption of these particles at the oil-water interface. The particles adsorbed at the interface form a dense physical barrier layer, effectively preventing the aggregation and merging of emulsion droplets, thereby significantly enhancing the kinetic stability of the emulsion. Based on this characteristic, the fine-tuning of emulsion microstructure and macroscopic properties can be achieved by scientifically screening parameters such as the type, size, and surface properties of the solid particles.

The hydrophobic dye DVP was covalently grafted onto the surface of GHS-protected AuNCs by EDC/NHS coupling to successfully prepare amphiphilic Au@DVP nanoclusters. Thanks to its amphiphilic emulsifying property, Au@DVP nanoclusters can form separated microdomain structures on the surface of emulsion droplets, significantly reducing interfacial energy and effectively inhibiting droplet coalescence, thereby ensuring the uniformity of particle size in the emulsion system during long-term storage. In the Cryo-TEM images, the spherical structure of the emulsion droplets can be clearly observed, and the high contrast at the edge is precisely the manifestation of the surface Au@DVP nanoclusters. When the boundary layer of emulsion droplets is exposed to the electron beam, the emulsion structure is disrupted, and uniformly dispersed gold nanoclusters are revealed, further confirming that amphiphilic gold nanoclusters are located at the emulsion boundary, forming micro-regions of the separation interface. In addition, by adjusting the grafting ratio of DVP, this emulsion system not only achieves multicolor fluorescence emission but also possesses an efficient photothermal conversion capability, laying a solid structural foundation for the multifunctional integration of materials (Figure 4.14) [22].

Similarly, amphiphilic gold nanoclusters AuNC@TA were successfully prepared by covalently bonding tryptamine to the surface of AuNC protected by GSH ligands. AuNC@TA is both hydrophilic and lipophilic and can be used as an excellent emulsifier for Pickering emulsions. Using styrene (St) monomer as the oil phase and

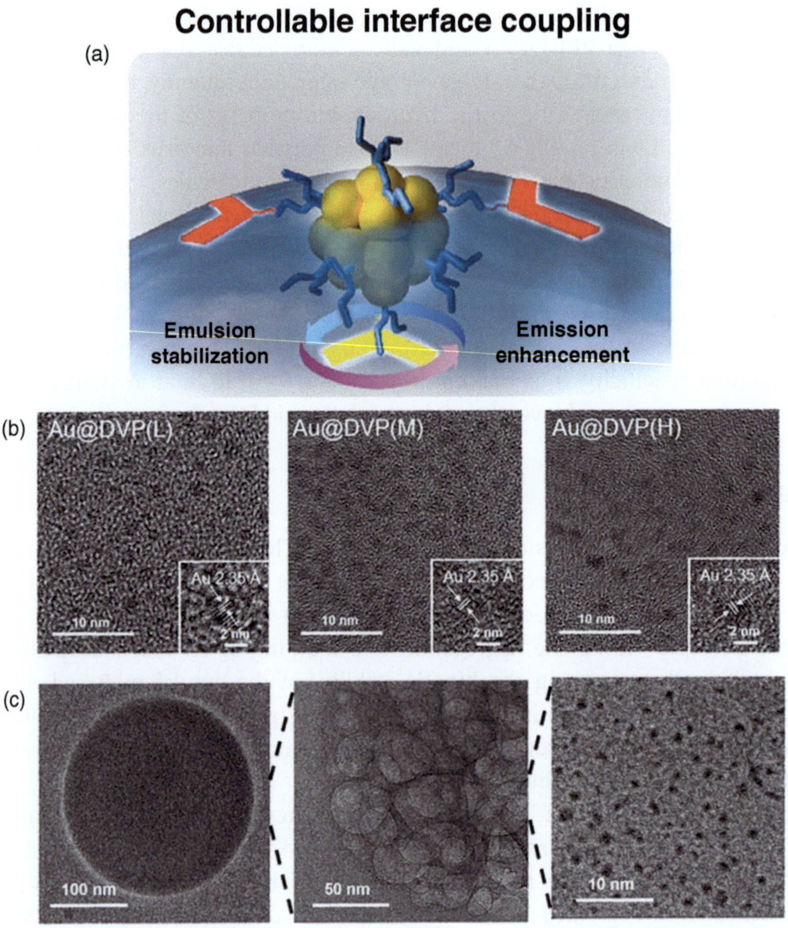

Figure 4.14 (a) The interaction between Au@DVP(H) and emulsion droplets. (b) TEM images of Au@DVPs with different graft ratios. (c) Cryo-TEM images of Au@DVP(L)-P. *Source:* Reproduced with permission from Liu et al. [22] / John Wiley & Sons.

aqueous solution as the water phase, this nanocluster can spontaneously adsorb at the styrene/water interface, and an O/W type Pickering emulsion was constructed. Based on the photothermal conversion performance of AuNC@TA, 660 nm red light can trigger the local photothermal effect of the emulsion. This further catalyzes the activation of the thermal initiator to achieve in situ polymerization of styrene monomer, with a polymerization efficiency of up to 87% within 3 h. This process not only demonstrates the efficient emulsification performance of AuNC@TA in emulsion systems but also paves the way for new applications in photothermal catalytic polymerization [23].

The two-dimensional lamellar structured g-C_3N_4, as a solid emulsifier, has demonstrated significant advantages in the preparation of photofunctional emulsions (Figure 4.15a). Styrene/water emulsion can be successfully prepared by simply mixing the water dispersion of g-C_3N_4 with styrene monomer

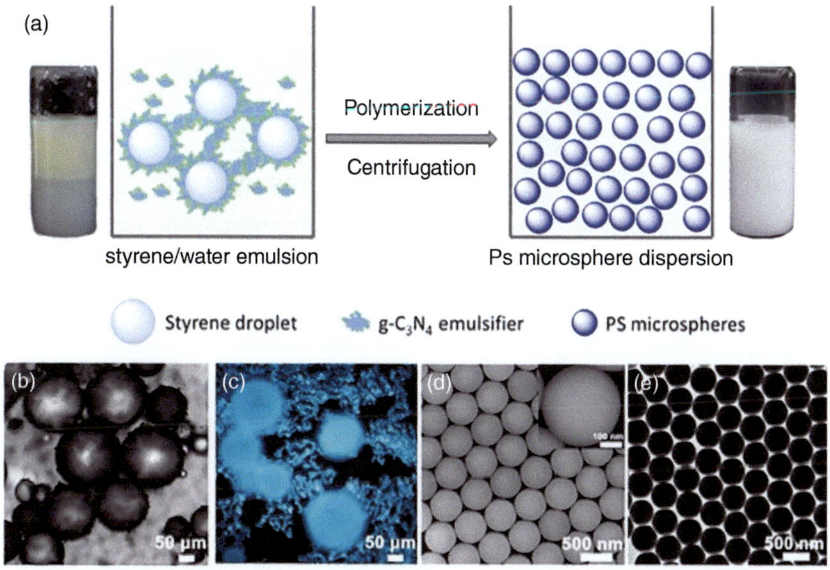

Figure 4.15 (a) Schematic of the Pickering emulsion polymerization process employing g-C_3N_4 as the emulsifier. (b) Microscopy and (c) fluorescence images of styrene/water emulsion droplets stabilized by g-C_3N_4. (d) SEM and (e) TEM images display the resulting PS microspheres. *Source:* Reproduced with permission from Cao et al. [24] / John Wiley & Sons.

and stirring. The microscopy images show that the average size of styrene droplets is approximately 200 μm. The fluorescence images further revealed that the emulsion droplets presented blue fluorescence, indicating that the fluorescent g-C_3N_4 was firmly anchored on the droplet surface, effectively preventing the coalescence of the emulsion (Figure 4.15b,c). Subsequently, the polymerization reaction was initiated by $K_2S_2O_8$. SEM and TEM images (Figure 4.15d,e) show the morphology of polystyrene (PS) microspheres, which have clear shapes, smooth surfaces, and uniform particle size distribution. These findings strongly show the superior performance of g-C_3N_4 in the emulsion polymerization process. In addition, the solid emulsifiers have a major impact on the nucleation stage of the emulsion polymerization process [24].

During the preparation of styrene-based emulsion, C_3N_4 was still used as the solid emulsifier, successfully encapsulating 4-vinylbenzeneboronic acid (VBA) and sialic acid (SA) molecules. Among them, 4-vinylphenylboronic acid is a functional copolymer monomer capable of forming a reversible boronic ester with SA molecules, thereby introducing SA template molecules into the complex system. Through polymerization reactions, complexes with optical functions are prepared. Subsequently, the boronic ester was dissociated by modifying the pH, leading to the release of SA molecules from the polymer nanoparticle surfaces, resulting in the formation of SA-imprinted fluorescent nanocomposites [25].

Montmorillonite (MMT) is a material with a layered structure. Its unique structural properties make it perform well in a variety of applications. Chitosan (CS), as a

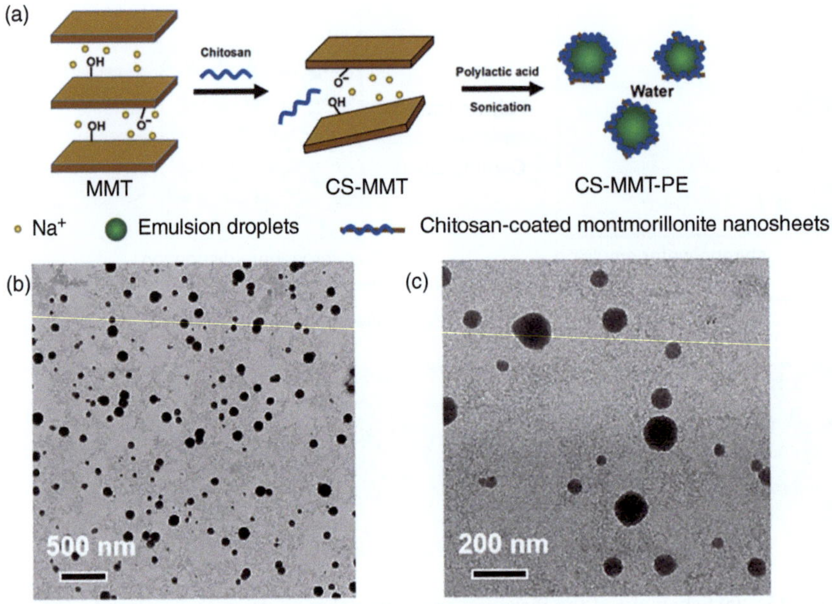

Figure 4.16 (a) The preparation process and (b, c) TEM images of CS-MMT-PE. *Source:* Reproduced with permission from Kang et al. [26] / Elsevier.

natural polymer with a positive charge, can easily embed into the layered structure of MMT, thereby increasing the interlayer spacing and promoting interlayer delamination. This embedding process not only improves the dispersibility of MMT but also enables CS to be combined with MMT nanosheets, forming CS-MMT composites with good water solubility. In emulsion preparation, with polylactic acid as the oil phase and aqueous solution as the water phase, CS-MMT can be used as an efficient emulsifier to construct a Pickering emulsion (Figure 4.16a). This emulsion has excellent stability. Its droplet size can be stabilized at approximately 200 nm, and no coalescence or macroscopic phase separation occurs within several weeks. By observing the microstructure of CS-MMT-PE, it can be found that there are a large number of nanoparticles in the image, and these nanoparticles correspond to obvious contrast, further confirming the existence of CS-MMT in the emulsion droplets (Figure 4.16b,c) [26]. Further, when preparing Pickering nanoemulsions from MMT nanosheets, the photothermal agent indocyanine green (ICG) was added to obtain PE@MMT-ICG. The droplet size of PE@MMT-ICG is smaller than that of PE@MMT, indicating that CS-MMT and ICG have a positive synergistic effect on the stability of the emulsion interface [27].

4.1.5 Liposome

Lipid nanoparticles (LNPs) are a class of nanoscale particles composed of lipid materials such as phospholipids and cholesterol. Lipid molecules spontaneously assemble in water to form monolayer or bilayer structures. The hydrophobic tails gather inward, while the hydrophilic heads face outward and come into contact with water, thus forming a stable inner cavity structure for encapsulating hydrophilic substances. Meanwhile, the hydrophobic regions in the lipid bilayer can also load hydrophobic substances, endowing it with the ability to simultaneously carry both hydrophilic and hydrophobic components.

By selecting the prodrug phospholipid TK-DOX, neutral lipid DOPE, and cationic lipid DOTAP molecules, the hydrophobic triazine-carbazole derivative (sCzTAZPO) fluorescent conjugated molecules can be self-assembled into stable positively charged liposome nanoparticles through hydrophobic and electrostatic interactions (Figure 4.17a) [28]. With increasing incorporation of sCzTAZPO,

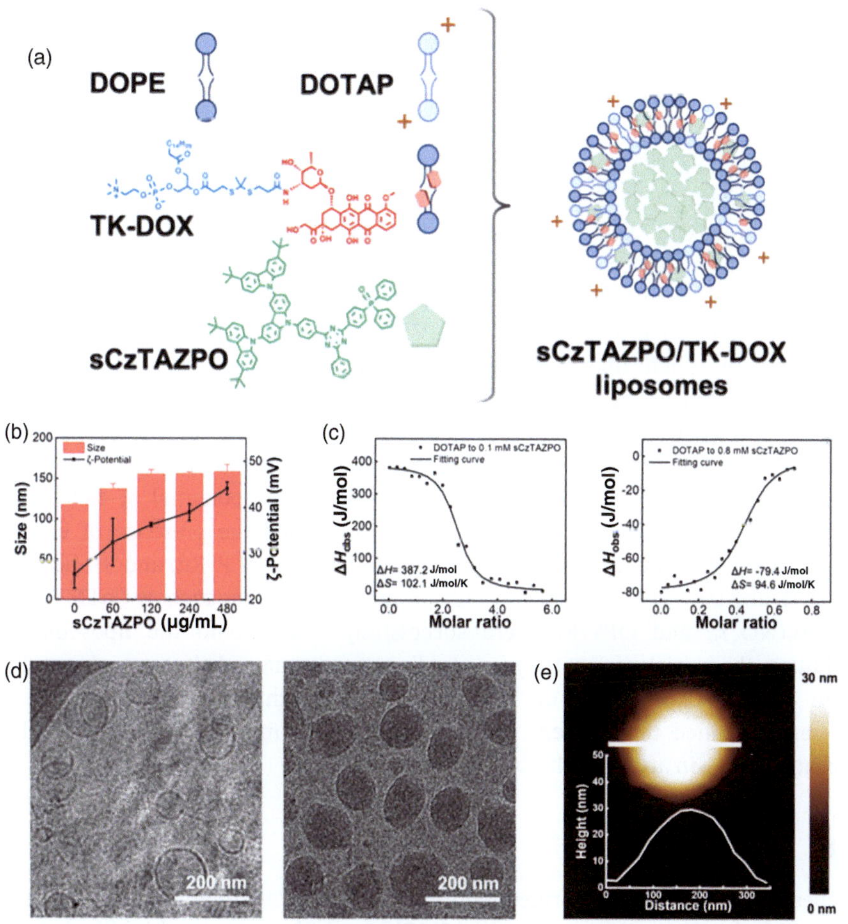

Figure 4.17 (a) The preparation of sCzTAZPO/TK-DOX liposomes. (b) Analysis of size and ζ-Potential of sCzTAZPO/TK-DOX liposomes with varying sCzTAZPO contents. (c) Measurements of ΔH upon titrating DOTAP into 0.1 and 0.8 mM sCzTAZPO solutions, respectively. (d) Cryo-TEM images comparing TK-DOX liposomes and sCzTAZPO/TK-DOX liposomes. (e) AFM image of sCzTAZPO/TK-DOX liposome. *Source:* Reproduced with permission from Liu et al. [28] / American Chemical Society.

the hydrodynamic diameter of the liposomes gradually increased, demonstrating interactions between the two components (Figure 4.17b). At low sCzTAZPO concentrations, its binding to phospholipids was primarily driven by hydrophobic effects. At higher concentrations, electrostatic interactions became significant. Under high-concentration conditions, sCzTAZPO forms aggregates, with its extended π-system of these aggregates attracts hydronium ions (H_3O^+), causing the aggregates to acquire a negative charge. This enables electrostatic interaction with the positively charged DOTAP lipids. Critically, this electrostatic attraction reduces the number of counterions associated with DOTAP, resulting in an increase in the net surface charge of DOTAP. Consequently, as the amount of sCzTAZPO increases, the surface potential of the liposomes shifts positively (Figure 4.17c). It was observed from the TEM image that in the absence of sCzTAZPO, TK-DOX liposomes presented a circular shape. However, the sCzTAZPO/TK-DOX liposomes loaded with sCzTAZPO maintained a complete spherical structure while showing significant electron-dense dark regions inside, confirming the successful encapsulation of sCzTAZPO (Figure 4.17d). The AFM image is shown in Figure 4.17e. The sCzTAZPO/TK-DOX liposome presents a dome shape with a height range of 4–30 nm, indicating its flexible structure.

Furthermore, POPC and cholesterol were selected as the primary components of the liposomes. By co-loading the photocatalyst CoTMPyP, photosensitizer $Ru(bpy)_3(NO_3)_2$, chemical catalyst oligo(*p*-phenylene vinylene)–rhodium complex (OPV-Rh), and biocatalyst creatine kinase (CK) during lipid membrane hydration, artificial photosynthetic liposomes were successfully fabricated. Through cryo-TEM observation, it can be clearly seen that the liposomes present a regular circular structure, and the clear outline of the phospholipid bilayer can be distinguished. Dynamic light scattering measurements indicate that the liposomes have an average particle size of approximately 620 nm and a Zeta potential of 48.8 mV. Furthermore, elemental analysis tests further confirmed the presence of each catalytic module in the liposomes, with the mass fractions of P, Ru, and Rh being 0.25, 0.04, and 0.07 wt%, respectively, indicating that molecules such as CoTMPyP, $Ru(bpy)_3(NO_3)_2$, and OPV-Rh were successfully loaded onto the liposomes. In addition, the particle size of the prepared liposomes remained almost unchanged within 7 days, demonstrating the excellent stability of this composite structure. This result verified the efficient construction of liposomes and their potential application value in artificial photosynthetic systems [29].

4.1.6 Micelle

Micelles are nanostructures formed by the self-assembly of amphiphilic polymers above critical micelle concentrations (CMC). Among them, amphiphilic polymers form polymer micelles through the directional arrangement of hydrophobic tails and hydrophilic heads. Pluronic F-127 (referred to as F127), as a commercialized triblock copolymer with a chemical structure of PEO_{100}-PPO_{65}-PEO_{100}, is widely used as a macromolecular stabilizer for the preparation of hydrophobic small molecules or conjugated polymer nanoparticles. When the concentration exceeds CMC, F127 is capable of self-assembling into micelles that possess a distinctive core-shell structure: hydrophobic PPO segments constitute the micelle core, while hydrophilic PEO segments form the outer shell [30]. It is worth noting that F127 micelles exhibit remarkable temperature response characteristics, which make them ideal carriers for regulating the fluorescence properties of dyes.

After the fluorescent molecule DPAC-Ty with VIE effect was rapidly added to the aqueous solution of F127 micelles, it could effectively enter the hydrophobic cavity of the micelles, forming a multicolored fluorescent micelle solution (Figure 4.18a). TEM images show that the formed spherical micelles have a diameter of less than 10 nm and are well dispersed (Figure 4.18b). Dynamic light scattering analysis revealed a notable difference in the size distribution of DPAC-Ty when present in the system of F127 solution or DMSO/H_2O mixture (Figure 4.18c). When there are no F127 micelles, DPAC-Ty will self-aggregate to form larger particles. When F127 micelles are present, DPAC-Ty is able to dissolve within the micelle core, resulting in a marked decrease in its aggregation. This confirms that F127 micelles can effectively promote the uniform dissolution and dispersion of hydrophobic

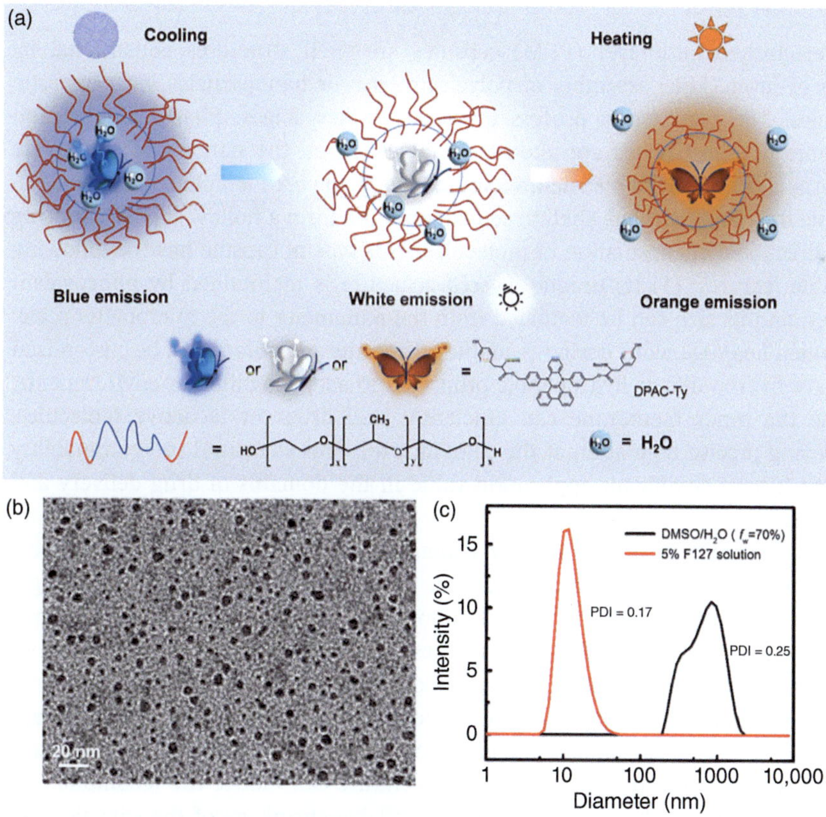

Figure 4.18 (a) The temperature-responsive emission of DPAC-Ty in F127 micelles. (b) TEM image of DPAC-Ty in F127 micelles. (c) DLS results of DPAC-Ty in DMSO/H_2O mixture and in F127 micelles. *Source:* Reproduced with permission from Liu et al. [31] / American Chemical Society.

DPAC-Ty molecules in aqueous solutions. When the temperature rises, the viscosity of the micelles increases, and their hydrophobicity enhances. The moisture in the core is gradually discharged, causing DPAC-Ty to vibrate freely and dissipate energy in a more hydrophobic environment, leading to a shift in fluorescence color to orange-red [31].

CB/F127-1-M micelles are formed by adding U-shaped CB DMSO solution to an F127 solution. These micelles exhibit a quasi-spherical or quasi-cubic shape, with an average diameter of around 146 nm and a relatively uniform size distribution. Upon increasing the temperature, the average size of CB/F127-1-M micelles gradually expands—from 160 at 15 to 240 nm at 55 °C. This size increase is attributed to the dehydration of the hydrophilic PEO segments at higher temperatures, which reduces the hydrophilicity of the micelle shell and promotes micelle aggregation. Therefore, the dynamic changes in the microstructure of micelles can provide favorable conditions for macro-environmental regulation of the conformation of fluorescent molecules, thereby achieving precise control of luminescence behavior [32].

4.1.7 Capsule

Polyelectrolyte multilayer (PEM) capsules are shell structures constructed by layer-by-layer (LbL) assembly of polyelectrolytes or nanoparticles with opposite charges. The preparation process involves two key stages: Firstly, anionic and cationic polyelectrolytes are alternately deposited on the nanoparticle template to form a multilayer film structure; Subsequently, the core template is dissolved to enable the polyelectrolyte shell to self-support and form a hollow cavity, achieving the directional encapsulation of molecules. This type of capsule has the following notable features: (1) Its unique layered structure is maintained by noncovalent forces, and its size can be regulated from the nanometer to the micrometer scale; (2) when heterogeneous design is adopted, the outer membrane can be customized to have hydrophilic or hydrophobic properties to adapt to different environments, while the inner membrane can efficiently load drugs or bioactive molecules, achieving precise regulation at the cellular level. This structural programmability makes it have significant application value in the domains of drug delivery and biosensing.

Using calcium carbonate ($CaCO_3$) nanoparticles as templates, the positively charged poly(diallyldimethylammonium chloride) (PDADMAC) and negatively charged poly(sodium 4-styrenesulfonate) (PSS) were alternately adsorbed through electrostatic interaction to synthesize a multilayer capsule (Figure 4.19). Subsequently, the $CaCO_3$ template was dissolved in ethylenediaminetetraacetic acid (EDTA) solution, resulting in the formation of hollow polyelectrolyte microcapsules. As the glass transition temperature of this polymer system is approximately 40 °C, heating the capsule above this temperature can induce the recombination of polymer chains, thereby achieving controllable shrinkage of the capsule size. During this process, the alkaline phosphatase (ALP) in the solution is actively encapsulated inside the capsule, forming an efficient heat-responsive enzyme encapsulation system [33].

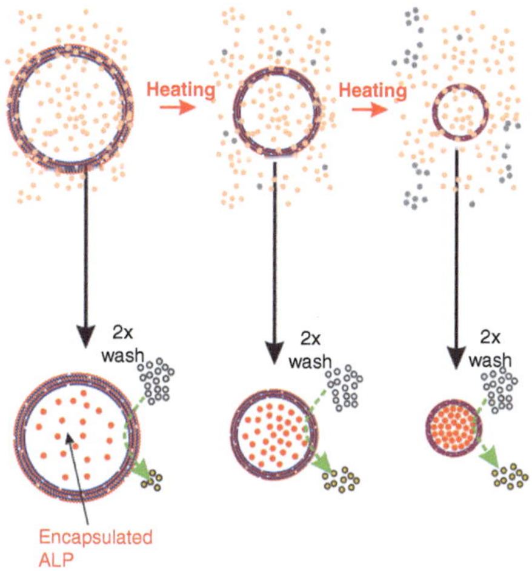

Figure 4.19 Schematic illustration of the ALP load within the multilayer capsule. *Source:* Van der Meeren et al. [33] / with permission of John Wiley & Sons.

In addition, inorganic gold nanoparticles and silver nanoparticles can also be used as building blocks to construct novel microcapsules. Firstly, calcium carbonate ($CaCO_3$) nanoparticles are mixed with silver ammonia solution, and the first layer of silver nanoparticles is formed on the core surface through the glucose reduction method. Subsequently, the gold nanoparticle solution was added, and the second layer of gold nanoparticles was formed through stirring and ultrasonic treatment. Repeat the above steps as needed, alternately depositing silver and gold nanoparticles until the required number of layers is reached. Finally, the calcium carbonate template is dissolved in EDTA solution, leaving a complete nanoparticle shell structure, and microcapsules can be obtained (Figure 4.20). The loaded precious metal

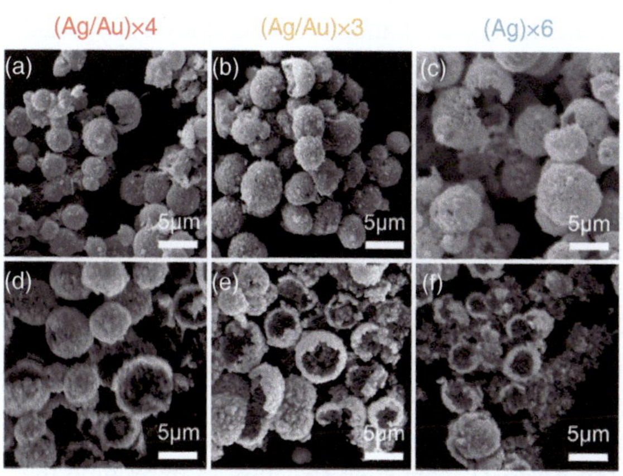

Figure 4.20 SEM images of the hollow microcapsules prepared with (a) (Ag/Au) × 4, (b) (Ag/Au) × 3, and (c) Ag × 6, and after low-intensity ultrasound treatment: (d) (Ag/Au) × 4, (e) (Ag/Au) × 3, and (f) Ag × 6. *Source:* Reproduced with permission from Li et al. [34] / Elsevier.

nanoparticles endow this microcapsule with unique optical functions, imparting broad potential for use in domains such as drug delivery and optical therapy [34].

4.2 Functions of Photofunctional Polymer Nanocomposites

4.2.1 Metal Enhanced Fluorescence Effect

When fluorescent substances approach a metal surface, the existence of surface plasmons on the metal can cause an augmentation of fluorescence emission. The separation between the fluorescent substances and the metal surface is a crucial factor affecting the degree of this metal-enhanced fluorescence (MEF) effect. If the fluorescent substances are overly close to the metal surface, a nonradiative energy transfer process may take place. In this process, dipole–dipole interactions allow excited-state fluorophores to transmit energy to the metal substrate, ultimately leading to fluorescence quenching. On the contrary, if the distance between the fluorophores and the metal surface is overly large, the interaction required for fluorescence enhancement decreases, thus weakening the MEF effect. Therefore, there is an optimal separation distance between the metal surface and the fluorescent substances, within which the maximum fluorescence enhancement can be attained.

Regarding the PFV/Ag@SiO_2 core-shell structured fluorescent nanoparticles, the dimension of the Ag nanoparticles is approximately 23 nm. The surface plasmon resonance (SPR) absorption peak is situated at 417 nm, which exhibits a high degree of consistency with the absorption spectrum of the fluorescent conjugated polymer. As a result, the excitation of the fluorescent conjugated polymer may efficiently trigger the plasmon resonance effect [35]. Furthermore, the fluorescence intensity is significantly affected by the interaction distance between the fluorophore and silver nanoparticles. In this system, the fluorescent nanoparticles employ a SiO_2 shell as both an isolation layer and a distance-regulating layer between the metallic surface and the fluorescent polymer PFV. Ag@SiO_2-2 nanoparticles feature a 4 nm SiO_2 shell layer encircling an Ag core. At an identical PFV concentration, the fluorescence intensity of these nanoparticles is approximately 1.3 times greater than that of pure SiO_2 nanoparticles. This phenomenon indicates that the SiO_2 barrier layer can effectively alleviate the fluorescence quenching effect exerted by the Ag nanoparticle surface on PFV. Significantly, the synergistic effect of the Ag core and the SiO_2 barrier layer gives rise to a certain level of MEF, which partially offsets the surface-induced self-quenching of PFV. Moreover, no significant change in the fluorescence lifetime of PFV is observed in PFV/Ag@SiO_2. Hence, it can be deduced that the primary mechanism of MEF in this system is the enhancement of the excitation. These results further validate the efficacy of the Ag@SiO_2 core-shell structure in improving the performance of fluorescent nanoparticles (Figure 4.21) [2].

Regarding the Ag@PLL/(FA/FAM-SE) fluorescent nanocomposites, a poly-L-lysine (PLL) shell layer is deposited on the surface of silver nanoparticles to regulate

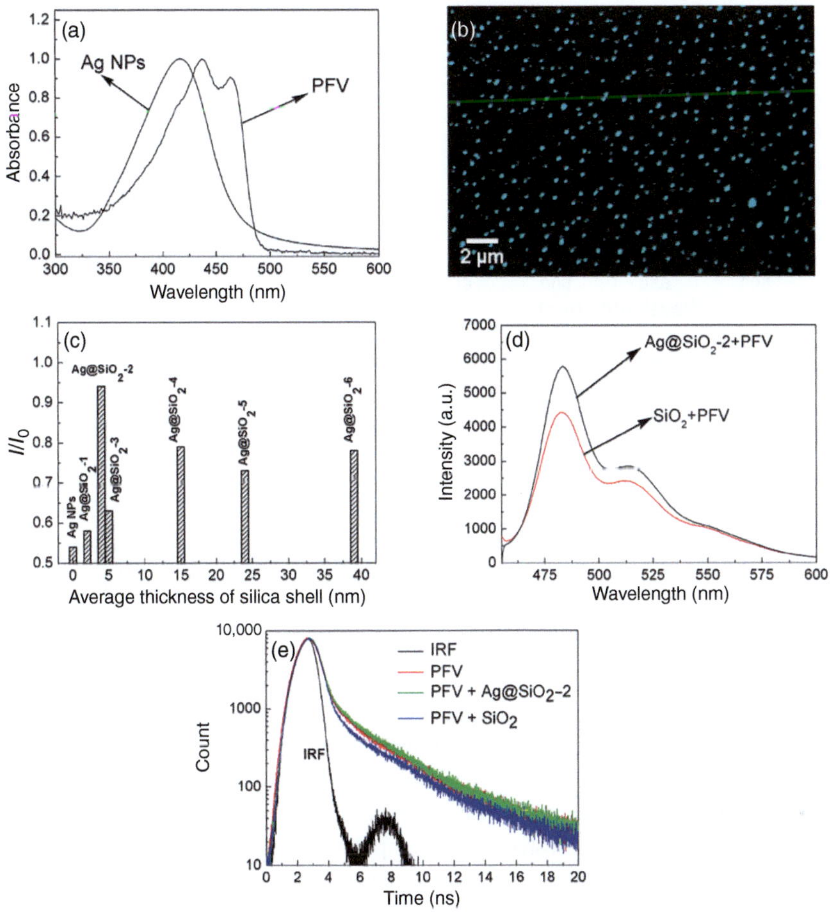

Figure 4.21 (a) Absorption spectra of silver nanoparticles and PFV. (b) Fluorescence image of the $Ag@SiO_2$–2 + PFV nanoparticles. (c) Ratio of fluorescence intensity for PFV after assembly on various nanoparticle surfaces compared to PFV in water. (d) Fluorescence emission of $Ag@SiO_2$–2 + PFV and SiO_2 + PFV nanoparticles. (e) Fluorescence decay profiles of PFV, $Ag@SiO_2$–2 + PFV, and SiO_2 + PFV. *Source:* Reproduced with permission from Tang et al. [2] / American Chemical Society.

the interaction distance between the silver core and the dye FAM. Through the modulation of the reaction concentration of PLL, diverse composite systems with shell thicknesses spanning from 0 to 13 nm are successfully synthesized. When excited at 470 nm, fluorescence intensity measurements at 515 nm indicate that the composite system with an 8 nm shell thickness demonstrates the strongest fluorescence signal, attaining an enhancement factor of 2.1 (Figure 4.22a). Moreover, after the immobilization of FAM-SE on the surface of the core-shell structure, a notable reduction in fluorescence lifetime is observed, decreasing from 3.87 to 1.28 ns (Figure 4.22b).

When an excited fluorophore is positioned in proximity to colloidal metallic structures, the electric field associated with its emitted light interacts with the SPR

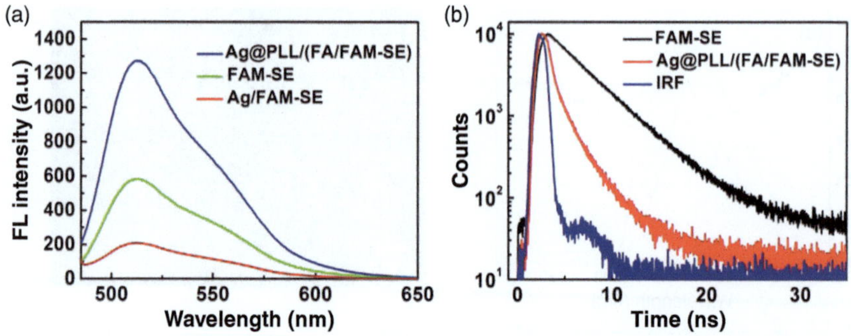

Figure 4.22 (a) Fluorescence spectra of Ag@PLL/(FA/FAM-SE), FAM-SE, and Ag/FAM-SE solutions. (b) Fluorescence decay of FAM-SE solution and Ag@PLL/(FA/FAM-SE) nanocomposites. *Source:* Tang et al. [7] / with permission of American Chemical Society.

of the metal. This interaction introduces an additional radiative decay rate, denoted as k_m. The quantum yield (Q) and fluorescence lifetime (τ) of the fluorophore can be determined by considering the radiative decay rate (k) and nonradiative decay rate (k_{nr}) intrinsic to the fluorophore, along with the contribution from k_m.

$$\tau = 1/(k + k_m + k_{nr})$$

$$Q = (k + k_m)/(k + k_m + k_{nr})$$

This decrease in fluorescence lifetime suggests that the radiative decay rate of FAM is notably enhanced under the influence of silver nanoparticles. This leads to an enhanced quantum yield and, consequently, increased fluorescence intensity. These discoveries provide valuable perspectives for the further optimization of the optical properties of core-shell structured composite materials [7].

In an aqueous solution, the polymer assumes a flexible chain conformation, the configuration and dimensions of which are capable of varying in accordance with changes in the surrounding environmental conditions. As a result, when employed as an intermediate barrier layer in the MEF system, responsive variations in distance can be triggered by adjusting external environmental factors, thus facilitating the effective regulation of the MEF effect.

A new type of hybrid nanoparticle with dual-responsive properties can be synthesized by cross-linking poly(3-acrylamidephenylboronic acid-co-acrylic acid) (PAPBA-PAA) with silver nanoparticles. This hybrid, referred to as Ag@PAPBA-PAA, combines the unique functionalities of two polymer segments, allowing precise control over both particle size and fluorescence intensity [36].

The APBA units within the polymer shell are highly sensitive to glucose and can reversibly bind glucose molecules (Figure 4.23a). Upon glucose binding, the APBA groups acquire additional charge, causing the shell to swell. Experiments show that Ag@PAPBA-PAA nanoparticles, initially averaging 38 nm in diameter, expand to 53 nm when exposed to glucose—a swelling rate of about 40%. Consequently, the expansion widens the distance separating Por^{4+} dye molecules from the silver nanoparticles' surface, leading to a marked enhancement in fluorescence intensity, up to 1.8 times the original value.

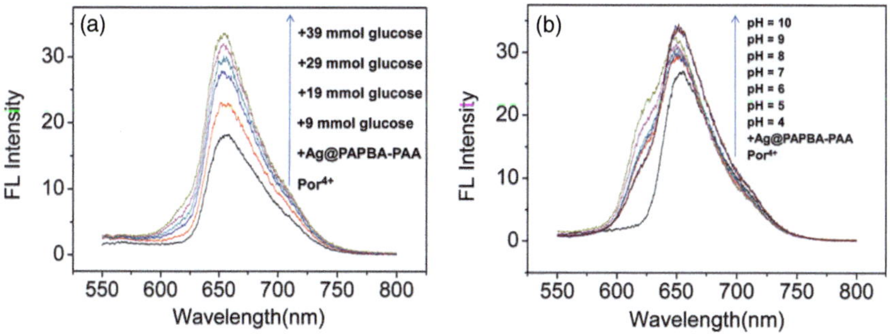

Figure 4.23 The fluorescence emission of the Por^{4+}/Ag@PAPBA-PAA nanocomposites regulated by (a) glucose concentrations and (b) pH value. *Source:* Zhang et al. [36] / with permission of American Chemical Society.

Poly(acrylic acid) (PAA) provides pH sensitivity (Figure 4.23b). Between pH 4 and 6, increasing the pH progressively ionizes the carboxyl groups along the PAA chain, resulting in larger nanoparticle sizes. As the pH rises from 6 to 8, ionization of PAA approaches saturation, stabilizing the nanoparticle diameter and indicating equilibrium in swelling. When the pH exceeds 8, the APBA moieties interact with hydroxide ions, becoming charged and increasing electrostatic repulsion, which causes further swelling of the nanoparticles. Conversely, lowering the pH contracts the PAPBA-PAA shell, reducing nanoparticle size. These pH-dependent changes are closely linked to variations in fluorescence intensity. Their unique swelling behavior provides a valuable platform for studying MEF and opens new avenues for the design of smart, stimuli-responsive materials.

There are significant differences in plasmon resonance properties between different metal nanostructures, which have a distinct impact on fluorescence enhancement. In this investigation, gold nanoflowers (AuNFs) were chosen as the core material, and a core-shell structured nanoparticle, AuNF@Gelatin, was synthesized by utilizing gelatin as the shell layer. The outer surface of AuNF@Gelatin has the ability to adsorb the positively charged conjugated polymer PFVCN. PFVCN displays a distinct absorption peak at approximately 450 nm, while the SPR absorption band of AuNFs is centered at 592 nm. The absorption spectra of these two components exhibit minimal overlap. Nevertheless, the fluorescence emission peak of PFVCN is situated at 550 nm, which shows substantial spectral overlapping with the absorption band of AuNFs. This spectral compatibility promotes efficient fluorescence enhancement. By adjusting the concentration of gelatin, the thickness of the shell layer in AuNF@Gelatin can be accurately controlled, thus enabling the optimization of the fluorescence enhancement performance. When the shell thickness reaches 10 nm (designated as AuNF@Gelatin-2), the fluorescence intensity of PFVCN is remarkably enhanced, doubling the initial intensity. These findings indicate that precise regulation of fluorescence enhancement can be accomplished through the rational design of the core-shell architecture and the corresponding shell thickness of gold nanoparticles (Figure 4.24) [37].

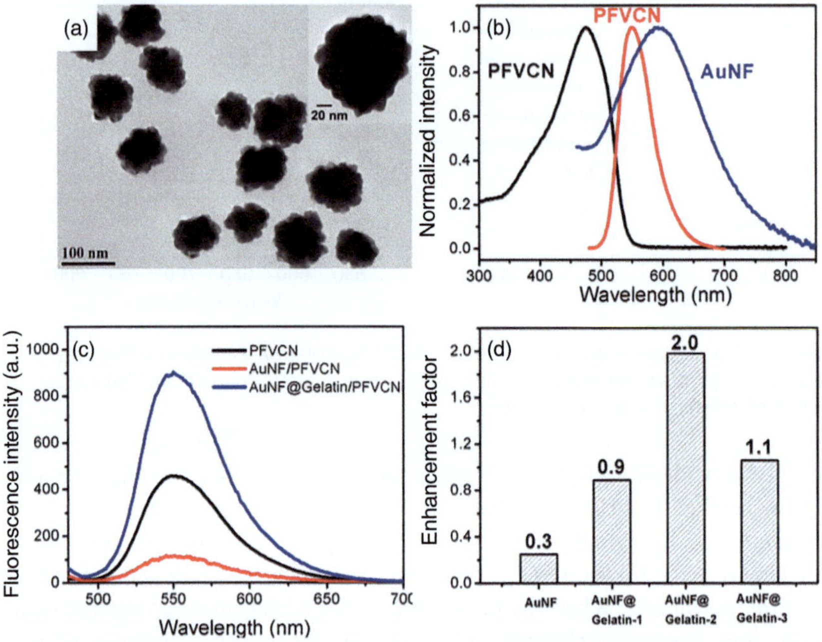

Figure 4.24 (a) TEM image showing the synthesized gold nanoflowers (AuNFs). (b) Normalized absorption spectrum (black) and emission spectrum (red) of PFVCN in water, along with the normalized extinction spectrum (blue) of AuNFs. (c) Fluorescence spectra of PFVCN measured in the presence of either AuNFs or AuNF@Gelatin-2. (d) Enhancement factor plotted as a function of AuNF@Gelatin nanoparticles with varying shell thicknesses. *Source:* Reproduced with permission from Cui et al. [37] / American Chemical Society.

Based on the principle of electrostatic adsorption, negatively charged gold nanospheres (AuNPs) can be effectively assembled with positively charged poly-L-lysine (PLL). Subsequently, via the EDC/NHS amidation reaction, the fluorescent dye Cy3 and the targeting peptide RGD were successfully conjugated to the Au@PLL composite structure. Spectral analysis indicates a significant overlap between the absorption spectrum of the AuNPs and Cy3, suggesting that this composite structure can substantially enhance the fluorescence emission of Cy3. Experimental findings further illustrate that when Cy3 is immobilized on the surface of gold nanospheres, its fluorescence intensity increases by approximately 1.5-fold. Nevertheless, fluorescence lifetime measurements suggest that the fluorescence lifetime of Cy3 remains nearly constant before and after modification, being 0.2 and 0.19 ns, respectively. This phenomenon is hypothesized to enhance the excitation efficiency of Cy3 molecules rather than accelerating the fluorescence decay rate, thus resulting in fluorescence enhancement [38].

The distribution pattern of noble metal nanoparticles exerts a significant influence on the MEF effect. For example, silver nanoparticles that are in situ reduced within PNIPAM-co-PAA microgels possess an average size of approximately 5 nm, with the maximum absorption peak at 380 nm. The pH-responsive behavior of PAA is an essential component of this system. Within the pH range of 5–10, the carboxyl

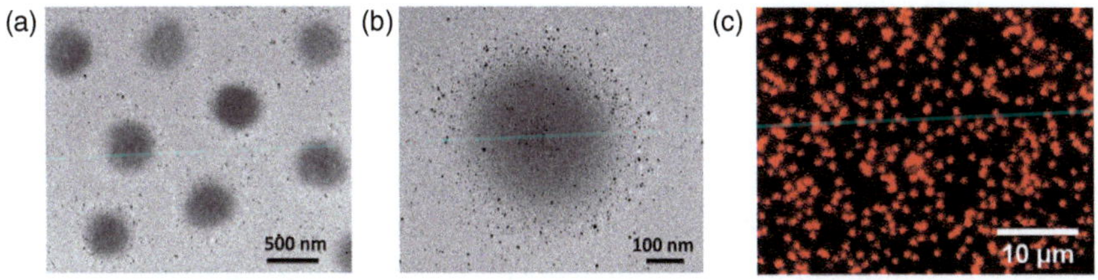

Figure 4.25 TEM images showing (a) hybrid PNIPAM-co-PAA microgel particles, (b) an individual microgel nanoparticle, and (c) a fluorescence image of microgels with adsorbed Por^{4+}. *Source:* Reproduced with permission from Tang et al. [16] / American Chemical Society.

groups along the PAA chains become ionized and hydrophilic, resulting in chain extension and microgel swelling. Simultaneously, the phase transition characteristics of PNIPAM cause the microgel to gradually contract as the temperature rises. A significant contraction occurs between 30 and 35 °C, and a stable, contracted state is maintained above 40 °C. By regulating the external pH and temperature conditions, the composite microgel can undergo reversible volume changes, thus enabling the tunability of the distance between silver nanoparticles and dyes. After the incorporation of the composite microgel, the fluorescence intensity of Por^{4+} increases by up to 2.1 times at 650 nm, clearly demonstrating a remarkable MEF effect. This finding provides novel perspectives on the application of noble metal nanoparticles in the development of functional materials (Figure 4.25) [16].

The silver nanoparticle solution, characterized by an average particle size of approximately 66 nm, was successfully synthesized via the citric acid reduction method. The nanoparticles display a distinct negative surface charge, as indicated by the zeta potential of −49.5 mV. To fulfill fluorescence enhancement, the fluorescent dye fluorescein isothiocyanate isomer (FITC) was chosen. The excitation and emission peaks of FITC are in close alignment with the maximum absorption peak of silver nanoparticles. In the mixture of silver nanoparticles and FITC, positively charged polyacrylamide was introduced to trigger nanoparticle aggregation. FITC molecules predominantly localize within the slit-like structures formed by the aggregated silver nanoparticles, where their fluorescence intensity is notably enhanced through the SPR effect [39].

4.2.2 Energy Transfer Mechanism

During the Förster resonance energy transfer (FRET) process, the donor molecule becomes excited upon absorbing a photon and subsequently transfers its energy to the acceptor molecule in a nonradiative manner within a spatial distance range of 1–10 nm. Upon the acceptor molecule receiving energy, it enters an excited state and emits fluorescence. The efficiency of FRET is primarily governed by two factors: the degree of spectral overlap between the emission spectrum of the donor fluorophore and the absorption spectrum of the acceptor, as well as the intermolecular distance separating the donor and acceptor molecules. Consequently, the appropriate selection of donor–acceptor pairs and the preparation of nanoscale aggregates are of great significance for the effective establishment of an energy transfer mechanism.

The energy donor conjugated oligomer OFBT-A and the energy acceptor DNA labeled with TR dye (TR-DNA) were efficiently assembled through a coprecipitation approach. This assembly notably diminished the distance between the donor and the acceptor, resulting in the successful fabrication of the nanocomposite. The fluorescence emission maximum of OFBT-A nanoparticles is observed at 525 nm, which exhibits substantial spectral overlap with the absorption band of TR, as illustrated in Figure 4.26a,b. This overlap facilitates efficient FRET from OFBT-A to TR. Upon incremental addition of TR-DNA, the fluorescence intensity of OFBT-A at 525 nm progressively decreases, whereas the fluorescence emission of TR at 610 nm correspondingly increases. These findings corroborate that FRET from OFBT-A to TR indeed takes place (Figure 4.26c). Moreover, the fluorescence intensity of TR exhibits an approximate 12-fold enhancement relative to direct excitation at 580 nm. As depicted in Figure 4.26d, the fluorescence color of the nanocomposites transforms from green to orange subsequent to TR-DNA loading. Notably, within the ionic strength range of 0–500 mM, only minimal fluorescence quenching was

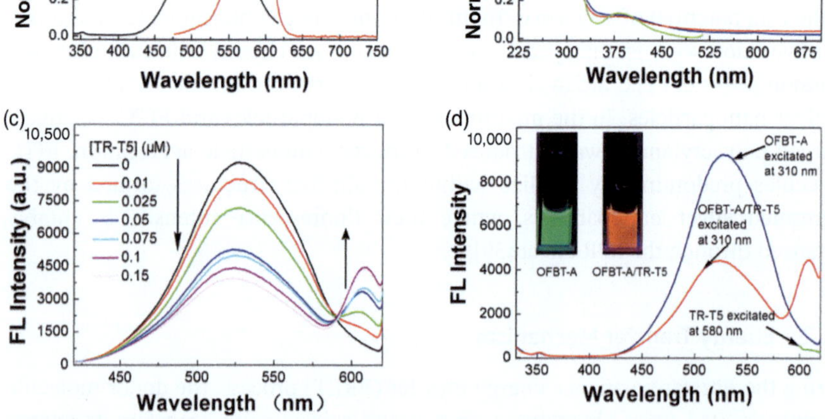

Figure 4.26 (a) Normalized emission spectrum of OFBT-A and absorption spectrum of TR. (b) Normalized absorption spectrum of OFBT-A solution, along with the absorption and excitation spectra of OFBT-A/TR-T5 nanocomposites. (c) Fluorescence spectra of OFBT-A/TR-T5 nanocomposites prepared via co-reprecipitation with different TR-T5 concentrations. (d) Fluorescence spectra and photographs of OFBT-A nanoparticles, OFBT-A/TR-T5 nanocomposites, and TR-T5. *Source:* Wang et al. [19] / with permission of American Chemical Society.

detected, and the FRET efficiency remained relatively stable. When the pH is in the range of 5.0–9.5, slight alterations in the emission spectrum are observed; however, at pH 3.0, fluorescence quenching reaches approximately 20%. Overall, the nanoparticles display good stability of FRET efficiency across the tested solution environments [19].

In the construction of a FRET system with the conjugated polymer PFO as the donor and $Ir(piq)_2(acac)$ as the acceptor, lipid molecules were introduced through the co-precipitation method. Consequently, nanoscale aggregates with uniform size and a more compact conformation were successfully prepared. The maximum emission wavelength of PFO is 438 nm, which exhibits an overlap with the broad absorption spectrum of $Ir(piq)_2(acac)$. This overlap facilitates efficient FRET. Within nanoscale aggregates, the reduced spatial separation between donor and acceptor molecules leads to an enhancement in FRET efficiency. Systematic variation of the concentration ratio of $Ir(piq)_2(acac)$ revealed that the emission intensity of $Ir(piq)_2(acac)$ at 625 nm increased progressively, while the emission intensity of PFO at 425 nm decreased in parallel. Notably, when the weight ratio of PFO to $Ir(piq)_2(acac)$ was adjusted to 20:5, the FRET efficiency, quantified as the ratio of emission intensities ($I_{625\,nm}/I_{425\,nm}$), attained a maximum value of 2.37. Moreover, upon excitation of PFO at 390 nm, the emission of $Ir(piq)_2(acac)$ was observed to be approximately 12-fold greater compared to direct excitation of $Ir(piq)_2(acac)$ at 477 nm. This indicates that PFO, as the donor, significantly enhances the emission of the acceptor. Time-resolved fluorescence spectroscopy further demonstrated that the average emission lifetime of PFO in PFO nanoparticles was 120 ps, whereas in the PFO/Ir nanocomposite, the emission lifetime of PFO was reduced to 15 ps. This reduction can be ascribed to an accelerated emission decay rate of PFO during the FRET process, thus confirming the high efficiency of the constructed FRET system (Figure 4.27) [40].

By introducing photoswitchable molecules, the energy transfer efficiency between them and conjugated fluorophores can be effectively modulated through the regulation of their local concentration. Specifically, eight dithienylethene (DTE) units are covalently linked to a cage-like polyhedral oligomeric silsesquioxane (POSS) molecule to construct a highly photo-switchable molecular structure, designated as POSS-DTE8. Subsequently, photofunctional nanoparticles are synthesized via a co-precipitation method using both POSS-DTE8 and conjugated polymers. Within polymer nanoparticles, the separation distance between the conjugated polymer and the DTE group generally measures less than 10 nm, a proximity that is adequately close to promote energy transfer.

DTE manifests two stable structural configurations: a noncolored open-ring form and a pigmented closed-ring form. Upon exposure to ultraviolet light irradiation, energy is transferred from the conjugated polymer to the closed form of the DTE group, leading to the efficient fluorescence quenching of the conjugated polymer. The incorporation of POSS-DTE8, which encompasses eight DTE groups, notably elevates the local concentration of DTE within the polymer nanoparticles. This guarantees efficient energy transfer with the surrounding conjugated polymers, thereby accomplishing effective fluorescence quenching. Additionally, under visible

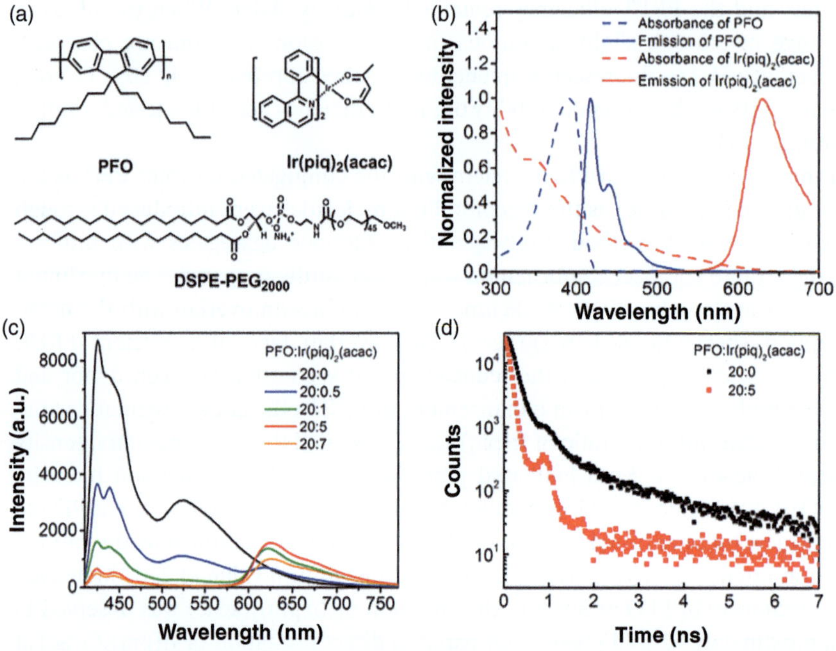

Figure 4.27 (a) Molecular structures of PFO, $Ir(piq)_2(acac)$, and DSPE-PEG2000. (b) Absorption and emission spectra of PFO (blue) and $Ir(piq)_2(acac)$ (red). (c) Emission spectra of PFO/Ir nanocomposites with different component ratios. (d) Fluorescence decay of PFO in both PFO nanoparticles and PFO/Ir nanocomposites. *Source:* Reprinted with permission from Guo et al. [40].

light irradiation, the fluorescence of the conjugated polymer can be restored to its initial intensity. Consequently, under alternating ultraviolet and visible light irradiation, the super photoswitch molecule POSS-DTE8 enables the precise regulation of the luminescent states of fluorescent conjugated polymer nanoparticles, facilitating reversible switching between "on" and "off" fluorescence states (Figure 4.28) [41]. This methodology not only enhances energy transfer efficiency but also provides novel insights for the development of light-controlled fluorescent materials.

Via hydrophobic and electrostatic interactions, fluorescent conjugated molecules sCzTAZPO can be systematically assembled with the liposomal prodrug TK-DOX to generate stable liposomal nanoparticles, sCzTAZPO/TK-DOX. The liposomal formulation presents two distinct absorption peaks corresponding to the characteristic absorption regions of sCzTAZPO and DOX, respectively, which validates the presence of both photofunctional components within the composite system. The fluorescence emission spectrum demonstrates that the liposome exhibits a characteristic emission peak of sCzTAZPO at 482 nm, along with detectable emission peaks of DOX at 554 and 598 nm. The spectral overlap between the emission band of sCzTAZPO and the absorption band of DOX implies that sCzTAZPO and DOX serve as energy donor and acceptor, respectively, which promotes FRET. Analysis of the decay curve indicates a more rapid decay of the sCzTAZPO exciton in the sCzTAZPO/TK-DOX liposome, with the lifetime decreasing from 13.5 to 1.9 ns,

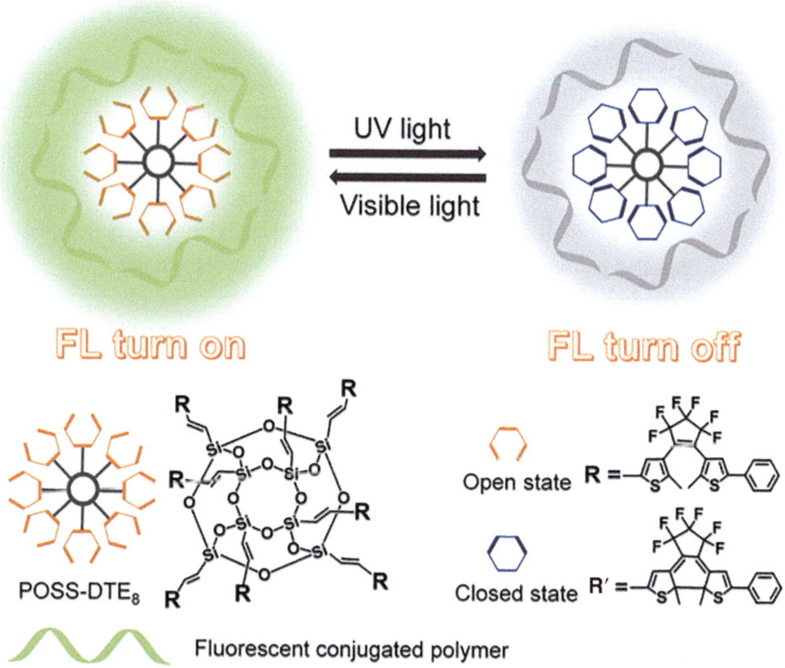

Figure 4.28 Schematic illustration of the mechanism by which the photoswitch molecule POSS-DTE8 modulates the fluorescence emission of conjugated polymers. *Source:* Liu et al. [41] / with permission of John Wiley & Sons.

thus confirming the occurrence of efficient FRET between sCzTAZPO and DOX. The FRET ratio ($I_{554\,nm}/I_{482\,nm}$) is determined to be 0.73 (Figure 4.29) [28].

Beyond liposomes, micellar systems can also function as effective platforms for establishing a FRET mechanism. Through the incorporation of donor and acceptor fluorescent molecules into the micellar structure, efficient energy transfer can be achieved. In the instance of the polymer poly[(9,9-bis(3′-((N,N-dimethyl)-N-ethylammonium)-propyl)-2,7-fluorene)-alt-2,7-(9,9-dioctylfluorene)] (PFN) micelle system, researchers encapsulated the hydrophobic fluorescent dyes C6 and NR separately within the micelles to construct a FRET-based system. The maximum emission peak of PFN is 433 nm, while the maximum absorption wavelengths of C6 and NR are 460 and 525 nm, respectively. Significantly, there is a substantial spectral overlap between the emission spectrum of PFN and the absorption spectrum of C6, which promotes energy transfer from PFN (donor) to C6 (acceptor). This spectral overlap leads to a notably higher energy transfer efficiency in the PFN/C6 system compared to the PFN/NR system. This finding further emphasizes the crucial role of spectral overlap in enabling efficient FRET [42].

To regulate the interaction distance more precisely between the energy donor and the acceptor, covalent bonding can be utilized instead of simple physical mixing. This method allows for a more stable fixation of the relative positions of the two components, thus improving both the efficiency and controllability of energy transfer.

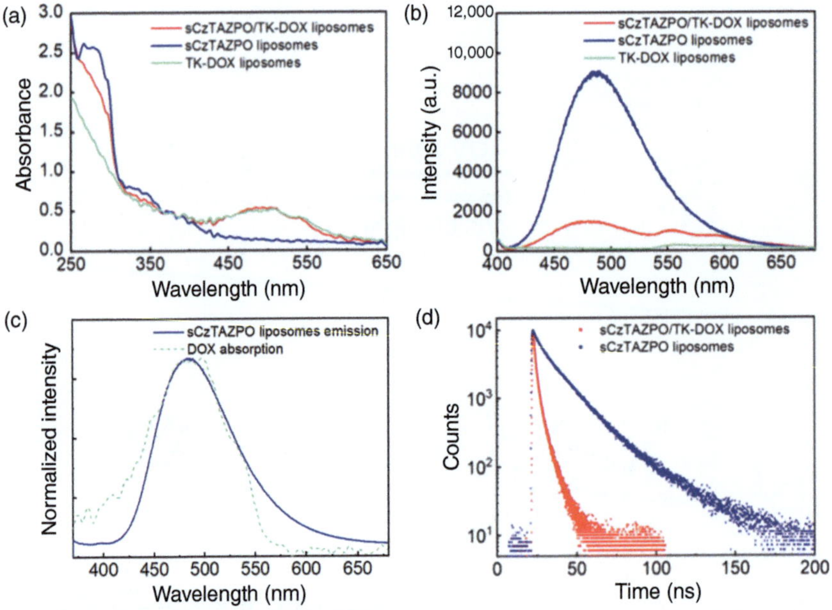

Figure 4.29 (a) Absorbance and (b) emission spectra of sCzTAZPO/TK-DOX liposomes, sCzTAZPO liposomes, and TK-DOX liposomes, respectively. (c) Normalized absorption spectrum of DOX and fluorescence spectrum of sCzTAZPO liposomes. (d) Fluorescence decay of sCzTAZPO/TK-DOX liposomes and sCzTAZPO liposomes. *Source:* Liu et al. [28] / with permission of American Chemical Society.

Via the thiolation reaction, spiropyran (SP) was effectively grafted onto the surface of gold nanoclusters (AuNCs) through Au–S bonds, leading to the formation of the AuNC–SP complex. AuNCs display a characteristic UV–Vis absorption peak at 370 nm. Upon exposure to ultraviolet radiation, a distinct absorption peak emerges at 545 nm, corresponding to the structural transformation of spiropyran from its closed-ring form (SH-SP) to the open-ring isomer (SH-MC). Notably, the emission spectrum of AuNCs at 527 nm demonstrates considerable spectral overlap with the absorption profile of SH-MC. This spectral overlap implies that AuNCs can serve as an energy donor, thus facilitating the establishment of a FRET system with SH-MC acting as the energy acceptor (Figure 4.30a). To further verify the occurrence of the FRET process, time-resolved fluorescence spectroscopy is utilized. The results demonstrate that after UV irradiation, the average fluorescence lifetime of the gold nanoclusters decreases from the initial 109.6–80.8 ns. This finding confirms that AuNCs serve as energy donors and that a FRET process indeed occurs within the AuNC-SP system (Figure 4.30b). This conversion is accompanied by remarkable alterations in fluorescence properties. Specifically, the emission at 527 nm gradually declines, and the quantum yield (QY) reduces to 0.27%. Meanwhile, a new emission peak corresponding to SH-MC emerges at 640 nm, with its intensity gradually increasing and the QY rising to 0.14%. Upon subsequent irradiation with visible light for 300 s, the emission at 640 nm markedly diminishes, whereas the emission at 527 nm exhibits partial recovery. This suggests that the spiropyran reverts

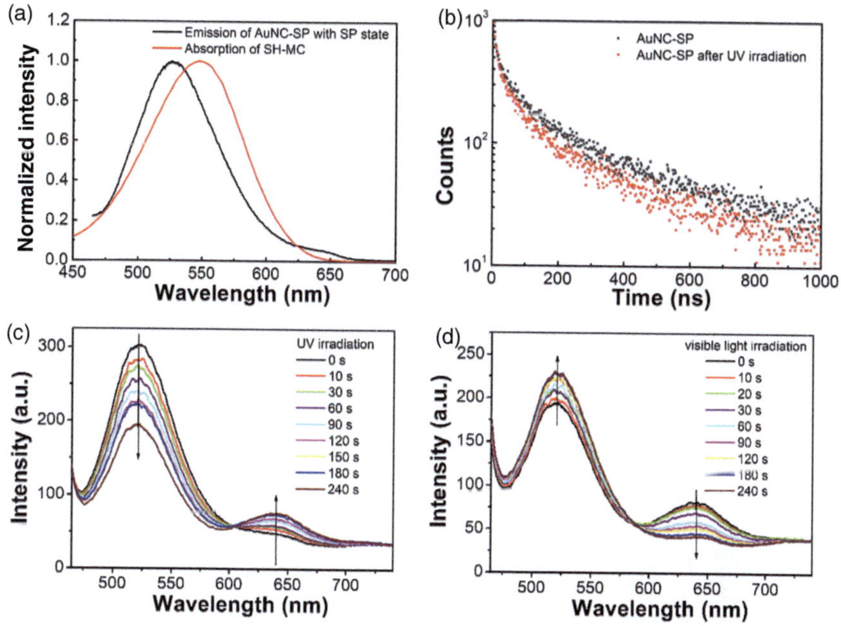

Figure 4.30 (a) Normalized fluorescence spectrum of AuNC-SP with SP state and normalized absorption spectrum of SH-MC. (b) Fluorescence decay profiles of AuNC-SP at 527 nm before and after UV light irradiation. Fluorescence spectra of AuNC-SP@CS under (c) UV light irradiation and (d) white light irradiation for varying durations. *Source:* Cong et al. [15] / with permission of American Chemical Society.

from the open-ring SH-MC to the closed-ring SH-SP structure. The FRET efficiency ($I_{640\,\text{nm}}/I_{527\,\text{nm}}$) of the AuNC-SP system is determined to be 37% (Figure 4.30c,d) [15].

The conjugated polymer PF-DBT-BIMEG is a material distinguished by its unique molecular structure and distinct physical characteristics. The polymer backbone is covalently linked by fluorene (PF), functioning as an energy donor, and dithienobenzothiadiazole (DBT), acting as an energy acceptor. On the side-chains, a multitude of oligoethylene glycol (EG) moieties and positively charged imidazole groups with high charge density (BIMEG) have been incorporated, which notably improve the polymer's hydrophilicity. Regarding its optical characteristics, PF-DBT-BIMEG displays a maximum absorption peak at 400 nm, mainly ascribed to the PF unit. Moreover, a wide absorption band between 450 and 600 nm is detected, corresponding to the DBT structure.

Because of the BIMEG side-chains' strong hydrophilia and the electrostatic repulsion among polymer chains, the polymer disperses as individual chains in aqueous solution. As a result, the donor (PF) and acceptor (DBT) parts are separated by a relatively large spatial distance, leading to a low energy transfer efficiency. Upon the addition of adenosine triphosphate (ATP), a remarkable alteration in the polymer's fluorescence response is observed. Specifically, while the emission at 655 nm rises noticeably, the emission at 423 nm sharply declines. Consequently, the emission intensity ratio I_{655}/I_{423} increases from 0.06 to 1.36, representing a

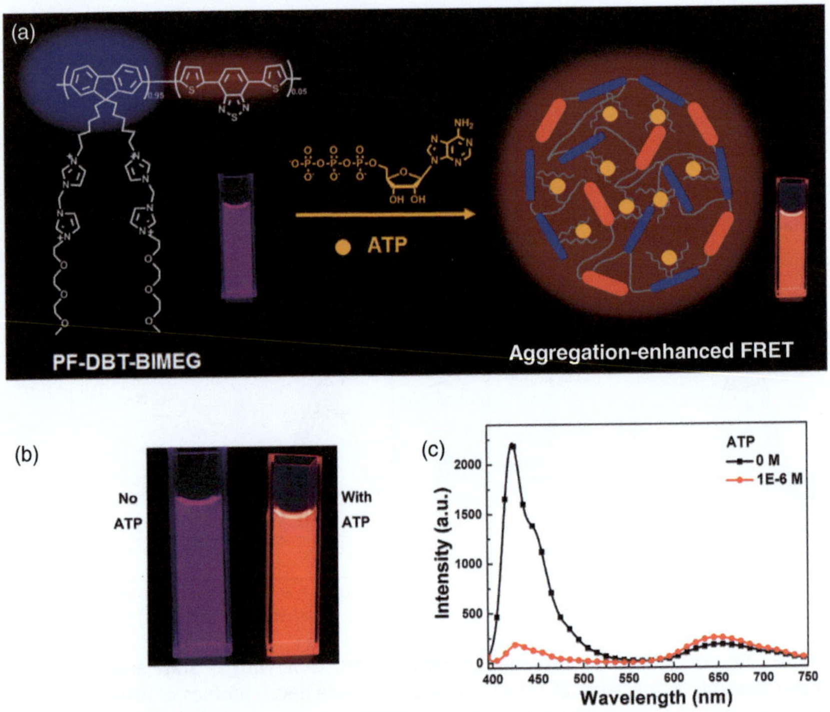

Figure 4.31 (a) Molecular structure of PF-DBT-BIMEG and ATP. (b) The photographs and (c) emission spectra of PF-DBT-BIMEG before and after aggregation by ATP. *Source:* Reproduced with permission from Cui et al. [43] / American Chemical Society.

22-fold enhancement (Figure 4.31) [43]. This phenomenon can be attributed to the following factors: (1) The bis-imidazole groups situated on the polymer chain demonstrate robust electrostatic interactions with the phosphate groups of ATP. (2) Hydrophobic interactions are established between the adenine moiety of ATP and the hydrophobic backbone of the polymer. (3) The resultant complex diminishes the distance in space between the acceptor DBT and the donor PF, thereby augmenting the efficiency of FRET and subsequently regulating the fluorescence signal.

The polymer PF-DBT-BIMEG is capable of forming an energy transfer system with the two-dimensional material PhCN as well. Owing to its negative surface charge (−33 mV), PhCN can strongly interact with the positively charged PF-DBT-BIMEG via electrostatic forces, resulting in the formation of stable colloidal complexes. This robust adsorption leads to a highly compact arrangement of WSCPs molecules on the surface of PhCN, thereby substantially decreasing the spatial separation between the polymer's donor and acceptor parts. As a consequence, the intermolecular FRET mechanism is activated. The initiation of FRET significantly improves the efficiency of energy transfer, which is reflected in a marked increase in the I_{655}/I_{420} intensity ratio [44].

The dual-emission fluorescent nanoparticle system fabricated using conjugated polymers PFO and MEH-PPV has successfully established a FRET mechanism.

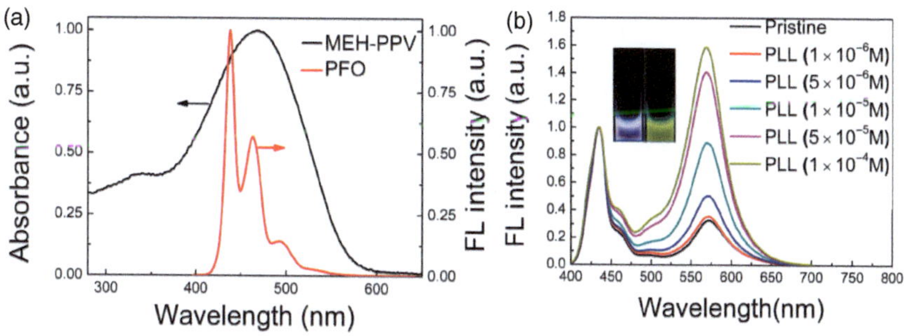

Figure 4.32 (a) Fluorescence spectrum of PFO and absorption spectrum of MEH-PPV. (b) Fluorescence spectra of the mixed nanoparticle solution after adding varying amounts of the cationic polyelectrolyte PLL. *Source:* Reproduced with permission from Wu et al. [45] / American Chemical Society.

The maximal emission peak of PFO's blue fluorescence is at 425 nm, in contrast to MEH-PPV, which emits orange fluorescence with a peak at 550 nm. As shown in Figure 4.32a, the UV–vis absorption spectrum of MEH-PPV nanoparticles covers a wide range from 300 to 580 nm, demonstrating substantial spectral overlap with the emission spectrum of PFO nanoparticles. This overlap acts as a crucial prerequisite for the efficient implementation of FRET [45].

To ensure that the distance between the mixed nanoparticles of PFO and MEH-PPV is within an appropriate range for the occurrence of the FRET effect, the researchers incorporated cationic PLL. As the concentration of PLL in the nanoparticle solution increased, the emission intensity of MEH-PPV increased gradually in comparison to that of PFO. Simultaneously, the luminescent color of the solution exhibited a notable shift from blue to yellow (Figure 4.32b). This phenomenon can be ascribed to the fact that PFO and MEH-PPV nanoparticles are characterized by negative surface charges, while the PLL chains are positively charged. The ensuing electrostatic interactions decrease the distance between the nanoparticles. Considering that FRET efficiency is highly reliant on inter-particle distance, the reduction in separation substantially facilitates more effective energy transfer from PFO to MEH-PPV, thus resulting in enhanced MEH-PPV emission intensity.

4.2.3 Electron Transfer Mechanism

Under photoexcitation circumstances, the electron transfer mechanism is generally initiated. This phenomenon occurs when a molecule absorbs a photon, resulting in the excitation of an electron from its ground state to a higher energy (excited) state. The excited electron can subsequently transfer to a neighboring molecule, thereby facilitating the process of electron transfer. The efficiency of this electron transfer is governed by several factors, including the spatial separation between the electron donor and acceptor species, the physicochemical characteristics of the surrounding medium (such as solvent polarity and viscosity), temperature, and the match between the donor and acceptor energy states.

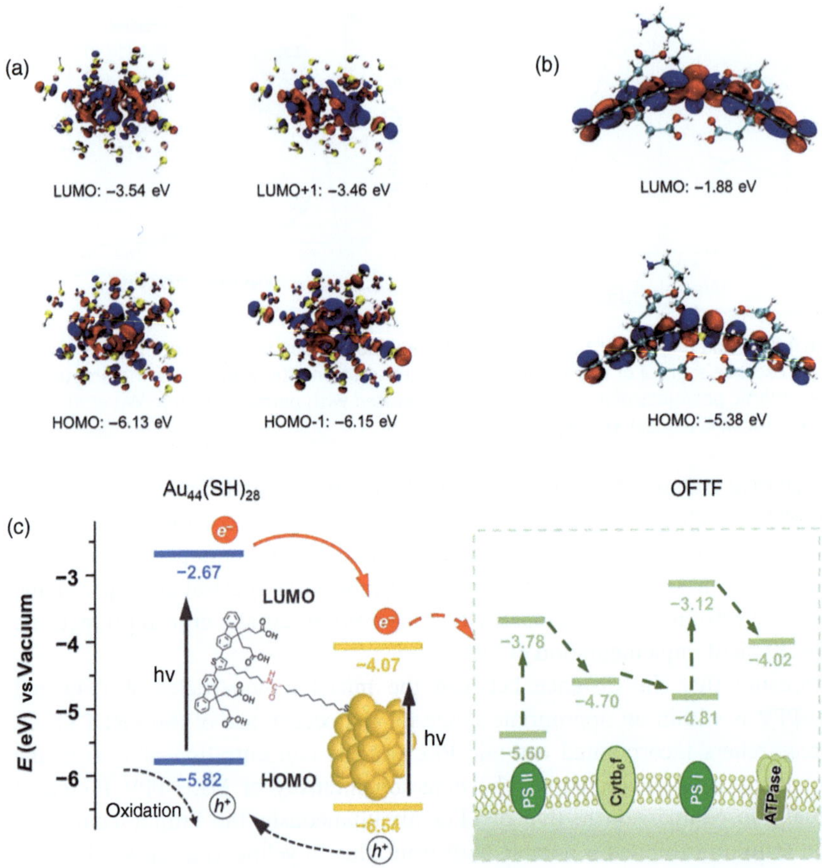

Figure 4.33 Stimulated the spatial distributions of frontier molecular orbitals for (a) AuNCs and (b) OFTF. (c) Schematic diagrams the electron transfer between OFTF and AuNCs. *Source:* Cong et al. [46] / with permission of John Wiley & Sons.

An ultrasmall AuNCs@OFTF heterojunction was successfully synthesized via the covalent bonding between gold nanoclusters (AuNCs) and conjugated oligomer OFTF. The simulated charge transfer process from OFTF to Au_{44} following photoexcitation suggests that the highest occupied molecular orbital (HOMO) is predominantly localized on the OFTF molecule, while the lowest unoccupied molecular orbital (LUMO) is mainly concentrated in the core region of Au44 (Figure 4.33a,b). This energy-level configuration provides a theoretical basis for the charge transfer from OFTF to Au44 under photoexcitation, thus directly corroborating the photoinduced electron transfer mechanism within the AuNCs@OFTF heterojunction. The computed band gaps (E_g) of OFTF and AuNCs are 3.15 and 2.47 eV, respectively. Further examination indicates that the conduction band (CB) of OFTF is situated at −2.67 eV, whereas that of AuNCs is positioned at −4.07 eV. These results validate the existence of a favorable energy level alignment within the AuNCs@OFTF heterojunction, which promotes efficient charge transfer from OFTF to AuNCs (Figure 4.33c,d) [46].

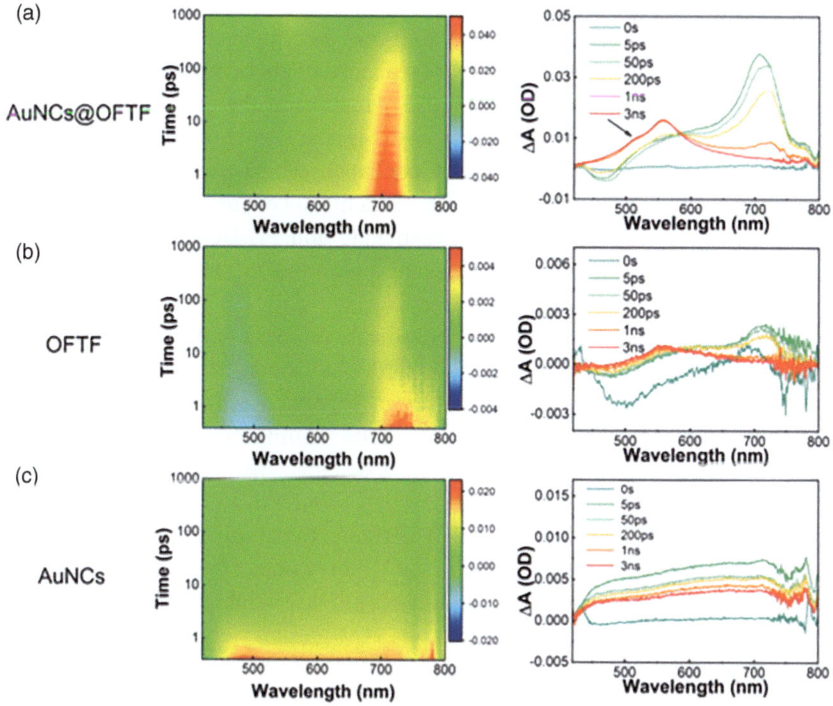

Figure 4.34 Transient absorption spectra and corresponding TA spectral signals of (a) AuNCs@OFTF, (b) OFTF, and (c) AuNCs. *Source:* Cong et al. [46] / with permission of John Wiley & Sons.

Under illumination conditions, AuNCs@OFTF absorb photons and generate electron-hole pairs. In femtosecond transient absorption spectroscopy (fs-TA), the negative signal observed at 480 nm corresponds to stimulated emission (SE), which represents the photon emission released when the molecule returns to its ground state following photoexcitation. The positive signals at 557 and 715 nm are attributed to triplet–triplet absorption (TTA) and excited-state absorption (ESA), respectively, reflecting the molecule's distinct absorption characteristics in its excited state. As illustrated in Figures 4.34a,b, compared to pristine OFTF, AuNCs@OFTF exhibits an enhanced signal at 509 nm, confirming the formation of a charge transfer state. This feature suggests that charge transfer occurs from OFTF to AuNCs upon their coupling. Furthermore, the persistence of the SE and ESA bands beyond 200 ps in AuNCs@OFTF provides additional evidence for the strong coupling between OFTF and AuNCs in the excited state (Figure 4.34c). Consequently, the AuNCs@OFTF heterojunction facilitates efficient photoinduced charge transfer through favorable energy level alignment and close interfacial contact [46].

The conjugated polymer PFO features positively charged quaternary ammonium salt side chains. Poly(L-glutamic acid) serves as a drug carrier and covalently bonds to the anticancer drug DOX via the EDC/NHS reaction. The quinone group in DOX serves as a highly effective quencher of the electron transfer process. Through electrostatic interactions, PFO and the negatively charged PG-DOX self-assemble into a nanocomposite with a mean size of about 50 nm. An increase in PG-DOX concentration leads to a gradual reduction in the emission intensity of PFO. This observation can be attributed to the interaction between the quinone group in DOX and the conjugated polymer backbone, which initiates an electron transfer process that results in fluorescence quenching. This mechanism not only elucidates the interfacial interaction between PFO and PG-DOX but also offers a crucial theoretical foundation for the design and optimization of drug delivery systems [47].

The conjugated molecule CB is composed of two cyclopenta-dithiophene (CDT) moieties connected to a BODIPY core through double bonds. To enhance solubility and processability, hexyl side chains were incorporated at the C-5 position of the CDT units. The optimized molecular structure assumes a stable "U"-shaped conformation (Figure 4.35a), with an interwing distance spanning approximately from 2.8 to 4.0 nm. This highly distorted geometry results in a decreased π-orbital overlap between the CDT units and the BODIPY core, thus limiting intramolecular charge transfer efficiency and giving rise to dual fluorescence emission characteristics. As depicted in the side-view representation of Figure 4.35b, the dihedral angle between the BODIPY core and the CDT plane in the ground state is approximately 50.8 °, which increases to around 74.0 ° in the excited state. Simultaneously, the distance between the two CDT planes extends to 5.7–8.6 Å (d_1). These computational results imply that CB experiences a conformational transition from a "folded" to an "unfolded" state upon photoexcitation.

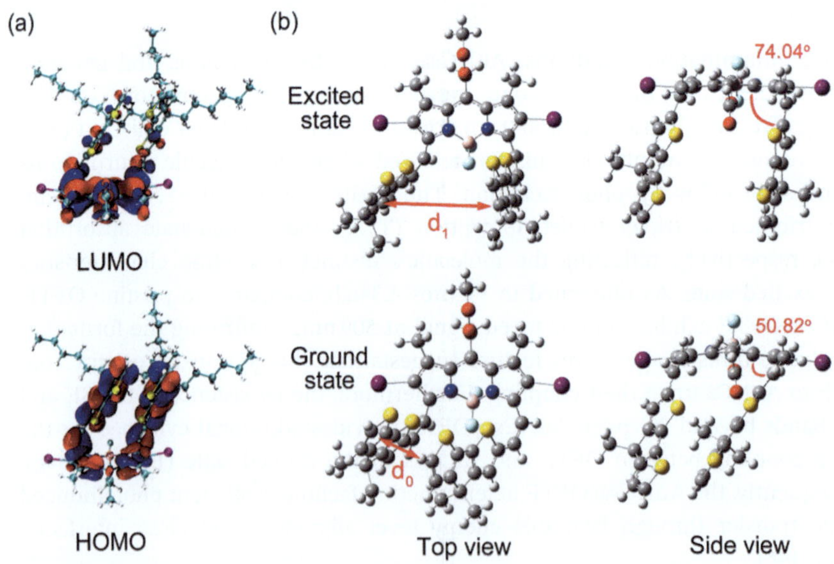

Figure 4.35 (a) The stimulated HOMO and LUMO orbitals of the CB. (b) Top and side views of the CB′ optimized molecular geometry in both its ground and excited states. *Source:* Peng et al. [32] / with permission of John Wiley & Sons.

In the CB structure, this rotational freedom heightens the probability of the molecule transitioning from a "folded" to an "unfolded" conformation upon excitation, thus facilitating twisted intramolecular charge transfer (TICT) from the CDT donor to the BODIPY acceptor. When CB forms a complex with F-127 micelles, the TICT emission peak of CB/F127-1-M emerges at 705 nm, indicating that CB is effectively dispersed within the micellar environment and readily undergoes the TICT process. The formation of a complex with F-127 micelles enhances the solubility and TICT characteristics of CB molecules, presenting a promising strategy for its application in responsive materials [32].

Through the encapsulation of electron acceptors within fluorescent polymer nanoparticles, the formation of exciplexes not only substantially alters the fluorescence emission characteristics but also presents novel approaches for nanoparticle functionalization. The formation of exciplex stems from the electron transfer between excited-state and ground-state molecules, leading to a remarkable red shift in the fluorescence emission. Inside the nanoparticles, the formation of exciplexes can be further regulated by controlling the spatial arrangement of electron donors (D) and electron acceptors (A). Such regulation facilitates tunable fluorescence color and alleviates fluorescence quenching induced by π-stacking interactions.

The molecular structure exerts a significant influence on the formation of exciplexes. As depicted in Figure 4.36, from the top perspective, two TPOTZ molecules manifest substantial spatial overlap, signifying strong π–π interactions

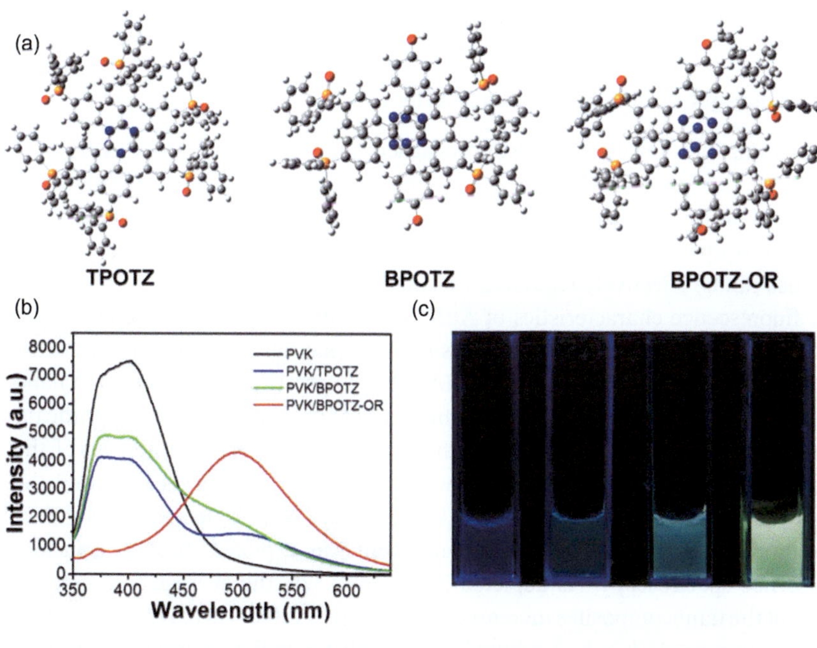

Figure 4.36 (a) Top view of the face-to-face π–π-stacking of TPOTZ, BPOTZ, and BPOTZ-OR, respectively. (b) Fluorescence emission spectra and (c) photographs of PVK, PVK/TPOTZ, PVK/BPOTZ, and PVK/BPOTZ-OR nanoparticles. *Source:* Reproduced with permission from Liu et al. [18] / American Chemical Society.

between them. Conversely, two BPOTZ molecules exhibit only a slight relative displacement when overlapping, suggesting the existence of considerable π–π stacking interactions. Nevertheless, BPOTZ-OR molecules demonstrate a notable displacement among themselves. This phenomenon can be ascribed to the incorporation of the alkyl chain structure, which enhances the molecular flexibility of BPOTZ-OR and reduces its rigidity. As a result, the π–π stacking between adjacent BPOTZ-OR molecules is attenuated. This decrease in intermolecular π–π interactions facilitates the interaction between BPOTZ-OR and PVK, thereby promoting the formation of an electron donor–acceptor system [18].

PVK nanoparticles exhibit intrinsic blue fluorescence. When TPOTZ or BPOTZ molecules are incorporated with PVK to form nanoparticles, the acceptor molecules show significant π–π stacking interactions. However, only a limited fraction of these acceptor molecules participates in intermolecular electron transfer with PVK, resulting in the formation of exciplexes with weak and restricted emission. As a result, the native fluorescence of PVK remains predominant. In contrast, BPOTZ-OR molecules exhibit weaker π–π stacking interactions, allowing PVK to interact with a greater number of BPOTZ-OR molecules. This facilitates the formation of more exciplexes through intermolecular electron transfer, producing strong exciplex emission and green fluorescence. The exciplex formation in PVK/BPOTZ-OR nanoparticles induces a red shift in fluorescence emission to 500 nm and a substantial Stokes shift of 170 nm. Background fluorescence interference is significantly reduced by this strong Stokes shift, highlighting the potential of these nanocomposites for bioimaging applications.

4.2.4 Aggregation-induced Emission Effect

Distinct from the prevalent aggregation-caused quenching (ACQ), the AIE effect pertains to the phenomenon in which molecular rotation and vibration are constrained during the aggregation process. Consequently, nonradiative energy dissipation is diminished, resulting in enhanced luminescence. Therefore, optical properties can be effectively regulated via controlled aggregation.

The fluorescence characteristics of AuNCs exhibit a substantial dependence on their surface ligand modification. Studies indicate that when AuNCs are protected by GSH from nanocomposites (AuNC-HA-PROT) with hyaluronic acid (HA) and protein (PROT) via robust interactions, their optical properties undergo significant alterations. As shown in Figure 4.37a, the fluorescence emission intensity of the composite system at 605 nm attains a level fivefold that of pure AuNCs, and the absolute quantum yield increases from 1.91% (for pure AuNCs) to 7.4%.

The mechanism of fluorescence enhancement was explored through transient fluorescence spectroscopy. As depicted in Figure 4.37b, the average fluorescence lifetime of the nanocomposites increased from 3.45 μs (AuNCs) to 5.57 μs. According to the equation $\tau = 1/(k_r + k_{nr})$, where k_r represents the radiative decay rate and k_{nr} represents the nonradiative decay rate, k_r is determined by the intrinsic molecular characteristics and remains invariant. Consequently, the extended fluorescence lifetime is mainly attributable to a decrease in k_{nr}. This implies that the HA-PROT

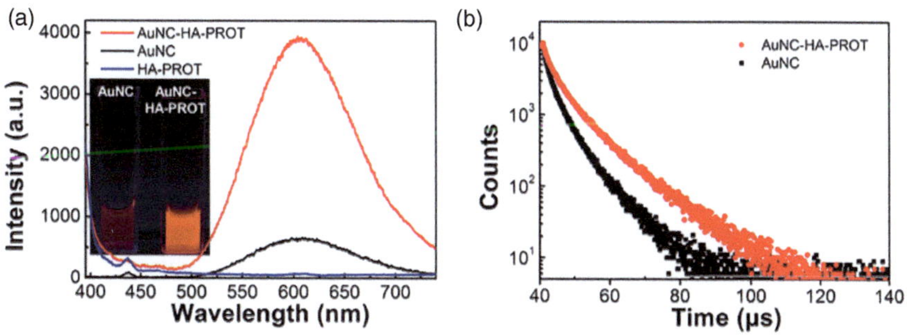

Figure 4.37 (a) Fluorescence spectra and (b) emission decay of AuNC-HA-PROT and AuNC. *Source:* Reproduced with permission from Xia et al. [14] / American Chemical Society.

cross-linked structure inhibits nonradiative transitions and thus improves radiative efficiency through the aggregation-induced emission effect [14].

Analogously, through the functionalization of the ligands of gold nanoclusters, amine-modified AuNC@ODA nanoclusters were successfully synthesized. These nanoclusters possess amphiphilic attributes and can selectively adsorb at the oil-water interface, thus forming stable emulsions (denoted as AuPE). The hydrophobic cavity within the emulsion can encapsulate the blue fluorescent dye tetraphenylethylene (TPE). During the interfacial polymerization process, AuNC@ODA reacted with hexamethylene diisocyanate (HDI), leading to the formation of AuPU composites with improved fluorescence characteristics. The emission of AuNC@ODA has an average fluorescence lifetime of 1.221 μs. However, the immobilization of AuNC@ODA within the AuPU matrix minimizes the nonradiative decay caused by the flexible ligands, resulting in enhanced fluorescence emission and an extended lifetime of 5.219 μs. Meanwhile, the measured fluorescence lifetime of TPE was approximately 2.41 ns, which is consistent with the characteristics of aggregation-induced emission enhancement. Collectively, these results imply that the AuPU-TPE integrated architecture not only spatially separates the two fluorescent components but also creates a confined microenvironment that significantly enhances the optical properties of both materials [48].

OF-Si-doped silica nanoparticles (OF/SiO_2 NPs) were synthesized by means of the reverse micelle method with ammonia serving as a catalyst, through the co-condensation reaction of OF/Si-containing trimethoxysilane side chains and tetraethyl orthosilicate (TEOS). Experimental findings indicate that when the doping concentration of the oligomer OF is as low as 0.025–0.05 mol%, the absolute fluorescence quantum yield of the resultant OF/SiO_2 NPs attains up to 97%. To conduct a more in-depth exploration of the high emission efficiency, the fluorescence lifetimes of diverse samples were measured. As shown in Table 4.1, the oligomer OF-Si demonstrates a quantum yield of 71.23% and an average fluorescence lifetime of 0.64 ns in chloroform, indicating remarkable luminescence. Its performance is similar to that of the conjugated polymer PFO. Nevertheless, the quantum yield of PFO nanoparticles declines to 51.84%, and the fluorescence lifetime is shortened to 0.36 ns, which can be mainly ascribed to

Table 4.1 Optical parameters of samples.

	λ_{abs} (nm)	λ_{emi} (nm)	Φ_F (%)	τ (ns)	Solvent
OF-Si	354	398,417	71.2	0.64	$CHCl_3$
OF/SiO_2 0.5% NPs	354	400,418	62.4		H_2O
OF/SiO_2 0.2% NPs	354	400,418	69.9	0.95	H_2O
OF/SiO_2 0.1% NPs	354	400,418	84.3		H_2O
OF/SiO_2 0.05% NPs	354	400,418	97.3	0.95	H_2O
OF/SiO_2 0.03% NPs	354	400,418	93.2		H_2O
OF/SiO_2 0.025% NPs	354	400,418	97.7		H_2O
OF/SiO_2 0.02% NPs	354	400,418	65.7	0.92	H_2O
PFO	392	418	63.1	0.52	$CHCl_3$
PFO NPs	401	438	51.8	0.36	H_2O

Source: Zhang et al. [49] / with permission of American Chemical Society.

the aggregation-induced quenching effect. Conversely, the average fluorescence lifetime of OF-Si–doped silica nanoparticles increases significantly from 0.64 to 0.95 ns, suggesting a substantial reduction in the nonradiative decay rate within this system. This phenomenon might be attributed to the rigid structure of the silica matrix, which effectively restrains the intramolecular motion and vibration of aggregated organic molecules, thus minimizing energy dissipation. This research provides novel perspectives and methodologies for the development of high-performance fluorescent nanomaterials.

Through the regulation of the aggregation form via host–guest assembly, the optical properties of materials can be effectively modulated. For example, DPBT was chosen as the luminescent core of the conjugated oligomer, and 4,4′-bipyridine (BPy) was introduced at both termini to act as binding sites for cyclodextrin (CB), thus forming the DPBT-BPy molecule. Subsequently, the terminal pyridine groups of DPBT-BPy were methylated to generate the DPBT-BPyMe molecule, which bears four positive charges. The region around the carbonyl portal of CB is negatively charged, as is the inner surface of its cavity. As a result, host–guest assemblies can be formed through the synergistic effects of ion-dipole interactions, hydrogen bonding, and hydrophobic effects involving the carbonyl portal and the rigid, hydrophobic cavity (Figure 4.38a) [50].

The experimental results demonstrate that the relative fluorescence quantum yields of both DPBT-BPy and DPBT-BPyMe increase by more than twofold upon complexation with CB[8]. For DPBT-BPy alone, the fluorescence lifetime exhibits a single-exponential decay with a lifetime of 1.69 ns. In contrast, the DPBT-BPy/CB[8] complex displays a double-exponential decay, with lifetimes of 1.00 and 4.01 ns, indicating the formation of a long-lived species within the system. The average fluorescence lifetime of the DPBT-BPy/CB[8] complex is 2.76 ns, representing a 1.6-fold increase compared to DPBT-BPy alone.

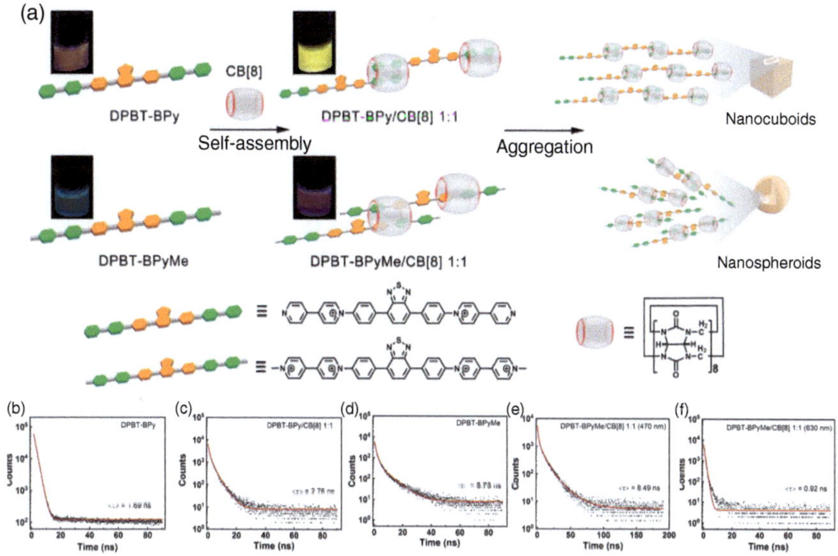

Figure 4.38 (a) Schematic diagram of the host–guest assembly of DPBT-BPy and DPBT-BPyMe with CB[8]. Emission decay of (b) DPBT-BPy and (c) DPBT-BPy/CB[8]. Emission decay of (d) DPBT-BPyMe and DPBT-BPyMe/CB[8] at (e) 470 nm and (f) 630 nm. *Source:* Zhang et al. [50] / with permission of American Chemical Society.

In aqueous solution, DPBT-BPyMe shows a double-exponential fluorescence decay with lifetimes of 1.26 and 7.24 ns, reflecting the presence of both monomeric and aggregated states. Upon forming DPBT-BPyMe/CB[8], the fluorescence lifetime fits a triple-exponential decay, with lifetimes of 1.30, 7.41, and 18.3 ns. The 18.3 ns component corresponds to a newly formed long-lived species. The average fluorescence lifetime of this complex at 470 nm is 8.49 ns, which is 2.2 times higher than that of DPBT-BPyMe alone. Additionally, the fluorescence lifetime measured at 630 nm for the DPBT-BPyMe/CB[8] complex suggests that encapsulation by CB[8] promotes intramolecular charge transfer from the DPBT unit to the pyridinium moiety (Figure 4.38b–f).

The principal factors responsible for the alteration in fluorescence color, the enhancement of emission efficiency, and the extension of fluorescence lifetime of such aggregates are as follows: (1) Host–guest interactions diminish the aggregation degree of conjugated oligomers in aqueous solutions; (2) the macrocyclic confinement effect restricts intramolecular rotational freedom, thereby suppressing nonradiative decay pathways and increasing emission efficiency. Moreover, this effect augments molecular rigidity and promotes intramolecular π-electron delocalization, resulting in an increase in fluorescence intensity; (3) the formation of host–guest complexes minimizes the quenching impacts of environmental factors like oxygen and solvent molecules, thereby prolonging the excited-state lifetime. Therefore, the host–guest assembly strategy facilitates the effective regulation of the optical properties of conjugated oligomers, providing novel perspectives and methodologies for the development of high-performance luminescent materials.

4.2.5 Photosensitization Effect

The essence of photosensitization resides in the role of the photosensitizer. When it absorbs light of an appropriate wavelength, it is promoted from its ground state to an excited state and then efficiently transfers energy (or an electron) to nearby oxygen molecules. This process produces reactive oxygen species (ROS), which drive the photosensitization effect.

Porphyrins belong to a category of photosensitizers that are extensively employed in antibacterial applications at present. Through the utilization of electrostatic adsorption between an anionic polythiophene derivative (PTP) and a cationic porphyrin photosensitizer, positively charged nanocomposites can be generated. In the composite system, the fluorescence amplification effect and energy transfer mechanism intrinsic to conjugated polymers can substantially enhance the fluorescence emission of porphyrins. This improvement encourages more effective energy transfer to nearby oxygen, which raises the amount of singlet oxygen produced [51].

Iridium (III) complexes facilitate efficient singlet-to-triplet intersystem crossing (ISC) as a direct consequence of the heavy atom effect, consequently demonstrating long-lived phosphorescence and large Stokes shifts. This characteristic endows them with considerable application potential in the domain of photosensitizers. Given that the phosphorescence of iridium (III) complexes is easily quenched by oxygen, their interaction with oxygen enables them to function as highly efficient photosensitizers. Incorporating iridium (III) complexes (e.g. $Ir(piq)_2(acac)$) into conjugated polymer (e.g. PFO) nanoparticles constitutes a simple method for fabricating nanocomposites containing photosensitizing agents.

In an aqueous solution saturated with air, the PFO/Ir-DSPE nanocomposite displays robust luminescence, and the solution presents a rose-red color. Nevertheless, when oxygen is bubbled into the solution, the phosphorescence from $Ir(piq)_2(acac)$ is notably weakened, while the emission from PFO remains relatively unaltered, causing a corresponding color transition to purple (Figure 4.39a). Analysis of the phosphorescence decay of the PFO/Ir-DSPE nanocomposite under varying environmental conditions reveals that the decay rate of $Ir(piq)_2(acac)$

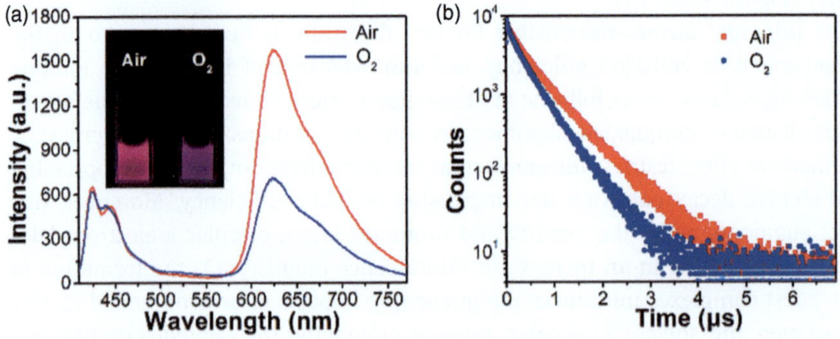

Figure 4.39 (a) Fluorescence spectra and (b) emission decay from PFO/Ir-DSPE nanocomposites under air or O_2 saturation. *Source:* Guo et al. [40] / with permission of American Chemical Society.

is markedly accelerated in oxygen-saturated aqueous solution compared to air-saturated aqueous solution. In particular, $Ir(piq)_2(acac)$ has an average lifetime of 357 ns in oxygen-saturated solution, which is much less than the 453.7 ns recorded in air-saturated solution (Figure 4.39b). This finding implies that the emission quenching of $Ir(piq)_2(acac)$ stems from collisions between its excited state and molecular oxygen, resulting in energy transfer to oxygen molecules and the subsequent generation of singlet oxygen. This mechanism further validates the efficacy of the iridium (III) complex as a photosensitizer [40].

Fluorescent conjugated polymer nanoparticles, functioning as distinctive photosensitizers, have garnered substantial research attention in the realm of photodynamic therapy owing to their outstanding photostability and low cytotoxicity. For example, the conjugated polymer PFPPBA is built from fluorene and phenyl units, resulting in a hydrophobic backbone, with side chains that bear hydrophilic quaternary ammonium salt and phenylboronic acid functionalities. The numerous hydroxyl groups existing in carbohydrate structures are capable of establishing covalent bonds with boronic acid, thus facilitating the formation of stable nanocomposites. Upon photon absorption by the PFPPBA conjugated polymer backbone, the transfer of energy from the excited triplet-state PFPPBA to molecular oxygen in the environment leads to the generation of singlet oxygen ($^1O_2$). The efficiency of PFPPBA nanoparticles in generating $^1O_2$ is predominantly determined by their triplet energy level (T_1). As shown in Figure 4.40a, the calculated T_1 energy level of the backbone is 2.58 eV, which notably surpasses the minimum energy requisite for $^1O_2$ generation (1.09 eV). As a result, PFPPBA nanoparticles display a robust capacity to sensitize molecular oxygen and efficiently generate singlet oxygen. Moreover, T_1 excitons are distributed across the conjugated structure, which leads to a substantial enlargement of the interaction surface between excitons and oxygen, thereby promoting the generation of $^1O_2$. To assess the singlet oxygen generation ability of PFPPBA nanoparticles, 2′,7′-dichlorofluorescein (DCFH) was utilized as an ROS probe. Experimental results indicate that after 10 min of irradiation, the emission intensity of DCF at 525 nm increases significantly, suggesting extensive

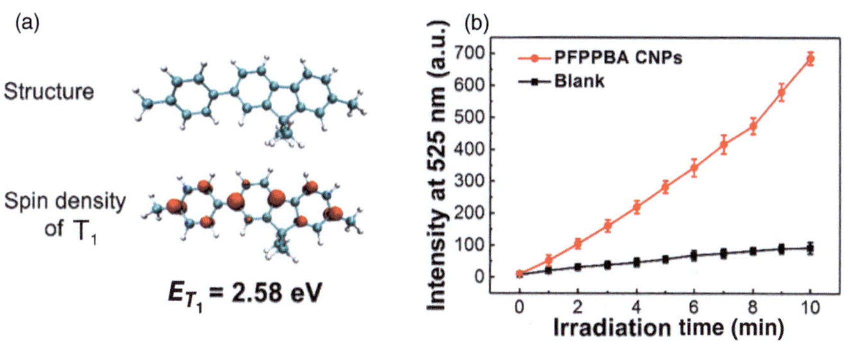

Figure 4.40 (a) The stimulated T_1 state and its energy level within the PFPPBA structure; (b) measurement of DCF fluorescence at 525 nm after the introduction of PFPPBA nanoparticles, followed by exposure to white light. *Source:* Liu et al. [52] / with permission of American Chemical Society.

oxidation of DCFH to DCF (Figure 4.40b). In comparison, the DCFH solution in the control group exhibits minimal fluorescence emission. This comparative result validates that PFPPBA nanoparticles possess a strong capacity for singlet oxygen generation. Due to their unique structural characteristics and excellent photosensitive properties, PFPPBA nanoparticles show considerable potential for application in the field of photodynamic therapy [52].

The AuNCs capped by glutathione (GSH) manifest exceptional photosensitization properties. Utilizing this attribute, the multifunctional nanocomposites, HA-PROT-AuNC, can be synthesized (Figure 4.41a). These nanoparticles effectively inhibit nonradiative transitions, thereby markedly augmenting fluorescence emission. Given the fixed ISC efficiency of AuNCs, the enhanced fluorescence of AuNCs also contributes to an increase in phosphorescence emission. At 77 K, the phosphorescence intensity of AuNC-HA-PROT was 1.96 times higher than that of pure AuNCs, according to a comparison of the phosphorescence spectra of AuNC-HA-PROT and HA-PROT. It confirms the enhancement of phosphorescence (Figure 4.41b). Furthermore, the generation of ROS can be accomplished through the interaction between a photosensitizer and the excited triplet state of oxygen. In this study, 2′,7′-dichlorofluorescein (DCFH) was utilized as a ROS probe.

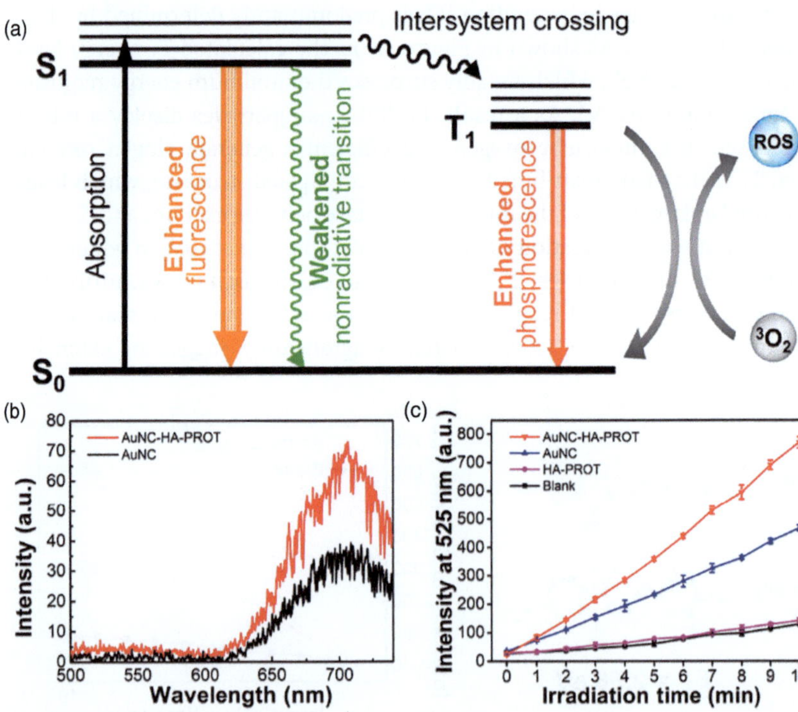

Figure 4.41 (a) Schematic illustration of enhanced singlet oxygen generation by the AuNC-HA-PROT nanocomposites. (b) Phosphorescence emission of AuNC-HA-PROT and AuNC at 77 K. (c) Fluorescence emission from DCF generated in the presence of AuNC-HA-PROT, AuNC, and HA-PROT upon light irradiation. *Source:* Xia et al. [14] / with permission of American Chemical Society.

As shown in Figure 4.41c, the experimental findings showed that the DCFH solution containing AuNCs gradually increased in emission of DCF at 525 nm under continuous irradiation, which was noticeably higher than that seen in the control group. Crucially, the DCF emission rate in the presence of AuNC-HA-PROT nanocomposites was higher than that in the presence of pure AuNCs. Meanwhile, the emission rate of the solution containing HA-PROT nanoparticles was comparable to that of the control. These findings clearly demonstrate that the modification significantly enhances the ROS generation capability of AuNCs within the AuNC-HA-PROT nanocomposites. Therefore, the GSH ligand-protected AuNCs and their nanocomposites exhibit excellent performance in photosensitization and ROS generation, providing significant references for applications in related fields [14].

Phenyl-modified carbon nitride (PhCN) is capable of absorbing visible light and generating electron-hole pairs. These electron-hole pairs can migrate to the surface of carbon nitride, where they react with molecular oxygen adsorbed on the surface, generating various ROS, including superoxide radicals ($\cdot O^{2-}$), hydroxyl radicals ($\cdot OH$), and singlet oxygen ($^1O_2$).

In diethyl phthalate, hydrogen peroxide (H_2O_2) is introduced to promote the decomposition of bis(2,4,6-trichlorophenyl) oxalate (TCPO), thus generating the high-energy intermediate dioxetanedione (C_2O_4). This intermediate is thoroughly blended with the 9,10-diphenylanthracene reaction system, enabling energy transfer from the unstable C_2O_4 to DPA. Consequently, CO_2 is produced, and 9,10-diphenylanthracene is promoted to an excited state. The 9,10-diphenylanthracene in the excited state is unstable and rapidly reverts to its ground state through radiative transition, emitting chemiluminescence with a central wavelength of 450 nm. The mixture is used as the organic phase. Phenyl-modified carbon nitride solution is used as the aqueous phase. Subsequently, the two phases are combined to form a functional emulsion, and its absorption spectrum exhibits favorable spectral overlap with the chemiluminescent light source based on 9,10-diphenylanthracene. The chemiluminescence generated in this way can effectively drive the photocatalytic activity of PhCN, resulting in the formation of electron-hole pairs and ROS (Figure 4.42a) [53].

5,5-Dimethyl-1-pyrroline-*N*-oxide (DMPO) is utilized as a spin-trapping agent for the detection of hydroxyl radicals ($\cdot OH$). When DMPO interacts with $\cdot OH$, it undergoes a chemical reaction to yield the DMPO-OH radical, which presents a characteristic signal in electron paramagnetic resonance (EPR) spectroscopy. The excitation of PhCN under a chemiluminescent light source results in the generation of a larger amount of $\cdot OH$ radicals (Figure 4.42b).

2,2,6,6-Tetramethyl-4-piperidone hydrochloride (TEMP) is employed for the detection of singlet oxygen ($^1O_2$). Upon reaction with $^1O_2$, TEMP undergoes oxidation to form TEMPO, resulting in a characteristic radical signal detectable via EPR. Following the addition of PhCN, chemiluminescence photons are effectively captured by PhCN. Consequently, a pronounced TEMPO signal is observed, indicating the generation of a substantial quantity of $^1O_2$ (Figure 4.42c).

5-tert-Butoxycarbonyl-5-methyl-1-pyrroline-*N*-oxide (BMPO) is utilized as a spin-trap agent for the detection of superoxide anion radicals ($\cdot O^{2-}$). When BMPO

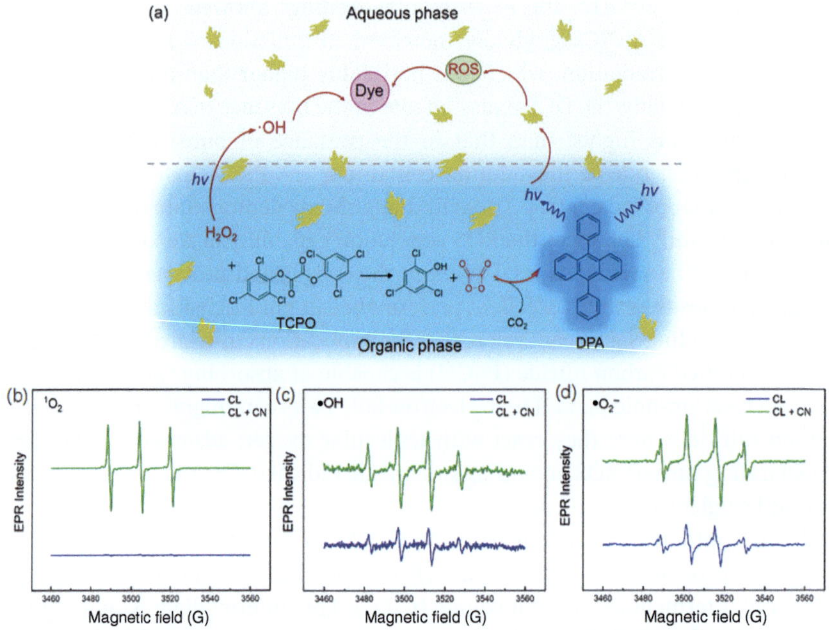

Figure 4.42 (a) Schematic illustration of chemiluminescence-driven ROS generation. Electron paramagnetic resonance (EPR) spectra of (b) $^1O_2$, (c) ·OH, and (d) $\cdot O_2^-$ in the emulsion containing PhCN (CL + CN) after H_2O_2 addition. *Source:* Luo et al. [53] / with permission of American Chemical Society.

comes into contact with $\cdot O^{2-}$, it rapidly reacts to generate BMPO-OOH, which yields a distinct signal in EPR spectroscopy. The subsequent reaction and decomposition of BMPO-OOH result in the formation of BMPO-OH, which also displays the same EPR signal. Upon the introduction of PhCN, the radical signal originating from BMPO is significantly enhanced, accompanied by a notable alteration in the peak pattern. This finding suggests that PhCN promotes the generation of $\cdot O^{2-}$ (Figure 4.42d).

The host–guest interaction markedly improves electron transfer efficiency in the supramolecular system, enabling conventional type II photosensitizers to function effectively as type I photosensitizers. In this research, traditional iodinated BODIPY (G) derivatives were utilized as guest molecules, whereas electron-rich bispillar[5]arene (BP5A) functioned as the host and electron donor, resulting in the formation of a supramolecular polymer complex. By decreasing the spatial separation between BODIPY and BP5A, the host–guest interaction facilitates more effective electron transfer from BP5A to BODIPY. Subsequently, the reaction with molecular oxygen generates $O^{2-\cdot}$ via a type I photochemical pathway, as depicted in Figure 4.43a. An analysis of the HOMO and LUMO energies of both host and guest molecules indicates that upon photoexcitation of molecule G, electrons from the HOMO orbital of BP5A can be transferred to the LUMO orbital of G* to generate $G^{-\cdot}$, which then transfers electrons to O_2, leading to the formation of

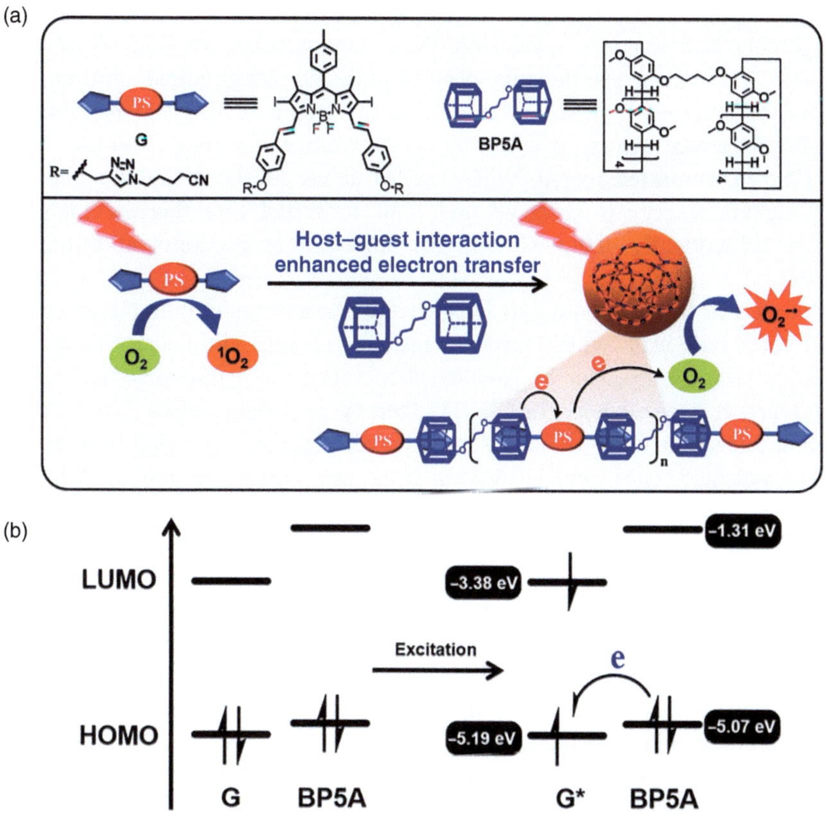

Figure 4.43 (a) The fabrication of HG nanocomposites and photoinduced generation of $O^{2-\cdot}$. (b) Schematic diagram of the electron transfer process between G and BP5A (Open Access).

$O^{2-\cdot}$ (Figure 4.43b) [54]. Therefore, the host–guest interaction can be employed to transform type II photosensitizers into type I photosensitizers, facilitating the effective regulation of both type I and type II pathways.

4.2.6 Photothermal Effect

The photothermal effect refers to the process wherein a material absorbs light, generally visible or infrared radiation, and transforms it into thermal energy. When exposed to a light source, the material absorbs photons, causing an augmentation in the kinetic energy of electrons or atoms, thereby giving rise to the generation of heat.

Noble metal nanostructures, including gold nanorods, have attracted substantial attention as near-infrared-activated nanomaterials due to their unique photothermal characteristics. These nanostructures typically exhibit two characteristic absorption peaks associated with transverse and longitudinal SPRs. The wavelength at which SPR occurs is predominantly governed by the morphology, dimensional parameters, and elemental composition of the metal nanoparticles. Through collective electron oscillations, these nanoparticles can achieve strong absorption of specific wavelengths. Significantly, the longitudinal absorption peak can be precisely adjusted by modifying the length of the gold nanorods, facilitating coverage across an extensive spectral range spanning from the visible to the near-infrared region. When excited at the localized SPR frequency, electromagnetic energy is absorbed and then converted into thermal energy, which is subsequently dissipated into the surrounding environment through diffusion.

The surface of the gold nanorods is positively charged, facilitating the adsorption of polyacrylic acid, which carries a negative charge. The polyacrylic acid, rich in carboxyl groups, can be conjugated to vancomycin through the conjugation reaction mediated by EDC/NHS, thereby obtaining AuNR@PAA-VAN nanocomposites. The longitudinal absorption peak of the gold nanorod composite was observed at 831 nm, and an 808 nm laser was selected as the irradiation source to evaluate its photothermal properties. As depicted in Figure 4.44a, upon irradiating ultrapure water with a laser, the temperature increased by merely 0.42 °C. In contrast, the temperature of the gold nanocomposite solution exhibited a substantial increase, reaching 20 °C under identical conditions. Further calculations indicated a photothermal conversion efficiency of 23.3% for the gold nanocomposite material. Additionally, after three cycles of irradiation with the 808 nm laser, the temperature variation demonstrated that the material exhibited excellent reproducibility during the heating process (Figure 4.44b). These results confirm that the gold nanocomposite material possesses outstanding photothermal performance, coupled with high reproducibility and stability under extended laser irradiation [55].

In addition, ultrasmall-sized AuNCs demonstrate a notable photothermal effect. By covalently grafting the hydrophobic dye DVP onto the surface of AuNCs possessing photothermal properties, amphiphilic Au@DVP nanoclusters were successfully synthesized. These nanoclusters are capable of interacting with the two-phase interface of the emulsion, thereby effectively stabilizing the emulsion droplets.

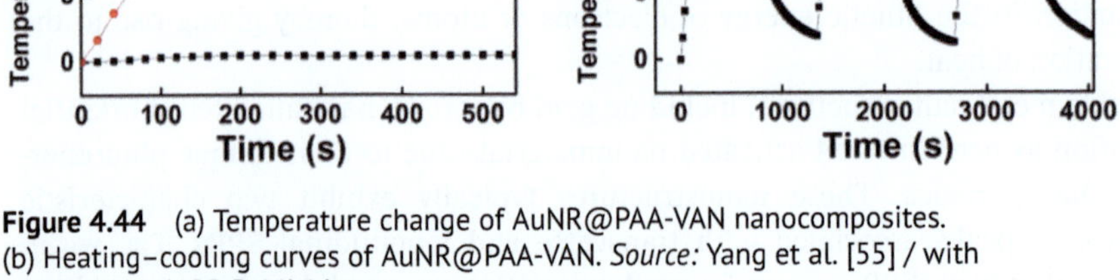

Figure 4.44 (a) Temperature change of AuNR@PAA-VAN nanocomposites. (b) Heating–cooling curves of AuNR@PAA-VAN. *Source:* Yang et al. [55] / with permission of IOP Publishing.

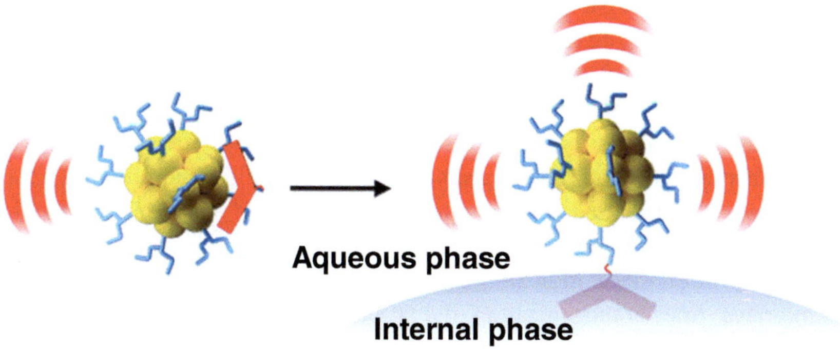

Figure 4.45 Schematic representation of the photothermal response of Au@DVP(L) in solution and at the interface of an emulsion droplet. *Source:* Yang et al. [22] / with permission of John Wiley & Sons.

Table 4.2 Photothermal conversion efficiencies of Au@DVP-Ps.

Sample	Photothermal conversion efficiency (%)
Au@DVP(L)-P	46.15
Au@DVP(M)-P	31.94
Au@DVP(H)-P	25.09

Source: Adapted from Yang et al. [22].

The nanoclusters exhibit excellent photothermal conversion performance within the emulsion system, efficiently converting absorbed light energy into thermal energy (Figure 4.45) [22].

However, the photothermal conversion efficiency varies significantly among emulsions prepared using gold nanoclusters with different grafting ratios. As depicted in Table 4.2, according to the photothermal experimental results, under 660 nm irradiation, Au@DVP(L)-P, which had a relatively low grafting ratio, manifested the highest photothermal conversion efficiency of 46.15%. Subsequently, Au@DVP(M)-P presented an efficiency of 31.94%. Conversely, Au@DVP(H)-P, which had the highest grafting ratio, demonstrated the lowest efficiency, merely reaching 25.09%. These results suggest that as the DVP grafting ratio declined, the photothermal conversion efficiency of Au@DVP-Ps increased notably. This tendency further elucidates the influence of DVP grafting on the photothermal conversion performance of the emulsion. Au@DVP(L) exhibited the highest loading capacity within the emulsion system. Moreover, the conformational alterations of DVP in Au@DVP(L) resulted in coverage of the AuNCs, thereby interfering with and weakening the energy conversion ability of the AuNCs core. Nevertheless, the DVP moieties' anchoring inside the Au@DVP(L)-P droplets alleviated the disruptions caused by DVP and promoted heat dissipation. This anchoring effect may gradually diminish with the increasing DVP content. Consequently, the high photothermal conversion efficiency observed in gold nanoparticles is closely related to their high loading quantity in the emulsion and the anchoring effect exerted by DVP.

In the investigation of the commercially available photothermal molecule ICG, it has been discovered that its spatial distribution has a major effect on the photophysical characteristics of Pickering emulsions. The amphiphilic photothermal agent ICG was incorporated into a montmorillonite-based Pickering emulsion (PE@MMT-ICG). Before homogenization, 59.32% of ICG was detected at the water–chloroform interface, indicating a spontaneous propensity of ICG to adsorb at this interface. After homogenization, the proportion of ICG located at the interface rose to 70.25%, which can be ascribed to the expansion of the interfacial area during the process. Upon laser irradiation, PE@MMT-ICG exhibited remarkable photothermal performance, manifested by a substantial temperature increase. As depicted in Figure 4.46, subsequent to 16 min of irradiation, the temperatures of the ICG aqueous solution and PE@MMT-ICG rose by 17.5 and 30.6 °C, respectively. Considering that the volumes of the two liquid specimens and the ICG concentration were equivalent, it can be deduced that the Pickering nanoemulsion, serving as a delivery system for ICG, demonstrates superior photothermal characteristics in comparison to the ICG aqueous solution. The augmented and more persistent photothermal performance of PE@MMT-ICG can be ascribed to the inhibition of ICG self-aggregation, which diminishes the proportion of absorbed light energy expended in endothermic decomposition and photodegradation processes [27].

The multifunctional nanocomposites, CNP@AMP-AuNCs, were synthesized by exploiting the photothermal conversion properties of the conjugated polymer PDPP-DBT [56]. These nanoparticles exhibit remarkable photothermal conversion capacity and structural stability. When irradiated with 808 nm near-infrared light at an intensity of 1 W/cm^2, the temperature of the material rapidly rises by 20 °C within 5 min and exceeds 50 °C within 15 min (Figure 4.47a). The photothermal conversion efficiency attains 34.5%, which is comparable to that of traditional photothermal conjugated polymers like polyaniline and polypyrrole [57, 58]. This efficient photothermal performance can be ascribed to the high-temperature

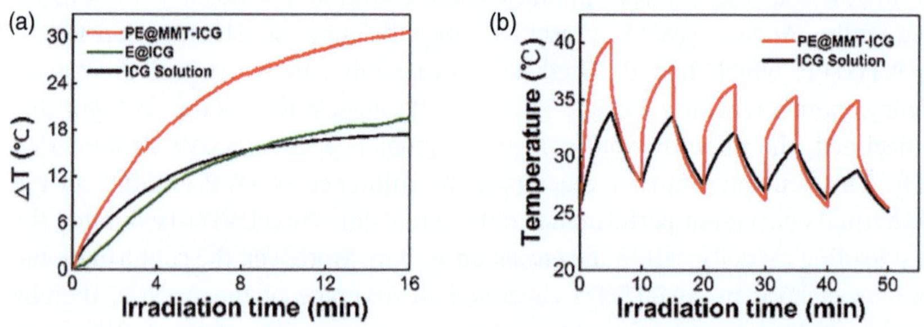

Figure 4.46 (a) Temperature changes of the ICG solution, E@ICG emulsion, and PE@MMT-ICG emulsion under 808 nm laser irradiation. (b) Heating–cooling curves of the ICG solution and PE@MMT-ICG. *Source:* Zan et al. [27] / with permission of American Chemical Society.

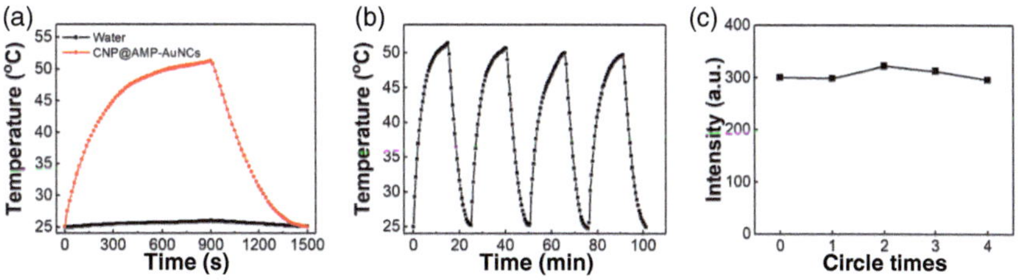

Figure 4.47 (a) Temperature changes of CNP@AMP-AuNCs with 808 nm NIR laser irradiation. (b) Four consecutive heating–cooling cycles and (c) the associated fluorescence emission for CNP@AMP-AuNCs. *Source:* Zhu et al. [9] / with permission of American Chemical Society.

environment generated under near-infrared irradiation. Such an environment not only promotes electron donation from thiol, phenolic, and amino groups to Au(III), thus accelerating the reduction of Au(0) (increasing the Au(0)/Au(I) ratio to 4.9) but also elevates the collision frequency among Au(0) atoms, facilitating the formation of a greater number of Au(0)NC@Au(I)-AMP core-shell nanostructures. The in situ reduced gold nanoclusters not only display superior fluorescence emission characteristics but also maintain stable fluorescence intensity over four cycles of near-infrared irradiation (Figure 4.47b,c). This indicates enhanced cycling stability in comparison to the clinically utilized photothermal dye ICG. These results imply that CNP@AMP-AuNCs further enhance thermal stability while retaining the original photothermal properties of CNPs, thereby offering a robust material platform for precise photothermal therapy (PTT) [9].

During the conversion of light energy to thermal energy, photothermal materials induce the production of acoustic signals in the surrounding tissues. These acoustic waves are then detected by a detector and reconstructed into photoacoustic images through computer processing. Due to the minimal attenuation of acoustic waves as they propagate through biological tissues, photoacoustic imaging is particularly well-suited for deep tissue visualization, effectively combining the high contrast of optical imaging with the substantial penetration depth of ultrasound. Furthermore, compared to the visible and NIR-I spectral windows, the NIR-II window is characterized by markedly reduced tissue autofluorescence, making it especially advantageous for high-resolution biological imaging applications [59].

For example, dopamine was polymerized on MoS_2 nanosheets (NSs), followed by covalent conjugation with polyethylene glycol (PEG) and a triphenylphosphonium (TPP)-terminated peptide (pep-TPP, CCKKGGKK-TPP), leading to the synthesis of mitochondria-targeted MoS_2@PDA-Fe@PEG/TPP nanosheets (MPFPT NS). As shown in Figure 4.48, the NIR-II photoacoustic (PA) signal intensity of MPFPT NS at 1210 nm increased proportionally to concentration, indicating a strong linear correlation within the 0–1.6 mg/mL range. These results suggest that MPFPT NS have considerable potential as NIR-II PA imaging nanotheranostic agents [60].

Cy5.5 and its quencher, QSY21, display strong absorption in the near-infrared range and are capable of producing photoacoustic signals. An alkyne-terminated peptide was linked to azide-functionalized Cy5.5 via a click chemistry reaction,

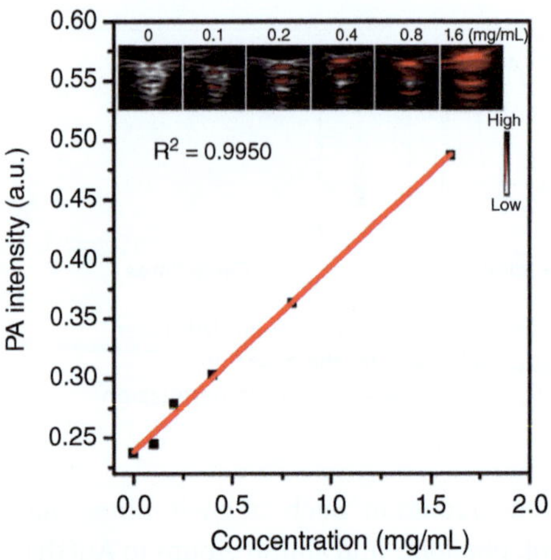

Figure 4.48 Variation in PA intensity with increasing concentrations of MPFPT NS aqueous dispersions. Inset: Corresponding PA images of MPFPT NS. *Source:* Reproduced with permission from Li et al. [60] / American Chemical Society.

and subsequently, the quencher QSY21 was attached via an NHS coupling reaction (Figure 4.49a). Because of the intrinsic hydrophobicity of these molecules, they spontaneously self-assemble into nanoparticles in an aqueous solution. The probe generates photoacoustic signals at two wavelengths: 680 nm (referred to as PAS_{680}) and 730 nm (referred to as PAS_{730}). Experimental findings suggest that the signal intensity of PAS680 is significantly and negatively correlated with the concentration of matrix metalloproteinase-2 (MMP-2), while the intensity of PAS730 remains relatively unaffected (Figure 4.49b). In the presence of MMP-2, the peptide substrate is cleaved, thus disrupting the FRET from Cy5.5 to QSY21. This phenomenon is ascribed to the enzymatic cleavage, which causes the quencher to be unable to effectively suppress the fluorescence of Cy5.5. Consequently, the increased photoacoustic activity of Cy5.5 leads to a reduction in photoacoustic signal intensity at 680 nm. This design effectively utilizes the enzyme-mediated regulation of FRET to facilitate the specific detection and imaging of MMP-2 activity [61].

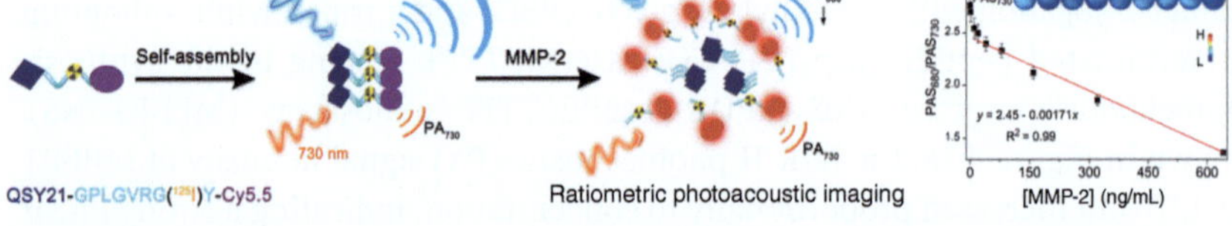

Figure 4.49 (a) The photoacoustic imaging mechanism of the nanocomposites. (b) Changes in PAS_{680}/PAS_{730} with elevated concentrations of MMP-2. The inset shows the corresponding PA images. *Source:* Yin et al. [61] / with permission of American Chemical Society.

4.3 Bioapplications of Photofunctional Polymer Nanocomposites

4.3.1 Cell Imaging

Cell imaging is essential to the research of cell and tissue functions, the pathogenesis of diseases, and clinical diagnosis [62]. Conventional cell imaging acquires information on the structure of biological tissues and the progress of biochemical reactions through fluorescence signals generated by organic small molecules or quantum dot materials. However, single imaging materials often have drawbacks such as easy photobleaching or high biological toxicity [63, 64]. Therefore, by leveraging the diverse and stable optical signals, high water solubility, and biocompatibility provided by the composite system, it is possible to design optical functional probes with specific functions, enabling precise and sensitive biological imaging at the subcellular, cellular, and animal levels.

Targeted imaging is a technique that enables high-precision imaging by combining specific molecules, cells, or tissues and using specific markers or probes. It significantly improves the resolution and specificity of imaging, thereby better facilitating the observation and analysis of biological processes or pathological changes. The core of this technology lies in the specific binding ability of the labeled molecules to the target.

The cyclic peptide RGD-modified nanoparticles can specifically recognize the cell surface integrin $\alpha_v\beta_3$ for targeted binding. The PVK/BPOTZ-OR nanoparticles are formed by a co-precipitating method. The amino groups of RGD can be modified onto the surface of the nanoparticles through the NHS-EDC reaction with the carboxylic groups on the surface of the nanoparticles. Human breast cancer cells MDA-MB-231 overexpress the integrin $\alpha_v\beta_3$. Human breast cancer cells MCF-7 and human normal liver cells L-02 have low expression of the integrin $\alpha_v\beta_3$ [65, 66]. Therefore, MCF-7 and L-02 cells are used as the control groups. After co-culture of the three types of cells with PVK/BPOTZ-OR@RGD nanoparticles, as shown in Figure 4.50a, significant green fluorescence was observed in human breast cancer cells MDA-MB-231, while MCF-7 cells showed less fluorescence, and human normal liver cells L-02 showed no fluorescence signal (Figure 4.50b). The analysis results indicate that, following the recognition of the RGD-modified nanoparticle surfaces by integrins expressed on MDA-MB-231 cells, a large number of the nanoparticles were taken into the cell interior. From the cell imaging results, it can be seen that this fluorescent nanoparticle can distinguish cancer cells with different expression levels of integrin $\alpha_v\beta_3$ and has the function of targeting cells [18].

Polysaccharides present on the cell surface are integral to mediating cellular communication, particularly during processes such as cell development and differentiation [67]. Aberrant expression patterns of these polysaccharides can serve as biomarkers for certain cancer cell types. Notably, sialic acid (SA), a critical constituent of cell-surface polysaccharides, is frequently overexpressed in malignant cells, and this heightened expression is widely recognized as a significant indicator of various cancers. Phenylboronic acid (PBA) can bind covalently to compounds containing a cis-diol group. Under normal physiological conditions,

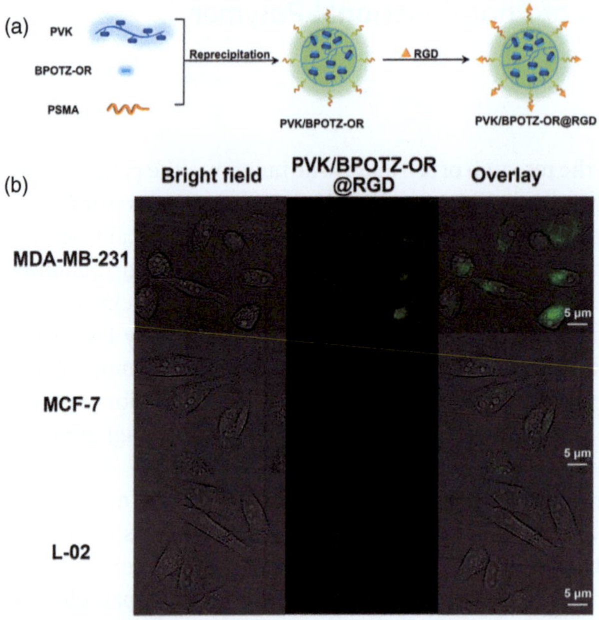

Figure 4.50 (a) The structure of PVK/BPOTZ-OR@RGD. (b) Fluorescence images of MDA-MB-231, MCF-7, and L-02 cells co-cultured with PVK/BPOTZ-OR@RGD. *Source:* Reproduced with permission from Liu et al. [18] / American Chemical Society.

the binding ability of phenylboronic acid to sialic acid is not significantly different from its binding ability to other monosaccharides. Molecular imprinting technology is a method that introduces template molecules during the preparation of polymers or nanoparticles, and then removes the template molecules. The pores and binding sites formed in the nanoparticles can specifically recognize the template molecules. Using sialic acid as the template molecule, through molecular imprinting technology, its specific recognition ability can be significantly enhanced.

For example, the method of stabilizing Pickering emulsion polymerization by using phenyl-modified carbon nitride (PhCN) was adopted, successfully preparing fluorescent nanocomplexes SVBA/PhCN-SA NPs with sialic acid (SA) imprinting (Figure 4.51a). During the experiment, PSVBA/PhCN-SA NPs were co-cultured with prostate cancer cells DU 145 and cervical cancer cells HeLa.

For PSVBA/PhCN nanoparticles lacking molecular imprinting capability, comparable green fluorescence was detected within the cytoplasm of both DU 145 and HeLa cells. This observation suggests that, despite the high density of phenylboronic acid moieties present on the surface of PSVBA/PhCN NPs, these nanoparticles do not exhibit selective binding to the SA targets that are overexpressed on the cellular membrane. In contrast, following co-incubation with PSVBA/PhCN + SA NPs, fluorescence signals in DU 145 and HeLa cells were nearly undetectable. This outcome may be attributed either to electrostatic repulsion between the SA-functionalized nanoparticle surfaces and the cell membrane or to the blockade of nanoparticle groups capable of membrane interaction by the SA complex. Following

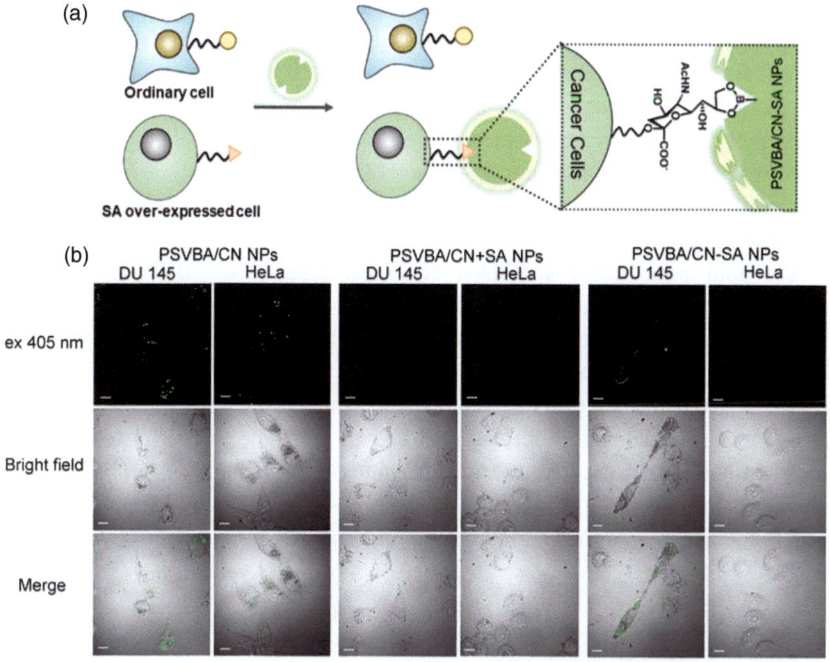

Figure 4.51 (a) Schematic illustration of the specific recognition and binding mechanism of PSVBA/PhCN-SA NPs with cancer cells. (b) Fluorescence imaging of DU 145 and HeLa cells following incubation with PSVBA/PhCN NPs, PSVBA/PhCN + SA NPs, and PSVBA/PhCN-SA NPs, respectively. *Source:* Reproduced with permission from Luo et al. [25] / American Chemical Society.

co-culture with PSVBA/PhCN-SA nanoparticles possessing molecular imprinting functionality, pronounced fluorescence signals were observed on the membranes of DU 145 cells. This finding indicates that the nanoparticles predominantly bind to SA expressed on the cell surface. Conversely, no significant fluorescence was detected in HeLa cells, providing further evidence that SA-imprinted nanoparticles exhibit specific recognition and binding affinity toward DU 145 cancer cells characterized by SA overexpression (Figure 4.51b). Therefore, this study successfully prepared SA-imprinted nanoparticles with targeting recognition ability through molecular imprinting technology, providing new ideas and methods for the specific diagnosis and treatment of prostate cancer [25].

The process by which nano-imaging materials enter cells mainly relies on the endocytosis mechanism. The endocytosis mechanism is a complex biological process involving the coordinated action of various cell membrane structures and related proteins [68]. Among them, membrane structures such as clathrin, actin, and microtubule proteins play crucial roles in the process of endocytosis. Additionally, micropinocytosis and the energy provided by sugars are also important components of the endocytosis mechanism. These factors work together to make the endocytosis process have a longer temporal characteristic, and its efficiency and properties are regulated and influenced by various factors. In contrast, the physical fusion method shows significant advantages in the process of nanomaterials entering cells. The physical fusion method usually has the characteristics of high efficiency, strong controllability, and simple operation. Physical methods allow for precise control over the observable properties of nanomaterials, including their size, morphology, and surface features, thereby achieving the regulatory effect on the process of nanomaterials entering cells. This method not only improves the efficiency of nanomaterials entering cells but also provides more possibilities for the application of nanoscale imaging materials.

For the gold nanocluster stabilized emulsion system Au@DVP(H)-P, after co-culturing with human osteosarcoma MG-63 cells for 15 min, uniform green and red fluorescence were detected within the cytoplasm (Figure 4.52a,b), suggesting rapid cellular uptake of the emulsion. Continuous cell imaging, as presented in Figure 4.52c, demonstrated that increasing cytoplasmic fluorescence was accompanied by a rapid enhancement of fluorescence signals along the cell periphery. These findings further indicate that the internalization of the emulsion by the cells occurs independently of the concentration of Au@DVP(H)-P. Additionally, no permanent damage to the cell membrane was observed in the experiment.

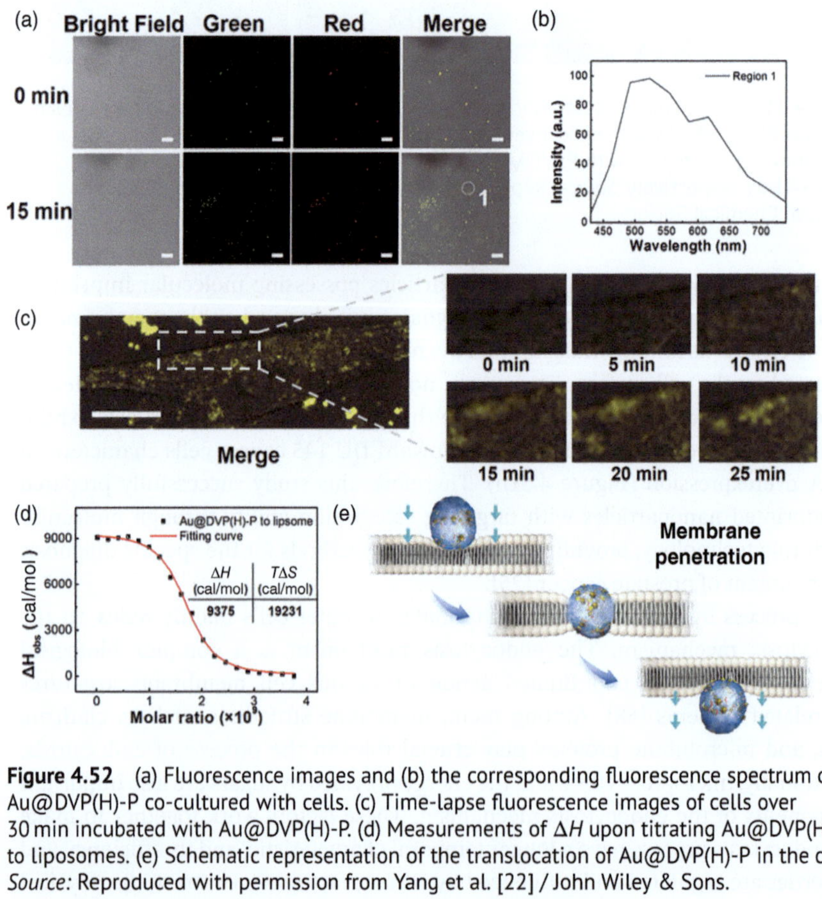

Figure 4.52 (a) Fluorescence images and (b) the corresponding fluorescence spectrum of Au@DVP(H)-P co-cultured with cells. (c) Time-lapse fluorescence images of cells over 30 min incubated with Au@DVP(H)-P. (d) Measurements of ΔH upon titrating Au@DVP(H)-P to liposomes. (e) Schematic representation of the translocation of Au@DVP(H)-P in the cell. *Source:* Reproduced with permission from Yang et al. [22] / John Wiley & Sons.

To assess the interfacial interactions, liposomes were employed to mimic the cell membrane structure for isothermal titration calorimetry (ITC) analysis. As depicted in Figure 4.52d, both the ΔH and $T\Delta S$ values associated with the interaction between Au@DVP(H)-P and liposomes are positive, indicative of hydrophobic interactions. Given the hydrophilic nature of the glutathione ligand on Au@DVPs and the hydrophobic character of PLGA, the surface of Au@DVP-Ps is composed of distinct hydrophilic and hydrophobic microdomains. During cellular interaction, the hydrophilic regions of Au@DVP-Ps preferentially associate with the hydrophilic head groups of the cell membrane, while the hydrophobic domains interact with the membrane's hydrophobic tail regions. This dual affinity facilitates the translocation of Au@DVP-Ps into the cell interior (Figure 4.52e). These findings offer new insights into the potential applications of emulsion-based systems for cell imaging [22].

To evaluate the suitability of Au@DVP(H)-P for long-term cell tracking, cells were continuously imaged over a period of 1 hour. The results demonstrated that Au@DVP(H)-P displayed distinct green and red fluorescence patterns depending on its location within the cells. As shown in the three-dimensional fluorescence overlay in Figure 4.53a, both fluorescence signals originated from the cytoplasm. This variation in fluorescence distribution was primarily due to differences in the polarity of the cellular microenvironment. At Au@DVP(H)-P, the densely grafted DVP may respond to intracellular polarity shifts during prolonged incubation.

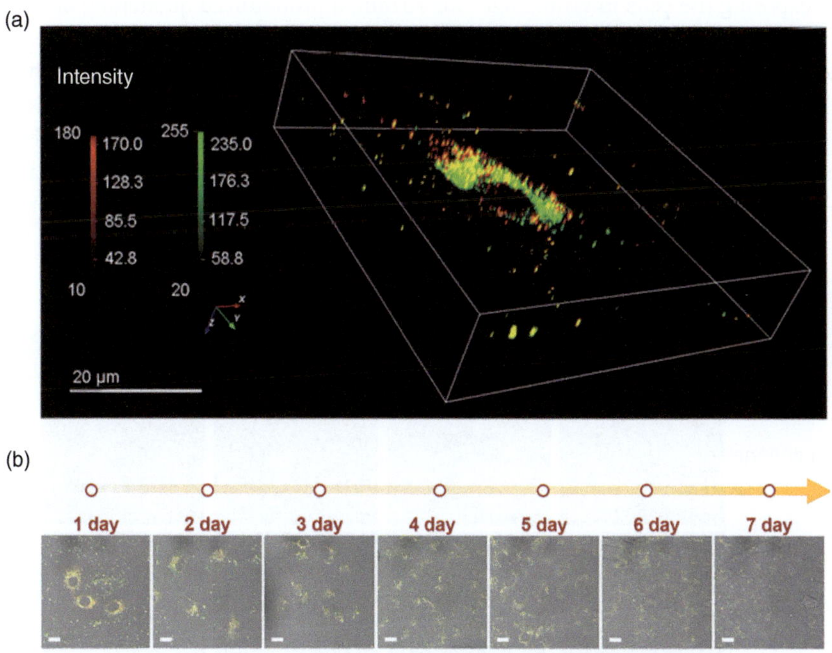

Figure 4.53 (a) 3D fluorescence reconstruction of one cell stained with Au@DVP(H)-P. (b) The fluorescence images of cells following 7 days of continuous culture after staining with Au@DVP(H)-P. *Source:* Reproduced with permission from Yang et al. [22] / John Wiley & Sons.

Additionally, even after 7 days of cell culture, Au@DVP(H)-P fluorescence remained detectable (Figure 4.53b). These results indicate that Au@DVP(H)-P can remain within the living cells and can be observed for cell division and proliferation, making it suitable for efficient cell tracking. Therefore, Au@DVP(H)-P shows good adaptability in long-term cell tracking, and its fluorescence properties provide a powerful tool for studying the polarity changes in the cell microenvironment and related physiological/pathological processes.

Utilizing fluorescent probes to achieve noninvasive real-time imaging of intracellular structures is of great significance for cell research [69]. The ultra-small gold nanoclusters (AuNCs) possess unique physical and chemical properties that enable them to penetrate natural barriers such as cell membranes and nuclear envelopes and enter the cell nucleus smoothly. By combining the photochromic AuNC-SP with the biological molecule chitosan (CS), a new fluorescent probe, AuNC-SP@CS, can be obtained. After co-culturing with MCF-7 cells, pronounced green fluorescence attributable to the nanocomposite was detected intracellularly, whereas red fluorescence was absent. When the cells that had taken up AuNC-SP@CS were irradiated with ultraviolet light, as shown in Figure 4.54, a bright red signal emitted from the SH-MC in AuNC-SP@CS was observed. This indicates that in the cells, spiropyran's closed loop state transforms into the open MC state upon exposure to ultraviolet light. Meanwhile, the weakening of the green fluorescence signal is attributed to FRET that occurs between AuNCs and the photochromic SH-SP ligand. After exposing the cells to visible light for 10 min, a pronounced quenching of the red signal, accompanied by an enhancement of the green signal, was observed. This is due to the restoration of the spiropyran from SP in AuNC-SP@CS to the MC state, thereby inhibiting the FRET process. In the second round of ultraviolet and visible light exposure, the reversible change in fluorescence color can still be detected. This demonstrates that AuNC-SP@CS maintains the photoisomerization of MC/SP during intracellular internalization and successfully achieves efficient and precise dual-color imaging of tumor cells [15].

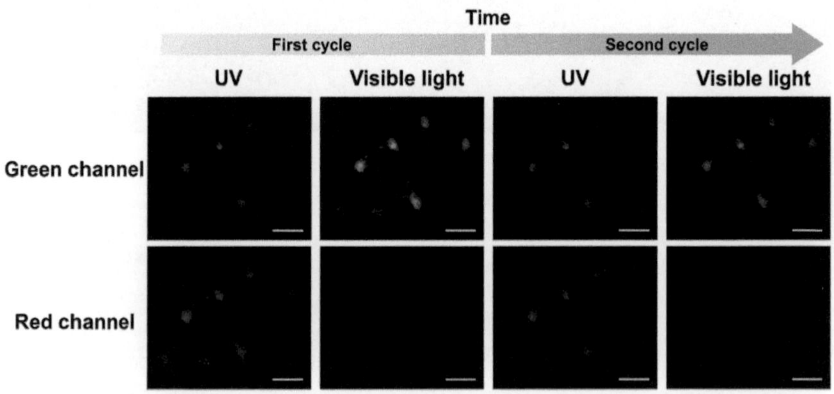

Figure 4.54 Fluorescence imaging of MCF-7 cells with AuNC-SP@CS during two consecutive UV–visible light irradiation cycles. *Source:* Reproduced with permission from Cong et al. [15] / American Chemical Society.

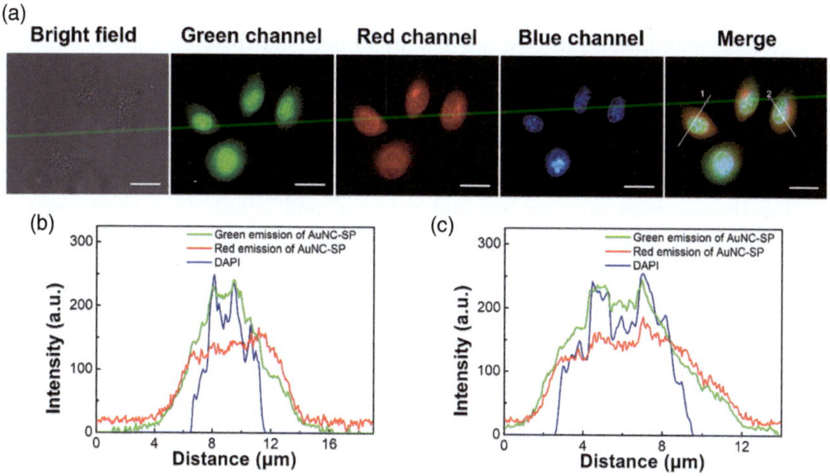

Figure 4.55 (a) Representative fluorescence images of MCF-7 cells incubated with AuNC-SP@CS and DAPI. The corresponding colocalization intensity distributions along arbitrary line 1 (b) and line 2 (c). *Source:* Reproduced with permission from Cong et al. [15] / American Chemical Society.

In staining MCF-7 cells using the commercial nucleic acid dye DAPI, the overlapping areas of green and red fluorescence signals appear orange in the superimposed fluorescence images, while the nucleus region appears cyan due to the effective superimposition of the blue fluorescence of the nucleic acid dye DAPI with the green and red fluorescence of AuNC-SP (Figure 4.55). This result not only confirmed the distribution of AuNC-SP in the cell nucleus but also revealed its unique fluorescence properties. Further analysis indicated that the AuNC-SP@CS nanocomposite could smoothly enter the cytoplasm, which was mainly attributed to the lysosomal degradability characteristic of chitosan (CS). Under enzymatic degradation, chitosan is decomposed, thereby releasing AuNC-SP. Since the size of AuNC-SP is significantly smaller than the diameter of the cell nucleus pores, it can easily transit the nuclear pores and accumulate within the nucleus. Therefore, the AuNC-SP@CS nanocomposite also possesses reversible light control properties, which enables it not only to be used for fluorescence imaging in the cytoplasm but also to further penetrate into the nucleus interior to achieve comprehensive imaging of tumor cells.

In fluorescence imaging, the self-luminescence of biological systems can generate background noise, which reduces the resolution and accuracy of probe fluorescence imaging. The absorption and scattering of light signals by biological tissues also lead to attenuation of excitation and emission light, affecting the effective imaging depth of the probe. Inorganic photofunctional materials have rich energy level structures, which facilitate the construction of various imaging systems, such as two-photon fluorescence imaging, fluorescence lifetime imaging, and photoacoustic imaging, and help to overcome the aforementioned problems.

The nonlinear optical properties of photonic functional materials have been developed for multimodal imaging. For instance, two-photon imaging has many unique advantages over single-photon imaging, mainly including: (1) The excitation wavelengths of two-photon imaging are usually in the near-infrared region, which enhances the tissue penetration ability; (2) the energy of the excitation wavelength of two-photon is low, which can avoid the fluorescence bleaching problem and also reduce the phototoxicity to biological tissues; (3) the long wavelength of two-photon can avoid the spontaneous fluorescence of biological tissues at low wavelengths; (4) two-photon fluorescence imaging has higher lateral and longitudinal resolutions, thus enabling the acquisition of higher-resolution three-dimensional images.

The dual-ligand–modified gold nanoclusters (AuNCs) prepared using amphipathic ligand (ZW) and 11-mercaptoundecanoic acid (MUA) have demonstrated significant potential in the field of biological imaging. Through the coupling reaction of EDC/NHS and MUA, streptavidin (SA) and biotin are covalently attached to the surface of AuNCs, forming AuNC-SA and AuNC-biotin (Figure 4.56a–d). These functionalized nanoclusters then bind to biotinylated antibodies and are used to process MCF-7 cells, thereby achieving specific assembly. The data demonstrate that AuNC-SA and AuNC-biotin undergo robust, specific membrane-associated interactions. The high two-photon absorption cross-section of AuNC@ZW/MUA enables its effective empolyment in two-photon microscopy, wherein the emitted green fluorescence is chiefly confined to the MCF-7 cell membrane. It effectively reduces the interference of background spontaneous fluorescence (Figure 4.56e,f) [70].

Furthermore, the long fluorescence lifetime characteristic of AuNCs further enhances their imaging effect (Figure 4.56g). By comparing the lifetimes of AuNCs

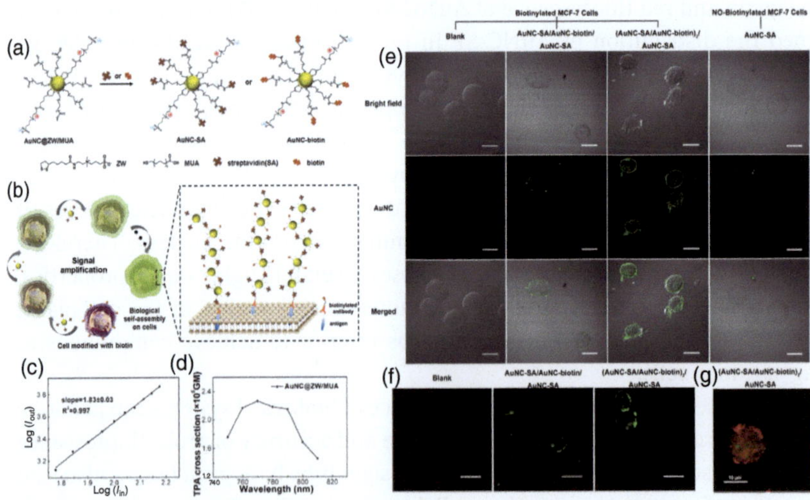

Figure 4.56 (a) The structure of AuNC-SA and AuNC-biotin, and (b) their biotin–streptavidin driven assembly. (c) Linear dependence of the integrated fluorescence intensity of AuNC@ZW/MUA on two-photon excitation power. (d) Two-photon absorption cross sections of AuNC@ZW/MUA. (e) Fluorescence images, (f) two-photon fluorescence images, and (g) fluorescence lifetime images of assembled AuNC-SA/AuNC-biotin nanocomplexes on MCF-7 cells. *Source:* Reproduced with permission from Jiang et al. [70] / American Chemical Society.

in the cell membrane and cytoplasm, it was found that the lifetime of AuNCs on the cell membrane (approximately 8 ns) was significantly longer than that of AuNCs in the cytoplasm (approximately 5 ns). The difference in the lifetimes of AuNCs in the cell membrane and cytoplasm reveals the significant differences in the internal and external environments of the cell. By merging two-photon imaging with fluorescence lifetime imaging and employing the bioconjugation-driven assembly of AuNC-SA and AuNC-biotin, detailed information about the EpCAM antigen on the MCF-7 cell membrane was obtained. This multimodal imaging technique not only improves the resolution and specificity of imaging but also provides a powerful tool for in-depth research on cell membrane–related biological processes.

Photoacoustic imaging converts light energy into heat through optical functional probes, causing instantaneous thermal elastic expansion of tissue structure, thereby generating ultrasonic waves to provide images with high contrast, high spatial resolution, and high penetration. Photoacoustic imaging synergistically integrates the high molecular specificity inherent to optical imaging with the superior tissue penetration capabilities of ultrasound. By exploiting the photoacoustic effect, this modality effectively circumvents the limitations imposed by optical scattering, thereby surpassing the conventional depth restriction of high-resolution optical imaging (approximately 1 mm). Consequently, photoacoustic imaging enables high-resolution visualization of deep-seated biological tissues in vivo [71, 72].

The series of materials constructed based on π-conjugated polymers of indigoid have an ultra-narrow bandgap characteristic of approximately 1.2 eV, which makes them ideal candidates for photoacoustic imaging. The conjugated polymer can be co-precipitated with saturated phospholipid DPPC to successfully prepare the nanocomposite SPN4. This material exhibits a significant absorption peak at 800 nm and shows an excellent extinction coefficient ($253\,L\,g^{-1}\,cm^{-1}$). In in vivo experiments, SPN4 was injected subcutaneously into the lateral abdominal region of mice for photoacoustic imaging. Figure 4.57 shows a comparison of the imaging results of five characteristic bands before and after injection in the same field of view ($14 \times 14 \times 6\,mm^3$): (1) The 600 nm band shows low contrast of SPN4 due to the strong absorption of blood; (2) the 800 nm band shows significant signal enhancement due to the dominant absorption of SPN4; (3) even in the 1000 nm long wavelength band, SPN4 can still maintain visible contrast in the background interference of blood and water. It is worth noting that the high contrast area at the injection site is due to the local aggregation effect of SPN4 during the subcutaneous injection process [73].

4.3.2 Optical Therapies

Photodynamic therapy has been clinically adopted for the killing of pathogenic bacteria and tumor treatment and has thus developed into a relatively complete disease treatment technology [74]. The mechanism of photodynamic therapy is that the photosensitizer molecules absorb light and undergo ISC to reach T_1. Then, through electronic transfer or energy transfer, the energy is delivered to the oxygen outside the molecule or the cellular components, generating ROS

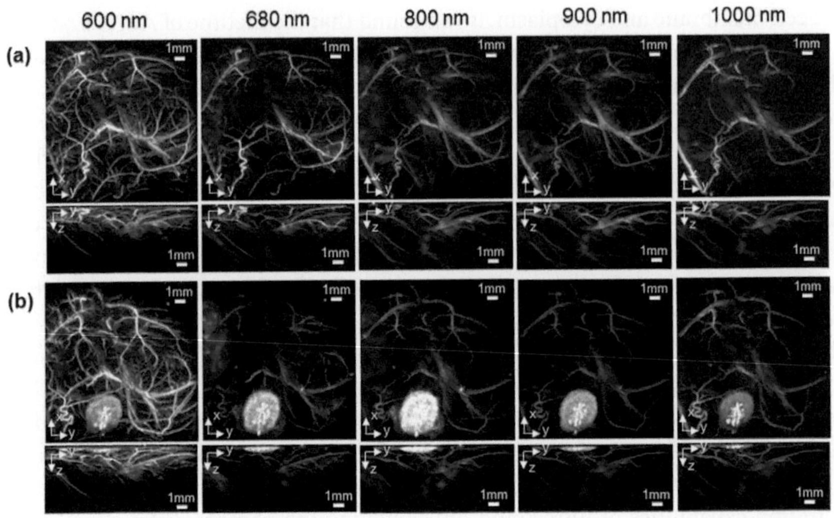

Figure 4.57 in vivo photoacoustic images of the vasculature, taken (a) before and (b) after the injection of SPN4 (Open access).

substances, including singlet oxygen, superoxide anion radicals, hydroxyl radicals, and hydrogen peroxide with high reactivity, which oxidize the cellular structure and ultimately cause the death of pathogenic bacteria or tumor cells. According to the different ways of generating ROS, photodynamic therapy is usually divided into two types. The first type is that the photosensitizer molecules in T_1 directly interact with biological molecules, generating free radicals through electron transfer or hydrogen transfer. These free radicals react rapidly and produce substances with stronger oxidizing properties to damage the cell structure. The second type of action is that the photosensitizer molecules in T_1 interact with ground-state oxygen molecules to produce singlet oxygen, and then the electrophilic singlet oxygen further oxidizes the biological molecules [75].

The D-A-D type isomer 6MNVDPP uses methoxy-naphthaldehyde as the electron donor and 3,6-dimethyl-substituted diketopyrrolopyrrole (DPP) as the electron acceptor. It is prepared as a nanocomposite through the co-precipitation method with the amphiphilic conjugated polymer DSPE-PEG2000. In this nanocomposite, the 6MNVDPP molecules exhibit a typical J aggregation. Neighboring 6MNVDPP molecules establish bimolecular D-A interactions through C—H···O hydrogen bonding between hydrogen atoms on the naphthalene ring and the carbonyl group on the planar ring of DPP. These multiple intermolecular D-A interactions work in synergy with the J aggregation pattern, giving rise to a compact π–π stacking of 6MNVDPP and ultimately forming a 3D spatial D-A interlocked network (Figure 4.58a–c) [76].

This structure of 6MNVDPP significantly enhances the electron coupling between molecules and promotes the separation of photogenerated electrons and holes. Upon photoexcitation, molecules in the excited state transfer energy to molecular oxygen via electron transfer processes, resulting in the formation of superoxide

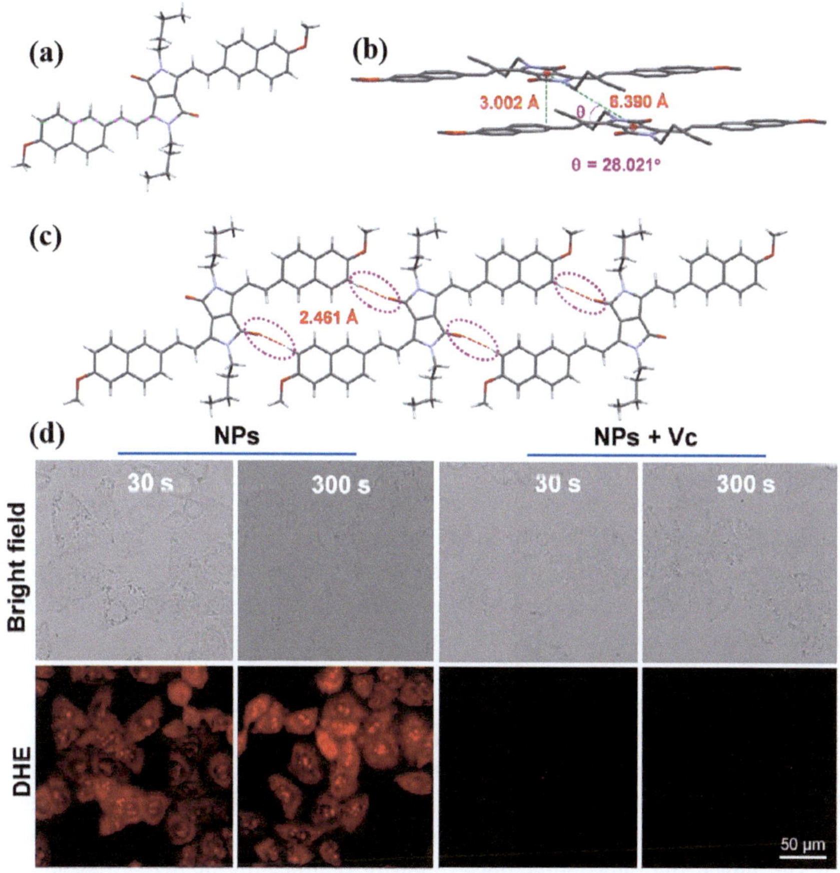

Figure 4.58 (a) Single crystal structure, (b) dimer structure, and (c) molecular interactions of 6MNVDPP. (d) Generation of $O_2^{\cdot-}$ by 6MNVDPP NPs with incubated with HeLa cells, using DHE as the indicator and Vc as a radical scavenger. *Source:* Feng et al. [76] / with permission of John Wiley & Sons.

anion radicals ($O_2^{\cdot-}$). The generation of $O_2^{\cdot-}$ at the cellular level was verified using dihydroethidium as a fluorescent probe. The experimental results showed that after 808 nm laser irradiation, adding vitamin C (Vc, a free radical inhibitor), the red emission signal was significantly quenched, further verifying the I-type photodynamic activity of the 6MNVDPP nanocomplex (Figure 4.58d). Therefore, by regulating the intermolecular interactions, 6MNVDPP exhibits excellent photophysical properties and photodynamic activity, providing a new idea for the development of efficient I-type photosensitizers.

The strategy of enhancing ROS has demonstrated significant application potential in the fields of nanomaterials and photodynamic therapy. Using the co-precipitation technique, nanoparticles were formed with conjugated polymer PFBT as the energy donor and TPP as the energy acceptor. Based on the energy transfer amplification system of the conjugated polymer, the emission intensity of TPP was significantly

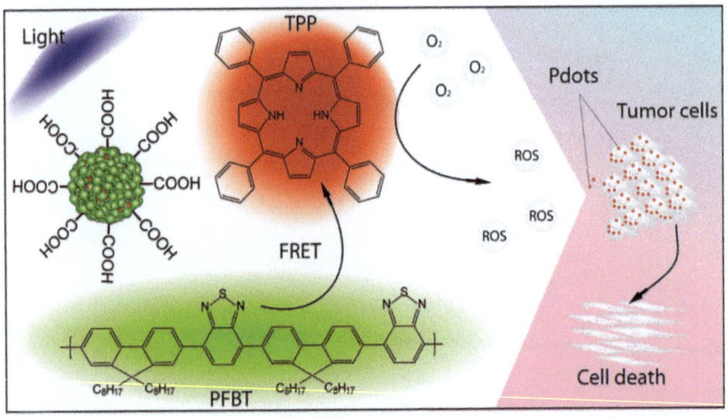

Figure 4.59 The enhanced ROS production based on the FRET effect between PFBT dots and TPP. *Source:* Li et al. [77] / with permission of American Chemical Society.

increased, and the number of excited-state molecules was also increased, thereby enhancing the singlet oxygen production capacity of TPP by 14 times compared to using TPP alone. After injecting these nanoparticles into tumor-bearing mice and undergoing photodynamic therapy, the tumors were completely suppressed (Figure 4.59). Moreover, Pdots showed no obvious side effects on the major organs of mice and exhibited good biocompatibility [77].

By regulating the surface modification of the materials, the photophysical properties, and the molecular assembly methods, the generation efficiency of ROS can be significantly enhanced, thereby achieving efficient antibacterial or antitumor effects. For instance, the MEF effect is applied to enhance the production of ROS. Taking the PAA/CS@Au/RB nanocomposite as an example, after the RB molecules are modified onto the surface of the nanocomposite, due to the influence of the SPR effect, the phosphorescence intensity significantly increases. Because the production of singlet oxygen is governed by the energy transfer from the triplet excited-state photosensitizer to oxygen, this nanocomposite can significantly increase the yield of ROS for killing bacteria [78].

In addition, by enhancing the generation of ROS from the material design aspect, the combination of multiple interactions and multivalent binding can achieve a close and specific binding of photosensitive materials to biological systems, while promoting the effective enrichment of photosensitive materials and improving the killing efficiency.

Research on fluorescent conjugated polymer PFPPBA nanoparticles has shown that those containing phenylboronic acid (PBA) and quaternary ammonium salt (QAS) can efficiently and specifically kill *P. aeruginosa* through a process known as the “sugar bridge” interaction. Figure 4.60a simulates the binding mode between the bacterial surface lectin LecA, lactulose, and the PBA groups on the surface of the nanoparticles. The oxygen atoms of the C(4)-OH and C(6)-OH groups form hydrogen bond networks with the His50, Gln53, and Asp100 residues of LecA, while the C(4)-hydroxyl group of the galactose moiety in lactulose coordinates with calcium ions in LecA. Additionally, the fructose portion of lactulose attaches to the PFPPBA nanoparticles due to the oxygen atom of the fructose hydroxyl group in

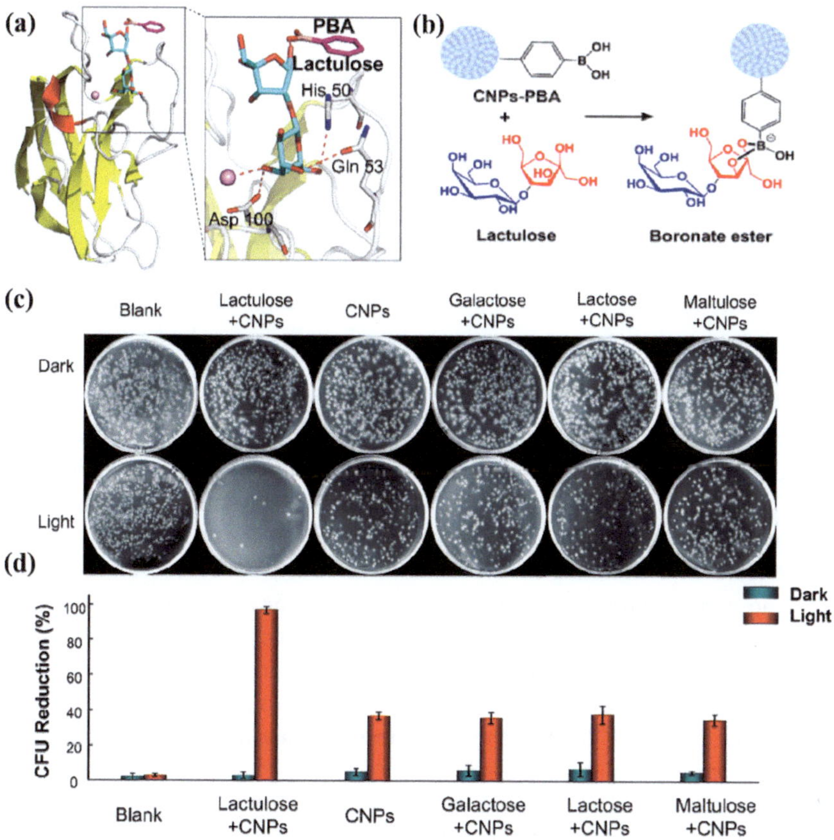

Figure 4.60 (a) The stimulated docking mode of lactulose to LecA and PBA; (b) the link between lactulose and PBA groups of PFPPBA nanoparticles; (c) bacterial colony counts; and (d) the killing efficiency of *P. aeruginosa* co-cultured with PFPPBA nanoparticles and different sugars upon irradiation. *Source:* Liu et al. [52] / with permission of American Chemical Society.

lactulose having a significant affinity for the PBA moiety of PFPPBA nanoparticles. Meanwhile, the galactose segment of lactulose successfully interacts with LecA (Figure 4.60b). Sugar molecules are introduced to the bacterial surface and then interact with the fluorescent conjugated polymer to enable controlled, targeted enrichment.

The PFPPBA nanoparticles can accumulate on the surface of bacteria through the "sugar bridge" mediated by lactulose and produce ROS to kill the bacteria. As shown in Figure 4.60c, the colony counts of untreated and different sugar-treated *P. aeruginosa* were essentially the same when co-cultured with PFPPBA nanoparticles under dark conditions. However, under white light irradiation, the colony count of *P. aeruginosa* treated with lactulose and co-cultured with PFPPBA nanoparticles decreased significantly, with a killing rate of 97% (Figure 4.60d). This indicates that lactulose, as the "sugar bridge", can effectively promote the enrichment of nanoparticles on the bacterial surface and efficiently kill the bacteria through the

ROS generated under light. Therefore, lactulose binds to LecA through its galactose portion and interacts with the PBA groups of PFPPBA nanoparticles through its fructose portion. This allows for the controlled and targeted accumulation of the PFPPBA nanoparticles on the bacterial surface. At the same time, the ROS generated by the nanoparticles under light further enhances the bactericidal effect, providing an efficient strategy for the targeted treatment of *P. aeruginosa* [52].

Based on this strategy, by controlling the enrichment of photosensitive nanoparticles on adherent bacteria, it is possible to effectively inhibit the adhesion of bacteria to host cells and achieve efficient sterilization effects. Taking human bladder epithelial cells as the host cells as an example, the glycoproteins on their cell surfaces contain a large number of mannose residues, which makes it easy for bacteria to recognize and adhere to them [79]. The FimH protein structure in the type I fimbriae of *E. coli* has the function of adhering to cells. In the experiment, after adding mannose trisaccharide and PFPPBA nanoparticles, the mannose at the terminus of the mannose trisaccharide can bind to the bacterial FimH lectin, thereby significantly inhibiting the bacteria's adhesion to cells. Moreover, the neighboring diol structure of the mannose trisaccharide also mediates the enrichment of more nanoparticles on the bacteria, further inhibiting the bacteria's invasive activity. This result indicates that mannose trisaccharide and nanoparticles have significant antiadhesion effects. Under light conditions, the killing efficiency of this strategy against bacteria is as high as 97%, demonstrating its great potential in antibacterial applications [80].

Thermal therapy utilizes materials with high photothermal conversion efficiency to convert the excitation light energy into heat, causing a local temperature to increase and disrupting the biological activity of large molecules and cellular structures within the cells. This leads to the necrosis of tumor cells within a few minutes. The use of gold nanoparticle shells to mediate PTT has been applied in the high-precision clinical treatment of prostate cancer [81]. Conjugated polymer materials with near-infrared absorption also exhibit excellent photothermal conversion performance, with an efficiency comparable to that of gold nanostructures. They show good photothermal therapeutic effects both in vitro and in vivo [82].

Nanocomposites (Pdots) were prepared by co-precipitation of conjugated polymers with photothermal conversion function and PSMA, and were used for intratumoral injection therapy. After laser irradiation, the temperature at the tumor site of the mice significantly increased, while there was no obvious temperature change at the nonirradiated site (Figure 4.61). This confirmed the precise photothermal effect of Pdots. Further analysis of tumor volume changes revealed that the tumors in the control group continued to grow. The tumor growth rates of the laser irradiation group and the group that received only Pdots injection were basically the same as those of the control group. The photothermal treatment group achieved complete tumor ablation on the second day, and no signs of tumor recurrence were observed in the subsequent 14 days, fully demonstrating the high efficiency of Pdots in photothermal tumor treatment. After photothermal treatment, only black ablation plaques remained at the tumor site (Figure 4.61), visually demonstrating the excellent performance of Pdots in tumor ablation. These results verified the significant effect of Pdots in photothermal tumor treatment, not only achieving

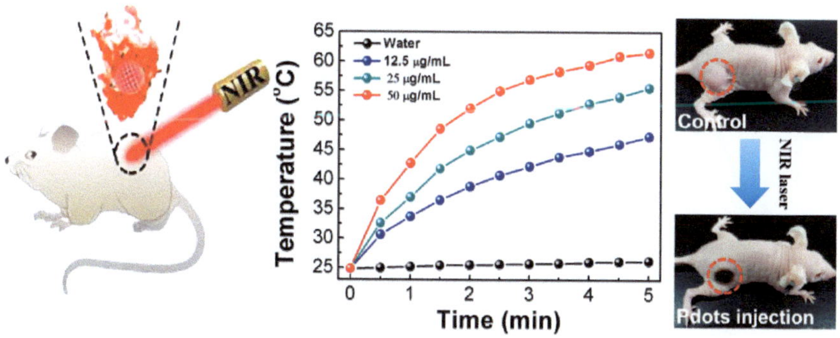

Figure 4.61 Schematic diagram of Pdots in photothermal tumor therapy in nude mice. *Source:* Reproduced with permission from Li et al. [56] / American Chemical Society.

rapid tumor ablation but also effectively prolonging the survival period of nude mice with tumors, providing a strong experimental basis for the clinical application of photothermal treatment techniques [56].

Conjugated polymers with photothermal effects demonstrate great potential for multifunctional applications in tumor treatment. They can not only be used for tumor ablation but also serve as remote activation tools to enhance the efficacy of cancer treatment [83]. Taking amphiphilic semiconductor polymers and bromelain as examples, used respectively as photothermal conversion materials and temperature-sensitive enzymes, the nanocomposite PCB1-Bro was prepared by their combination (Figure 4.62). Bromelain is an efficient proteinase that digests collagen effectively, and its optimal activity temperature is 45 °C. The experiment showed that under light conditions, the activity of bromelain in PCB1-Bro was 2.3 times that under nonlight conditions. This photothermal-triggered enzymatic activation mechanism can effectively digest collagen in the extracellular matrix, thereby significantly enhancing the tissue permeability and retention (EPR) effect of PCB1-Bro. During the subsequent photothermal treatment process, the average tumor temperature induced by a large accumulation of PCB1-Bro reached 50.5 °C, which was much higher than the threshold temperature for inducing cell apoptosis and necrosis (43 °C), thereby significantly improving the effect of PTT. This scientific design provides new strategies and possibilities for tumor treatment [84].

By combining conjugated polymer nanoparticles with photothermal conversion function and heat-activated IFN-γ plasmids, a novel strategy for thermally initiating tumor immunotherapy was developed. This plasmid employs the heat shock promoter HSP70. When the cell temperature reaches above 39 °C, the promoter is activated and initiates the expression of the IFN-γ gene. The expressed IFN-γ cytokine is secreted extracellularly and binds to the IFN-γ receptor on the surface of macrophages, thereby activating related signaling pathways and inducing the polarization of macrophages to the pro-inflammatory M1 subtype. M1-type macrophages will secrete a series of inflammatory factors, which can directly induce the death of cancer cells.

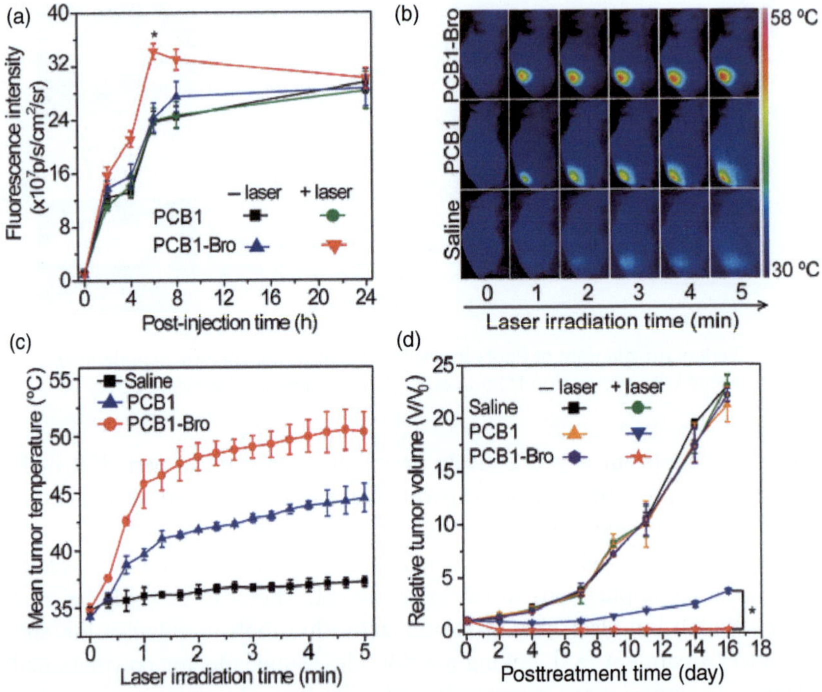

Figure 4.62 (a) Fluorescence intensity and (b) infrared thermal imaging of tumors after incubation with PCB1 and PCB1-Bro upon irradiation. (c) Tumor temperature and (d) tumor volume after incubation with PCB1 and PCB1-Bro upon irradiation. *Source:* Li et al. [84] / with permission of John Wiley & Sons.

The experimental results show that IFN-γ protein can be detected in the system that has been transfected with the plasmid (Figure 4.63a,b). After laser activation, the content of IFN-γ protein increased by three times. Compared with the control group, the proportion of M1/M2 type activated by light significantly increased. Compared with the group that simply transfected the IFN-γ plasmid, the proportion of M1/M2 increased by two times; compared with the untransfected group and the tumor-associated macrophage group, the proportion of M1/M2 increased by four times (Figure 4.63c). This near-infrared light-activated optogenetic technique can effectively re-activate the tumor-associated macrophages around cancer cells, promoting their transformation into the pro-inflammatory M1 subtype, laying the foundation for inhibiting tumor growth. Further experiments co-cultured macrophages with breast cancer cells 4T1. Only the 4T1 cells co-cultured with the experimental group (G6) exhibited a marked reduction in cell viability, according to the results (Figure 4.63d–f). The expression level of the proliferation protein Ki-67 in 4T1 cells was significantly lower than that of the control group, proving that this system can regulate the communication between macrophages and breast cancer cells and inhibit cell proliferation. This optogenetic platform is triggered by near-infrared laser, providing a promising strategy for noninvasive precision medicine in cancer immunotherapy. Therefore, by combining photothermal

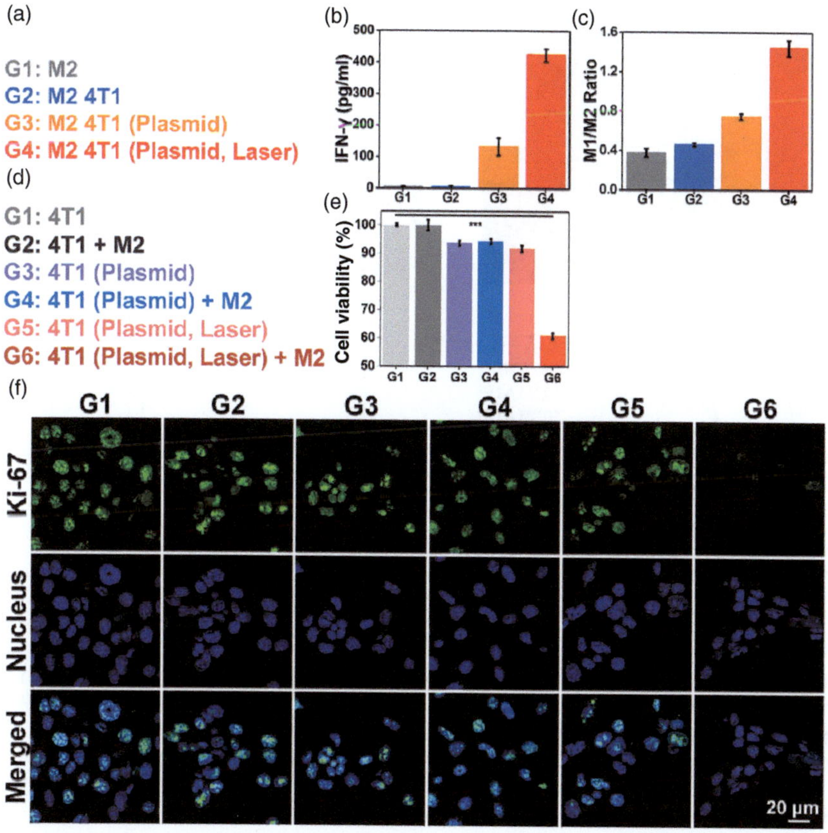

Figure 4.63 (a) A list of the different treatments. (b) The level of IFN-γ expression and (c) the M1/M2 ratio following the treatments listed in (a). (d) The treatment process list. (e) 4T1 cell viability following the treatments listed in (d). (f) Immunofluorescence images of 4T1 cells stained with Ki-67 (green) and Hoechst 33342 (blue). *Source:* Fu et al. [85] / with permission of John Wiley & Sons.

conversion and gene expression regulation, a new type of tumor immunotherapy strategy has been successfully developed, providing new ideas and methods for improving the tumor microenvironment and inhibiting tumor growth [85].

4.3.3 Optical Detections

Based on the excellent binding properties of optical functional polymer nanocomposites with biological molecules and structures, this material has seen rapid development in the field of optical detection. Its advantages lie in three main areas: Firstly, it enables real-time dynamic monitoring and can respond quickly to biomolecular interactions. Secondly, the operating process is straightforward and does not require complex pretreatment steps. Thirdly, it is extremely sensitive and can detect trace biomarkers with precision. These characteristics make this material an important technical platform for optical detection.

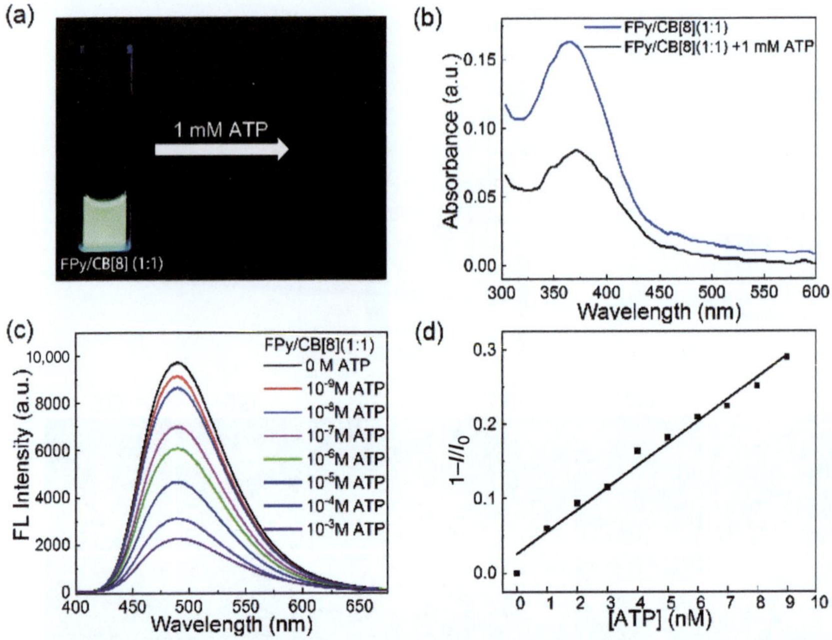

Figure 4.64 (a) Fluorescence images and (b) absorption spectra of the FPy/CB[8] solution without and with ATP. (c) Fluorescence intensities of the FPy/CB[8] solution at varying ATP concentrations. (d) Relationship between fluorescence intensity ($1-I/I_0$) and ATP concentration. *Source:* Ye et al. [88] / with permission of Royal Society of Chemistry.

Adenosine triphosphate (ATP), as an important biological molecule, provides energy for numerous life processes. Therefore, the detection of ATP holds significant research value [86, 87]. Based on the FPy/CB[8] (1:1) host–guest assembly, where the two positively charged pyridine groups are completely exposed in the solvent molecules, they possess binding sites rich in anions. The ATP molecule contains three negatively charged phosphate groups, which can interact with positively charged substances (such as fluorene derivatives with two pyridine units, FPy molecules) through electrostatic interactions, promoting the aggregation of the FPy/CB[8] (1:1) assembly. During this process, hydrophobic interaction and π–π stacking enhance the nonradiative decay of excitons, leading to reduced fluorescence intensity of the FPy molecule when aggregated, which exhibits a clear aggregation-induced quenching behavior (Figure 4.64a,b).

As the ATP concentration gradually increases, the fluorescence quenching degree of the FPy/CB[8] (1:1) system also increases. Even at an ATP concentration of 1 nM, some degree of fluorescence quenching can be observed (Figure 4.64c). Within the range of ATP concentrations from 1 to 9 nM, the change in fluorescence intensity ($1-I/I_0$) shows a linear relationship with the ATP concentration, and the linear relationship can be expressed as: $1-I/I_0 = 0.0297[\text{ATP}] + 0.0255$ (Figure 4.64d). With its low detection limit, this assay serves as a powerful tool for highly sensitive ATP detection [88].

In fingerprint research, apart from the overall pattern, which can provide basic information about an individual, a large amount of detailed information is also hidden in the microscopic structure of the fingerprint. Fingers contain various chemical components, mainly including sweat secreted by the human body, sebum-like substances, and various environmental pollutants. Due to the amphiphilic nature of fingerprints, when designing the developer, it is necessary to ensure that it has a high affinity with the components in the fingerprint to achieve an effective development effect. Moreover, common fluorescent background base materials in daily life or overlapping fingerprints found at crime scenes may cause difficulties in identifying the fingerprint pattern.

To address these issues, researchers incorporated red light-conjugated oligomers into silica nanoparticles through covalent bonds, thereby preparing OFDBT/SiO_2-Epoxy NPs. Compared with pure conjugated oligomer nanoparticles, the fluorescence quantum yield of this composite nanoparticle increased to 37.7%, and its photostability was significantly enhanced. Experiments demonstrated that the fingerprint patterns prepared from OFDBT/SiO_2-Epoxy NPs could be clearly observed with the naked eye on various substrates such as glass, stainless steel, plastic bags, and leaves, and were minimally affected by the green background fluorescence (Figure 4.65a). This is because the red emission's ability to effectively minimize background signal interference, while the developer can fully retain the detailed features of the fingerprint, including sweat glands, folds, triangular points, spine terminations, islands, bifurcations, core points, eyes, bridging, path deviations, and intersections (Figure 4.65b). The high-resolution imaging effect proved that OFDBT/SiO_2-Epoxy NPs have excellent fingerprint imaging capabilities, because the introduction of the epoxy group can enhance the hydrophobicity of the particle surface and improve the affinity of the nanoparticles to fingerprints. High-resolution local area details are very useful for obtaining personal identification information. OFDBT/SiO_2-epoxy NPs can be directly applied to latent fingerprint detection without any surface treatment. Meanwhile, Silica nanoparticles and conjugated oligomers are widely recognized for their nontoxic and biocompatible properties. Therefore, OFDBT/SiO_2-epoxy NPs not only serve as a safe detection agent but also show great potential in latent fingerprint development applications. Its excellent fingerprint imaging ability can provide rich, detailed feature information, providing important technical support for forensic science and identity recognition fields [89].

The enhanced fluorescence signal technology not only significantly improves the detection sensitivity but also can be used for precise bacterial identification. This technology is highly valuable for identifying bacterial infections, particularly different pathogenic bacteria, as it can effectively enhance diagnostic accuracy and provide a scientific basis for formulating more targeted treatment plans for patients. Attached to oligo-polyethylene glycol, AuNC-MUA and AuNC-MUA/OEG exhibit significant differences in their biointerfacial interaction mechanisms with *Candida* due to variations in surface ligands. Furthermore, the architecture of the cell walls of different species of *Candida* varies considerably, as do their surface hydrophobic states. Taking advantage of the diversity of hydrogen bonds

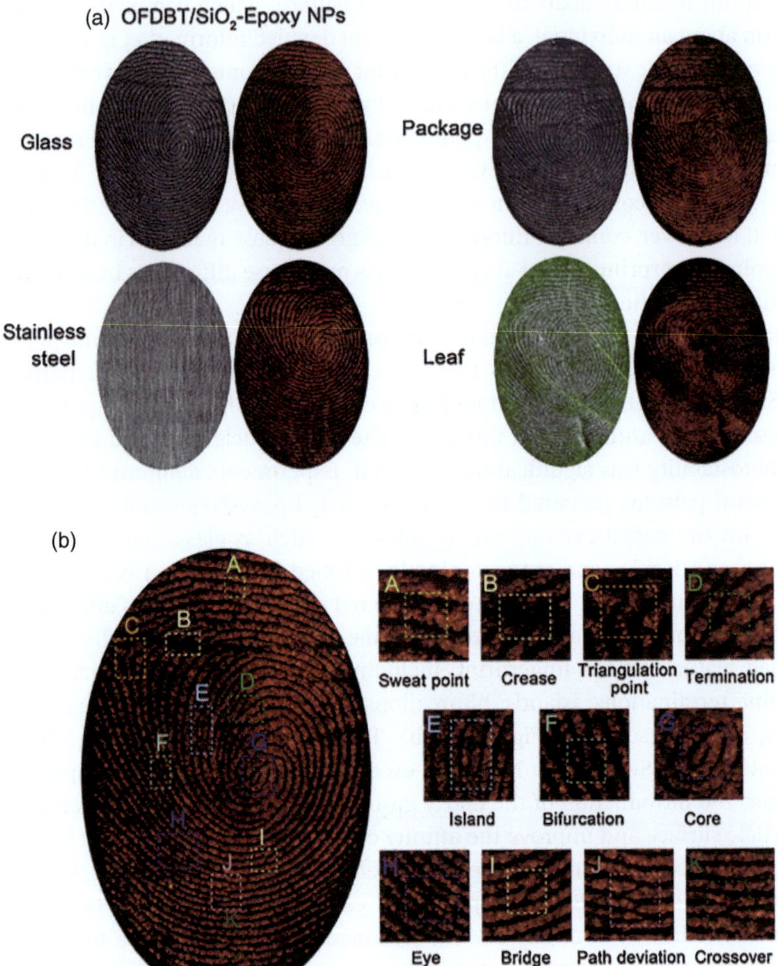

Figure 4.65 (a) Bright-field and fluorescence images of latent fingerprints (LFPs) developed with OFDBT/SiO_2-epoxy nanoparticles on various substrates. (b) Fluorescence images of LFPs developed by OFDBT/SiO_2-epoxy NPs on glass and the magnified regional images with specific features. *Source:* Reproduced with permission from Yang et al. [89] / Elsevier.

and hydrophobic interactions present at the biological interface, after 30 min of interaction, varying degrees of fluorescence enhancement were observed in these two gold nanoclusters. Additionally, these two gold nanoclusters can specifically bind to *Candida* cells, providing a powerful tool for reliably identifying *Candida* species [90].

Unlike single-color fluorescence signals, dual-color fluorescence signals can be directly compared and analyzed, thereby efficiently revealing the interrelationships between target molecules. Through simultaneous detection of two types of fluorescence, not only can the precise distinction of signal sources (such as through different wavelengths or fluorescent dyes) be achieved but also the background noise interference can be significantly reduced. This design simultaneously enhances the sensitivity and specificity of the detection.

Reactive oxygen species are essential for signal transduction processes in living organisms. Among them, hypochlorite ion (ClO^-), a potent ROS, can regulate inflammatory-related pathological processes [91]. Two oligomers (1,4-bis((E)-2-(10-hexyl-10H-phenothiazin-3-yl)vinyl)benzene) and (4,7-bis((E)-2-(10-hexyl-10H-phenothiazin-3-yl)vinyl)benzo[c][1,2,5]thiadiazole) were synthesized. The phenothiazine groups in their structures can react specifically with ClO^-, generating different fluorescence colors in response. By co-precipitating the green-light–emitting oligomer with the morpholine-modified PEG block polymer, a nanoparticle LNP targeting lysosomes was successfully prepared. This nanoparticle can achieve a fluorescence transition from green to blue in the presence of ClO^- (Figure 4.66a). By co-precipitating the red-light-emitting oligomer with the triphenylphosphine-modified PEG block polymer, a nanoparticle MNP targeting mitochondria was successfully prepared. This nanoparticle can achieve a fluorescence transition from red to orange in the presence of ClO^- (Figure 4.66b). LNP was further mixed with MNP nanoprobes and co-cultured with macrophages. To mimic the inflammatory environment, the cells were pretreated with bacterial cell wall lipopolysaccharide (LPS) and cytokine IFN-γ to induce ClO^- production by macrophages. It was observed that in the lysosomes and mitochondria of stimulated macrophages, the nanoprobes were activated by ClO^-, presenting blue and orange fluorescence signals, respectively. After using *N*-acetyl-L-cysteine (NAC) to clear ROS, the signals in the blue and orange channels in the cells were significantly weakened, which further verified the specific response mechanism of the nanoprobes (Figure 4.66c) [92].

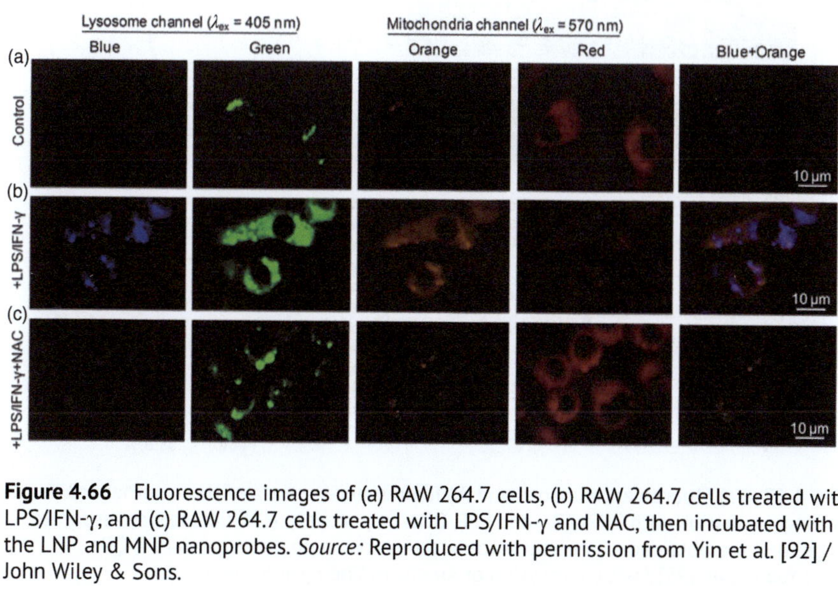

Figure 4.66 Fluorescence images of (a) RAW 264.7 cells, (b) RAW 264.7 cells treated with LPS/IFN-γ, and (c) RAW 264.7 cells treated with LPS/IFN-γ and NAC, then incubated with the LNP and MNP nanoprobes. *Source:* Reproduced with permission from Yin et al. [92] / John Wiley & Sons.

While eliminating bacterial microorganisms, the use of fluorescent labeling and other methods for identifying and diagnosing microbial pathogens can provide effective guidance for antibacterial treatment [93, 94]. Based on the development of conjugated polymer-quantum dot composite materials, the dual functions of identifying and killing microbial pathogens have been achieved. The water-soluble CdSe/ZnS quantum dots have a negatively charged surface, and through electrostatic interaction with cationic polyfluorene derivatives (PFP), they form a complex, thereby enabling efficient FRET between PFP and quantum dots. When negatively charged bacteria were introduced, the increased affinity between PFP and the bacteria caused PFP to dissociate from the quantum dots, consequently diminishing the energy transfer effect. The experimental results showed that the three pathogenic bacteria treated with PFP all exhibited blue fluorescence, indicating that PFP could effectively coat the surface of the pathogens. In contrast, pathogens exposed to quantum dots did not exhibit detectable fluorescence, likely as a result of electrostatic repulsion between the quantum dots and the pathogen surfaces. Furthermore, a notably higher intensity of PFP fluorescence was observed in *S. aureus*, indicating that the electrostatic complex formed between PFP and *S. aureus* was more effective than those formed with *E. coli* or *C. albicans* (Figure 4.67a). Across different pathogens, the FRET interaction between PFP and quantum dots was attenuated to varying extents, as demonstrated by the restoration of fluorescence intensity at 426 nm and the corresponding decrease at 620 nm (Figure 4.67b).

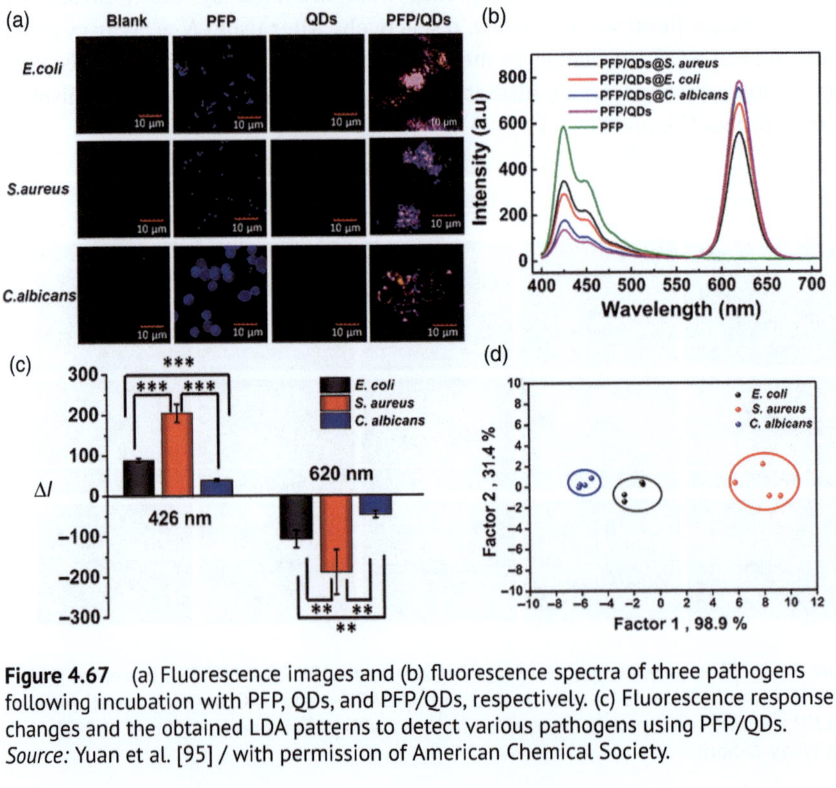

Figure 4.67 (a) Fluorescence images and (b) fluorescence spectra of three pathogens following incubation with PFP, QDs, and PFP/QDs, respectively. (c) Fluorescence response changes and the obtained LDA patterns to detect various pathogens using PFP/QDs. *Source:* Yuan et al. [95] / with permission of American Chemical Society.

Further, the pathogen was identified by measuring the fluorescence intensity changes ΔI_{426} and ΔI_{620}, and applying linear discriminant analysis (LDA). LDA transformed the fluorescence intensity changes into standard scores and classified them into three categories. The contribution rates of the two typical factors were 98.9% and 31.4% respectively (Figure 4.67c). This study achieved a detection accuracy of over 95%, capable of effectively distinguishing *S. aureus*, *E. coli*, and *C. albicans*, fully demonstrating the convenience and reliability of this method (Figure 4.67d). This research provides a new technical means for the rapid identification of pathogens and precise antibacterial treatment [95].

During the refrigeration of red blood cells (RBCs), their structural and biochemical characteristics undergo significant changes, resulting in the release of free hemoglobin and thereby reducing the oxygen-carrying capacity of RBCs [96]. The iridium (III) complex in the PFO/Ir-DSPE nanocomposite materials exhibits long-lived red phosphorescence emission, which can be used to monitor the oxygen-carrying status within cells through fluorescence lifetime imaging technology. As shown in Figure 4.68, the PFO/Ir-DSPE nanocomposite exhibits an average lifetime of 1.3 ns in fresh RBCs, whereas its lifetime is markedly prolonged to 12 ns in RBCs subjected to 14 days of storage. This indicates that fresh RBCs have a strong oxygen-carrying capacity. It is worth noting that the fresh RBC membrane shows a green signal; the PFO/Ir-DSPE nanocomposite exhibits a longer lifetime when localized within the membrane compared to its residence in the cytoplasm (Figure 4.68a). This result further confirms the integrity of the fresh RBC membrane and the lower oxygen content. However, in refrigerated RBCs following 14 days of storage, the boundary between the membrane and the cytoplasm becomes blurred, making it difficult to distinguish (Figure 4.68b). This indicates that long-term refrigeration significantly alters the structure of the RBC membrane, thereby affecting its function. The release of hemoglobin further weakens the oxygen-carrying capacity of the RBCs. Therefore, PFO/Ir-DSPE nanocomposite materials, as an efficient probe, can not only detect the oxygen content but also effectively evaluate the state

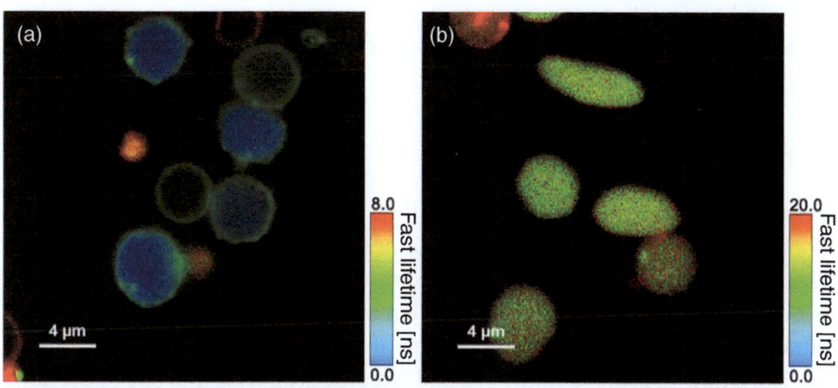

Figure 4.68 Fluorescence lifetime imaging of RBCs incubated with PFO/Ir-DSPE nanocomposites at (a) 0 day and (b) 14 days. *Source:* Reproduced with permission from Guo et al. [40] / Elsevier.

of RBCs. This technology provides an important tool and method for the study of RBC storage and function [40].

Fluorescent signals also have significant application value in gas molecule detection. *N*,*N*-Dimethyl-1-dodecanamine ($C_{12}A$) exhibits CO_2 responsiveness. By mixing it with water and DMSO solvents in a certain proportion, an emulsion system was successfully prepared. By introducing DPAC-Ty fluorescent molecules with vibration-induced emission (VIE) effect, this system achieved visual monitoring of CO_2. In the emulsion system, the DPAC-Ty fluorescent molecules initially emit a pink signal. When CO_2 is present, the $C_{12}A$ molecules undergo protonation reactions, forming amphiphilic molecules $C_{12}AH^+$ with positive charges, and further self-assemble into spherical micelles. These micelles can carry the fluorescent molecules, and their cores contain a certain proportion of water, thereby suppressing the free vibration of the hydrophobic DPAC-Ty molecules and emitting intense blue fluorescence.

During the germination process of plant seeds, a large amount of oxygen is consumed for aerobic respiration. This process involves organisms utilizing oxygen to completely oxidize and decompose organic substances, releasing energy and generating CO_2 and water. Therefore, by detecting CO_2, the progress of seed germination can be monitored in real time. For example, in the experiment of mung bean seed germination, the following phenomena can be observed: In the first 12 h, the bean seeds showed no significant changes, and the fluorescence signal remained pink, indicating that the seeds had not yet entered a significant germination stage. However, during the second and third 12-h periods, the mung bean seeds broke through the seed coat and grew roots in large numbers, indicating that the seeds were rapidly germinating and their respiration was significantly enhanced, producing a large amount of CO_2. At this time, the fluorescence signal changed from pink to blue, which directly reflected the increase in carbon dioxide concentration. In the subsequent 12 h, the growth state of the mung bean seeds tended to be stable, possibly because the low oxygen content inhibited further germination of the seeds (Figure 4.69). This system not only demonstrates the efficiency and sensitivity of fluorescence signals in gas detection but also provides a new technical means for the real-time monitoring of the plant seed germination process [97].

4.3.4 Drug Delivery

Fluorescent polymer nanoparticles, due to their unique structural tunability and functional designability, have become important materials for constructing ideal drug delivery carriers. Moreover, the photophysical properties of the light-functional components they carry enable the visualization and monitoring of the drug release and administration process through real-time imaging techniques. In the field of drug delivery, how to achieve controlled and targeted drug release is one of the key breakthroughs in this research direction.

The use of light-controlled release technology to achieve precise drug delivery is a highly regarded research direction. The sCzTAZPO fluorescent conjugated molecule

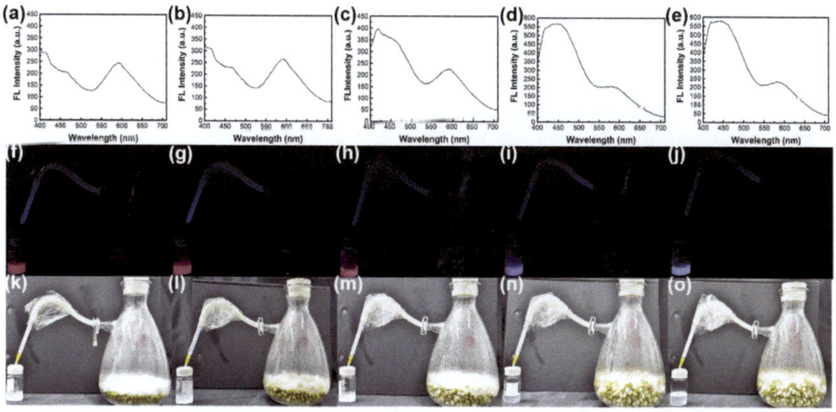

Figure 4.69 (a–e) Fluorescence spectra and (f–j) corresponding fluorescence images of the CO_2-responsive system recorded at 12 h intervals. (k–o) Photographs of mung bean seeds at different stages of germination. *Source:* Yan et al. [97] / with permission of Elsevier.

was combined with TK-DOX to self-assemble into stable lipid nanocomposites. These composites were then combined with RBCs to prepare an optical functional system, sCzTAZPO/TK-DOX@RBCs, which has the ability for light-responsive drug release. In this system, through electrostatic and hydrophobic interactions, sCzTAZPO is distributed in the cytoplasm of RBCs, presenting random punctate fluorescence signals, while the lipophilic TK-DOX is inserted into the membrane layer of RBCs (Figure 4.70a). This modular assembly strategy fully utilizes the natural oxygen transport ability of RBCs, enabling sCzTAZPO to maintain its photosensitizing activity in the tumor hypoxic microenvironment and continuously produce ROS. The generated ROS can specifically break the TK bond, enabling the prodrug to be activated and released on demand at the tumor site. Fluorescence imaging shows that before illumination, due to the strong FRET effect between sCzTAZPO and DOX, the blue fluorescence signal was weak; after illumination, as the TK bond was broken, the FRET effect weakened, and the blue fluorescence of sCzTAZPO significantly increased, while the red fluorescence of DOX showed a dispersed distribution, confirming the effect of light-controlled release. in vivo experiments verified the significant antitumor effect of sCzTAZPO/TK-DOX@RBCs under light conditions. The experimental results demonstrated that the average tumor weight in the illuminated group was 85% lower than that in the control group, fully demonstrating the efficient antitumor activity of this system under light activation (Figure 4.70b–d) [28].

Conjugated polymer PPV-ST was synthesized by covalently linking visible light-responsive Stenhouse adduct (DASA) with folic acid in the PPV side chain. Under the irradiation of 550 nm visible light, DASA, as a light-responsive unit, can transform from a hydrophobic, conjugated, colored triene to a closed-ring, zwitterionic, colorless cyclopentenone. Folic acid provides targeting ability, enabling the drug to act more precisely on the target cells (Figure 4.71a). PPV-ST demonstrates amphiphilic characteristics, facilitating its self-assembly into nanoparticles

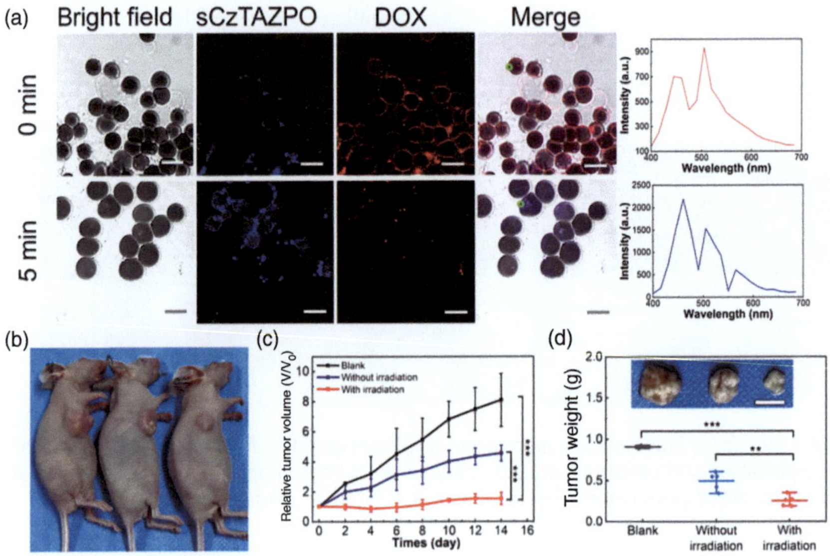

Figure 4.70 (a) Fluorescence images of released DOX from sCzTAZPO/TK-DOX@RBCs composites after white light irradiation. (b) Photographs of HeLa tumor-bearing mice; (c) changes in relative tumor volume; and (d) the tumor weights of mice treated with sCzTAZPO/TK-DOX@RBCs upon irradiation. *Source:* Reproduced with permission from Liu et al. [28] / American Chemical Society.

(PPV-ST-NP) in aqueous environments and enabling the co-encapsulation of the anticancer agents DOX and CPT. Fluorescence images revealed that CPT is effectively released from PPV-ST-NP, as indicated by its blue fluorescence, and subsequently internalized into both the cytoplasm and nucleus of the cells. Notably, the PPV-ST-NPs themselves were predominantly retained within lysosomal compartments, as evidenced by the distinct green fluorescence emission (Figure 4.71b). These observations provide compelling evidence that CPT is liberated from the nanoparticles and is able to penetrate intracellular compartments, whereas the carrier nanoparticles remain sequestered within lysosomes. Based on these photo isomerizing molecules with hydrophobic-hydrophilic change properties, the drug delivery system has significant guiding significance in avoiding unnecessary toxic side effects of drugs on normal tissues and constructing ultra-long-lasting formulations. This design not only improves the targeting and release efficiency of the drug but also reduces potential damage to healthy tissues, providing a new idea for the development of light-responsive drug delivery systems [98].

4.3.5 Photosynthesis

Optical functional polymer nanocomposites can efficiently absorb sunlight and convert it into electrical energy or chemical energy, thereby supporting the development of green energy. In terms of photosynthesis, these materials can simulate the photosynthesis process of plants, using solar energy to drive chemical reactions and

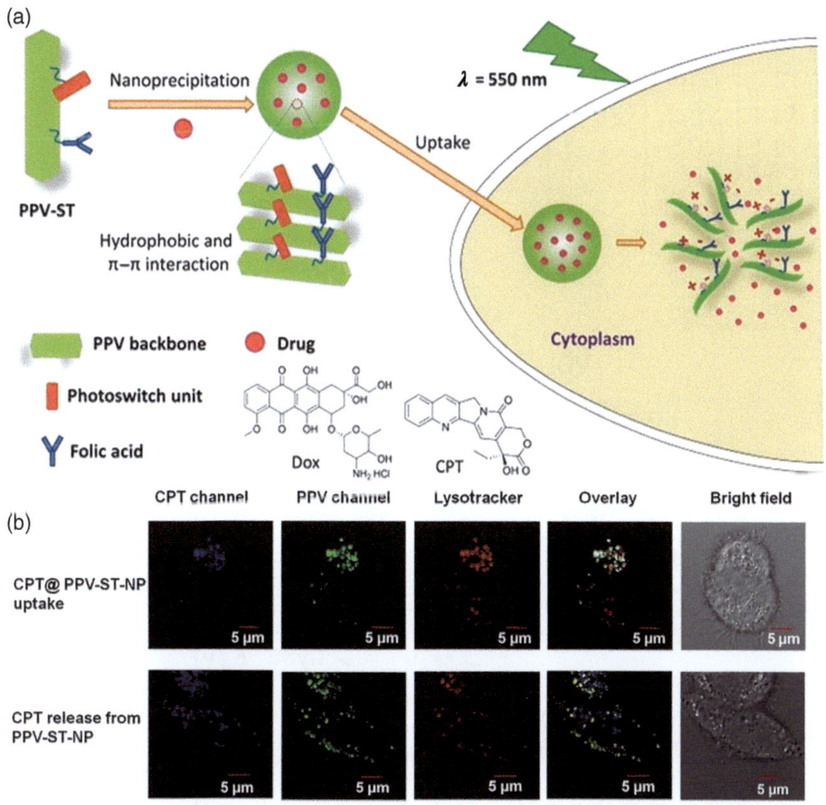

Figure 4.71 (a) Mechanism of light-triggered drug release from targeted PPV-ST nanoparticles. (b) Fluorescence images showing the cellular uptake of CPT-loaded PPV-ST nanoparticles, and the intracellular release of CPT when exposed to light.
Source: Senthilkumar et al. [98] / with permission of John Wiley & Sons.

converting carbon dioxide and water into organic compounds. By optimizing the processes of light absorption and energy transfer, the efficiency of photosynthesis can be improved.

The artificial photosynthetic system based on conjugated polymer heterojunction and biological hybridization can efficiently reduce CO_2 into CH_3COOH. As shown in Figure 4.72a,b, both conjugated PDI and conjugated polymer PFP carry positive charges, serving as n-type and p-type semiconductors, respectively. They are combined with the surface of the *M. thermoacetica* through electrostatic interaction and form a *p*–*n* junction on the bacterial surface. The energy level offsets of HOMO and LUMO between these two molecules are both over 0.3 eV (ΔE_{LUMO} is 1.29 eV, ΔE_{HOMO} is 0.48 eV), which provides sufficient driving force for the separation of holes/electrons. Therefore, after PFP/PDI is excited, the generated electron-hole pairs undergo effective migration, resulting in a significant photocurrent.

On the surface structure of bacteria, the redox potentials of the oxidoreductase flavoprotein (F_p) and the redox protein (R_d) match that of PFP/PDI (Figure 4.72c,d).

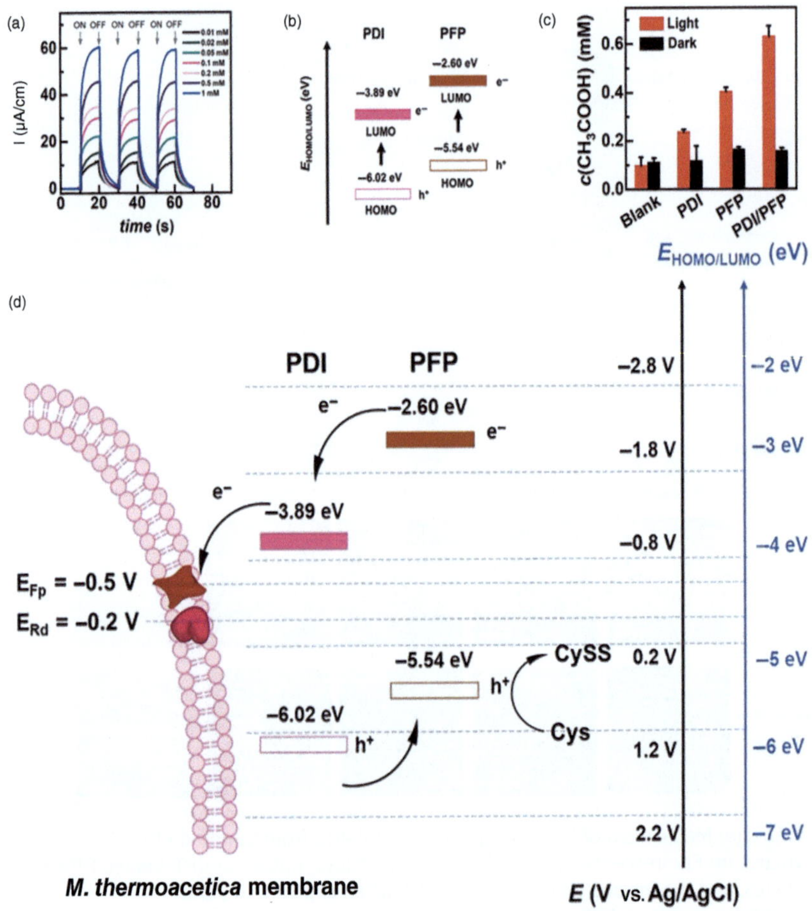

Figure 4.72 (a) Transient photocurrent responses and (b) energy level diagram of the PDI and PFP molecules. (c) Quantity of CH_3COOH produced, and (d) the underlying mechanism of the PDI/PFP/*M. thermoacetica* biohybrid system. *Source:* Gai et al. [99] / with permission of John Wiley & Sons.

This enables *M. thermoacetica* to effectively capture photogenerated electrons and synthesize CH_3COOH through the Wood-Ljungdahl pathway. Under dark conditions, the concentration of CH_3COOH produced by the combination of conjugated molecules with *M. thermoacetica* is only around 0.1 mM, while under light conditions, *M. thermoacetica* also produces the same concentration of CH_3COOH. However, after 12 h of light exposure, the PFP/PDI complex of *M. thermoacetica* can produce 0.63 mM of CH_3COOH, significantly higher than the yield under dark conditions. This result indicates that photofunctional materials can modify non-photosynthetic bacteria and endow them with the ability to enhance photosynthesis, thereby significantly improving the efficiency of CH_3COOH synthesis. This artificial photosynthetic system, based on the heterojunction of conjugated polymers and bacterial biological hybridization, demonstrates the potential application of photofunctional materials in biological systems. It also provides an innovative and efficient approach for the efficient reduction of CO_2 and the synthesis of CH_3COOH [99].

By integrating photocatalytic, chemical catalytic, and biological catalytic modules to construct an artificial photosynthesis system, the process of natural photosynthesis can be simulated. In this system, the photocatalytic functional module is integrated with the chemical and biological catalytic modules through liposomes, achieving efficient energy conversion and substance synthesis. In the photocatalytic module, electron transfer occurs between the photocatalyst CoTMPyP and the photosensitizer $Ru(bpy)_3(NO_3)_2$, using $Na_2S_2O_8$ as a sacrificial agent, to realize the water splitting reaction. This process simulates the light reaction part of photosynthesis and can efficiently produce O_2. In the chemical catalysis module, NAD^+ is reduced to NADH in the presence of the oligomeric styrene-rhodium complex (OPV-Rh) and sodium formate (HCOONa). This step provides important reducing power for the subsequent biological catalysis reactions. In the biological catalysis module, creatine kinase (CK) facilitates the conversion of ADP to ATP using phosphocreatine as a phosphate donor (Figure 4.73). This process simulates ATP synthesis in photosynthesis and provides energy for cellular metabolism. Through the integration of liposomes, this artificial photosynthesis system has increased the production of O_2, NADH, and ATP compared to the solution system. Moreover, in the presence of alcohol dehydrogenase (ADH), NADH in the system can be used to produce methanol, which is 37% higher than in the solution system. The successful construction of this system provides a new approach for the research of artificial photosynthesis and demonstrates the potential of integrating multiple catalytic modules to better utilize solar energy resources. In the future, it is anticipated that this technology will be crucial in energy conversion, environmental protection, and biomanufacturing [29].

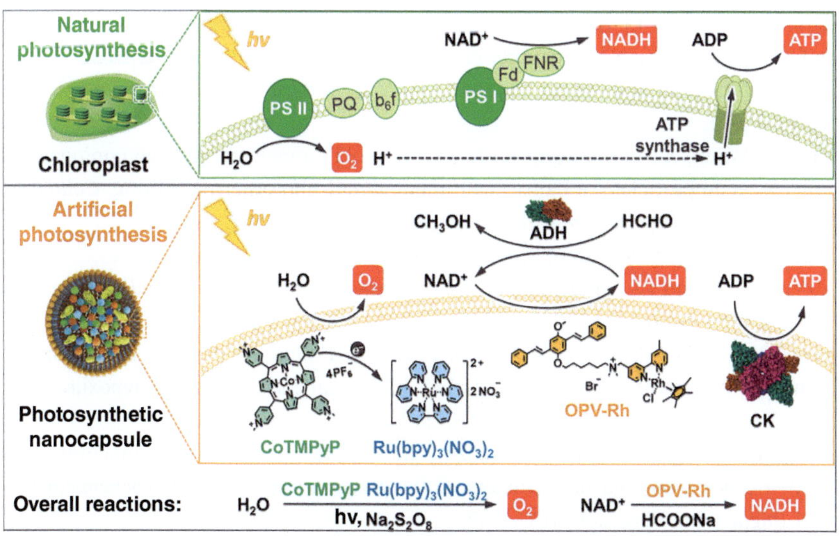

Figure 4.73 Photosynthetic mechanism of optical functional polymer nanocapsules. *Source:* Zhu et al. [29] / with permission of John Wiley & Sons.

References

1 Li, J.-F., Zhang, Y.-J., Ding, S.-Y. et al. (2017). Core-shell nanoparticle-enhanced raman spectroscopy. *Chemical Reviews* 117 (7): 5002–5069.

2 Tang, F., He, F., Cheng, H., and Li, L. (2010). Self-assembly of conjugated polymer-Ag@SiO_2 hybrid fluorescent nanoparticles for application to cellular imaging. *Langmuir* 26 (14): 11774–11778.

3 Wang, X., He, F., Tang, F. et al. (2011). Preparation of hybrid fluorescent-magnetic nanoparticles for application to cellular imaging by self-assembly. *Colloids and Surfaces a-Physicochemical and Engineering Aspects* 392 (1): 103–109.

4 Tang, F., Ma, N., Wang, X. et al. (2011). Hybrid conjugated polymer-Ag@PNIPAM fluorescent nanoparticles with metal-enhanced fluorescence. *Journal of Materials Chemistry* 21 (42): 16943–16948.

5 Kim, D.J., Kang, S.M., Kong, B. et al. (2005). Formation of thermoresponsive gold nanoparticle/PNIPAAm hybrids by surface-initiated, atom transfer radical polymerization in aqueous media. *Macromolecular Chemistry and Physics* 206 (19): 1941–1946.

6 Wang, C., Tang, F., Wang, X., and Li, L. (2015). Self-assembly of fluorescent hybrid core-shell nanoparticles and their application. *ACS Applied Materials & Interfaces* 7 (24): 13653–13658.

7 Tang, F., Wang, C., Wang, X., and Li, L. (2015). Facile synthesis of biocompatible fluorescent nanoparticles for cellular imaging and targeted detection of cancer cells. *ACS Applied Materials & Interfaces* 7 (45): 25077–25083.

8 Xia, B., Wang, X., He, F. et al. (2012). Self-assembly of conjugated polymer on hybrid nanospheres for cellular imaging applications. *ACS Applied Materials & Interfaces* 4 (11): 6332–6337.

9 Zhu, S., Wang, X., Li, S. et al. (2020). Near-infrared-light-assisted in situ reduction of antimicrobial peptide-protected gold nanoclusters for stepwise killing of bacteria and cancer cells. *ACS Applied Materials & Interfaces* 12 (9): 11063–11071.

10 Zhu, S., Wang, X., Liu, L., and Li, L. (2020). Gold nanocluster grafted conjugated polymer nanoparticles for cancer cell imaging and photothermal killing. *Colloids and Surfaces a-Physicochemical and Engineering Aspects* 597: 124764.

11 Wan, X.-K., Xu, W.W., Yuan, S.-F. et al. (2015). A near-infrared-emissive alkynyl-protected Au_{24} nanocluster. *Angewandte Chemie, International Edition* 54 (33): 9683–9686.

12 Peng, R., Luo, Y., Cui, Q. et al. (2022). Conjugated oligomer-directed formation of hollow nanoparticles for targeted photokilling cancer cells under hypoxia. *Advanced Optical Materials* 10 (5): 2102377.

13 Kang, Y., Zhang, J., Wu, Z. et al. (2024). Thermal-responsive gated nanocomposite of antibacterial gold nanoclusters. *Colloids and Surfaces A: Physicochemical and Engineering Aspects* 684: 133169.

14 Xia, J., Wang, X., Zhu, S. et al. (2019). Gold nanocluster-decorated nanocomposites with enhanced emission and reactive oxygen species generation. *ACS Applied Materials & Interfaces* 11 (7): 7369–7378.

15 Cong, Y., Wang, X., Zhu, S. et al. (2021). Spiropyran-functionalized gold nanoclusters with photochromic ability for light-controlled fluorescence bioimaging. *ACS Applied Bio Materials* 4 (3): 2790–2797.

16 Tang, F., Ma, N., Tong, L. et al. (2012). Control of metal-enhanced fluorescence with pH- and thermoresponsive hybrid microgels. *Langmuir* 28 (1): 883–888.

17 Kang, Y., Zhang, Y., Li, X. et al. (2023). Protein-assisted molybdenum disulfide as biomimetic nanozyme for antibacterial application. *ACS Applied Bio Materials* 6 (7): 2898–2904.

18 Liu, L., Liu, R., Wang, X. et al. (2018). Facile preparation of fluorescent nanoparticles with tunable exciplex emission and their application to targeted cellular imaging. *ACS Applied Bio Materials* 1 (1): 185–192.

19 Wang, X., He, F., Li, L. et al. (2013). Conjugated oligomer-based fluorescent nanoparticles as functional nanocarriers for nucleic acids delivery. *ACS Applied Materials & Interfaces* 5 (12): 5700–5708.

20 Wang, J., Cheng, Y., Peng, R. et al. (2020). Co-precipitation method to prepare molecularly imprinted fluorescent polymer nanoparticles for paracetamol sensing. *Colloids and Surfaces a-Physicochemical and Engineering Aspects* 587: 124342.

21 Liu, R., Cui, Q., Wang, C. et al. (2017). Preparation of sialic acid-imprinted fluorescent conjugated nanoparticles and their application for targeted cancer cell imaging. *ACS Applied Materials & Interfaces* 9 (3): 3006–3015.

22 Yang, N., Kang, Y., Cong, Y. et al. (2023). Controllable gold nanocluster-emulsion interface for direct cell penetration and photothermal killing. *Advanced Materials* 35 (50): e2208349.

23 Yang, N., Kang, Y., Liu, J. et al. (2025). Photothermal miniemulsion polymerization by amphiphilic gold nanoclusters. *Chemistry – An Asian Journal* 20 (5): e202401194.

24 Cao, Q., Cui, Q., Yang, Y. et al. (2018). Graphitic carbon nitride as a distinct solid stabilizer for emulsion polymerization. *Chemistry – A European Journal* 24 (9): 2286–2291.

25 Luo, Y., Yang, Y., Cui, Q. et al. (2019). Fluorescent nanoparticles synthesized by carbon-nitride-stabilized pickering emulsion polymerization for targeted cancer cell imaging. *ACS Applied Bio Materials* 2 (11): 5127–5135.

26 Kang, Y., Zan, Y., Cong, Y. et al. (2024). A clay-based pickering nanoemulsion with antibacterial activity. *Colloids and Surfaces A: Physicochemical and Engineering Aspects* 686: 133337.

27 Zan, Y., Liu, J., Zhao, Z. et al. (2025). A montmorillonite-based pickering nanoemulsion for the integration of photothermal therapy and NIR-responsive drug delivery. *ACS Applied Bio Materials* 8 (1): 652–660.

28 Liu, J., Cong, Y., Wang, X. et al. (2024). Modular assembly of photoactive lipid nanoparticles on red blood cells toward enhanced phototherapy efficacy. *ACS Applied Materials & Interfaces* 16 (46): 63306–63316.

29 Zhu, C., Gao, Z., Yu, W. et al. (2024). Conjugated molecules based multi-component artificial photosynthesis system for producing multi-objective products. *Small* 20 (8): e2306440.

30 Picheth, G.F., Marini, T.C., Taladriz-Blanco, P. et al. (2019). Influence of pluronic F127 microenvironments on the photochemical nitric oxide release from S-nitrosoglutathione. *Journal of Colloid and Interface Science* 544: 217–229.

31 Liu, S., Wang, J., Tang, F. et al. (2020). Aqueous systems with tunable fluorescence including white-light emission for anti-counterfeiting fluorescent inks and hydrogels. *ACS Applied Materials & Interfaces* 12 (49): 55269–55277.

32 Peng, R., Zhang, H., Yao, C. et al. (2023). U-shaped conjugated oligomer-based micelles as a ratiometric fluorescent nanothermometer for diverse cellular applications. *Advanced Optical Materials* 11 (14): 2300222.

33 Van der Meeren, L., Li, J., Konrad, M. et al. (2020). Temperature window for encapsulation of an enzyme into thermally shrunk, $CaCO_3$ templated polyelectrolyte multilayer capsules. *Macromolecular Bioscience* 20 (7): e2000081.

34 Li, J., Van der Meeren, L., Verduijn, J. et al. (2022). New all-nanoparticle microcapsules for ultrasound release and laser remote killing of cancer cells. *Materials Today Communications* 33: 104287.

35 Chen, Y., Munechika, K., and Ginger, D.S. (2007). Dependence of fluorescence intensity on the spectral overlap between fluorophores and plasmon resonant single silver nanoparticles. *Nano Letters* 7 (3): 690–696.

36 Zhang, J., Ma, N., Tang, F. et al. (2012). pH- and Glucose-responsive core-shell hybrid nanoparticles with controllable metal-enhanced fluorescence effects. *ACS Applied Materials & Interfaces* 4 (3): 1747–1751.

37 Cui, Q., He, F., Wang, X. et al. (2013). Gold nanoflower@gelatin core-shell nanoparticles loaded with conjugated polymer applied for cellular imaging. *ACS Applied Materials & Interfaces* 5 (1): 213–219.

38 Wang, C., Tang, F., Wang, X., and Li, L. (2016). Synthesis and application of biocompatible gold core-poly-(L-Lysine) shell nanoparticles. *Colloids and Surfaces a-Physicochemical and Engineering Aspects* 506: 425–430.

39 Xia, B., He, F., and Li, L. (2014). Metal-enhanced fluorescence using aggregated silver nanoparticles. *Colloids and Surfaces a-Physicochemical and Engineering Aspects* 444: 9–14.

40 Guo, J., Wang, X., Zhu, S. et al. (2019). Dual-emitting nanocomposites for oxygen-carrying capacity analysis and boosted singlet oxygen generation in stored red blood cells. *Dyes and Pigments* 171: 107751.

41 Liu, R., Yang, Y., Cui, Q. et al. (2018). A diarylethene-based photoswitch and its photomodulation of the fluorescence of conjugated polymers. *Chemistry – A European Journal* 24 (67): 17756–17766.

42 Wang, J., Peng, R., Luo, Y. et al. (2021). Preparation of fluorescent conjugated polymer micelles with multi-color emission for latent fingerprint imaging. *Colloids and Surfaces A: Physicochemical and Engineering Aspects* 615: 126192.

43 Cui, Q., Yang, Y., Yao, C. et al. (2016). Aggregation-induced energy transfer of conjugated polymer materials for ATP sensing. *ACS Applied Materials & Interfaces* 8 (51): 35578–35586.

44 Cui, Q., Xu, J., Shen, G. et al. (2017). Hybridizing carbon nitride colloids with a shell of water-soluble conjugated polymers for tunable full-color emission and synergistic cell imaging. *ACS Applied Materials & Interfaces* 9 (50): 43966–43974.

45 Wu, M., Xu, X., Wang, J., and Li, L. (2015). Fluorescence resonance energy transfer in a binary organic nanoparticle system and its application. *ACS Applied Materials & Interfaces* 7 (15): 8243–8250.

46 Cong, Y., Wang, X., Bai, H. et al. (2024). Intracellular gold nanocluster/organic semiconductor heterostructure for enhancing photosynthesis. *Angewandte Chemie, International Edition* 63 (48): e202406527.

47 Feng, X., Lv, F., Liu, L. et al. (2010). Conjugated polymer nanoparticles for drug delivery and imaging. *ACS Applied Materials & Interfaces* 2 (8): 2429–2435.

48 Zhao, X., Qi, D., Zan, Y. et al. (2025; https://doi.org/10.1002/chem.202501597). Gold nanocluster-based luminescence microcomposites by interfacial polymerization. *Chemistry: —A European Journal* 31: e01597.

49 Zhang, S., Liu, R., Cui, Q. et al. (2017). Ultrabright fluorescent silica nanoparticles embedded with conjugated oligomers and their application in latent fingerprint detection. *ACS Applied Materials & Interfaces* 9 (50): 44134–44145.

50 Zhang, Q., Li, H., Kang, Y. et al. (2024). Tunable fluorescence, morphology, and antibacterial behaviors of conjugated oligomers via host-guest supramolecular self-assembly. *ACS Applied Bio Materials* 7 (4): 2533–2543.

51 Xing, C., Xu, Q., Tang, H. et al. (2009). Conjugated polymer/porphyrin complexes for efficient energy transfer and improving light-activated antibacterial activity. *Journal of the American Chemical Society* 131 (36): 13117–13124.

52 Liu, L., Wang, X., Zhu, S. et al. (2020). Controllable targeted accumulation of fluorescent conjugated polymers on bacteria mediated by a saccharide bridge. *Chemistry of Materials* 32 (1): 438–447.

53 Luo, Y., Peng, R., Cui, Q. et al. (2021). Internal chemiluminescence light-driven photocatalysis. *ACS Applied Materials & Interfaces* 13 (50): 60471–60477.

54 Teng, K.-X., Niu, L.-Y., and Yang, Q.-Z. (2022). A host-guest strategy for converting the photodynamic agents from a singlet oxygen generator to a superoxide radical generator. *Chemical Science* 13 (20): 5951–5956.

55 Yang, N., Wang, C., Wang, X., and Li, L. (2018). Synthesis of photothermal nanocomposites and their application to antibacterial assays. *Nanotechnology* 29 (17): 175601.

56 Li, S., Wang, X., Hu, R. et al. (2016). Near-infrared (NIR)-absorbing conjugated polymer dots as highly effective photothermal materials for *in vivo* cancer therapy. *Chemistry of Materials* 28 (23): 8669–8675.

57 Yang, J., Choi, J., Bang, D. et al. (2011). Convertible organic nanoparticles for near-infrared photothermal ablation of cancer cells. *Angewandte Chemie, International Edition* 50 (2): 441–444.

58 Zha, Z., Yue, X., Ren, Q., and Dai, Z. (2013). Uniform polypyrrole nanoparticles with high photothermal conversion efficiency for photothermal ablation of cancer cells. *Advanced Materials* 25 (5): 777–782.

59 Shi, T., Huang, C., Li, Y. et al. (2022). NIR-II phototherapy agents with aggregation-induced emission characteristics for tumor imaging and therapy. *Biomaterials* 285: 121535.

60 Li, X., Xiao, H., Xiu, W. et al. (2021). Mitochondria-targeting MoS2-based nanoagents for enhanced NIR-II photothermal-chemodynamic synergistic oncotherapy. *ACS Applied Materials & Interfaces* 13 (47): 55928–55938.

61 Yin, L., Sun, H., Zhang, H. et al. (2019). Quantitatively visualizing tumor-related protease activity in vivo using a ratiometric photoacoustic probe. *Journal of the American Chemical Society* 141 (7): 3265–3273.

62 Specht, E.A., Braselmann, E., and Palmer, A.E. (2017). A critical and comparative review of fluorescent tools for live-cell imaging. *Annual Review of Physiology* 79: 93–117.

63 Li, K. and Liu, B. (2014). Polymer-encapsulated organic nanoparticles for fluorescence and photoacoustic imaging. *Chemical Society Reviews* 43 (18): 6570–6597.

64 Liu, H., Zhang, X., Wu, X. et al. (2011). Rapid sonochemical synthesis of highly luminescent non-toxic AuNCs and Au@AgNCs and Cu (II) sensing. *Chemical Communications* 47 (14): 4237–4239.

65 Feng, T., Ai, X., Ong, H., and Zhao, Y. (2016). Dual-responsive carbon dots for tumor extracellular microenvironment triggered targeting and enhanced anticancer drug delivery. *ACS Applied Materials & Interfaces* 8 (29): 18732–18740.

66 Ma, Y., Ai, G., Zhang, C. et al. (2017). Novel linear peptides with high affinity to $\alpha v\beta_3$ integrin for precise tumor identification. *Theranostics* 7 (6): 1511–1523.

67 Raman, R., Raguram, S., Venkataraman, G. et al. (2005). Glycomics: an integrated systems approach to structure-function relationships of glycans. *Nature Methods* 2 (11): 817–824.

68 Rennick, J.J., Johnston, A.P.R., and Parton, R.G. (2021). Key principles and methods for studying the endocytosis of biological and nanoparticle therapeutics. *Nature Nanotechnology* 16 (3): 266–276.

69 Hinde, E., Thammasiraphop, K., Duong, H.T.T. et al. (2017). Pair correlation microscopy reveals the role of nanoparticle shape in intracellular transport and site of drug release. *Nature Nanotechnology* 12 (1): 81–89.

70 Jiang, X., Wang, X., Yao, C. et al. (2019). Surface-engineered gold nanoclusters with biological assembly-amplified emission for multimode imaging. *Journal of Physical Chemistry Letters* 10 (17): 5237–5243.

71 Pu, K., Shuhendler, A.J., Jokerst, J.V. et al. (2014). Semiconducting polymer nanoparticles as photoacoustic molecular imaging probes in living mice. *Nature Nanotechnology* 9 (3): 233–239.

72 Abelha, T.F., Dreiss, C.A., Green, M.A., and Dailey, L.A. (2020). Conjugated polymers as nanoparticle probes for fluorescence and photoacoustic imaging. *Journal of Materials Chemistry B* 8 (4): 592–606.

73 Stahl, T., Bofinger, R., Lam, I. et al. (2017). Tunable semiconducting polymer nanoparticles with INDT-based conjugated polymers for photoacoustic molecular imaging. *Bioconjugate Chemistry* 28 (6): 1734–1740.

74 Huang, Z. (2005). A review of progress in clinical photodynamic therapy. *Technology in Cancer Research & Treatment* 4 (3): 283–293.

75 Nosaka, Y. and Nosaka, A.Y. (2017). Generation and detection of reactive oxygen species in photocatalysis. *Chemical Reviews* 117 (17): 11302–11336.

76 Feng, L., Li, C., Liu, L. et al. (2022). A Facile structural isomerization-induced 3D spatial D-A interlocked network for achieving NIR-II phototheranostic agents. *Angewandte Chemie, International Edition* 61 (50): e202212673.

77 Li, S., Chang, K., Sun, K. et al. (2016). Amplified singlet oxygen generation in semiconductor polymer dots for photodynamic cancer therapy. *ACS Applied Materials & Interfaces* 8 (6): 3624–3634.

78 Wang, C., Cui, Q., Wang, X., and Li, L. (2016). Preparation of hybrid gold/polymer nanocomposites and their application in a controlled antibacterial assay. *ACS Applied Materials & Interfaces* 8 (42): 29101–29109.

79 Feenstra, T., Thogersen, M.S., Wieser, E. et al. (2017). Adhesion of *Escherichia coli* under flow conditions reveals potential novel effects of FimH mutations. *European Journal of Clinical Microbiology & Infectious DIiseases* 36 (3): 467–478.

80 Liu, L., Wang, X., Ban, D. et al. (2020). Controllable accumulation of conjugated polymer nanoparticles on the surface of adhesive bacteria. *Colloids and Surfaces A: Physicochemical and Engineering Aspects* 591: 124569.

81 Rastinehad, A.R., Anastos, H., Wajswol, E. et al. (2019). Gold nanoshell-localized photothermal ablation of prostate tumors in a clinical pilot device study. *Proceedings of the National Academy of Sciences of the United States of America* 116 (37): 18590–18596.

82 Li, S., Deng, Q., Zhang, Y. et al. (2020). Rational design of conjugated small molecules for superior photothermal theranostics in the NIR-II biowindow. *Advanced Materials* 32 (33): e2001146.

83 Wang, Y., Li, S., Zhang, P. et al. (2018). Photothermal-responsive conjugated polymer nanoparticles for remote control of gene expression in living cells. *Advanced Materials* 30 (8): 1705418.

84 Li, J., Xie, C., Huang, J. et al. (2018). Semiconducting polymer nanoenzymes with photothermic activity for enhanced cancer therapy. *Angewandte Chemie, International Edition* 57 (15): 3995–3998.

85 Fu, X., Huang, Y., Zhao, H. et al. (2021). Near-infrared-light remote-controlled activation of cancer immunotherapy using photothermal conjugated polymer nanoparticles. *Advanced Materials* 33 (34): e2102570.

86 Knowles, J.R. (1980). Enzyme-catalyzed phosphoryl transfer reactions. *Annual Review of Biochemistry* 49: 877–919.

87 Ghosh, A., Ronner, P., Cheong, E. et al. (1991). The role of ATP and free ADP in metabolic coupling during fuel-stimulated insulin release from islet beta-cells in the isolated perfused rat pancreas. *The Journal of Biological Chemistry* 266 (34): 22887–22892.

88 Ye, R., Cui, Q., Yao, C. et al. (2017). Tunable fluorescence behaviors of a supramolecular system based on a fluorene derivative and cucurbit[8]uril and its application for ATP sensing. *Physical Chemistry Chemical Physics* 19 (46): 31306–31315.

89 Yang, Y., Liu, R., Cui, Q. et al. (2019). Red-emissive conjugated oligomer/silica hybrid nanoparticles with high affinity and application for latent fingerprint detection. *Colloids and Surfaces a-Physicochemical and Engineering Aspects* 565: 118–130.

90 Cong, Y., Wang, X., Yao, C. et al. (2022). Controlling the interaction between fluorescent gold nanoclusters and biointerfaces for rapid discrimination of fungal pathogens. *ACS Applied Materials & Interfaces* 14 (3): 4532–4541.

91 Milkovic, L., Gasparovic, A.C., Cindric, M. et al. (2019). Short overview of ROS as cell function regulators and their implications in therapy concepts. *Cells* 8 (8): 793.

92 Yin, C., Zhu, H., Xie, C. et al. (2017). Organic nanoprobe cocktails for multilocal and multicolor fluorescence imaging of reactive oxygen species. *Advanced Functional Materials* 27 (23): 1700493.

93 Wang, X., Li, H., Yang, J. et al. (2024). Chemical nose strategy with metabolic labeling and "antibiotic-responsive spectrum" enables accurate and rapid pathogen identification. *Analytical Chemistry* 96 (1): 427–436.

94 Yuan, H., Liu, Z., Liu, L. et al. (2014). Cationic conjugated polymers for discrimination of microbial pathogens. *Advanced Materials* 26 (25): 4333–4338.

95 Yuan, H., Zhao, H., Peng, K. et al. (2020). Conjugated polymer-quantum dot hybrid materials for pathogen discrimination and disinfection. *ACS Applied Materials & Interfaces* 12 (19): 21263–21269.

96 Thanh Duc, M., d'Orlye, F., Menager, C. et al. (2013). Red blood cells decorated with functionalized core-shell magnetic nanoparticles: elucidation of the adsorption mechanism. *Chemical Communications* 49 (47): 5393–5395.

97 Yan, Y., Li, Y., Wang, J. et al. (2022). A carbon dioxide responsive fluorescent system based on micellar transformation. *Colloids and Surfaces a-Physicochemical and Engineering Aspects* 641: 128457.

98 Senthilkumar, T., Zhou, L., Gu, Q. et al. (2018). Conjugated polymer nanoparticles with appended photo-responsive units for controlled drug delivery, release, and imaging. *Angewandte Chemie, International Edition* 57 (40): 13114–13119.

99 Gai, P., Yu, W., Zhao, H. et al. (2020). Solar-powered organic semiconductor-bacteria biohybrids for CO_2 reduction into acetic acid. *Angewandte Chemie, International Edition* 59 (18): 7224–7229.

5

Photofunctional Polymer Fiber Composites

At present, there are two major limitations in the application of traditional phototherapy: insufficient depth of tissue penetration and lack of spatial and temporal accuracy. On the one hand, short-wavelength visible light suffers exponential attenuation when transmitted in vivo, and it is difficult to reach deep tissues effectively. On the other hand, photosensitizers that are systemically administered lack the ability to locate lesions, which easily causes nontargeting effects. The emergence of photofunctional polymer fibers provides a new idea and technical path to solve these problems.

Photofunctional polymer fibers are changing the way photons interact with living things. Such materials precisely regulate the structure of polymer chains (such as β-sheet nanocrystals, semicrystalline regions, or dynamic cross-linked networks) through molecular-level optical design, which can be applied to the design of polymer chains, integrating photon transmission, energy conversion, and biological response functions into micron-scale flexible carriers. This innovative design has not only achieved a leap in performance but also brought new possibilities to the biomedical field [1].

First, the polymer's long-chain molecular structure, as a substrate material, provides good protection for the photoactive groups, which enables it to resist the erosion of the complex environment in the organism. The stability and reliability of the photofunctional material in the physiological system are significantly improved. Second, the polymer itself can also be used as a light-emitting unit to directly respond to external light stimulation and produce light emission at a specific wavelength. This property does not rely on additional fluorescent labels or dyes, thus effectively avoiding the risk of biotoxicity and immune responses that may be triggered by exogenous photofunctional materials. In addition, the rich chemical modification sites and diverse bonding modes of polymers enable them to accurately match and combine with biomolecules to achieve seamless docking of photofunctional materials and biological systems. This seamless docking provides the possibility of in situ, real-time, dynamic monitoring and regulation of cell signal transduction, metabolic networks, and physiological and pathological processes, and opens up a new way for biomedical research and application.

Photofunctional polymer fibers can be used to construct endogenous therapeutic platforms, breaking through the limitations of traditional phototherapy.

Photofunctional Polymer Composites for Bioapplications, First Edition. Xiaoyu Wang and Lidong Li.

For example, implantable hydrogel optical fibers can be seamlessly integrated into soft tissues through mechanical adaptation, achieving a high degree of compatibility with living organisms. Their porous network structure can not only monitor the changes in metabolites in real time but also accurately activate photosensitizers, so as to achieve precise treatment of lesions. This design integrates diagnosis and treatment functions, which significantly improves the efficiency and safety of phototherapy. The application of biodegradable fiber scaffolds further expands the potential of photofunctional polymer fibers. This kind of scaffold can realize the functional transformation from mechanical support and photodynamic sterilization to guided tissue regeneration in a preset period.

In general, photofunctional polymer fibers are a class of polymer composites that has made important breakthroughs in materials science thanks to their unique design and functions. They also provide a powerful tool for life science and medical research and promote the in-depth study of the interaction between photons and living organisms.

5.1 Preparation and Structures of Photofunctional Fiber Composites

The shape of the photofunctional fiber composites is significantly different from that of other photofunctional polymer composites. It is a slender object with a large ratio of length to cross-sectional diameter, as well as flexibility and strength. This unique macrostructure (such as fiber orientation, crystallinity, and porosity) is achieved through the selection of specific molecular weights, molecular chain structures, or functional groups, and is combined with efficient and stable processing technology. Molecular screening provides the basic properties for the materials, and the processing technology further regulates the microscopic arrangement and overall morphology of the fibers, so that they show excellent performance in photofunctional applications. Here are some of the processes used to make the photofunctional fiber composites.

5.1.1 Electrospinning

Electrospinning is an advanced technique for fabricating continuous polymer fibers, and has emerged as a principal method for the production of nanofiber materials, particularly with the recent advancements in nanotechnology. This prominence is attributed to its cost-effectiveness, relative simplicity of apparatus, process controllability, and broad applicability across various fields.

The electrospinning process utilizes electrostatic forces to eject and stretch polymer solutions, resulting in fibers with diameters ranging from the micrometer scale down to the nanometer scale. The standard electrospinning setup comprises four essential items: a high-voltage direct current (DC) power supply, a collector (commonly a grounded conductive metal plate), a syringe fitted with a metallic needle, and a syringe pump to ensure a consistent polymer solution flow rate.

Initially, the polymer is either dissolved in a volatile, conductive organic solvent or melted through the application of heat. The resultant melt or polymer solution is put into a syringe, and the syringe pump precisely regulates the flow rate. Upon application of a high-voltage potential—typically extending from several kilovolts to tens of kilovolts—to the needle tip, a powerful electric field is applied to the polymer solution at the needle's tip. This field exerts a force that counteracts the surface tension of the droplet. As the electric field strength increases, the droplet at the needle tip transitions from a spherical to a conical shape, forming what is known as a Taylor cone. The tip of the Taylor cone emits a charged jet of polymer solution when the electrostatic force exceeds the surface tension. As the jet's diameter reduces, the combined actions of the electric field and the rising charge density cause it to continue stretching and thinning. Concurrently, solvent evaporation or polymer solidification occurs during the flight of the jet, resulting in the deposition of continuous polymer fibers onto the collector. This process enables the fabrication of nanofibers with controlled morphology and properties, making electrospinning a versatile and powerful tool in the field of nanomaterials science.

By controlling the process conditions and changing the collection device, various morphologies of nanofiber structures can be obtained, such as disorderly arrangement, directional arrangement, patterned, spider-web-like, bead-like, ribbon, spiral, porous, necklace-like, firecracker-like, rice-granular, and core-shell structures (Figure 5.1). These diverse structures make electrospinning technology have a wide range of applications in materials science, biomedicine, environmental engineering, and other fields [2].

For example, PAN-CQDs nanofibers were successfully prepared by electrospinning using polyacrylonitrile (PAN) dissolved in dimethylformamide solution, followed by the addition of photosensitizer carbon quantum dots (CQDs) (Figure 5.2). The surface of PAN-CQDs nanofibers is slightly rough compared to the smooth surface of pure PAN nanofibers. It was observed that the diameter of all fibers was evenly distributed, but increased in the presence of CQDs. This is because the addition of nanodots affects the viscosity and surface tension of the polymer solution, thus changing the fiber formation process. TEM images showed that CQDs exhibited a uniform distribution within the fiber structure, with no evidence of aggregation. Therefore, the diameter of the final NF can be effectively controlled by adjusting the mass ratio of PAN to CQD. Furthermore, the uniformly distributed CQD introduced in NF has less effect on the fiber surface, indicating that the method is able to achieve functionalization without significantly changing the fiber surface properties [3].

The electrospinning method can also achieve high-efficiency preparation by simultaneously spraying two different solutions in a syringe device using a double-nozzle assembly. As shown in Figure 5.3a, the spinneret diameter of the PVP-coated SnO_2 solution is 0.4 mm, while that of the PVP-coated TiO_2 solution is 0.7 mm. Through this "two-channel" design, TiO_2/SnO_2 heterojunction nanofibers can be successfully prepared after calcination and applied in the field of photocatalysis. As can be seen from the scanning electron microscopy (SEM) images (Figure 5.3b), the structure of the prepared fibers is consistent with that of the fibers

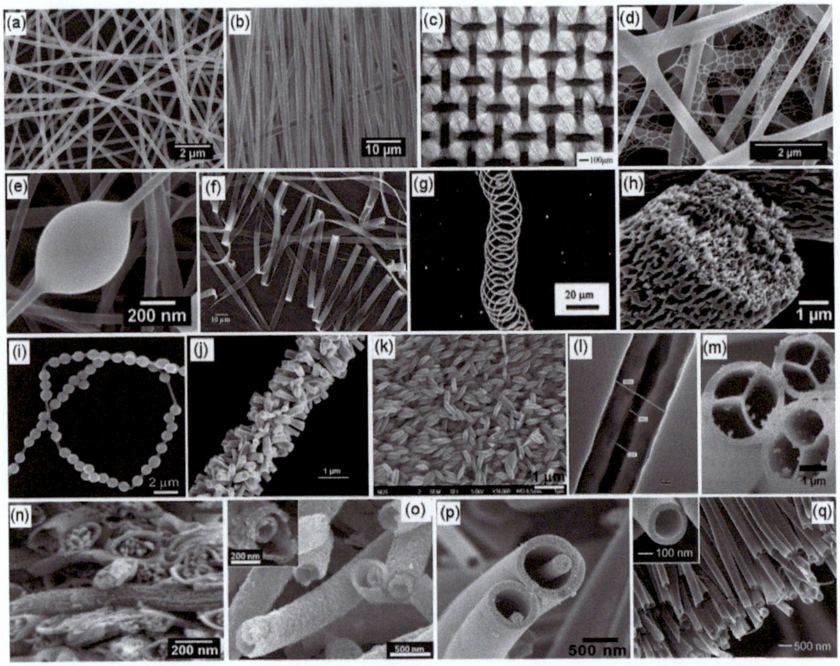

Figure 5.1 Various micro- and nanostructures of fiber produced by electrospinning. (a) Random orientation, (b) direction, (c) pattern, (d) spider's web, (e) stringing beads, (f) ribbon, (g) helical, (h) porous, (i) necklace, (j) firecracker, (k) rice grain, (l) core–shell, (m) multi-channel tube, (n) multi-core cable, (o) tube-in-tube, (p) nanowire-in-microtube and (q) hollow structures. *Source:* Reproduced with permission from Wang et al. [2] / Elsevier.

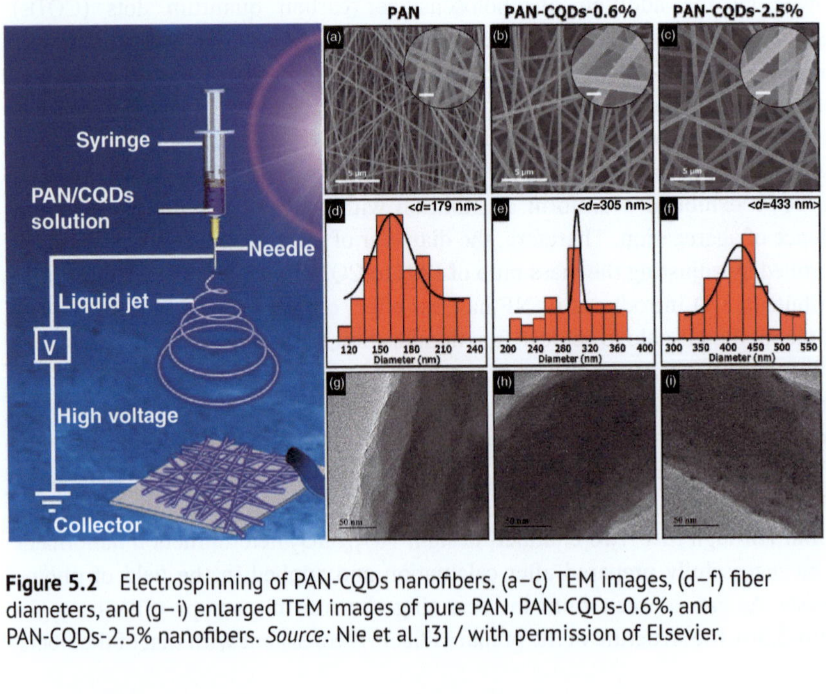

Figure 5.2 Electrospinning of PAN-CQDs nanofibers. (a–c) TEM images, (d–f) fiber diameters, and (g–i) enlarged TEM images of pure PAN, PAN-CQDs-0.6%, and PAN-CQDs-2.5% nanofibers. *Source:* Nie et al. [3] / with permission of Elsevier.

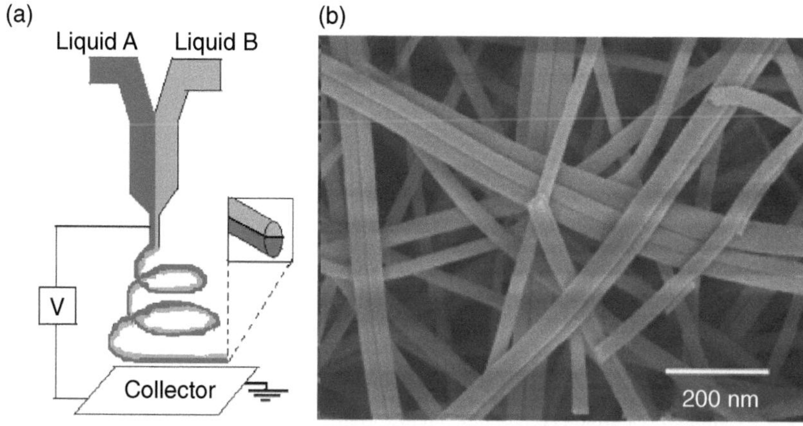

Figure 5.3 (a) electrospinning of the bicomponent TiO_2/SnO_2 nanofibers and (b) representative SEM image of the resulting nanofibers. *Source:* Reproduced with permission from Liu et al. [4] / American Chemical Society.

prepared by the traditional electrospinning method. With an average diameter of roughly 100 nm, the bicomponent nanofibers are continuous and can reach lengths on the order of centimeters. In addition, there is a clear boundary between the two nanofibers, indicating that two parallel jets of solution fuse to form a complete fiber during electrospinning. Each small nanofiber has a diameter of about 50 nm. This method provides a new possibility for the preparation and application of nanocomposites [4].

Electrospinning can also be combined with atomic layer deposition (ALD) technology to produce photofunctional polymer fiber composites. ALD is a deposition method based on repeated reactions of gaseous precursors with solid substrates at low pressures [5]. The nanoscale and macroscopic structure of the electrospun fibers can provide reaction sites for the precursors, and ALD vacuum conditions provide a sufficient mean free path for gas molecules to diffuse into the interior of nanofibers and undergo surface reactions. At the specified time, the precursor reacts completely with the substrate surface when pulsed into the ALD chamber under vacuum conditions. After each step of the surface reaction, inert gases are applied to sweep away any remaining precursors and volatile products, as shown in Figure 5.4 [6]. In this way, the thickness of the product on the substrate surface can be precisely adjusted according to the number of cycles. The ALD technique provides good shape retention for the as-grown materials, which is a key factor in the selection of ALD to be combined with electrospinning.

For example, Microemulsions comprising polyacrylonitrile (PAN), polystyrene (PS), and polyvinyl pyrrolidone (PVP) were synthesized and subsequently utilized as precursors for fiber formation via the electrospinning technique. The resulting fibers were subjected to calcination at an elevated temperature of 800 °C, yielding porous carbon fibers (MPCF). $Pd(C_5HF_6O_2)_2$ supported by nitrogen was then used as a palladium source to deposit palladium on MPCF by the ALD technique. In the early stage of deposition, palladium atoms gradually aggregate and nucleate to

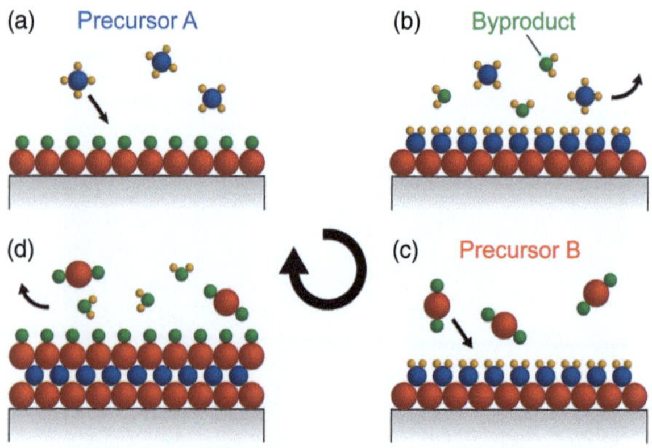

Figure 5.4 Schematic of an ALD technique: (a) precursor A is introduced and reacts on the surface; (b) excess chemicals are removed from the chamber; (c) precursor B is introduced and reacts on the modified surface; (d) excess chemicals are removed from stringing beads the chamber. *Source:* Asundi et al. [6] / with permission of American Chemical Society.

form atomic clusters. After a certain number of cycles, palladium nanoparticles (Pd-NPs) are finally formed, which can be used for the preparation of Pd/MPCF fibers with photothermal catalytic properties (Figure 5.5a). Figure 5.5b shows the macroporous structure of 14-Pd/MPCF with channel diameters ranging from 50 to 100 nm. Figure 5.5c further highlights the parallel channel-like pore structure in the fibers, with a slightly rougher surface. Figure 5.5d shows the loading of palladium

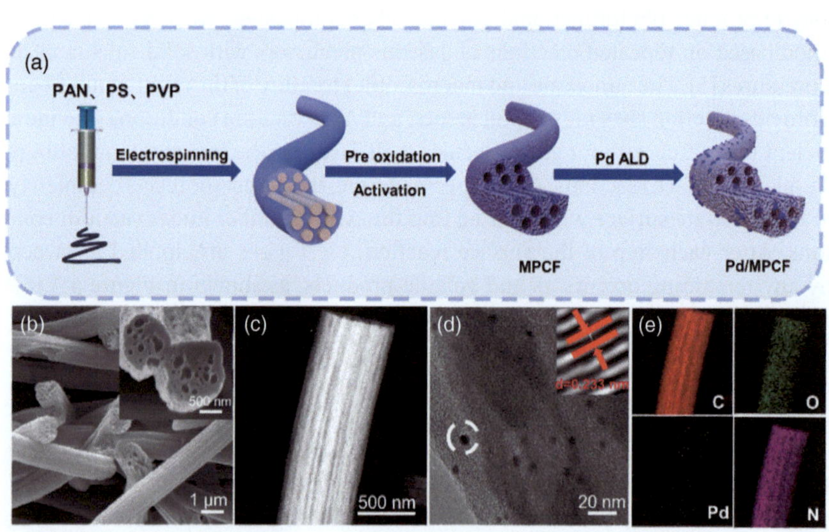

Figure 5.5 (a) The preparation of Pd/MPCF using electrospinning combined with ALD technology. (b) SEM image and (c) TEM image of Pd/MPCF. (d) HRTEM image of Pd-NPs on the walls of MPCF. (e) Elemental mapping images of C, O, Pd, and N in Pd/MPCF. *Source:* Reproduced with permission from Zhang et al. [7] / Elsevier.

nanoparticles with a lattice fringe spacing of 0.233 nm corresponding to the (111) face of palladium. Furthermore, EDX spectrum analysis showed that palladium (Pd), carbon (C), oxygen (O), and nitrogen (N) elements were uniformly distributed throughout the fiber (Figure 5.5e), further confirming the efficient loading of palladium nanoparticles on porous carbon fibers and providing new ideas for the development of photothermal catalytic materials [7].

5.1.2 3D Printing

The core of fabricating photofunctional polymer fibers by 3D printing technology lies in the precise control of the viscoelasticity of the precursor solution. The process is mainly realized by the following steps: firstly, the pre-gel solution is extruded through a nozzle, and the pre-gel solution is molded by its self-assembly characteristics or subsequent cross-linking. For example, cross-linking reactions can be initiated using ultraviolet (UV) light. The coaxial nozzle utilized in the 3D printing system features a dual-channel architecture. The sheath channel is perfused with a mixture comprising acrylamide (AA) monomer, photoinitiator, cross-linking agent, and alginate, while the core channel is supplied with a catalyst and an alginate ionic crosslinker. When UV light is irradiated onto the pre-gel solution, the polymerization reaction of the monomers in the irradiated area is selectively activated, forming a polyacrylic acid (PAA) covalent network. This network then crosslinks with the alginate network of Ca^{2+} ions to form a double interpenetrating structure, resulting in hydrogel fibers with excellent mechanical properties (Figure 5.6) [8]. The gradient design of the mechanical properties of fiber

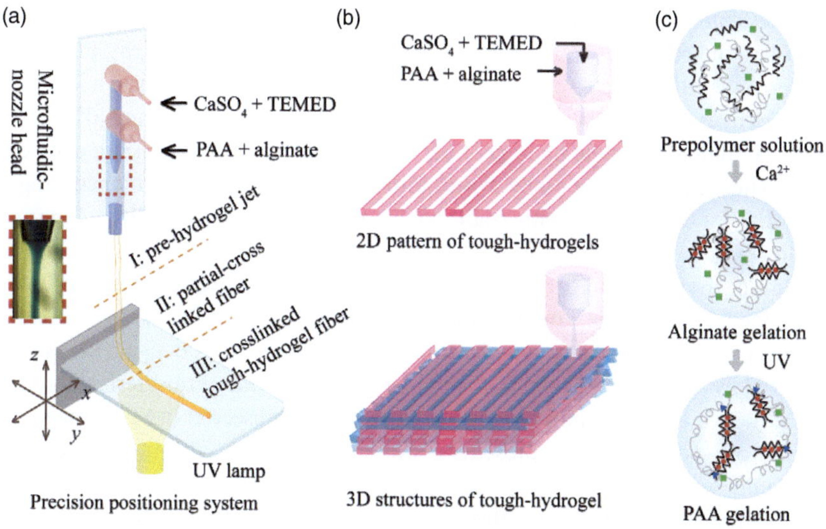

Figure 5.6 (a) 3D printing system and (b) the process for the preparation of hydrogel fiber. (c) Schematic illustrating of the double interpenetrating structure of hydrogel fibers. *Source:* Wei et al. [8] / with permission of American Chemical Society.

materials can also be realized by precisely controlling the gradient distribution of light intensity.

Similarly, an 80 °C gelatin solution can be mixed with the drug DOX, and sodium alginate can be added for 3D printing. Subsequently, the fiber scaffolds are formed by cross-linking with Ca^{2+}. Then, core/shell fibers with photothermal properties are prepared by orderly immersion in polycaprolactone (PCL) and dopamine solutions [9].

Among them, the critical control points include maintaining the fiber morphology stability of the solution during extrusion and maintaining the structural shape integrity during the covalent cross-linking pretreatment stage. The precise implementation of these steps and control points ensures that the final prepared fiber materials have the required optical functions and mechanical properties. In this way, 3D printing technology can not only efficiently fabricate photofunctional polymer fibers but also achieve precise regulation of their properties, opening up new possibilities for the development of materials science in the future.

5.1.3 Wet Spinning Process

Wet spinning is one of the most common methods of constructing fibrous materials, which can be produced on a large scale due to its mature and controllable operating process. The formation mechanism is that the precursor solution is injected into the coagulation bath, and the spinning is carried out by cooling coagulation or a nonsolvent-induced phase separation process.

For example, hydrogel precursors containing alginate and end-functionalized poly(ethylene glycol) prepolymer (PEGDA), as well as a hydrogel precursor solution containing photoinitiator I2959, are injected into a coagulation bath containing Ca^{2+} by wet spinning to form nascent hydrogel fibers. The ionic cross-linked network ensures the continuity of the spinning process, while under UV light, PEGDA is polymerized with 2-hydroxyethyl acrylate to form a covalent cross-linked network that endows hydrogel fibers with excellent mechanical properties. Through this wet spinning technology, a single hydrogel fiber with a length of more than 6 m can be prepared. These fibers can be easily woven into textiles (Figure 5.7), demonstrating the potential for integration with traditional textiles and the feasibility of mass production [10].

In addition, the fibers prepared by traditional wet spinning can also be further functionalized. For example, after making textiles from wool/acrylic (W/A) blended fibers, they can be dyed by sequentially immersing them in Rose Bengal (RB) and Cationic Yellow X-8GL dyes at 100 °C to produce photodynamic antibacterial fabrics (Figure 5.8). Two different morphologies of fibers can be observed in SEM images: the surface of wool yarn fibers presents a scale-like structure with a diameter of $21.8 \pm 6.8\,\mu m$, while the surface of acrylic fibers is smoother with a diameter of $16.8 \pm 7.1\,\mu m$. In addition, the fabric fibers treated with RB and RB/CY3 have rougher surfaces than the pristine W/A fibers. This phenomenon can be attributed to both dye loading and the side effects of dyeing and soaping processes. These functional treatments not only enrich the properties of fibers but also provide more possibilities for their application in the field of textiles [11].

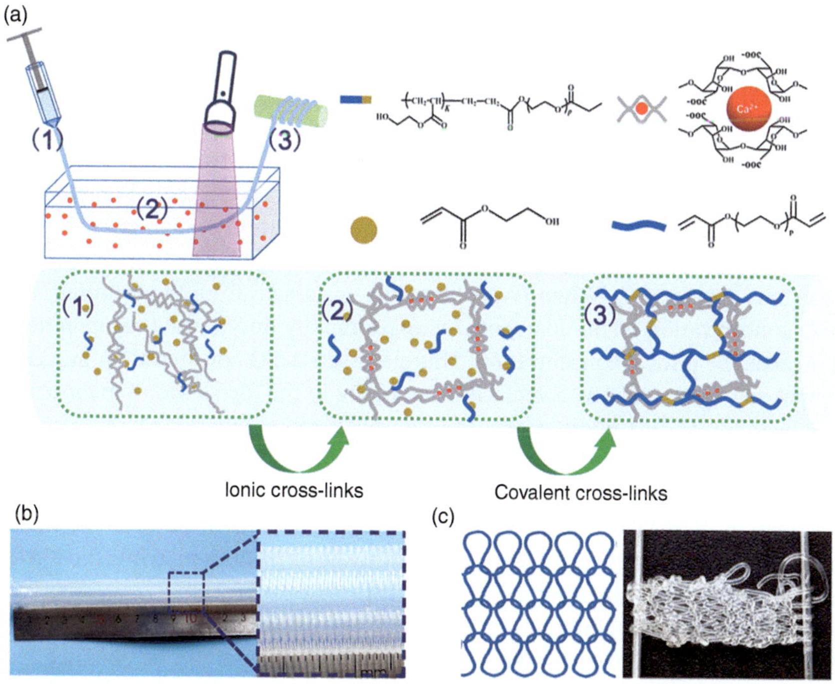

Figure 5.7 (a) Schematic illustrating the wet spinning process for hydrogel fiber fabrication. (b) Photograph of the collected single hydrogel fiber on a spool. (c) Photograph of knitted hydrogel fibers. *Source:* Song et al. [10] / with permission of John Wiley & Sons.

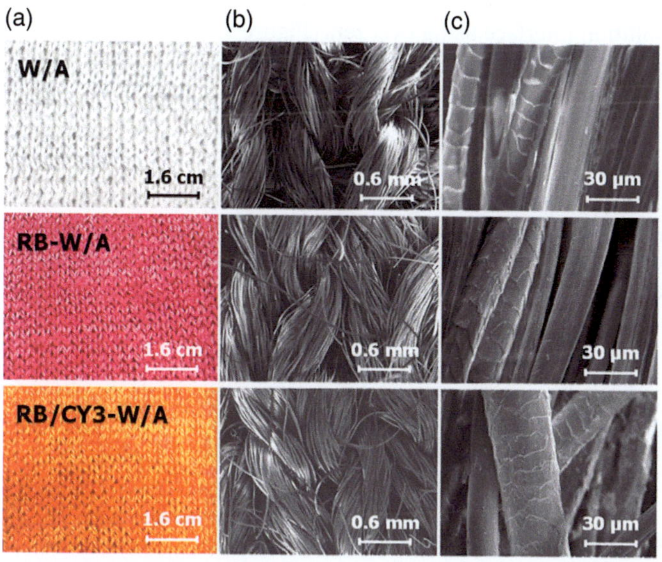

Figure 5.8 (a) Photographs, (b) SEM images, and (c) enlarged SEM images of W/A fibers, RB-W/A fibers and RB/CY3-W/A fibers. *Source:* Reproduced with permission from Chen et al. [11] / American Chemical Society.

5.1.4 Stretching Process

The stretching process is based on the entanglement interaction between molecular chains to prepare photofunctional polymer fibers by artificial drawing. The mechanism is similar to the formation of spider silk: spiders secrete liquid proteins from their glands, forming a gel mixture that, when it leaves the body, quickly transforms into solid fibers. For example, in the spinning process based on sodium polyacrylate (PAAS), PAAS chains form concentrated solutions in a mixed solution of water and dimethyl sulfoxide (DMSO) through winding and hydrogen bonding interactions (Figure 5.9). When PAAS filaments are drawn from the solution, the DMSO concentration in the filaments increases rapidly due to water evaporation and exceeds the critical concentration. This elevated DMSO concentration induces the rapid aggregation of PAAS molecules, resulting in the formation of microscale PAAS hydrogel (PAH) fibers. Concurrently, the surplus aqueous phase spontaneously forms droplets, driven by surface tension effects. The resulting PAH fibers are subsequently immersed in a polymethyl acrylate (PMA) solution, facilitating the formation of MAPAH fibers with a distinct core-shell architecture. The fiber diameters can be precisely regulated within the range of 20–300 μm, while the PMA shell thickness remains between 1 and 2 μm [12].

Based on the process of interfacial polyelectrolyte complexation (IPC) at the oppositely charged polyelectrolyte interface, supramolecular fibers can be obtained by stretching. For instance, four positively charged conjugated polymers (CCP)—PFP, PPE, PT, and PBF—exhibiting distinct fluorescence emissions (blue, green, orange, and red, respectively) were utilized. Supramolecular fibers were successfully fabricated by combining negatively charged DNA with different conjugated polymers and positively charged histone proteins (H). This was followed by the application of an upward force at the solution interface using tweezers (Figure 5.10a) [13].

Observation through an optical fiber microscope revealed that these fibers exhibit a bead-like structure, which is distributed periodically along the fiber axis (Figure 5.10b). The diameter of these fibers varies from several micrometers to tens of micrometers, with the specific size possibly depending on the composition and stretching rate. SEM images further show that the roughness of the fiber increases after the addition of the conjugated polymer, which is explained by the surface tension change of the polyelectrolyte solution (Figure 5.10c). In addition, the fluorescence images showed that the conjugated polymers were uniformly

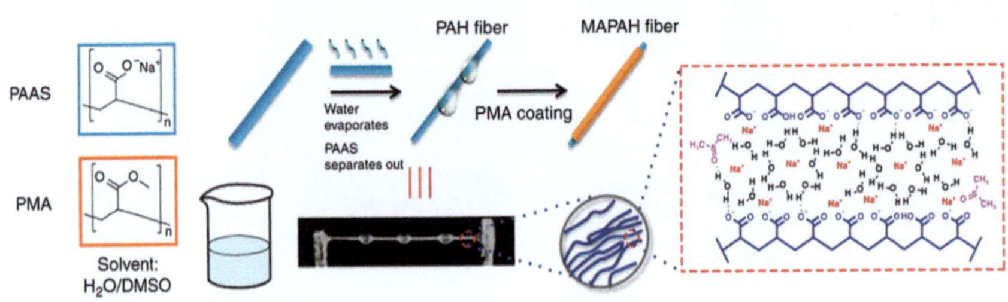

Figure 5.9 The scheme of synthesis of PAH and MAPAH fibers. *Source:* Zhao et al. [12] / Springer Nature / CC BY 4.0.

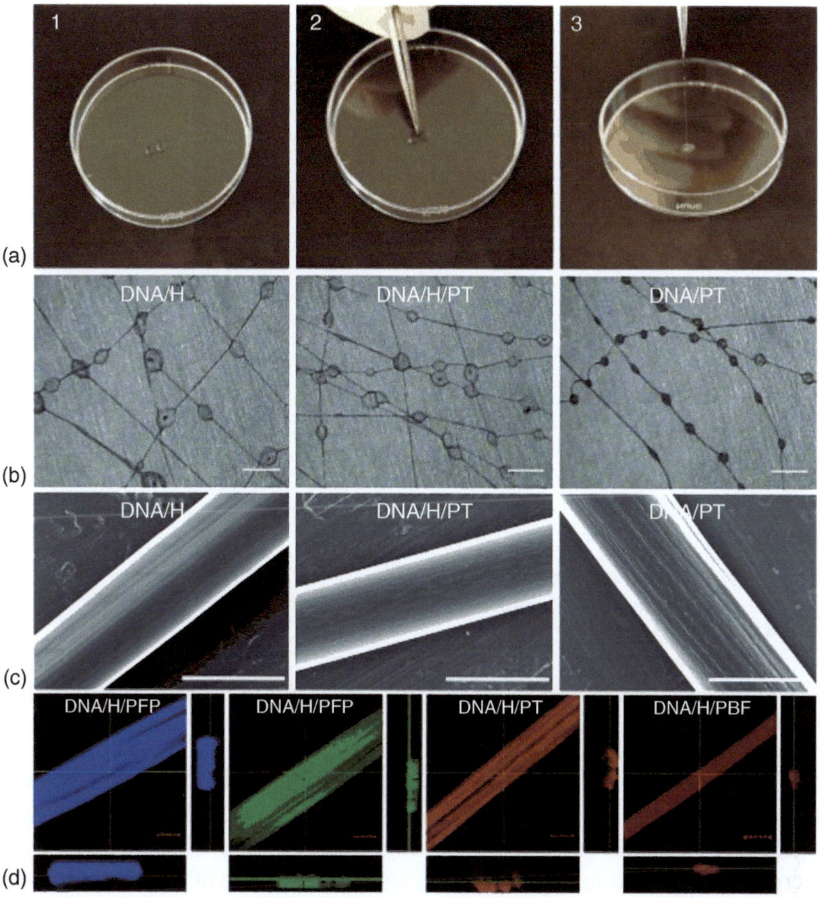

Figure 5.10 (a) Stretching process used to prepare DNA/H/CCP fibers. (b) Microscope images and (c) SEM images of DNA/H, DNA/PT, and DNA/H/PT fibers. (d) Fluorescence images of DNA/H/PFP, DNA/H/PPE, DNA/H/PT and DNA/H/PBF, each displaying distinct colors. *Source:* Reproduced with permission from Wang et al. [13] / John Wiley & Sons.

dispersed in the fibers, and DNA/H/CCP fibers emit blue, green, and red light (Figure 5.10d). This study not only demonstrates the application potential of IPC technology in the preparation of supramolecular fibers but also provides a new idea for the development of optical-functional fiber materials through the regulation of fluorescence color.

5.1.5 Capillary-templated Polymerization

The capillary-templated polymerization is an advanced technique for in situ polymerization, wherein hydrogel precursor solutions are injected into molds, such as capillaries or tubes, under various controlled conditions, including UV irradiation, thermal effects, and physical interactions. This approach facilitates the fabrication of hydrogel fibers with precisely regulated diameters and limited

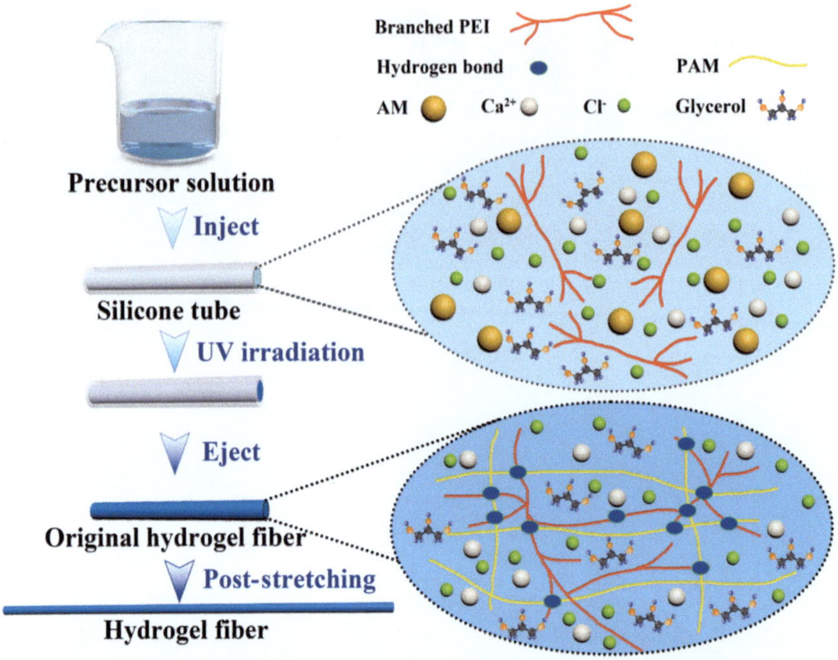

Figure 5.11 Schematic illustration of hydrogel fiber preparation via polymerization within a templated capillary.

lengths. Due to its simplicity, cost-effectiveness, and high efficiency, this method has garnered significant scientific interest. The molds utilized in this process often exhibit tunable mechanical properties, elasticity, transparency, biocompatibility, and micro- or nanostructural features, enabling easy separation of the fibers. These attributes further enhance the versatility and application potential of the hydrogel fibers in diverse fields.

For example, Figure 5.11 shows the manufacturing process for a hydrogel fiber. In the operation, the hydrogel precursor liquid is injected into the silica gel tube by a syringe. The precursor solution consists of water, glycerol, calcium chloride ($CaCl_2$), acrylamide (AM), branched PEI, *N,N′*-methylenebis(acrylamide) (MBA), and photoinitiator-2960. Subsequently, the gelation process is triggered by exposing the silicone tube filled with the precursor solution to UV irradiation for a fixed duration. In this process, MBA acts as a cross-linking agent to covalently link PAM chains, while branched PEI and PAM work synergistically through noncovalent interactions such as hydrogen bonds to form a three-dimensional cross-linked hydrogel network structure. Finally, the formed hydrogel fiber is pushed out of the mold by applying air pressure from one end of the silicone tube. This method is not only easy to operate but can also accurately control the properties of hydrogel fibers by adjusting the composition of the precursor solution and polymerization conditions [14].

Similarly, acrylamide-modified polyethylene glycol was combined with a glucose-responsive fluorescent monomer, acrylamide, and *N,N′*-methylenebis(acrylamide).

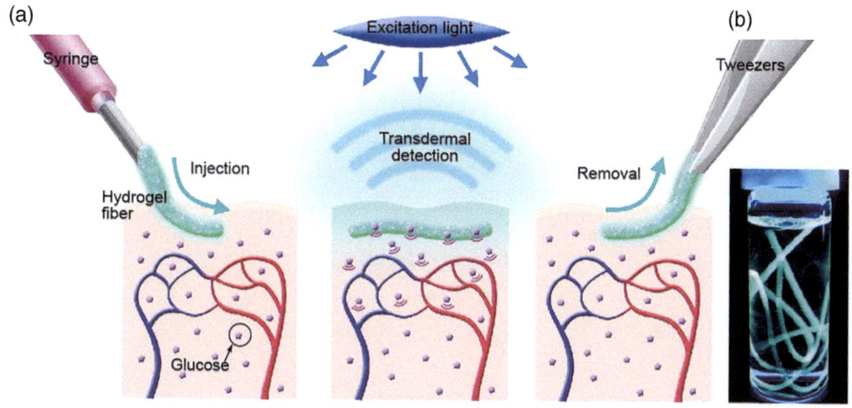

Figure 5.12 (a) Schematic illustration of a fluorescent hydrogel fiber used in vivo for glucose detection. (b) Photograph of hydrogel fibers exhibiting green fluorescence under ultraviolet illumination. *Source:* Heo et al. [15] / PNAS.

Subsequently, sodium persulfate and *N,N,N′,N′*-tetramethylethylenediamine were added to the mixture [15]. The resulting solution was injected into polyolefin microcapillaries and subjected to cross-linking at 37 °C. Upon completion of the cross-linking reaction, the fluorescent hydrogel microfibers were extracted from the capillaries. These fibers exhibited a diameter of $956 \pm 9\,\mu m$, which is smaller than that of a standard needle, thereby facilitating their injection into biological tissues (Figure 5.12). The fabricated fluorescent hydrogel microfibers demonstrated excellent biocompatibility and mechanical integrity, and were capable of real-time monitoring of glucose concentration fluctuations in vivo. This approach offers promising opportunities for advanced biomedical applications.

5.1.6 Microfluidic Method

The microfluidic method is a technique that enables the controlled fabrication of fibers by precisely manipulating the dimensions of fluids within microchannels. This approach is characterized by minimal sample consumption, operational convenience, high precision, and rapid processing. A notable advantage of microchannel systems is the capability to introduce two or more coaxial streams, facilitating complex fiber architectures. The underlying mechanism of microfluidic spinning involves the injection of a precursor solution into the inlet of a microfluidic chip, where it is conveyed through microchannels under laminar flow conditions. Alternatively, the precursor may be extruded directly from the tip of an embedded syringe or the outlet of a capillary glass tube. Upon exiting the microchannel, the precursor solution undergoes rapid cross-linking or solidification, which may be induced by UV irradiation, ionic or chemical cross-linking agents, or solvent exchange processes. This results in the formation of hydrogel fibers within a short time frame [16].

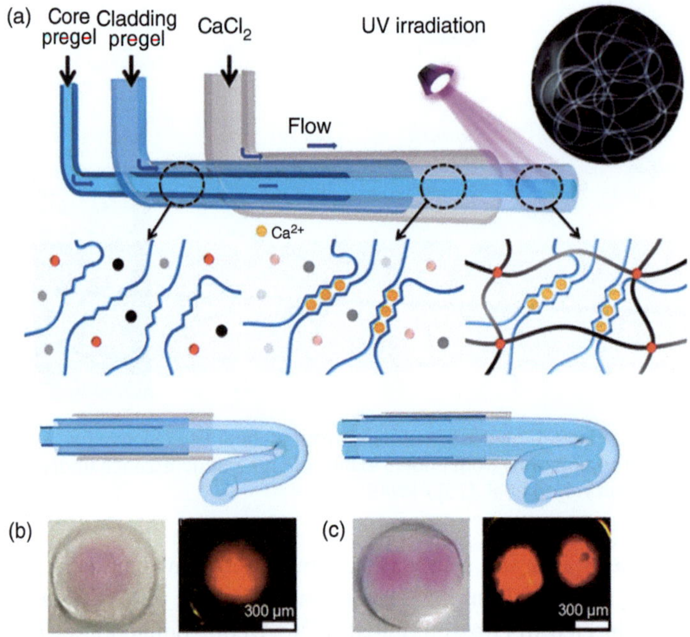

Figure 5.13 (a) Microfluidic method for preparing optical fibers with double networks. Schematic diagrams, optical micrographs, and fluorescence images of optical fibers with (b) a single core and (c) two cores. *Source:* Fitria et al. [17] / with permission of John Wiley & Sons.

As shown in Figure 5.13, three capillaries with different diameters are arranged coaxially to form a microchannel system. The fiber core was injected with a mixture of AAm, *N,N′*-methylenebis(acrylamide) (bisAA), and alginate in aqueous solution, while the cladding was filled with DMSO solution of AAm, bisAA, dopamine methacrylamide (DMA), and alginate. Lithium phenyl-2,4,6-trimethylbenzoyl phosphate was then added as an initiator to activate the radical polymerization. Calcium chloride ($CaCl_2$) solution was injected into the outermost capillary, facilitating the rapid diffusion of Ca^{2+} ions into the adjacent inner and middle capillaries. Subsequently, these Ca^{2+} ions interact with alginate ions, inducing ionic cross-linking and resulting in the formation of a solidified Ca-Alg network. Because of the continuous effect of this process, the solution successfully maintains the shape of the microfibril during the flow process. Then, the photopolymerization of the AAm, bisAA, and DMA components is initiated under the irradiation of UV light. Finally, the covalently crosslinked network and the ionically crosslinked network cross each other to form a dual network structure. To make the hydrogel optical fiber's core and cladding easily distinguishable, Rhodamine B was added to the solution. By observing the red signal of Rhodamine B, the diameters of the core and cladding were measured to be around 821–1290 μm, respectively. This design demonstrates the precise control of microchannel technology [17].

5.2 Functions of Photofunctional Fiber Composites

5.2.1 Energy/Electron Transfer Effect

Luminescent fiber fabrication commonly entails the incorporation of luminescent materials into fibers via physical doping or chemical coupling, either prior to or following fiber formation [18]. These luminescent agents—encompassing inorganic quantum dots, organic fluorescent dyes, luminescent carbon dots, and rare earth metal complexes—are extensively utilized in the synthesis of electrospun nanofibers. The optical characteristics of these luminescent materials are modulated by various parameters, including fiber architecture, morphology, wettability, and polarity. Consequently, luminescent nanofibers have attracted considerable interest for both fundamental research and practical applications. Furthermore, by controlling energy or electron transfer processes among chromophores, the luminescence properties of fibers can be finely tuned, resulting in enhanced efficiency and stability of luminescent output.

The complex of europium ions (Eu^{3+}) with organic ligands has attracted considerable attention due to its unique optical properties. Fluorescent nanofibers based on the $Eu(TTA)_3Phen$ complex, prepared by electrospinning, exhibit excellent luminescent properties. First, Eu^{3+} were complexed with 4,4′,4-trimethyltriphenylamine (TTA) and 1,10-phenanthroline (Phen) to form the $Eu(TTA)_3Phen$ complex. This step resulted in strong binding between Eu^{3+} and the organic ligand via the formation of coordination bonds. Subsequently, the $Eu(TTA)_3Phen$ complex was mixed with SiO_2. During this process, Si—O—Eu and O → Eu bonds were formed, immobilizing the $Eu(TTA)_3Phen$ complex on the SiO_2 surface to create the $SiO_2@Eu^{3+}$ complex. This immobilization significantly enhanced the stability of the complex. Next, $SiO_2@Eu^{3+}$ was blended with PAN and electrospun into a fluorescent nanofiber membrane, named FPNM. As an efficient method for nanofiber preparation, electrospinning ensures the uniform distribution of the fluorescent complex within the fiber structure. In the nanofiber membrane, C—N bonds of PAN react with SiO_2 to form O=C—N bonds, which participate in Eu^{3+} coordination, thereby significantly improving the efficiency of energy transfer from ligands to Eu^{3+}.

Optical performance assessments reveal that the $Eu(TTA)_3Phen$ complex exhibits absorption in the 300–420 nm range with a maximum peak of 370 nm (Figure 5.14a). In the nanofiber membrane, the excitation intensity shows significant enhancement, attributed to the energy transfer facilitation caused by O=C—N bonds formed through the reaction between PAN and SiO_2. Correspondingly, fluorescence emission corresponding to the $5D_0 \rightarrow 7F_2$ transition of 4f electrons was observed at 612 nm, with its intensity markedly increased in the membrane (Figure 5.14b). Further analysis reveals that the fluorescence lifetime of the $Eu(TTA)_3Phen$ complex in the nanofiber membrane extends from 595.4 to 692.9 μs (Figure 5.14c). This phenomenon indicates that the integration of PAN with SiO_2 not only extends

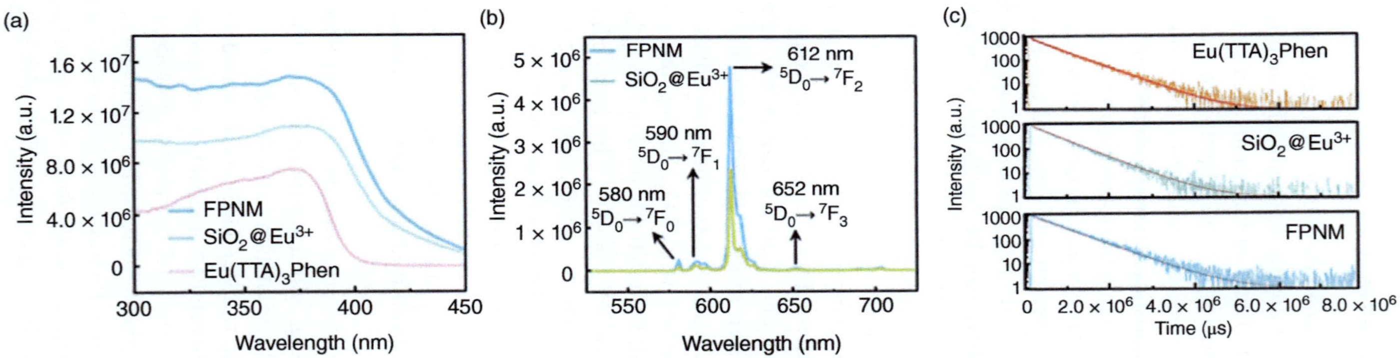

Figure 5.14 (a) Excitation spectra, (b) emission spectra, and (c) fluorescence decay profiles of $Eu(TTA)_3Phen$, $SiO_2@Eu^{3+}$, and FPNM, respectively. *Source:* Yan et al. [19] / with permission of Elsevier.

the fluorescence lifetime but also increases the efficiency of energy transfer from the ligands to Eu^{3+}. Consequently, the excellent compatibility between PAN and $Eu(TTA)_3Phen$ in the nanofiber structure significantly enhances its optical properties [19].

In the mixed solution of pyrene and polystyrene (PS), the addition of tetrabutylammonium hexafluorophosphate (TBAH) enables the fabrication of fluorescence-amplified nanofibers via electrospinning [20]. The phenyl groups on the PS side chains can form a sandwich-like π–π stacking structure with pyrene molecules, facilitating electron transport. Additionally, the scaffold-like architecture of PS acts as a spacer to prevent excessive aggregation of pyrene molecules, effectively reducing self-quenching effects (Figure 5.15a).

The highly interconnected porous structure of electrospun pyrene/PS/TBAH films significantly reduces diffusional resistance for target molecules while enhancing intermolecular interactions with explosive analytes, thereby improving detection sensitivity. For all nitroaromatic explosives, their low-lying LUMO energy levels facilitate photoinduced electron transfer (PET) from the excited state of pyrene/PS/TBAH films to these electron-deficient nitro compounds (Figure 5.15b). This "molecular linear energy transfer + amplified quenching effect" mechanism substantially boosts detection sensitivity.

As a common solid-state light source, white light typically requires doping with multiple luminescent materials. However, the energy transfer effects between multiple dyes often affect the color purity of white light materials. By controlling the gap that separates donor and acceptor molecules, the efficiency of energy transfer can be effectively controlled, thereby achieving efficient white light emission [21].

For instance, core–shell structured porous white-light-emitting nanofibers (WLENFs) based on polylactide (PLA) were fabricated using a coaxial electrospinning technique. Blue-light-emitting dye 1,8-naphthalic anhydride (B) was incorporated into the shell layer as the energy donor, while the orange-light-emitting

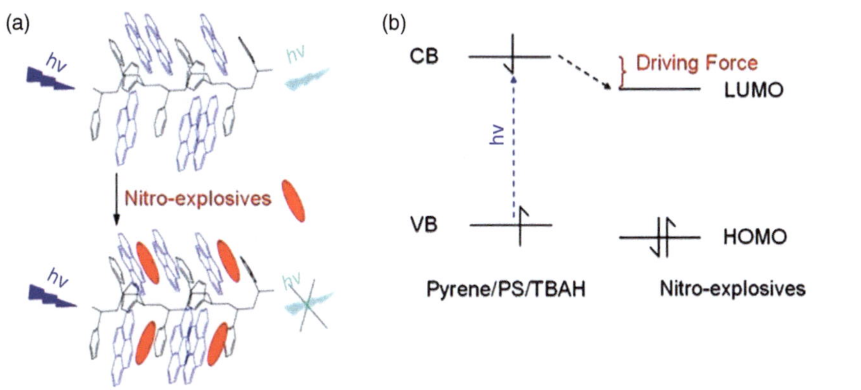

Figure 5.15 (a) Schematic representation of the interactions between pyrene and PS in the presence and absence of nitroaromatic explosives. (b) The PET process occurring between pyrene/PS/TBAH fibers and nitroaromatic explosives. *Source:* Wang et al. [20] / with permission of John Wiley & Sons.

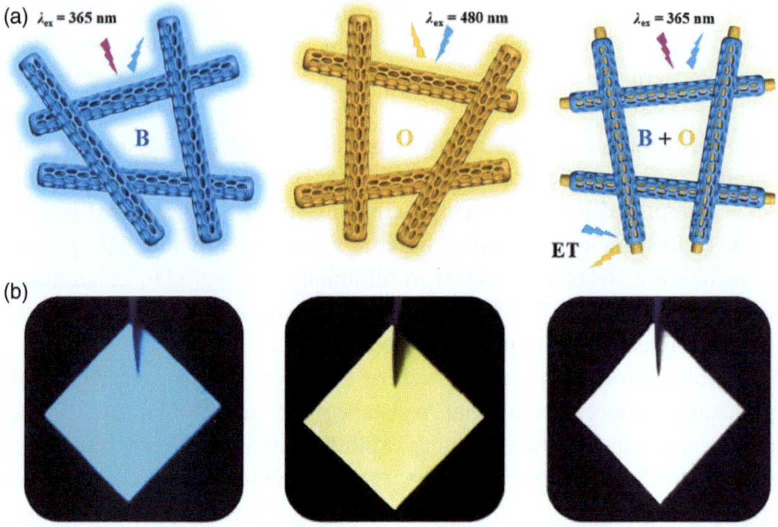

Figure 5.16 (a) Schematic representation of emission and (b) fluorescence images for B porous fibers, O porous fibers, and B + O porous fibers, respectively. *Source:* Reproduced with permission from Qin et al. [22] / Elsevier.

acceptor dye perylene mono-anhydride (O) was embedded in the core (Figure 5.16a). With the shell thickness exceeding 200 nm and the donor-acceptor distance surpassing 10 nm, this spatial separation effectively blocks energy transfer between B and O. Computational analysis of fluorescence decay time in different fibers revealed an energy transfer efficiency of only 18.06% (Table 5.1), further validating the structural effectiveness. Additionally, the porous architecture reduces interference from the nanofiber shell on the core's fluorescent molecules, enhancing overall fluorescence efficiency. Under UV irradiation, blue fluorescence from the shell layer transfers to the core, exciting orange emission and producing a combined light of orange and blue, thus achieving white-light emission (Figure 5.16b). In summary, the coaxial electrospinning strategy for porous WLENFs not only precisely controls energy transfer but also boosts fluorescence efficiency through its porous structure, providing a viable solution for achieving high-performance white-light emission [22].

Table 5.1 Fluorescence lifetimes of O in various fibers measured at 570 nm.

Shape	Method	Sample	τ_1 (ns)	τ_2 (ns)	τ_{avg} (ns)	Efficiency (%)
Porous fiber	Single nozzle	SOF	6.06	13.56	6.69	0
Porous fiber	Single nozzle	S2F	0.24	8.87	1.44	78.43
Porous fiber	Coaxial nozzle	C2F	2.83	6.26	5.48	18.06

Source: Qin et al. [22] / with permission of Elsevier.

5.2.2 Photochromic Effect

By incorporating photochromic molecules (such as spiropyran) into fiber materials through physical blending or chemical modification, these composite materials exhibit rapid and reversible color switching (coloration/decoloration) under specific wavelength light stimuli. This occurs via photoinduced reversible isomerization reactions, ultimately enabling the fabrication of fiber materials with dynamically responsive optical properties.

Using the wet spinning method with PAN as the spinning matrix, long-persistent red-emitting $Y_2O_2S{:}Eu^{3+}$, Mg^{2+}, Ti^{4+} phosphors were blended with a spiropyran derivative (YOS) that changes to red under UV irradiation to prepare red photochromic fibers (RPLF). Similarly, $SrAl_2O_4{:}Eu^{2+}$, Dy^{3+} phosphors were mixed with SAOD to produce green photochromic fibers (GPLF); blue-emitting $Sr_2MgSi_2O_7{:}Eu^{2+}$, Dy^{3+} phosphors were blended with SMSO to obtain blue photochromic fibers (BPLF) (Figure 5.17a). After exposure to UV light, the colorless spiropyran transforms into a colored merocyanine structure, causing the three groups of white fibers to display red, green, and blue colors (Figure 5.17b). This process is reversible; under visible light irradiation, they can revert to white. Although the open-ring merocyanine exhibits some luminescence, its efficiency is relatively low. Therefore, upon UV excitation, these fibers can emit corresponding red, green, and blue light, which matches the emission of rare-earth long afterglow luminescent materials. As a result, the fibers show dual responses in color and fluorescence under UV light [23].

Based on the above fiber fabrication, the incorporation of thermochromic crystal violet lactone (CVL) enables enriched color modulation (Figure 5.18). Below 38 °C, CVL adopts a quinoid structure with extended π-conjugation, absorbing blue-green light to exhibit red coloration. When photoexcited by UV irradiation at temperatures exceeding 38 °C, CVL undergoes a structural transition to its

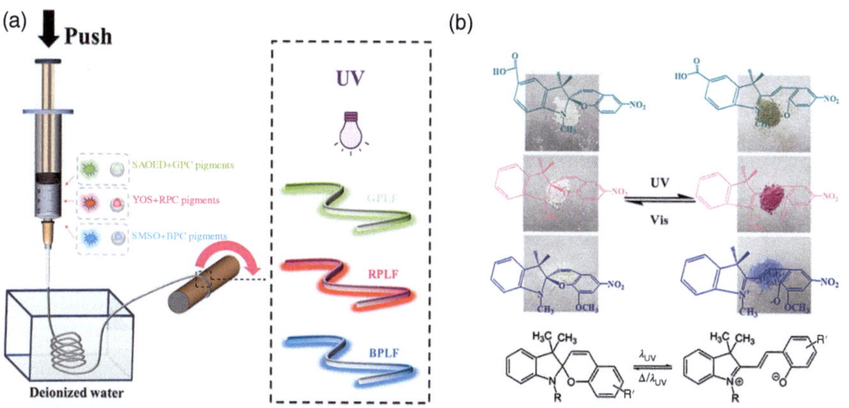

Figure 5.17 (a) Schematic illustration of the wet spinning for stimuli-responsive RPLF, GPLF, and BRLF fibers. (b) Photochromic response of various spiropyran derivatives. *Source:* Shen et al. [23] / with permission of Elsevier.

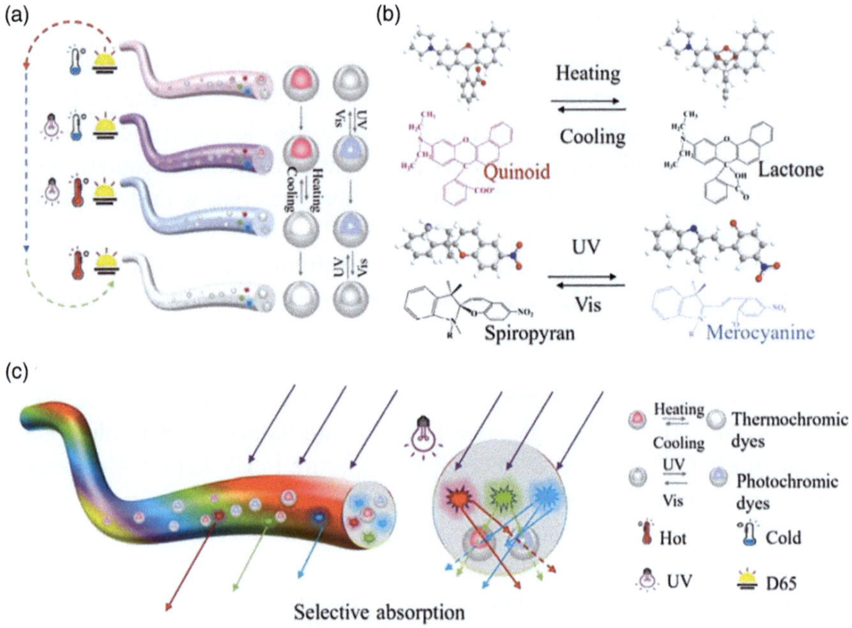

Figure 5.18 (a) Schematic representations of the chromic process, (b) chromic mechanism, and (c) the light response of polymer fiber composites under varying conditions. *Source:* Shen et al. [24] / with permission of Elsevier.

lactone form, diminishing conjugation and shifting absorption to the UV region. This dramatically reduces visible-light absorption, causing a color shift from red to white. Under UV excitation, the spiropyran (SP) component undergoes ring-opening isomerization to merocyanine, further transitioning from white to blue. Importantly, the SP photochromic dye and the CVL thermochromic dye exhibit selective absorption of specific emission bands from the rare-earth complex, contingent upon their respective isomeric configurations. This wavelength-selective absorption allows the fiber to modulate specific emission bands of hybrid luminescent materials, thereby dynamically tuning output colors. Such sophisticated optical response mechanisms significantly expand the chromatic expressiveness of fibrous materials [24].

5.2.3 Photosensitization Effect

The combination of photosensitizers with fibers enables the preparation of functionalized fibers for photodynamic therapy (PDT). By simply mixing photosensitizers with polymer solutions and processing them into polymer nanofibers through spinning techniques, these agents can be effectively incorporated into the polymer matrix. These photofunctional polymer fibers efficiently deliver targeted wavelengths of light to tumor tissues or lesions, thereby activating the photosensitizers to generate reactive oxygen species (ROS), achieving precise therapeutic effects on affected areas.

Using TTVB as a dopant and electropositive PVDF-HFP as the polymer matrix, a nanofiber membrane, TTVB@NM, was successfully fabricated through the electrospinning technique. The TTVB molecule features a characteristic D-π-A structure, where the vinylthiophene group serves as a π-bridge connecting the electron donor triphenylamine fragment and the electron acceptor 3-ethylbenzo[d]thiazol-3-ium group. This donor-acceptor molecular configuration not only effectively reduces the S_1–T_1 energy gap but also enhances intersystem crossing (ISC) through the introduction of heteroatoms S and N, thereby endowing TTVB with excellent photocatalytic activity for generating ROS.

Experimental observations using DCFH as a ROS probe revealed that TTVB generates substantial ROS under white light irradiation, significantly enhancing DCF fluorescence (Figure 5.19). Furthermore, singlet oxygen ($^1O_2$), superoxide anion ($O_2^{\cdot -}$), and hydroxyl radical (·OH) were detected using indicators 9,10-anthracenediyl-bis(methylene) dimalonic acid (ABDA), dihydrorhodamine 123 (DHR 123), and hydroxyphenyl fluorescein (HPF), respectively. Results demonstrated that while TTVB produced less $^1O_2$ under sunlight than commercial photosensitizer RB, it generated significantly more $O_2^{\cdot -}$ and ·OH compared to RB. Consequently, the fabricated TTVB@NM nanofibers exhibit remarkable photosensitization properties, demonstrating potential for antimicrobial applications [25].

The integration of photosensitizer protoporphyrin IX (PpIX) with cellulose diacetate (CDA) enables the fabrication of an antibacterial fiber via electrospinning (Figure 5.20a). Using TEMPO as a spin-trapping agent for $^1O_2$, distinct characteristic triplet electron paramagnetic resonance (EPR) signals were observed under white-light irradiation (Figure 5.20b). To further validate the Type II mechanism of $^1O_2$ generation during photodynamic processes in PpIX/CDA microfibers, the addition of $^1O_2$ quencher 1,4-diazabicyclo[2.2.2]octane (DABCO) resulted in significant signal attenuation. This outcome provides compelling evidence for the pivotal photosensitizing role of PpIX within the fibrous matrix [26].

5.2.4 Photocatalytic Effect

Under light illumination, semiconductors generate electron-hole pairs (excitons), which can diffuse to the semiconductor surface and initiate redox reactions there. This process constitutes the fundamental mechanism of photocatalytic reactions. To enhance photocatalytic performance, photocatalytic materials can be physically incorporated into polymer solutions to create nanofibers with high catalytic efficiency. These nanofibers not only significantly increase the material's specific surface area but also improve contact efficiency with reactants, thereby substantially boosting photocatalytic performance.

ZnO and TiO_2 are the most commonly used photocatalytic materials, widely applied in environmental purification and energy conversion due to their excellent photocatalytic performance and stability. ZnO-based nanofibers can be fabricated through the electrospinning technique. The procedure entails the preparation of a homogeneous solution by combining polyvinyl alcohol (PVA) and konjac glucomannan, followed by the incorporation of ZnO nanoparticles. Citric acid is

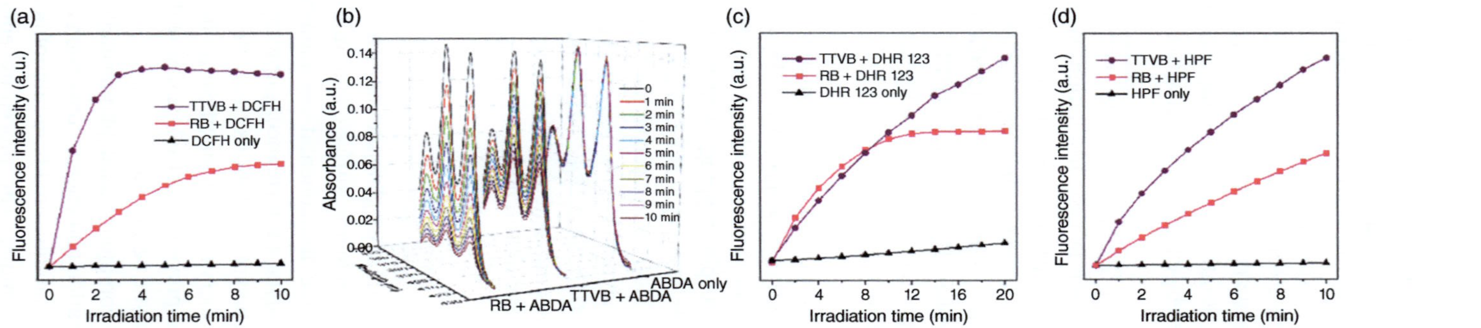

Figure 5.19 ROS generation by TTVB and RB in aqueous solution under light irradiation, evaluated using: (a) DCFH as a probe for ROS; (b) ABDA as the probe for $^1O_2$; (c) DHR 123 as the probe for $O_2^{\cdot -}$; (d) HPF as the probe for ·OH. *Source:* Li et al. [25] / with permission of Elsevier.

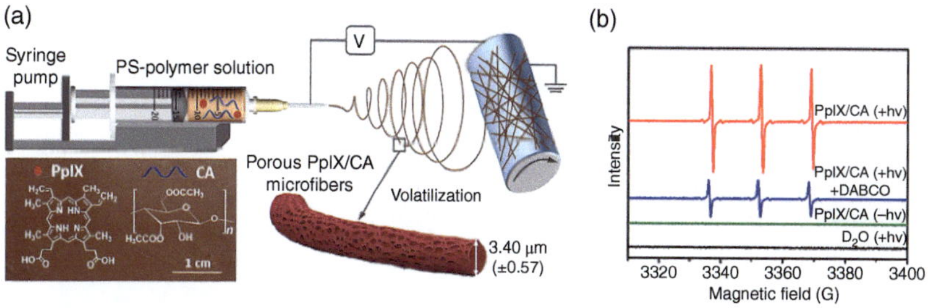

Figure 5.20 (a) The diagram of electrospinning used to prepare PpIX/CA fibers. (b) EPR spectra of TEMPO measured with and without PpIX/CA or DABCO, both under irradiation and in the dark. *Source:* Wang et al. [26] / with permission of Elsevier.

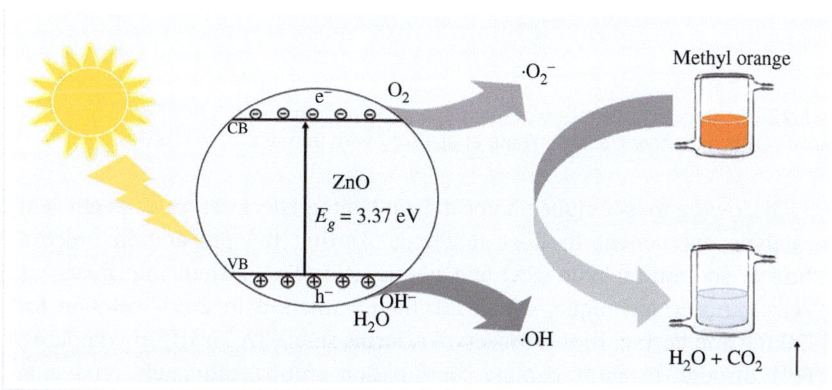

Figure 5.21 The schematic illustration of the photocatalytic effect of ZnO nanoparticles. *Source:* Lv et al. [27] / with permission of American Chemical Society.

subsequently introduced as a cross-linking agent. The resulting mixture is then subjected to electrospinning to fabricate nanofibers (Figure 5.21). In the prepared nanofibers, ZnO exhibits a bandgap of 3.37 eV. When exposed to light, ZnO absorbs photons, which results in e^- being excited from the valence band to the conduction band. This process also generates h^+ within the valence band. Specifically, conduction band e^- interact with O_2 adsorbed on the ZnO surface, leading to the formation of $\cdot O_2^-$. These $\cdot O_2^-$ can then react with water molecules to produce H_2O_2 and hydroxide ions, OH^-. The h^+ in the valence band directly interacts with H_2O or OH^-, generating ·OH. The generated ·OH and h^+ possess strong oxidizing properties, enabling direct oxidation of dye molecules into smaller molecules or mineralization into harmless substances, thereby achieving dye degradation. The nanofiber membrane prepared by electrospinning technology not only boasts a high specific surface area and well-structured pore morphology but also effectively disperses ZnO nanoparticles to enhance photocatalytic efficiency. Furthermore, cross-linking agent citric acid significantly improves the mechanical properties and stability of the fiber membranes, making it more promising for practical applications [27].

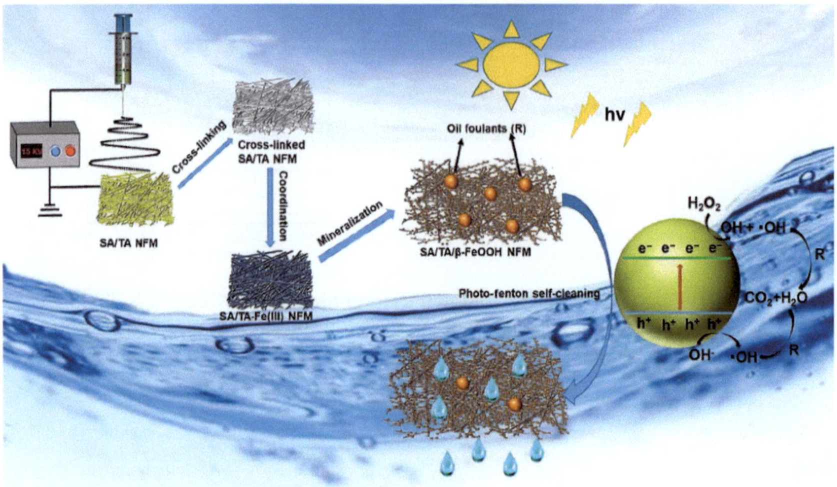

Figure 5.22 Schematic illustration of the preparation and photocatalytic effect of SA/TA/β-FeOOH nanofibers. *Source:* Wang et al. [28] / with permission of Elsevier.

β-FeOOH, owing to its unique photocatalytic-Fenton effect, serves as an efficient photocatalytic component in fiber materials. During the preparation process, a mixture of sodium alginate (SA) and tannic acid (TA) is spun into fibers via the electrospinning technique, which are then immersed in $CaCl_2$ solution for cross-linking and curing. In this process, TA forms stable TA-Fe(III) tri-complexes with Fe^{3+} through its multi-dentate coordination groups, ultimately producing SA/TA/β-FeOOH composite nanofibers via mineralization (Figure 5.22). This design not only enhances the mechanical qualities and stability of the material but also significantly improves its photocatalytic activity.

The photocatalytic-Fenton reactions on this membrane are as follows:

$$\beta - \mathrm{FeOOH} + \mathrm{hv} \rightarrow \mathrm{e}^- + \mathrm{h}^+$$

$$\mathrm{e}^- + \mathrm{H_2O_2} \rightarrow \cdot\mathrm{OH} + \mathrm{OH}^-$$

$$\mathrm{h}^+ + \mathrm{OH}^- \rightarrow \cdot\mathrm{OH}$$

$$\cdot\mathrm{OH} + \mathrm{R} \rightarrow \mathrm{CO_2} + \mathrm{H_2O}$$

Firstly, β-FeOOH is excited under visible light irradiation to generate e^- and h^+. The electrons react with H_2O_2 to produce ·OH and OH^-. The holes directly interact with OH^-, further generating additional ·OH. These high-concentration ·OH can completely oxidize organic pollutants (R) into harmless CO_2 and H_2O. This process endows the nanofiber with excellent self-cleaning properties [28].

Heterojunctions can also be constructed to further enhance photocatalytic activity. A mixed solution of cellulose acetate and PVP was electrospun and subsequently deacetylated to obtain CeP nanofibers. The MIL-100(Fe) structure was in situ synthesized on these fibers, with β-FeOOH immobilized within the MIL-100(Fe) framework, ultimately yielding a core-shell structured fiber composite

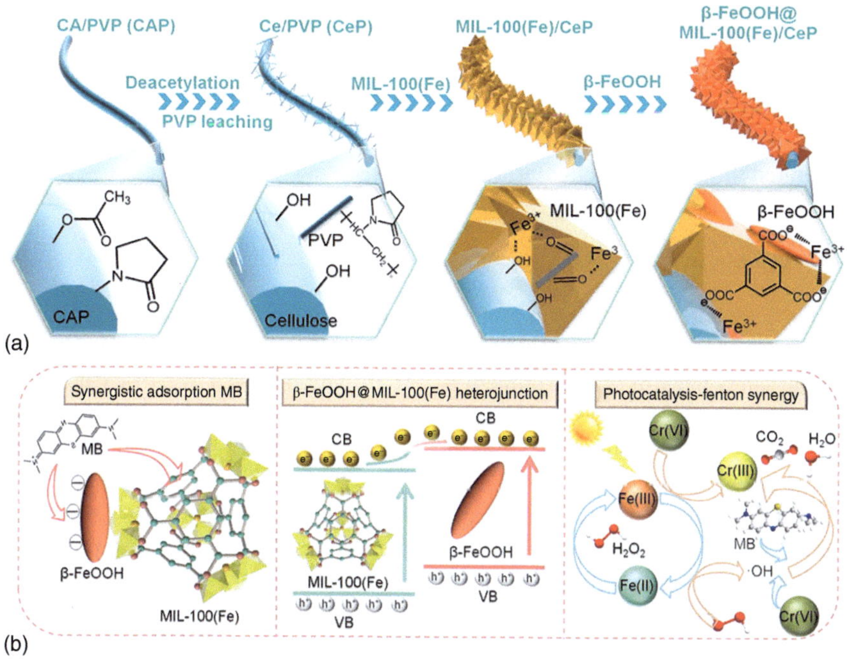

Figure 5.23 (a) Schematic illustration of the core-shell β-FeOOH@MIL-100(Fe)/CeP nanofiber morphologies. (b) The photocatalytic mechanism of β-FeOOH@MIL-100(Fe)/CeP nanofiber. *Source:* Lu et al. [29] / with permission of Elsevier.

β-FeOOH@MIL-100(Fe)/CeP (Figure 5.23a). The reaction processes on this membrane are as follows:

$$Cr(VI) + Fe^{2+} \rightarrow Cr(III) + Fe^{3+}$$

$$H_2O_2 + Fe^{3+} \rightarrow HO_2^{\cdot} + H^+ + O_2^{\cdot-} + Fe^{2+}$$

$$H_2O_2 + O_2 + e^- \rightarrow OH^- + OH^{\cdot} + O_2^{\cdot-}$$

$$R + Cr(VI) + OH^{\cdot} + HO_2^{\cdot} + O_2^{\cdot-} \rightarrow H_2O + CO_2 + Cr(III)$$

The reaction mechanism involves a synergistic cycle: Initially, Fe^{2+} reduces highly toxic Cr(VI) to less toxic Cr(III), while being oxidized to Fe^{3+}. Subsequently, Fe^{3+} undergoes a Fenton reaction with H_2O_2, regenerating Fe^{2+} and producing hydroperoxyl radicals ($HO_2^{\cdot}$) and $O_2^{\cdot-}$ to sustain the iron catalytic cycle. Concurrently, photogenerated electrons react with O_2 and H_2O_2, generating hydroxyl radicals (·OH), hydroxide ions (OH^-), and additional $O_2^{\cdot-}$, establishing a self-perpetuating oxidative chain. Ultimately, ·OH, $HO_2^{\cdot}$, and $O_2^{\cdot-}$ radicals degrade organic chemicals (R) to CO_2 and H_2O, while Cr(III) forms stable precipitates via hydrolysis or complexation. This synergistic system demonstrates exceptional efficacy for treating chromium-laden organic wastewater.

In this system, the negatively charged β-FeOOH facilitates dye adsorption. Furthermore, the β-FeOOH@MIL-100(Fe) heterojunction structure significantly enhances visible light absorption and utilization efficiency while promoting the effective separation of photogenerated electron-hole pairs. Meanwhile, iron ions in β-FeOOH@MIL-100(Fe) generate a photo-Fenton synergistic effect upon visible light exposure, accelerating the degradation of the pollutant (Figure 5.23b). Consequently, this fiber composite exhibits excellent self-cleaning properties and can be regenerated through visible light irradiation [29].

5.2.5 Photothermal Conversion Effect

Nanoparticles with photothermal properties can be securely embedded within fiber structures, enabling stable heat generation under illumination while preventing potential safety hazards or environmental contamination caused by particle detachment or ion release. This design not only enhances material stability and safety but also provides a more reliable solution for practical applications of photothermal conversion technology.

Cu_2S nanoparticles are extensively utilized in composites due to their superior photothermal conversion properties. In a study of photothermal fibers, polymer PDLLA and PCL were blended and dissolved, followed by the addition of a Cu_2S solution to prepare the spinning solution. Then the CS-PLA/PCL composite fiber was successfully fabricated via electrospinning. Experimental results demonstrate that the 30CS-PLA/PCL membrane containing Cu_2S exhibits significant photothermal effects under 808 nm near-infrared (NIR) irradiation: In dry conditions, the surface temperature rapidly rises to approximately 61 °C within 5 min; in humid conditions, the temperature reaches 52 °C within 10 min. In contrast, the 0CS-PLA/PCL membrane without Cu_2S shows no noticeable heating effect under identical conditions (Figure 5.24). Additionally, the 30CS-PLA/PCL membrane maintains excellent structural stability during heating, with no significant shape

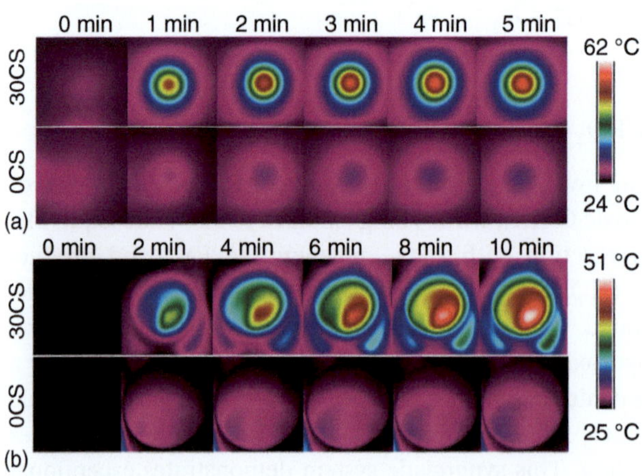

Figure 5.24 Photothermal performance of 30CS-PLA/PCL and 0CS-PLA/PCL fibers in the (a) dry and (b) humid environments under continuous light irradiation. *Source:* Reproduced with permission from Wang et al. [30] / American Chemical Society.

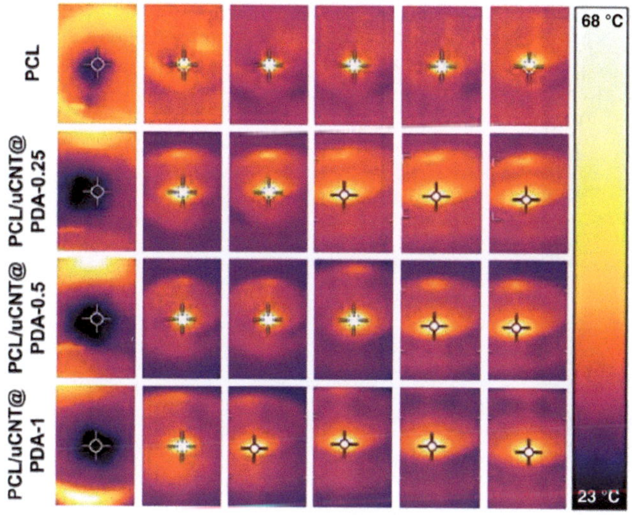

Figure 5.25 Infrared thermal images of PCL/uCNT@PDA fibers with varying PDA content under irradiation. *Source:* Reproduced with permission from Patil et al. [31] / Elsevier.

deformation. These findings indicate that the fiber membrane demonstrates controllable and outstanding photothermal behavior under NIR irradiation, demonstrating potential for photothermal tumor therapy and wound healing applications [30].

In another study, unzipped multiwalled carbon nanotubes (uCNT) were prepared, and polydopamine (PDA) was in situ polymerized on their surfaces to form the uCNT@PDA composite. PCL was then mixed with the uCNT@PDA solution to create a spinning solution, which was subsequently processed into a film through spinning techniques. To evaluate its photothermal conversion performance, NIR light was irradiated onto the PCL/uCNT@PDA film (Figure 5.25). Experimental results showed that under continuous 0.5 W/cm^2 NIR light exposure for 5 min, the temperature of PCL/uCNT@PDA-0.25 nanofibers rapidly increased from 24 to 52 °C. Meanwhile, under identical conditions, PCL/uCNT@PDA-1 reached a temperature of 67.4 °C. These findings indicate that the prepared nanofibers exhibit significant photothermal effects, with their performance being dependent on the loading amount of uCNT@PDA [31].

5.2.6 Light-guiding Properties

Different from other forms of photofunctional composites, photofunctional polymer fiber materials with excellent optical properties, mechanical properties, and biocompatibility have become the key breakthrough to promote the development of optical waveguide technology to miniaturization, intelligence, and multi-scenario applications.

By encapsulating silica-coated upconversion nanoparticles (UCNPs) with poly(ethylene glycol) diacrylate (PEGDA) and photoinitiator Irgacure, the obtained

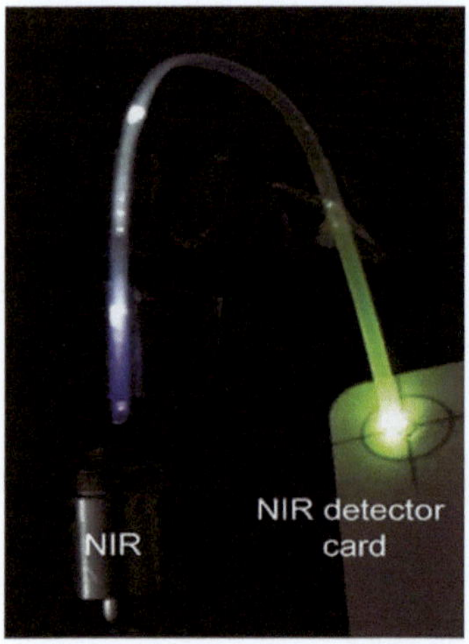

Figure 5.26 UCNPs based fiber for near-infrared light conduction. *Source:* Reprinted with permission from Teh et al. [32].

composite was then loaded into fluorinated ethylene propylene (FEP) tubes. UV-initiated cross-linking further yielded an implantable optical device, which functions as a light source capable of converting NIR radiation to visible light and transmitting it to deep-seated tissues (Figure 5.26). PEGDA's transparency and flexibility make it an ideal optical conductor, while the FEP coating effectively prevents UCNP leakage, ensuring stability and biocompatibility. The fiber-optic–pigment combination enables wireless PDT when activated by external NIR light, offering a novel, noninvasive approach for treating deep-seated tumors [32].

Silk protein, renowned for its excellent biocompatibility, can be utilized to fabricate optical fibers with superior waveguide performance. The manufacturing process begins with casting and drying silk protein solutions to form a film, which is then encapsulated within a silk protein gel to create the fiber material. As the cladding layer of the optical waveguide, the hydrogel has a refractive index of approximately 1.34, lower than that of the core silk protein film (1.54). This creates a step-index waveguide structure, enabling effective guidance of light. As shown in Figure 5.27, when the fiber is inserted into biological tissue, light can be observed simultaneously in both the cladding and core before bending. However, after bending, light becomes confined exclusively within the silk protein film core, demonstrating successful light confinement. Experimental measurements reveal

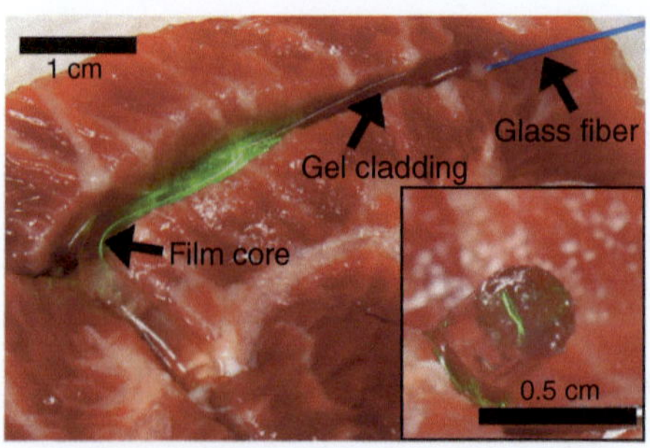

Figure 5.27 Green light from a multimode fiber confined within the silk film core of the waveguide. *Source:* Applegate et al. [33] / Optica Publishing Group.

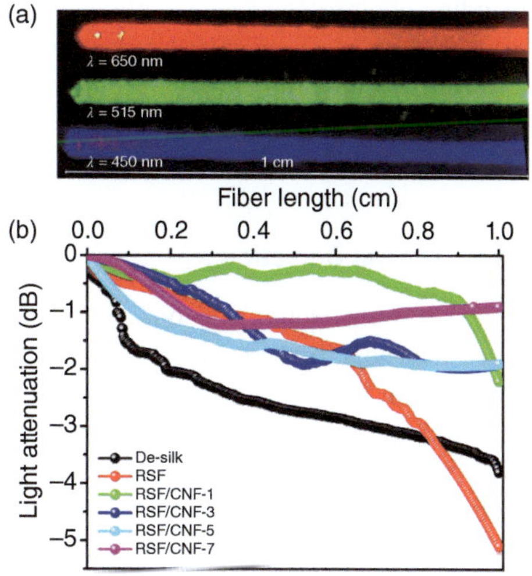

Figure 5.28 (a) Optical images of RSF/CNF-7 hybrid waveguides demonstrating guided laser light at wavelengths of 650, 515, and 450 nm. (b) Light attenuation curves of RSF/CNF hybrid fibers with varying lengths, compared to degummed silk fiber and pure RSF fiber. *Source:* Lu et al. [34] / with permission of Elsevier.

an optical loss of 2.0 dB/cm, comparable to other polymer waveguides, primarily due to surface scattering at the film's edges [33].

The combination of silk protein and cellulose nanofibers (CNF) enables the preparation of regenerated silk protein (RSF)/CNF fibers with exceptional optical waveguide properties. The process begins with preparing solutions of RSF and CNF, which are then uniformly blended to form a spinning solution. Composite fibers are subsequently fabricated using bio-inspired microfluidic spinning technology. The incorporation of CNF facilitates hydrogen bond formation with RSF molecules, not only significantly enhancing the dispersion stability of CNF within the protein matrix but also inducing orderly axial alignment of CNF fibers under high-speed shear flow. This structural regulation endows RSF/CNF composites with superior molecular orientation and high crystallinity. Optical performance assessments (Figure 5.28a) reveal efficient laser transmission at wavelengths of 650, 515, and 450 nm. Specifically, the RSF/CNF-7 composite, incorporating 7 wt% CNF, demonstrates exceptionally low optical loss, measured at 1.0 dB/cm. This value is significantly lower than that observed for natural degummed silk fiber (De-silk fiber) (28 dB/cm) and RSF fiber (11 dB/cm), as depicted in Figure 5.28b. These superior waveguide characteristics originate from both the highly ordered internal microstructure of the fibers and the synergistic interaction between CNF and RSF molecules [34].

5.3 Bioapplications of Photofunctional Fiber Composites

Optical polymer fiber composites efficiently integrate the functional attributes of photoactive materials with the structural merits of fibrous materials. This category of materials not only preserves the flexibility, lightweight nature, and high mechanical strength typical of conventional fibers but also attains novel functionalities, including photoresponsiveness, photocatalytic activity, and optical sensing, via the incorporation of photoactive components. By merging structural integrity with multifunctionality, optical polymer composite fibers exhibit substantial potential for applications in advanced domains such as sensor technology, biomedical engineering, and optogenetics.

5.3.1 Optical Detection

Optical sensing technology has demonstrated remarkable advantages, such as noninvasiveness, high sensitivity, and real-time monitoring, in glucose detection, providing an efficient and secure method for blood glucose management. This methodology involves the capillary-templated polymerization of a p(AM-*co*-PEGDA-*co*-3-APBA) fiber core, wherein acrylamide (AM), poly(ethylene glycol) diacrylate (PEGDA), and 3-(acrylamido)phenylboronic acid (3-APBA) are covalently cross-linked to form the copolymeric structure. Following core formation, the fiber is further encapsulated with an alginate outer layer, which is subsequently cross-linked via calcium ions (Figure 5.29a). In the sensing process, glucose molecules diffuse into the fiber core and bind specifically to 3-APBA's cis-diol groups, resulting in an elevation of the system's Donnan osmotic pressure. As a consequence, the fiber expands, causing a change in its refractive index and a reduction in light loss within

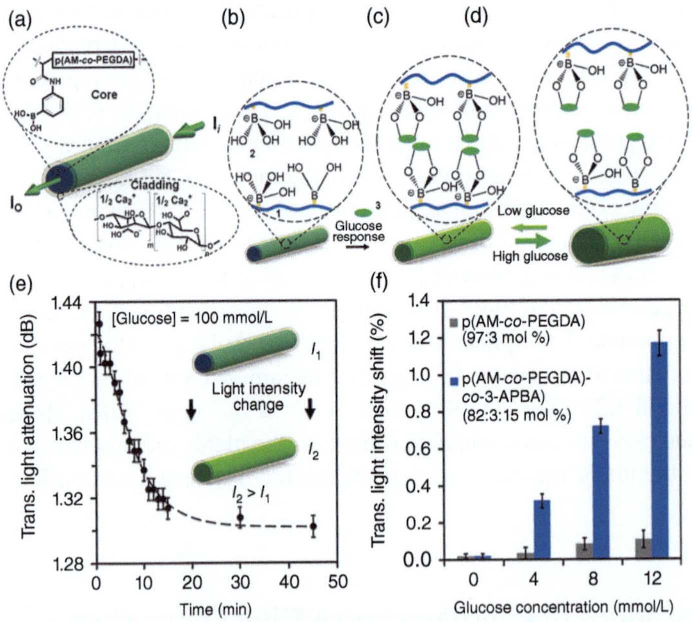

Figure 5.29 (a–d) Schematic illustration of the structure and glucose detection mechanism of p(AM-*co*-PEGDA-*co*-3-APBA) fiber coated with alginate. (e) Transmitted light attenuation through the fiber composites over time of glucose interaction. (f) Transmitted light intensity through the p(AM-*co*-PEGDA-*co*-3-APBA) and p(AM-*co*-PEGDA) fibers measured at various glucose concentrations. *Source:* Yetisen et al. [35] / John Wiley & Sons / CC BY 4.0.

the fiber (Figure 5.29b–d). Experimental findings suggest that an increase in glucose concentration results in a change in the transmitted light intensity of the hydrogel fiber, with a detection sensitivity of 1.2 mmol/L (Figure 5.29e,f).

In comparison to traditional continuous glucose monitoring devices, the hydrogel optical fiber sensor presents several remarkable advantages. Firstly, its flexibility and biocompatibility render it highly suitable for subcutaneous implantation, thus facilitating real-time monitoring of glucose levels in interstitial fluid. Secondly, the sensor showcases outstanding reusability and an extended continuous sensing duration, which substantially diminishes the requirement for frequent device replacement. Moreover, the sensor manifests a reliable linear response within the physiological glucose concentration range and demonstrates robust resistance to potential interfering substances, including fructose, lactate, and metal ions [35].

Investigating temperature fluctuations at the single-cell level is crucial for advancing our understanding of cellular physiology and for enabling targeted studies of disease mechanisms. In this research, a photofunctional fiber was developed by immobilizing (UCNPs, specifically $NaYF_4$:Yb^{3+}/Er^{3+}) onto the surface of natural spider silk. Within these UCNPs, the $^2H_{11/2} \rightarrow {}^4I_{15/2}$ electronic transition is responsible for emission at 525 nm, while the $^4S_{3/2} \rightarrow {}^4I_{15/2}$ transition yields emission at 548 nm. The energy difference between the thermally coupled $^2H_{11/2}$ and $^4S_{3/2}$ states is minimal, allowing these levels to maintain thermal equilibrium. Upon increasing temperature, a larger population of electrons is thermally promoted from the $^4S_{3/2}$ state to the $^2H_{11/2}$ state, resulting in an enhanced emission intensity at 525 nm and a concomitant decrease at 548 nm. Consequently, this system exhibits a ratiometric fluorescence response that is sensitive to temperature variations.

In the experiment, 980 nm light was employed to excite the fiber, and the up-conversion fluorescence spectra from diverse regions of C127 cells were documented. Figure 5.30a presents the spectral profiles obtained from the membrane region adjacent to the cell nucleus (R_1) and from an extracellular location proximal to the cell (R_2). Based on the fluorescence intensity ratios computed at R_1 and R_2 (Figure 5.30b), the corresponding values are $FIR_1 = 0.291$ and $FIR_2 = 0.246$, with the corresponding temperatures being $T_1 = 313$ K and $T_2 = 298$ K, respectively. These results imply that the temperature of the membrane near the cell nucleus is elevated compared to that in the extracellular location. To conduct a more in-depth exploration of the spatial temperature distribution, the FIR values around the C127 cells were systematically mapped (Figure 5.30c). The results demonstrate that the membrane temperature rises as the measurement location nears the nucleus. This observation offers valuable perspectives on intracellular temperature fluctuations and their potential impacts on cellular physiological processes [18].

A single-filament microfiber exhibiting both photoconductive and luminescent properties can be obtained by physically isolating natural lotus root filaments. This microfiber serves as an optical fiber for biosensing applications. In this study, *Helicobacter pylori*, a bacterium linked to gastric diseases, was selected as the target analyte. A 440 nm light source with a power of 10 nW was coupled into the microfiber, and the transmitted optical signal was collected at the distal end (Figure 5.31a). During the experiment, the microfiber was immersed in a buffer

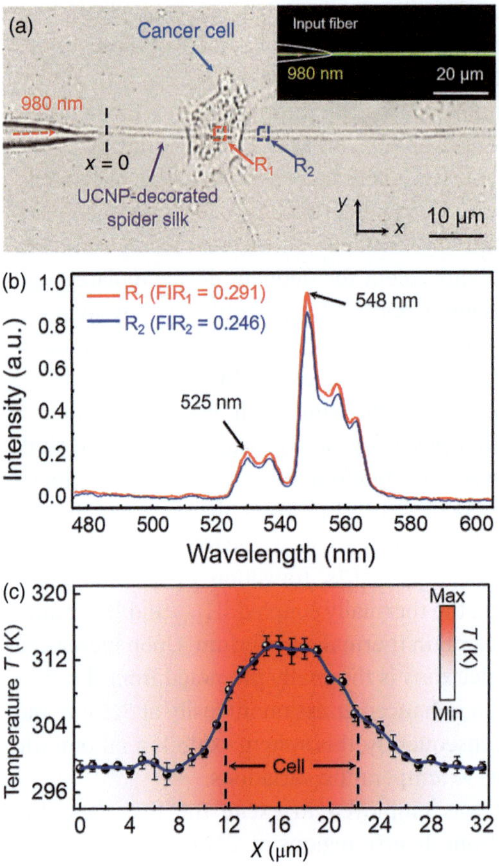

Figure 5.30 (a) Optical microscope image and (b) spectral intensity plots of UCNP-decorated spider silk fiber detecting temperatures at R_1 and R_2. (c) The fiber measures temperatures in the vicinity of the cell. *Source:* Gong et al. [18] / with permission of American Chemical Society.

solution containing *H. pylori*, followed by the addition of H_2O_2 to induce bacterial apoptosis (Figure 5.31b). After 25 min (Figure 5.31c), a gradual decrease in bacterial activity was observed. By 50 min (Figure 5.31d), the bacteria had undergone apoptosis and adhered to the microfiber surface, resulting in increased light scattering and a corresponding reduction in transmitted light intensity. Thus, the apoptotic state of *H. pylori* can be monitored by tracking changes in transmitted light intensity (Figure 5.31e).

Throughout the apoptosis process, a time-dependent decrease in output light intensity was recorded, indicating that apoptotic bacteria adhere to the microfiber and enhance light scattering. After 45 min, the light intensity stabilized, suggesting that nearly all *H. pylori* cells in the solution had entered apoptosis. In control experiments, bacterial samples without H_2O_2 maintained high activity for up to 50 min, with only minor fluctuations in light intensity. Additionally, the introduction of H_2O_2 alone, in the absence of bacteria, did not significantly affect transmitted light intensity, confirming that H_2O_2 does not interfere with optical transmission (Figure 5.31f). By measuring changes in transmitted light and leveraging the high environmental sensitivity of microfibers, the real-time monitoring of the bacterial

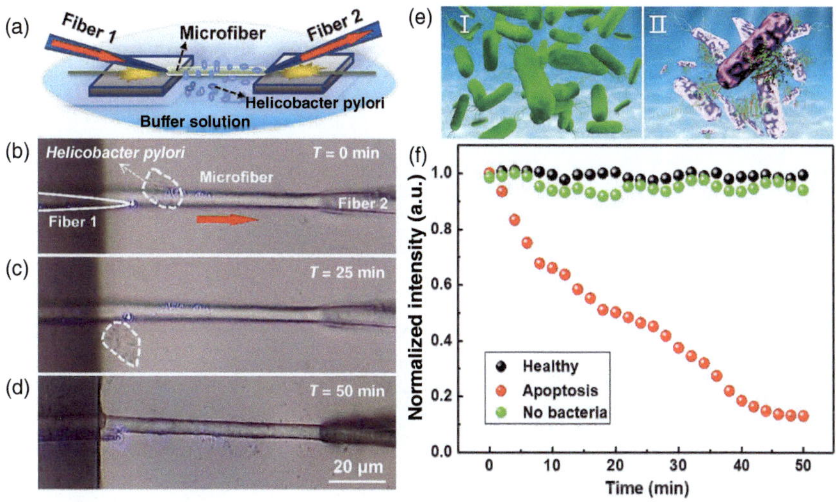

Figure 5.31 (a) Schematic diagram of the microfiber used for *Helicobacter pylori* detection. (b–d) Optical micrographs showing the interaction between microfiber with *H. pylori* at 0, 25, and 50 min, respectively. (e) Schematic diagrams of (I) healthy and (II) apoptotic *H. pylori*. (f) Light intensity of the microfiber in the prescence of apoptotic versus heathy *H. pylori*. *Source:* Yang et al. [36] / with permission of American Chemical Society.

apoptosis process has been successfully realized. This approach provides novel opportunities for the development of biosensors and holds significant potential for application, particularly in the fields of biomedicine and environmental monitoring [36].

5.3.2 Optical Therapy

Photofunctional fibers loaded with photosensitizers can overcome the limitation of inadequate light penetration depth in traditional PDT, thus serving as an ideal carrier for deep tissue treatment. Due to their unique design, these fibers ensure the complete activation of photosensitizers within deep tissue layers. This technology not only improves the precision and efficacy of therapeutic interventions but also minimizes damage to surrounding healthy tissues, suggesting extensive application potential. For example, in oncology treatments, photofunctional fibers can penetrate deeply into tumor masses, enabling the targeted destruction of cancer cells while preserving adjacent normal tissues. Moreover, this innovative approach can be applied to the management of various other conditions, including infectious and dermatological diseases, demonstrating its multidisciplinary applicability [37].

The PCL was blended with a gelatin solution, and then the PG fiber scaffolds were successfully fabricated via electrospinning. Subsequently, by means of the layer-by-layer self-assembly technique, negatively charged graphene oxide (GO) and positively charged g-C_3N_4 loaded with the photosensitizer Ce6 were alternately deposited on the surface of the scaffold. Finally, the photofunctional polymer fiber composite Ce6@pCN-GO-PG was synthesized. This fiber composite integrates the

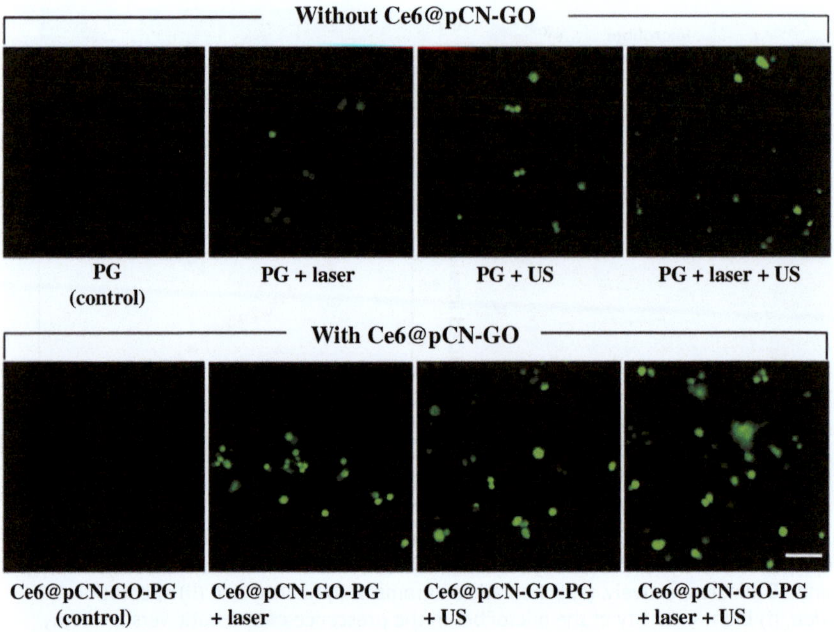

Figure 5.32 Fluorescence images of DCFH-DA in cells cultured on Ce6@pCN-GO-PG fibers under laser or US and combined laser + US irradiation conditions. *Source:* Reproduced with permission from Sun et al. [38] / American Chemical Society.

unique properties of the photosensitizer, graphene, and g-C_3N_4, thus acting as an efficient platform for PDT.

In the cellular experiment, 2′,7′-Dichlorofluorescein diacetate (DCFH-DA) was utilized as a fluorescent probe for ROS to assess the generation of ROS within SKBR3 cells. The findings revealed that when the Ce6@pCN-GO-PG scaffold was combined with either ultrasound (US) or laser irradiation, the cultured cells exhibited the highest intensity of green fluorescence, indicating that the generation of ROS in this system was the greatest (Figure 5.32). Comparative analysis suggested that the primary mechanism underlying US-enhanced cell death is its mechanical disruption effect. Specifically, US induces physical damage to the cell membrane via the formation and collapse of cavitation bubbles. This mechanical action enables the sensitizer to approach the cells more closely, thereby promoting extensive ROS generation and ultimately resulting in effective cell death [38].

The AIEgen photosensitizer can be effectively integrated into the electroactive polymer PVDF-HFP through an electrospinning approach, leading to the creation of the photofunctional polymer fiber composite TTVB@NM. These fibers demonstrate the capacity to produce diverse ROS, including singlet oxygen ($^1O_2$), superoxide radicals ($O_2^{\cdot -}$), and hydroxyl radicals (·OH), thus presenting excellent photosensitization properties. In the transmission of respiratory diseases, in addition to infectious droplets, infectious aerosols also constitute a significant transmission pathway. Due to their relatively small particle size (typically less than 10 μm), aerosols are capable of remaining airborne for prolonged durations and can be transported across considerable distances [39]. To assess the bacterial interception capacity of TTVB@NM, an aerosol incorporating *E. coli* was utilized as a model within the experimental framework. The findings indicated that after the nebulization of the *E. coli* aerosol, a substantial amount of bacteria was captured on the surface of the TTVB@NM fiber, while no bacterial presence was identified on the underlying medical mask (Figure 5.33a).

The effective pathogen interception ability of TTVB@NM was further verified via SEM images (Figure 5.33b). Additionally, TTVB@NM manifested a remarkable antibacterial effect subsequent to irradiation under sunlight for 5 or 10 min. The pristine fiber film merely exhibited partial inhibitory activity against *E. coli* and M13 bacteriophage under sunlight, with inhibition rates of 37% and 42%, respectively (Figure 5.33c). Conversely, TTVB@NM demonstrated prominent

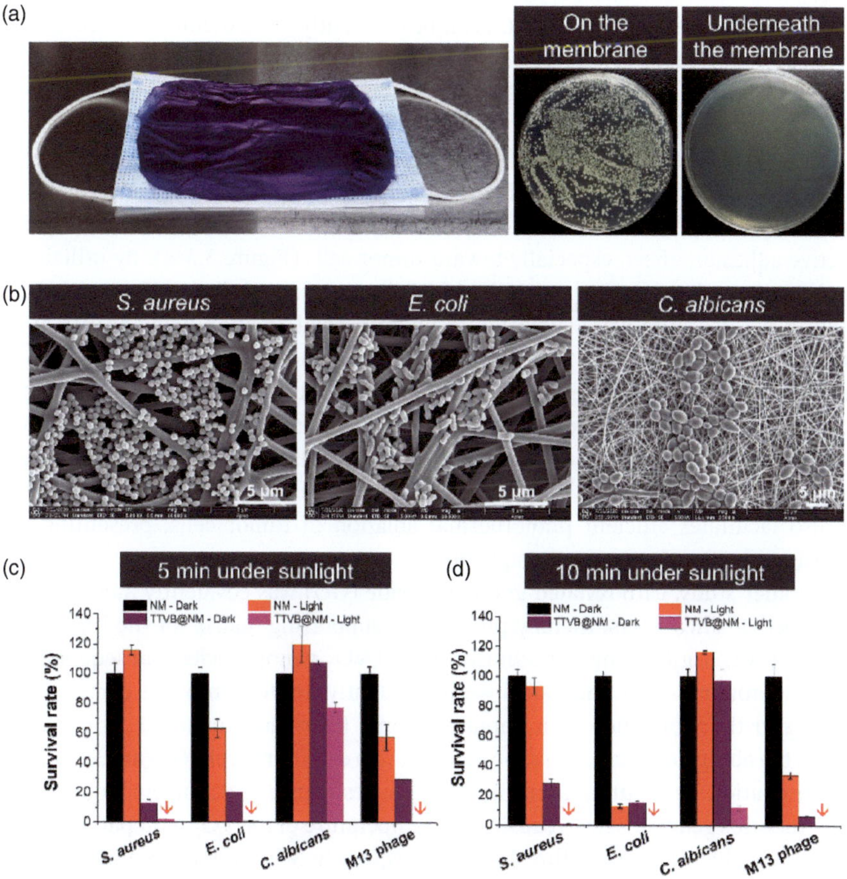

Figure 5.33 (a) TTVB@NM fibers coated on a medical mask after exposure to *E. coli* aerosols, with bacterial quantification on both fibers and mask using the plate-counting method. (b) SEM images showing various pathogens on the surface of TTVB@NM fibers. Antibacterial efficiency of TTVB@NM fibers after (c) 1 min and (d) 5 min of sunlight irradiation. *Source:* Li et al. [25] / with permission of Elsevier.

bactericidal activity in the absence of light, which was predominantly ascribed to TTVB's positive charge characteristics. More crucially, TTVB@NM was capable of inactivating over 99% of *Staphylococcus aureus*, *E. coli*, and M13 after merely 5 min of sunlight exposure. For *Candida albicans*, an irradiation duration of 10 min attained an inhibition efficiency of 90% (Figure 5.33d). These results imply that TTVB@NM not only efficiently captures microorganisms but also expeditiously eradicates pathogens in droplets and aerosols upon sunlight activation, suggesting its potential as a surface material for purification applications [25].

Apart from inducing cell death via the photodynamic effect, nanomaterials possessing photothermal conversion capabilities, such as gold nanoparticles and carbon-based materials, can also be integrated into fibrous matrices. When excited by NIR light, these materials generate localized hyperthermia, thus accomplishing photothermal cell ablation. This method not only improves therapeutic efficacy but also reduces damage to the surrounding normal tissues.

Photothermal graphene oxide (GO) can be covalently grafted onto the surface of electrospun poly(caprolactone) (PCL) fibers, leading to the successful fabrication of a PCL_{MF}-GO fibrous film. The surface of GO is replete with abundant functional groups, such as hydroxyl and carboxyl groups, which can adsorb proteins from serum to form a bioactive layer. This bioactive layer significantly enhances cell adhesion. Additionally, the high hydrophilicity and adsorption capacity of GO improve the interaction between cells and the material surface, manifesting a selective adhesion effect, especially toward tumor cells (Figure 5.34a). By utilizing the photothermal conversion capability of GO, breast cancer cells (MCF-7) adhered to the PCL_{MF}-GO film were subjected to photothermal ablation using an 810 nm laser for 200 s. As shown in Figure 5.34b,c, fluorescence microscopy verified the effective elimination of cells on the PCL_{MF}-GO film, with only cell debris remaining and large cell-free regions being clearly discernible. In contrast, individual cells persisted on the surface of the unmodified PCL_{MF} film. These results indicate that the PCL_{MF}-GO film not only demonstrates excellent cell adhesion properties but also facilitates efficient photothermal ablation of tumor cells, presenting a promising platform for tumor therapy [40].

In another study, carboxylated graphene oxide (GO) was covalently attached to the surface of silica fibers, thereby providing a supporting matrix for the immobilization of Au@Ag_2S triangular nanosheets and VO_2 nanoparticles. Subsequently, alpha-fetoprotein (AFP) antibodies were covalently grafted onto the composite, resulting in the fabrication of a photofunctional fiber-based material. This fiber exhibits the ability to specifically capture hepatoma carcinoma cells (Figure 5.35a,b). Upon excitation by a pump laser, the plasmon resonance effect of the Au@Ag_2S nanostructures generates heat, causing the temperature of the VO_2 nanoparticles to exceed their phase-transition threshold. Consequently, VO_2 undergoes a transition from an insulating state to a metallic state, thereby enhancing the plasmonic effect and amplifying the photothermal response. As depicted in Figure 5.35c, after irradiation with a 130 mW pump laser, the surface temperature of the microfiber increased rapidly: it reached 63 °C within 2 min and 77 °C within 6 min. This swift thermal elevation facilitates efficient ablation and destruction of cancer

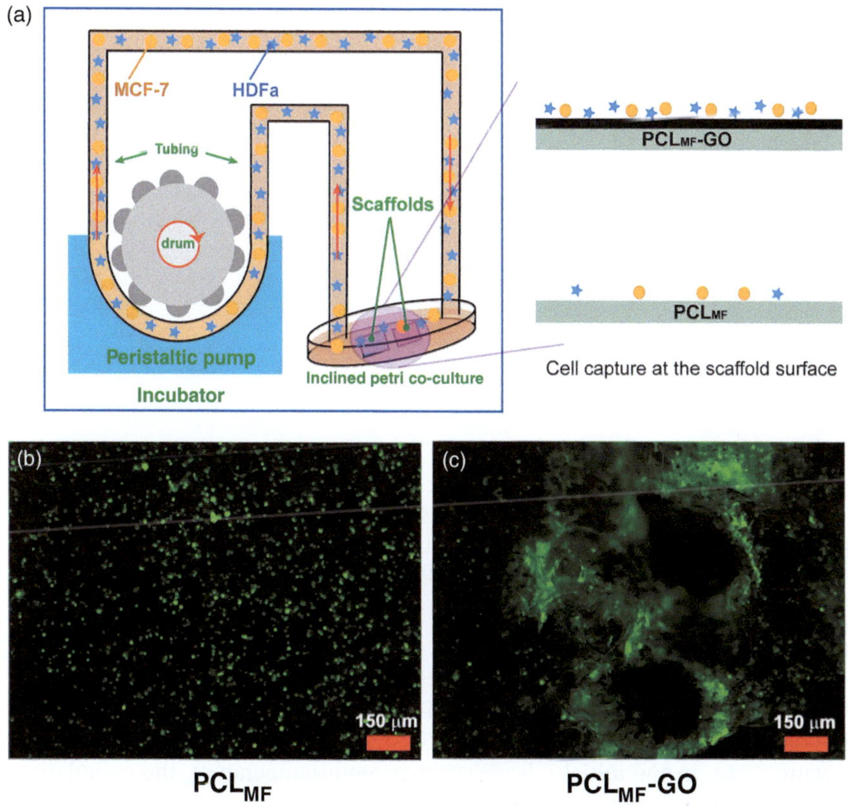

Figure 5.34 (a) Schematic illustration of the PCL_{MF}-GO subtract used to capture cells. Fluorescence images of MCF-7 cells cultured on (b) PCL_{MF} and (c) PCL_{MF}-GO substrates following laser irradiation. *Source:* Reproduced with permission from Mauro et al. [40] / Elsevier.

cells adhered to the fiber surface. Furthermore, the temperature gradient formed between the microfiber and its surrounding environment can attract nearby free cancer cells to the heated zone, where they are also destroyed. Therefore, this photofunctional composite fiber demonstrates significant potential for applications in plasmon-enhanced photothermal regulation, especially in the fields of optical fiber sensing and photothermal therapy [41].

The integration of the photothermal effect and light-controlled release of drugs can significantly improve the efficiency of cell ablation. This method utilizes the localized hyperthermia generated by the photothermal effect to directly damage cellular structures. Simultaneously, it integrates light-controlled release mechanisms to accurately deliver drugs or therapeutic agents under specific light stimuli, thus implementing a dual-targeting strategy against the targeted cells.

Doxorubicin (DOX) was chosen as a model drug and loaded onto mesoporous silica-coated gold nanorods (Au@SiO_2-DOX). The resultant complex was then combined with poly(*N*-isopropylacrylamide) (PNIPAM) and octaglycidyl

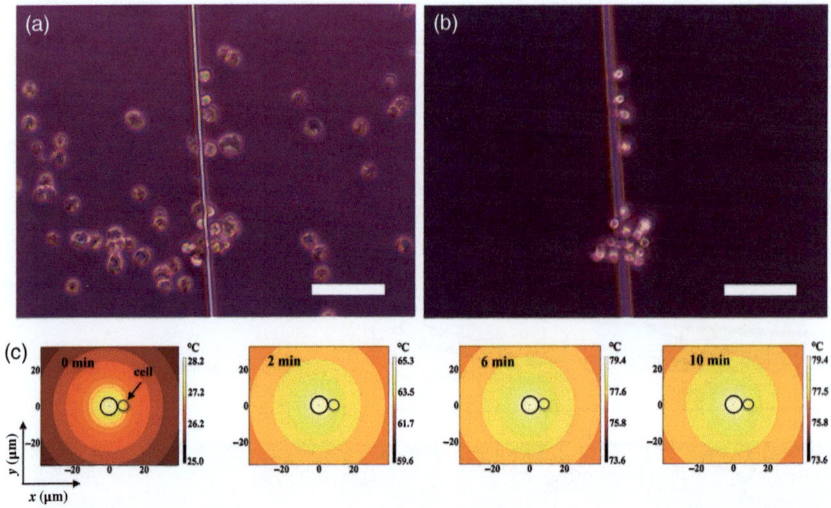

Figure 5.35 Optical images of photofunctional fiber composite (a) captured hepatoma carcinoma cells and (b) then washed with PBS. (c) Thermal distribution maps of the fiber composites under different irradiation duration. *Source:* Xiao et al. [41] / with permission of John Wiley & Sons.

polyhedral oligomeric silsesquioxane (OpePOSS), followed by electrospinning and cross-linking to prepare PNIPAM–POSS–Au@SiO_2 nanofibers. As shown in Figure 5.36a, in the initial release cycle at room temperature, the cumulative release rate of DOX reached 21.3%, corresponding to a kinetic release constant of 0.36%/min. Upon exposure to NIR irradiation, the solution temperature increased to 39.5 °C within 5 min (Figure 5.36b), resulting in an accelerated DOX release rate of 1.32%/min, which was approximately 13 times higher than the initial rate. In the subsequent second, third, and fourth cycles, the release constants at room temperature remained consistently around 0.1%/min.

The pulsed release mechanism of DOX is activated when the temperature surpasses the lower critical solution temperature of PNIPAM, leading to the contraction of the fibers. This phenomenon enhances drug release and promotes water convection. After 24 h of treatment with the released DOX, a significant reduction in the quantity of HeLa cells was noted, and almost no viable cells remained after 48 h. Conversely, the untreated cells continued to proliferate (Figure 5.36c,d). Therefore, the drug delivery system based on PNIPAM–POSS–Au@SiO_2 nanofibers not only enables controlled and efficient DOX release but also demonstrates promising potential for application in tumor therapy [42].

In addition, light-responsive systems encapsulating photofunctional components can be developed to enable on-demand activation of such photofunctions. For example, PCL nanofibers were prepared by means of the electrospinning technique as a matrix. Subsequently, silver nanoparticles (AgNPs) were synthesized through an in situ reduction method within a chitosan solution and encapsulated into nanogels (AgNPs-nanogels). Thereafter, the AgNPs-nanogels were stably

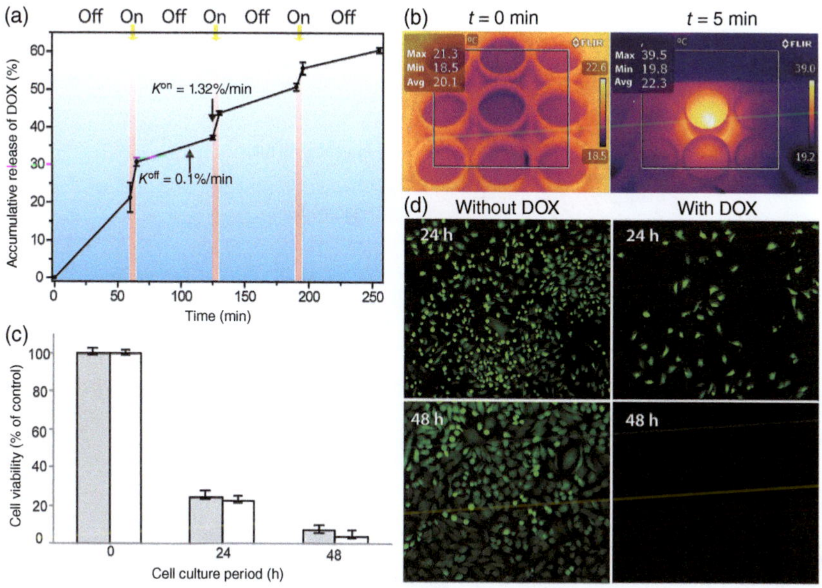

Figure 5.36 (a) Pulsed release of DOX from PNIPAM–POSS–Au@SiO$_2$ nanofibers under sequential NIR light on–off cycles. (b) Infrared images of PNIPAM-POSS-Au@SiO$_2$-DOX nanofibers upon NIR irradiation. (c) Cell viability of HeLa cells incubated with released DOX (gray) and ree DOX (white). (d) Live staining of HeLa cells with green signals after incubated with released DOX for 24 h and 48 h. *Source:* Li et al. [42] / with permission of John Wiley & Sons.

immobilized on the surface of the nanofibers via intermolecular hydrogen bonding, along with the electrostatic interaction between the negatively charged nanofibers and the positively charged AgNPs-nanogels.

Subsequently, the nanofiber mats were exposed to a 405 nm laser source for 150 s to trigger the release of AgNPs (Figure 5.37a). After irradiation, the fiber surface presented a smoother morphology, indicating that the laser excitation induced the surface plasmon resonance of AgNPs, thus disrupting the cross-linked structure of the nanogel and promoting the release of AgNPs (Figure 5.37b,c). As per the inhibition zone assay, the PCL nanofibers functionalized with AgNPs-loaded nanogels manifested superior antibacterial performance. This light-responsive release mechanism enables the controlled delivery of AgNPs and substantially enhances the antimicrobial efficacy of the material, presenting promising potential for advanced biomedical applications, such as in smart wound dressings [43].

5.3.3 Optical Display

Photofunctional polymer fibers can provide crucial support in surgical and diagnostic applications by facilitating high-resolution in vivo imaging through the transmission of optical signals. For example, high-strength fibers serving as surgical sutures can be synthesized through free radical polymerization,

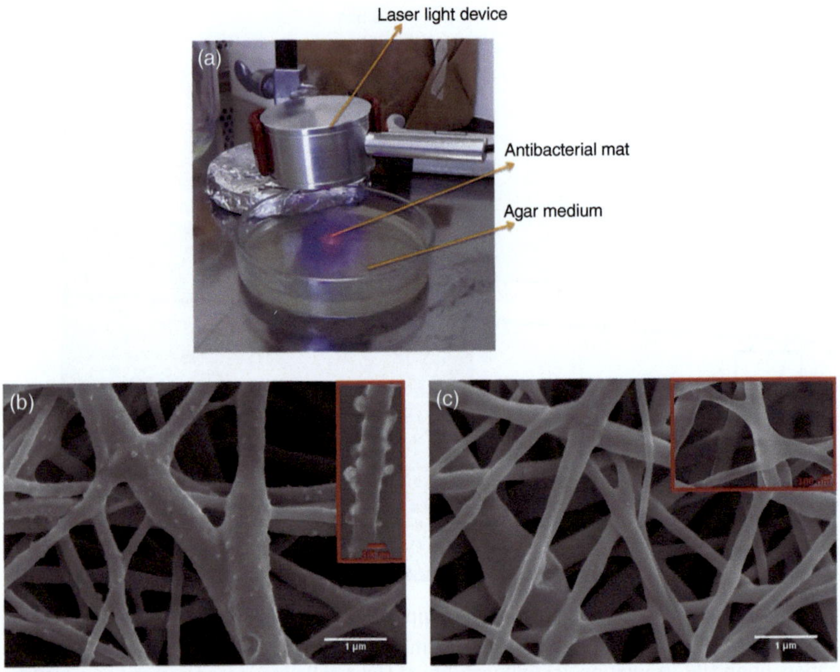

Figure 5.37 (a) The photograph of antibacterial mat composed of PCL nanofiber containing AgNPs-nanogels under laser irradiated. SEM images of the nanofibers (b) before and (c) after irradiation. *Source:* Reproduced with permission from Ballesteros et al. [43] / Elsevier.

with 2-acrylamido-2-methylpropanesulfonic acid (AMPS) being the primary monomer, together with other hydrogen-bond-containing monomers such as (*N*-(2-hydroxyethyl)crylamide, *N*-[di(hydroxymethyl)methyl]acrylamide, and (*N*-[tris (hydroxymethyl)methyl]acrylamide). The polymerization process also integrates the cross-linking agent *N,N′*-methylenebis(acrylamide) and Zr^{4+} to improve mechanical properties, resulting in $AT_3MBAA_{0.01}Zr^{4+}{}_{0.2}$@PLA (LSS) fibers.

Owing to the aggregation-induced emission effect of sulfonic acid groups within the fiber, the LSS fiber demonstrates intense fluorescence when irradiated with 365 nm UV light. The unique luminescent characteristics of the suture enable direct visualization and assessment of its integrity following surgical procedures, thereby providing clinicians with a straightforward and effective diagnostic tool. In murine models, the suture is typically difficult to discern under ambient light conditions. However, when exposed to UV light, the fluorescent properties of the fiber allow for precise localization and evaluation of the suture's continuity, independent of the animal's skin pigmentation (Figure 5.38). This characteristic expedites the timely assessment of the necessity for suture replacement, thus facilitating optimal wound healing. In general, this photofunctional polymer fiber improves the precision and visibility of surgical suturing and offers a valuable asset for postoperative wound surveillance [44].

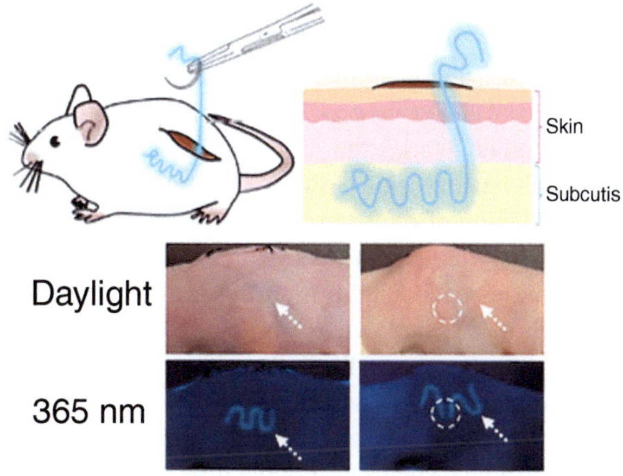

Figure 5.38 Photographs and fluorescence images of LSS sutures implanted subcutaneously in mice, visualized under ultraviolet irradiation. *Source:* Wen et al. [44] / with permission of John Wiley & Sons.

Photochromic fibers demonstrate promising application prospects in the domains of smart wearables and advanced textile technologies. For example, biobased bacterial cellulose (BC) is treated with TEMPO oxidation, converted into a spinning solution, and covalently bonded to spiropyran-based photochromic microcapsules (PM) via EDC/NHS chemistry. Finally, photochromic fibers are produced using wet spinning technology. Utilizing these fibers, elaborate flower patterns and the "JNU" logo with diverse hues and color intensities can be embroidered, effectively demonstrating their photochromic characteristics (Figure 5.39). Experimental findings indicate remarkable color alterations in the patterns before and after UV irradiation at ambient temperature. In the absence of UV irradiation, the PM embedded within the fibers persist in the colorless spirocyclic form. Nevertheless, upon exposure to UV light, the C—O bond in the spiropyran molecule experiences heterolytic cleavage, initiating an electronic rearrangement from the closed-ring to the open-ring structure. This structural change modifies the spatial configuration of the two ring systems linked by the spirocarbon atom, transitioning from a perpendicular to a

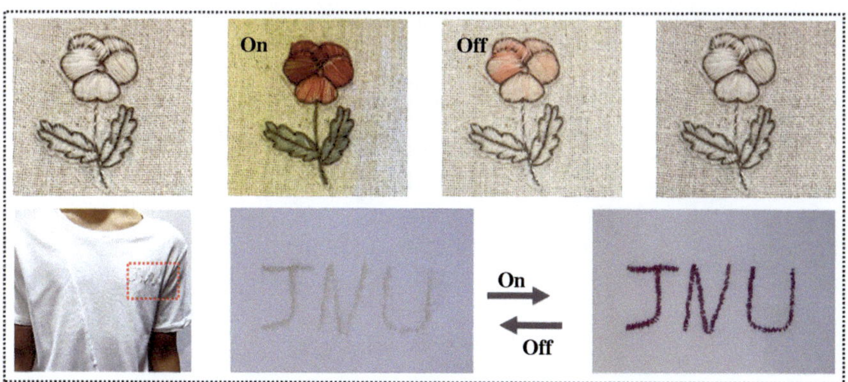

Figure 5.39 Fabric patterns formed by the photochromic fibers under UV irradiation. *Source:* Reproduced with permission from Xie et al. [45] / John Wiley & Sons.

coplanar arrangement. As a result, an extended conjugated system is established, which absorbs visible light and displays distinct color variations. This research not only emphasizes the promising applications of photochromic fibers in textile design but also provides novel perspectives for the development of smart materials [45].

In another work, the integration of inorganic fluorescent powders with polyvinylidene fluoride (PVDF) and polymethyl methacrylate (PMMA) facilitates the production of photochromic fibers featuring outstanding performance. These fibers display homogeneous light emission when subjected to UV excitation, and high luminescence stability is preserved along the transmission direction. Moreover, the brightness variation remains below 1% across diverse bending radii, suggesting remarkable mechanical flexibility and optical stability.

In the experiment, researchers precisely controlled the switching and brightness of light sources connected to each optical fiber segment, successfully achieving the desired overall display effect of the fiber pattern. This technology not only enriches the visual content inherent to the fabric but also presents novel prospects for the progress of smart textiles. For example, by integrating existing capacitance-sensing yarns with stable performance and photochromic fibers into daily-use textiles, the on/off states of three differently colored fibers on a wristband can be controlled in real-time via hand gestures (Figure 5.40). This interactive design enhances the user experience and paves new paths for the application of wearable devices. Overall, this type of photochromic fiber, which is based on inorganic phosphors, PVDF, and PMMA, combines outstanding luminescent characteristics, mechanical stability, and interactive capabilities. Consequently, it offers crucial technical support for the development of smart textiles and wearable devices [46].

5.3.4 Optogenetics

Optogenetics represents an innovative interdisciplinary approach that amalgamates genetic and optical techniques to precisely manipulate cellular activities, especially those of neurons, through the utilization of light. The cornerstone of this technology lies in the employment of light-sensitive proteins, such as microbial opsins. These proteins have the capacity to regulate ion channels in the cell membrane

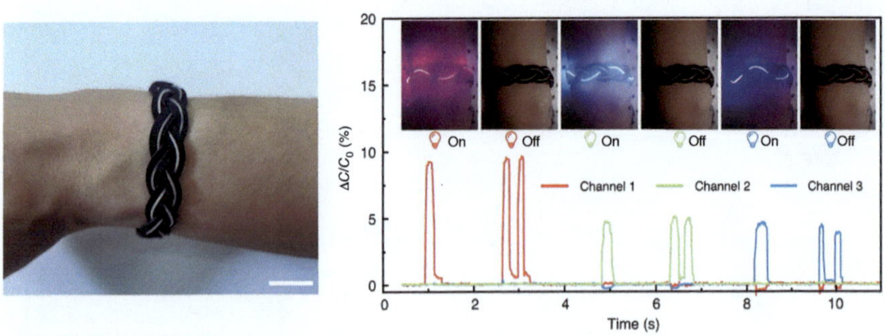

Figure 5.40 Capacitive-triggered multicolor emission of fibers in response to hand gestures. *Source:* Li et al. [46] / Springer Nature / CC BY 4.0.

when exposed to light of specific wavelengths, consequently triggering either neuronal excitation or inhibition [47]. This mechanism offers exceptional precision and controllability, thereby advancing the field of neuroscience research.

For example, researchers transfected the photosensitive protein ChR2 (channelrhodopsin-2), originally derived from green algae, into neurons using a viral vector, thereby successfully achieving neuronal photoactivation. This milestone signifies a major advancement in the field of optogenetics. However, due to the limited tissue penetration capability of light, the practical application of optogenetics remains heavily reliant on the synergistic interaction between optical-functional fibers and photosensitive proteins [48].

Photofunctional polymer fibers assume a crucial position in this technology as they allow for the integration of diverse multifunctional components, encompassing light sources, microelectrodes, microfluidic channels, and thermal sensors, into a unified system. Once implanted into a living organism, these fibers exhibit long-term stability and outstanding biocompatibility, thus promoting the coordinated execution of multiple functions, such as optogenetic stimulation, neural signal recording, and targeted drug delivery. This integrated multifunctional design functions as a potent platform for the advancement of neuroscience research and clinical applications, making a substantial contribution to the extensive development and implementation of optogenetic technologies [49].

A degradable optical fiber based on poly(L-lactic acid) (PLLA) can be manufactured via a simple hot stretching procedure. This kind of optical fiber shows considerable potential for utilization as a neural interface in the domains of optogenetics and neuroscience research. As depicted in Figure 5.41a–c, PLLA fibers were

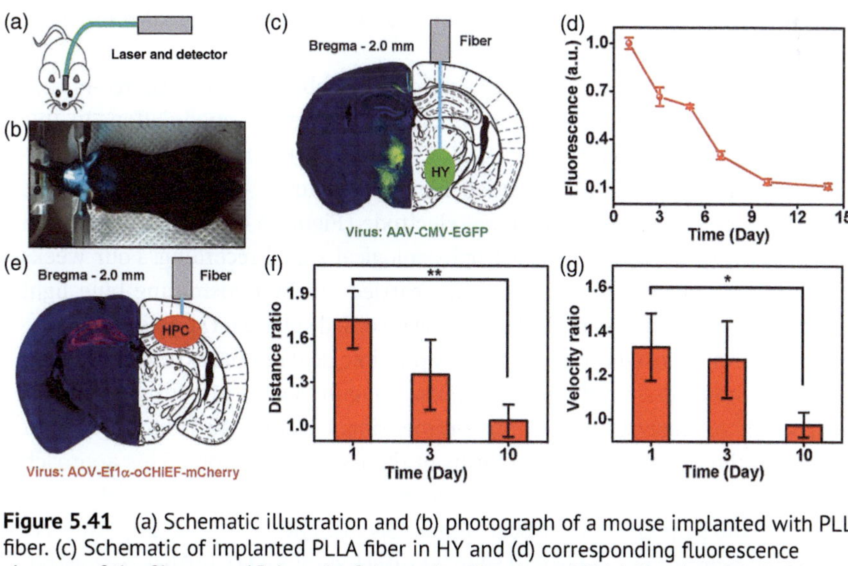

Figure 5.41 (a) Schematic illustration and (b) photograph of a mouse implanted with PLLA fiber. (c) Schematic of implanted PLLA fiber in HY and (d) corresponding fluorescence changes of the fiber over 15 days. (e) Schematic of implanted PLLA fiber in HPC. (f) Distance and (g) velocity ratios of movement of mice on days 1, 3, and 10 with the laser switched on and off. *Source:* Fu et al. [50] / with permission of John Wiley & Sons.

implanted into the mouse brain, and the AAV-CMV-EGFP virus was then injected into the hypothalamus (HY). The virus efficiently infected diverse cell types, resulting in strong expression of EGFP. After the complete expression of EGFP, in vivo fluorescence intensity was collected through the PLLA fibers. The outcomes indicated a progressive decrease in fluorescence intensity over time, implying that the PLLA fibers display favorable biodegradability and optical performance in vivo (Figure 5.41d).

Subsequently, the researchers carried out optogenetic experiments by implanting PLLA fibers into the mouse brain and inducing epileptic seizures via the overactivation of hippocampal (HPC) neurons, with results presented in Figure 5.41e–g. Specifically, the AAV-DIO-oCHiEF-mCherry virus, which selectively infects excitatory neurons, was bilaterally injected into the HPC of mice. After the full expression of the oCHiEF protein, PLLA fibers were implanted, and a 473 nm laser was employed to deliver light stimulation. The experimental results indicated that on the first day, light transmitted through the PLLA fiber effectively activated the HPC, leading to mild epileptic seizures. These seizures were characterized by enhanced locomotor activity, including an increased traveled distance and a higher movement speed. However, once the laser was deactivated, the mice promptly reverted to their baseline behavior. The same experimental protocol was replicated on the 3rd and 10th days with the same laser power. The results showed a significant decline in both the traveled distance and movement speed over time. By the 10th day, no discernible behavioral differences were detected regardless of laser activation [50].

This study not only confirmed the efficacy and biodegradability of PLLA fibers in optogenetics but also highlighted their promising application potential in neuroscience research, offering a valuable reference for the development of safer and more efficient neural interfaces.

The alginate-polyacrylamide hydrogel optical fiber fabricated through free-radical polymerization exhibits applicability for in vivo optogenetic modulation. In this research, adeno-associated virus (AAV)-CaMKIIα-ChR2-mCherry was administered via intra-hippocampal injection in mice (Figure 5.42a), followed by the implantation of a customized hydrogel electrode (Figure 5.42b) to facilitate both in vivo optical stimulation and electrophysiological signal recording. Four weeks after injection, optogenetic activation was carried out by transmitting blue light pulses through the hydrogel optical fiber. The recorded spike data suggested that the activated neurons evoked significant electrophysiological responses, thus validating the effective transmission of blue light pulses (Figure 5.42c). Moreover, the study incorporated an analysis of the three-dimensional spatial distribution of unit clusters and the average spike waveforms of the recorded neurons. This approach enabled the identification of two distinct neuronal populations that were selectively activated by the hydrogel optical fiber (Figure 5.42d). The inter-peak amplitudes of the sorted waveforms ranged from 150 to 200 μV (Figure 5.42e), further corroborating the ability of the hydrogel optical fiber to serve as a biocompatible waveguide for in vivo optogenetic stimulation.

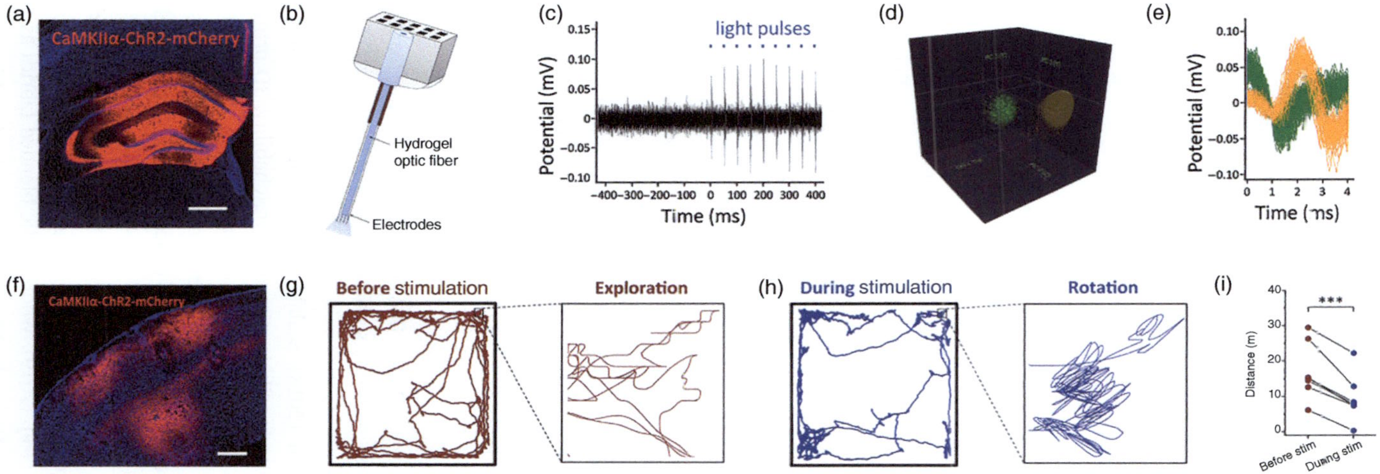

Figure 5.42 (a) Fluorescence image of expressed CaMKIIα-ChR2-mCherry in hippocampal neurons. (b) Schematic illustration of a hydrogel optical fiber-intergrated electrode array. (c) Electrophysiological signals obtained prior to and during light pulse stimulation. (d) Principal component analysis and (e) recorded potentials of two neuronal units shown in (c). (f) Fluorescence image of expressed CaMKIIα-ChR2-mCherry in M1. Representative movement traces before (g) and during (h) stimulation with the fiber. (i) Quantification of distance traveled before and after stimulation. *Source:* Wang et al. [51] / with permission of John Wiley & Sons.

To further verify the long-term functionality of the hydrogel optical fiber, the fiber was implanted into the primary motor cortex (M1) of mice (Figure 5.42f). In the absence of light stimulation, the mice were permitted to engage in free exploration for a duration of 10 min (Figure 5.42g). When exposed to light, the mice manifested an immediate elevation in right-turning and rotational behaviors, while their total locomotor activity was notably diminished (Figure 5.42h,i). These results indicate that blue light pulses transmitted through the hydrogel optical fiber can effectively activate M1 glutamatergic neurons, thus modulating behavioral responses [51].

References

1 Babazadeh-Mamaqani, M., Razzaghi, D., Roghani-Mamaqani, H. et al. (2024). Photo-responsive electrospun polymer nanofibers: mechanisms, properties, and applications. *Progress in Materials Science* 146: 101312.

2 Wang, X., Ding, B., Yu, J., and Wang, M. (2011). Engineering biomimetic superhydrophobic surfaces of electrospun nanomaterials. *Nano Today* 6 (5): 510–530.

3 Nie, X., Wu, S., Mensah, A. et al. (2020). Carbon quantum dots embedded electrospun nanofibers for efficient antibacterial photodynamic inactivation. *Materials Science & Engineering, C: Materials for Biological Applications* 108: 110377.

4 Liu, Z., Sun, D.D., Guo, P., and Leckie, J.O. (2007). An efficient bicomponent TiO_2/SnO_2 nanofiber photocatalyst fabricated by electrospinning with a side-by-side dual spinneret method. *Nano Letters* 7 (4): 1081–1085.

5 Johnson, R.W., Hultqvist, A., and Bent, S.F. (2014). A brief review of atomic layer deposition: from fundamentals to applications. *Materials Today* 17 (5): 236–246.

6 Asundi, A.S., Raiford, J.A., and Bent, S.F. (2019). Opportunities for atomic layer deposition in emerging energy technologies. *ACS Energy Letters* 4 (4): 908–925.

7 Zhang, X., Yue, G., Li, H. et al. (2024). Boosting Suzuki coupling reaction by multiscale porous Pd/carbon nanofibers photothermal catalyst. *Carbon* 230: 119658.

8 Wei, S., Qu, G., Luo, G. et al. (2018). Scalable and automated fabrication of conductive tough-hydrogel microfibers with ultrastretchability, 3D printability, and stress sensitivity. *ACS Applied Materials & Interfaces* 10 (13): 11204–11212.

9 Liu, C., Wang, Z., Wei, X. et al. (2021). 3D printed hydrogel/PCL core/shell fiber scaffolds with NIR-triggered drug release for cancer therapy and wound healing. *Acta Biomaterialia* 131: 314–325.

10 Song, J., Chen, S., Sun, L. et al. (2020). Mechanically and electronically robust transparent organohydrogel fibers. *Advanced Materials* 32 (8): 1906994.

11 Chen, W., Chen, J., Li, L. et al. (2019). Wool/acrylic blended fabrics as next-generation photodynamic antimicrobial materials. *ACS Applied Materials & Interfaces* 11 (33): 29557–29568.

12 Zhao, X., Chen, F., Li, Y. et al. (2018). Bioinspired ultra-stretchable and anti-freezing conductive hydrogel fibers with ordered and reversible polymer chain alignment. *Nature Communications* 9: 3479.

13 Wang, F., Liu, Z., Wang, B. et al. (2014). Multi-colored fibers by self-assembly of DNA, histone proteins, and cationic conjugated polymers. *Angewandte Chemie International Edition* 53 (2): 424–428.

14 Xu, S., Yan, Y., Zhao, Y. et al. (2021). Spinnable adhesive functional-hydrogel fibers for sensing and perception applications. *Journal of Materials Chemistry C* 9 (16): 5554–5564.

15 Heo, Y.J., Shibata, H., Okitsu, T. et al. (2011). Long-term in vivo glucose monitoring using fluorescent hydrogel fibers. *Proceedings of the National Academy of Sciences of the United States of America* 108 (33): 13399–13403.

16 Duan, X., Yu, J., Zhu, Y. et al. (2020). Large-scale spinning approach to engineering knittable hydrogel fiber for soft robots. *ACS Nano* 14 (11): 14929–14938.

17 Fitria, G., Kwon, M., Lee, H. et al. (2023). Microfluidic fabrication of highly efficient hydrogel optical fibers for in vivo fiber-optic applications. *Advanced Optical Materials* 11 (18): 2300453.

18 Gong, Z., Wu, T., Chen, X. et al. (2021). Upconversion nanoparticle decorated spider silks as single-cell thermometers. *Nano Letters* 21 (3): 1469–1476.

19 Yan, X., Wang, Y., Li, B. et al. (2022). Polyacrylonitrile fluorescent nanofibers for selective and reversible copper detection in aqueous solutions. *Applied Surface Science* 602: 154302.

20 Wang, Y., La, A., Ding, Y. et al. (2012). Novel signal-amplifying fluorescent nanofibers for naked-eye-based ultrasensitive detection of buried explosives and explosive vapors. *Advanced Functional Materials* 22 (17): 3547–3555.

21 Zhao, S., Qin, Z., Gong, S. et al. (2021). White light emitting nanofibers constructed by electrospinning: advances and perspectives. *Journal of Materials Chemistry C* 9 (15): 5036–5054.

22 Qin, Z., Zhang, P., Wu, Z. et al. (2018). Coaxial electrospinning for flexible uniform white-light-emitting porous fibrous membrane. *Materials & Design* 147: 175–181.

23 Shen, X., Akbarzadeh, A., Shi, C. et al. (2021). Preparation and characterization of photo-stimuli-responsive fibers based on lanthanide-activated phosphors and spiropyran dye. *Journal of Materials Research and Technology* 13: 1374–1387.

24 Shen, X., Hu, Q., Zhai, S. et al. (2023). Independent dual-responsive luminous composite fibers with controllable full-color emissions. *Journal of Luminescence* 255: 119550.

25 Li, M., Wen, H., Li, H. et al. (2021). AIEgen-loaded nanofibrous membrane as photodynamic/photothermal antimicrobial surface for sunlight-triggered bioprotection. *Biomaterials* 276: 121007.

26 Wang, T., Ke, H., Chen, S. et al. (2021). Porous protoporphyrin IX-embedded cellulose diacetate electrospun microfibers in antimicrobial photodynamic inactivation. *Materials Science & Engineering, C: Materials for Biological Applications* 118: 111502.

27 Lv, D., Wang, R., Tang, G. et al. (2019). Ecofriendly electrospun membranes loaded with visible-light responding nanoparticles for multifunctional usages: highly efficient air filtration, dye scavenging, and bactericidal activity. *ACS Applied Materials & Interfaces* 11 (13): 12880–12889.

28 Wang, Y., He, Y., Yu, J. et al. (2022). Alginate-based nanofibrous membrane with robust photo-Fenton self-cleaning property for efficient crude oil/water emulsion separation. *Separation and Purification Technology* 287: 120569.

29 Lu, W., Duan, C., Zhang, Y. et al. (2021). Cellulose-based electrospun nanofiber membrane with core-sheath structure and robust photocatalytic activity for simultaneous and efficient oil emulsions separation, dye degradation and Cr(VI) reduction. *Carbohydrate Polymers* 258: 117676.

30 Wang, X., Lv, F., Li, T. et al. (2017). Electrospun micropatterned nanocomposites incorporated with Cu_2S nanoflowers for skin tumor therapy and wound healing. *ACS Nano* 11 (11): 11337–11349.

31 Patil, T.V., Dutta, S.D., Patel, D.K. et al. (2023). Electrospinning near infra-red light-responsive unzipped CNT/PDA nanofibrous membrane for enhanced antibacterial effect and rapid drug release. *Applied Surface Science* 612: 155949.

32 Teh, D.B.L., Bansal, A., Chai, C. et al. (2020). A flexi-PEGDA upconversion implant for wireless brain photodynamic therapy. *Advanced Materials* 32 (29): 2001459.

33 Applegate, M.B., Perotto, G., Kaplan, D.L., and Omenetto, F.G. (2015). Biocompatible silk step-index optical waveguides. *Biomedical Optics Express* 6 (11): 4221–4227.

34 Lu, L., Fan, S., Geng, L. et al. (2021). Low-loss light-guiding, strong silk generated by a bioinspired microfluidic chip. *Chemical Engineering Journal* 405: 126793.

35 Yetisen, A.K., Jiang, N., Fallahi, A. et al. (2017). Glucose-sensitive hydrogel optical fibers functionalized with phenylboronic acid. *Advanced Materials* 29 (15): 1606380.

36 Yang, X., Xu, L., Xiong, S. et al. (2024). Light-emitting microfibers from lotus root for eco-friendly optical waveguides and biosensing. *Nano Letters* 24 (2): 566–575.

37 Costa, S.M., Fangueiro, R., and Ferreira, D.P. (2022). Drug delivery systems for photodynamic therapy: the potentiality and versatility of electrospun nanofibers. *Macromolecular Bioscience* 22 (5): e2100512.

38 Sun, D., Zhang, Z., Chen, M. et al. (2020). Co-immobilization of Ce6 sono/photosensitizer and protonated graphitic carbon nitride on PCL/gelation fibrous scaffolds for combined sono-photodynamic cancer therapy. *ACS Applied Materials & Interfaces* 12 (36): 40728–40739.

39 Gralton, J., Tovey, E., McLaws, M.-L., and Rawlinson, W.D. (2011). The role of particle size in aerosolised pathogen transmission: a review. *Journal of Infection* 62 (1): 1–13.

40 Mauro, N., Scialabba, C., Pitarresi, G., and Giammona, G. (2017). Enhanced adhesion and in situ photothermal ablation of cancer cells in surface-functionalized electrospun microfiber scaffold with graphene oxide. *International Journal of Pharmaceutics* 526 (1–2): 167–177.

41 Xiao, A., Zheng, J., Wu, X. et al. (2023). Ultrasensitive detection and cellular photothermal therapy via a self-photothermal modulation biosensor. *Advanced Optical Materials* 11 (8): 2202711.

42 Li, Y.-F., Slemming-Adamsen, P., Wang, J. et al. (2017). Light responsive hybrid nanofibres for on-demand therapeutic drug and cell delivery. *Journal of Tissue Engineering and Regenerative Medicine* 11 (8): 2411–2420.

43 Ballesteros, C.A.S., Correa, D.S., and Zucolotto, V. (2020). Polycaprolactone nanofiber mats decorated with photoresponsive nanogels and silver nanoparticles: slow release for antibacterial control. *Materials Science & Engineering, C: Materials for Biological Applications* 107: 110334.

44 Wen, K., Zhang, C., Zhang, G. et al. (2024). Jellyfish-inspired artificial spider silk for luminous surgical sutures. *Advanced Materials* 36 (36): e2314158.

45 Xie, L., Yu, Y., Wu, D. et al. (2025). UV-induced photochromic macrofibers derived from bacterial cellulose. *Small* 21 (20): e2408097.

46 Li, P., Wang, Y., He, X. et al. (2024). Wearable and interactive multicolored photochromic fiber display. *Light: Science & Applications* 13 (3): 48.

47 Deisseroth, K. (2011). Optogenetics. *Nature Methods* 8 (1): 26–29.

48 Boyden, E.S., Zhang, F., Bamberg, E. et al. (2005). Millisecond-timescale, genetically targeted optical control of neural activity. *Nature Neuroscience* 8 (9): 1263–1268.

49 Sridharan, A., Rajan, S.D., and Muthuswamy, J. (2013). Long-term changes in the material properties of brain tissue at the implant-tissue interface. *Journal of Neural Engineering* 10 (6): 066001.

50 Fu, R., Luo, W., Nazempour, R. et al. (2018). Implantable and biodegradable poly(L-lactic acid) fibers for optical neural interfaces. *Advanced Optical Materials* 6 (3): 1700941.

51 Wang, L., Zhong, C., Ke, D. et al. (2018). Ultrasoft and highly stretchable hydrogel optical fibers for in vivo optogenetic modulations. *Advanced Optical Materials* 6 (16): 1800427.

6

Photofunctional Polymer Composite Films

Photofunctional polymer composite films are multifunctional materials fabricated through the integration of a polymer matrix with photoactive components. According to their structural support forms, these films can be categorized into two types: self-supported films and substrate-supported films. Both types are capable of facilitating light absorption, emission, conversion, or modulation, thereby enabling a wide spectrum of optical functionalities. This is achieved through the integration of optical functional components, including fluorescent dyes, photosensitive molecules, quantum dots, or nanoparticles. For example, these films can display properties including enhanced fluorescence, photoresponsiveness, and light energy conversion, and their performance can be dynamically regulated in response to external light stimuli.

The highly adaptable and tunable architecture, combined with the exceptional optical properties of photofunctional polymer composite films, endows them with significant potential for applications in fields such as precision medicine, advanced biomaterials, and biosensing. For example, in biological detection, they can serve as highly sensitive optical sensors for monitoring alterations in biomolecules or cellular activities. In the realm of precision medicine, these films can be employed to develop intelligent drug delivery systems that are capable of realizing precise drug release in response to light signals. Furthermore, photofunctional polymer composite films demonstrate substantial potential in numerous other areas, including optoelectronic devices, flexible displays, and photocatalysis. As a result, these films present innovative solutions for the progress of biotechnology. Their multifunctionality and wide-ranging applicability have rendered them a research focus in the fields of biomaterials and optical materials.

6.1 Structures of Photofunctional Polymer Composite Films

The properties of photofunctional polymer composite films are not solely determined by their constituent components; instead, they are also significantly influenced by the multilevel morphological structures both within the films and on their surfaces. These structural features dictate the mechanisms of light–matter

Photofunctional Polymer Composites for Bioapplications, First Edition. Xiaoyu Wang and Lidong Li.

interactions, as well as the pathways for energy and mass transfer, thereby determining the overall functional performance of the device. In the following section, the discussion will be organized based on the macroscopic support structures of the films.

Self-supported films possess the ability to preserve structural integrity without the requirement of a substrate. Their autonomy and remarkable mechanical strength are generally realized via the design of cross-linked structures or the integration of reinforcing elements. These attributes empower such films not only to display outstanding flexibility and processability but also to adapt to intricate environments, thus broadening their potential applications in the biomedical domain. Conversely, substrate-supported films are affixed to a substrate, such as glass, silicon wafers, or flexible plastics, which offers mechanical support and protection. These films manifest exceptional stability and durability, rendering them especially appropriate for utilization in optical sensors and optoelectronic apparatuses.

6.1.1 Self-Supported Film

Natural polymeric materials exhibit substantial potential in the advancement of photofunctional composite films, attributable to their advantageous biocompatibility and abundant availability. These materials contain a multitude of functional groups within their molecular architectures, such as amino, carboxyl, and hydroxyl groups. These functional groups can promote multiple intermolecular interactions and play a vital part in the formation of self-supporting films.

Chitosan is a natural polysaccharide characterized by its excellent film-forming properties. It is a cationic aminopolysaccharide composed primarily of D-glucosamine units interconnected by β-(1 → 4) glycosidic bonds, with a defined proportion of *N*-acetyl-D-glucosamine residues incorporated within its polymeric structure [1]. Its structural unit is composed of one primary amino group (—NH_2) and two free hydroxyl groups (—OH) (Figure 6.1). The existence of these functional groups allows chitosan to easily form robust intramolecular and intermolecular hydrogen bonds. These hydrogen bonds not only promote a regular alignment of the chitosan chains but also contribute to a dense crystalline structure, thus endowing chitosan with stable physicochemical properties. Acetic acid is commonly utilized as a solvent in the preparation of chitosan-based membranes. During the dissolution process, the chitosan chains experience breakage and reorganization. Through methods such as solvent evaporation or extrusion, intermolecular forces, including hydrogen bonds and hydrophilic interactions, are established between the macromolecular chains, resulting in the formation of a translucent film with specific mechanical strength and selective gas permeability [2].

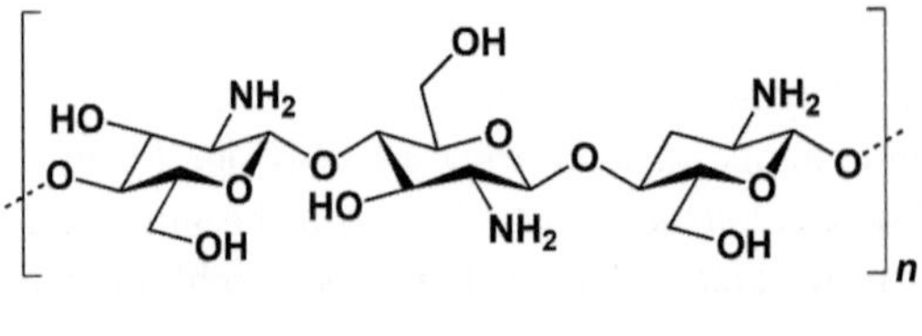

Figure 6.1 Molecular structure of chitosan.

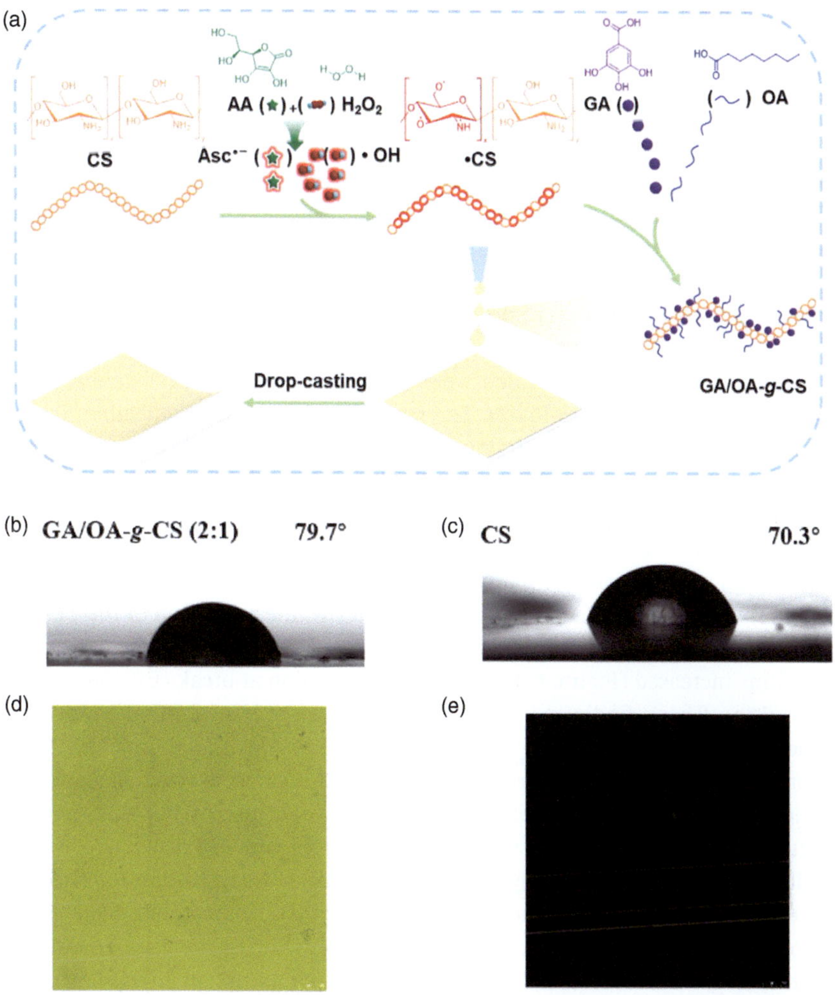

Figure 6.2 (a) The preparation procedures of GA/OA-*g*-CS film. Water contact angles of (b) GA/OA-*g*-CS and (c) CS films. Fluorescence images of GA/OA-*g*-CS (d) with and (e) without curcumin. *Source:* Zhang et al. [3] / with permission of Elsevier.

For example, gallic acid (GA), serving as a photosensitizer, can generate reactive oxygen species (ROS) upon exposure to blue light irradiation. Through a free radical–induced grafting reaction, GA and octanoic acid (OA) are covalently grafted onto chitosan, resulting in the synthesis of the GA/OA-g-CS polymer (Figure 6.2a). Both CS and GA/OA-g-CS are dissolved in an acetic acid solution to prepare the film-forming solution, which is subsequently cast onto glass slides and dried under controlled oven conditions. This process yielded CS and GA/OA-g-CS polymer films. Experimental results demonstrate that the contact angle of the GA/OA-g-CS film is 79.7° (Figure 6.2b), significantly higher than that of the CS film (70.3°) (Figure 6.2c), indicating enhanced hydrophobicity of the GA/OA-g-CS film. Following this, the films are stained with curcumin. Observations reveal that the surface of the GA/OA-g-CS film exhibits yellow fluorescence (Figure 6.2d), whereas the pure CS film does not display any fluorescence (Figure 6.2e). Hydrophobic chains on the GA/OA-g-CS film's surface are responsible for this increased fluorescence because they facilitate efficient hydrophobic interactions with curcumin [3].

Moreover, nanoparticles function not only as photofunctional components but also possess the capacity to penetrate the hydrogen-bond-reinforced network structure of chitosan when used as fillers. This creates a "tortuous path," which restricts the diffusion of water molecules and gases. Consequently, the mechanical strength and barrier properties of the film improve [4].

TP-CS nanocomposite films were fabricated by incorporating photofunctional TiO_2 nanoparticles (TP-NPs), which were surface-modified with polytannic acid (PTA), into a chitosan (CS) solution. The resulting mixture was then subjected to solution casting and subsequent drying to obtain the final films. The cross-sectional morphology of the obtained films maintained uniformity and density, similar to that of the pure CS films (Figure 6.3a). Nevertheless, the film thickness increased because of an increase in TP-NP loading. Atomic force microscopy (AFM) analysis indicated the existence of nanoscale surface protrusions, accompanied by a substantial increase in surface roughness. Specifically, the roughness of the TP-CS film increased compared to the pure CS film. Moreover, the tensile strength (TS) of TP-CS films increased (Figure 6.3b), while the elongation at break (EB) decreased (Figure 6.3c). These findings imply that the incorporation of TP-NPs notably enhances the mechanical strength of the CS film but impairs its ductility. This phenomenon can be mainly ascribed to the formation of robust hydrogen bonds and electrostatic interactions between the TP-NPs and chitosan chains, which strengthen the internal network structure of the film (Figure 6.3d) [5].

Sodium alginate is a naturally occurring anionic polysaccharide constituted by β-D-mannuronic acid (M) and α-L-guluronic acid (G) monomers, which are interconnected via $(1 \rightarrow 4)$ glycosidic bonds, as depicted in Figure 6.4 [6]. Sodium alginate has garnered a lot of interest because of its exceptional film-forming, hydrophilicity, and biocompatibility. Despite the fact that sodium alginate films display pronounced hydrophilic properties, their mechanical strength and water resistance are still restricted. To overcome these limitations, ion cross-linking is frequently utilized to increase the films' structural integrity and water resistance. This method not only enhances the functional performance of the material but also expands its potential applications in the biomedical domain [7].

For instance, optical functional graphene oxide (GO) was integrated with a sodium alginate (Na-alg) solution. This solution was then poured into a mold and dried to fabricate Na-alg/GO nanocomposite films. Subsequently, the film was immersed in a polyvalent cationic solution to trigger cross-linking. SEM images clearly demonstrated the microstructural differences among the diverse films. The surface of the pure Na-alg film presented a smooth appearance, accompanied by a compact internal structure (Figure 6.5a,b). In contrast, the Na-alg/GO film exhibited distinct banded surface structures, while the interior formed a layered structure attributable to the incorporation of GO (Figure 6.5c,d). The ion-cross-linked M-alg/GO film displayed a surface morphology analogous to that of the Na-alg/GO film, yet with

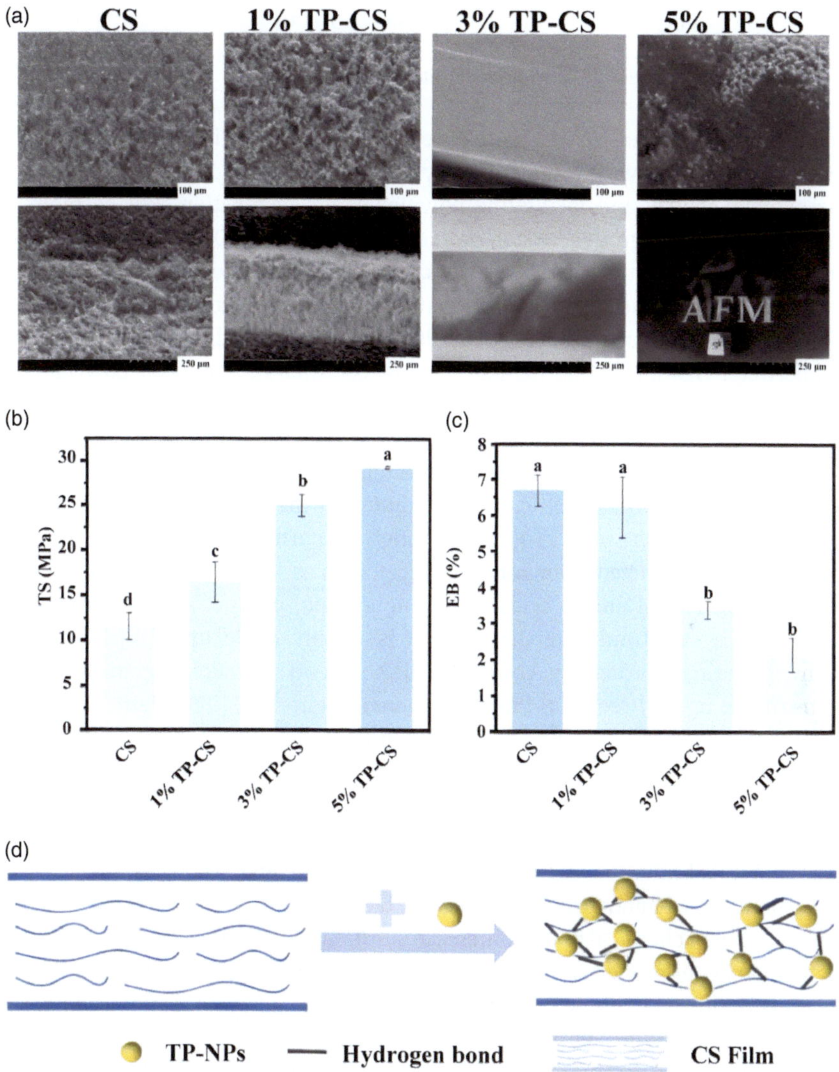

Figure 6.3 (a) SEM images showing the surface and cross-section morphologies of pure CS and TP-CS films with varying TP-NP contents. (b) TS and (c) EB of pure CS and TP-CS films with varying TP-NP contents. (d)Schematic illustration of the film formation process of TP-CS film. *Source:* Ding et al. [5] / with permission of Elsevier.

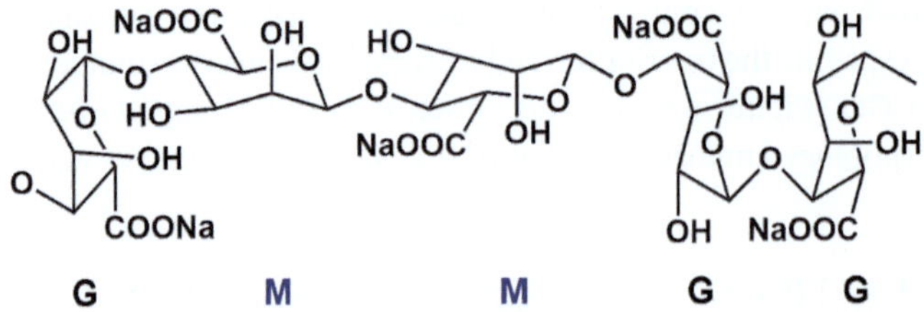

Figure 6.4 Molecular structure of sodium alginate.

a more compact internal structure (Figure 6.5e–j). This implies that polyvalent cations can coordinate with Na-alg, thereby modifying the structural organization of the alginate matrix and facilitating polymer chain aggregation. Furthermore, the oxygen-containing groups located on the basal planes of GO, together with the carboxyl groups present at the sheet edges, are capable of interacting with polyvalent metal ions. Consequently, the formation of these nanocomposite films is a result of a synergistic combination of multiple interactions, including hydrogen bonds between GO sheets and alginate chains, coordination between polyvalent cations and alginate, and the interaction between polyvalent cations and GO sheets (Figure 6.5k) [8].

In addition to natural polysaccharides, certain proteins also demonstrate notable film-forming capabilities, which can be attributed to their diverse array of functional groups. Gelatin is a protein derived through the partial hydrolysis of collagen, which is typically extracted from animal sources such as skin, bones, and tendons. It contains a variety of amino acids, including glycine, proline, and hydroxyproline, and has a structural composition that is mainly made up of hydrophilic α-chains, β-chains (formed by covalently linking two α-chains together), and γ-chains (made up of three cross-linked α-chains) (Figure 6.6) [9]. Gelatin exhibits distinctive film-forming capabilities, accompanied by commendable biocompatibility and nutritional characteristics. Nevertheless, its intrinsic hydrophilic nature renders gelatin-based materials prone to moisture absorption and softening under high-humidity conditions, consequently constraining their practical utilization. To overcome this limitation, researchers have adopted diverse strategies, including the incorporation of nanoparticle composites, blending with plant extracts, or the implementation of chemical modifications, to substantially enhance the mechanical strength and barrier properties of gelatin-based films. The structural integrity of the materials is improved by these strategies, and the resulting films are endowed with supplementary functional characteristics as a result.

For example, optically functional TiO_2 nanoparticles were incorporated into a gelatin solution, followed by the addition of grapefruit seed extract (GSE). Subsequently, the resultant mixture was cast onto a template and dried to fabricate a Gel/GSE/TiO_2 composite film. FE-SEM analysis of the microstructure of the gelatin-based film demonstrated that the surface of the pure gelatin film was smooth and dense (Figure 6.7a), while the film integrated with GSE presented

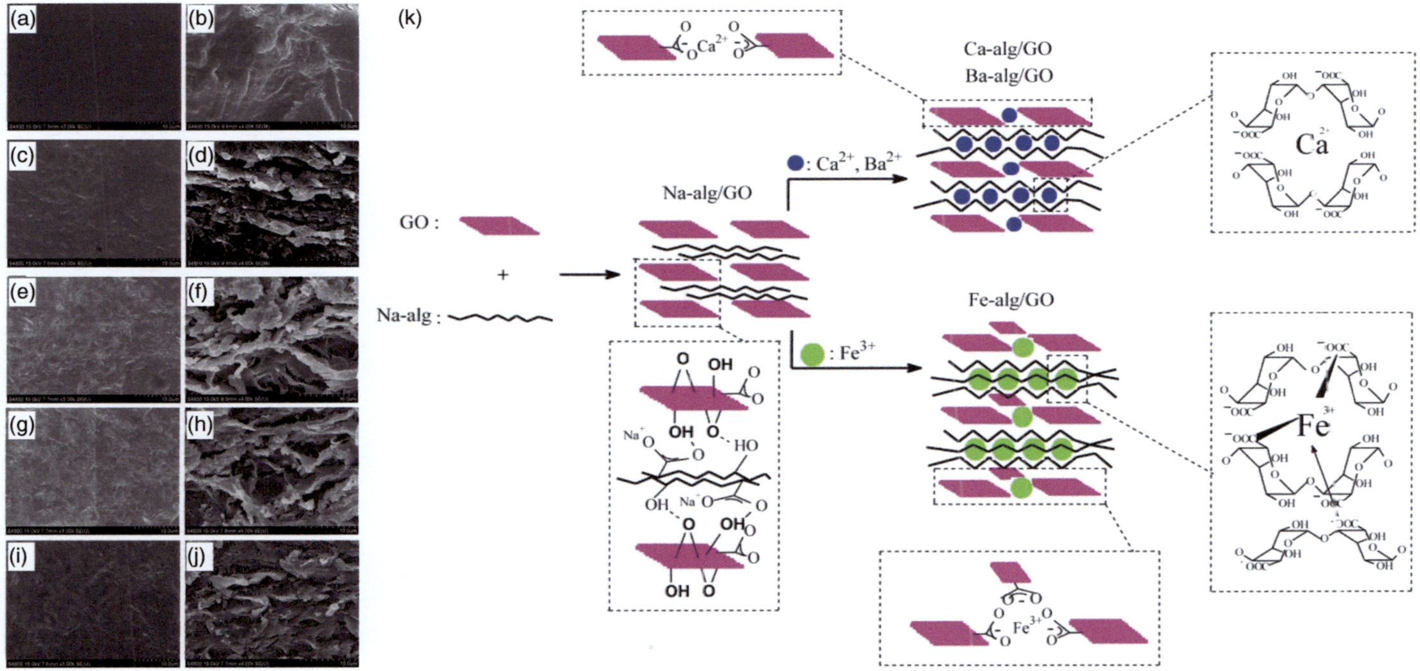

Figure 6.5 SEM images showing the surface and cross-section morphologies of (a, b) Na-alg film, (c, d) Na-alg/GO film, (e, f) Ca-alg/GO film, (g, h) Ba-alg/GO film, and (i, j) Fe-alg/GO film. (k) Schematic illustration of the film formation process of M-alg/GO film with different metal ions. *Source:* Zheng et al. [8] / with permission of John Wiley & Sons.

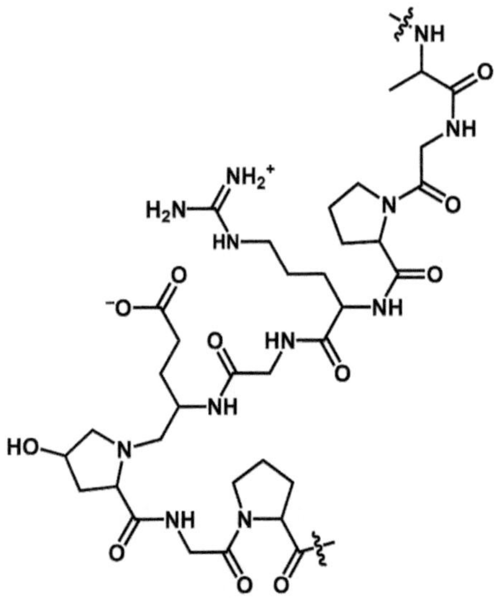

Figure 6.6 Molecular structure of gelatin.

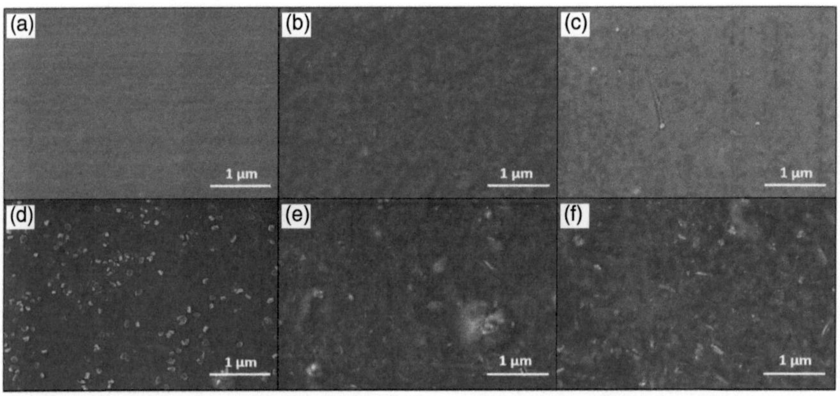

Figure 6.7 SEM images of surface of (a) Gel, (b) Gel/GSE, (c) Gel/GSE/$TiO_2^{0.5\%}$, (d) Gel/GSE/$TiO_2^{1\%}$, (e) Gel/GSE/$TiO_2^{3\%}$, and (f) Gel/GSE/$TiO_2^{5\%}$ films. *Source:* Reproduced with permission from Riahi et al. [10] / Elsevier.

a slightly irregular morphology (Figure 6.7b). As the content of TiO_2 increased, more nanostructures gradually appeared within the film (Figure 6.7c–f). When the TiO_2 concentration reached 3 wt%, nanoaggregates were detected (Figure 6.7e). The experimental findings indicated that a TiO_2 concentration of 1 wt% achieved the optimal dispersion within the gelatin matrix embedded with GSE, offering a valuable reference for improving the performance of gelatin-based films [10].

Silk fibroin, a natural protein sourced from silk, exhibits a molecular architecture consisting of a heavy chain (390 kDa) and a light chain (26 kDa), covalently linked through disulfide bonds. The molecular chain is replete with numerous active amino acid residues, providing an abundance of functional sites for subsequent chemical modifications. In regard to its secondary structure, silk fibroin predominantly comprises α-helix, β-sheet, and random coil structures (Figure 6.8).

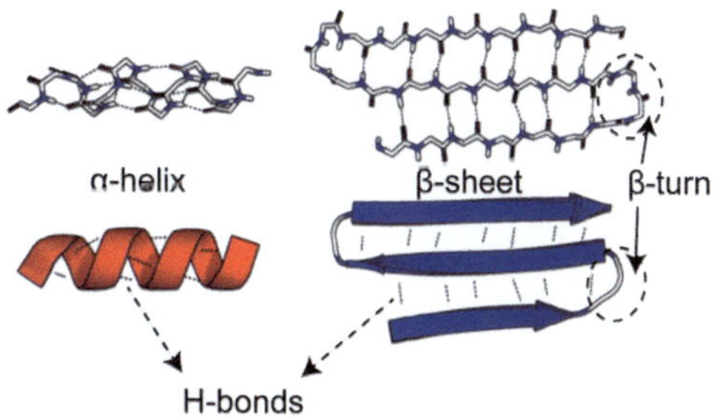

Figure 6.8 Schematic illustration of α-helix and β-sheet of silk fibroin. *Source:* Qiu et al. [11] / with permission of John Wiley & Sons.

Among them, the β-sheet structure forms densely packed crystalline regions via intermolecular hydrogen bonds. These crystalline domains endow silk fibroin with high mechanical strength, enabling it to endure substantial external forces. Simultaneously, the presence of random coil and α-helix structures confers a certain degree of elasticity, allowing the material to deform under stress without readily fracturing. This distinctive combination of rigidity and flexibility renders silk fibroin an excellent multifunctional biomaterial, well-suited for a diverse array of high-performance applications [11].

Prodigiosin demonstrates a pH-responsive discoloration characteristic and can be mixed with silk fibroin solution as a photofunctional component. Glycerol and 1-ethyl-3-(3-dimethylaminopropyl) carbodiimide (EDC) are added as plasticizers and cross-linking agents, respectively, to enhance the mechanical qualities of the resultant film. Subsequently, the obtained mixture is poured into a mold and dried to acquire the SEGP composite film (Figure 6.9a). The thickness of the pure silk fibroin film (SF) is 69 ± 4 μm, while that of the SEGP composite film is 98 ± 3 μm.

SEM images reveal that the SF film features a dense and smooth surface (Figure 6.9b). When glycerol is added, the film structure becomes porous, with notably larger pore diameters (Figure 6.9c). This is because the oxygen atoms in glycerol are highly electronegative, and the O—H bonds are strongly polar. These properties cause hydrogen bonds to be formed with the side groups of the silk fibroin protein chains. This results in the reorganization of hydrophilic and hydrophobic regions within the protein chains, creating numerous pores on the film's surface. Conversely, the film with only EDC added demonstrates a smoother surface morphology (Figure 6.9d), which is a consequence of the condensation reaction between carboxyl and amino groups. When both glycerol and EDC are incorporated, the resultant silk fibroin-based film presents a more uneven surface with smaller pore sizes (Figure 6.9e). Moreover, the surface roughness of the the silk fibroin-based film with 20.0 wt% glycerol, 20.0 wt% EDC, and 0.33 wt% prodigiosins (SEGP) composite film doped with prodigiosin is significantly reduced, suggesting strong compatibility between silk fibroin and prodigiosin (Figure 6.9f) [12].

Cellulose, a natural polymer, consists of D-glucose units connected through 1,4-β-glycosidic bonds (Figure 6.10). Its molecular chain encompasses a substantial number of hydroxyl groups (—OH), which can form both intramolecular and intermolecular hydrogen bonds. These hydrogen bonds, in conjunction with van der Waals forces among adjacent glucose units, promote the parallel alignment and

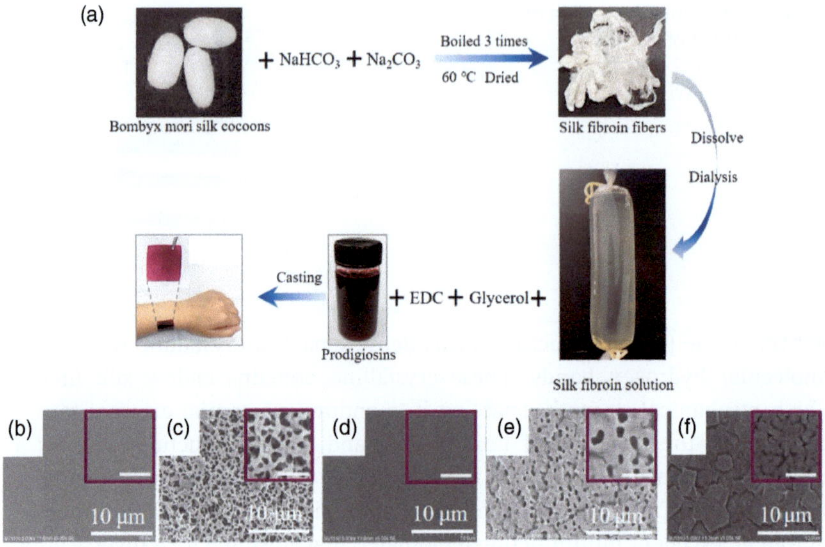

Figure 6.9 (a) Illustration of the film formation process of SEGP film. SEM images (b) SF film, (c) SG film with glycerol, (d) SE film with EDC, (e) SEG film with both glycerol and EDC, and (f) SEGP film. *Source:* Reproduced with permission from Liu et al. [12] / Elsevier.

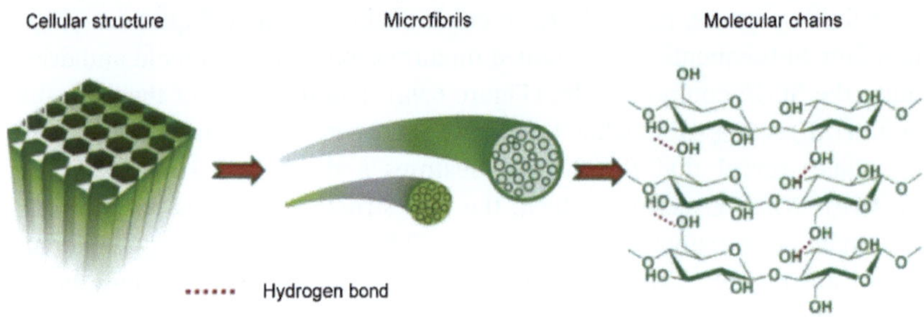

Figure 6.10 Molecular structure and microstructure of cellulose. *Source:* Long et al. [13] / with permission of Elsevier.

stacking of cellulose chains, resulting in the formation of a nanofibrillar structure containing both ordered (crystalline) and disordered (amorphous) domains. Owing to its semi-rigid backbone and superior film-forming properties, cellulose is extensively employed in membrane technology. Nevertheless, pure cellulose-based membranes frequently display drawbacks such as insufficient mechanical strength, poor water resistance, and limited antimicrobial performance. Consequently, current research endeavors are predominantly focused on the development of cellulose composite membranes through organic/inorganic hybridization methods to improve their overall performance [13].

Lignocellulose was extracted from dissolving wood pulp and then combined with corncob residue obtained from xylitol production to fabricate a lignocellulose-based film (PR20). Subsequently, a photofunctional composite film (HPR20) was prepared

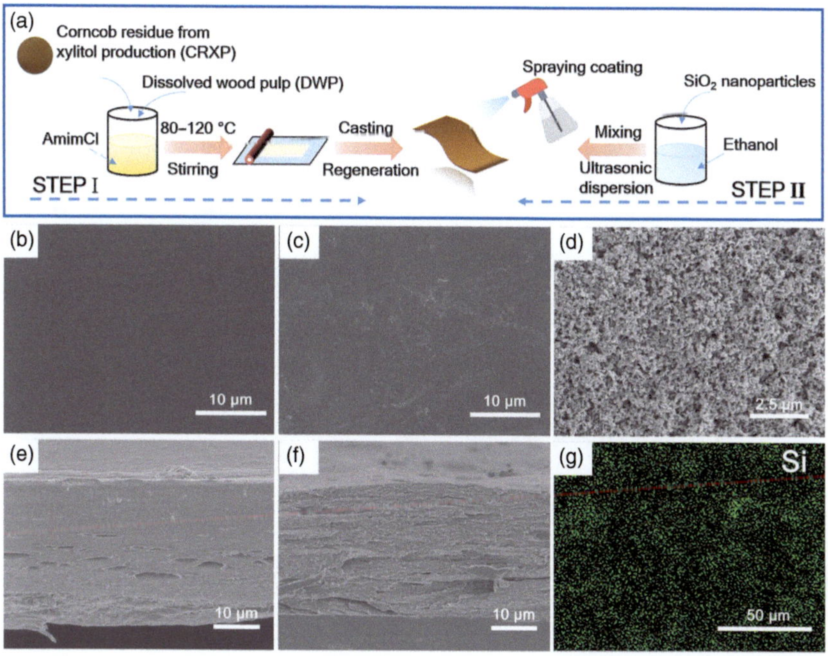

Figure 6.11 (a) Illustration of the film formation process of lignocellulose-based film. SEM images of the (b) surface and (c) cross-section of the pure cellulose film. SEM images of the (d) surface and (e) cross-section of PR20 films. (f) SEM images of the surface of HPR20 film. (g) Elemental map of the Si element of HPR20 film. *Source:* Song et al. [14] / with permission of American Chemical Society.

by depositing hydrophobic SiO_2 nanoparticles onto the PR20 film (Figure 6.11a). Experimental findings demonstrated that the surface of the pure lignin film was smooth, featuring a compact internal structure (Figure 6.11b,c). In contrast, the PR20 film exhibited a relatively rough surface and a layered microstructure, which could be ascribed to the incorporation of corncob residue (Figure 6.11d,e). After the application of SiO_2 nanoparticles, the surface roughness of the HPR20-3 film increased further (Figure 6.11f). Elemental mapping of silicon indicated a uniform distribution of SiO_2 nanoparticles across the film surface (Figure 6.11g) [14].

Moreover, the cellulose film can be further functionalized via the incorporation of organic molecules. For example, the prepared microfibrillated cellulose (MFC) film was immersed in a benzoxazine (BR) solution, followed by thermal curing to acquire an MFC/BR composite film. As the BR content rises, the transparency of the MFC/BR film enhances notably, accompanied by a decrease in haze (Figure 6.12a). Microstructural analysis indicates that the surface of the MFC/BR composite film is smoother in comparison to that of the pure MFC film (Figure 6.12b,c). Additionally, most of the internal voids within the MFC/BR film are effectively filled with BR (Figure 6.12d,e). Elemental mapping of *N* validates the uniform distribution of BR on both the surface and section of the composite film (Figure 6.12f,g). The compact internal structure of the MFC/BR film is due to the layered cellulose acting

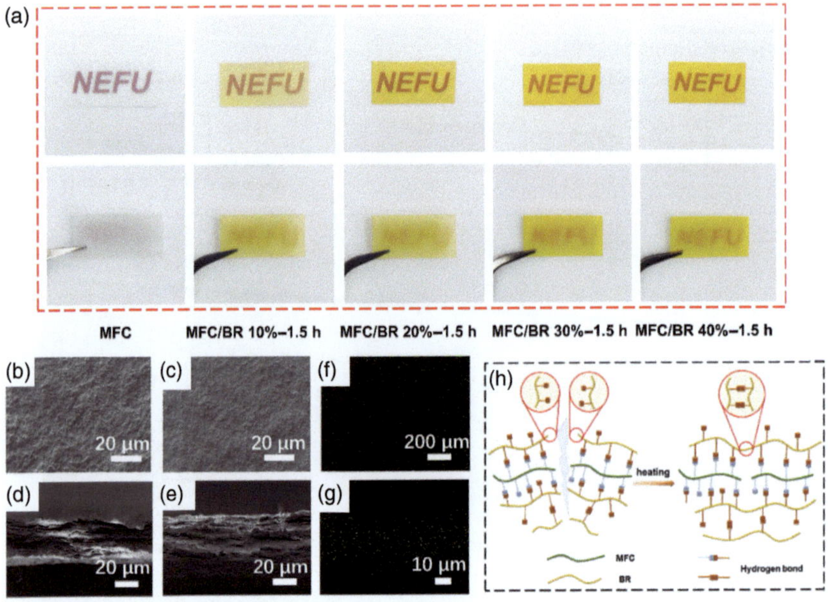

Figure 6.12 (a) Photographs of patterns covered by MFC and MFC/BR films with different BR contents. SEM image of surface of (b) MFC and (c) MFC/BR films. SEM image of cross-section of (d) MFC and (e) MFC/BR films. Elemental mapping of *N* on the (f) surface and (g) cross-section of MFC/BR films. (h) Illustration of reversible hydrogen bonding in MFC/BR films. *Source:* Li et al. [15] / with permission of Elsevier.

as the "bricks" in the structural framework, and the BR filling the spaces between them, acting as the "mortar" and contributing to mechanical reinforcement. Furthermore, hydrogen bonding is a crucial factor underlying the high mechanical strength of cellulose-based composite films. The numerous hydrogen bonds formed between cellulose and BR promote strong interfacial adhesion (Figure 6.12h) [15].

6.1.2 Substrate-Supported Film

In addition to self-supported films, substrate-supported films can also be fabricated, which depend on the physical and chemical interactions between the film materials and the substrate surface. Thus, the selection of appropriate substrate materials and the control of interfacial interactions are crucial factors for ensuring the structural and property stability of the films during the fabrication process.

Physical coating is a widely used technique for preparing thin films, including methods such as dip coating, spin coating, drop coating, and spray coating. Each method exhibits unique characteristics and is applicable to specific circumstances. Among these, dip coating is the most straightforward approach. It involves immersing the substrate in a preformulated solution containing the active substance. Through the uniform lifting of the substrate one or multiple times, the active substance is uniformly coated on its surface. Subsequently, appropriate posttreatment and drying procedures are implemented to acquire the desired

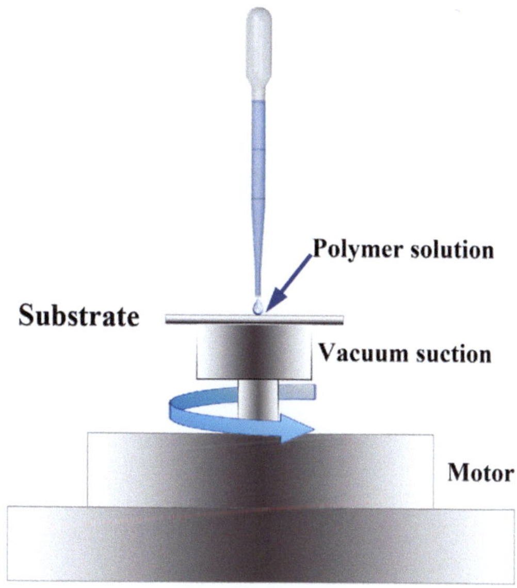

Figure 6.13 Schematic illustration of film preparation by spin coating.

functional thin film. Spin coating involves the application of a polymer solution onto a substrate, which is then rotated at a high velocity. The centrifugal force disperses the solution uniformly across the surface, forming a thin film while expelling the excess material (Figure 6.13). This method is particularly well-suited for applications demanding uniform thin films. Dip coating and spray coating methods are similar to spin coating but are more convenient to use. Dip coating involves directly dropping the polymer solution onto the substrate, while spray coating involves spraying the solution onto the substrate. The solvent then naturally evaporates, and the polymer precipitates to form a film. These two methods are suitable for large-area coating or situations requiring rapid film formation.

For example, when a quartz plate is immersed in a dopamine solution, the dopamine undergoes a process called self-polymerization, generating polydopamine particles that cover the surface of the substrate uniformly. Subsequently, the quartz plate is immersed in a glutaraldehyde solution, and then in a solution containing BSA-protected AuNCs, facilitating covalent bond formation between the AuNCs and the polydopamine layer. This procedure leads to the formation of a photofunctional PDA/AuNCs film (Figure 6.14a). SEM imaging indicates that polydopamine nanoparticles with an average diameter of approximately 100 nm are spontaneously formed on the quartz surface (Figure 6.14b), acting as an adhesive layer for the immobilization of BSA-AuNCs. After immobilization, the surface of the film becomes significantly smoother (Figure 6.14c) [16].

The multilayer composite film can be fabricated utilizing the layer-by-layer (LbL) self-assembly technique [17]. This technique is predicated on the sequential deposition of individual layers, with each layer held together by specific intermolecular interactions. These interactions may include electrostatic forces, hydrogen bonding, halogen bonding, coordination bonding, and, in some cases, covalent bonding. Through the sequential and alternate deposition of distinct molecular

Figure 6.14 (a) The preparation process of PDA/AuNCs film. SEM images of the (b) PDA film and (c) PDA/AuNCs film. *Source:* Li et al. [16] / MDPI / CC BY 4.0.

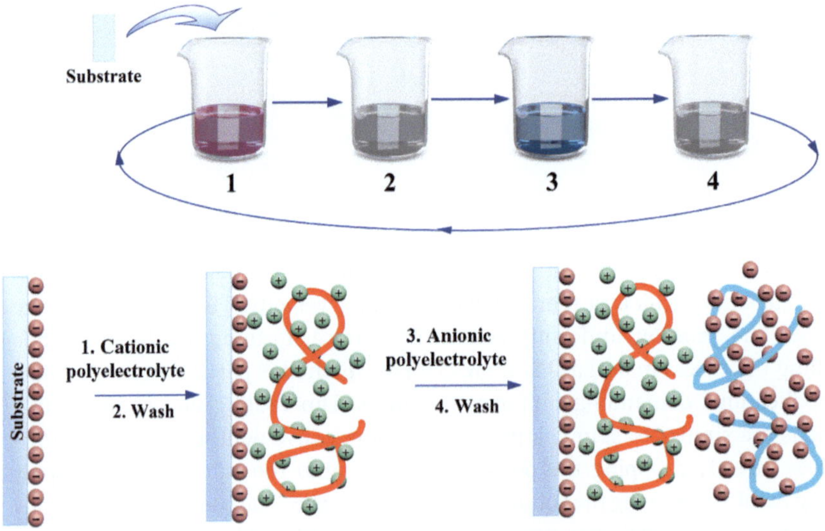

Figure 6.15 Schematic illustration of film preparation by LbL technology.

layers, a structurally well defined and stable thin film can be spontaneously generated. Among these, the assembly based on electrostatic interaction is the simplest and most extensively utilized approach. As depicted in Figure 6.15, the general procedure for LbL technology is as follows: Initially, the substrate surface undergoes chemical treatment to introduce negative charges. Subsequently, the substrate is submerged in a solution of cationic polyelectrolyte and left to stand for a specific duration, during which the polyelectrolyte is adsorbed onto the surface through electrostatic attraction. After that, the substrate is washed with ultrapure water and then dried with nitrogen gas before being immersed in an anionic polyelectrolyte solution to facilitate the adsorption of a second layer. By repeatedly executing this sequence, a multilayered self-assembled thin film can be fabricated. This method also permits the incorporation of photofunctional components, facilitating the development of thin films with enhanced and diverse functionalities.

The PAA/PDDA multilayer film was successfully prepared via the alternate deposition of the positively charged polyelectrolyte PDDA and the negatively charged polyelectrolyte PAA. Subsequently, silver nanoseeds were synthesized by

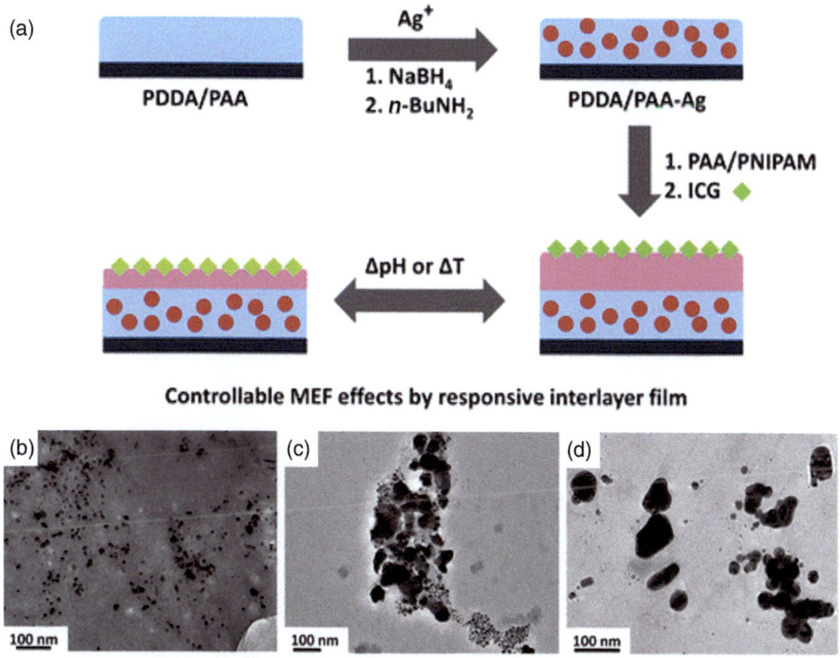

Figure 6.16 (a) The preparation process of a composite film incorporating Ag nanoparticles. (b) TEM images of Ag seeds, Ag nanoparticles after (c) 1 h and (d) 2 h of reduction in the film, respectively. *Source:* Tong [18] / Royal Society of Chemistry.

alternately immersing the film into an $AgNO_3$ solution and a reducing agent, the $NaBH_4$ solution. To increase the size of the silver nanoparticles further, the film was subjected to alternating immersions in $AgNO_3$ and *n*-butylamine solutions. Ultimately, a composite film with photofunctions was acquired by alternately depositing PAA and PNIPAM layers (Figure 6.16a). TEM images suggest that silver nanoparticles can grow in situ within the multilayer matrix, and continuous growth takes place as the reduction time prolongs (Figure 6.16b–d) [18].

Langmuir–Blodgett (LB) film technology involves a process in which amphiphilic molecules, characterized by distinct hydrophobic and hydrophilic segments, are dispersed in a volatile organic solvent. The resulting solution is subsequently spread at the air–water interface [19]. Following solvent evaporation, the amphiphilic molecules spontaneously organize into an ordered monolayer on the surface of the water through self-assembly. Utilizing a specialized device, this monolayer can be transferred onto a solid substrate. Although the LB technique features mild technical requirements, ease of operation, and the ability to produce uniform, ordered thin films with easily controlled thickness, it relies on specialized equipment.

The amphiphilic conjugated polymer poly(phenylene)ethynylene (PPE) is well-suited for the fabrication of thin films through the LB technique. As shown in Figure 6.17a, this polymer contains both hydrophilic triethylene glycol side chains and hydrophobic ethyl-protected carboxyl groups, which endow the essential amphiphilic characteristics required for film formation via the LB technique

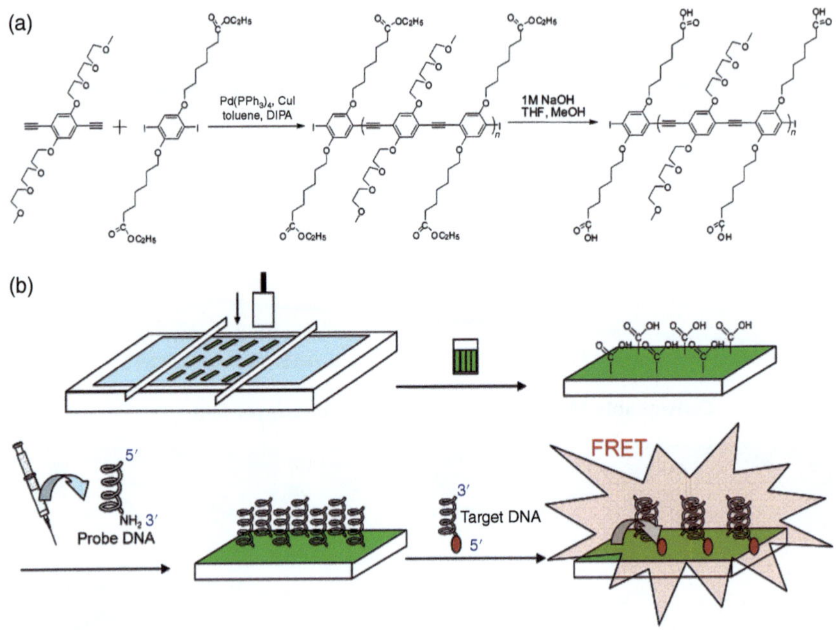

Figure 6.17 (a) Molecular structure of PPE. (b) The preparation process of a photofunctional composite film by the LB technique. *Source:* Pun et al. [20] / with permission of American Chemical Society.

and for subsequent transfer onto a planar substrate (Figure 6.17b). After the removal of the ethyl protecting groups, the carboxyl groups can form covalent bonds with amino-modified single-stranded DNA, thereby enabling DNA to be immobilized on the substrate surface. When the target DNA, labeled with the fluorescent dye HEX, is introduced, it hybridizes with the immobilized DNA to form a double-stranded helical structure, thus promoting fluorescence resonance energy transfer (FRET) [20].

Although physical methods for preparing photofunctional thin films offer many outstanding advantages, the stability of the resulting films is often limited. To enhance the stability of the films, chemical modification has emerged as a crucial approach for the surface functionalization of materials. Among the existing techniques, self-assembled monolayer (SAM) technology has demonstrated exceptional efficacy in improving the stability of thin films. This method introduces reactive groups onto the substrate surface, facilitating the spontaneous formation of a structurally ordered thin film through chemical bonding.

The silylation reaction is one of the most common and important methods for the monolayer modification of oxide surfaces [21]. The essence of the process lies in the condensation reaction that occurs between the silylation reagent and the hydroxyl groups present on the substrate's surface, forming stable silicon-oxygen bonds. Via this chemical procedure, alkyl groups or terminally functionalized alkyl groups are efficiently bonded to the substrate surface. This modification approach not only enhances the structural stability of the thin film but also constructs a solid platform for subsequent functionalization.

The glass slide was immersed in an aminopropyl trimethoxysilane solution for silanization modification, which led to the introduction of amino groups on its surface. Subsequently, 1,4-phenylenediisothiocyanate was used to activate amino groups, followed by covalent coupling of amine-terminated peptide nucleic acid (PNA), resulting in the successful preparation of a monolayer molecular film. Next, a target single-stranded DNA modified with the fluorescent dye C* and possessing a complementary base sequence was introduced, enabling specific hybridization with the PNA (Figure 6.18a). Owing to the exposed phosphate backbone of the DNA, the film surface acquired a negative charge. Eventually, the cationic conjugated polymer PFBT was added and adsorbed onto the negatively charged surface through electrostatic interactions (Figure 6.18b). This configuration promoted the establishment of a FRET system between PFBT and the dye C*, thus enabling efficient signal transduction and detection [22].

In addition, the indium tin oxide (ITO) surface can be treated with NH_4OH/H_2O_2 to introduce hydroxyl groups, followed by photochemical grafting of 5-hexen-1-ol. Subsequently, an initiator for atom transfer radical polymerization containing bromine is attached to the ITO surface through a reaction with 2-bromoisobutyryl bromide. Finally, a zwitterionic polymer was synthesized through atom transfer radical polymerization (Figure 6.19a). Employing gold as a photomask template, the carbon-bromine (C—Br) bonds in the exposed regions were selectively cleaved by ultraviolet irradiation, resulting in the degradation of the initiator in these

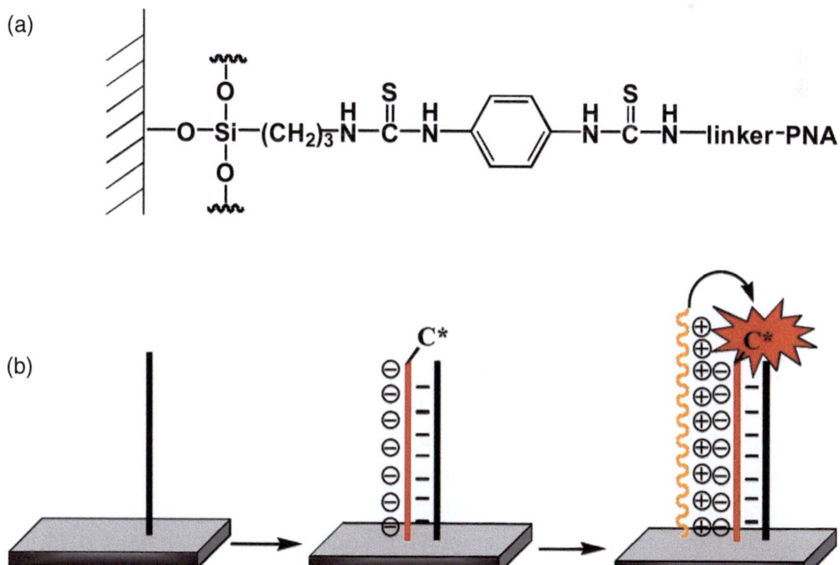

Figure 6.18 (a) PNA probe immobilized on substrate via SAM technology. (b) Self-assembly of DNA and PFBT for detection. *Source:* Liu and Bazan [22] / National Academy of Sciences. Copyright (2005) National Academy of Sciences, U.S.A.

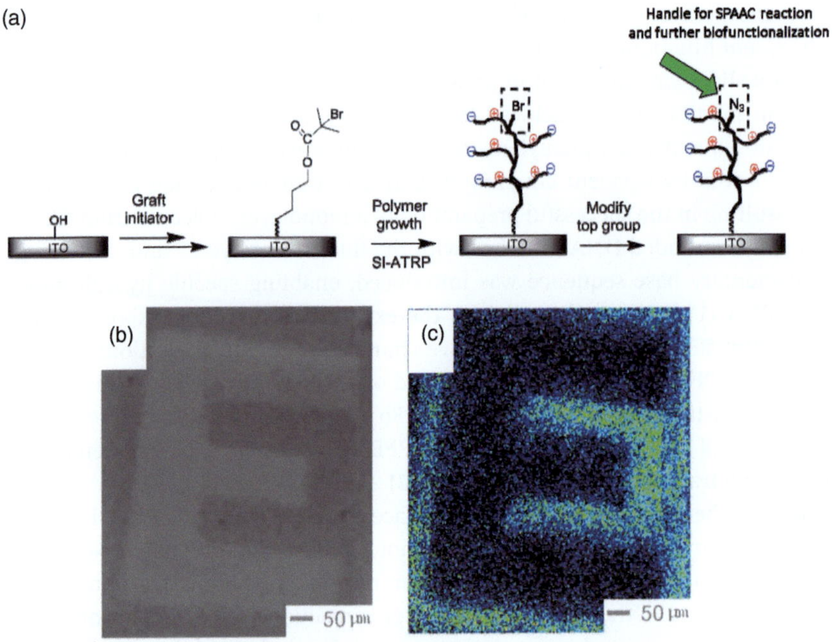

Figure 6.19 (a) The preparation process for zwitterionic polymer brushes on an ITO substrate. (b) SEM image and (c) elemental mapping of *N* in the polymer film on the ITO substrate. *Source:* Li et al. [23] / with permission of American Chemical Society.

areas and facilitating surface patterning. The resultant patterned zwitterionic polymer on the ITO substrate was verified by an SEM image (Figure 6.19b). The blue intensity in the SEM image corresponds to the surface nitrogen density, and the blue regions coincide with the areas containing the polymer brush structure (Figure 6.19c) [23].

6.2 Functions of Photofunctional Polymer Composite Films

6.2.1 Fluorescence Enhancement Effect

The enhancement of fluorescent molecule emission by metal plasmons primarily depends on the interaction distance between them. As a macroscopic composite material, photofunctional composite polymer films can precisely control this distance at the nanoscale, thereby demonstrating significant potential in the regulation of metal-enhanced fluorescence (MEF) effects.

A composite film was prepared by employing a quartz plate functionalized with silver nanoprisms (Ag NPRs) as the substrate, involving the alternating assembly of the cationic polyelectrolyte polyethyleneimine (PEI) and the anionic polyelectrolyte poly(acrylic acid) (PAA). The antibody was immobilized by conjugating its amino groups with the carboxyl groups of PAA via carbodiimide (EDC) and

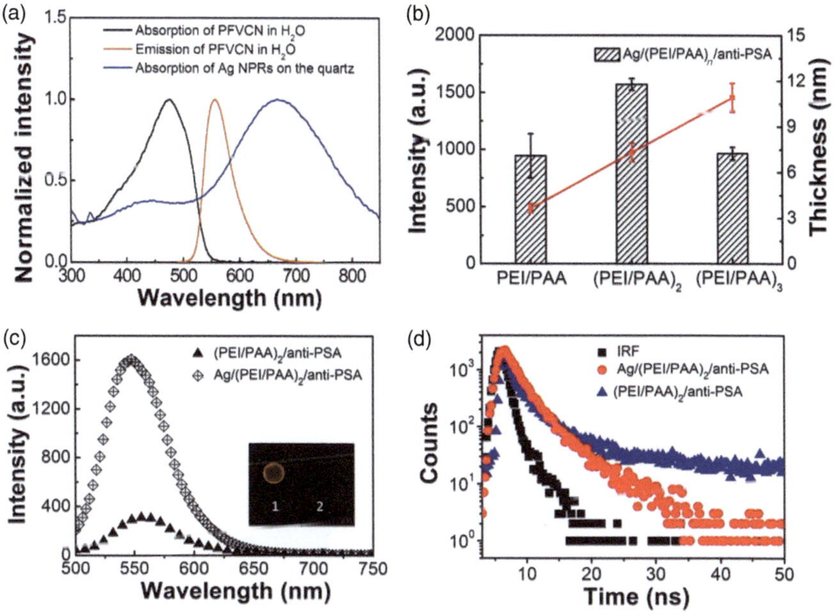

Figure 6.20 (a) Absorption and emission spectra of PFVCN solution along with the absorption spectrum of Ag NPRs. (b) Intensity changes of PFVCN emission on Ag/(PEI/PAA)$_n$/anti-PSA film with varying interlayer. (c) Fluorescence spectra and images of PFVCN on Ag/(PEI/PAA)$_2$/anti-PSA and (PEI/PAA)$_2$/anti-PSA films. (d) Emission decay profiles of PFVCN on (PEI/PAA)$_2$/anti-PSA and on Ag/(PEI/PAA)$_2$/anti-PSA films. *Source:* Wang et al. [24] / with permission of John Wiley & Sons.

N-hydroxysuccinimide (NHS) facilitated coupling chemistry. Considering the negatively charged surface of the antibody, it can further adsorb the cationic conjugated polymer poly[9,9′-bis(6,6′-(N,N,N-trimethylaminium)fluorene-2,7-ylenevinylene-co-alt-2,5-dicyano-1,4-phenylene] (PFVCN), which contains quaternary ammonium salt groups in its side chains. As depicted in Figure 6.20a, there is a significant overlap between the absorption and emission spectra of PFVCN and those of Ag NPRs, which provides a basis for the occurrence of the MEF effect. Three types of multilayer films, designated as (PEI/PAA)$_n$/anti-prostate-specific antigen (anti-PSA), with different interlayers, were constructed using the LbL technique. Experimental results indicate that the emission intensity of PFVCN reaches its maximum when $n = 2$ (Figure 6.20b). This result verifies that the separation between PFVCN and the silver nanostructures is a crucial factor influencing the MEF effect. The enhanced orange fluorescence can be visually detected (Figure 6.20c). Further mechanistic analysis shows that the fluorescence decay rate of PFVCN is significantly faster in the composite film containing Ag NPRs (1.49 ns) than in the film without Ag NPRs (3.44 ns), suggesting that the fluorescence enhancement mainly stems from an increase in the radiative decay rate of PFVCN (Figure 6.20d) [24].

Likewise, the plasmon absorption peak of the silver substrate demonstrates a favorable overlap with both the absorption and emission peaks of the conjugated polymer PFDBT-BIMEG, thus establishing the basis for the implementation of

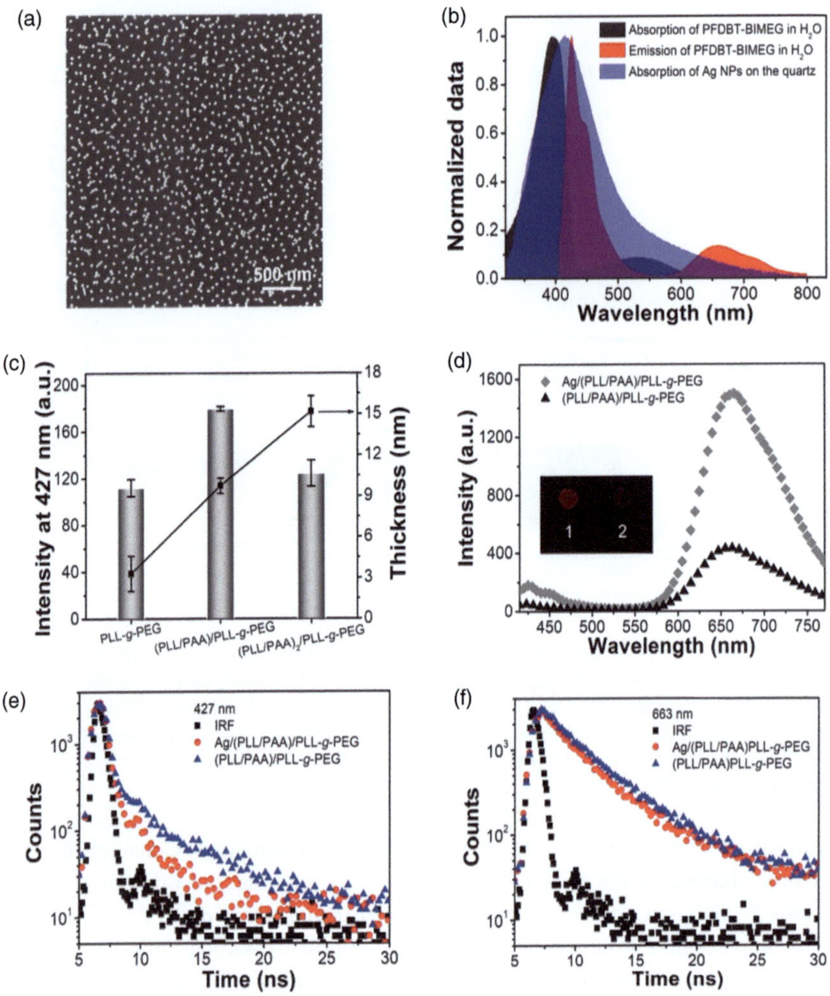

Figure 6.21 (a) SEM image of Ag film. (b) Absorption and emission spectra of PFDBT-BIMEG, along with absorption spectrum of Ag film. (c) The changes of 427 nm emission of PFDBT-BIMEG on Ag/(PLL/PAA)$_n$/PLL-*g*-PEG with varying interlayer. (d) Fluorescence spectra of PFDBT-BIMEG on Ag/(PLL/PAA)/PLL-*g*-PEG and (PLL/PAA)/PLL-*g*-PEG film. Fluorescence decay of (e) 427 nm emission and (f) 663 nm emission of PFDBT-BIMEG on (PLL/PAA)/PLL-*g*-PEG and Ag/(PLL/PAA)/PLL-*g*-PEG films. *Source:* Wang et al. [25] / with permission of John Wiley & Sons.

the MEF effect (Figure 6.21a,b). On the silver substrate, polylysine (PLL) and polyacrylic acid (PAA) were alternately deposited through LbL assembly, with the molecular brush PLL-g-PEG adsorbed on the outermost layer. Subsequently, the conjugated polymer PFDBT-BIMEG, which contains bis-imidazole groups in its side chains, was deposited via drop-casting. The MEF effect can be accurately regulated by adjusting the quantity of bilayer films.

The experimental results demonstrate that the 427 nm emission of PFDBT-BIMEG reaches its peak when the thickness of the (PLL/PAA)/PLL-g-PEG film is precisely

9.7 nm (Figure 6.21c). At this optimal thickness, the fluorescence signal exhibits an enhancement factor of 3.8 relative to the reference sample. Similarly, at 663 nm, the fluorescence intensity is increased by a factor of 3.5 (Figure 6.21d). Furthermore, upon exposure to ultraviolet illumination, a pronounced enhancement in the red fluorescence of PFDBT-BIMEG is readily observable. It is noteworthy that PFDBT-BIMEG emits blue fluorescence in the solution phase, yet exhibits red fluorescence in the film state. This fluorescence shift is attributed to the pronounced aggregation of PFDBT-BIMEG molecules within the thin film, which leads to an increased FRET efficiency through both intramolecular and intermolecular pathways.

To further explore the fluorescence enhancement mechanism, fluorescence lifetime measurements were carried out. At the emission wavelength of 427 nm, the fluorescence decay rate is markedly accelerated in the presence of silver nanoparticles. The measured fluorescence lifetimes are 0.5 ns (with silver nanoparticles) and 1.06 ns (without silver nanoparticles), respectively (Figure 6.21e). Conversely, the fluorescence lifetimes at 663 nm remain relatively comparable, being 2.80 and 2.88 ns, respectively (Figure 6.21f). These findings imply that the fluorescence amplification at 427 nm is mostly caused by the accelerated radiative decay generated by the surface plasmon effect of silver nanoparticles, while the amplification at 663 nm is predominantly regulated by the FRET effect [25].

Unlike in the solution phase, the photofunctional components within the thin film exist in an aggregated state, facilitating the realization of aggregation-enhanced effects. In this aggregated state, the internal rotation and vibration of the photofunctional molecules are constrained, leading to a reduction in nonradiative energy dissipation and consequently resulting in enhanced luminescence.

Copper nanoclusters are generated via in situ reduction within a mixed solution containing polyvinylpyrrolidone, polyvinyl alcohol, divalent copper salts, and glutathione, followed by an aging treatment. In this process, dehydration of the precursors induces the aggregation of copper nanoclusters. Such aggregation results in an enhancement of photoluminescence (PL), accompanied by a certain degree of blue shift and an extension of the fluorescence lifetime (Figure 6.22a,b). Ultimately, a Cu NC/polymer composite film with fluorescence enhancement characteristics is obtained. This aggregation process is closely associated with intermolecular interactions, including π–π stacking and hydrogen bonding. These interactions impede the flexibility of the ligands on the copper nanoclusters and suppress nonradiative decay caused by intramolecular motion. Moreover, forming Cu(I)···Cu(I) bonds strengthens the internanocluster interactions, endowing the film with greater rigidity and further restricting nonradiative transitions. Consequently, the luminescence intensity is enhanced. Simultaneously, weakened Cu(I)···Cu(I) bonds within the nanoclusters cause a shift of the PL toward higher energies, leading to a blue shift of the emitted light. Upon photoexcitation, the excited state rapidly relaxes to the S_1 state, followed by an intersystem crossing (ISC) transition to the T_1 state. Therefore, the nanoclusters possess a relatively long fluorescence lifetime. Both the S_1 and T_1 states have notable charge-transfer properties. Prior to film formation, the intramolecular vibration and rotation of the ligands on the

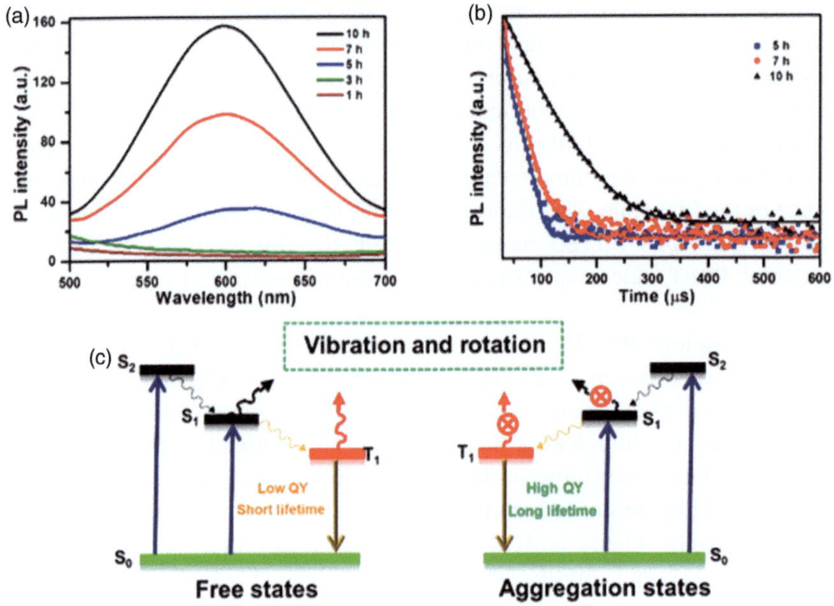

Figure 6.22 (a) PL spectra and (b) PL decay profiles of Cu NC/polymer composite films with varying dehydration times. (c) The PL enhancement mechanism in Cu NC/polymer composite films. *Source:* Wang et al. [26] / with permission of American Chemical Society.

copper nanoclusters result in nonradiative relaxation. However, after forming the composite film, the ligands' intramolecular motion is blocked, reducing the probability of nonradiative relaxation and thereby significantly increasing the PL quantum yield, which ultimately attains 30% (Figure 6.22c) [26].

The tetraphenylethylene (TPE) molecule, which demonstrates the aggregation-induced emission (AIE) effect, is chemically integrated into chiral helical substituted polyacetylenes through a copolymerization procedure. Subsequently, the obtained copolymer is blended with a polymethyl methacrylate (PMMA) solution and cast into a film. After drying, a film with chiral luminescence characteristics is acquired. This system demonstrates the regulation of circularly polarized luminescence (CPL) signals from nothing to emergence. In solution or dispersed states, the chiral conformation of the polymer backbone cannot be effectively transmitted to the fluorophore, leading to the lack of a distinguishable CPL signal. Nevertheless, upon transitioning to the solid-state film, the polymer chains assume a long-range ordered supramolecular structure via organized arrangement or self-assembly (Figure 6.23). This structural organization generates a distinct Cotton effect that overlaps with the fluorescence emission band, thus effectively activating the CPL signal [27].

6.2.2 Energy/Electron Transfer Effect

For the establishment of an effective energy transfer mechanism, it is of paramount importance to select an appropriate energy donor and to precisely control the

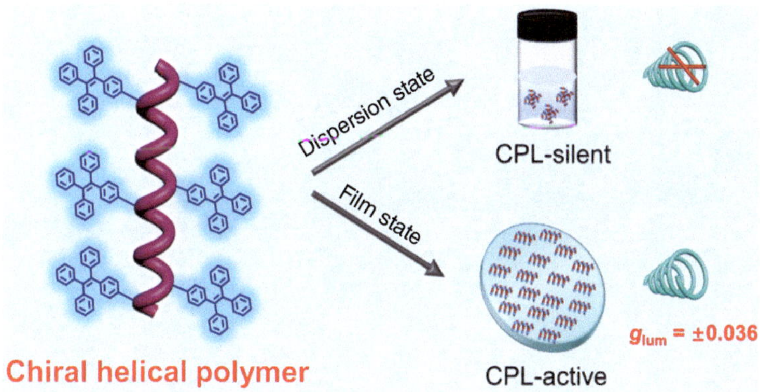

Figure 6.23 Illustration of the CPL-active film. *Source:* Lu et al. [27] / with permission of American Chemical Society.

spacing between the donor and the acceptor through a thin film as a separator. For example, a polyelectrolyte multilayer film can be prepared by means of the LbL technique. Using a quartz plate as the substrate, the negatively charged surface hydroxyl groups of the quartz are exploited to adsorb positively charged PEI. Subsequently, the negatively charged polyelectrolyte polystyrene sulfonate (PSS) and the positively charged polyelectrolyte poly(diallyldimethylammonium chloride) (PDDA) are deposited alternately. A conjugated polymer, PFV, containing quaternary ammonium salt side chains, is then adsorbed onto the PSS layer, followed by the deposition of a pH-responsive polyelectrolyte, poly(allylamine hydrochloride) (PAH). The positive charges on the PAH surface further promote the adsorption of negatively charged DNA, into which the EB dye can specifically intercalate within the double-stranded DNA (dsDNA) structure. As a result, a composite film incorporating two chromophores can be successfully constructed (Figure 6.24a).

From the spectral analysis, it can be seen that the emission spectrum of PFV and the absorption spectrum of EB have a high degree of overlap, which provides a basis for the energy transfer mechanism between the PFV donor and the EB acceptor (Figure 6.24b). In addition, by regulating the thickness of the spacer layer PAH, the FRET efficiency can be optimized. Experiments show that the thickness of the PAH is 2.6, 3.5, and 5.5 nm, respectively, at pH values of 3.0, 4.2, and 5.0; while at pH 7.0, the thickness is approximately 1.7 nm and the FRET efficiency is at its highest (Figure 6.24c). This indicates that the closer the distance between donor and acceptor, the more conducive it is for the occurrence of FRET [28].

Conjugated polymers not only possess excellent luminescent properties but also can act as photosensitizers. By establishing an energy transfer mechanism with oxygen, they can sensitize the production of ROS from oxygen. Taking the flexible polydimethylsiloxane (PDMS) membrane as an example, silver nanoparticles were prepared by in situ reduction, and positively charged poly(L-lysine) (PLL) and negatively charged poly(acrylic acid) (PAA) were alternately deposited. Finally,

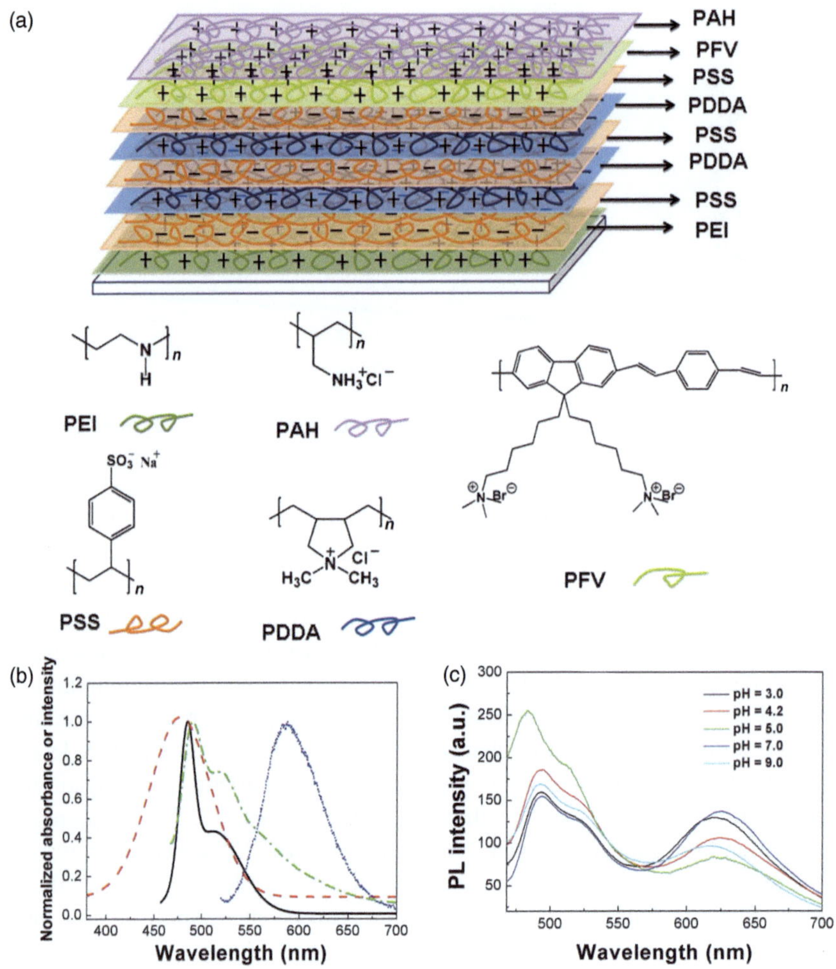

Figure 6.24 (a) The film preparation process exhibiting the FRET effect. (b) Normalized emission spectra of PFV in water (black) and on film (green), along with normalized absorption (red) and emission (blue) spectra of EB. (c) Emission spectra of a PEs/PFV/PAH/dsDNA(EB) film with PAH under various pH conditions. *Source:* Wang et al. [28] / Royal Society of Chemistry.

the conjugated polymer PFEMO with side chains of quaternary ammonium salts was adsorbed on the surface (Figure 6.25a).

According to the Jablonski energy level diagram, electrons are excited to the S_1 state under illumination, and then through ISC, they reach the T_1 state, transferring their energy to oxygen, thereby sensitizing oxygen (Figure 6.25b). During this process, the T_1 is crucial for the production of ROS. Simulation results show that the energy level of PFEMO is 1.96 eV, which is much higher than the 1.09 eV required for sensitizing oxygen. At the same time, the T_1 exciton is distributed in most areas of the main chain, increasing the chance of contact with oxygen, thereby significantly promoting the generation of ROS (Figure 6.25c). The highest triplet-triplet

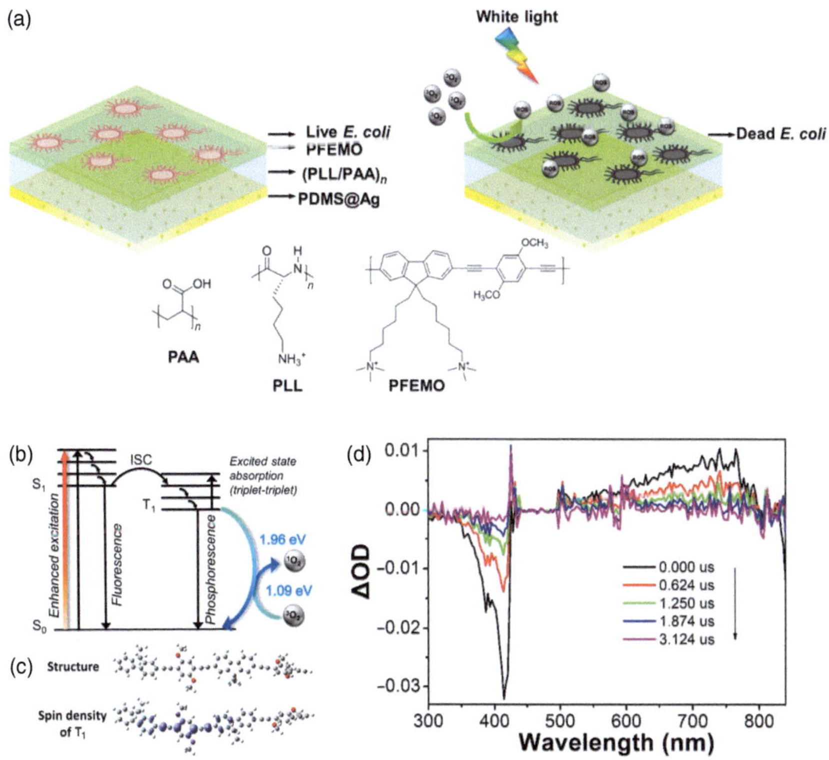

Figure 6.25 (a) The prepared flexible film with photosensitization effect. (b) Stimulated energy level and (c) spin density plot of the T_1 state of PFEMO. (d) Transient absorption spectra of PFEMO. *Source:* Wang et al. [29] / with permission of American Chemical Society.

absorption occurs at 730 nm and lasts roughly 1.52 μs, showing the presence of the T_1 in the conjugated polymer (Figure 6.25d). Therefore, PFEMO can efficiently sensitize oxygen through the energy transfer mechanism [29].

By increasing the conjugation degree of conjugated molecules, the binding ability with target molecules can be effectively enhanced. For instance, pyrene can easily combine with nitroaromatics through electron donor–acceptor interactions. Researchers introduced pyrene into the conjugated polymer framework and synthesized conjugated polymers PyPE-1 and PyPE-2. They cast them onto glass plates, respectively, to prepare photofunctional films 1 and 2. As a control, the conjugated polymer film without pyrene molecules was labeled as PPE.

By incorporating nitroaromatics, the photoinduced electron transfer mechanism can be established (Figure 6.26a). In this process, the electrons on the lowest unoccupied molecular orbital (LUMO) of the donor polymer (PPE or PyPE) will transfer to the adjacent electron acceptor molecule (such as trinitrotoluene [TNT]). Subsequently, the electrons on the LUMO of the acceptor will recombine with the holes on the highest occupied molecular orbital (HOMO) of the donor through non-radiative processes, resulting in fluorescence quenching (Figure 6.26b). Through theoretical simulation, it was found that the LUMO energies of PPE and PyPE

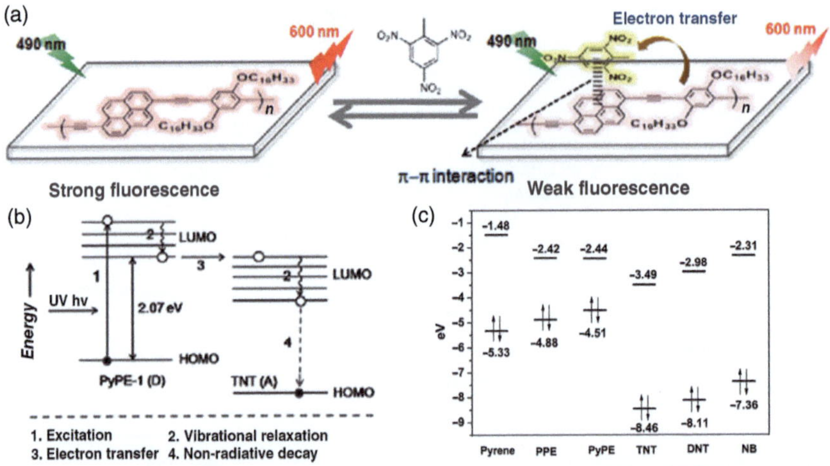

Figure 6.26 (a) The structure of the film with electron transfer effect. (b) The photoinduced electron-transfer mechanism of the film. (c) The calculated HOMO and LUMO energies of PPE and PyPE. *Source:* He et al. [30] / with permission of American Chemical Society.

are extremely similar: −2.42 and −2.44 eV. Thus, their driving forces for the same analyte should be similar (Figure 6.26c). In addition, the internal recombination energies (λ_i) of OPE2 + TNT and Py + TNT were calculated, with values of 0.230 and 0.226 eV, respectively, which are consistent with the typical π–π stacked electron donor–acceptor (D–A) systems. This indicates that the PPE-TNT and PyPE-TNT systems should have similar internal recombination energies. Therefore, the photoinduced electron transfer mechanism enables films 1 and 2 to be used for the high-sensitivity detection of TNT, mainly attributed to the enhancement of electron coupling. In the films, introducing the pyrene to PPE films causes the polymer layer to become more disordered, allowing acceptor molecules to diffuse and form π–π stacked D–A aggregates. This discovery provides important theoretical basis and experimental support for the design and development of highly sensitive chemical sensors [30].

Furthermore, in the thin film system, the separation between the electron donor and the electron acceptor is relatively small. When the electron donor (or acceptor) is in an excited state, it may collide with the ground-state acceptor (or donor) molecule, potentially resulting in transient charge transfer complexes, or even generating an excimer or exciplex.

The researchers used the reprecipitation method to prepare nanoparticles from fluorene or carbazole-based compounds and the conjugated polymer MEH-PPV. The nanoparticles composed of fluorene or carbazole-based compounds (NP) emitted purple or blue fluorescence, serving as the X component; while the MEH-PPV nanoparticles emitted light with a peak at 575 nm in the orange color under light excitation, serving as the Y component (Figure 6.27a,b). By dropping solutions of different volume ratios (X:Y) onto a quartz plate and drying them, a series of binary nanoparticle films was obtained.

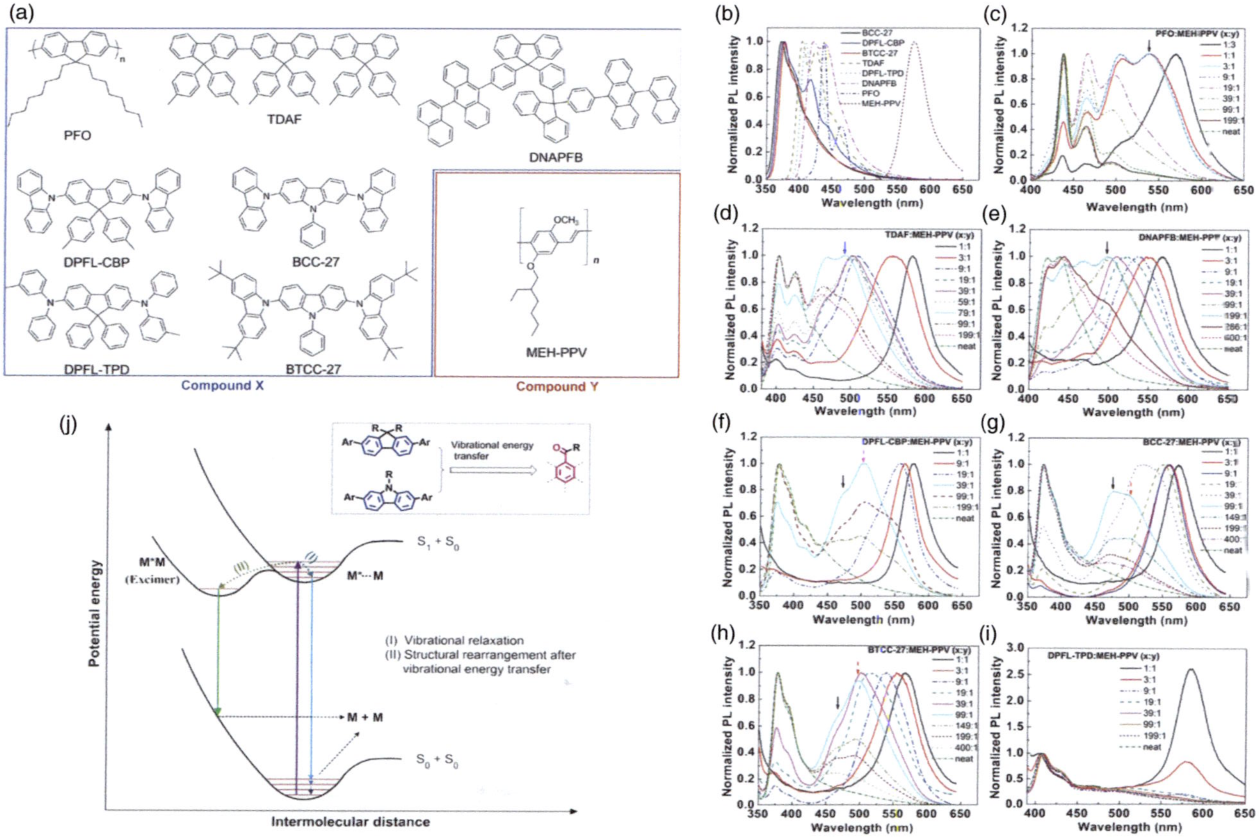

Figure 6.27 (a) The molecular structures and (b) photoluminescence (PL) spectra of compound X (PFO, TDAF, DNAPFB, DPFL-CBP, BCC-27, BTCC-27, and DPFL-TPD) and compound Y (MEH-PPV). PL spectra of X/Y binary nanoparticle films, where X is: (c) PFO, (d) TDAF, (e) DNAPFB, (f) DPFL-CBP, (g) BCC-27, (h) BTCC-27, and (i) DPFL-TPD. (j) The PL mechanism in the binary nanoparticle films. *Source:* Xu et al. [31] / with permission of John Wiley & Sons.

In these composite films, a strong long-wave emission band (>500 nm) can be observed, but its position is not fixed. Instead, it gradually shifts toward the blue as the proportion of X component increases (Figure 6.27c–i). The reason for this phenomenon lies in the fact that after being excited by light, the excited molecules in X nanoparticles can release their excess vibrational energy through two pathways: one is vibration relaxation (I), and the other is the transfer of molecular vibrational energy (II). Specifically, if there are no Y nanoparticles nearby, only the natural emission of the X compound (I) can be observed. However, if Y nanoparticles are present, the excited molecules within X nanoparticles will experience structural rearrangement through long-range molecular inter-vibrational energy transfer (II), thereby promoting the formation of an excimer (Figure 6.27j). Therefore, in addition to DPFL-TPD, excimer emission can also be observed from X nanoparticles.

In the case of DPFL-TPD, due to the nonplanar diphenylamino groups in the molecule, the resulting steric hindrance prevents the excited molecules from rearranging their structures to adapt to the configuration of the excimer. Therefore, no excimer was observed in the binary nanoparticle films containing DPFL-TPD. This study not only reveals the influence of molecular structure and energy transfer mechanism on the luminescence behavior but also provides an important theoretical basis for the design of new optical functional materials [31].

6.2.3 Photoisomerization Effect

Photoisomeric molecules (such as spiropyran and azobenzene) can be introduced into the solid-state film substrate through physical adsorption or chemical bonding, thereby endowing the film with dynamic response functions. For example, by covalently connecting carboxyl-functionalized spiropyran molecules with the GAGAGAGY polypeptide, the SP-8Y complex was successfully prepared. Subsequently, SP-8Y was added to the silk fibroin solution, fully mixed, and then poured into a polystyrene mold. After drying treatment, the SP-8Y/RSF film was obtained.

This film exhibits rapid photochromic properties. Under ultraviolet light, the SP-8Y component rapidly transforms from a closed structure to an open form. It achieves maximum absorption within 5 s, and the color of the film changes to red (Figure 6.28a,b). Furthermore, the film exhibits excellent stability in an aqueous environment, retaining its original color even after 24 h of immersion (Figure 6.28c). By contrast, the SP-COOH/RSF composite film without the polypeptide becomes colorless and transparent due to SP-COOH leakage, which further demonstrates the crucial role of the GAGAGAGY polypeptide in stability [32].

Additionally, the conjugated polymer PAA-C6-Azo, which contains azobenzene and has a negative charge, can be adsorbed onto the aminated substrate's surface through electrostatic interaction. Through the LbL technique, PAA-C_6-Azo and positively charged polyelectrolyte PDAC are alternately deposited to form a multilayer film structure. This film utilizes the host–guest interaction between azobenzene (Azo) and α-CD-RhB to achieve the regulation of optical functions (Figure 6.29).

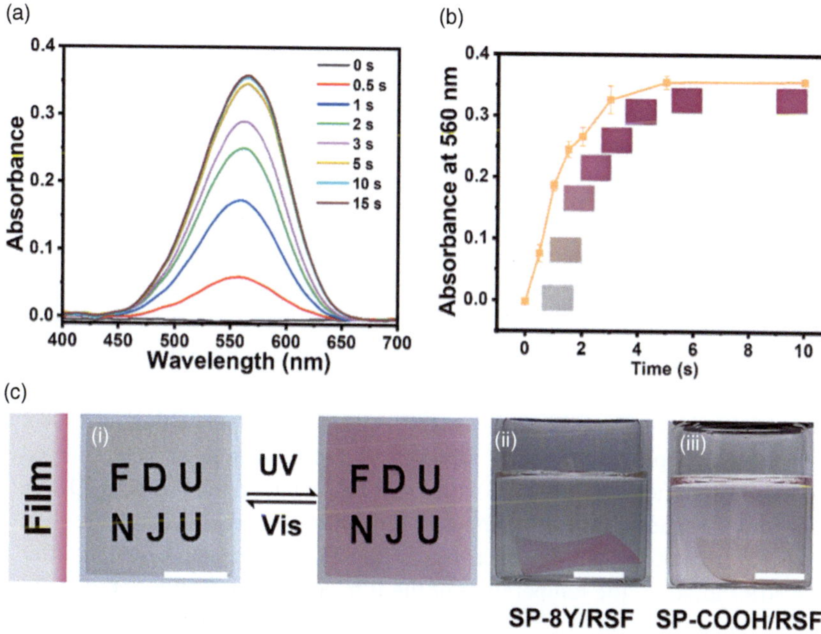

Figure 6.28 (a) The absorption spectra and (b) absorption intensity at 560 nm of the SP-8Y/RSF composite films under different durations of UV light irradiation. (c) (i) Photoisomerization effect of SP-8Y/RSF composite film. (ii) SP-8Y/RSF composite film and (iii) SP-COOH/RSF composite film after immersion in water. *Source:* Chen et al. [32] / with permission of John Wiley & Sons.

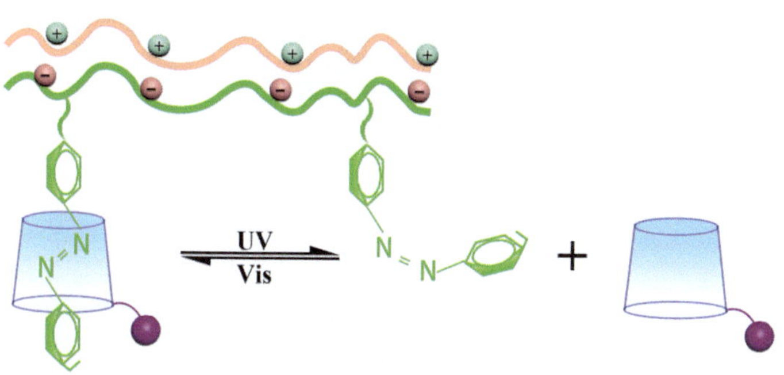

Figure 6.29 Illustration of host–guest interaction mediated by photoisomeric azobenzene (Azo) in the film.

Based on the analysis results of the absorption spectra, as the number of assembly layers increases, the absorption peaks corresponding to the π–π^* transition of the azobenzene group (300–370 nm) and the absorption peak of RhB (566 nm) all show a linear growth trend. This indicates that the deposition process of the multilayer film is regular and controllable. Moreover, the multilayer film exhibits a red fluorescence similar to RhB, further confirming the successful loading of RhB. When the film is exposed to ultraviolet light, the azobenzene molecule changes from the hydrophobic trans configuration to the polar cis configuration, resulting in the disruption of the host–guest interaction between Azo and α-CD-RhB, and RhB is released from the film. This process causes the fluorescence signal to gradually disappear, achieving the function of light-controlled release [33].

6.2.4 Photothermal Effect

Embedding components with photothermal conversion functions in the film enables controllable heat generation. This technology utilizes the unique optical and thermal properties of nanoparticles or specific molecules to efficiently convert light energy into heat energy, thereby achieving localized heating effects within the film.

After adsorbing positively charged dendritic PEI on the silicon substrate, gold nanorods (GNRs) with photothermal conversion properties were further adsorbed. This process was repeated 10 times, followed by assembling negatively charged poly(sodium-4-styrenesulfonate) (PSS) and positively charged doxorubicin (DOX), also repeating 10 times. Finally, a composite film with photothermal function was obtained (Figure 6.30a). Through near-infrared (NIR) imaging observation, after 5 min of NIR radiation, the temperature of the film containing GNRs significantly increased, reaching up to 60.7 °C; when the light source was turned off, the

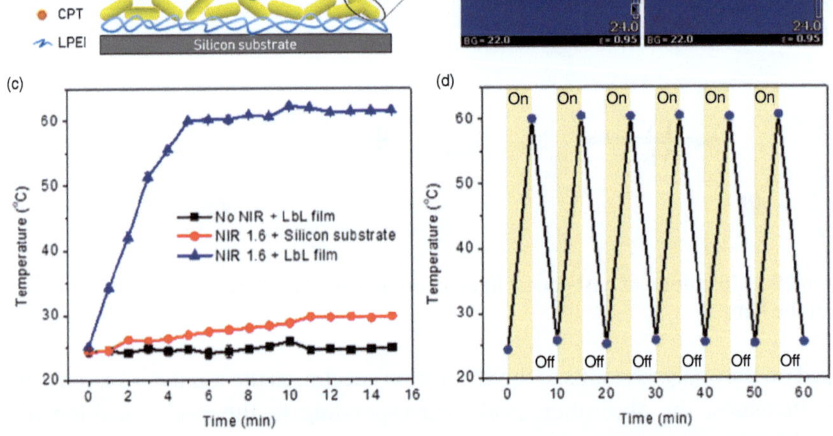

Figure 6.30 (a) The preparation of LbL films containing GNRs. (b) Infrared image of the LbL film before and after irradiation. (c) Temperature change of the LbL film and silicon substrate with and without irradiation. (d) Temperature response of the LbL film under on–off NIR irradiation. *Source:* Singh et al. [34] / with permission of American Chemical Society.

film temperature decreased by 25.7 °C (Figure 6.30b). Without irradiation, the temperature of the film did not change. In contrast, the temperature of the silicon substrate only reached 29 °C after 15 min irradiation (Figure 6.30c). This indicates that the photothermal effect mainly originates from the gold nanorods in the film. During the five cycles of NIR irradiation and switching, the temperature increase and decrease amplitudes of the film remained consistent, demonstrating the excellent photothermal stability of the film (Figure 6.30d) [34].

GSH-Au NCs with photothermal conversion ability were added to the gellan gum (GG) solution, and then poured into the mold for drying. Subsequently, they were cross-linked with Ga^{2+} to prepare the fibrous-Au-GG film. When this fiber film was exposed to 660 nm laser radiation, the infrared imaging showed that the irradiated area of the composite film rapidly increased by 35 °C within 3 min, demonstrating excellent photothermal responsiveness (Figure 6.31a). Subsequently, the temperature remained stable, as a thermal balance was created through ambient heat dissipation and photothermal conversion. After turning off the laser, the temperature gradually decreased. In contrast, the fibrous-GG film without GSH-Au NCs did not change in temperature under the same irradiation conditions. To verify its performance further, the fibrous-Au-GG film was immersed in water, and the temperature was recorded. The photothermal conversion efficiency was calculated to be as high as 31.2% based on the cooling data (Figure 6.31b,c). Additionally, the fibrous Au-GG film was subjected to six photothermal cycle tests, with almost no

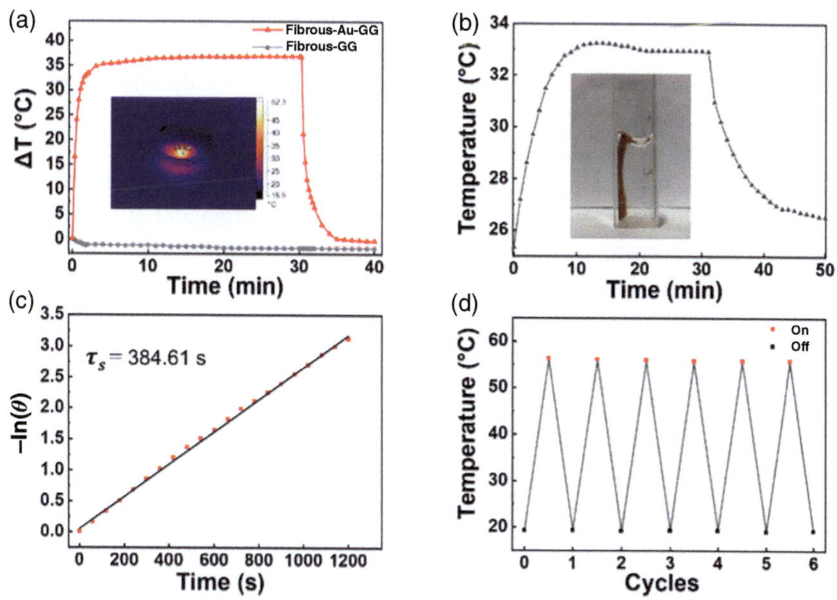

Figure 6.31 (a) Temperature change and thermal image of fibrous-Au-GG and fibrous-GG with irradiation. (b) Temperature change recorded under radiation in water containing fibrous Au-GG. (c) The change of negative natural logarithm of temperature as a function of cooling time. (d) Temperature response of the fibrous-Au-GG film under on–off NIR irradiation. *Source:* Li et al. [35] / Royal Society of Chemistry.

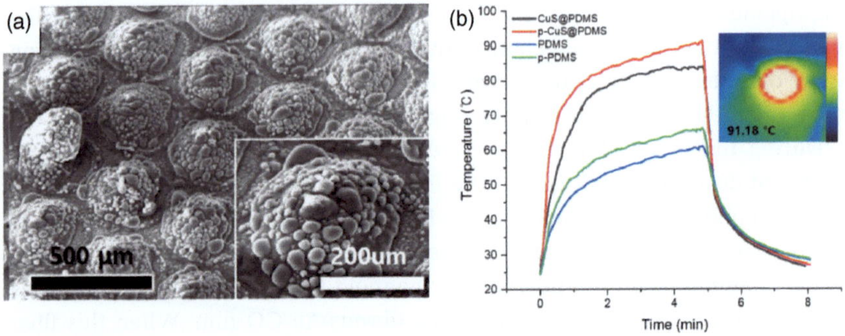

Figure 6.32 (a) SEM image of p-PDMS film. (b) Temperature changes of CuS@PDMS, p-CuS@PDMS, PDMS, and p-PDMS films under NIR irradiation, accompanied by the infrared image of p-CuS@PDMS film. *Source:* Chung et al. [36] / with permission of Elsevier.

significant change observed in the peak temperature (Figure 6.31d). This indicates that the film has excellent photothermal stability [35].

In addition, a patterned film can be prepared using a template. For example, a patterned PU template can be fabricated, and a mixture solution of photothermal CuS nanoparticles, PDMS, and perfluorodecyltrichlorosilane is poured into the PU mold. After casting and drying, it is finally solidified, and a p-CuS@PDMS film is obtained. By observing the SEM images, the surface of the p-PDMS film presents a honeycomb-like arrangement of protrusions, which are created by the aggregation of smaller particles and have a dome shape (Figure 6.32a). Their height is $91.7 \pm 4.9\,\mu m$, and the diameter is $236.1 \pm 15.8\,\mu m$. Under NIR radiation, the temperature changes of different films are measured. The results show that the temperature of the ordinary PDMS film rises to 61 °C, while the patterned p-PDMS film rises to 66 °C. The temperature of the CuS@PDMS film increases significantly to 84 °C, and the patterned p-CuS@PDMS film has the maximum temperature of 91 °C (Figure 6.32b). This phenomenon may be linked to a greater surface area of the patterned structure, thereby enhancing the light absorption efficiency. This research is an important reference in the field of micro-nano structure design and photothermal applications [36].

6.2.5 Other Photofunctions

Room-temperature phosphorescence (RTP) is a photophysical phenomenon that enables long-lasting phosphorescence emission at room temperature and normal pressure. Unlike traditional phosphorescence materials, RTP materials do not require low temperatures or inert gas protection. They can achieve stable luminescence simply through molecular structure design and material optimization. In the film material system, a rigid environment can effectively suppress nonradiative transitions through the regulation of aggregation state, thereby constructing high-performance RTP functional films.

The researchers synthesized a variety of imidazole derivatives and copolymerized them with the acrylamide (AM) monomer to successfully prepare three polymers:

PAMI, PAMMI, and PAMBI (Figure 6.33a). Subsequently, by using the injection molding and drying processes, films of PAMMI-2, PAMI-2, and PAMBI-2 were obtained. The experimental findings showed that the fluorescence and phosphorescence intensities of the films exhibited a consistent change pattern, with the specific order being PAMMI-2 > PAMBI-2 > PAMI-2 (Figure 6.33b).

The transient emission spectroscopy analysis revealed that the average excited-state lifetimes (τ_p) of the PAMI-2, PAMMI-2, and PAMBI-2 films were 2.4, 136.6, and 88.7 ms, respectively. These values all reached the millisecond level, confirming the phosphorescent properties of the films (Figure 6.33c). Under ultraviolet light irradiation, the PAMI-2 film exhibited blue luminescence, while the PAMMI-2 and PAMBI-2 films showed intense cyan luminescence. After the lights were turned off, green phosphorescent signals could still be observed in the PAMMI-2 and PAMBI-2 films (Figure 6.33d) [37].

The phosphorescence emission mechanism is primarily attributed to amide groups forming a rigid environment through hydrogen bonding on the PAM chain. This effectively reduces nonradiative transitions of luminescent groups, stabilizing the triplet state and protecting it from interference by external factors, such as oxygen and water. Moreover, compared with PAMI-2, the heavy atom effects of the imidazole cations and iodine ions in PAMMI-2 and PAMBI-2 significantly enhance the spin-orbit coupling (SOC) effect and promote the ISC process. However, the benzyl rotation in PAMBI-2 increases the energy dissipation of nonradiative transitions, thus PAMMI-2 exhibits the strongest phosphorescence performance at room temperature. Through molecular design and polymer engineering, functional films with excellent RTP properties have been successfully achieved.

Optical functional polymer films have extensive applications in the field of electroluminescence (EL), especially when used to regulate the fluorescence–phosphorescence emission system. They can be utilized to develop white organic light-emitting diodes (WOLEDs). In the research, two conjugated polymers with blue emission, PTPATPP and PTPATPPO, were synthesized. The organic layer of the ITO/PEDOT:PSS/conjugated polymer:[Ir(2-PhQ)$_2$acac]/SPPO13/Al device was prepared using the spin coating method (Figure 6.34a). In the structures of these two polymers, triphenylamine acts as the electron donor, and triphenylphosphine oxide/triphenylphosphine acts as the electron acceptor, forming a typical D–A (donor–acceptor) structure. At the same time, triphenylamine and triphenylphosphine oxide/triphenylphosphine can respectively serve as the hole transport unit and electron transport unit, further optimizing the performance of the device.

Through calculation, it was found that the ΔE_{ST} (singlet-triplet energy level difference) of PTPATPP and PTPATPPO was 0.49 and 0.47 eV, respectively. This relatively small energy level difference is conducive to achieving fluorescence-phosphorescence (F—P) hybrid luminescence. In the fabricated devices, the observed white light spectrum range covered 420–750 nm and consisted of two parts. One part came from the emission of the polymer, and the other part came from the emission of Ir(2-phq)$_2$acac (Figure 6.34b,c). This indicates that in the emission layer, the polymer not only successfully played the role of a fluorescent

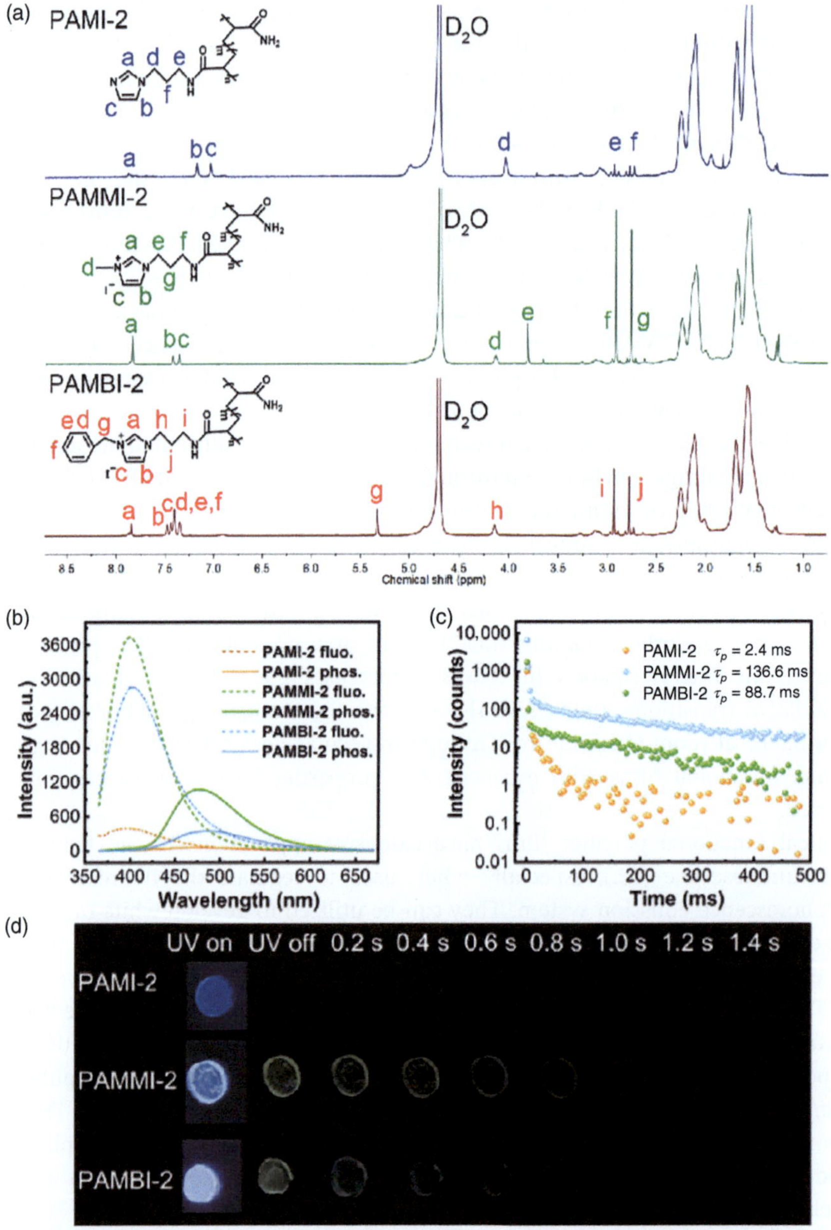

Figure 6.33 (a) ^{1}H NMR spectra and molecular structures of PAMI-2, PAMMI-2, and PAMBI-2 copolymers. (b) Fluorescence and phosphorescence emissions of PAMI-2, PAMMI-2, and PAMBI-2 copolymer films. (c) Emission decay profiles of PAMI-2, PAMMI-2, and PAMBI-2 copolymer films. (d) Photographs of PAMI-2, PAMMI-2, and PAMBI-2 copolymer films with the UV light (on and off). *Source:* Li et al. [37] / with permission of Elsevier.

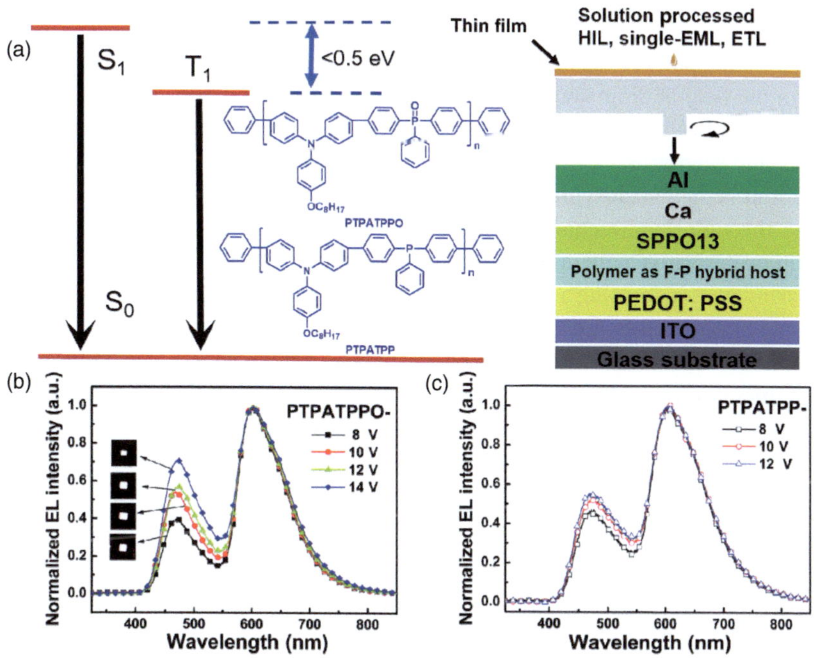

Figure 6.34 (a) The structure and energy levels of an F–P device based on PTPATPPO or PTPATPP. EL spectra of (b) PTPATPPO and (c) PTPATPP based F–P devices at different voltages. *Source:* Wang et al. [38] / Royal Society of Chemistry.

emission material but also effectively promoted phosphorescence emission as a main material. This design provides a feasible strategy for achieving efficient WOLEDs [38].

6.3 Bioapplications of Photofunctional Polymer Composite Films

Photofunctional polymer composite films, with their adjustable structure and diverse optical functions, have demonstrated significant advantages in the biomedical field. In the detection field, their high sensitivity and rapid response characteristics make them an ideal material for manufacturing high-precision sensors. In the treatment field, their photodynamic and photothermal conversion properties open up new avenues for precise treatment. In the field of energy conversion, photofunctional films are often used as light-absorbing layers or electron transport layers to effectively enhance the photoelectric conversion efficiency, providing important support for the innovation of electronic devices. In the packaging field, the intelligent design and high barrier performance of optical functional films effectively extend the shelf life of food and medicine, while enhancing the functionality and practicality of packaging.

6.3.1 Optical Detection

Photofunctional polymer composite films are an extremely important material system in the field of biomedicine. They can be used for highly sensitive detection of metal ions, small molecules, biological macromolecules (such as nucleotides and proteins), and even cells.

For example, 3-glycidoxypropyltrimethoxysilane is used to functionalize the substrate through SAM technology. This is followed by the coupling of a novel polyfluorene derivative with amino-modified side chains at specific locations using a microarray spotter, forming a sensing microarray. Cu^{2+} has an empty d orbital and can strongly quench the fluorescence of nearby fluorophores through electron or energy transfer. In the experiment, when the film was immersed in a Cu^{2+} solution, the blue fluorescence of the conjugated polymer was significantly quenched, with an intensity reduction of 50% (Figure 6.35a). The reason for this phenomenon is that after Cu^{2+} binds to the conjugated polymer, the excited conjugated polymer undergoes electron transfer, resulting in the quenching of the fluorescence. It is worth noting that the fluorescence of the film in the mixed metal ion (such as Hg^{2+}, Zn^{2+}, and Ni^{2+}) solution did not decrease (Figure 6.35b), indicating that this detection system has high selectivity for Cu^{2+} [39].

A dual-emissive film can be fabricated by integrating gold nanoclusters stabilized with histidine (AuNC@histidine), which exhibit blue fluorescence, with the polymer poly-BrNpA, characterized by orange RTP emission. The preparation process of this film included mixing the PVA solution with the above two materials and forming the final sensing film through casting and drying. In this system, hydrogen bonds connect the polymer chains, helping to fix poly-BrNpA and suppress the

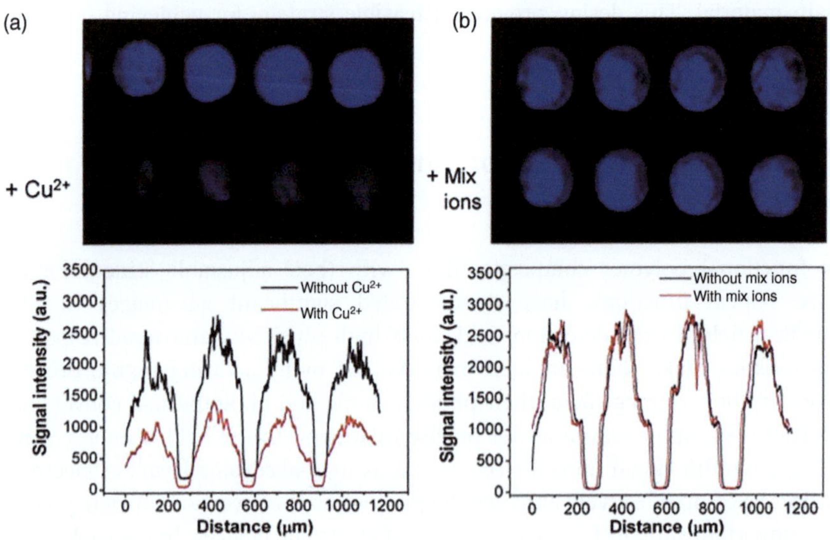

Figure 6.35 Fluorescence images and the signal intensities of polyfluorene modified microarrays before and after treating with (a) Cu^{2+} ions and (b) mixed ions. *Source:* Lv et al. [39] / with permission of John Wiley & Sons.

nonradiative relaxation process. This forms a highly cross-linked network structure, thereby enhancing the copolymer system's RTP emission.

However, the hydrogen bonds between these molecules are susceptible to the influence of environmental humidity. As the humidity increases, the hydrogen bonds may be disrupted, causing the phosphorescence emission intensity of poly-BrNpA to gradually decrease and eventually drop to one-quarter of the original level. In contrast, the fluorescence emission of AuNC@histidine is not affected by humidity and remains stable (Figure 6.36a). This change is visible to the naked eye when illuminated with a 365 nm light source. The change in Commission Internationale de I'Eclairage (CIE) chromaticity coordinates accurately reflects the change in color of the emission, which changes from a red signal (humidity of 10%) to a white signal (humidity of 40%) and finally to a green signal (humidity of 90%) (Figure 6.36b,c). Moreover, during the cyclic alternation process of humidity between 10% and 90%, the RTP signal of the film exhibits dynamic response characteristics (Figure 6.36d). Therefore, this light-functional polymer composite film not only has high sensitivity to humidity but also can be reused, demonstrating its potential application value in optical sensing [40].

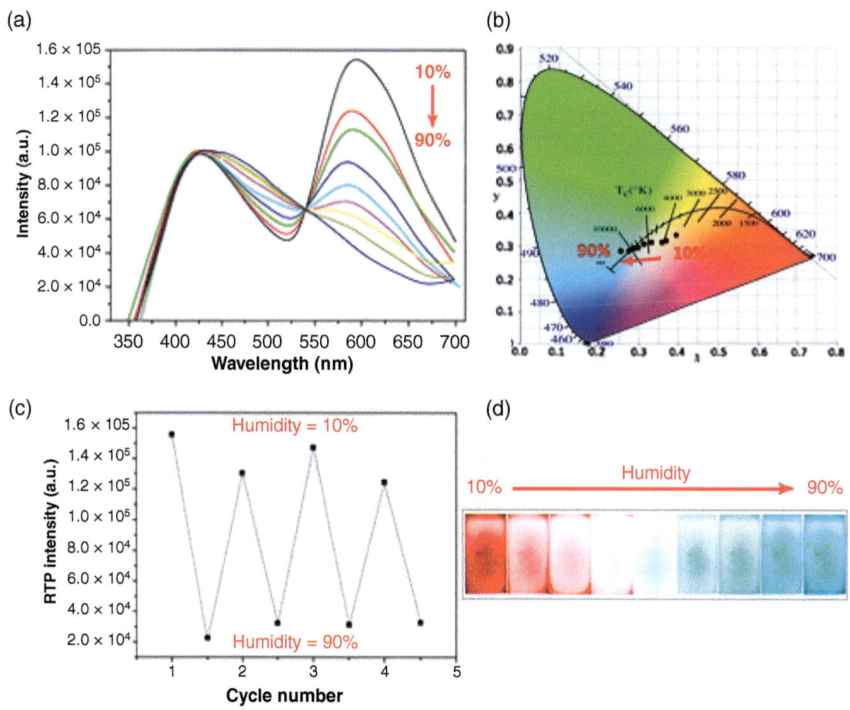

Figure 6.36 (a) Variation of emission intensity of the film at humidity ranging from 10% to 90%. (b) CIE coordinates derived from the data in (a). (c) Photoluminescence images of the film at humidity ranging from 10% to 90%. (d) Reversibility of the film's RTP intensity in response to humidity. *Source:* Wang et al. [40] / with permission of American Chemical Society.

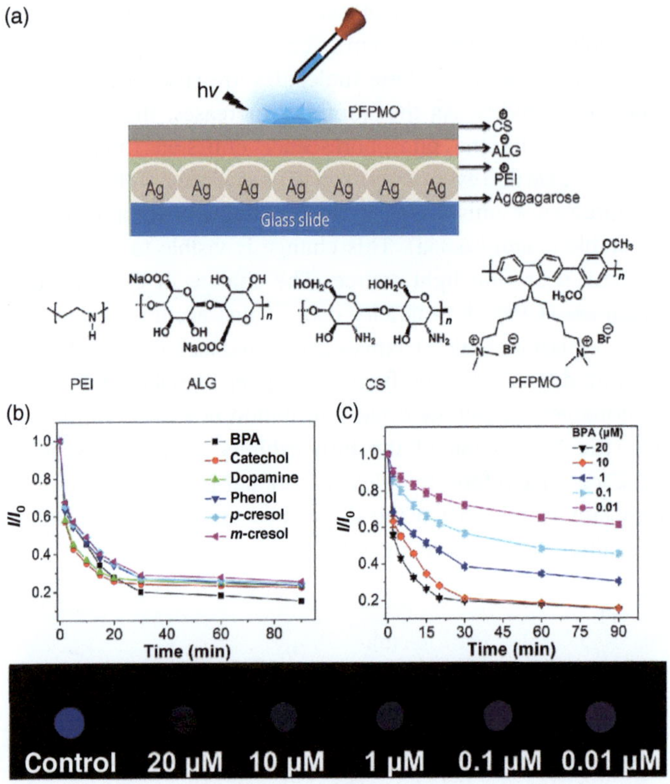

Figure 6.37 (a) The structure of the LbL film. Variation of I/I_0 over time with (b) different phenolic compounds and (c) varying concentrations of BPA. (d) Fluorescence images of PFPMO under UV light at different BPA concentrations. *Source:* Wang et al. [41] / Springer Nature / CC BY-NC-SA 3.0.

Using silver nanoparticles as the substrate, LbL technology was applied to construct intermediate layers of polyelectrolyte (PEI), chitosan, and sodium alginate. By precisely controlling the distance between the silver substrate and the surface-positively charged conjugated polymer PFPMO, the fluorescence signal was enhanced by 8.5 times through the MEF effect (Figure 6.37a). Subsequently, the negatively charged tyrosinase was introduced into the nanocomposite film. Quinones are created when phenolic compounds undergo oxidation, which is catalyzed by tyrosinase. Then the generated quinones quench the fluorescence of PFPMO through an electron transfer mechanism.

It can be observed from Figure 6.37b that after adding different phenolic compounds, with the extension of reaction time, quinones gradually form, and the fluorescence emission intensity of PFPMO decreases accordingly. Specifically, for catechol and dopamine, fluorescence quenching reaches its maximum after 20 min. For bisphenol A (BPA), phenol, *p*-cresol, and *m*-cresol, the quenching degree reaches its maximum approximately 30 min later. Among them, the quenching efficiency of BPA is the highest, reaching 85%. By comparing the structures of these phenolic compounds, it can be found that the reaction time between catechol and dopamine is relatively short. This is because their reaction process only involves a single-step enzymatic reaction directly converting catechol into quinone. The reactions of other phenolic compounds are relatively slow because their reaction process involves multiple enzymatic reactions: first, the phenolic substrate is converted into catechol, and then quinones are formed. In addition, the uniqueness of BPA lies in the fact that it contains two phenolic rings in its molecule, thus enabling the formation of two quinone structures. In contrast, other phenolic compounds can only generate one quinone structure, which gives BPA a stronger fluorescence quenching effect. Further tests were conducted on different concentrations of BPA, and it was found that the emission of PFPMO gradually decreased with the increase of reaction time (Figure 6.37c). Under the UV light, it can be directly observed that the fluorescence intensity of the tested sample demonstrates a notable alteration compared with the blank sample (Figure 6.37d). Therefore, this system can sensitively detect BPA, with a detection limit as low as 0.01 μM. This research provides an effective method for the high-sensitivity detection of phenolic compounds [41].

Cellulose nanofibers modified with fluorescein isothiocyanate (FITC) and proporphyrin IX (PpIX) were prepared into a thin film, BAsCNF. Among them, FITC in the film serves as the reactive substance for biogenic amines. With the increase in the concentration of biogenic amines, the fluorescence color gradually intensifies. CA-PpIX does not react with biogenic amines, so it can be used as a background color. When the film is sealed in contact with fresh shrimp, a gradual color change is observed over time, transitioning from red to orange and ultimately shifting to a yellowish-green color. At 25 °C, BAsCNF turns yellowish-green the next day. However, in a low-temperature environment of −18 °C, it does not turn orange until the sixth day (Figure 6.38). This color change can effectively reflect the freshness of shrimp, providing an intuitive and sensitive method for food preservation detection [42].

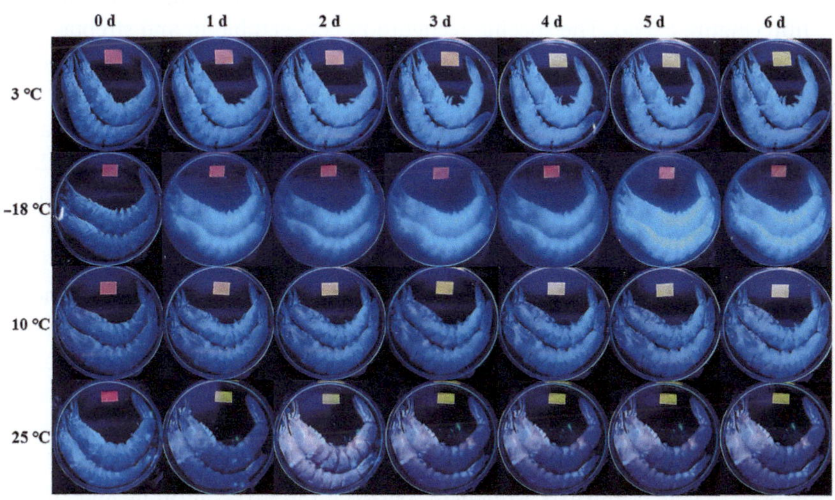

Figure 6.38 The fluorescence images of BAsCNF films on shrimp stored at 3, −18, 25, and 10 °C. *Source:* Reproduced with permission from Quan et al. [42] / Elsevier.

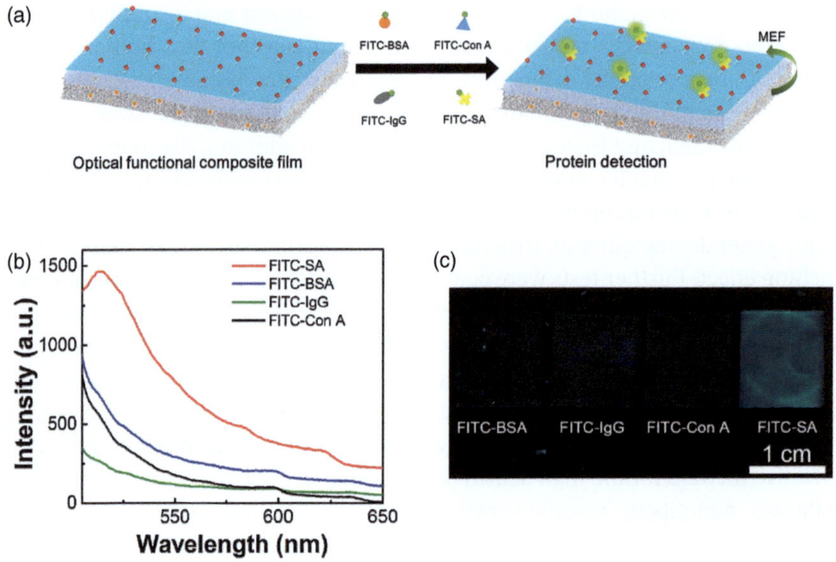

Figure 6.39 (a) Protein detection mechanism of the photofunctional composite film. (b) FITC emission intensity and (c) fluorescence images of the films after interaction with different proteins. *Source:* Wang et al. [43] / with permission of Elsevier.

Furthermore, the fabrication of multilayer films on filter paper represents an efficient and versatile approach. The inherent porous architecture of the filter paper facilitates effective adsorption of $HAuCl_4$, followed by the in situ generation of gold nanoparticles (AuNPs) via reduction with ascorbic acid. On this basis, by using self-assembly technology, polyethyleneimine (PEI), poly(glycolic acid) (PGA), and poly(L-lysine)-graft-poly(ethylene glycol)-biotin (PLL-g-PEG-biotin) can be adsorbed successively on the surface of gold nanostructures to form a stable multilayer film (Figure 6.39a). The specific recognition between biotin and streptavidin (SA) was utilized to capture FITC-labeled SA. Through spectral analysis, the FITC fluorescence signal of the target protein SA can be clearly observed, while other proteins do not show any fluorescence signal (Figure 6.39b,c). This result indicates that optical functional composite films can selectively capture and detect SA. Moreover, the fluorescence decay rate of FITC has markedly increased owing to the MEF effect, while the fluorescence signal intensity has been amplified by a factor of 3.5. This significantly improves the sensitivity and accuracy of the detection [43].

The detection of DNA sequences can be achieved through a similar method. Specifically, multilayer films of chitosan (CS) and sodium alginate (ALG) were used to construct the detection system. Among them, ALG not only serves as a reducing agent but also as a stabilizer, capable of in situ reducing silver ammonia solution, thereby preparing silver (Ag) nanoparticles. In addition, the fluorescent probe acriflavine was introduced into the system, which can be embedded in DNA to detect changes in DNA sequences. Between the silver substrate and DNA, a multilayer film of CS and poly(acrylic acid) (PAA) was assembled to precisely control the distance between Ag nanoparticles and acriflavine embedded in DNA.

An increase in the DNA sequence results in a greater number of acriflavine being implanted, significantly enhancing the fluorescence signal. This characteristic enables the system to effectively monitor the DNA elongation process catalyzed by TdT (terminal deoxynucleotide transferase) [44].

Introducing specific antibodies on the film surface can achieve the capture of specific antigens and even tumor cells. As shown in Figure 6.40a, Ag NPRs are used as the base and are covered with self-assembled multilayer films. When the antigen undergoes biocoupling with the carboxyl group on PAA, the recognition between the antigen and the antibody alters the distance between the outermost adsorbed PFVCN and the bottom silver nanostructure, thereby regulating the MEF effect and further controlling the PFVCN fluorescence signal. The anti-PSA was selected in the experiment to capture the PSA antigen. The results showed that the fluorescence intensity decreased significantly. However, after interacting with albumin, the fluorescence intensity remained almost unchanged. Furthermore, the film without Ag NPR did not show a fluorescence signal response (Figure 6.40b). Further research indicates that the PFVCN's fluorescence intensity falls as the

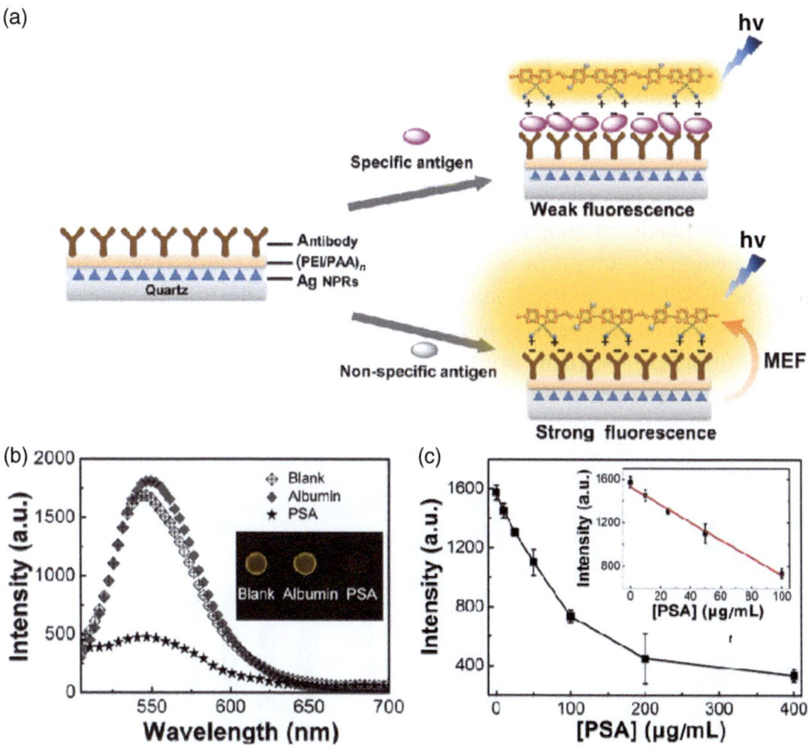

Figure 6.40 (a) The structure and detection mechanism for a special antigen of the photofunctional composite film with MEF effect. (b) Emission spectra and fluorescence images of the film before and after exposure to PSA and albumin. (c) Changes in PFVCN intensity on the film with varying concentrations of PSA. *Source:* Wang et al. [24] / with permission of John Wiley & Sons.

PSA concentration rises, and the detection limit of this composite film for PSA is 9.7 μg/mL (Figure 6.40c). The reason for the decrease in fluorescence is that when the anti-PSA antibody binds to the target antigen PSA, the separation between Ag NPR and PFVCN grows, resulting in a weakened MEF effect and thus generating a weaker fluorescence signal. However, when interacting with nontarget proteins, the distance between the metal and the fluorescent substance remains unaffected, the MEF effect is maintained, and the fluorescence intensity remains unchanged [24].

Furthermore, when the antiepithelial cell adhesion molecule (anti-EpCAM) is conjugated to the film surface, EpCAM-positive cell lines can be captured, such as the human breast cancer cell line MCF-7, the triple-negative breast cancer cell line MDA-MB-231, and the colonic adenocarcinoma cell line HT-29 (Figure 6.41a). Experiments show that after interacting with MCF-7, HT-29, and MDA-MB-231 cells, the fluorescence intensity of PFVCN decreased by 86%, 80%, and 79% respectively. However, for EpCAM-negative cells, HeLa and Jurkat cells showed only a 28% decrease, and no fluorescence intensity was observed after interaction with 293T cells (Figure 6.41b). Therefore, this composite film based on distance-dependent MEF effect shows good application potential in the specific and selective recognition of cells. Therefore, by introducing specific antibodies on the film surface and combining them with the MEF effect, highly sensitive and selective detection of target antigens and tumor cells can be achieved. This technology provides new ideas and methods for biomedical detection and cancer diagnosis.

In addition to cells, bacteria can also be detected in films. The substrate of this film is composed of silver nanoparticles, and the molecular brush structure on its surface can effectively prevent the adhesion of proteins or bacteria (Figure 6.42a). By adding the positively charged conjugated polymer PFDBT-BIMEG, negatively charged bacteria can be captured and immobilized on the film surface. Furthermore, the two-color fluorescence arising from FRET within PFDBT-BIMEG was observed (Figure 6.42b). The MEF effect of the thin film substantially enhances this dual-color fluorescence, thereby improving both the sensitivity and accuracy of the detection.

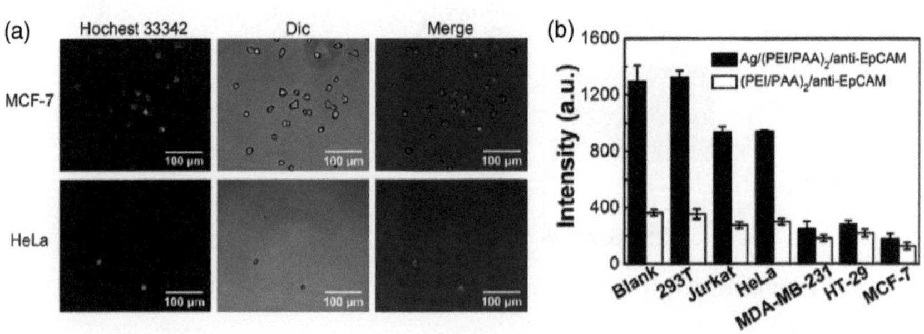

Figure 6.41 (a) Fluorescence images of Hochest 33,342-stained MCF-7 and HeLa cells captured on Ag/$(PEI/PAA)_2$/anti-EpCAM film. (b) Changes in PFVCN intensity on Ag/$(PEI/PAA)_2$/anti-EpCAM and $(PEI/PAA)_2$/anti-EpCAM films exposure to six distinct cells. *Source:* Wang et al. [24] / with permission of John Wiley & Sons.

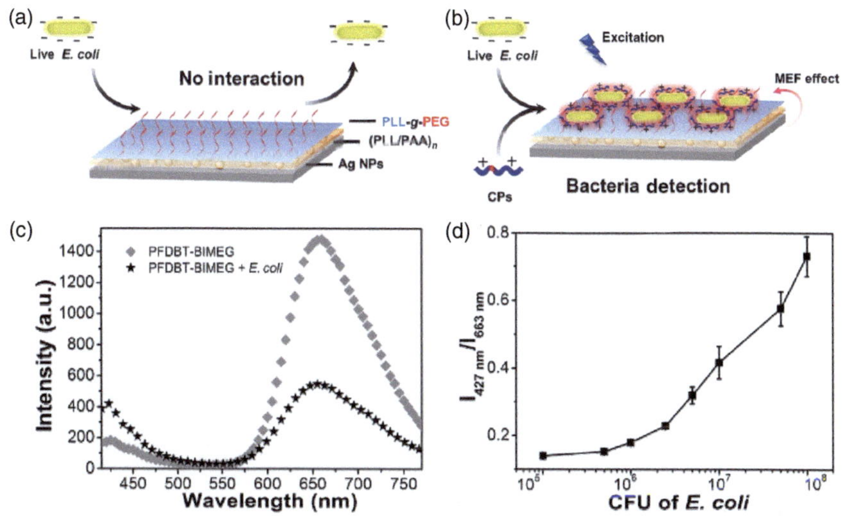

Figure 6.42 (a) The structure and (b) the mechanism of bacteria detection of the photofunctional composite film. (c) Emission spectra of PFDBT-BIMEG on the film, both before and after capturing *E. coli*. (d) Ratio of $I_{427\,nm}$ to $I_{663\,nm}$ of PFDBT-BIMEG on the film at varying *E. coli* concentrations. *Source:* Wang et al. [25] / with permission of John Wiley & Sons.

After interacting with *Escherichia coli*, the 427 nm emission of PFDBT-BIMEG was significantly enhanced, while the emission at 663 nm decreased (Figure 6.42c). This alteration of the signal indicates that after interacting with bacteria, the molecular aggregation degree of PFDBT-BIMEG has changed. Meanwhile, its interaction with bacteria also leads to a weakened self-aggregation of PFDBT-BIMEG on the planar substrate and a decrease in energy transfer efficiency. When the concentration of *E. coli* increased, the FRET ratio ($I_{427\,nm}/I_{663\,nm}$) increased linearly (Figure 6.42d). Therefore, by simply monitoring the changes in FRET efficiency, this biological detection system can quickly, sensitively, and efficiently detect and count the number of bacteria [25].

6.3.2 Optical Therapy

Photodynamic therapy utilizes photofunctional polymer composite films as implantable or topical platforms. Under specific light irradiation, the photosensitizers loaded on the films are stimulated to generate ROS, especially singlet oxygen $^1O_2$, achieving precise killing of pathogenic microorganisms or tumor cells. Compared with systemic administration, film carriers offer spatially controllable therapeutic effects and significantly reduce systemic toxicity. Its designable morphology (such as flexible patches) ensures efficient oxygen penetration to support ROS generation, while the polymer matrix (such as biocompatible PDMS) optimizes the stability, local retention, and adhesion to biological tissues of the photosensitizer.

For instance, when FDA-approved photosensitizer verteporfin (VP) is added to a PDMS film and exposed to 690 nm light, the film can generate $^1O_2$ (Figure 6.43a).

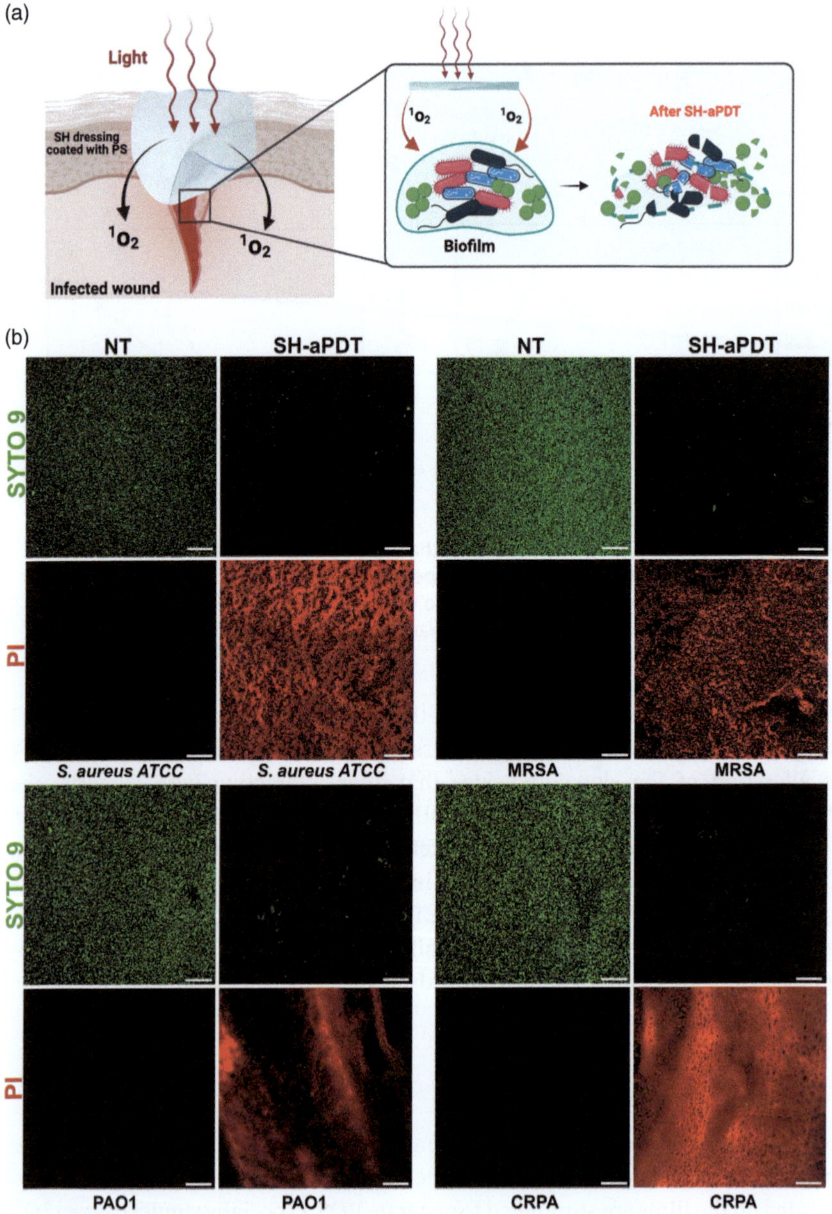

Figure 6.43 (a) Mechanism of the SH-aPDT film for disrupting biofilm under irradiation. (b) Fluorescence images of ATCC, MRSA, PAO1, and CRPA biofilms after treatment with SH-aPDT under irradiation and staining with SYTO 9/PI. *Source:* Cabral et al. [45] / with permission of American Chemical Society.

Researchers interacted the obtained SH-aPDT film with the biofilms of different bacteria and evaluated the survival rate of the biofilms through a fluorescence confocal microscope. In the experiment, green fluorescence was released by living bacteria that were stained with SYTO 9, while red fluorescence was released by dead bacteria that were stained with PI. The results showed that, regardless of whether it was Gram-negative bacteria (such as *Pseudomonas aeruginosa* PAO1 and carbapenem-resistant *P. aeruginosa* CRPA) or Gram-positive bacteria (such as *Staphylococcus aureus* ATCC and methicillin-resistant *S. aureus* MRSA), most biofilms died under light irradiation and emitted red fluorescence. In contrast, the control group did not show any killing effect on the bacteria (Figure 6.43b). In addition, SH-aPDT films can induce the rupture of biofilms, causing their surfaces to become irregular and the porosity to increase [45].

Gold nanocluster (GSH-AuNC) with photosensitive properties was encapsulated on the surface of SiO_2 and further fixed on PDMS films to form SiO_2-AuNC@PDMS composite films (Figure 6.44a). Based on the AIE effect of GSH-AuNC, the fluorescence quantum yield of the system was significantly increased, and the ability to generate ROS was enhanced. Upon exposure of the *E. coli* suspension to the films, experimental results indicated that the PDMS film in the control group exhibited no bactericidal activity. In contrast, the survival rate of *E. coli* on the SiO_2-AuNC@PDMS film decreased to approximately 1% following 20 min of irradiation (Figure 6.44b). Notably, irradiation alone did not result in bacterial inactivation. These findings demonstrate that the integration of the SiO_2-AuNC coating markedly enhances the antibacterial efficacy of the PDMS film (Figure 6.44c). This provides an important idea for the development of new antibacterial film materials [46].

Photofunctional polymer composite films serve as a versatile platform for heat-mediated biomedical applications by integrating efficient photothermal conversion properties with biocompatible carrier matrices. The photothermal conversion agent embedded within the film demonstrates high efficiency in transducing NIR light—characterized by its deep tissue penetration—into thermal energy, thereby inducing a controlled and localized temperature elevation. This precise thermal modulation is particularly advantageous for tumor thermal ablation, enabling selective cytotoxicity. Furthermore, NIR irradiation can induce structural changes within the polymer matrix, modulating membrane permeability and facilitating on-demand,

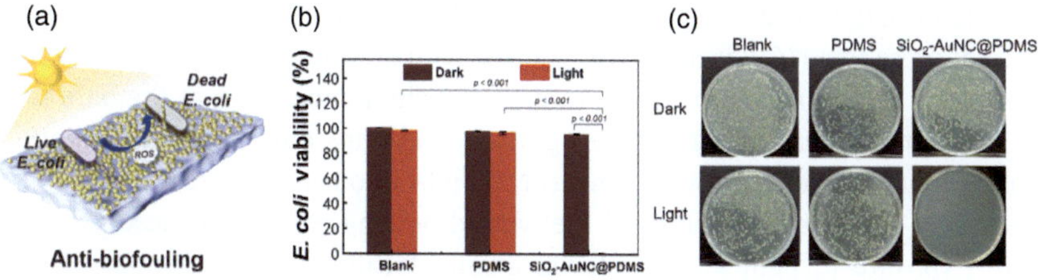

Figure 6.44 (a) The antibacterial mechanism of the SiO_2-AuNC@PDMS film. (b) Viability of *E. coli* on PDMS and SiO_2-AuNC@PDMS with and without irradiation. (c) Colony-forming units of *E. coli* after incubation on PDMS and SiO_2-AuNC@PDMS films with and without irradiation. *Source:* Zhang et al. [46] / with permission of Elsevier.

site-specific drug release. The film carrier not only safeguards labile therapeutic molecules but also features tunable response thresholds, release kinetics, and biodegradability, which can be tailored to align with specific clinical requirements, thereby optimizing therapeutic efficacy and minimizing off-target effects. Additionally, the inherent flexibility, adhesiveness, and engineered thermal management of the polymer film contribute to efficient confinement of heat within the targeted region, thus preserving adjacent healthy tissues. Collectively, this design paradigm enhances treatment safety and mitigates adverse side effects.

Using an intraocular lens (IOL) as the substrate, by alternately depositing positively charged PEI and negatively charged GO, leveraging the photothermal properties of GO. The resulting rGO@IOL composite films were subsequently implanted into rabbits for experimental evaluation. In the fifth week of observation, the experimental group with rGO@IOL combined with NIR light irradiation showed significant advantages: the posterior capsular kept smooth and clear, and the severity score of posterior capsular opacification (PCO) was only 0.6, significantly lower than that of the other control groups (Figure 6.45a,b). Although the rGO/PEI layer improved the surface hydrophilicity of the IOL and potentially reduced lens epithelial cell (LEC) adherence, it was unable to prevent PCO from developing without NIR irradiation. However, rGO@IOL combined with NIR radiation significantly prevented the occurrence of PCO by successfully eliminating residual LEC in the capsule. In the rGO@IOL + NIR group, almost no fibrin tissue was found in the optical area of the IOL, and the field of vision in the pupil area remained clear. In contrast, the posterior capsule of the other groups showed clear accumulation of proliferative tissue, which caused the pupil region to become opaque (Figure 6.45c).

This study shows that rGO@IOL combined with NIR light irradiation provides an effective method for preventing PCO. The mechanism may include the removal of LEC induced by photothermal effects and the improvement of the surface properties of the film [47].

In the preparation process of chitosan-tannic acid complex (CT) films, the loading of photothermal agent indocyanine green (ICG) and drug neomycin can be achieved through simple mixing, thereby developing photofunctional composite films CT-I/N. Skin patches were fabricated from CT, CT-I, and CT-I/N films and applied to the skin of rats for experiments. The experimental results show that under NIR irradiation, the surface temperature of the CT and CT-I/N patches loaded with ICG rapidly rose to 40 °C within 2 min, and the temperature continued to rise with the extension of the irradiation time. In contrast, the temperature of the bare CT patch without ICG and under simple lighting conditions only increased slightly, indicating that ICG can effectively convert NIR light into heat (Figure 6.46A).

Subsequent experiments evaluated the wound healing efficacy of these patches in vivo. Exposure to light alone induced a mild localized hyperthermia, which exhibited modest soothing and moisturizing effects and marginally facilitated wound repair. The CT-I/N patch, owing to the intrinsic antibacterial and hemostatic properties of chitosan as well as the antioxidant activity of tannic acid, demonstrated a baseline capacity to promote wound healing. Notably,

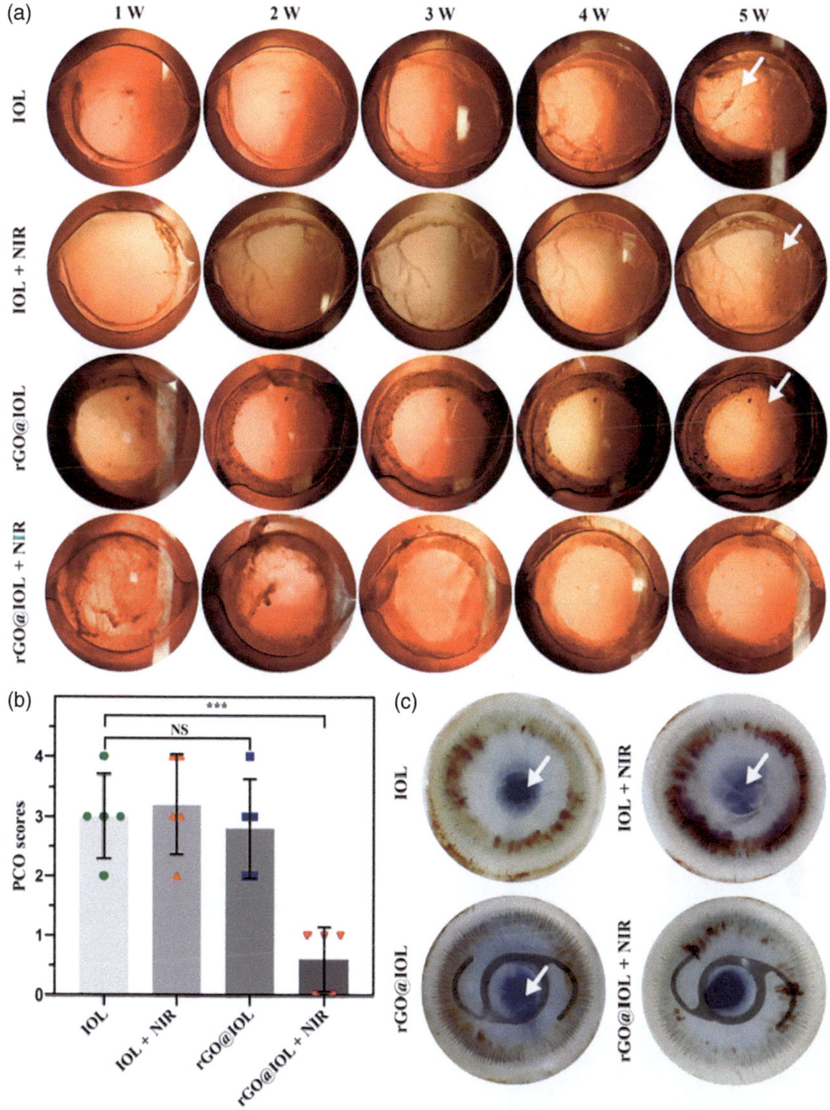

Figure 6.45 (a) Slit-lamp microscope images, (b) PCO scores, and (c) Miyake-Apple views of the eyes of posterior capsular before and after implanted rGO@IOL, with and without NIR irradiation. *Source:* Zhang et al. [47] / with permission of Elsevier.

application of the CT-I/N dressing in conjunction with NIR irradiation resulted in the most pronounced therapeutic benefit, with complete wound closure observed within 6–7 days (Figure 6.46B). Therefore, the CT-I/N composite film combines the dual mechanisms of NIR photothermal conversion and drug release, becoming an efficient photofunctional wound dressing [48].

Polydopamine with photothermal conversion properties can also be combined with silk fibroin to prepare SD nanoparticles. Subsequently, curcumin and bovine

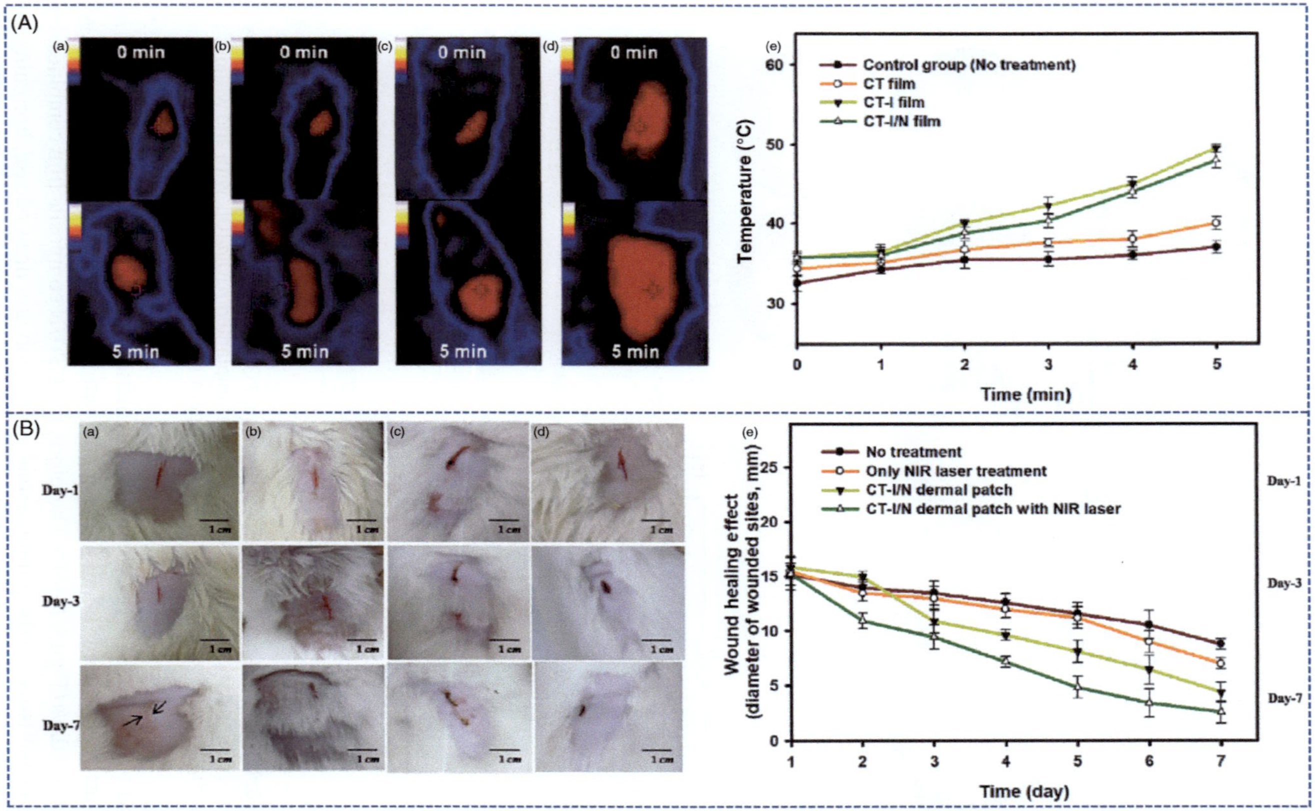

Figure 6.46 (A) Thermal images of wound treated with (a) no intervention, (b) CT film, (c) CT/I film, and (d) CT-I/N film. (e) Corresponding temperature changes for each treatment. (B) Photographs of wounds treated with (a) no intervention, (b) NIR laser alone, (c) CT-I/N film, and (d) CT-I/N film with NIR laser. (e) The diameter of wounds from days 1 to 7 of each treatment. *Source:* Rahman et al. [48] / with permission of Elsevier.

serum albumin (BSA) were respectively incorporated into SD nanoparticles to obtain two types of nanoparticles: SD/curcumin and SD/BSA. These nanoparticles can be further used to prepare SD films and fixed with heparin, ultimately obtaining SDH films. This kind of film has potential application value in the field of drug delivery. Under the irradiation of NIR light, the effective release of BSA and curcumin can be monitored. However, the experimental results show that the release rate of BSA is faster than that of curcumin. This is mainly because curcumin has strong hydrophobicity, which leads to a tighter interaction between it and the film system, thereby slowing down its release rate [49].

6.3.3 Energy Storage

Photofunctional polymer films can combine photocatalytic and electrical technologies to directly convert light energy into electrical or chemical energy and store it, demonstrating broad application prospects. This technology not only makes efficient use of solar energy but also provides an innovative solution for the development and storage of clean energy.

Inspired by the organic photovoltaic (OPV) structure, researchers have developed an artificial photosynthesis system based on multilayer film devices. Poly(3-hexylthiophene) (P3HT) films are fabricated via spin coating to construct a multilayer architecture consisting of ITO/ZnO/P3HT:PCBM/MoO_3. Subsequently, the nonphotosynthetic bacterium *S. ovata* is introduced onto the film surface. The incorporation of a P3HT:PCBM heterojunction is designed to suppress exciton recombination, thereby improving charge separation and transfer efficiency. Upon illumination, P3HT generates photoexcitons. The resulting electrons can be transferred in two ways: directly to the active sites of the bacterial membrane through the film-bacteria interface, or indirectly via electron mediators such as potassium ferricyanide. Concurrently, photogenerated holes oxidize cysteine, yielding cystine and protons (H^+). The generated H^+ ions are subsequently utilized by hydrogenase enzymes in the Wood–Ljungdahl pathway (WLP), facilitating the bioconversion of CO_2 to acetic acid (Figure 6.47a).

The experimental results show that when the prepared film is placed as artificial leaves in a pond, the artificial leaves with an area of 400 cm^2 can produce 8.2 g of acetic acid per week (Figure 6.47b). It is worth noting that multilayer semiconductor films can be recovered from bioartificial photosynthetic systems. After recovery, they can be reloaded with bacteria to carry out a new round of photosynthesis. In three cycle tests, the system maintained a stable acetic acid production, demonstrating superior long-term stability and recyclability (Figure 6.47c). This indicates that the bioartificial photosynthetic system has good sustainability in reducing carbon dioxide, providing a new idea for the application of future carbon reduction technologies [50].

The water-soluble conjugated polymer PFP can be adsorbed on the surface of carbon paper and combine with negatively charged thylakoids through the quaternary ammonium salt groups on the side chains, successfully preparing biohybrid PFP/thylakoid electrode films (Figure 6.48a). PFP has excellent ultraviolet

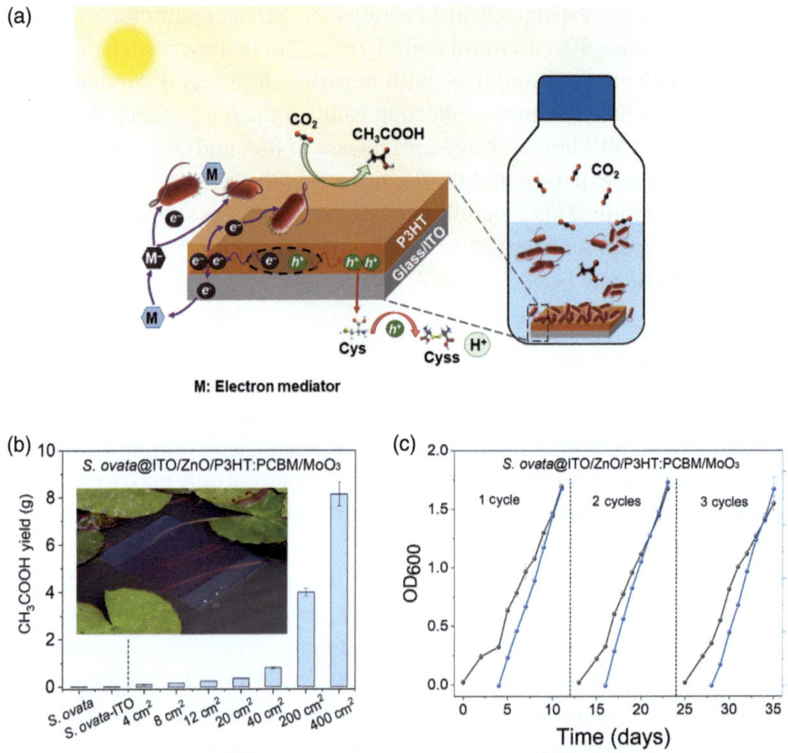

Figure 6.47 (a) The working mechanism of the *S. ovata*@ITO/ZnO/P3HT:PCBM/MoO_3 multilayer. (b) Total acetate yield as a function of increasing film area. (c) OD_{600} and the acetate yield of the multilayer film over three consecutive cycles. *Source:* Wen et al. [50] / American Association for the Advancement of Science / CC BY-NC-SA 3.0.

light absorption capacity and can be used as an artificial antenna to expand the light capture range of thylakoids, thereby accelerating the light reaction process of thylakoids. Under white light irradiation, PFP absorbs ultraviolet light and converts it into blue light that thylakoids can absorb, significantly improving the light energy capture efficiency of thylakoids. As the thylakoid photoreaction intensifies, more electrons are transferred to the carbon paper electrode through the electron transport chain, generating a stronger photocurrent. In addition, after the addition of PFP, the resistance of the thylakoid electrode significantly decreased from 2869 to 660 Ω (Figure 6.48b). PFP accelerates the electron transfer rate at the interface by improving the interface properties. Due to the synergistic effect of these two aspects, the photocurrent of the PFP/thylakoid electrode reached 1245.7 nA/cm^2, which was threefold more than that of the thylakoid electrode at 316.6 nA/cm^2 (Figure 6.48c). This result indicates that the introduction of PFP significantly enhances the photoelectric performance of the electrode, providing an important reference for the application of biohybrid electrodes [51].

The photocatalyst CdS can be embedded in the pore structure of regenerated cellulose (RC) films to form a stable CdS/RC composite film, and the film with

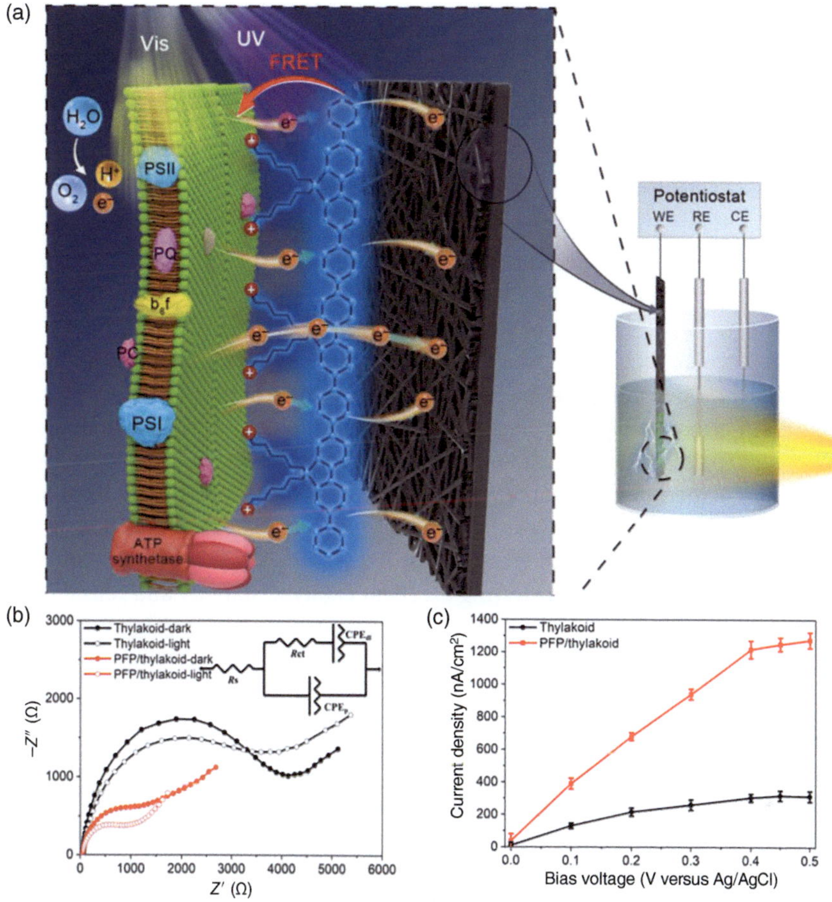

Figure 6.48 (a) The working mechanism of the PFP/thylakoid electrode film. (b) Electrochemical impedance spectroscopy analysis of thylakoid and PFP/thylakoid electrodes under dark and illuminated conditions. (c) Current density of thylakoid and PFP/thylakoid electrode as a function of applied electrode potential. *Source:* Zhou et al. [51] / with permission of John Wiley & Sons.

photocatalytic functions can be prepared by loading Pt nanoparticles. Under light conditions, CdS nanoparticles generate photogenerated excitons and migrate to the nanoparticle surface. Among these components, Pt nanoparticles serve as highly effective co-catalysts, facilitating interfacial charge separation via the Schottky barrier established at the Pt/CdS interface. Photogenerated electrons from CdS migrate to the Pt nanoparticles, where they take part in proton reduction reactions on the Pt surface, leading to the evolution of hydrogen gas. Concurrently, photogenerated holes are scavenged through their reaction with sacrificial agents present in the system.

The unique architecture of this nanocomposite film confers notable advantages: the high dispersion of CdS within the RC matrix results in an increased specific

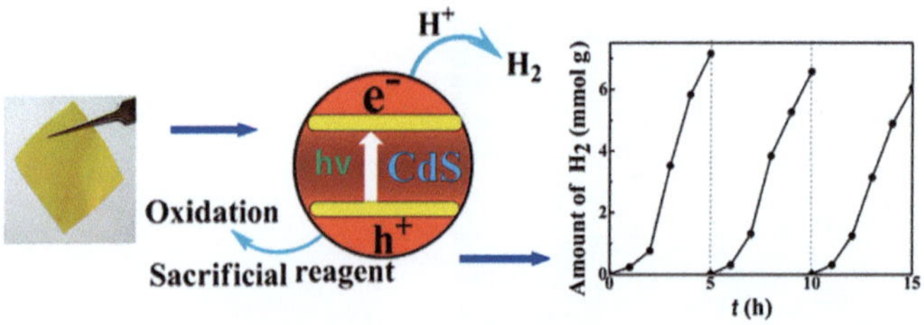

Figure 6.49 Photocatalytic H_2 generation mechanism and stability of the CdS/Pt-loaded RC film. *Source:* Ke et al. [52] / with permission of American Chemical Society.

surface area, which facilitates more efficient transfer and separation of interfacial charge carriers. This structural feature not only accelerates the kinetics of charge transport but also markedly suppresses the bulk recombination of photogenerated electron-hole pairs, thereby substantially enhancing the overall photocatalytic performance. Experimental data show that after two consecutive photochemical reactions, the catalytic activity of the CdS/RC film remains stable. In the third cycle, its hydrogen production efficiency still reached 93% of that of the first reaction (6.721 mmol/g), demonstrating excellent catalytic stability and reusability (Figure 6.49) [52].

6.3.4 Food Packaging

The application of photofunctional films in the field of food packaging is gradually becoming a hot topic in the cross-disciplinary research of food science and materials engineering. This type of film endows packaging materials with innovative functions such as antibacterial and preservation through mechanisms like photocatalysis and light regulation, providing a more efficient solution for the storage and transportation of food.

Taking the film prepared by compounding gelatin, GSE, and TiO_2 nanoparticles as an example, the addition of TiO_2 significantly improved the mechanical properties and water vapor barrier performance of the film. Meanwhile, TiO_2 can absorb ultraviolet light. Without sacrificing the transparency of the film, it almost completely blocks the transmission of ultraviolet light, endowing the film with excellent ultraviolet blocking ability. This characteristic not only helps to extend the shelf life of food but also effectively prevents ultraviolet rays from damaging the nutritional components of food [10].

Furthermore, TiO_2 exhibits photocatalytic activity that facilitates the degradation of organic compounds. Notably, when TiO_2 is immobilized within a cellulose matrix and exposed to UV irradiation, it effectively catalyzes the breakdown of organic matter [53]. Researchers mixed TiO_2 with low-density polyethylene (LDPE) to prepare different types of plastic films. In the experiment, strawberries were preserved in three different films (Figure 6.50): LDPE film (A), a film containing TiO_2 and titanate coupling agent TC-114 (B), and a film containing TiO_2 and titanate coupling agent TC-27 (C).

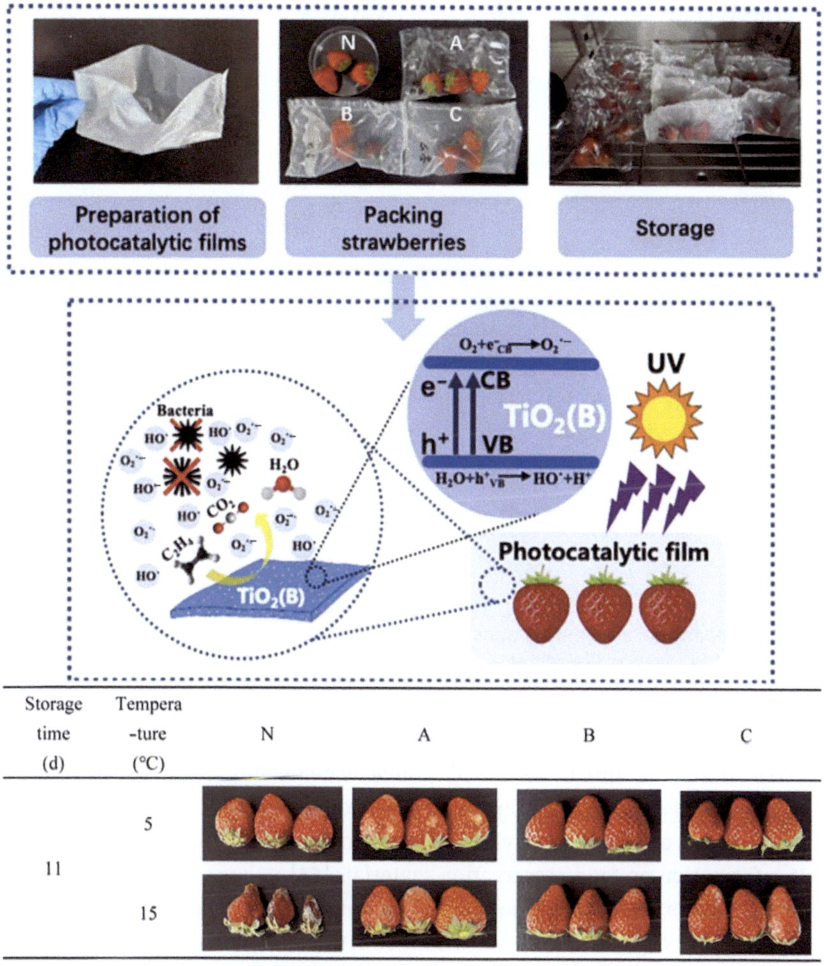

Figure 6.50 Mechanism of organic compound degradation and strawberry preservation effects of the photocatalytic film. *Source:* Yin et al. [54] / with permission of Elsevier.

Under light conditions, electrons in the TiO_2 valence band are excited to the conduction band, creating highly reactive electrons (e^-), while leaving the valence band with positively charged holes (h^+). The e^- and O_2 can combine to generate superoxide radicals ($\cdot O_2^-$), while the holes combine with hydroxyl ions (OH^-) and water molecules (H_2O) adsorbed on the surface of TiO_2 to produce hydroxyl radicals ($\cdot OH$). These ROS possess powerful oxidizing capabilities and can break common organic bonds such as C—C, C—H, C—O, and N—H, ultimately converting ethylene into water and carbon dioxide. The experimental data indicate that by the 11th day of storage, unpackaged strawberries exhibited pronounced wilting at both 5 and 15 °C. Strawberries stored with film A also showed substantial deterioration, characterized by the appearance of white spots on the fruit surface. In contrast, strawberries preserved using films B and C retained favorable color

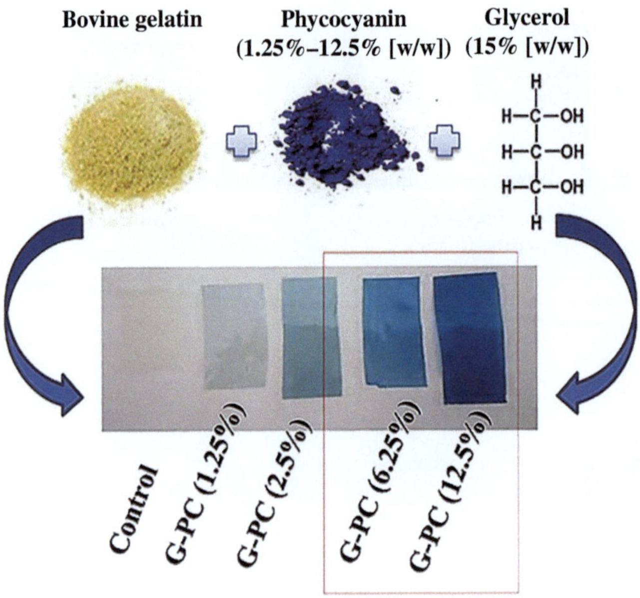

Figure 6.51 Preparation and photographs of gelatin films containing different concentrations of PC. *Source:* Chentir et al. [55] / with permission of Elsevier.

and overall condition at 5 and 15 °C, respectively. It is noteworthy, however, that strawberries stored in film C displayed visible water stains. Among the tested packaging materials, film B demonstrated superior performance in maintaining strawberry quality, effectively delaying the onset of spoilage. This research offers valuable information on food preservation materials [54].

In addition, the preparation of food packaging materials by mixing phycocyanin, PC, with gelatin and glycerol has significant functional advantages. Phycocyanin can effectively absorb ultraviolet light and visible light, enhance the light barrier performance of materials, and thereby reduce the oxidation effect of light on food (Figure 6.51). In addition, both apoprotein and phycobilin (chromophore tetrapyrrole) in phycocyanin can react with free radicals, significantly enhancing the antioxidant activity of the material. Therefore, films containing phycocyanin not only provide excellent light barrier protection but also further extend the storage time of food through their antioxidant effect, thus having broad application prospects [55].

References

1 Sutharsan, J. and Zhao, J. (2023). Physicochemical and biological properties of chitosan based edible films. *Food Reviews International* 39 (9): 6296–6323.

2 Yu, D., Yu, Z., Zhao, W. et al. (2022). Advances in the application of chitosan as a sustainable bioactive material in food preservation. *Critical Reviews in Food Science and Nutrition* 62 (14): 3782–3797.

3 Zhang, K., Jiang, L., Chen, R.-K. et al. (2024). A versatile food packaging film fabricated by gallic acid/octanoic acid grafted chitosan: preparation, characterization, high-efficient cells capture and photodynamic eradication of bacteria and biofilms. *Food Hydrocolloids* 152: 109914.

4 Souza, V.G.L., Pires, J.R.A., Rodrigues, C. et al. (2020). Chitosan composites in packaging industry-current trends and future challenges. *Polymers* 12 (2): 417.

5 Ding, K., Xie, Y., Xu, H. et al. (2024). Visible light-responsive TiO_2-based hybrid nanofiller reinforced multifunctional chitosan film for effective fruit preservation. *Food Chemistry* 460: 140539.

6 Shao, H., Wu, X., Xiao, Y. et al. (2024). Recent research advances on polysaccharide-, peptide-, and protein-based hemostatic materials: a review. *International Journal of Biological Macromolecules* 261: 129752.

7 Pacheco, M.S., Barbieri, D., da Silva, C.F., and de Moraes, M.A. (2021). A review on orally disintegrating films (ODFs) made from natural polymers such as pullulan, maltodextrin, starch, and others. *International Journal of Biological Macromolecules* 178: 504–513.

8 Zheng, H., Yang, J., and Han, S. (2016). The synthesis and characteristics of sodium alginate/graphene oxide composite films crosslinked with multivalent cations. *Journal of Applied Polymer Science* 133 (27): 43616.

9 Tongnuanchan, P., Benjakul, S., Prodpran, T., and Nilsuwan, K. (2015). Emulsion film based on fish skin gelatin and palm oil: physical, structural and thermal properties. *Food Hydrocolloids* 48: 248–259.

10 Riahi, Z., Priyadarshi, R., Rhim, J.-W., and Bagheri, R. (2021). Gelatin-based functional films integrated with grapefruit seed extract and TiO_2 for active food packaging applications. *Food Hydrocolloids* 112: 106314.

11 Qiu, W., Patil, A., Hu, F., and Liu, X.Y. (2019). Hierarchical structure of silk materials versus mechanical performance and mesoscopic engineering principles. *Small* 15 (51): e1903948.

12 Liu, J., Yang, M., Tan, J. et al. (2022). pH-responsive discoloration silk fibroin films based on prodigiosin from microbial fermentation. *Dyes and Pigments* 198: 109994.

13 Long, Q., Jiang, G., Zhou, J. et al. (2024). Cellulose ionic gel and its sustainable thermoelectric devices – design, applications and prospects. *Nano Energy* 120: 109130.

14 Song, Y., Xu, Y., Li, D. et al. (2021). Sustainable and superhydrophobic lignocellulose-based transparent films with efficient light management and self-cleaning. *ACS Applied Materials & Interfaces* 13 (41): 49340–49347.

15 Li, S., Cui, B., Jia, X. et al. (2024). A cellulose-based film with self-healing performance for light management. *Industrial Crops and Products* 212: 118312.

16 Li, Z., Xiao, W., Huang, R. et al. (2019). A gold nanoclusters film supported on polydopamine for fluorescent sensing of free bilirubin. *Sensors (Basel)* 19 (7): 1726.

17 Decher, G. (1997). Fuzzy nanoassemblies: toward layered polymeric multicomposites. *Science* 277 (5330): 1232–1237.

18 Tong, L., Ma, N., Tang, F. et al. (2012). pH and thermoresponsive Ag/polyelectrolyte hybrid thin films for tunable metal-enhanced fluorescence. *Journal of Materials Chemistry* 22 (18): 8988–8993.

19 Roberts, G.G. (1985). An applied science perspective of Langmuir-Blodgett films. *Advances in Physics* 34 (4): 475–512.

20 Pun, C.-C., Lee, K., Kim, H.-J., and Kim, J. (2006). Signal amplifying conjugated polymer-based solid-state DNA sensors. *Macromolecules* 39 (22): 7461–7463.

21 Ulman, A. (1996). Formation and structure of self-assembled monolayers. *Chemical Reviews* 96 (4): 1533–1554.

22 Liu, B. and Bazan, G.C. (2005). Methods for strand-specific DNA detection with cationic conjugated polymers suitable for incorporation into DNA chips and microarrays. *Proceedings of the National Academy of Sciences of the United States of America* 102 (3): 589–593.

23 Li, Y., Giesbers, M., Gerth, M., and Zuilhof, H. (2012). Generic top-functionalization of patterned antifouling Zwitterionic polymers on indium tin oxide. *Langmuir* 28 (34): 12509–12517.

24 Wang, X., Li, S., Zhang, P. et al. (2015). An optical nanoruler based on a conjugated polymer-silver nanoprism pair for label-free protein detection. *Advanced Materials* 27 (39): 6040–6045.

25 Wang, X., Cui, Q., Yao, C. et al. (2017). Conjugated polyelectrolyte-silver nanostructure pair for detection and killing of bacteria. *Advanced Materials Technologies* 2 (7): 1700033.

26 Wang, Z., Xiong, Y., Kershaw, S.V. et al. (2017). In situ fabrication of flexible, thermally stable, large-area, strongly luminescent copper nanocluster/polymer composite films. *Chemistry of Materials* 29 (23): 10206–10211.

27 Lu, N., Gao, X., Pan, M. et al. (2020). Aggregation-induced emission-active chiral helical polymers show strong circularly polarized luminescence in thin films. *Macromolecules* 53 (18): 8041–8049.

28 Wang, X., He, F., Tang, F., and Li, L. (2012). Optically amplified DNA detection on self-assembled solid films using conjugated polyelectrolytes. *Journal of Materials Chemistry* 22 (30): 15303–15308.

29 Wang, X., Zhu, S., Liu, L., and Li, L. (2017). Flexible antibacterial film based on conjugated polyelectrolyte/silver nanocomposites. *ACS Applied Materials & Interfaces* 9 (10): 9051–9058.

30 He, G., Yan, N., Yang, J. et al. (2011). Pyrene-containing conjugated polymer-based fluorescent films for highly sensitive and selective sensing of TNT in aqueous medium. *Macromolecules* 44 (12): 4759–4766.

31 Xu, X., Xu, Z., Ye, S. et al. (2015). Investigation of abnormal long-wavelength fluorescence emissions occurring in binary organic nanoparticle films. *Particle & Particle Systems Characterization* 32 (10): 962–969.

32 Chen, X., Wen, H., Yao, J. et al. (2025). Highly water-dispersible spiropyran-octapeptide supramolecules: efficient, multi-processable, and versatile photoswitches for time-dependent dual-mode encryption. *Advanced Functional Materials* 35 (6): 2414865.

33 Li, J., He, L., Wang, J. et al. (2014). Light-controlled drug releasing polymer films combining LbL self-assembly and host-guest interactions. *Express Polymer Letters* 8 (3): 143–153.

34 Singh, B., Chejara, M.R., and Park, M.-H. (2023). Light-responsive layer-by-layer film containing gold nanorods for sequential drug release. *ACS Omega* 8 (50): 48405–48412.

35 Li, B., Kang, Y., Wang, X. et al. (2023). Gold nanocluster decorated fibrous substrate for photo-modulated cellular growth. *Journal of Materials Chemistry C* 11 (7): 2600–2607.

36 Chung, W., Lee, B., and Kim, Y. (2025). Micropatterned CuS@PDMS films with anti-adhesive property for photothermal sterilization. *Progress in Organic Coatings* 198: 108909.

37 Li, H., Zhang, Q., Zhang, H. et al. (2024). Organic room-temperature phosphorescence of imidazole-based films by self-assembly enhancement strategies. *Colloids and Surfaces A: Physicochemical and Engineering Aspects* 689: 133760.

38 Wang, J., Xu, X., Tian, Y. et al. (2015). A novel blue fluorescent polymer for solution-processed fluorescent-phosphorescent hybrid WOLEDs. *Journal of Materials Chemistry C* 3 (12): 2856–2864.

39 Lv, F., Feng, X., Tang, H. et al. (2011). Development of film sensors based on conjugated polymers for copper (II) ion detection. *Advanced Functional Materials* 21 (5): 845–850.

40 Wang, X., Xu, Y., Ma, X., and Tian, H. (2018). Multicolor photoluminescence of a hybrid film via the dual-emitting strategy of an inorganic fluorescent au nanocluster and an organic room-temperature phosphorescent copolymer. *Industrial & Engineering Chemistry Research* 57 (8): 2866–2872.

41 Wang, X., He, F., Zhu, X. et al. (2014). Hybrid silver nanoparticle/conjugated polyelectrolyte nanocomposites exhibiting controllable metal-enhanced fluorescence. *Scientific Reports* 4: 4406.

42 Quan, Z., He, H., Zhou, H. et al. (2021). Designing an intelligent nanofiber ratiometric fluorescent sensor sensitive to biogenic amines for detecting the freshness of shrimp and pork. *Sensors and Actuators B: Chemical* 333: 129535.

43 Wang, X., Jiang, X., Zhu, S. et al. (2017). Preparation of optical functional composite films and their application in protein detection. *Colloids and Surfaces A: Physicochemical and Engineering Aspects* 535: 69–74.

44 Zhu, X., Wang, X., He, F. et al. (2015). Self-assembled nanocomposite film with tunable enhanced fluorescence for the detection of DNA. *ACS Applied Materials & Interfaces* 7 (2): 1334–1339.

45 Cabral, F.V., Xu, Q., Greer, A. et al. (2024). Superhydrophobic dressing for singlet oxygen delivery in antimicrobial photodynamic therapy against multidrug-resistant bacterial biofilms. *ACS Applied Bio Materials* 7 (9): 6175–6185.

46 Zhang, J., Wu, Z., Wang, X. et al. (2024). Photofunctional gold nanocluster-based nanocomposite coating for enhancing anti-biofouling and anti-icing properties of flexible films. *Surfaces and Interfaces* 54: 105161.

47 Zhang, C., Guo, Q., Tong, Z. et al. (2022). Thin film nanoarchitectonics of layer-by-layer assembly with reduced graphene oxide on intraocular lens for photothermal therapy of posterior capsular opacification. *Journal of Colloid and Interface Science* 619: 348–358.

48 Rahman, M., Chowdhury, F., Uddin, K. et al. (2023). Nanostructured chitosan-polyphenolic patch for remote NIR-photothermal controlled dermal drug delivery. *International Journal of Biological Macromolecules* 241: 124701.

49 Chung, T.-W., Tyan, Y.-C., Lin, S.-W. et al. (2021). Developing photothermal-responsive and anti-oxidative silk/dopamine nanoparticles decorated with drugs which were incorporated into silk films as a depot-based drug delivery. *International Journal of Biological Macromolecules* 185: 122–133.

50 Wen, N., Jiang, Q., and Liu, D. (2024). Polymer semiconductor films and bacteria hybrid artificial bio-leaves. *Science Advances* 10 (44): eadp8567.

51 Zhou, X., Zhou, L., Zhang, P. et al. (2019). Conducting polymers-thylakoid hybrid materials for water oxidation and photoelectric conversion. *Advanced Electronic Materials* 5 (3): 1800789.

52 Ke, D., Liu, S., Dai, K. et al. (2009). CdS/regenerated cellulose nanocomposite films for highly efficient photocatalytic H_2 production under visible light irradiation. *The Journal of Physical Chemistry C* 113 (36): 16021–16026.

53 Zeng, J., Liu, S., Cai, J., and Zhang, L. (2010). TiO_2 immobilized in cellulose matrix for photocatalytic degradation of phenol under weak UV light irradiation. *The Journal of Physical Chemistry C* 114 (17): 7806–7811.

54 Yin, C., Qian, J., Cai, M. et al. (2024). Photocatalytic films with high ethylene degradation efficiency for strawberry preservation. *Food Packaging and Shelf Life* 44: 101317.

55 Chentir, I., Kchaou, H., Hamdi, M. et al. (2019). Biofunctional gelatin-based films incorporated with food grade phycocyanin extracted from the Saharian cyanobacterium *Arthrospira* sp. *Food Hydrocolloids* 89: 715–725.

7

Photofunctional Polymer Hydrogel Composites

Photofunctional polymer composite hydrogels are a type of high-performance material developed by integrating photofunctional components with hydrogel networks. The hydrogel itself is a three-dimensional network structure formed by polymer chains through chemical cross-linking or physical entanglement. This architecture enables the hydrogel to absorb and retain substantial quantities of water (typically >90%) while maintaining structural integrity, elasticity, and effective energy dissipation. By precisely adjusting polymer concentrations, cross-linking density, and types, the mechanical characteristics of hydrogels can be precisely adjusted to fit the needs of different tissue environments.

The core feature of photofunctional polymer composite hydrogels lies in their ability to incorporate photofunctional components (such as fluorescent dyes, photothermal conversion materials, and photosensitive molecules) into the polymer chains through chemical modification or physical embedding. These functional components can stably bind to the hydrogel's numerous functional groups, such as carboxyl, amino, and hydroxyl moieties, while maintaining high water content and excellent biocompatibility of the hydrogel. This integration endows hydrogels with diverse photoresponsive capabilities, including fluorescence emission, photothermal conversion, and photosensitive degradation.

Photofunctional composite hydrogels demonstrate extensive application potential in biomedical fields. Their high water content and tunable pore size allow the diffusion and transport of oxygen, nutrients, metabolic wastes, and bioactive molecules, creating an ideal microenvironment for cell behavior regulation, tissue engineering, and drug delivery. Furthermore, the incorporation of photofunctional components enables fluorescence imaging, photothermal therapy (PTT), and light-controlled drug release, thereby showcasing unique advantages in antimicrobial and antitumor applications, tissue repair, and regenerative medicine.

7.1 Structures of Photofunctional Polymer Hydrogel Composites

Photofunctional polymer hydrogel composites are a new type of functional material, and their structure mainly consists of the following three parts: (1) hydrophilic

Photofunctional Polymer Composites for Bioapplications, First Edition. Xiaoyu Wang and Lidong Li.

polymer network skeleton; (2) functional optical components (such as fluorescent molecules, quantum dots, and upconversion nanoparticles); (3) mesoscopic-scale interaction interfaces. Among them, the structural design of the polymer network largely determines the mechanical properties, environmental responsiveness, and the expression effect of photofunctions of the material, while the cross-linking method directly affects the microstructure characteristics of the hydrogel.

7.1.1 The Network Structure of Chemical Cross-linking Hydrogels

Traditional chemical cross-linking methods usually form covalent cross-linking networks through the polymerization reaction of monomers. This cross-linking method can provide stable mechanical properties, endowing the hydrogel with high strength and durability. However, a significant drawback of this covalent cross-linked network is its irreversibility: once the network is damaged, it is extremely difficult to repair, thereby affecting the performance and service life of the hydrogel. To overcome the limitations of traditional cross-linking, dynamic chemical bonds can be introduced into the hydrogel network. For instance, borate ester bonds are reversible and can self-repair under specific conditions. Disulfide bonds can also endow hydrogels with self-healing capabilities through dynamic cleavage and recombination. By combining functional components such as fluorescent molecules and inorganic photofunctional nanoparticles with cross-linked networks, the hydrogel not only maintains structural stability but also possesses optical properties and intelligent response characteristics.

7.1.1.1 Permanent Covalent Bond

By using initiators to generate free radicals and initiate the chain growth and molecular chain cross-linking of alkene-based monomers, a polymer network structure with a C—C bond configuration can be formed. The photoinitiated cross-linking system requires a series of photoinitiators that can generate active free radicals under the irradiation of light of different wavelengths. These generate highly active free radicals, which in turn trigger the polymerization reaction of alkene-based monomers, ultimately forming polymers with three-dimensional network structures. In the traditional free radical photopolymerization process, organic small molecule photoinitiators are usually UV-responsive, require high energy, and the initiators will decompose. The development of new types of photoinitiators is of great significance for constructing stable hydrogel networks.

Take imine-type COF (COF-V) as an example, it has a conjugated structure and vinyl groups. Under light conditions, the COF-V promotes the excitation of electrons from the valence band to the conduction band, resulting in the generation of electron–hole pairs. The holes remaining in the valence band can subsequently interact with water molecules, facilitating the formation of ·OH radicals. These ·OH groups will attack the double bonds of acrylamide (AM) molecules to form adducts with free radicals, thereby triggering the chain growth reaction of polyacrylamide (PAM) (Figure 7.1a). This reaction needs to be carried out in an oxygen-free environment to avoid the inactivation of macromolecular

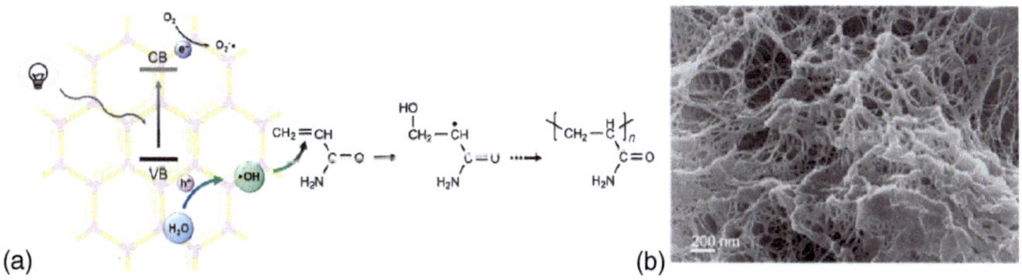

Figure 7.1 (a) The mechanism of photocatalytic polymerization of AM by using COF-V. (b) SEM image of the PAM/COF-V hydrogel. *Source:* Peng et al. [1] / with permission of American Chemical Society.

free radicals by oxygen. It is notable that the double bonds on COF-V are more likely to capture photogenerated free radicals than monomer molecules, thereby preferentially initiating surface polymerization. Owing to the porous architecture of COF-V, which permits the free diffusion of AM molecules smaller than 1 nm into its internal channels, PAM chains can react not only with the vinyl groups present on the surface but also within the internal channels of COF-V. This facilitates the formation of a multivalent cross-linked network that extends throughout both the interior and exterior of the carrier. SEM characterization confirmed that the PAM/COF-V hydrogel prepared in this manner exhibits a continuous porous architecture, characterized by pore sizes that vary from a few nanometers to several hundred nanometers (Figure 7.1b). Since no cross-linking agent was used, the hydrogel's cross-linking behavior is entirely driven by the vinyl units of COF-V. This unique design enables COF-V to have the dual functions of an efficient cross-linking agent and photoinitiator simultaneously, markedly boosting the mechanical performance of the hydrogel and simplifying the polymerization procedure [1].

Gold nanoclusters (AuNCs) can also be used as photosensitizers to efficiently generate free radicals in photocatalytic reactions. For example, the gold nanocluster AuNC@ImPAA contains imidazole as ligands and acrylamide groups. Under white light irradiation, some excited electrons transition from the HOMO energy level to the LUMO energy level, while leaving holes in the HOMO energy level. These holes react with water molecules to form hydroxyl radicals (·OH), which in turn trigger the polymerization reaction of acrylamide monomers, forming a covalent cross-linked network of polyacrylamide (PAM). Experiments show that when the concentration of AuNC@ImPAA reaches 4 mmol/L, the system can complete gelation within 5 min of light exposure, while the control group lacking AuNC@ImPAA remains in a liquid state even after 15 min. SEM analysis shows that the AuNC/PAM hydrogel presents a characteristic porous network architecture, with an average pore diameter of approximately 3.68 μm.

This photocatalytically induced covalent cross-linking network can achieve precise control of cross-linking density by modifying the AuNC@ImPAA concentration. As a multifunctional cross-linking point, multiple acrylamide groups on the surface of AuNCs can simultaneously connect multiple PAM chains, significantly enhancing the structural strength of the network. Covalent cross-linking endows hydrogels with excellent chemical and physical stability, enabling them to perform well in a variety of application scenarios [2].

Graphitic carbon nitride (g-C_3N_4) has become an ideal photocatalytic initiator due to its visible light responsiveness and high stability, and it can remain stable without degradation after initiating polymerization reactions. Under visible light irradiation, g-C_3N_4 can generate hole-electron pairs and further sensitize oxygen under aerobic conditions, converting into various reactive oxygen species (ROS), including ·OH, ·O_2, and $^1O_2$. Among them, ·OH can initiate the polymerization of *n*-isopropylacrylamide (NIPAM) monomers, thereby preparing g-C_3N_4/PNIPAM hydrogel. After the hydrogel adsorbs the aniline monomers, g-C_3N_4 in the system continues to act as a photocatalytic initiator, initiating the oxidative polymerization of aniline through the ROS generated, forming polyaniline (PAN), and ultimately constructing the g-C_3N_4/PNIPAM/PAN composite hydrogel (Figure 7.2a).

SEM analysis revealed that the g-C_3N_4/PNIPAM hydrogel has a loosely packed, porous architecture characterized by pore diameters spanning from several

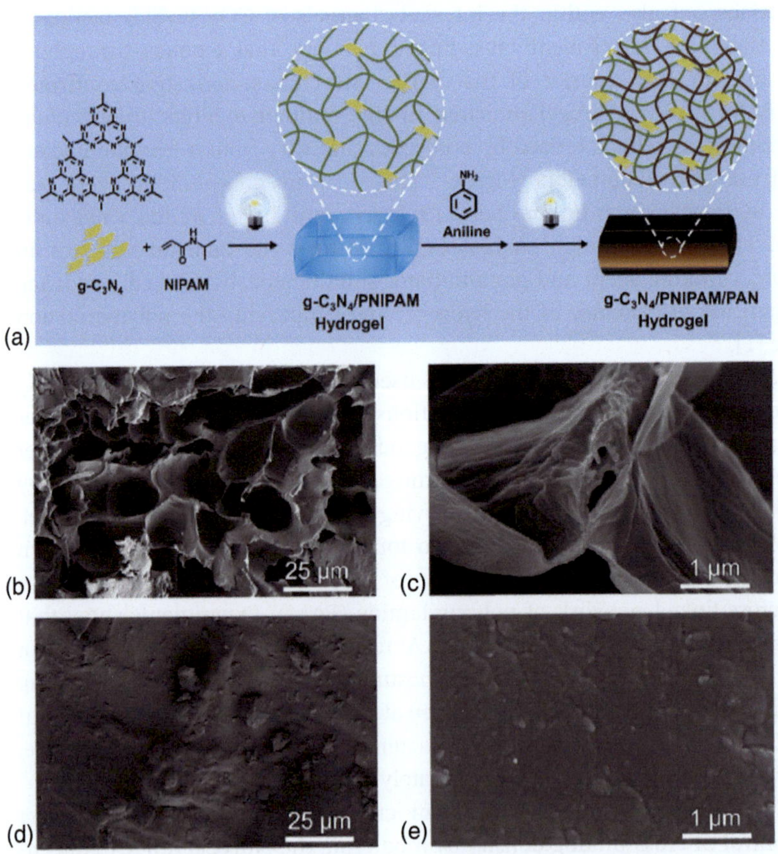

Figure 7.2 (a) The preparation process for the g-C_3N_4/PNIPAM/PAN hydrogel. SEM images of (b, c) g-C_3N_4/PNIPAM hydrogel and (d, e) g-C_3N_4/PNIPAM/PAN hydrogel. *Source:* Luo et al. [3] / with permission of Elsevier.

micrometers up to 20 μm. This large-pore structure is conducive to the adsorption of aniline monomers (Figure 7.2b,c). After the polymer was generated, the micropores originally existing in g-C_3N_4/PNIPAM disappeared, and the pore diameters were filled by the newly generated polymer, which proved that PAN was successfully incorporated into the composite hydrogel (Figure 7.2d,e). This hierarchical cross-linking strategy driven by photocatalysis forms an interpenetrating network structure, and the mechanical properties are improved threefold [3].

Since polysaccharides are found in the body as a component of the extracellular matrix (ECM), they are frequently employed to create biocompatible hydrogels. Materials based on chitosan have garnered a lot of interest due to their minimal toxicity, excellent biocompatibility, and inherent biodegradability. Chitosan, which is derived from the naturally occurring biopolymer chitin, is known for its amino group. The cross-linking of the chitosan chain is accomplished via covalent bonding between the carboxyl groups of the photosensitizer silicon phthalocyanines (SiPc) and the amino groups of chitosan, facilitated by a carbodiimide-mediated coupling reaction (Figure 7.3a). This process ultimately forms a three-dimensional uniform microporous structure of SiPcCh hydrogel, with the diameters of the interconnected pores ranging from 10 to 100 μm (Figure 7.3b), providing the material with excellent mechanical properties and biological activity [4].

Biological macromolecules such as gelatin are rich in functional groups like carboxyl and amino groups. These groups can react with cross-linking agents through EDC/NHS-mediated amidation reactions to form amide bonds, thereby constructing hydrogel systems. For example, when synthesizing glutathione (GSH)-modified gold nanoclusters using the one-pot method, silver ions are introduced to induce the formation of gold/silver clusters (Au/Ag clusters), which exhibit photosensitization properties. Subsequently, the GSH ligands of Au/Ag clusters were covalently cross-linked with the carboxyl/amino groups of gelatin using EDC/NHS-mediated amination reaction, and Au/Ag@gelatin sponge was finally prepared. The microstructure was characterized by SEM, and it was found that pure gelatin exhibited a disordered surface, a cross-linked network, and relatively few pores. In contrast, Au/Ag@gelatin exhibited a more uniform surface topology, characterized by closely packed and regularly distributed pores, indicative of a well-defined porous architecture. This structural optimization can be attributed to the interaction established by Au/Ag clusters as cross-linking points with gelatin, which significantly improves the overall performance of the material. This research not only demonstrates the possibility of combining biological macromolecules with nanomaterials but also provides new ideas for the preparation of functional hydrogels [5].

7.1.1.2 Dynamic Covalent Bonds

The Schiff base reaction is a chemical process in which a reversible covalent imine bond is formed through the condensation of a primary amine and an aldehyde functional group. This reaction is mild and can be carried out under physiological conditions, such as in an environment containing cells, tissues, and bioactive molecules. This process can form a hydrogel network without affecting cell or

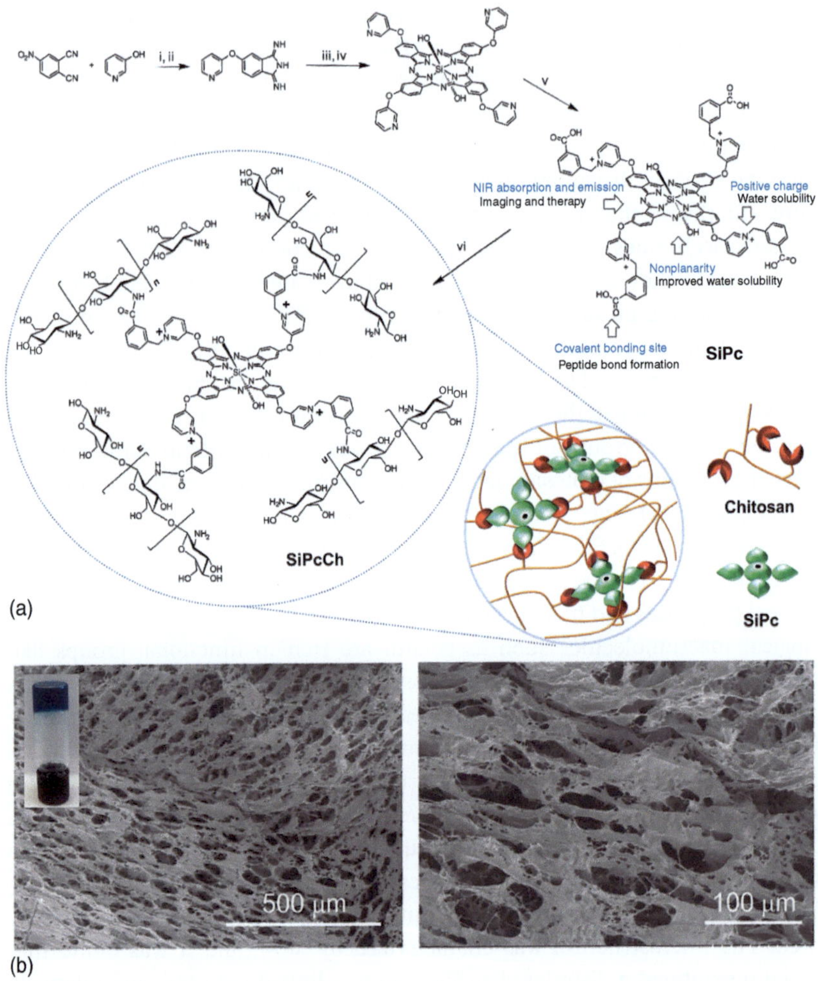

Figure 7.3 (a) The preparation process and (b) the obtained SEM images of SiPcCh hydrogel. *Source:* Strokov and Galstyan [4] / John Wiley & Sons / CC BY-NC-ND 4.0.

molecule activity, and is highly biocompatible. In addition, imine Schiff base bonds are pH-responsive, and the cross-linked network established through such dynamic covalent bonds can impart self-healing properties to hydrogels [6].

Using oxidized hyaluronic acid (OHA) and hydroxypropyl chitosan (HPCS) as substrates, aldehyde-modified OHA and aminated HPCS can spontaneously form imine bond cross-linked hydrogel systems through the Schiff base reaction (Figure 7.4a). The polydopamine nanoparticles loaded in the hydrogel can act as highly efficient photothermal converters and adsorb the anticancer drug DOX through π–π interactions. Meanwhile, cyclodextrin derivatives (NH_2-CD) increased the solubility of the antiinflammatory drug curcumin through host–guest interactions (CUR@NH_2-CD). Under pH 7.4 conditions, imine bonds can be formed by mixing HPCS solution with OHA solution for 60 s. FTIR spectroscopic analysis

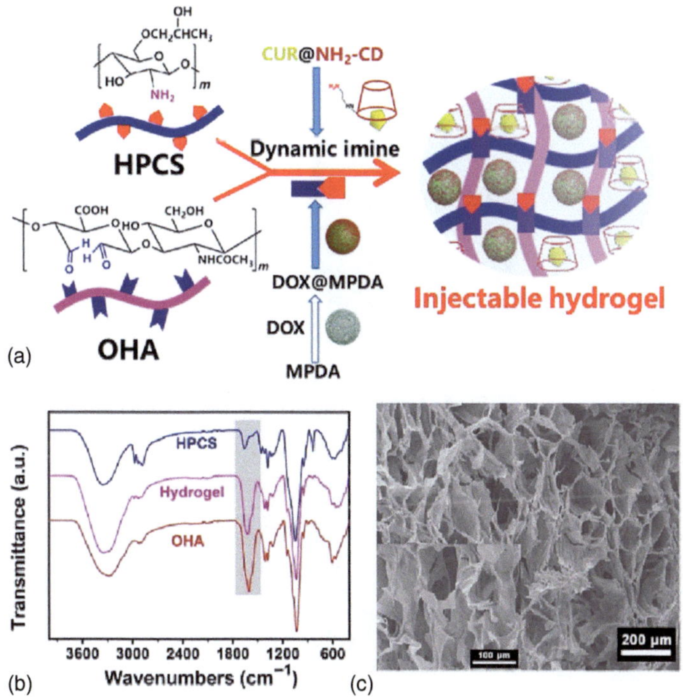

Figure 7.4 (a) Schematic illustration of the preparation process for the injectable hydrogel. (b) FTIR spectra of HPCS, OHA, and the resulting hydrogel. (c) The SEM images of the hydrogel. *Source:* Rong et al. [7] / with permission of Elsevier.

confirmed the presence of all characteristic absorption peaks associated with the gel matrix constituents, OHA and HPCS. Importantly, the absorption band at 1735/cm, indicative of the aldehyde groups on OHA, was absent in the spectrum, while a distinct peak emerged at 1640/cm, corresponding to the stretching vibration of the imine bond (—CH=N—) (Figure 7.4b). SEM images further revealed that the hydrogel possesses a network architecture, characterized by a honeycomb-like morphology (Figure 7.4c). Therefore, the structure of the hydrogel was confirmed [7].

ε-Poly(L-lysine) was grafted onto the HPCS chain, and the conjugated polymer HPCS-g-EPL containing amino groups was successfully prepared. Oxidized dextran (OD), attributable to the abundance of aldehyde groups present, can react with the amino groups in HPCS-g-EPL to form cross-linked structures (Figure 7.5a). Iron-doped carbon dots (Fe-CD) are used as antibacterial agents. Through physical mixing with functional substrates, the antibacterial performance of the material is further enhanced. Fe-CD demonstrates pronounced absorbance in the near-infrared (NIR) region, enabling its application as an effective photothermal agent for PTT. Additionally, Fe-CD functions as a catalyst, facilitating the conversion of H_2O_2 into highly reactive ·OH, thereby exerting antibacterial effects.

During the process of preparing the hydrogel, the pre-prepared oxidized glucan and HPCS-g-EPL solution were uniformly mixed in a certain proportion and

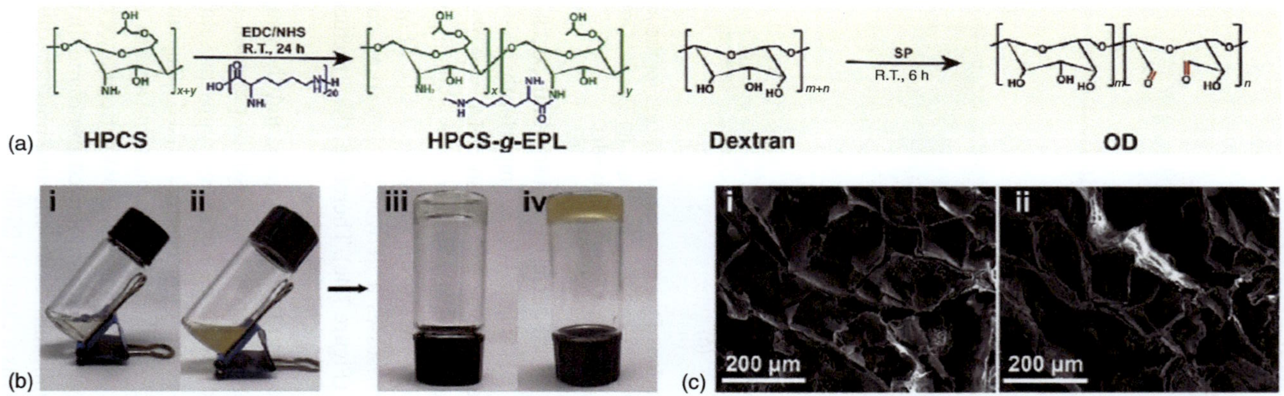

Figure 7.5 (a) The structures of HPCS-*g*-EPL and OD. (b) The photographs of gelation process of OD/HPCS-EPL with and without Fe-CDs. (c) SEM images of OD/HPCS-EPL hydrogel with and without Fe-CDs. *Source:* Dong et al. [8] / with permission of American Chemical Society.

left standing at room temperature for a period. Eventually, a stable hydrogel OD/HPCS-EPL was formed. The Fe-CD–loaded hydrogel appears yellow, indicating that Fe-CD has been successfully embedded into the hydrogel (Figure 7.5b). SEM images show that the hydrogel presents a continuous porous network structure regardless of whether they contain Fe-CD or not (Figure 7.5c), which further confirms the successful preparation of the hydrogel. This porous structure not only facilitates the loading and release of drugs but also provides a favorable microenvironment for cell growth and tissue regeneration [8].

Borate ester bonds have important applications in the preparation of hydrogels. As a Lewis acid, boric acid can accept electron pairs [9] and form complexes with Lewis bases (such as hydroxide ions or electron donor groups containing O and N atoms). In aqueous solutions, boric acid exists in a neutral state or combines with hydroxide ions, and its sp^2 and sp^3 hybrid orbitals, respectively, present triangular plane and square geometric structures. The relative balance between these two forms depends on the Lewis acidity and the pKa value of the boric acid derivative.

Polymers containing boric acid can form hydrogels by cross-linking with 1,2- or 1,3-diols. Depending on the pH value, boric acid in the hydrogel can separate or combine with 1,2- or 1,3-diols, forming a dynamic equilibrium (Figure 7.6). Therefore, borate ester bonds are not permanent bonds but are dynamic. In addition, the free boric acid and diols formed by hydrolysis can interact with adjacent free diols and boric acid to form new borate ester bonds. Based on this rearrangement mechanism, borate ester bond hydrogels can exhibit self-healing properties without external intervention.

Guar gum (GG) is a water-soluble natural polysaccharide extracted from the endosperm of guar beans, a leguminous plant. It is a nonionic galactomannan and has excellent biological safety and biodegradability [10]. Its structure is mainly composed of pyranose linked at the (1,4) site as the main chain, and the side chain is mainly α-D-galactose at the 6 site [11]. The ortho-diol groups in this structure can form dynamic borate ester bonds with borate ions, endowing the material

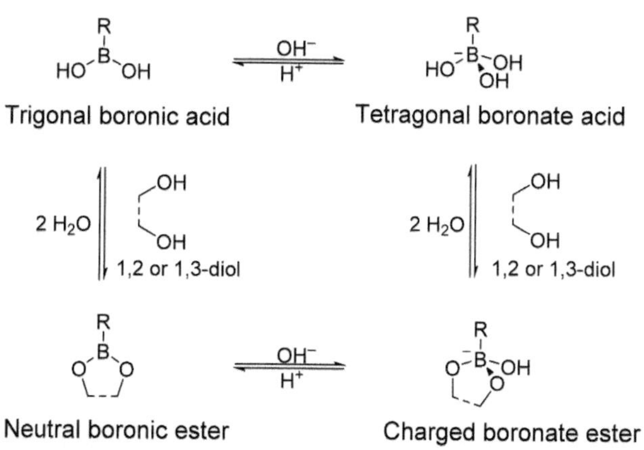

Figure 7.6 Formation mechanism of dynamic borate ester bonds. *Source:* Cho et al. [9] / Royal Society of Chemistry.

with self-healing properties. Borate ester bonds exhibit high reversibility through double decomposition, transesterification, or hydrolysis/re-esterification, enabling hydrogels to re-form after damage. This property holds significant application value in chemical and biological processes.

Fe^{3+} is a trace element in the human body, which can rapidly chelate with catechol and catechol groups in tannic acid (TA), demonstrating wide NIR absorption and excellent photothermal efficiency. The TA/Fe^{3+} nanocomposites were mixed with the GG solution. After adding borax solution, the mixture changed from a flowing liquid to a solid hydrogel within 40 s, completing the sol–gel transformation (Figure 7.7a). This rapid gelation process is attributed to the highly efficient formation of dynamic borate ester bonds between the diol groups in GG and the borate ions. Experiments show that, regardless of whether nanocomposites (NCs) are loaded or not, hydrogels exhibit a similar interconnected porous network structure (Figure 7.7b). The elemental mapping images further confirmed that carbon (C), oxygen (O), and iron (Fe) elements were uniformly distributed in the GBTF 0.2 hydrogel, indicating that TA/Fe^{3+} NCs successfully combined with the GG hydrogel (Figure 7.7c) [12].

Similarly, the GG solution was mixed with Cu nanoparticles that have photothermal effects, then the sodium tetraborate aqueous solution was added and shaken for about 30 s. Then, the system was left to stand for 15 min, and Cu-nanoparticle hydrogel composites could be obtained. The hydrogel is brown in color, in contrast to the color of the sol. At this point, the maximum surface plasmon absorption peak of the Cu nanoparticles is located at 617 nm. Compared with the Cu nanoparticles in the aqueous solution, the absorption peak has undergone a redshift (Figure 7.8a). This is because, once embedded in the hydrogel, the refractive index of the surrounding hydrogel is higher than that of water. Additionally, the SEM image shows that Cu nanoparticles are uniformly dispersed in the network structure, with almost no agglomeration phenomenon (Figure 7.8b). When a piece of the composite hydrogel was cut off and left for just 5 min, the fracture in the hydrogel had almost disappeared, leaving only a relatively shallow mark. This phenomenon proves that the composite hydrogel itself has remarkable self-healing properties (Figure 7.8c) [13].

In biomolecules, the protein containing cysteine can form stable cross-linked network structures through disulfide bonds (—S—S—) generated by terminal sulfhydryl groups (—SH). The cleavage and remodeling of disulfide bond networks primarily depend on thiol-disulfide exchange reactions. These reactions can trigger chemical cross-linking under ultraviolet (UV) radiation, while simultaneously decomposing cross-linked bonds in the presence of reducing agents or reduction potentials. Due to the reversibility of disulfide bonds, disulfide networks exhibit dynamic characteristics and endow materials with self-healing capabilities.

The recovery mechanism of dynamic disulfide bonds under UV light irradiation involves photochemical reactions and changes in molecular structure. The high energy of UV light can directly act on disulfide bonds (—S—S—), causing them to break and generate sulfur radicals (—S$^\cdot$). After the irradiation stops, these radicals recombine under thermodynamic drive and revert to disulfide bonds. For instance, using water-soluble galactoglucomannan (GGM) as the substrate, thiolated GGM

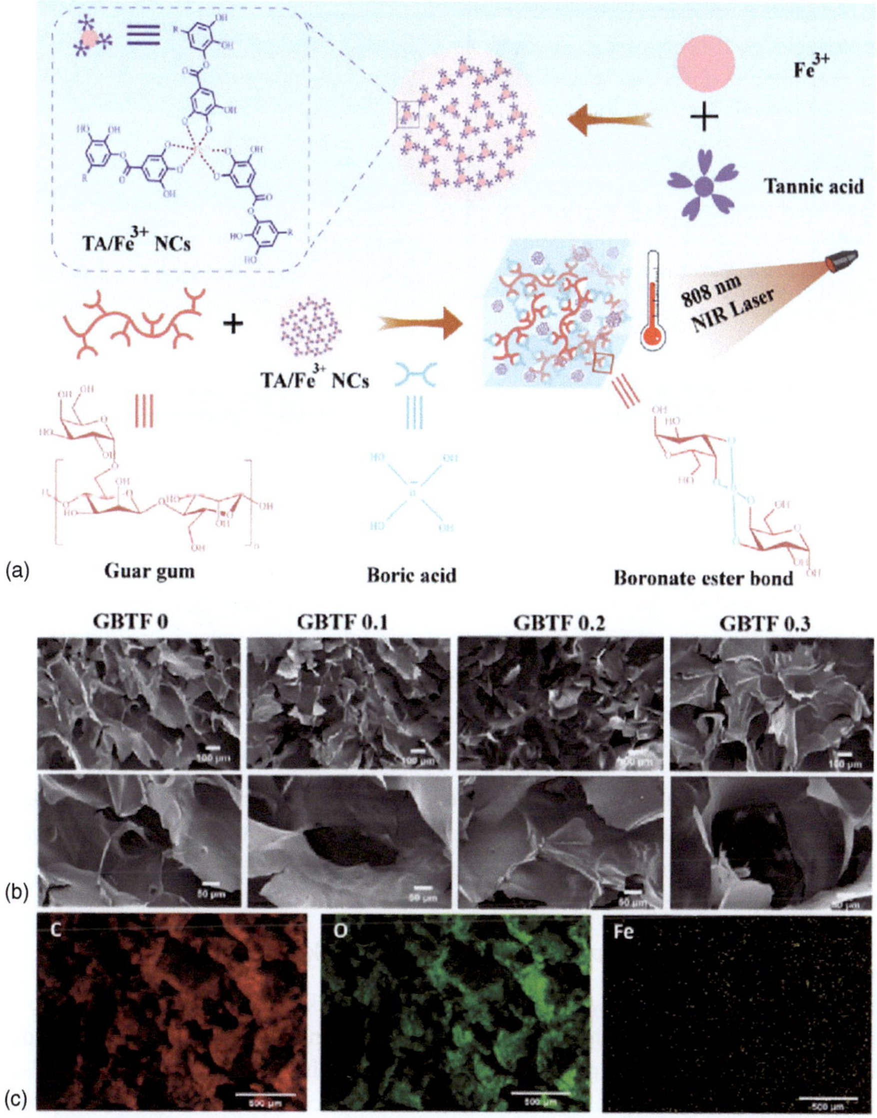

Figure 7.7 (a) The preparation process for GBTF hydrogel. (b) SEM images of different GBTF hydrogels at different magnifications. (c) Elemental mapping images of C, O, and Fe elements in GBTF 0.2 hydrogel. *Source:* Yang et al. [12] / with permission of American Chemical Society.

(GGM-SH) was synthesized via ring-opening reaction of γ-thiobutyrolactone. Under UV irradiation, the photoinitiator (IR2959) generates carbon radicals that attack the thiol groups in GGM-SH to form thiol radicals, thereby forming a disulfide bond cross-linked network. This dynamic cross-linked network can not only be compounded with TEMPO-oxidized cellulose nanofibrils (T-CNF) to form a high-strength hydrogel but also exhibits excellent self-healing performance,

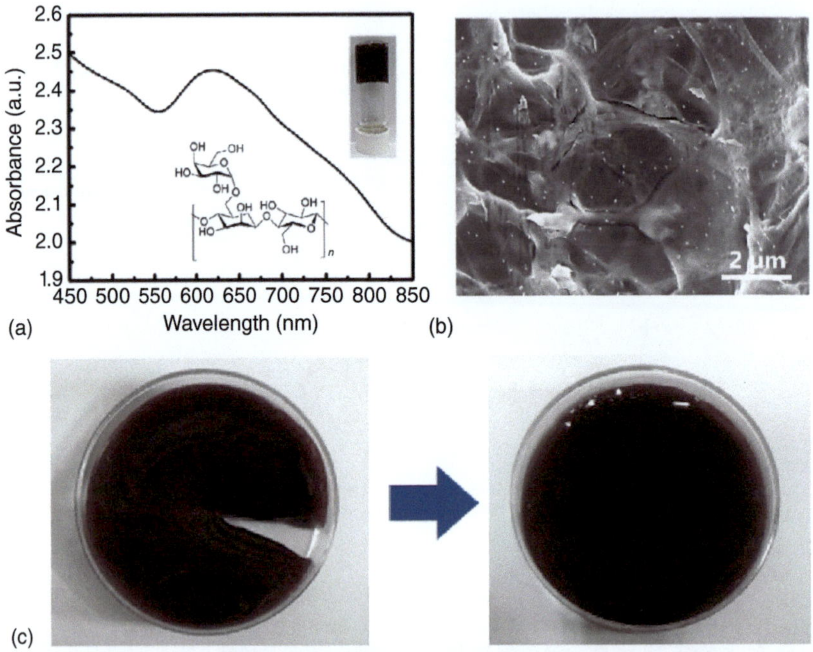

Figure 7.8 (a) UV-vis absorption spectrum, (b) SEM image, and (c) self-healing effect of Cu-nanoparticle hydrogel. *Source:* Chen et al. [13] / with permission of American Chemical Society.

such as restoring integrity after 160 s of UV irradiation (Figure 7.9). This dynamic material design based on disulfide bonds offers new ideas and broad applications for biomedical and smart materials [14].

7.1.2 The Network Structure of Physical Cross-linking

For physical cross-linked hydrogels, the main cross-linking methods include hydrogen bonds, metal coordination, hydrophobic association, host–guest interaction, and crystallization cross-linking. These cross-linking methods are noncovalent bonds and do not involve changes in chemical valence. Physical cross-linking methods are generally dynamic and reversible, meaning that physically cross-linked hydrogels have self-healing capabilities.

The physical cross-linking method of photofunctional polymer hydrogels, while retaining the characteristics of traditional noncovalent cross-linking, achieves synergistic regulation of optical functions through special design. Not only do noncovalent interactions endow materials with self-repairing capabilities, but they also provide a stable microenvironment for photofunctional components. For example, the hydrophobic microarea protects fluorescent molecules from being quenched by the aqueous phase, while the coordination effect regulates the luminescence properties of metal nanoclusters. Additionally, the crystalline domains function as physical cross-linking sites, thereby enhancing the stability

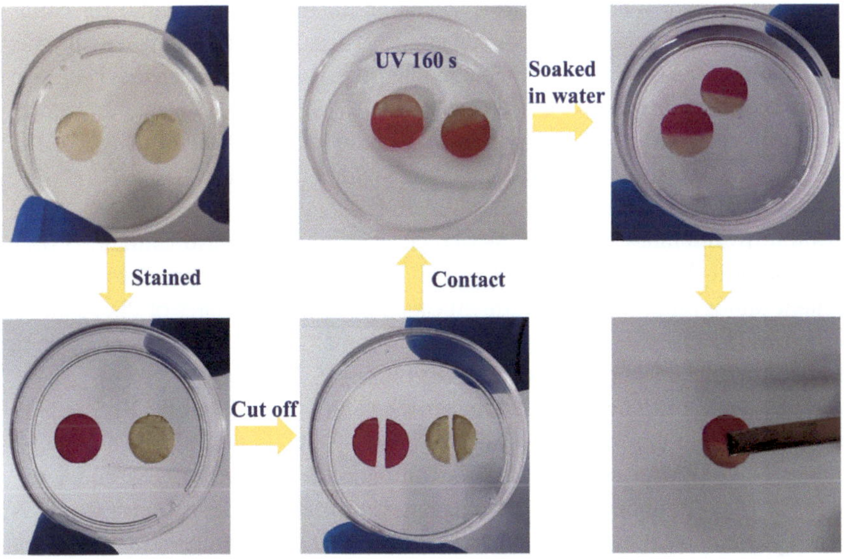

Figure 7.9 The self-healing effect of composite hydrogel facilitated by disulfide bonds. *Source:* Bi et al. [14] / Elsevier / CC BY 4.0.

of the encapsulated luminescent material. By precisely controlling the strength and density of these physical cross-links, researchers have successfully developed photofunctional hydrogels with high elasticity, rapid responsiveness, and long-term stability, providing an ideal material platform for flexible optoelectronic devices and biosensors.

7.1.2.1 Hydrogen Bond Cross-Linking

Hydrogen bonds (X—H...Y) are the most common and directional saturated interactions in physical cross-linking. They are essentially dynamic, reversible connections formed by electron-deficient hydrogen atoms (X—H) and lone pairs of electrons on electronegative atoms (O, N, F). The bond energy ranges from 5 to 30 kJ/mol. This interaction plays a central role in biological activities [15]. For example, DNA achieves the transmission of genetic information through precise pairing of A—T stabilized by two hydrogen bonds and C—G stabilized by three hydrogen bonds. Similarly, proteins maintain secondary structures through main-chain amide hydrogen bonds (such as C=O···H—N in α helices). In the field of materials, the wide binding constants of hydrogen bonds (10^2–10^6/M) [16] can provide unique design flexibility for hydrogels. By adjusting the number of hydrogen bond donors and acceptors, the solvent environment (e.g. DMSO/water competition) or dynamic response units (e.g. pH sensitivity), advanced hydrogel materials with self-repairing properties, high strength (MPa level), and environmental responsiveness can be produced. These materials have already demonstrated significant value in fields such as biosensing and tissue engineering [17–19].

Konjac glucomannan (KGM) is a natural, food-grade, nonionic polysaccharide composed of glucose and mannose. It is highly soluble in water and biocompatible [20]. When KGM dissolves in an alkaline solution, its ester bonds undergo hydrolysis, generating hydroxyl groups. These hydroxyl groups can interact through intramolecular and intermolecular hydrogen bonds to form a rich hydrogen bond network. Due to the presence of these hydrogen bonds, KGM will transform into a thermally irreversible hydrogel in an alkaline solution. When KGM is used as the functional base and mixed with dopamine hydrochloric acid (DA · HCl) under alkaline conditions, KGM will undergo hydrolysis and gelation. Additionally, the dopamine hydrochloric acid will self-polymerize to form polydopamine (PDA). This composite hydrogel system, due to its unique chemical structure, exhibits excellent tumor photothermal treatment capabilities. The presence of PDA enables the hydrogel to absorb light energy and convert it into heat energy, thereby exerting the photothermal effect in tumor treatment [21].

After blending the vinyl-modified spiropyran (SPMA) with Pluronic F127, a pink transparent aqueous solution was formed initially. Subsequently, 2-hydroxyethyl methacrylate (HEMA) monomer, initiator ammonium persulfate (APS), and accelerator *N,N,N′,N′*-tetramethyl ethylenediamine (TEMED) were added successively. A 10-hour free radical polymerization was conducted at room temperature, resulting in a milky opaque HEMA-SPMA-F127 composite hydrogel. In this system, the SPMA molecules undergo copolymerization with HEMA through the acrylate groups and integrate into the polymer main chain, effectively preventing the free migration of small molecules. Since no chemical cross-linking agent was added during the reaction process, the formation of the hydrogel network mainly relies on physical cross-linking [22]. Specifically, the hydroxyl/carbonyl groups of HEMA are capable of forming hydrogen bonds with the hydroxyl groups present on F127 micelles, while the methyl groups of HEMA can generate hydrophobic interactions with the hydrophobic PPO segments of F127 (Figure 7.10a). This enables F127 micelles to serve as a physical cross-linking point for PHEMA chains, facilitating the formation of a three-dimensional network.

FITR spectroscopic analysis (Figure 7.10b) shows that the C=O characteristic peak shifts from 1725 to 1706/cm, and the O—H stretching vibration peak moves from 3423 to 3370/cm. These redshift phenomena directly confirm the formation of hydrogen bond networks [24]. This intermolecular force decreases the degree of hydration of the polymer chains, promoting inter-chain aggregation into dense clusters and giving the hydrogel a macroscopically opaque appearance. SEM image further shows that this hydrogel has a tightly three-dimensional network structure with pore diameters of only 1–3 μm, significantly smaller than the pore diameter range of 10–100 μm of conventional hydrogels (Figure 7.10c). This extremely small pore size is the direct evidence of the highly aggregated and cross-linked polymer chains [23].

By modifying the surface of nanoparticles with polymer coatings, hydrogen bonding can be introduced. A reaction was carried out using sodium hypophosphite as the reducing agent and polyacrylamide (PAM) as the template under conditions of pH = 2 and 90 °C, successfully producing highly crystalline copper nanoparticles coated with PAM. Subsequently, this nanoparticle solution with photothermal properties was mixed with polyvinyl alcohol (PVA), forming a Cu-PVA hydrogel. As

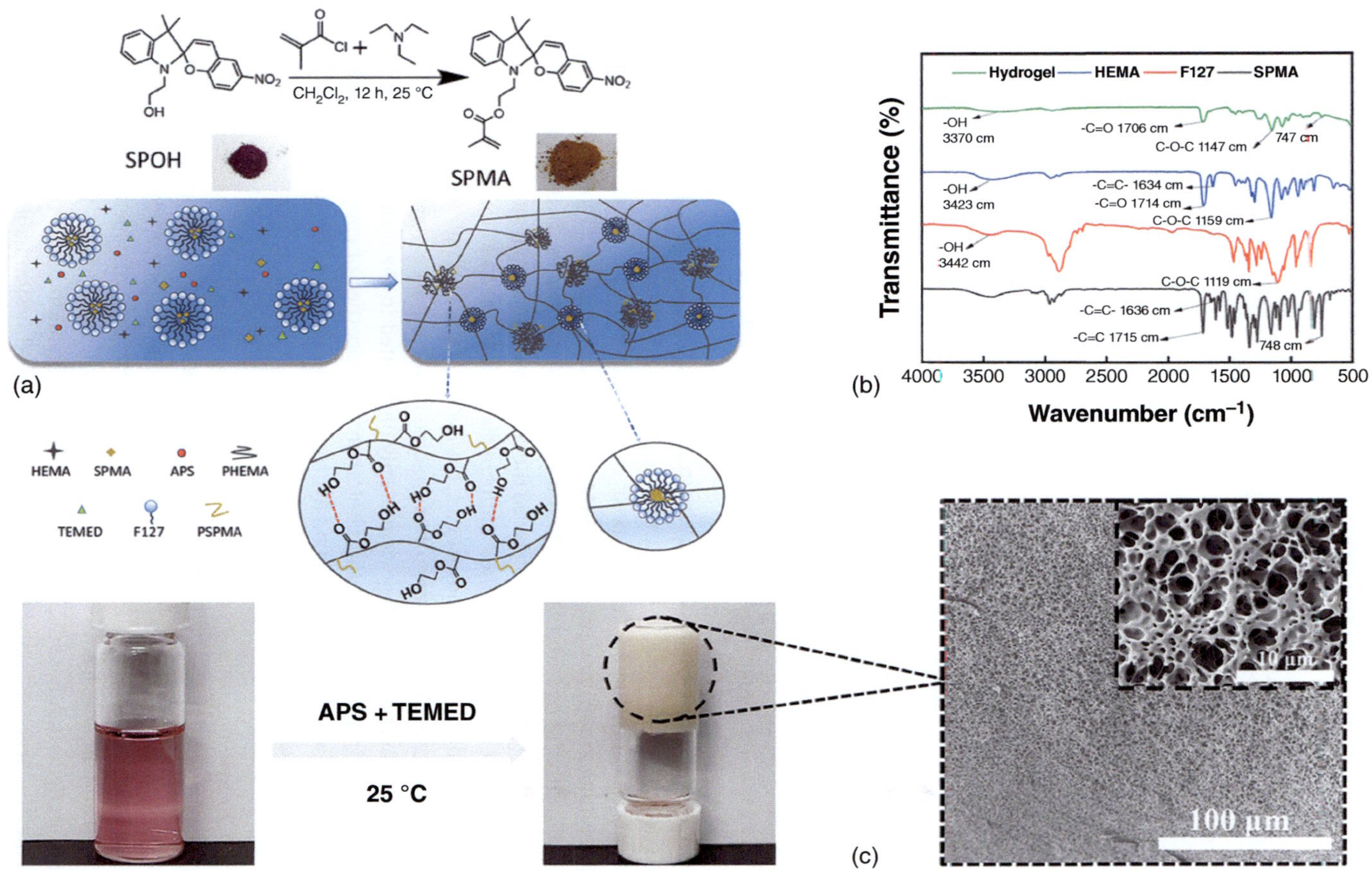

Figure 7.10 (a) Scheme illustration of the preparation process of HEMA-SPMA-F127 hydrogel. (b) FTIR spectra of HEMA-SPMA-F127 hydrogel, HEMA, F127, and SPMA, respectively. (c) Photograph and SEM image of the HEMA-SPMA-F127 hydrogel with different magnifications. *Source:* Gao et al. [23] / with permission of John Wiley & Sons.

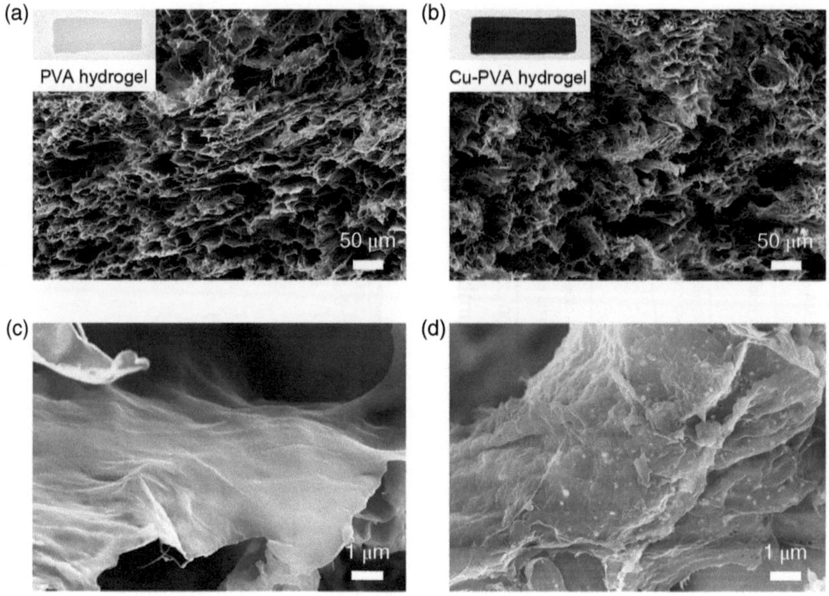

Figure 7.11 The photographs and SEM images of (a) pure PVA hydrogel and (b) Cu-PVA hydrogel; enlarged SEM views of (c) PVA hydrogel and (d) Cu-PVA hydrogel. *Source:* Reproduced with permission from Wang et al. [25] / Elsevier.

shown in Figure 7.11, the Cu-PVA hydrogel displayed the distinctive brick-red hue of copper nanoparticles in contrast to the pure PVA hydrogel.

The PAM coating on the surface of copper nanoparticles played a crucial role. The amine groups present in the PAM molecules are capable of forming hydrogen bonds with the hydroxyl groups found in the PVA molecules. This hydrogen bond cross-linking effect not only enhanced the stability of the copper nanoparticles in the hydrogel but also promoted their uniform dispersion. The SEM images showed that after embedding the Cu nanoparticles, the hydrogel still maintained a complete network structure. From the magnified images, high-contrast Cu nanoparticles could be clearly observed, and they were evenly distributed in the hydrogel matrix. This Cu-PVA hydrogel combines the photothermal properties of copper nanoparticles and the flexibility of hydrogen bond hydrogels, demonstrating excellent comprehensive performance [25].

7.1.2.2 Metal Coordination Effect

The coordination interaction between metal ions and organic ligands can be used to create a functional hydrogel. The key strength of this type of metal coordination hydrogel is its high level of designability. By selecting specific metal ions (e.g. Fe^{3+}, Zn^{2+}) and organic ligand structures (e.g. carboxylic acid groups, pyridine groups), the strength and dynamic properties of the coordination bonds can be controlled, achieving structural stability and mechanical tunability simultaneously. More noteworthy is the fact that these hydrogels are sensitive to environmental stimuli like variations in pH, temperature, and light levels, and can undergo a reversible sol–gel

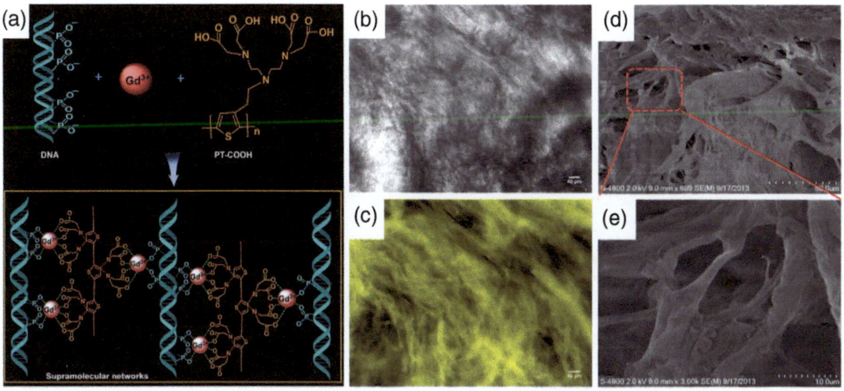

Figure 7.12 (a) Illustration of Gd^{3+} cross-linked DNA and PT-COOH for preparing hydrogel; (b) optical image and (c) fluorescence image of conjugated polymer-DNA hydrogel. (d, e) SEM images of the conjugated polymer-DNA hydrogel. *Source:* Hu et al. [26] / with permission of American Chemical Society.

phase transition. This unique dynamic characteristic enables it to adapt to the differentiated mechanical strength requirements of different applications.

The interaction between carboxylates and metal ions is the reason for the gelation of some hydrophilic polymer solutions. Conjugated polymers with carboxy side chains were designed, synthesized, and used to prepare conjugated polymer-DNA hydrogels. Gadolinium ions were employed to mediate the self-assembly process between the carboxylated conjugated polymers and the phosphorylated DNA (Figure 7.12a). Fluorescence images showed that PT-COOH emitted a bright yellow fluorescence (Figure 7.12b,c). SEM images revealed the presence of a distinct cross-linked network structure within the hydrogels (Figure 7.12d,e) [26].

The thiol–metal interaction is a common feature of metal-ligand binding. Its dynamic characteristics provide a unique opportunity for designing self-healing networks. Upon introduction of *N,N′*-bis(acryloyl)cystamine (BACA), which contains disulfide bonds, into a solution of gold nanoparticles (Au NPs), BACA exhibits specific adsorption onto the Au NP surfaces through the formation of RS—Au bonds. By mixing these functionalized nanoparticles with NIPAm monomers and conducting in situ free radical polymerization under the conditions of potassium persulfate (KPS) initiation and TEMED acceleration, hydrogels with Au NPs (GNP-15) as cross-linking points can be prepared (Figure 7.13a–c). SEM images show that this hydrogel has a uniform honeycomb-like structure, demonstrating its uniform morphology and structure (Figure 7.13d).

The abundance of dynamic RS—Au bonds within the hydrogel imparts self-healing capabilities. The figure demonstrates the self-healing process, wherein two separated hydrogel blocks rejoin upon exposure to NIR light. The healing efficiency can reach 96% within 1 min. This is because each cross-section contains molecular chains of RS—Au bonds. When irradiated by an NIR laser, the Au NPs act as a photothermal conversion element, generating localized heating and inducing the on/off switching of surface ligands, as well as subsequent dynamic

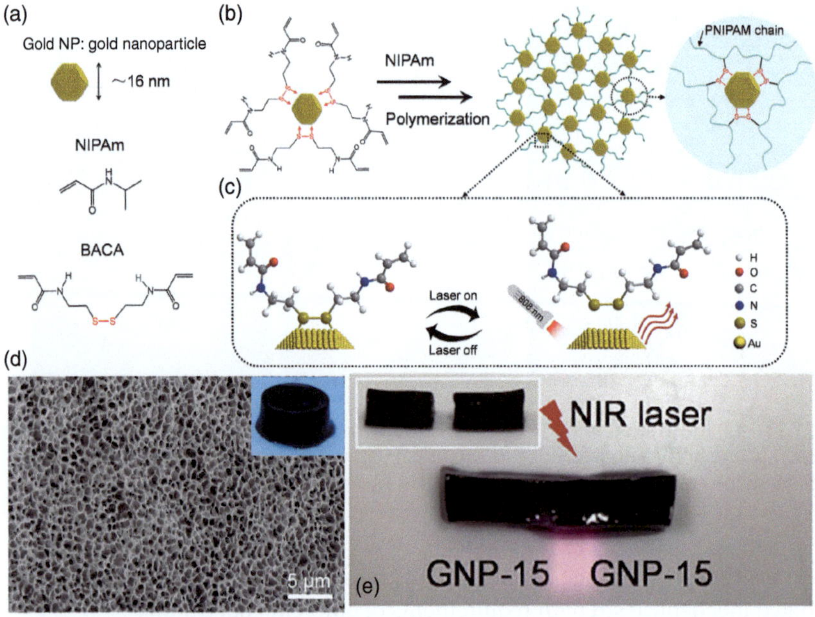

Figure 7.13 (a) The structures of Au NPs, NIPAm, and BCAC. (b) Preparation mechanism and (c) self-healing mechanism in GNP-15 hydrogel; (d) SEM image of the GNP-15 hydrogel; (e) photographs demonstrating the self-healing process of the GNP-15 hydrogel under NIR laser irradiation. *Source:* Qin et al. [27] / with permission of Elsevier.

migration. This promotes the reconstruction of RS—Au bonds, ultimately achieving self-healing (Figure 7.13e) [27].

7.1.2.3 Hydrophobic Interaction

Hydrophobic associative hydrogel refers to a physical hydrogel in which the hydrophobic regions of the molecular chains serve as the cross-linking points. Micelles are an important component of hydrophobic associative hydrogels. Hydrophobic monomers have poor solubility in water, while micelles formed by surfactants can encapsulate these hydrophobic monomers. In the micelles, the hydrophobic monomers undergo copolymerization with the hydrophilic monomers in the aqueous environment, and the hydrophobic segments are tightly wound around the interior of the micelles [28]. On one hand, micelles act as dynamic cross-linking points, causing hydrophilic molecular chains to cross-link here and form a network structure; on the other hand, when the hydrogel undergoes deformation under external force, the hydrophobic molecular chains entangled within the micelles will be stretched and unentangled, thereby generating a large amount of dissipated energy. Therefore, hydrophobic associative hydrogels generally have high toughness. The incorporation of hydrophobic segments into hydrophilic polymer chains induces the spontaneous aggregation of these hydrophobic domains via hydrophobic interactions. These aggregated regions serve as physical cross-linking points within the hydrophilic polymer network. Owing

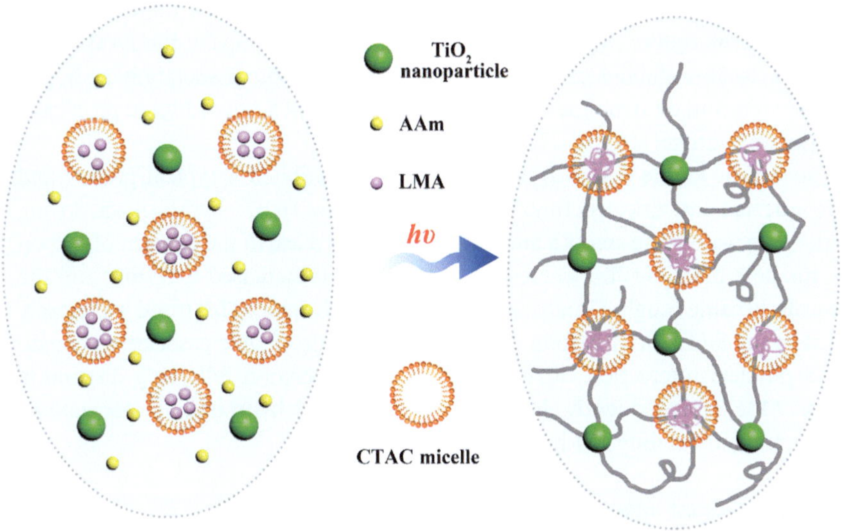

Figure 7.14 Schematic illustration of the TiO_2/HyA composite hydrogels.

to their dynamic and reversible nature, such hydrophobic associations can impart self-healing capabilities to the resultant hydrogel.

Using TiO_2 as the photoinitiator and cross-linking point, a TiO_2-assisted hydrophobic associative hydrogel (TiO_2/HyA) can be prepared through in situ free radical polymerization (Figure 7.14). When cetyltrimethylammonium chloride (CTAC) and NaCl are dissolved in water, lauryl methacrylate (LMA) is added as a hydrophobic monomer to the CTAC/NaCl solution to form stable micelles. Subsequently, acrylamide (AAm) and a solution containing TiO_2 nanoparticles are added. TiO_2 acts as the photoinitiator and initiates the in situ polymerization of AAm under UV light. After polymerization, the formation of a large number of hydrophobic associative domains can serve as cross-linking points. The polyacrylamide chains will form hydrogen bonds with the abundant hydroxyl groups present on the surface of TiO_2 nanoparticles, resulting in cross-linking. Therefore, the TiO_2/HyA hydrogel is prepared through the synergistic effect of hydrophobic association and TiO_2 nanoparticles [29].

Similarly, carbon dots prepared by the hydrothermal method using β-cyclodextrin and hydrochloric acid as raw materials have a fluorescence emission maximum of 535 nm and contain a large number of hydroxyl and carboxyl groups. Stearyl methacrylate and sodium dodecyl benzene sulfonate form stable micelles when mixed with an acrylamide solution and carbon dots. This triggers a polymerization reaction that generates polyacrylamide (PAM). During this process, pronounced hydrophobic interactions occur between the hydrophobic side chains of stearyl methacrylate and sodium dodecyl benzene sulfonate, resulting in the formation of a large number of hydrophobic association domains and stable dispersion. These hydrophobic side chains are connected to a part of the PAM chain after polymerization, thereby forming some cross-linking points. At the same time, the abundant hydroxyl and carboxyl groups present on the surface of the carbon dots enable the formation of hydrogen bonds with the amide groups of the PAM chain. Therefore, by introducing carbon dots into a hydrophobic association hydrogel, a double cross-linked structure was formed, constructing a novel double physically cross-linked hydrogel (HCD).

As shown in Figure 7.15a–d, all the hydrogels exhibited a typical porous and interconnected network structure. The pore diameter of HCD 0 hydrogel was 50 μm, and the pore walls were smooth and flat. With the increase in the content of carbon dots, the pore diameter first decreased to 15 μm, then increased to 35 μm, and the pore walls became rough. Finally, they became smooth again. This trend was closely related to the addition of carbon dots. Moreover, hydrophobic association constitutes a physical interaction capable of rapid reorganization following disruption (Figure 7.15e). Consequently, hydrogels incorporating hydrophobic associations demonstrate outstanding self-healing properties [30].

7.1.2.4 Host–Guest Interaction

Hydrogels constructed through physical cross-linking mediated by host–guest interactions represent a crucial element of the hydrogel system. Host–guest supramolecular hydrogels possess dynamic characteristics due to their noncovalent supramolecular interactions (Figure 7.16). Compared with other physical cross-linking methods, such as electrostatic or hydrophobic interactions, host–guest interaction has unique advantages. The well-defined stoichiometric ratio and directional nature of the system, such as the formation of a single host–guest complex at each cross-linking site, facilitate the accurate prediction and rational design of material properties. This property helps to achieve greater homogeneity across length scales in the network, thereby improving the overall performance of the hydrogel [31].

For example, carboxymethyl cellulose (CEC) was grafted with β-cyclodextrin (β-CD) and adipic dihydrazide (ADH), respectively, resulting in two derivatives: CEC-CD and CEC-ADH. Subsequently, at 60 °C, CEC-CD was mixed with AD-CHO in a PBS solution at a molar ratio of [CD:AD = 1.5:1] for 2 h. Since the host in CEC-CD has a hydrophobic cavity that can encapsulate 1-adamantane (AD), the mixture remained in a transparent solution state. In contrast, due to the lack of formation of the CD/AD complex, AD-CHO could not be completely dissolved in CEC-ADH, which shows the significance of host–guest interactions at the supramolecular level.

Next, CEC-CD/AD-CHO was combined with CEC-ADH, resulting in the rapid formation of the CAA hydrogel within seconds to minutes. This swift gelation is attributable to the accelerated kinetics of acylhydrazone bond formation, in conjunction with supramolecular host–guest interactions, which collectively facilitate efficient cross-linking within the system (Figure 7.17a,b). Through SEM images, it can be observed that the CAA hydrogel has an interconnected porous network structure (Figure 7.17c). This hydrogel can load the anticancer drug DOX with red luminescence. When two heart-shaped CAA hydrogels were cut into halves and closely contacted, a complete hydrogel was formed and appeared

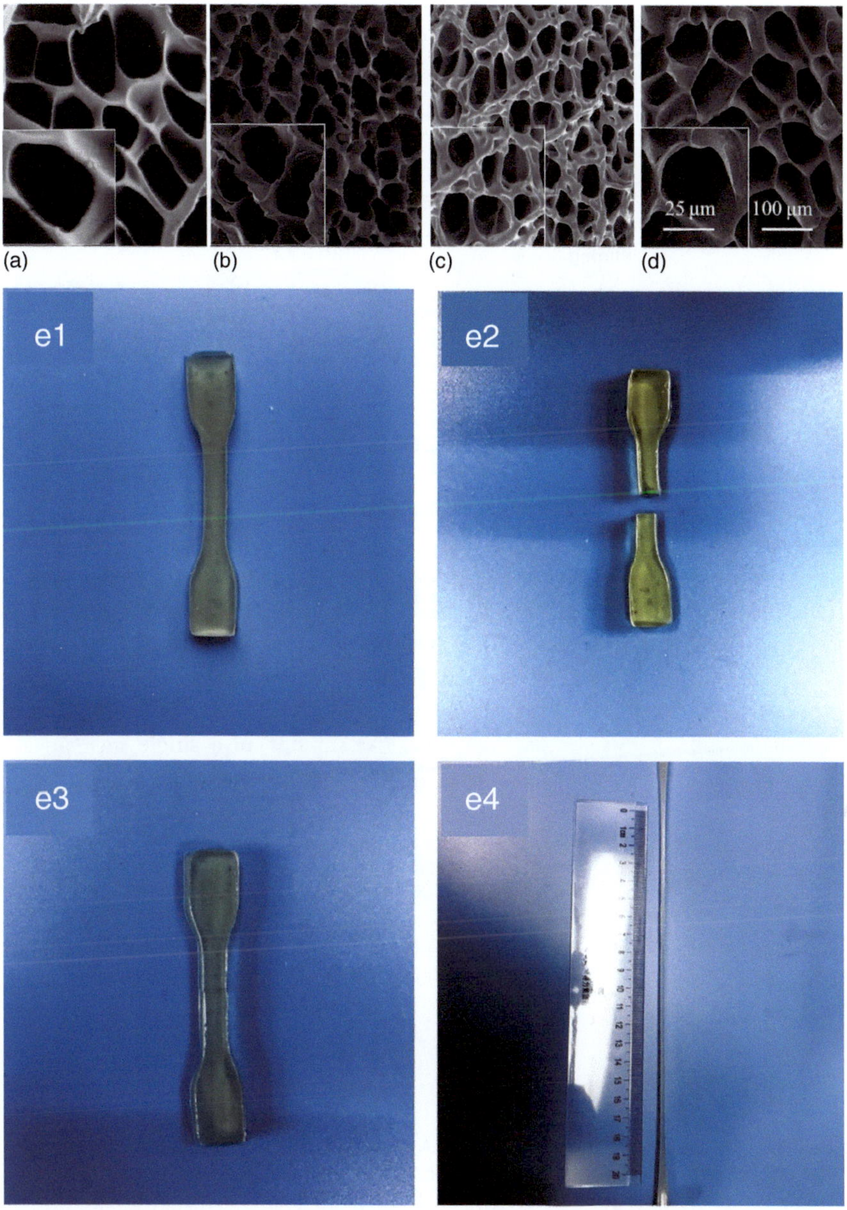

Figure 7.15 SEM images of (a) HCD 0, (b) HCD 2, (c) HCD 4, (d) HCD 8 hydrogels with different carbon dot content. (e) The self-healing effect of HCD 4 hydrogel. *Source:* Reproduced with permission from Zhang et al. [30] / John Wiley & Sons.

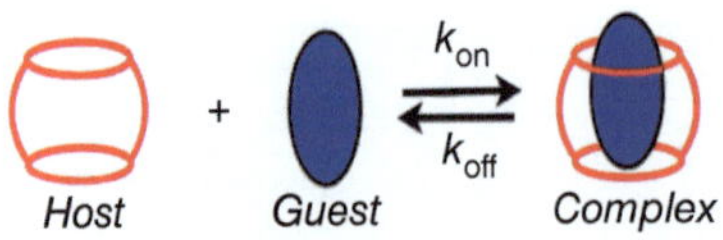

Figure 7.16 Schematic illustration of dynamic host–guest interaction. *Source:* Mantooth et al. [31] / with permission of John Wiley & Sons.

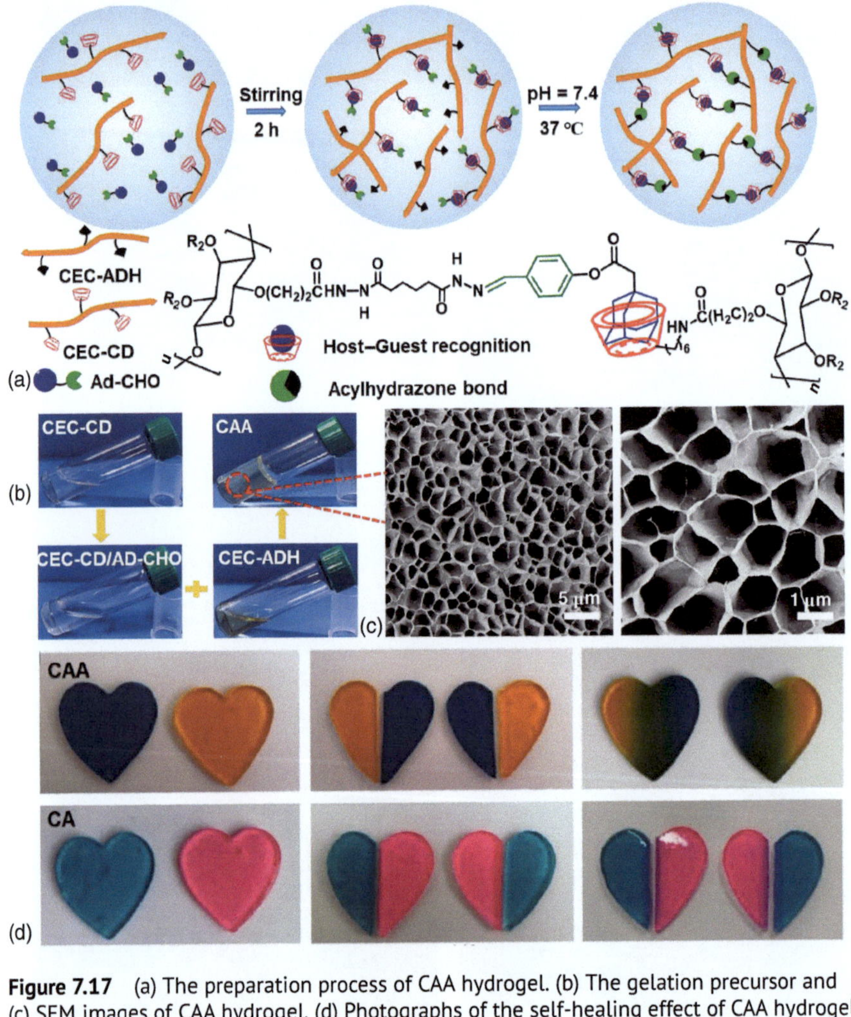

Figure 7.17 (a) The preparation process of CAA hydrogel. (b) The gelation precursor and (c) SEM images of CAA hydrogel. (d) Photographs of the self-healing effect of CAA hydrogel in comparison to CA hydrogel. *Source:* Jiang et al. [32] / with permission of Elsevier.

green (Figure 7.17d). However, the CA hydrogel without AD did not show any healing phenomenon even after several days. This result indicates that the host–guest interactions play an indispensable role in the self-healing process of the hydrogel [32].

The conjugated polymer PPG with photothermal effects is modified by an amideation reaction to link polyethylene glycol (PEG) segments to the side chains, thereby enhancing its water solubility. After adding the host molecule α-cyclodextrin

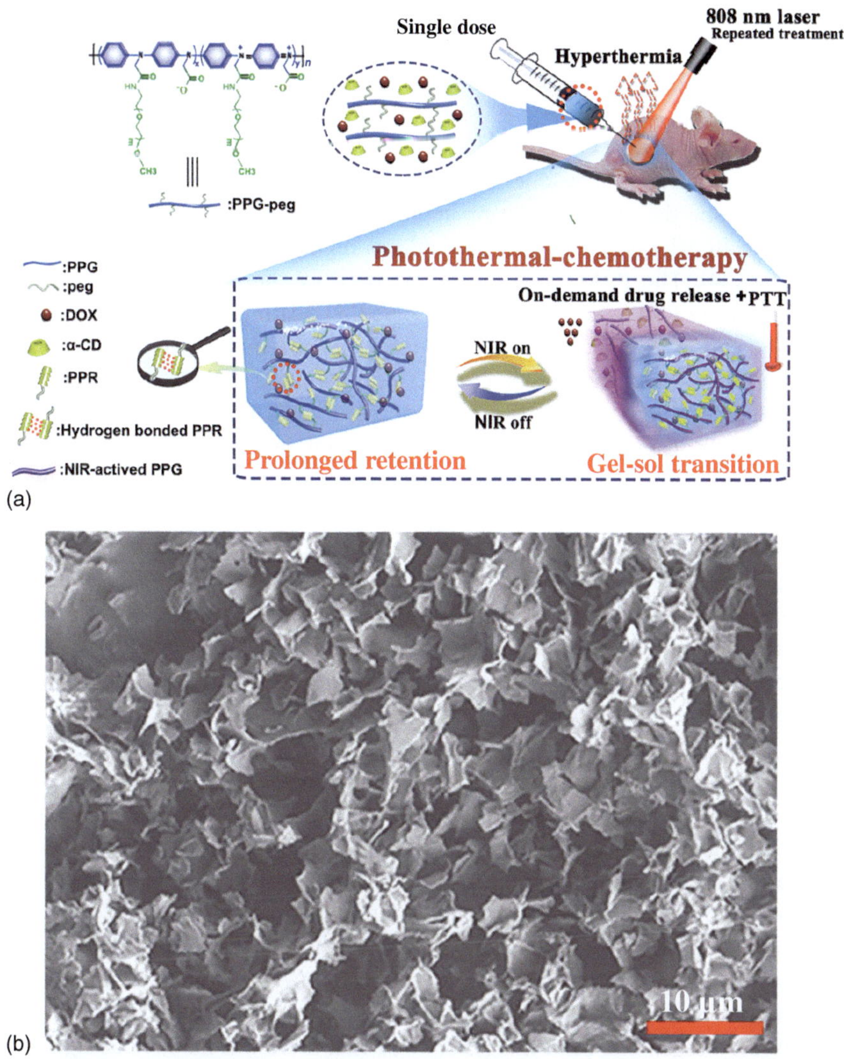

Figure 7.18 (a) The preparation process and applications of the photothermal hydrogel. (b) SEM image of the photothermal hydrogel. *Source:* Liu et al. [33] / with permission of Elsevier.

(α-CD) to the aqueous solution, based on the host–guest interaction, α-CDs and the PEG segments assemble into a polyrotaxane (PPR) (Figure 7.18a). Utilizing the strong hydrogen bonding between the PPRs, microcrystalline aggregates are formed as physical cross-linking points, ultimately resulting in a supramolecular hydrogel. The hydrogel exhibits a distinct porous internal architecture (Figure 7.18b). In control experiments, a simple physical mixture of PPG, PEG, and α-CD failed to yield hydrogel formation, underscoring the critical role of covalent linkage between PPG and PEG chains in hydrogel assembly. Upon exposure to low-intensity 808-nm laser irradiation, the hydrogel efficiently transduces incident laser energy into localized thermal elevation, resulting in a terminal temperature of approximately 50 °C. At this temperature, the α-CDs and the PEG segments dissociate, causing the hydrogel to transform into a solution [33].

7.1.2.5 Crystallization and Cross-Linking

The crystallization technology, due to its simple gelation conditions and excellent mechanical properties, has significant application value in the polyvinyl alcohol (PVA) hydrogel system, and is particularly highly regarded in the field of biomedicine [34]. The PVA hydrogel was prepared by the repeated freeze-thaw method. First, the PVA was completely dissolved by heating. Then, the homogeneous PVA solution was placed in a −20 °C freezer for several hours and thawed at room temperature. This process was repeated multiple times [35].

During the freezing process, the growth of ice crystals pushes aside the PVA molecular chains, causing the system to undergo phase separation and form a PVA-rich phase. Owing to the intrinsic propensity of PVA for crystallization, the molecular chains in these regions form crystalline domains through orderly arrangement, eventually resulting in a three-dimensional network structure (Figure 7.19) [36]. This physical cross-linking network endows the hydrogel with unique mechanical properties while avoiding the potential toxicity introduced by chemical cross-linking agents. Compared with other hydrogels, the crystalline cross-linked PVA hydrogel possesses core characteristics such as controllable performance, ultra-high purity, biocompatibility, and outstanding mechanical strength.

Adding photofunctional components to the PVA hydrogel system can enable the development of shape memory functionality. Since the crystals of PVA can melt at higher temperatures (50–60 °C), they exhibit temperature responsiveness. Therefore, thermal stimulation can be used to disintegrate this network and restore the original shape [37]. The specific implementation method is as follows:

Firstly, poly(dopamine) nanoparticles (PDAPs) with photothermal properties were mixed with PVA solution at 90 °C, then frozen at −15 °C, and subsequently restored to room temperature. This process was repeated 1–3 times to obtain PDAPs/PVA hydrogel. The SEM images showed that there was no obvious sign of

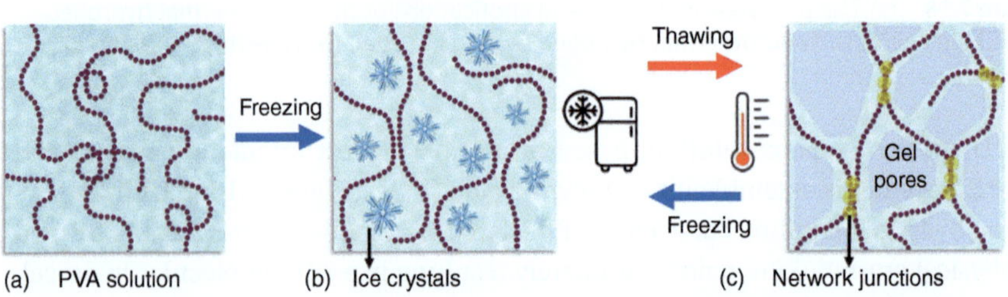

Figure 7.19 The preparation process of PVA hydrogels by the freeze-thaw method. Schematic illustration of (a) PVA chains dispersed in solution; (b) freezing induces phase separation, entrapping PVA chains between ice crystals; and (c) thawing transforms ice crystals into hydrogel pores, forming the gel network. *Source:* Adelnia et al. [36] / with permission of Elsevier.

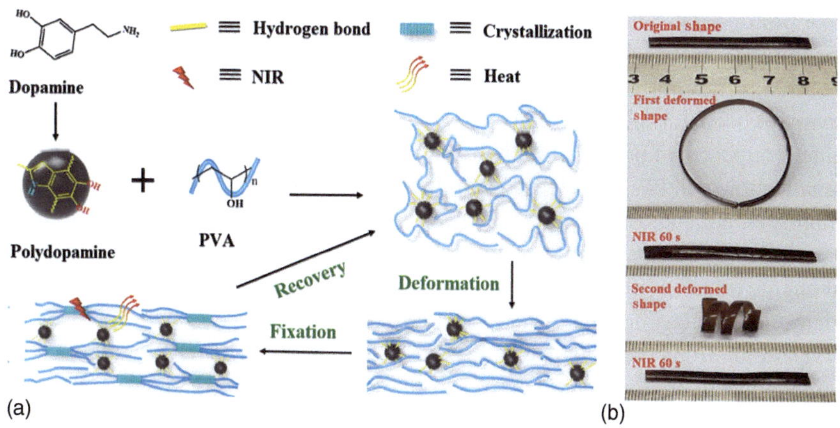

Figure 7.20 (a) Schematic representation of the shape-memory property of PVA hydrogel; (b) shape-memory process of PVA hydrogel activated by infrared laser. *Source:* Yang et al. [38] / with permission of John Wiley & Sons.

aggregation, and PDAPs were evenly spread throughout the composite hydrogel. In this system, PDAPs and PVA chains form a stable network through a large number of hydrogen bonds as the permanent phase, while the PVA crystallites generated through repeated freezing/thawing cycles serve as the reversible phase [38].

The process of shape memory involves first stretching the hydrogel to a certain point of deformation, after which the temporary shape is set under the strain through a freezing/thawing cycle to form PVA crystals (Figure 7.20a). The photothermal effect of PDAPs converts NIR light into thermal energy, increasing the local temperature until it reaches the melting point of the PVA crystals (50–60 °C). This triggers the recovery of the shape (Figure 7.20b). This process not only achieves the shape memory function but also demonstrates the efficient responsiveness of the hydrogel under photothermal stimulation.

7.2 Functions of Photofunctional Polymer Hydrogel Composites

7.2.1 Electron Transfer Mechanism

In photofunctional polymer hydrogel composites, the electron transfer mechanism is affected by several crucial factors, including the separation between the electron donor and acceptor, the alignment of their energy levels, the surrounding medium (e.g. solvent characteristics), and temperature. The network architecture of the hydrogel can immobilize molecules at specific spatial positions, thereby effectively regulating electron transfer efficiency. Additionally, the hydrogel generally has a high-water content exceeding 90%, establishing a microenvironment similar to the liquid phase. This characteristic, combined with the temperature-responsive property of the hydrogel matrix, allows for the dynamic regulation of molecular mobility, which subsequently impacts the electron transfer rate.

The protonation-sensitive fluorescent molecule PPBEN was integrated into the temperature-sensitive F127 micellar hydrogel. By introducing H_2SO_4 solutions of diverse concentrations to regulate the $[H^+]$ within the hydrogel system, a series of responsive behaviors could be detected. Under acidic conditions, both the primary amino group located on the alkyl chain and the tertiary amino group present in the aromatic moiety of PPBEN undergo protonation, which significantly improves their solubility in aqueous solutions. Moreover, the protonation of these tertiary amino groups weakens their electron-donating capacities, thus decreasing the intramolecular charge transfer efficiency within PPBEN. This effect manifests as a blue shift in both the absorption and emission spectra [39]. The fluorescence emission of PPBEN mainly stems from the radiative transition of $S_1 \rightarrow S_0$, while the emission of the protonated form predominantly derives from the transition between LUMO+1 and HOMO−1, accounting for approximately 95% of the total emission (Figure 7.21a).

The F127-PPBEN hydrogel presents a characteristic three-dimensional network structure. When the H^+ concentration is 5 mmol/L, the hydrogel demonstrates a

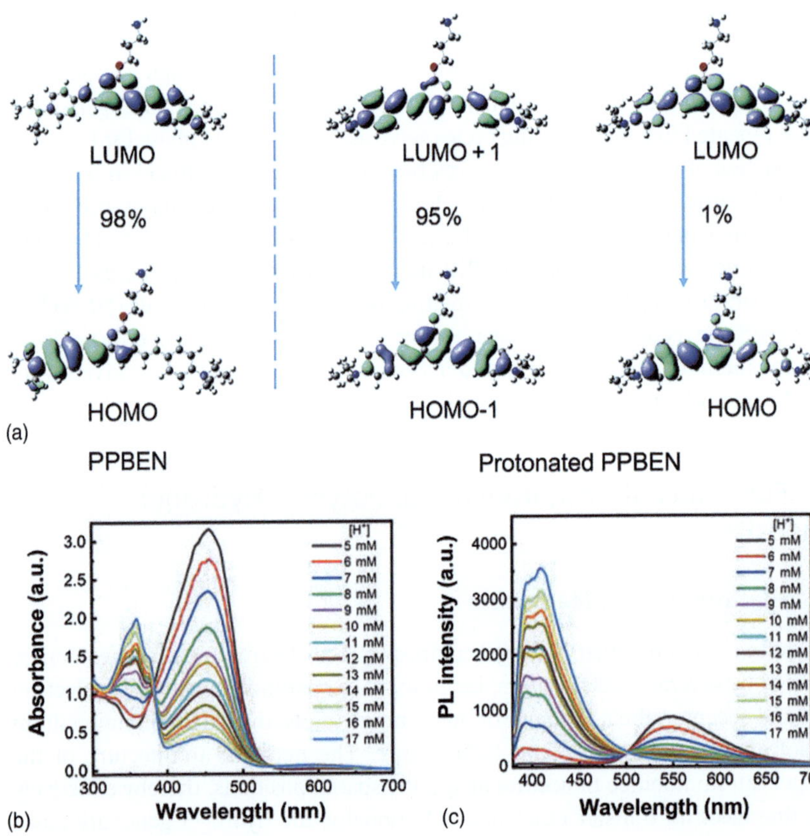

Figure 7.21 (a) Calculated $S_1 \rightarrow S_0$ emission of PPBEN and protonated PPBEN. *Source:* Reprinted from our published Tang et al. [39]. (b) Absorption and (c) emission spectra of F127-PPBEN hydrogel under varying $[H^+]$. *Source:* Wang et al. [40] / with permission of John Wiley & Sons.

singular absorption maximum at 450 nm (Figure 7.21b) and a distinct emission maximum at 550 nm (Figure 7.21c). This implies that, under these conditions, the majority of PPBEN molecules exist in a deprotonated state. This phenomenon can be ascribed to the shielding effect offered by the F127 micelles, which facilitate the retention of poorly water-soluble PPBEN molecules in a deprotonated state within the hydrophobic core of the micelles. As the H^+ concentration rises, the 550 nm emission gradually declines, while a 411 nm emission emerges, with its intensity increasing steadily. Simultaneously, alterations in the absorption spectrum are detected: the absorption peak at 450 nm gradually decreases, and a new peak at 350 nm appears and strengthens. These spectral variations indicate that, with the increase in H^+ concentration, a larger number of PPBEN molecules undergo protonation and dissolve in the acidic aqueous environment outside the F127 micelles. Notably, even when the H^+ concentration reaches as high as 17 mmol/L, the absorption peak at 450 nm still persists. This further validates that a portion of the PPBEN molecules remains encapsulated within the F127 micelles and is thus protected from protonation by the hydrophobic microenvironment. Consequently, the F127 micelles in the hydrogel exert a significant shielding effect on PPBEN molecules, and the aqueous environment within the hydrogel matrix allows for the precise regulation of the protonation state and optical properties of PPBEN through the modulation of the H^+ concentration [40].

An efficient electron transfer mechanism can be realized by incorporating electron donors (ZnO) and acceptors (viologen derivatives V_1^{2+}, V_2^{2+}, V_3^{2+}) into the polyvinyl alcohol (PVA) hydrogel system. The specific preparation process is as follows: PVA is dissolved in a water/glycerol mixed solvent containing ZnO and viologen under high-temperature conditions, and then cooled at low temperature to form a physically cross-linked organohydrogel (OHG). The cross-linked network structure of the OHG is stabilized by hydrogen bonds and PVA crystalline domains.

The observed color change phenomenon in viologens arises from an intermolecular electron transfer process, which occurs between an electron donor and viologen molecules acting as electron acceptors, triggered by light-induced excitation (Figure 7.22a). In this hydrogel system, the conduction band energy level of ZnO (−3.98 eV) is higher than the reduction potentials of three viologen derivatives (V_1^{2+}: −4.11 eV, V_2^{2+}: −4.31 eV, V_3^{2+}: −4.40 eV) (Figure 7.22b). As a result, photogenerated electrons produced by ZnO upon excitation can spontaneously transfer to viologen molecules, thereby initiating a coloration reaction (Figure 7.22c). The experimental results demonstrate that the PVA/V_1^{2+}/0.52 wt% ZnO hydrogel exhibits a light blue color, with an absorbance of 0.60 at 606 nm, after 1 s of irradiation, and gradually transitions to a dark blue hue after 300 s. Samples containing V_2^{2+} and V_3^{2+} also exhibit significantly enhanced color change responses, which validates the general applicability of ZnO nanoparticles in photosensitizing viologens. In contrast, the PVA/V_1^{2+}/0 wt% ZnO OHG sample only shows a weak color change (with an absorbance of 0.38) after 300 s of irradiation. This can be ascribed to the fact that, in the absence of ZnO, only counter anions act as electron donors, and the presence of water molecules impedes the electron transfer process. The incorporation of ZnO nanoparticles improves the photochromic efficiency

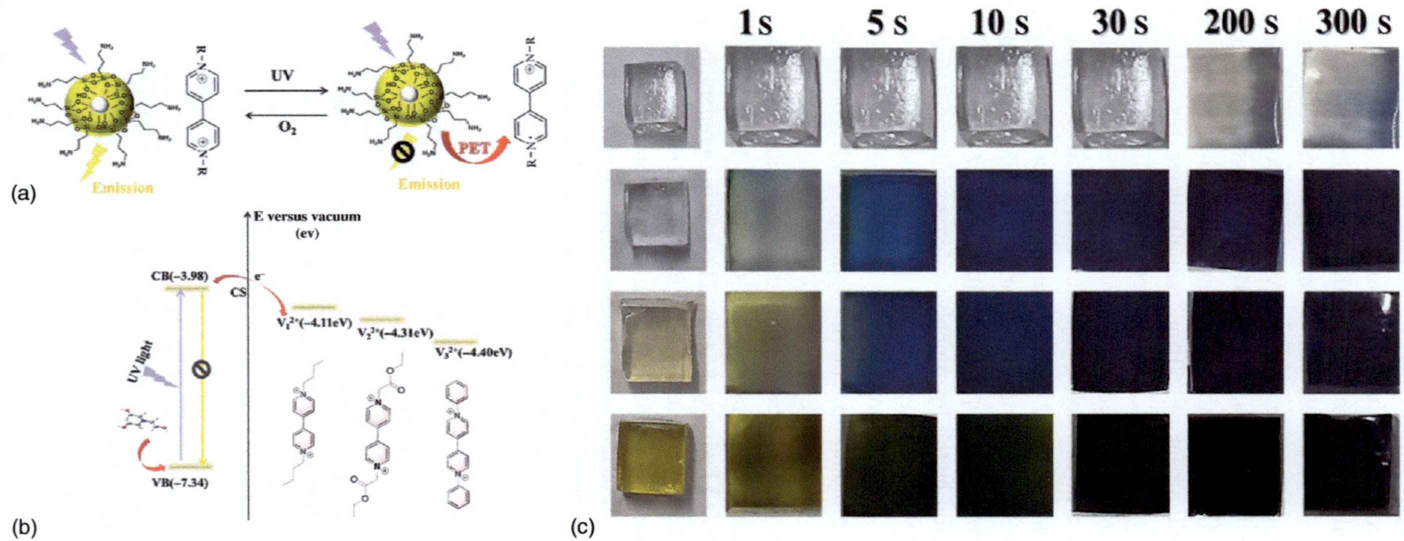

Figure 7.22 (a) Schematic illustration of photoinduced electron transfer between ZnO and viologen molecules. (b) Energy level diagram of ZnO and V_1^{2+}, V_2^{2+}, V_3^{2+}. (c) Photographs of PVA/V_1^{2+}/0 wt% ZnO, PVA/V_1^{2+}/0.52 wt% ZnO, PVA/V_2^{2+}/0.52 wt% ZnO and PVA/V_3^{2+}/0.52 wt% ZnO hydrogels under varying irradiation durations. *Source:* Chen et al. [41] / with permission of Elsevier.

considerably, increasing it by two orders of magnitude. The addition of ZnO not only provides an efficient electron transfer pathway, replacing the slower electron donation mechanism mediated by counter anions but also enhances the contact probability between electrons and acceptor molecules through the microstructural characteristics of the hydrogel. Therefore, the inclusion of ZnO notably improves the photochromic performance of viologen-based hydrogels [41].

7.2.2 Energy Transfer Mechanism

As a multifunctional matrix, the hydrogel is capable of effectively immobilizing both donor and acceptor molecules. Through the precise regulation of the spatial configuration of these molecules within the hydrogel, the separation between the donor and the acceptor can be meticulously adjusted to fall within the optimal range (usually 1–10 nm) necessary for fluorescence resonance energy transfer (FRET), thus improving the efficiency of energy transfer.

For instance, the fluorescent conjugated oligomer $OFBTCO_2Na$ forms bonds with PNAGA via hydrogen bonding and hydrophobic interactions, facilitating the successful construction of a three-dimensional microporous scaffold characterized by high tensile strength and robust fluorescent properties. This scaffold can effectively encapsulate fibroblasts and showcase excellent biocompatibility and mechanical stability. Owing to its high emission intensity and favorable photostability, $OFBTCO_2Na$ can serve as an energy donor in a FRET system. Its emission spectrum shows a high degree of overlap with the absorption spectrum of the cell membrane dye DiD, which meets one of the crucial prerequisites for FRET. When DiD is gradually introduced into the $OFBTCO_2Na$ solution, a progressive decrease in the emission intensity of $OFBTCO_2Na$ at 540 nm is witnessed, accompanied by a corresponding increase in the emission intensity of DiD at 670 nm. This indicates that efficient energy transfer has taken place between the two molecules. Significantly, compared to DiD under the same excitation conditions, the fluorescence signal is amplified by nearly three times, further validating the high efficiency of the FRET process (Figure 7.23a,b) [42].

Upon encapsulation of DiD-stained cells within the hydrogel matrix, an efficient energy transfer system was successfully established. When the hydrogel underwent mechanical stretching, the local contact area between the cells and the scaffold expanded, promoting more intimate interaction between the cells and a higher local concentration of $OFBTCO_2Na$. Simultaneously, the closer proximity between the hydrogel scaffold and the cells diminished the distance between DiD and $OFBTCO_2Na$, thus enabling effective energy transfer. This process was accompanied by a significant enhancement of the red fluorescence signal (Figure 7.23c). These findings not only emphasize the potential of hydrogels in biosensing applications but also highlight their value in cellular imaging [42].

The establishment of a cascading energy transfer mechanism within the hydrogel system is an effective strategy for luminescence regulation. Layered rare earth hydroxides (LRH) containing Gd^{3+}, Tb^{3+}, or Eu^{3+} ions can be synthesized and then combined with sodium salicylate (SA) and acrylamide monomers to fabricate

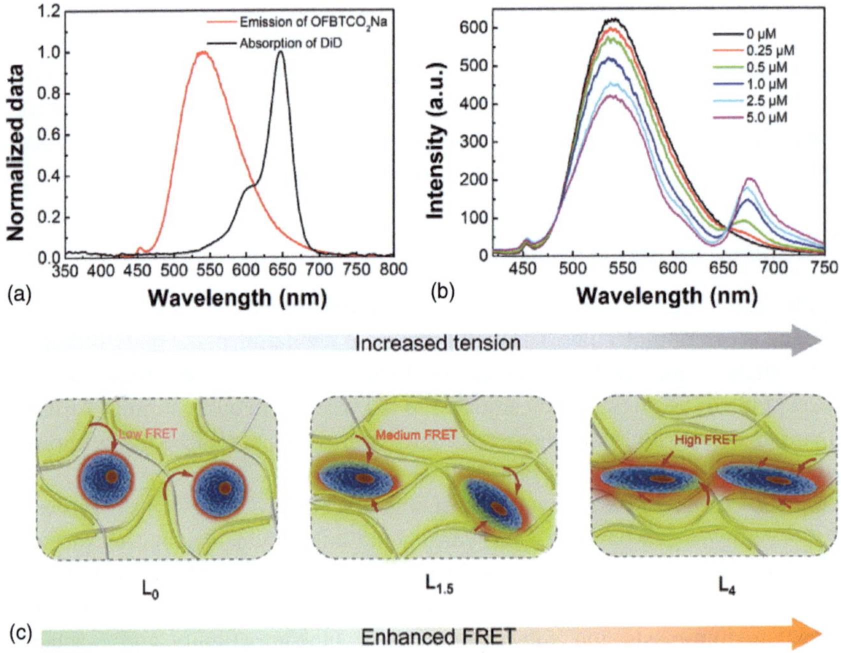

Figure 7.23 (a) Normalized emission spectrum of $OFBTCO_2Na$ alongside the UV-visible absorption spectrum of DiD. (b) Fluorescence spectra of the $OFBTCO_2Na$/DiD system after sequential additions of DiD. (c) Schematic representation of the FRET effect of $OFBTCO_2Na$/DiD system in L_0, $L_{1.5}$, and L_4 hydrogels at varying stretched lengths. *Source:* Zhang et al. [42] / with permission of American Chemical Society.

LRH/PAM hydrogels through in situ radical polymerization. In this system, LRH acts as the host matrix and promotes energy transfer to the sensitizer SA, thereby enhancing its phosphorescent emission. SA functions as a highly efficient sensitizer for Tb^{3+} due to the proximity of its lowest triplet state (23,300/cm) to the resonance energy level of Tb^{3+} ($^5D_4$: 20,500/cm), facilitating effective energy transfer (Figure 7.24). This energy transfer enables Tb^{3+} to emit green light at a wavelength of 544 nm during the $^5D_4 \rightarrow {}^7F_5$ transition. Additionally, the resonance energy level of Tb^{3+} ($^5D_4$: 20,500/cm) is closely aligned with the resonance energy levels of Eu^{3+} ($^5D_1$: 19,100/cm and $^5D_0$: 17,300/cm), allowing further energy transfer from Tb^{3+} to Eu^{3+}. As a result, Eu^{3+} emits red light at a wavelength of 614 nm during the $^5D_0 \rightarrow {}^7F_2$ transition. Experimental results indicate a relatively high energy transfer efficiency between Tb^{3+} and Eu^{3+}. Specifically, when the mole fraction of Eu^{3+} reaches 0.2, the energy transfer efficiency reaches 40.3%. Consequently, by adjusting the mole fraction of Eu^{3+}, the photoluminescent color of the hydrogel can be continuously modulated, ranging from green to yellow-green, yellow, orange-red, pink, and blue-violet. This cascading energy transfer mechanism not only provides novel perspectives for the development of luminescent materials but also offers new opportunities for the design and application of multifunctional hydrogel systems [43].

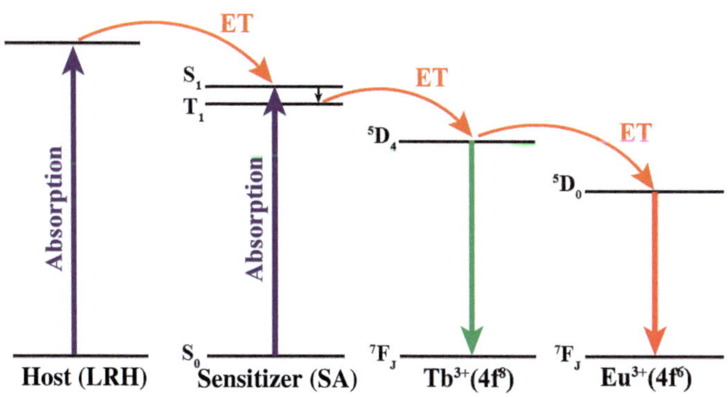

Figure 7.24 Schematic diagram of the energy-transfer process of $LGd_{0.5}Tb_{0.5-x}Eu_xH$/PAM hydrogels ($x = 0$–0.5).

7.2.3 Fluorescence Enhancement Effect

The three-dimensional network architecture of hydrogels is capable of not only immobilizing the spatial locations of molecules but also facilitating the aggregation of photofunctional molecules, constraining intramolecular rotation and vibration. Consequently, this reduces nonradiative energy dissipation and realizes aggregation-induced fluorescence enhancement.

Gold nanoclusters (AuNCs) with glutathione (GSH) serving as a ligand were successfully utilized in the synthesis of AuNCs@ZIF nanocomposites through coordination interactions between Zn(II) ions and either the carboxyl group of glutathione or the nitrogen atoms of 2-methylimidazole (Figure 7.25a). In comparison to free AuNCs, the emission intensity of AuNCs@ZIF was significantly elevated, exhibiting a fourfold increment (Figure 7.25b). Furthermore, the fluorescence lifetime of AuNCs@ZIF (16.8 μs) was substantially longer than that of free AuNCs (12.8 μs), indicating that ZIF-8 effectively diminished nonradiative relaxation pathways and prolonged the fluorescence lifetime (Figure 7.25c). The ZIF-8 framework offers a confined space for the AuNCs, which constrains the intramolecular vibration and rotation of the ligands without changing the charge state of gold. By markedly suppressing nonradiative relaxation processes, the aggregation-induced emission (AIE) effect is significantly enhanced.

Moreover, the AuNCs@ZIF composite nanoparticles were blended with bovine serum albumin (BSA) and sodium alginate (SA). Via glutaraldehyde-induced cross-linking of BSA and calcium (II)-mediated cross-linking of SA, a self-sorting process occurred, resulting in the formation of a hydrogel featuring an interpenetrating double-network structure with a thickness of 0.35 mm (Figure 7.25d,e). SEM images indicated that the double-network hydrogel was composed of entangled fibers and cross-linked junctions, constructing a dense three-dimensional network structure (Figure 7.25f). Remarkably, even under high-temperature conditions (55 and 75 °C), the hydrogel maintained stable red emission. This phenomenon could be ascribed to the protective functions of both ZIF-8 and the hydrogel matrix on the AuNCs, thus preserving the stability of their fluorescent properties [44].

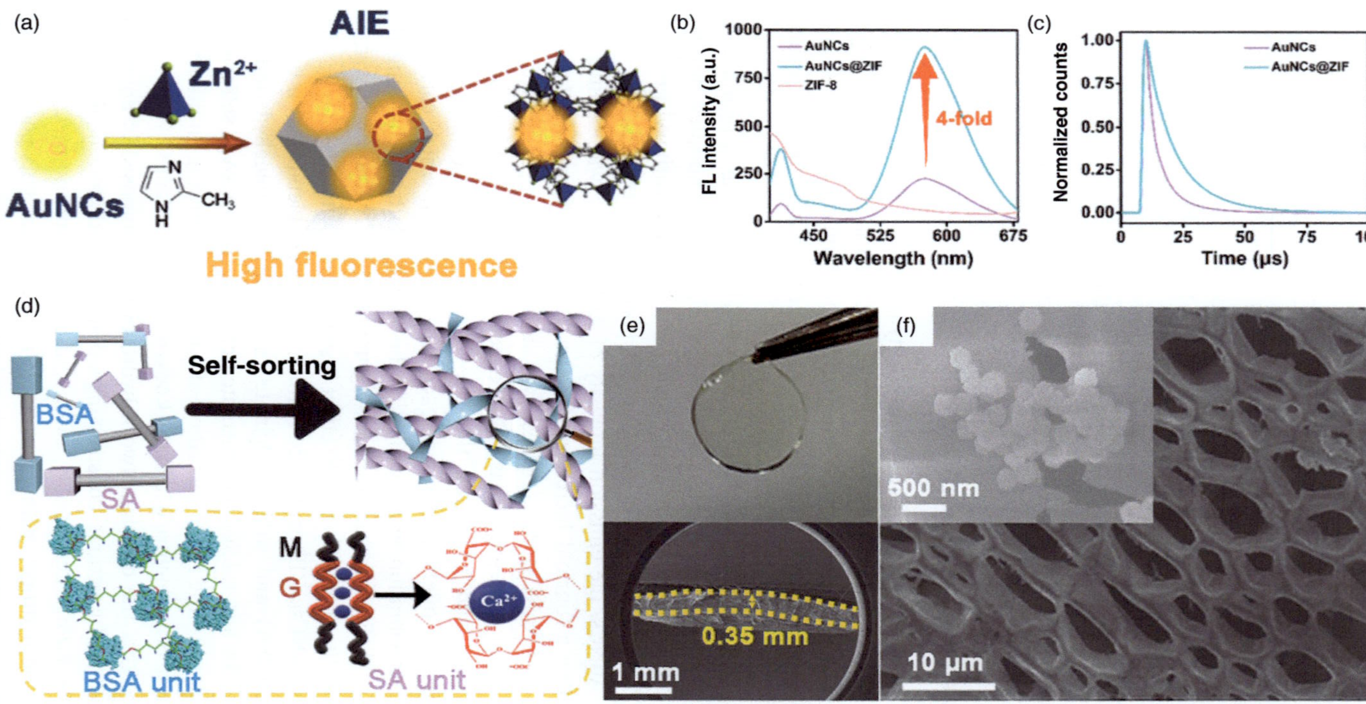

Figure 7.25 (a) Schematic representation of the preparation of AuNCs@ZIF nanocomposites. (b) Fluorescence emission spectra of AuNCs, AuNCs@ZIF, and ZIF-8. (c) Fluorescence decay profiles of AuNCs and AuNCs@ZIF. (d) Schematic illustration of hydrogel formation. (e) Photographs and (f) SEM images of the hydrogel. *Source:* Yan et al. [44] / with permission of American Chemical Society.

The photofunctional molecule TPE-4TA was co-dissolved with poly(vinyl alcohol) (PVA) in dimethyl sulfoxide. Subsequently, the linker molecule benzene-1,4-diboronic acid (BDA) was introduced, which facilitated the cross-linking of the PVA polymer through borate ester bonds. Thereafter, the solvent was replaced with deionized water, leading to the formation of a highly transparent dye-doped PVA hydrogel matrix (Figure 7.26a). The resultant hydrogel film displayed negligible background fluorescence, indicating that the molecular probes were probably well-dispersed or dissolved with high mobility within the polyhydroxy matrix. The fluorescence emission of TPE-4TA was significantly suppressed owing to its AIE characteristic. Moreover, due to the strong affinity of tetrazole/tetrazole ester groups for the polyhydroxy alcohol groups on the PVA backbone, these AIE molecules are confined within the hydrogel network, preventing leakage. Meanwhile, various metal ions could freely diffuse into the hydrogel and interact with the embedded molecular probes. Overall, the PVA hydrogel matrix provides a favorable chemical and physical environment for fluorescence sensing applications based on tetrazole-functionalized AIE molecules.

Time-resolved fluorescence decay demonstrated that upon the introduction of Hg^{2+}, the decay half-life of TPE-4TA notably increased from 3.06 to 4.28 ns (Figure 7.26b), implying the emergence of a fluorescence mechanism regulated by AIE. When the hydrogel was submerged in diverse Hg^{2+}-containing solutions, the emission intensity of the film augmented with the elevation of Hg^{2+} concentration, and the fluorescence signal was homogeneously distributed, suggesting that a stable AIE process transpired across the entire spatial configuration (Figure 7.26c). These results further corroborate the efficacy and structural stability of the PVA hydrogel matrix in promoting fluorescence-based detection [45].

The surface plasmon resonance (SPR) effect manifested by noble metal nanoparticles, like silver nanoparticles, can notably augment fluorescence signals. The network architecture of the hydrogel substrate facilitates the in situ synthesis and homogeneous dispersion of these nanoparticles (Figure 7.27). During the preparation procedure, polyvinyl alcohol (PVA) is initially blended with a borax solution, followed by the gradual dropwise addition of a silver nitrate solution. Subsequently, the resultant mixture is allowed to stand undisturbed in a dark environment at room temperature for 24 h, leading to the production of a vivid yellow product. This phenomenon is ascribed to the formation of a hydrogel by borax in an alkaline medium, which is replete with hydroxyl groups that can reduce Ag^{+} in situ. As a result, the PBH@AgD hydrogel is successfully synthesized.

This three-dimensional network structure not only guarantees the homogeneous dispersion of silver nanoparticles but also establishes an optimal microenvironment for fluorescence enhancement. The fluorophores within insulin proteins can elevate the local concentration of insulin around silver nanoparticles via coordination interactions involving sulfur or nitrogen atoms. This interaction results in a substantial enhancement of the fluorescence emission peak at 300 nm within 1 min, thereby attaining a highly efficient fluorescence amplification effect [46].

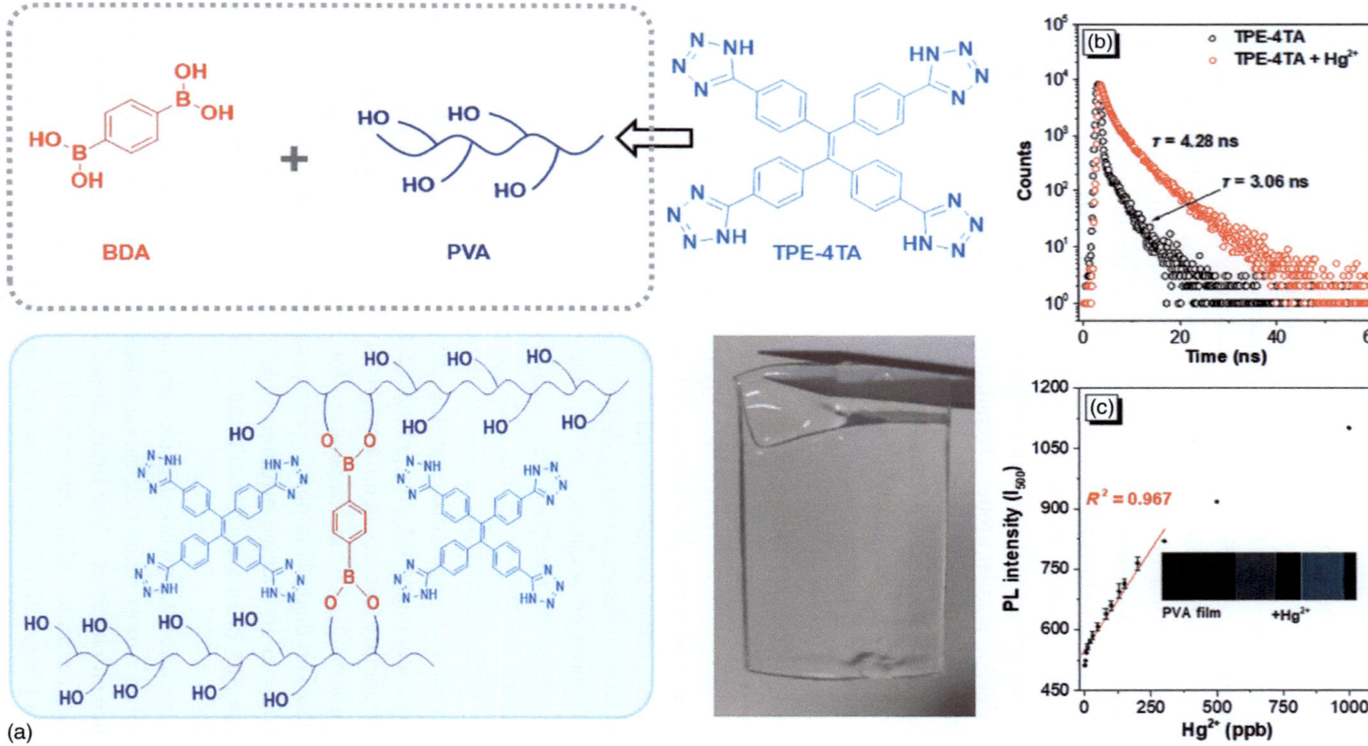

Figure 7.26 (a) The structures and photograph of TPE-4TA-doped PVA hydrogel. (b) Fluorescence decay profiles of TPE-4TA with and without Hg^{2+}. (c) Fluorescence intensities and photographs of TPE-4TA-doped PVA hydrogel with and without Hg^{2+}. *Source:* Reproduced from Wu et al. [45] / John Wiley & Sons / CC BY 4.0.

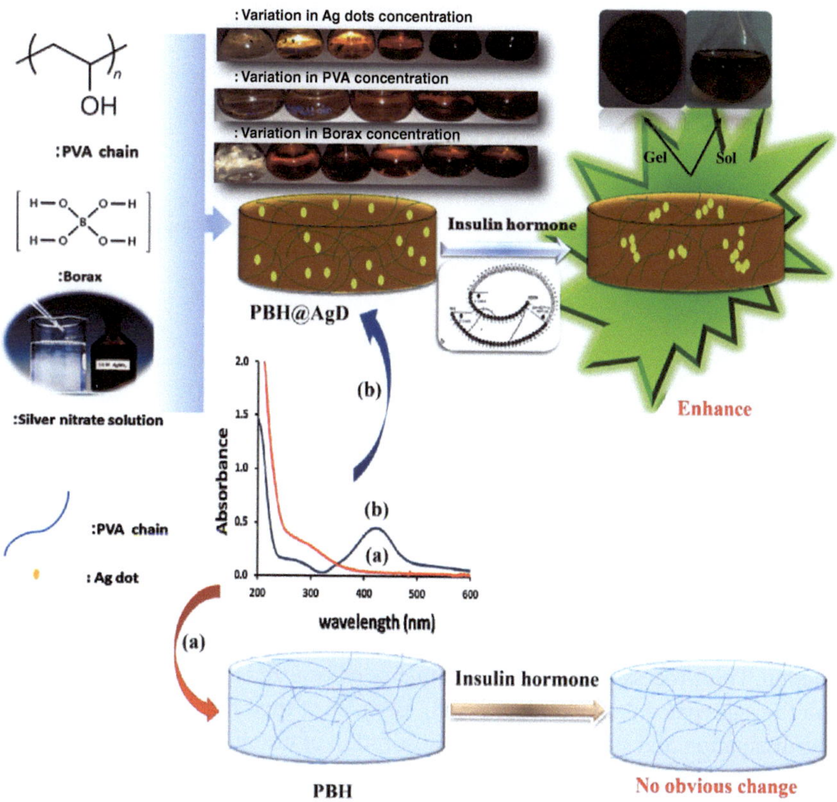

Figure 7.27 The structure of PBH@AgD hydrogel with enhanced emission. *Source:* Pourreza et al. [46] / with permission of Elsevier.

7.2.4 Photothermal Conversion Effect

Hydrogels with photothermal conversion capabilities are typically composed of a three-dimensional cross-linked polymeric network, such as poly(*N*-isopropylacrylamide), which is integrated with photothermal functional components, including gold nanorods, carbon-based materials, or polydopamine nanoparticles. Upon exposure to light, the photothermal agents rapidly convert light energy into thermal energy through mechanisms such as SPR or nonradiative relaxation of excited electrons. This conversion occurs on the picosecond timescale, resulting in localized temperature elevation within the hydrogel matrix [47].

Metal nanoparticles possess the ability to absorb visible or NIR light within a specific wavelength spectrum. When light is absorbed, the SPR phenomenon results in a significant augmentation of the quantity of "hot electrons" on the nanoparticle surface. These hot electrons promptly transfer their energy to the metal lattice via coupling with phonons. Subsequently, the lattice dissipates this energy into the surrounding environment in the form of thermal radiation [13].

For example, a polyacrylamide hydrogel embedded with gold nanoclusters (AuNC/PAM) has been observed to facilitate the transformation of gold

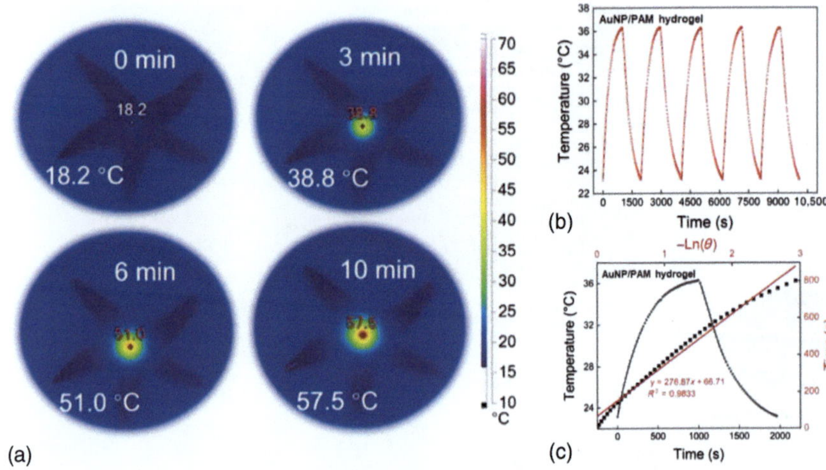

Figure 7.28 (a) Infrared thermal imaging and (b) the photothermal cycling stability of AuNP/PAM hydrogel; (c) calculated photothermal conversion efficiency from a single irradiation cycle. *Source:* Yang et al. [2] / with permission of Elsevier.

nanoclusters into larger gold nanoparticles (AuNPs) upon exposure to UV light irradiation. This process results in the emergence of localized SPR effects, characteristic of the aggregated AuNPs (Figure 7.28). Because of their effective nonradiative energy dissipation and significant optical absorption of visible light, AuNPs display excellent photothermal conversion performance. When irradiated with a 660 nm laser, the temperature of the hydrogel increases gradually as the irradiation time is extended. Infrared thermal imaging offers a more intuitive visualization of the photothermal effect of the AuNP/PAM hydrogel. As the irradiation duration lengthens, the temperature at the center of the laser spot rises steadily, reaching an increment of 39.3 °C after 10 min. Over five irradiation cycles, the temperature variations remainstable, indicating the outstanding photothermal stability of the AuNP/PAM hydrogel. Through the analysis of the temperature variation within a single irradiation cycle, the photothermal conversion efficiency of the AuNP/PAM hydrogel is determined to be 14.14%, further validating its effective photothermal functionality [2].

Polyaniline (PANI) exhibits a broad spectral absorption capability, efficiently absorbing both visible and NIR light. This characteristic allows the π-electron system of PANI to become photoexcited upon illumination. The resulting excited-state electrons dissipate energy in the form of heat through nonradiative relaxation pathways, such as electron–phonon interactions, thereby demonstrating notable photothermal conversion efficiency. This mechanism underscores the significant potential of polyaniline in photothermal material applications.

Using the composite hydrogel g-C_3N_4/PNIPAM/PAN, which incorporates polyaniline, as an example, the temperature of the material increased significantly under 808-nm laser irradiation, and this heating effect was closely associated with the laser power density. During a 10-min irradiation at different power densities,

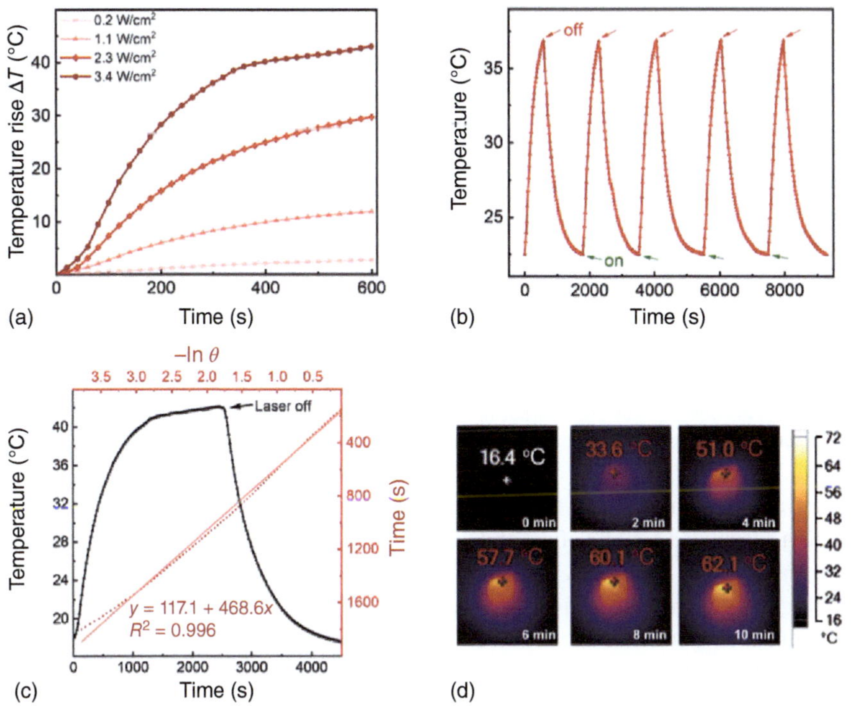

Figure 7.29 (a) The temperature rise profiles of g-C_3N_4/PNIPAM/PAN hydrogel under irradiation with varying power densities. (b) The photothermal cycling stability of the g-C_3N_4/PNIPAM/PAN hydrogel. (c) Calculated photothermal conversion efficiency from a single irradiation cycle. (d) Infrared thermal imaging of the g-C_3N_4/PNIPAM/PAN hydrogel. *Source:* Luo et al. [3] / with permission of Elsevier.

the temperature of the hydrogel rose by 2.7, 11.9, 29.6, and 43.0 °C, respectively, suggesting its highly efficient photothermal response capacity (Figure 7.29a). Moreover, the g-C_3N_4/PNIPAM/PAN hydrogel exhibited exceptional photothermal stability during temperature cycling tests comprising five laser on/off cycles. When the laser was activated, the temperature increased rapidly, declining gradually after it was switched off. The maximum temperature rise remained approximately constant throughout the five cycles, validating the reliability and durability of this composite hydrogel in photothermal conversion (Figure 7.29b). Its photothermal conversion efficiency was calculated to be 35.5% (Figure 7.29c). In the infrared thermal imaging experiment, the temperature at the center of the hydrogel increased rapidly over irradiation time, ultimately rising by 45.7 °C (Figure 7.29d). This result further verified the outstanding photothermal performance of the g-C_3N_4/PNIPAM/PAN hydrogel in practical applications [3].

The photothermal conversion mechanism of semiconductor materials can be elaborated as follows: upon the incidence of light on the material surface, free electrons within the valence band go into the conduction band after absorbing light, leading to the generation of electron-hole pairs. Subsequently, these excited electrons release energy through nonradiative relaxation processes. The energy is

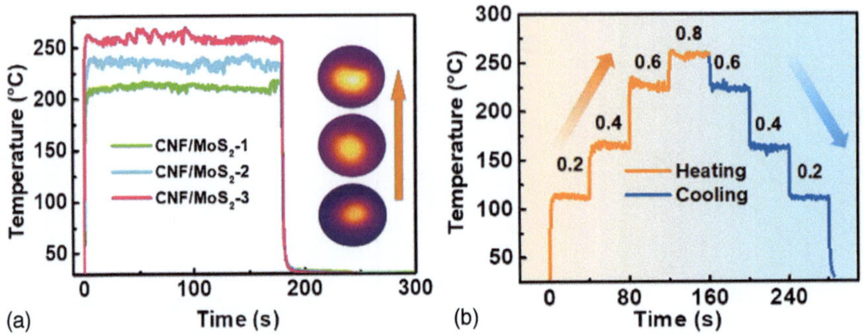

Figure 7.30 (a) Temperature rise profiles of CNF/MoS_2-1, CNF/MoS_2-2, and CNF/MoS_2-3 aerogels with increased MoS_2 doping under irradiation. (b) Temperature changes in CNF/MoS_2-3 aerogels at various irradiation powers. *Source:* Yuan et al. [48] / with permission of Elsevier.

predominantly transferred to the material lattice in the form of phonons, thereby inducing a localized heating effect and ultimately elevating the surface temperature of the material.

Taking molybdenum disulfide (MoS_2) as an example, its outstanding photothermal conversion performance allows it to be combined with cellulose nanofibers (CNF) for the fabrication of functional hydrogels. Experimental results demonstrate that composite aerogels with varying MoS_2 doping exhibit pronounced temperature gradients under NIR laser irradiation, with temperature increases ranging from 213.7 to 238.1 °C and up to 260.0 °C (Figure 7.30a). Infrared thermal imaging clearly illustrates the spatial distribution of the thermal field. Cyclic irradiation tests of CNF/MoS_2-3 aerogel conducted at different light intensities reveal consistent and stable heating and cooling behavior (Figure 7.30b). These findings confirm the composite hydrogel's excellent structural stability and reliable photothermal performance over multiple irradiation cycles [48].

7.2.5 Photosensitization Effect

The hydrogel has demonstrated a remarkable capability as a carrier in the application of photosensitizers. It can effectively protect photosensitizers from environmental degradation, thereby prolonging their functional duration and ensuring stability during photodynamic therapy or photocatalytic reactions. The photofunctionality of the PNAGA supramolecular hydrogel can be further optimized through the integration of the water-soluble conjugated oligomer OFBT-COOH (Figure 7.31a). Under white light irradiation, the excited-state electrons of OFBTCOOH undergo intersystem crossing (ISC) to transition to the triplet excited state (T_1). The resulting excitons are predominantly distributed along the entire polymer backbone, thereby increasing the interaction interface with adjacent oxygen molecules. The energy of the T_1 state is calculated to be 1.62 eV, which surpasses the minimum energy threshold (1.09 eV) necessary for the conversion of triplet oxygen ($^3O_2$) into singlet oxygen ($^1O_2$).

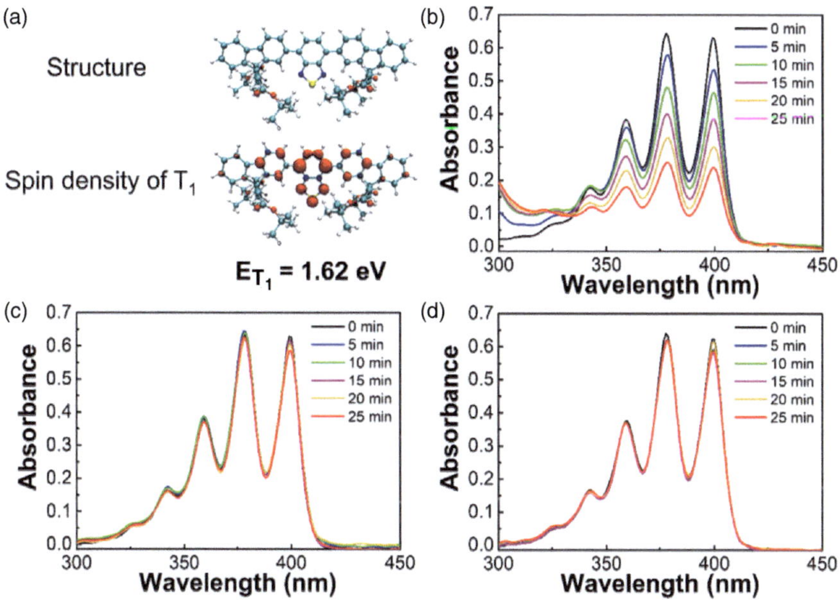

Figure 7.31 (a) The calculated T_1 energy of the OFBTCOOH. The absorption spectra of the ADPA solution under continuous light irradiation (b) with and (c) without OFBT/PNAGA hydrogel, and (d) in comparison with PNAGA hydrogel. *Source:* Zhang et al. [49] / with permission of American Chemical Society.

To assess the ability of the OFBT/PNAGA hydrogel to generate singlet oxygen ($^1O_2$), 9,10-anthracenediyl-bis(methylene)dimalonic acid (ADPA) was utilized as a specific $^1O_2$ probe. ADPA reacts selectively with $^1O_2$ to produce an endoperoxide compound, which is associated with a decrease in absorbance. The experimental outcomes indicated that the absorbance of ADPA gradually declined under continuous irradiation in the presence of the OFBT/PNAGA hydrogel (Figure 7.31b). Conversely, no substantial change in absorbance was detected in the pure ADPA solution or in the ADPA solution containing the PNAGA hydrogel (Figure 7.31c,d). As a result, the OFBT/PNAGA hydrogel not only demonstrates effective photosensitizer-loading characteristics but also holds significant potential for antibacterial applications [49].

The small-molecule photosensitizer chlorin e6 (Ce6) is capable of yielding a photosensitization effect when integrated into a hydrogel system. This is accomplished through the covalent conjugation of Ce6 with catalase (CAT) and the subsequent incorporation of the complex into the hydrogel matrix. CAT promotes the catalytic decomposition of hydrogen peroxide (H_2O_2) into oxygen (O_2), thus providing additional molecular oxygen to sustain the photosensitization reaction. The generation of ROS by Ce6-CAT upon photoactivation was confirmed via electron spin resonance (ESR) spectroscopy. In the experimental protocol, 2,2,6,6-tetramethylpiperidine (TEMP) and dimethylpyridine *N*-oxide (DMPO) were utilized as probes for the specific detection of $^1O_2$ and superoxide anion radicals ($O_2\cdot-$), respectively. The findings showed that Ce6-CAT effectively produced both

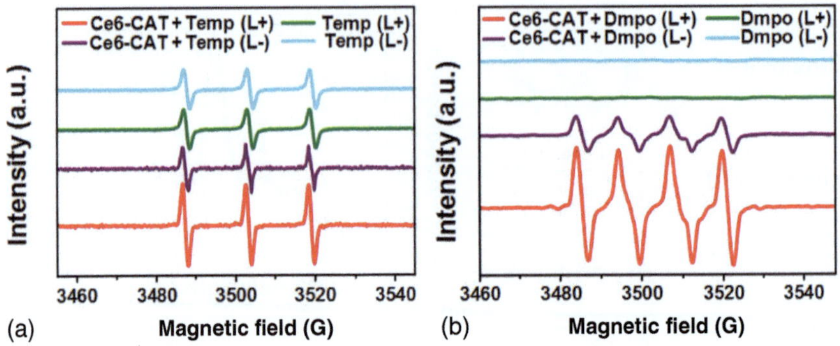

Figure 7.32 (a) The ESR spectra of TEMP and (b) DMPO with and without Ce6-CAT under light irradiation (L+) and without irradiation (L−). *Source:* Meng et al. [50] / with permission of John Wiley & Sons.

$^1O_2$ and $O_2{\cdot}-$ under 660 nm light irradiation, achieving comparable results to free Ce6 (Figure 7.32). Furthermore, the addition of H_2O_2 further augmented the ability of Ce6-CAT to produce $^1O_2$ upon photoactivation. This enhancement was predominantly attributed to the catalytic activity of CAT, which decomposed H_2O_2 to generate O_2, thereby facilitating the light-induced production of ROS [50].

Metal-organic frameworks (MOFs) have emerged as the optimal carrier material in photocatalysis due to their highly adjustable structural and porous characteristics. Incorporating MOFs into a hydrogel matrix that capitalizes on their responsiveness to visible light and efficient separation of photogenerated charge carriers can effectively realize the photocatalytic generation of ROS. For example, an anionic MOF framework can be synthesized using 1,3,5-benzenetriazole carboxylate (H_3BTT) ligands. Subsequently, silver nanoparticles (AgNPs) are introduced into the MOF pores via an in situ reduction approach, leading to the formation of an MOFs@Ag composite. Finally, a functional hydrogel material is fabricated through cross-linking with polysaccharide molecules. This photocatalytic system presents multiple advantages (Figure 7.33a): (1) The conduction band (−0.57 V versus NHE) and valence band (2.24 V versus NHE) of the MOF facilitate visible light absorption. Upon irradiation, photogenerated electrons rapidly migrate from the MOF surface to the AgNPs, while the corresponding holes remain in the valence band, thus attaining spatial separation of the charge carriers. (2) The surface SPR effect of AgNPs enhances the local electromagnetic field, which not only elevates the generation efficiency of electron-hole pairs within the MOF but also increases the concentration of active carriers through hot electron injection. (3) The Schottky barrier formed at the MOF–AgNP interface effectively inhibits electron-hole recombination, extends the carrier lifetime, and offers kinetic advantages for continuous ROS generation. Experimental results have confirmed that the system is capable of generating multiple ROS, including ·OH, $\cdot O_2^-$, and $^1O_2$ under visible light irradiation (Figure 7.33b–d). This synergistic multi-ROS generation mechanism contributes to a significant enhancement of the hydrogel's photocatalytic performance [51].

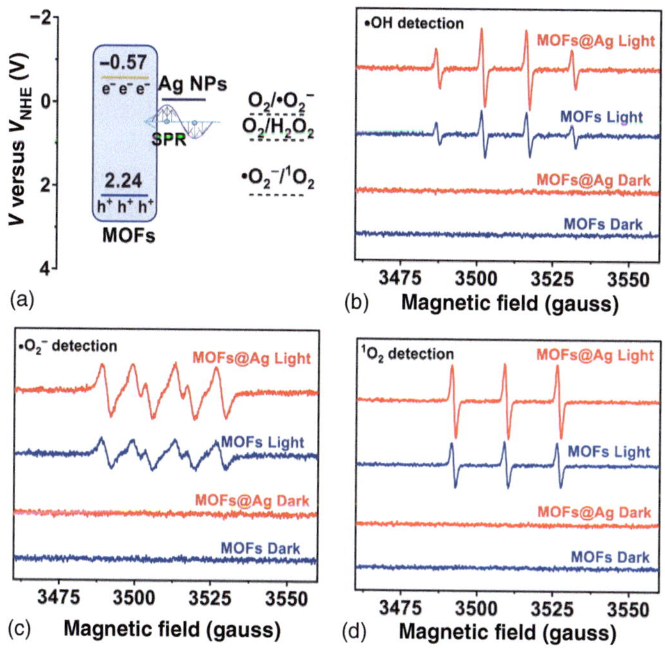

Figure 7.33 (a) Illustration of the energy levels of MOFs@Ag nanocomposites. The ESR signals of (b) ·OH, (c) $\cdot O_2^{-\cdot}$ and (d) $^1O_2$ with MOFs@Ag or MOFs under light irradiation and in the dark. *Source:* He et al. [51] / with permission of Elsevier.

7.2.6 Photoisomerization Effect

Photoisomerization is the reversible conversion of molecular geometric configurations when the molecules are irradiated with light of a specific wavelength. For example, azobenzene is excited upon exposure to UV light, transitioning from a trans-configuration to a cis-structure, and can return to the original state under visible light irradiation or through thermal relaxation. These structural alterations are closely related to the dynamic changes in molecular polarity, steric hindrance, and electronic energy levels. Integrating the photoisomerization effect into hydrogel systems can overcome the limitations of traditional liquid or solid carriers. The three-dimensional, cross-linked hydrogel network facilitates the immobilization of photosensitive molecules via covalent bonding or host–guest interactions. This prevents deactivation caused by aggregation or migration. It also enables precise regulation of isomerization efficiency and response speed by adjusting the hydrogel's dynamic microenvironment.

An ionogel was effectively synthesized via the combination of azobenzene (Azo), poly(*N*-isopropylacrylamide) (PNIPAM), and 1-ethyl-3-methylimidazolium bis(trifluoromethylsulfonyl)imide ($[C_2mim][NTf_2]$). Within this system, light acts as an external stimulus, azobenzene functions as a photosensitive element, and the sol–gel transition arising from the phase separation behavior of PNIPAm offers an amplification effect. Ultimately, actuation is achieved through alterations in ion conductivity. The ionogel generates a light-responsive electrical signal, which can be output to regulate the switching of an external circuit.

In the specific mechanism, under dark conditions and UV light irradiation, the mixtures of [C_2mim][NTf_2] and azobenzene are denoted as trans-IL and cis-IL, respectively. Exposure to white light results in trans-IL displaying a low affinity for PNIPAm, causing PNIPAm to aggregate. In contrast, under UV light, cis-IL exhibits a high affinity for PNIPAm, allowing more cis-IL molecules to permeate the PNIPAm aggregates. This permeation weakens the effect of the cross-linker, consequently inducing the gel-to-sol transition. Photoisomerization of azobenzene significantly alters the interaction between the ionic liquid (IL) and PNIPAm. Specifically, cis-azobenzene strengthens the interaction between the polymer and IL, while also inhibiting intramolecular polymer interactions. This light-regulated sol–gel transition not only facilitates a distinct switching behavior in ion conduction (Figure 7.34) but also provides a simple and effective strategy for regulating ion transport in optoelectronic devices [52].

Spiropyran is a photoisomerizable molecule that can undergo a structural transformation from a closed-ring, colorless, and nonfluorescent spiropyran (SP) form to an open-ring, colored, and red-fluorescent merocyanine (MC) form when irradiated with light. In this research, a multiresponsive fluorescent hydrogel, denoted as HEMA-SPMA-F127, was synthesized through the radical polymerization of 2-hydroxyethyl methacrylate (HEMA) in the presence of spiropyran-functionalized methacrylic acid (SPMA) and F127 micelles. During the experimental process, the hydrogel was irradiated with 365 nm light, and the photochromic response as a function of time was monitored. As shown in Figure 7.35, after 300 s of irradiation, the photochromic conversion rate of the hydrogel containing F127 was significantly improved. Specifically, the hydrogel containing 6.6 wt% F127 exhibited a fluorescence growth rate approximately 2.5-fold greater than that observed for the hydrogel prepared in the absence of F127. Furthermore, the color change observed in the hydrogel containing F127 was more distinct, suggesting a greater conversion of SPMA molecules to the MC state [23, 24].

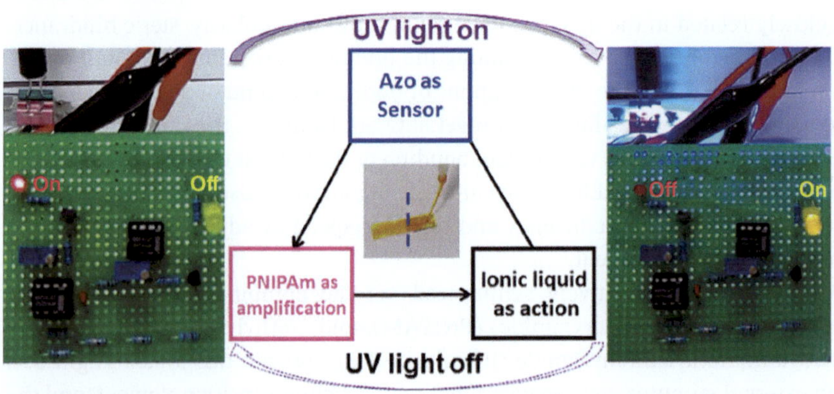

Figure 7.34 Ion-conductive switch based on light-responsive sol–gel transition of hydrogel. *Source:* Wang et al. [52] / with permission of American Chemical Society.

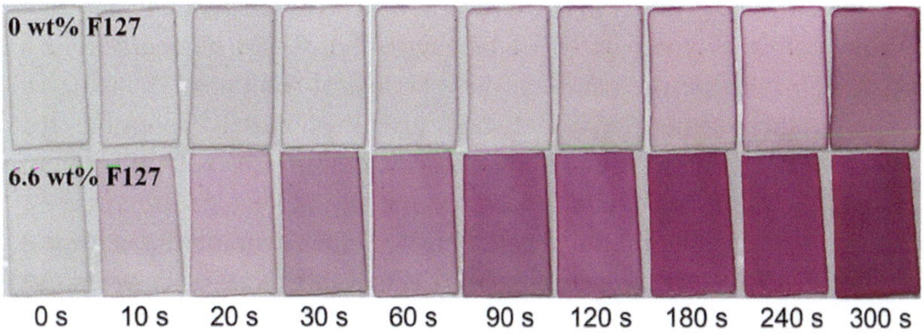

Figure 7.35 Photochromic response of spiropyran-based hydrogels with varying F127 concentrations under irradiation. *Source:* Reproduced with permission from Gao et al. [23] / John Wiley & Sons.

This phenomenon suggests an increase in the population of activatable SPMA molecules as well as an enhancement in the photoisomerization rate. Upon analysis, it is evident that F127 micelles generate a hydrophobic microenvironment within the hydrogel, which facilitates the conformational transition of SPMA molecules. The hydrophobic nature of this environment minimizes SPMA aggregation, thereby promoting the occurrence of photoisomerization reactions. Additionally, F127 micelles are capable of solubilizing SPMA molecules, further reducing their aggregation and consequently increasing the quantity of activatable SPMA species in the hydrogel system. This research offers novel insights into the design of photofunctional materials and highlights the critical function of micelles in modulating molecular conformation and reaction kinetics.

7.3 Bioapplications of Photofunctional Polymer Hydrogel Composites

The biomedical applications of hydrogels have been developed significantly since their first application in ophthalmic treatments in 1960 [53]. As previously discussed, polymer hydrogels, with their unique three-dimensional cross-linking network structure and tunable cross-linking forces, have laid down their basic structural characteristics as carriers of photofunctional materials. By incorporating photofunctional components and utilizing mechanisms like electron/energy transfer, photothermal conversion, and photosensitization, these hydrogels acquire diverse light-responsive functionalities. This organic integration of structure and function endows photofunctional polymer hydrogel composites with significant advantages in biomedical applications.

The polymer network matrix of hydrogels exhibits exceptional properties as a composite platform, facilitating the effective dispersion or chemical integration of diverse organic and inorganic photofunctional components, such as photosensitive dyes, noble metal nanoparticles, and conjugated polymers, within its three-dimensional structure. This capability enables the development of multifunctional composite materials with tailored functionalities. Furthermore, polymer matrices like polyethylene glycol (PEG), hyaluronic acid (HA), gelatin, and chitosan (CS) are distinguished by their excellent biocompatibility and biodegradability, which substantially mitigate the risk of adverse biological reactions. Their highly hydrated, porous three-dimensional network structures closely resemble the microenvironment of the natural ECM, allowing for the unrestricted diffusion of nutrients, metabolic byproducts, and bioactive signaling molecules. Importantly, these biomimetic structures provide a conducive microenvironment for critical cellular functions, which are fundamental to the advancement of regenerative medicine.

Moreover, the network architecture of hydrogels serves as an optimal platform for the precise design of photosensitive functionalities. By incorporating photosensitizing groups—such as azobenzene, spiropyran, or cinnamate esters—or through molecular modifications at polymer chains or cross-linking sites, these materials can acquire advanced smart-responsive capabilities. Such capabilities include remote, noncontact, and spatiotemporal control over photoinduced deformation, drug release, or modulation of mechanical properties. As a result, functional polymer hydrogel composites that combine customizable photofunctionality, superior biocompatibility, biomimetic ECM-like structures, and intelligent photoresponsiveness are emerging as highly promising advanced material platforms in cutting-edge biomedical applications. These applications span drug delivery, tissue engineering and regenerative medicine, biosensing and diagnostics, photodynamic and PTT, and the regulation of cellular behavior.

This section will focus on recent advancements, innovative design strategies, and future directions for these hydrogel composites in specific biological applications, supported by representative studies from the literature.

7.3.1 Optical Detection and Sensing

Photofunctional polymer hydrogel composites are used in biological detection and sensing because they utilize the three-dimensional network structure of hydrogels as a carrier and reaction platform. Photoresponsive polymers or nanomaterial probes can be immobilized or integrated into this structure. When target molecules or microorganisms interact specifically with these probes, it triggers significant changes in their optical signals, enabling highly sensitive and selective detection of analytes. As a carrier for photoresponsive probes, the three-dimensional, cross-linked network of a hydrogel provides an enormous specific surface area and numerous sites for the modification of functional groups. This facilitates functionalization and the dense integration of recognition elements and signal-reporting units. This significantly enhances the intensity of the sensing signal and detection sensitivity. Its porous structure and high water content enable the rapid diffusion and transport of target analytes, substrates, and buffer ions within the hydrogel, thereby reducing response times and improving detection efficiency. Additionally, the physicochemical properties of hydrogels closely resemble those of natural ECM, creating a biomimetic microenvironment that preserves the native conformations and biological activity of biomolecules, reducing inactivation and enhancing the reliability of detection. These characteristics make hydrogels ideal matrix materials for constructing next-generation, high-performance, portable, and even implantable biosensors.

Diabetes, a major metabolic disorder, presents a serious risk to public health worldwide. Statistics show that nearly 828 million adults worldwide were diagnosed with diabetes in 2022. This massive patient population has driven rapid growth in the glucose sensor market due to growing demand for blood glucose monitoring [54]. In recent years, hydrogel-based fluorescence sensors have emerged as a research hotspot in blood glucose detection due to their high sensitivity and rapid response characteristics.

A ratiometric fluorescent sensor was developed by incorporating rhodamine B and fluorescein as responsive fluorescent groups into the side chains of the monomers of polyacrylamide-based hydrogels and by integrating glucose oxidase (GOx) via simple embedding or covalent grafting. Under neutral pH conditions without glucose, the hydrogel demonstrated weak fluorescence emissions at 518 nm (green) and 598 nm (red). When glucose was added, the loaded GOx oxidized glucose into D-glucono-δ-lactone and H_2O_2. The subsequent oxidation of H_2O_2 generated oxygen and protons, reducing the pH of the hydrogel medium. Under these conditions, the fluorescein underwent a ring-opening reaction, significantly decreasing the green signal, while rhodamine B underwent a ring-opening reaction, enhancing the red signal. By monitoring the ratio changes between green and red fluorescence, quantitative glucose detection was achieved. Compared to single-fluorescence sensors, this ratiometric fluorescence sensor not only demonstrates rapid response and high sensitivity but also effectively reduces false-positive results and background interference, improving detection accuracy and reliability [55].

H_2O_2 is an essential biomarker in biomedical research since it is an active oxygen species in biological metabolism. It has two functions in living things: it is a crucial cellular signaling molecule that controls physiological functions like apoptosis, migration, differentiation, and cell proliferation [56]. Furthermore, when H_2O_2 builds up in cells at excessive levels, it can destroy macromolecules like DNA and proteins, causing oxidative stress and upsetting cellular equilibrium. Inflammatory disorders, neurological diseases, cardiovascular diseases, cancer, diabetes, and aging processes are all greatly influenced by this process [57].

To achieve high-sensitivity in situ detection of H_2O_2, researchers developed an electrochemiluminescence (ECL) sensor (Figure 7.36a). First, ABEI-Ag ECL luminescent material was synthesized by reacting luminol derivative ABEI with $AgNO_3$. Subsequently, PAni-PA hydrogel with high conductivity was prepared by mixing aniline and phytic acid, followed by polymerization initiated by $(NH_4)_2S_2O_8$. The ABEI-Ag solution was then applied to a glassy carbon electrode modified with PAni-PA conductive hydrogel to form the ABEI-Ag@PAni-PA composite sensor.

In the experiment, cells were induced to secrete H_2O_2 by stimulation with phorbol 12-myristate 13-acetate (PMA). The AgNPs in the hydrogel catalyzed H_2O_2 decomposition into $OH^{\bullet}$, significantly accelerating the ECL reaction of ABEI and generating intense fluorescence signals (Figure 7.36b). Results demonstrated that PMA-induced cell signals were markedly enhanced. Notably, cells adhered to the

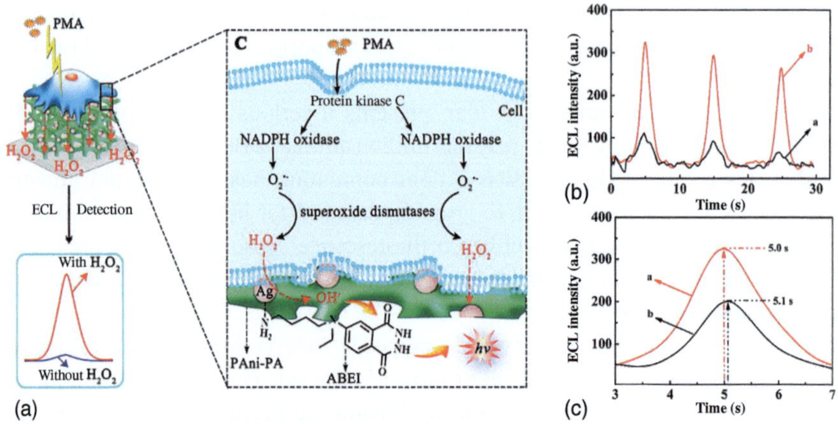

Figure 7.36 (a) The detection mechanism of released H_2O_2 from cells using ABEI-Ag@PAni-PA composite sensor. (b) ECL intensities of cells bound on ABEI-Ag@PAni-PA composite hydrogel before and after PMA stimulation. (c) Comparison of ECL intensities of cells bound on the ABEI-Ag@PAni-PA composite hydrogel and in solution. *Source:* Jiang et al. [58] / with permission of American Chemical Society.

hydrogel emitted stronger H_2O_2 signals than those in solution (Figure 7.36c). This occurs because the H_2O_2 released by hydrogel-bound cells, with its extremely short diffusion range, can be instantly captured by ABEI-Ag, enabling highly sensitive in situ detection of H_2O_2. The long π-conjugated backbone of the PAni-PA hydrogel matrix not only demonstrates excellent charge transport properties but also significantly enhances conductivity, thereby improving the detection sensitivity. This conductive hydrogel-based ECL sensor shows significant potential applications in biomedical research and disease diagnosis [58].

DNA detection holds profound significance in biomedical research, with applications spanning critical fields such as genetic mutation analysis, diagnosis, and treatment of hereditary diseases. To enhance detection efficiency, researchers have developed a novel oligonucleotide hybridization platform capable of achieving rapid and sensitive detection of cancer DNA without amplification. The innovation lies in utilizing hydrogel formation as a signal amplification strategy, which significantly improves both the accuracy and sensitivity of the detection process.

Specifically, this detection method first involves the binding of target cancer DNA with a polyethylene glycol (PEG) conjugated with single-stranded DNA (ssDNA) capture probes. Subsequently, the detection probes are linked to P1 peptides at their ends. When target DNA is present, the probes specifically bind to the DNA, releasing P1 peptides that self-assemble into a hydrogel network in solution. During this process, the fluorescent dye Cy3.5 is encapsulated within the hydrogel's hydrophobic micropores, which weakens quenching effects and significantly enhances fluorescence intensity (Figure 7.37). Leveraging amplified fluorescence signals, this method achieves high-sensitivity detection of cancer DNA. The successful development of this technology not only provides a powerful tool for early cancer diagnosis but also opens new avenues for DNA detection in biomedical research. Its high

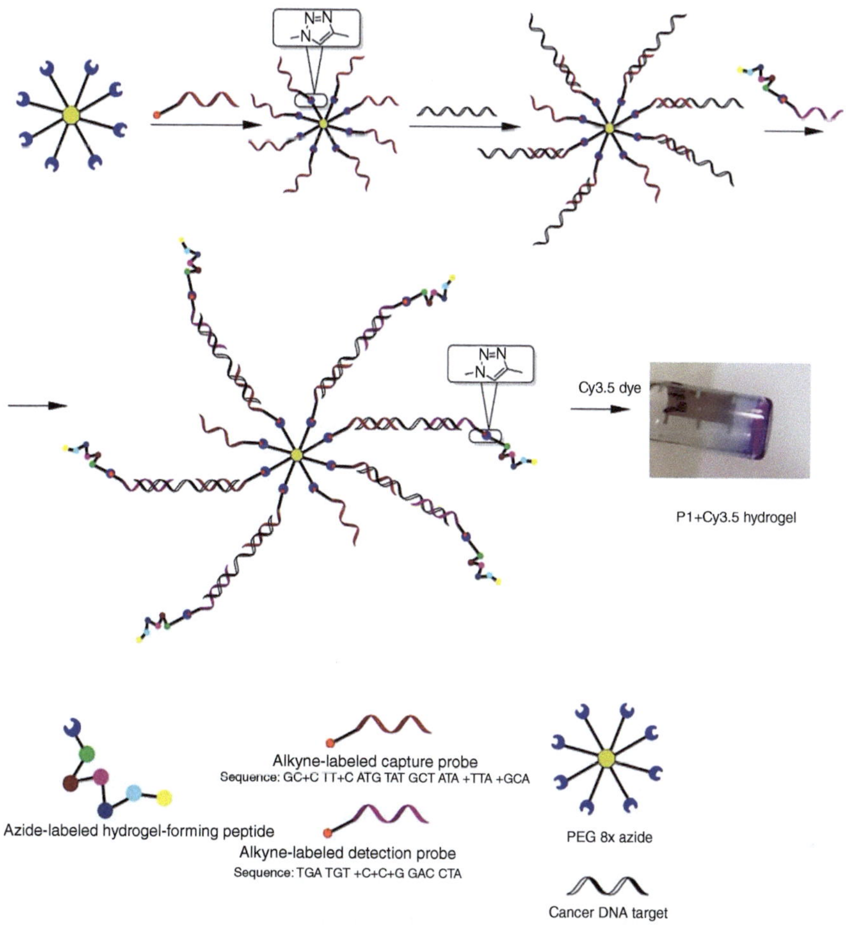

Figure 7.37 DNA detection mechanism of P1 + Cy3.5 hydrogel. *Source:* Jandi et al. [59] / with permission of American Chemical Society.

efficiency and sensitivity make it widely applicable in both clinical diagnosis and basic research [59].

Tetracycline (TC), a broad-spectrum antibiotic, is widely used in animal husbandry to control microbial diseases due to its excellent oral bioavailability and low cost. However, excessive use of tetracycline has led to its accumulation in the environment and animal products, posing a major risk to public health. Therefore, it is crucial to evaluate and precisely detect antibiotic levels in both field environments and samples.

A method for encapsulating blue fluorescent carbon dots in a sodium alginate hydrogel system enables quantitative detection of tetracycline. The carboxyl groups on both the alginate and carbon dot surfaces cross-link with Ca^{2+} to form a stable hydrogel network. When tetracycline is present in the environment, its phenolic hydroxyl or enol groups bind to these surface carboxyl groups, forming electron donor–acceptor complexes. This process facilitates nonradiative exciton

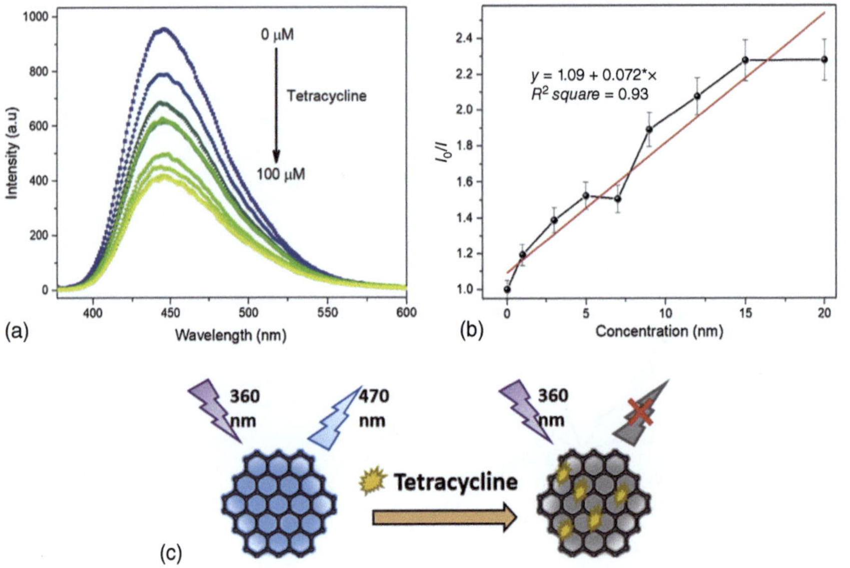

Figure 7.38 (a) Fluorescence intensity changes of CDs at varying tetracycline concentrations. (b) Detection limit of tetracycline was determined by the hydrogel. (c) Schematic diagram of the quenching mechanism of CDs by tetracycline. *Source:* Ehtesabi et al. [60] / with permission of Elsevier.

recombination, resulting in fluorescence quenching of the carbon dots (Figure 7.38). Experimental results demonstrate a strong linear relationship between fluorescence emission intensity at 470 nm and tetracycline concentration. This discovery provides a simple and sensitive method for quantifying tetracycline [60].

Photofunctional polymer hydrogel composites not only find extensive applications in molecular detection but also demonstrate significant potential for pathogen detection, which is crucial for early disease diagnosis and infectious disease prevention. Based on the specific interaction between sialic acid and hemagglutinin (HA) or influenza A virus (IAV) [61], a detection system can be developed through a sequential fabrication process as follows: Initially, IAV nanoparticles are employed as a biological template and incubated with sialic acid-AuNP to facilitate specific recognition. Subsequently, these functionalized nanoparticles are incorporated into a hydrogel matrix formed by the click reaction between polyethylene glycol diazide and cyclooctyne-functionalized PEG. Following hydrogel formation, the IAV template is selectively removed using HCl, thereby generating imprinted binding sites within the hydrogel network that are complementary to the viral morphology. Throughout this process, the AuNPs remain immobilized on the substrate via a combination of electrostatic interactions and strong Au—S bonds, ensuring structural stability and retention of functional properties.

When the hydrogel contacts IAV, the HA binds to sialic acid-AuNP, causing aggregation of AuNP within the hydrogel. This triggers the SPR peak redshift, transforming the hydrogel color from pink to purple (Figure 7.39). At the same time, the specific binding increases the additional cross-linking point, resulting in

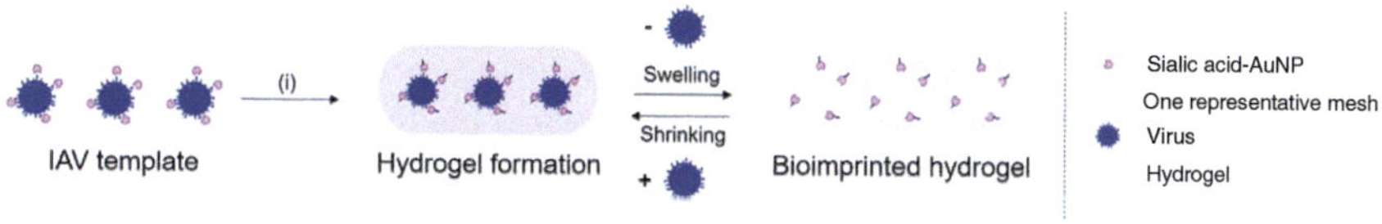

Figure 7.39 Formation process and IAV detection using the bioimprinted hydrogel. *Source:* Randriantsilefisoa et al. [62] / with permission of Royal Society of Chemistry.

a reduction of the hydrogel diameter by about 30%. This dual optical/mechanical response of color change and volume contraction is visible, enabling the rapid, convenient, and specific detection of viruses [62].

The detection and identification of bacterial species can be accomplished by selectively recognizing distinct metabolites or biomolecules, such as enzymes, that are secreted by specific bacterial strains. Taking *E. coli* as an example, its secreted β-glucuronidase (β-GUS) serves as a key detection marker [63]. This assay achieves specific detection of β-GUS by covalently coupling fluorescent 4-methylumbellone-β-D-glucuronide (MUG) or chromogenic 4-nitrophenyl-β-D-glucuronide (PNPG) to chitosan hydrogel via EDC/NHS reaction (Figure 7.40). When the target bacterial strain's β-GUS interacts with the hydrogel, it catalyzes hydrolysis reactions, releasing protonated coumarin derivatives such as 4-methylumbellone (4-MU) or 4-nitrophenol (4-NP). Notably, 4-MU emits fluorescence at 445 nm, while 4-NP produces an absorption peak at 405 nm, causing the solution to change from colorless to yellow. This enzymatic reaction-to-optical signal conversion not only demonstrates high specificity but also enables rapid and low-limit bacterial detection [64].

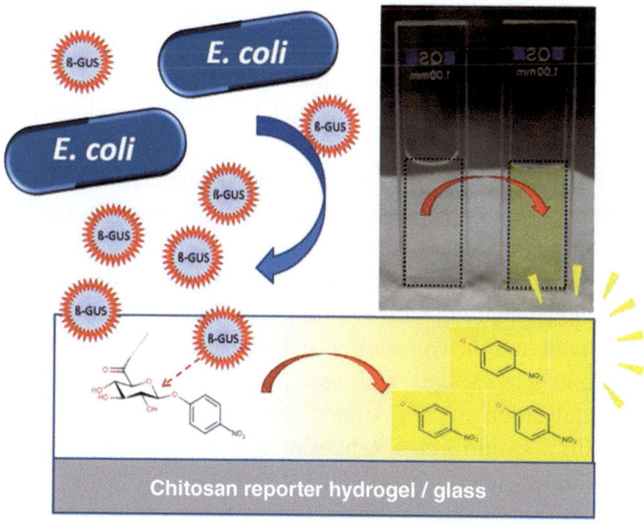

Figure 7.40 Detection mechanism of bacterial enzymes using the photofunctional hydrogel composite. *Source:* Ebrahimi et al. [64] / with permission of American Chemical Society.

The capture and release of specific tumor cells have demonstrated broad applications in early cancer diagnosis, dynamic monitoring, personalized treatment, and metastasis research. Notably, intelligent biological platforms constructed with biocompatible hydrogels that integrate photothermal and temperature-sensitive properties have provided innovative solutions for efficient tumor cell capture and controlled release.

At the heart of this platform lies a composite hydrogel p(NAm-*co*-AuNC-*co*-HA) developed using ligand-modified gold nanoclusters (AuNCs), which are co-polymerized with organic monomers such as acrylamide, *N*-acetylglucosamine (NAGA), and hyaluronic acid via free radical polymerization. Specifically, allylamine-modified AuNCs form a stable cross-linked network with polymer chains through covalent cross-linking and hydrogen bonds. As a key component of the hydrogel, hyaluronic acid (HA) specifically binds to CD44 receptors, enabling targeted capture of CD44-expressing tumor cells such as MDA-MB-231 cells. The AuNCs exhibit excellent photothermal conversion properties, converting light energy into heat when exposed to 660 nm irradiation, which raises the local temperature. This thermal effect disrupts hydrogen bonds between hydrogel molecules, reducing mechanical properties to achieve gentle cell release. This functionality is absent in hydrogels without AuNCs (Figure 7.41a,b). Notably, released cells retain viability and can be directly used for subsequent biological analyses [65].

7.3.2 Phototherapy

The photothermal sterilization and therapeutic effects of photofunctional polymer hydrogel composites are achieved by incorporating photothermal agents such as metal nanomaterials, graphene-based carbon materials, or conjugated polymers into the hydrogel. These materials absorb incident light energy and convert it into thermal energy through nonradiative transitions, thereby increasing local temperature. This process induces protein denaturation in bacteria or tumor cells, ultimately leading to apoptosis and achieving bacterial inactivation and tumor ablation.

The SPR effect in metal nanomaterials endows composite hydrogels with excellent photothermal properties. Among various photothermal conversion materials, gold nanorods (GNRs) exhibit two SPR absorption peaks due to their anisotropic rod-like structure. By adjusting the aspect ratio, these peaks can be precisely tuned from the visible to NIR regions, aligning with the "optical window" of biological tissues to maximize tissue penetration. This makes GNRs ideal materials for PTT. When embedded in biodegradable chitosan hydrogels, the GNR-hydrogel system exhibits effective biofilm eradication via PTT. Under 830 nm laser irradiation, the temperature increases by 10 °C within 2 min, exhibiting significant photothermal bactericidal and biofilm inhibitory effects, effectively managing antibiotic-resistant oral infections [66].

Silver nanoparticles (AgNPs) exhibit photothermal properties similar to gold nanoparticles. Using mesoporous silica nanorods (MSN) as carriers, AgNPs were synthesized within channels through an in situ reduction method. By adjusting the

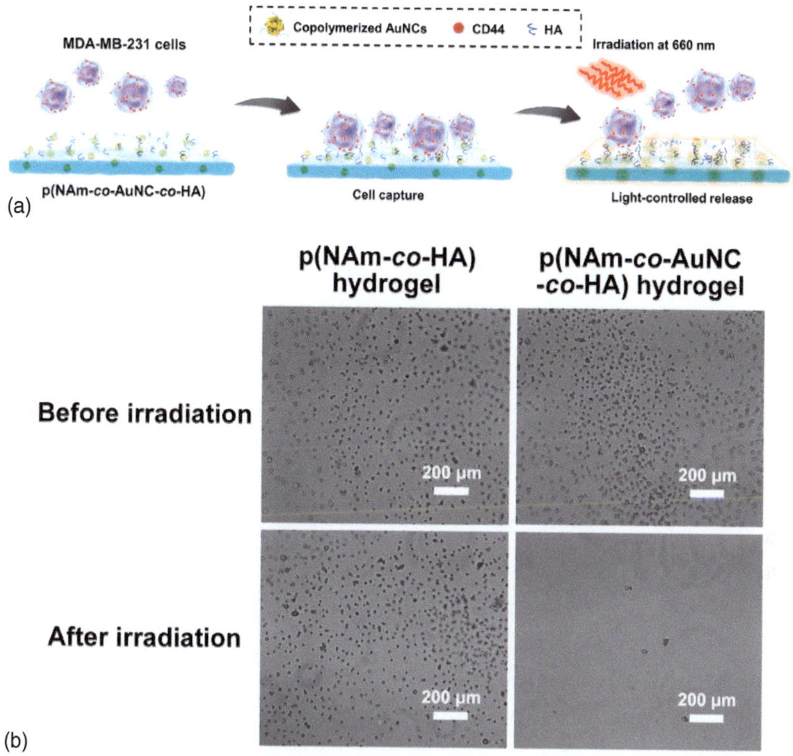

Figure 7.41 (a) Mechanism of cancer cell capture and light-triggered release via p(NAm-*co*-AuNC-*co*-HA) hydrogel. (b) Optical images of captured and released cells on p(NAm-*co*-AuNC-*co*-HA) and p(NAm-*co*-HA) hydrogels. *Source:* Zhu et al. [65] / with permission of American Chemical Society.

morphology, the SPR peak of AgNPs can be tuned to the NIR region. The MSN/Ag nanocomposite was mixed with gelatin to form an MSN/Ag-Gel hydrogel. After light irradiation, the temperature of the MSN/Ag-Gel hydrogel increased by 25.6 °C. The synergistic effects of photothermal killing and silver ion release mechanisms resulted in a bactericidal efficiency exceeding 99% against *E. coli* [67].

Hydrogel materials based on guar gum and borate ester bonds can effectively encapsulate Cu nanoparticles and inject them into mice. In experiments, the injection site was irradiated with a 660 nm laser for 5 min. Infrared thermal imaging revealed a significant temperature increase in the composite hydrogel within the body, rising from 29.2 to 32.6 °C (Figure 7.42a,b). Additionally, when mixed with bacterial solution, this composite hydrogel demonstrated excellent bactericidal efficacy by eliminating 99% of *E. coli* after just 10 min of irradiation. This research provides crucial scientific evidence for developing novel antibacterial materials [13].

In addition to inorganic materials and small organic molecules, conjugated polymers demonstrate higher photothermal conversion efficiency and excellent biocompatibility. To fully leverage these properties, researchers self-assembled

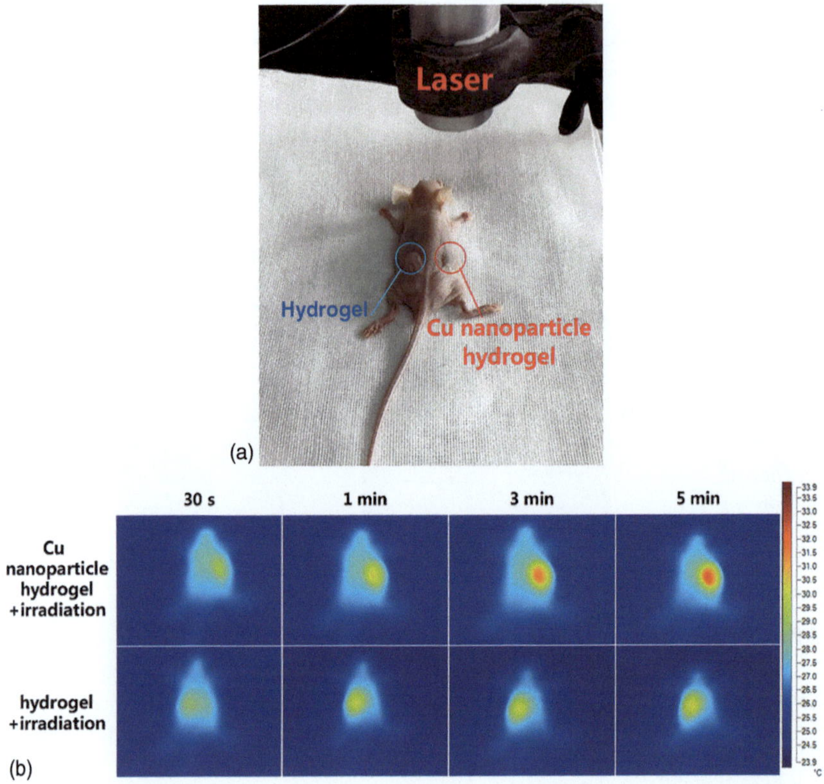

Figure 7.42 (a) Photograph of the in vivo photothermal performance setup for Cu nanoparticle hydrogel. (b) in vivo thermal images of Cu nanoparticle hydrogel and pure hydrogel under irradiation. *Source:* Reproduced with permission from Chen et al. [13] / American Chemical Society.

pH-sensitive bromothymol blue (BTB), β-glycerophosphate (β-GP), and an NIR-absorbing conjugated polymer PTDBD into chitosan (CS) hydrogels. The β-GP imparted thermal sensitivity to the hydrogel, enabling sol–gel transition behavior when the temperature increases. Through electrostatic interactions, the hydrogel maintains close contact with wound bacteria, thereby enhancing antibacterial efficacy [68].

Experimental results demonstrated that areas treated with BTB/PTDBD/CS hydrogel exhibited significantly higher temperatures under laser irradiation compared to untreated regions (Figure 7.43a,b). Despite the temperatures generated by PTDBD exceeding 55 °C during PTT—which may cause damage to adjacent healthy tissues—the infected wounds treated with BTB/PTDBD/CS hydrogels combined with laser irradiation exhibited the most rapid healing (Figure 7.43c). This demonstrates the hydrogel system's significant potential in photothermal applications for effectively promoting wound healing while minimizing collateral tissue damage.

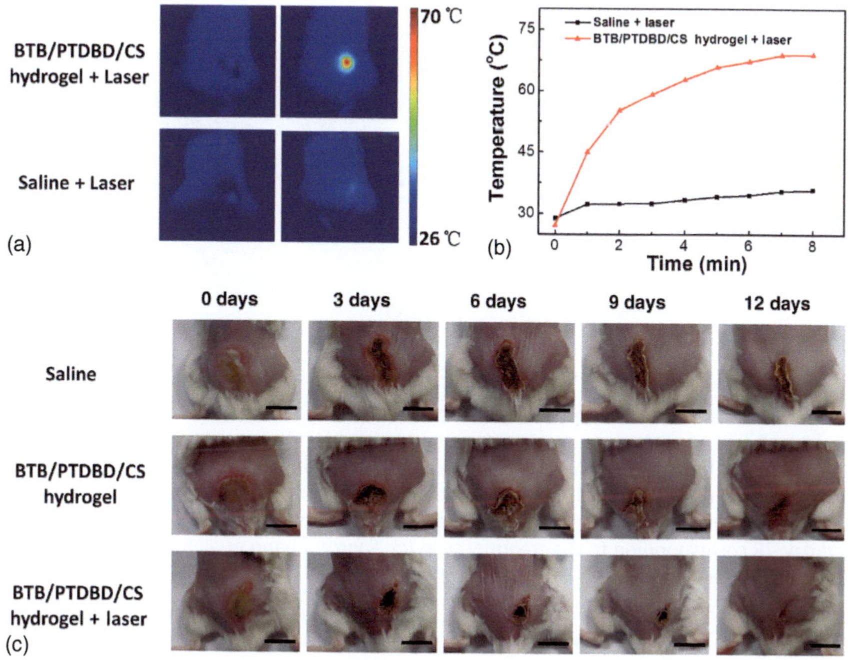

Figure 7.43 (a) In vivo thermal images and (b) corresponding temperature profiles of BTB/PTDBD/CS hydrogel. (c) Photographs of *Staphylococcus aureus*-infected mice wounds treated with saline, BTB/PTDBD/CS hydrogel with and without irradiation. *Source:* Wang et al. [68] / with permission of American Chemical Society.

The π-conjugated backbone of polydopamine (PDA) contains numerous delocalized electrons that can absorb NIR radiation and efficiently transduce it into thermal energy, making it highly applicable in PTT. In this study, carboxymethyl cellulose sodium (CMC-DA) modified with dopamine was used as the matrix. The introduction of the oxidizing agent sodium periodate ($NaIO_4$) into the CMC-DA solution initiated hydrogel formation. Throughout this process, dopamine underwent oxidation to form PDA, whereas $NaIO_4$ was concomitantly reduced to sodium iodide (NaI). Experimental results showed that after implanting this hydrogel at tumor sites in vivo, significant temperature elevation was observed when irradiated with an 808 nm laser for 1 min. The tumor's temperature rose by 27 °C within 5 min of irradiation, compared to no significant temperature rise in the control group (Figure 7.44a,b). Furthermore, post-photothermal treatment, the recurrence tumor volume in mice was significantly reduced, demonstrating the hydrogel's remarkable efficacy in suppressing tumor recurrence (Figure 7.44c,d) [69].

Similarly, the photodynamic bactericidal and therapeutic effects of photofunctional polymer hydrogel composites are achieved by incorporating photosensitizers into the hydrogel. These agents absorb photon energy and generate ROS through electron transfer or energy transfer pathways. This process damages biomolecules and cellular structures, leading to apoptosis. Consequently, the system effectively inactivates bacteria while promoting wound healing.

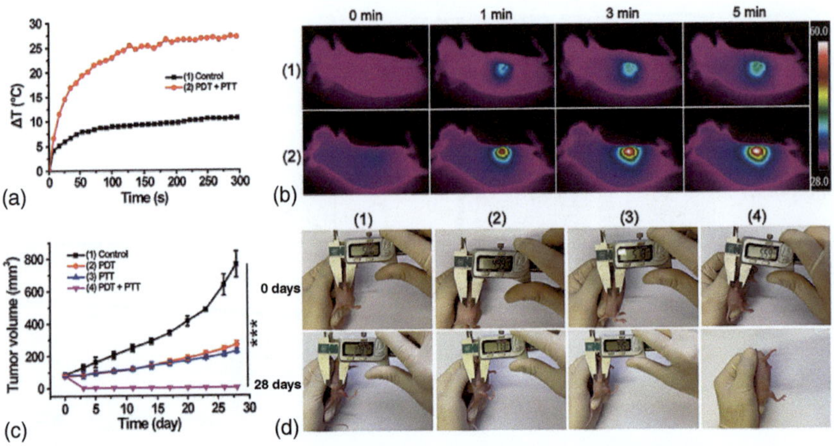

Figure 7.44 (a) Temperature variation of tumors implanted in the hydrogel during 5 min of laser irradiation; (b) corresponding infrared thermal images of the tumor; (c) changes in relative tumor volume following different treatments; (d) photographs of tumor size on day 0 and day 28 corresponding to the treatments in (c). *Source:* Zhang et al. [69] / with permission of Elsevier.

The commonly used photosensitizer tetrakis(4-carboxyphenyl)porphyrin (TCPP) can be covalently linked to ε-polylysine and subsequently copolymerized with vinyl-modified ε-polylysine to form a hydrogel network (mPL-Pr-TCPP). Under light exposure, TCPP generates various ROS, including $\cdot O_2^-$ and $^1O_2$. The abundant lysine residues in the hydrogel adsorb bacteria through electrostatic interactions, thereby enhancing the antibacterial efficacy of ROS. This bacterial capture mechanism effectively addresses the issues of short ROS lifespan and limited diffusion range. Experimental results demonstrate that applying mPL-Pr-TCPP hydrogel to wounds and exposing them to light for 15 min significantly promotes wound healing. As shown in Figure 7.45, after 12 days, the wound area was reduced to $0.1\ mm^2$. In contrast, wounds treated with PBS exhibited marked suppuration and ulceration within 2 days, accompanied by significant pus accumulation by day 4, indicative of severe tissue damage resulting from bacterial infection. Although the photosensitizer alone exhibits some antibacterial effects, its wound area ($21.3\ mm^2$) is comparable to commercial 3M dressings, making it less effective than the mPL-Pr-TCPP hydrogel [70].

In another work, the conjugated oligomer photosensitizer OFBTCOOH physically cross-links with poly(*N*-acrylamide glycolate) (PNAGA) monomers through hydrophobic interactions, which is incorporated into a hydrogel network during the free radical polymerization process, ultimately forming an OFBT/PNAGA supramolecular hydrogel (Figure 7.46a). The backbone of OFBTCOOH incorporates benzothiadiazole (a strong electron-withdrawing unit) and fluorene (an electron-donating unit). Such a push-pull electron system effectively facilitates ISC after photoexcitation, thereby efficiently generating triplet excitons and sensitizing oxygen to produce $^1O_2$. Under light irradiation for 20 min, the hydrogel composite demonstrated excellent photodynamic bactericidal efficacy

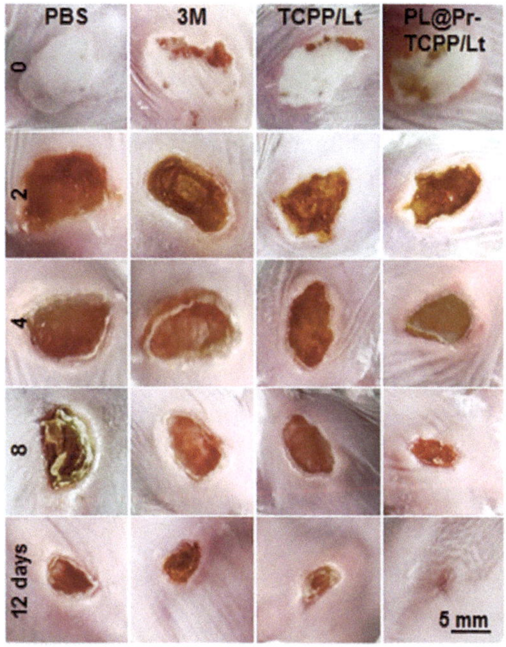

Figure 7.45 Photographs of wounds after treatment with PL@Pr-TCPP hydrogel, TCPP, 3M dressing, and PBS under light irradiation over a 12-day period. *Source:* Reproduced with permission from Ran et al. [70] / Elsevier.

by eliminating 99% of *E. coli* (Figure 7.46b). In contrast, the hydrogel without OFBTCOOH exhibited no such photodynamic killing capability. Additionally, the amide groups of PNAGA form stable hydration barrier layers on the hydrogel surface through hydrogen bonding with abundant water molecules. This structure not only resists biofouling but also prevents bacterial accumulation, thereby maintaining long-term antibacterial performance. These dual mechanisms endow the OFBT/PNAGA supramolecular hydrogel with broad application prospects in photodynamic antimicrobial therapy [49].

In addition, the photocatalytic activity of Au/Ag can be utilized to achieve photodynamic sterilization. Specifically, the method involved adding silver ions to glutathione (GSH)-protected gold nanoclusters to induce their aggregation into Au/Ag clusters, which enhanced fluorescence and photodynamic performance through the AIE effect. Subsequently, the Au/Ag clusters were covalently immobilized in a gelatin network via amidation reactions, forming Au/Ag@gelatin sponges. In the pig skin infection model experiment, the bacterial biofilm on the infected wound completely disappeared after treatment with Au/Ag@gelatin under irradiation (Figure 7.47). The results demonstrated that this bactericidal effect was entirely dependent on oxidative damage caused by ROS generated by light excitation, rather than cell membrane disruption from silver ion release. This highlights the potential application value of Au/Ag@gelatin sponges in photodynamic sterilization [5].

7.3.3 Light-Controlled Drug Release

Efficient and safe drug delivery remains one of the core challenges in modern disease treatment. Conventional methods often face issues such as low drug

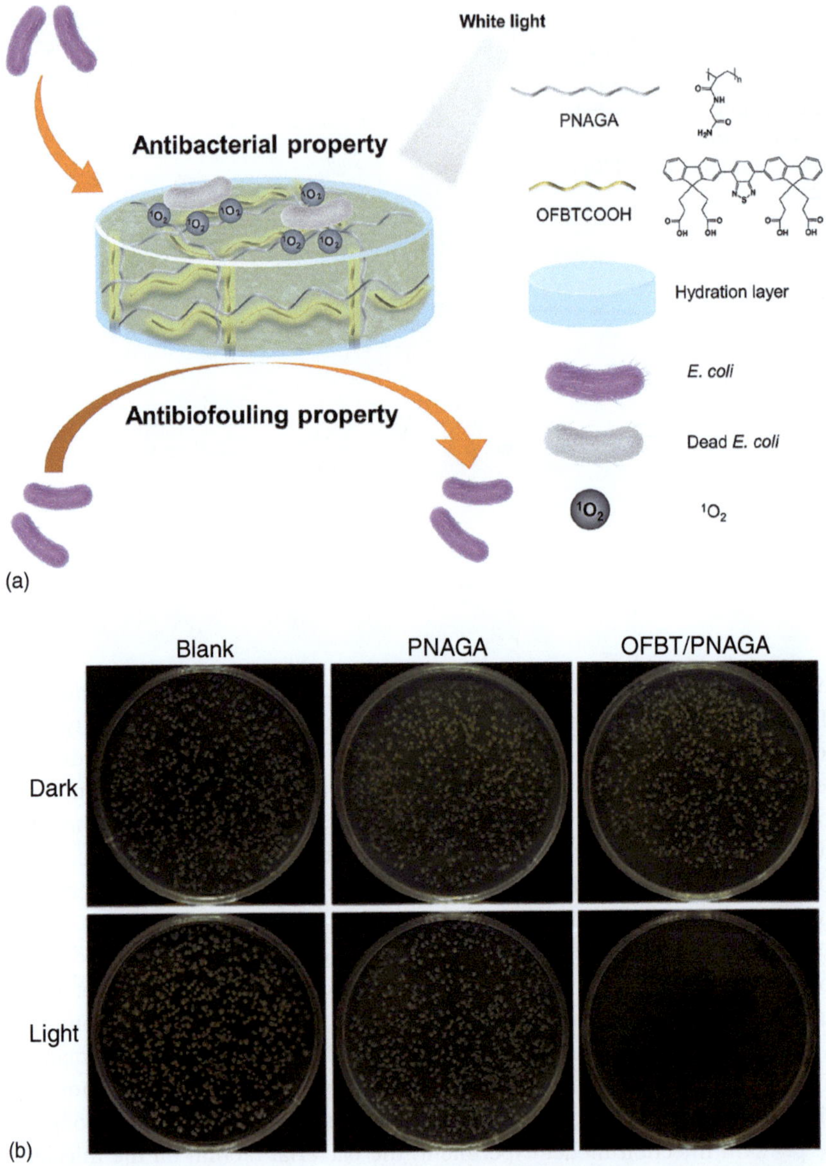

Figure 7.46 (a) Multiple antibacterial mechanisms of OFBT/PNAGA hydrogel. (b) Killing efficacy of *E. coli* of OFBT/PNAGA and PNAGA hydrogel under light irradiation. *Source:* Zhang et al. [49] / with permission of American Chemical Society.

bioavailability, significant off-target effects, and resulting toxic side effects. Particularly for treatments with a narrow therapeutic window, specific targets, or localized lesions (such as solid tumors, inflammatory sites, or specific neural regions), achieving precise spatiotemporal control of drug delivery and on-demand release at the target site is crucial. This not only significantly improves therapeutic

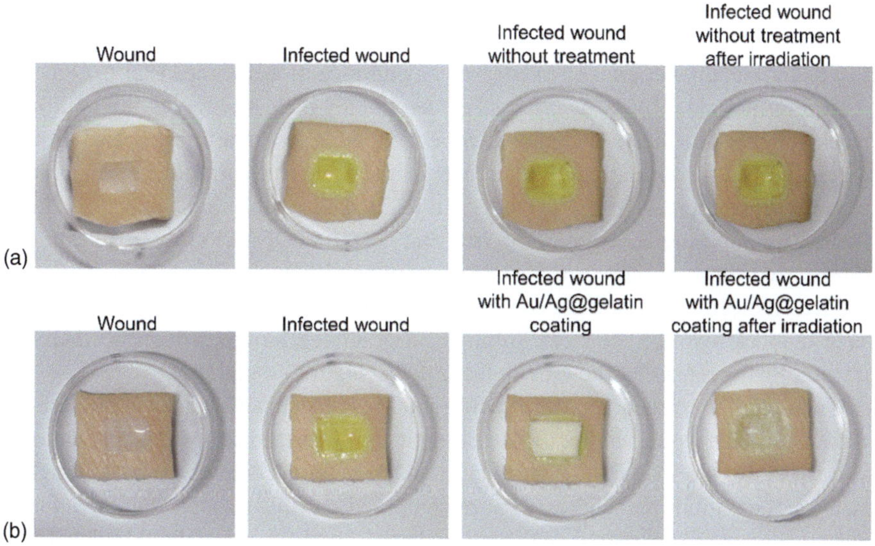

Figure 7.47 Images of *Pseudomonas aeruginosa*-infected wounds: (a) untreated control and (b) treated with Au/Ag@gelatin under light irradiation. *Source:* Reproduced with permission from Wang et al. [5] / IOP Publishing.

efficacy and reduces systemic toxicity but also decreases medication dosage and frequency while enhancing patient compliance. Hydrogels demonstrate exceptional applicability in drug delivery systems. Firstly, hydrogels consist of abundant water and cross-linked polymers. Their high water content (70–99%) closely mimics biological tissues, endowing them with excellent biocompatibility and hydrophilic drug-carrying capacity. Furthermore, formed in aqueous environments, hydrogels avoid drug denaturation and degradation in organic solvents. The cross-linking framework of hydrogel matrices gives them quasi-solid properties, maintaining mechanical strength spanning from 0.5 kPa to 5 MPa. This adaptability allows their application across various human tissues [71].

Photofunctional polymer hydrogel composites demonstrate exceptional potential in drug delivery systems thanks to their ability to achieve precise, on-demand, and spatiotemporally controlled drug release. This intelligent, controlled-release capability stems from the integration of photosensitive molecules, photoinitiators, or photoactive nanocomponents into the hydrogel's three-dimensional network. Light acts as an external stimulus that triggers controllable physicochemical changes in the hydrogel's photosensitive structures or components, such as volume expansion, reduced cross-link density, or cleavage of chemical bonds. These transformations effectively "activate" drug release pathways or "disrupt" the carrier's drug-binding capacity, enabling remote, noninvasive, and highly precise control of drug delivery behavior.

Azobenzene, one of the most common photoisomerization groups, can be converted from the E-form (trans) to the Z-form (cis) under UV irradiation, and restored to its original state through visible light exposure or heating. When in cis configuration, azobenzene exhibits higher polarity than the trans forms, a property that enables regulation of hydrophobic interactions. As early as 1967, this phenomenon was applied to construct a photoreactive hydrogel [72].

Hyaluronic acid (HA) is an essential element of the ECM and serves as a building block for this polymeric structure. Azobenzene acts as a photoresponsive group that exhibits a conformational preference for the trans configuration (Azo), which interacts with the cavity of β-cyclodextrin (CD). This enables host–guest interactions. By combining HA containing Azo and CD, the EGF@PR-S hydrogel can be synthesized through these interfacial interactions. When exposed to 365 nm light, the Azo molecules adopt a cis conformation, preventing interaction with the CD. This induces hydrogel de-cross-linking and gradual strength reduction, thereby facilitating the release of the growth factor EGF (Figure 7.48a) [73].

When applied to wounds, the experimental results showed that new epidermis grew toward the wound center in all treatment groups, significantly reducing wound size. Notably, the EGF@PR-S hydrogel treatment group demonstrated markedly faster regeneration after light exposure compared to other groups (Figure 7.48b). Specifically, wounds treated with this hydrogel nearly completely closed within 10 days, while wounds subjected to alternative treatments retained significant residual areas. The accelerated wound healing observed may be attributed to the UV-triggered sustained release of EGF at the wound site, in conjunction with the hydrogel's microenvironment that supports cellular growth. This photoresponsive supramolecular hydrogel not only exhibits excellent biocompatibility but also promotes cell proliferation, enhances EGF retention in skin tissues, thereby accelerating granulation tissue formation and angiogenesis [73].

To address the potential cytotoxicity issues of most UV-responsive Azo/CD-based hydrogels, the photoisomerization response wavelength of azobenzene can be red-shifted to create red-light–responsive hydrogels, making them suitable for more and broader clinical applications. For instance, hydrogels made from β-CD and methoxy-substituted azo (mAzo)-functionalized hyaluronic acid (HA) exhibit red-light responsiveness after extensive electron-rich group modifications. Upon red light irradiation, trans-mAzo undergoes photoisomerization to its cis-isomer, inducing host–guest complex dissociation. This process decreases the cross-linking density within the supramolecular hydrogel network, thereby facilitating controlled drug release (Figure 7.49a). FITC-BSA was encapsulated in the hydrogel in order to investigate the light-triggered release of the hydrogel. As shown in Figure 7.49b, hydrogels that were irradiated freely demonstrated significantly accelerated release during the first 3 days (81% versus 45% under dark conditions), which was almost double that of the dark group. Barrier-irradiated hydrogels exhibited a release rate of $65\% \pm 3\%$, which was reduced compared to the irradiated group, yet elevated relative to the dark control, due to reduced radiation attenuation by skin tissue. This phenomenon demonstrates that host–guest dissociation induced by 625 nm light effectively enhances the rate of protein release from the light-responsive hydrogels [74].

Ortho-nitrobenzyl is one of the most common photocleavable groups in photoresponsive hydrogel systems for biological applications. Its photo-degradation

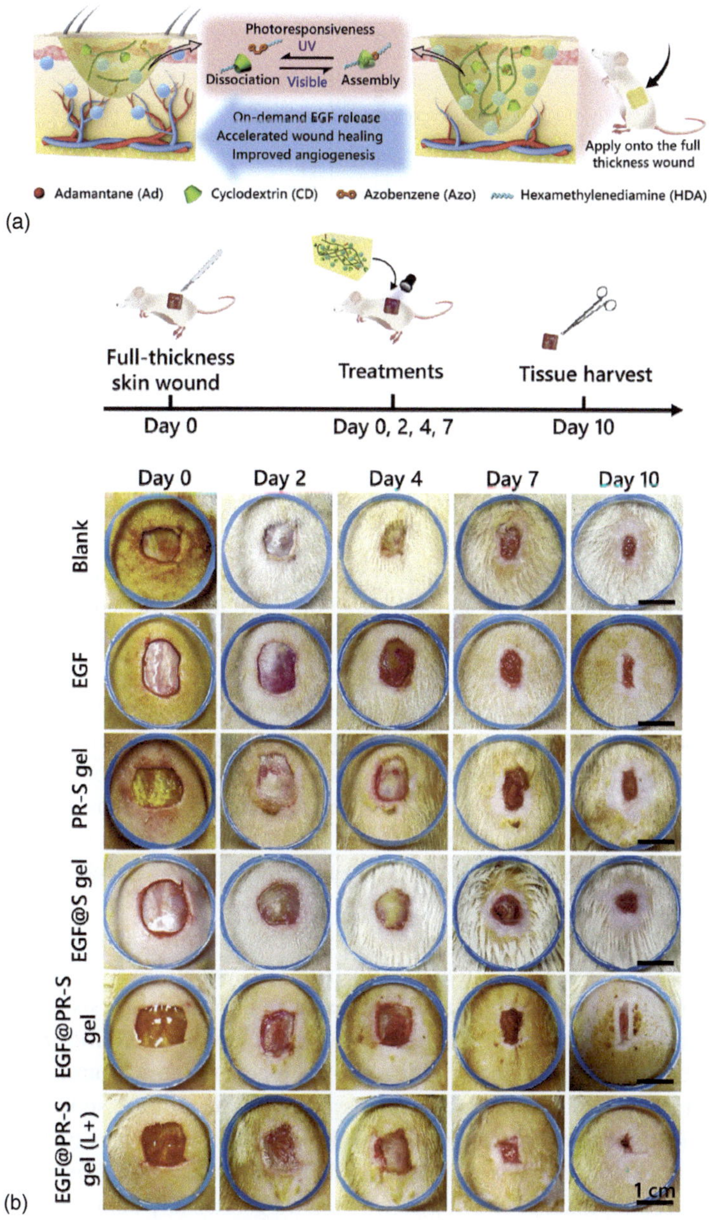

Figure 7.48 (a) The mechanism of light-controlled EGF release from EGF@PR-S hydrogel. (b) Photographs of wounds treated with EGF, PR-S gel, EGF@S gel, and EGF@PR-S, with and without light irradiation, captured on days 0, 2, 4, 7, and 10. *Source:* Zhao et al. [73] / with permission of Elsevier.

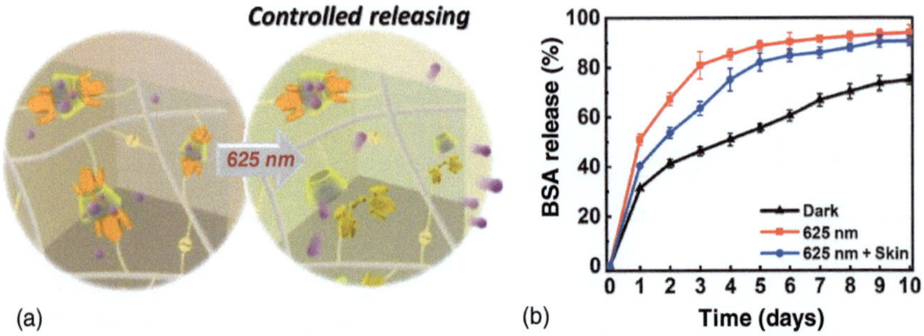

Figure 7.49 (a) The mechanism of red-light-controlled protein release from the mAzo-HA/CD-HA hydrogel. (b) BSA release profiles from mAzo-HA/CD-HA hydrogel under the conditions of darkness, 625 nm irradiation with and without skin. *Source:* Wu et al. [74] / with permission of American Chemical Society.

products and degradation process remain inert to biomolecules such as proteins and DNA, making it a promising candidate for biomedical applications. This photodegradation mechanism has been utilized to trigger drug release in hydrogels by disrupting polymer networks, enabling precise control over medication delivery. A prime example involves the four-arm star-shaped poly(ethylene glycol)-*b*-poly(γ-o-nitrobenzyl-L-glutamate) hydrogel, where hydrophobic domains are formed through intermolecular interactions between poly(γ-o-nitrobenzyl-L-glutamate) segments. These hydrophobic domains serve dual purposes: acting as cross-linking points and encapsulating hydrophobic drugs like doxorubicin (DOX). Upon light exposure, the ortho-nitrobenzyl ester group undergoes complete cleavage, disrupting the hydrophobic cross-links and releasing the encapsulated drug. Experimental results demonstrate that this photo-triggered drug release significantly enhances HeLa cell apoptosis rates, highlighting its potential value in cancer therapy [75].

In addition to the aforementioned drug release mechanisms, the highly efficient photothermal conversion agents integrated into the hydrogel system can effectively absorb light energy and convert it into localized heat when irradiated by specific wavelengths, particularly NIR light within the biological tissue "optical window." This rapid temperature rise occurs both within the hydrogel and its surrounding microenvironment. When the temperature surpasses the lower critical solution temperature (LCST), the thermoresponsive hydrogel network experiences a phase transition characterized by a change in its affinity for water, transitioning from a hydrophilic to a hydrophobic state. This transformation triggers a sharp contraction (deswelling) of the hydrogel, thereby releasing the encapsulated drugs.

For example, multiwalled carbon nanotubes (CNTs) were surface-functionalized to introduce amino and phenylboronic acid groups. These groups formed dynamic covalent bonds, such as imine bonds and borate ester bonds, with alginate dialdehyde (ADA), constructing the ADA/PEI/PP-CNT (AP/PCNT) hydrogel. Furthermore, poly(*N*-isopropylacrylamide) (PNIPAM) was directly incorporated into the hydrogel to create ADA/PEI/PNIPAM/PP-CNT (APN/PP-CNT) hydrogel. As a thermoresponsive polymer, poly(*N*-isopropylacrylamide) (PNIPAM) exhibits

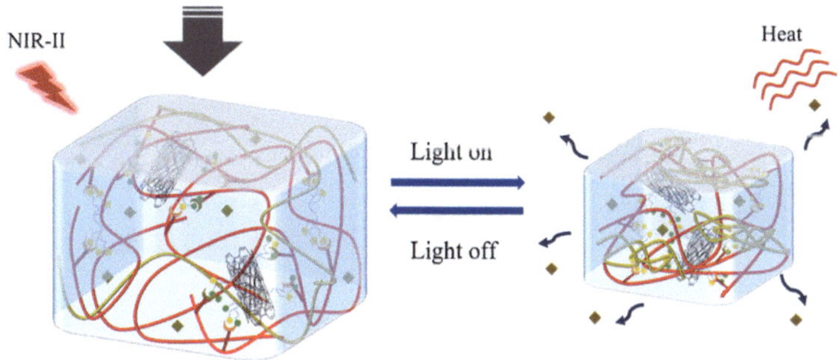

Figure 7.50 Mechanism of the NIR-II-controlled drug release from the APN/PP-CNT hydrogel. *Source:* Li et al. [76] / with permission of John Wiley & Sons.

contraction behavior when exposed to high temperatures generated by NIR-II light irradiation. This facilitates the release of the antibiotic neomycin from the hydrogel network (Figure 7.50) [76].

In another work, the antimicrobial peptide IK8 encapsulated in liposomes and the photothermal agent gold nanorods were co-loaded into the PEG hydrogel matrix. The lipid vesicles protect IK8 from degradation by proteases in vivo and provide temperature-controlled release capability. Under irradiation with an 860 nm laser, the hydrogel temperature rose to 55 °C within 10 min, triggering the phase transition of the lipid vesicles and releasing IK8. This released IK8 exerted bactericidal activity against *S. aureus* and *P. aeruginosa*. Furthermore, by increasing the laser intensity, the local high temperature reached 60 °C, directly damaging the bacterial cell membranes. The photothermal killing synergistically enhanced the bactericidal effect of IK8, ultimately achieving complete bacterial eradication (Figure 7.51) [77].

Using polydopamine (PDA) nanoparticles as a cross-linking agent, the system effectively cross-links thiol-terminated tetra-arm polyethylene glycol (PEG), forming PDA/PEG hydrogels. This hydrogel system not only demonstrates excellent biocompatibility but also enables the loading of anticancer drug SN38 onto PDA nanoparticles through π–π interactions and hydrogen bonding. This loading method ensures stable drug retention under physiological conditions with minimal leakage.

When PDA/PEG hydrogel is exposed to NIR light, PDA nanoparticles are capable of transducing light energy into thermal energy, resulting in a pronounced increase in local temperature. This thermal effect disrupts the π–π interaction between SN38 and PDA nanoparticles, enabling on-demand drug release from the hydrogel. The drug release exhibits an "on-off" switching characteristic with high temporal and spatial control capabilities. In experiments, after NIR irradiation, the local temperature in the gel-SN38 + NIR group reached approximately 44 °C (Figure 7.52a,b). This photothermal effect not only triggers drug release but also directly kills tumor cells through high temperatures. Experimental results show that tumor growth was entirely suppressed in the group treated with gel-SN38 + NIR irradiation, while tumor growth in other control groups (such as the drug-only group or photothermal-only group) was only partially suppressed (Figure 7.52c–e).

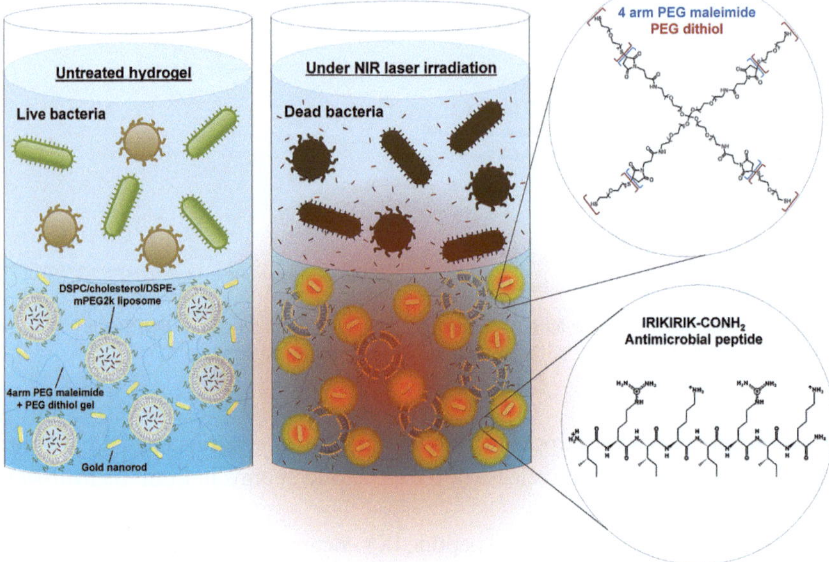

Figure 7.51 Mechanism of photothermal-triggered release of antimicrobial peptides from PEG-based hydrogel. *Source:* Moorcroft et al. [77] / with permission of American Chemical Society.

This outcome demonstrates the significant efficacy of PDA/PEG hydrogel in photo-controlled drug release and chemotherapeutic-PTT synergy [78].

7.3.4 Cell Culture

In complex biological tissues, cells reside within the ECM. As a vital component of human tissues, the natural ECM forms a hydrogel-like network structure composed of biological macromolecules. This structure not only provides physical support for cells but also actively participates in vital biological processes such as growth, differentiation, and signal transduction [79]. Hydrogel, an artificial material, closely mimics the structure and function of natural ECM, creating a microenvironment that resembles the in vivo environment. This structure facilitates cell adhesion, migration, and proliferation while supplying essential nutrients and oxygen. Furthermore, hydrogel's physical properties (e.g. hardness, elasticity, and roughness) and chemical composition (including functional groups and growth factor loading) can be precisely regulated to induce specific cellular behaviors. Compared with traditional two-dimensional cell culture, hydrogel supports three-dimensional cell cultivation, which more closely approximates the structural and functional characteristics of in vivo tissues. Three-dimensional culture not only aids cells in forming more complex tissue structures but also provides a more realistic model for tissue engineering, demonstrating broad application potential in regenerative medicine, drug screening, and disease research [80].

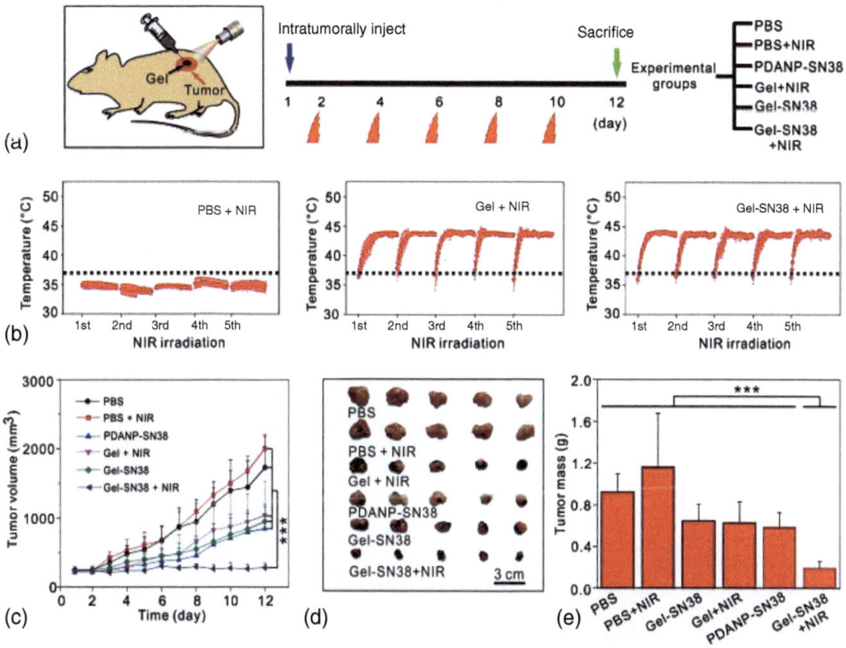

Figure 7.52 (a) in vivo chemo-photothermal tumor treatment groups: PBS, PBS + NIR, PDANP-SN38, Gel + NIR, Gel-SN38, and Gel-SN38 + NIR. (b) Tumor temperature variation during the 5th NIR irradiation with PBS, Gel, and Gel-SN38. (c) The volume, (d) photographs, and (e) mass of tumors following different treatments. *Source:* Wang et al. [78] / with permission of American Chemical Society.

In living organisms, cells generate physicochemical signals by sensing dynamic microenvironmental changes such as mechanical stress in the ECM, thereby guiding vital processes like growth, proliferation, and migration. Compared to traditional hydrogels, photoresponsive hydrogels demonstrate core advantages as cell culture scaffolds. They allow the chemical composition and physical properties to be modified in a noninvasive way with high spatiotemporal precision using light stimulation. This enables the simulation of ECM's dynamic microenvironment, significantly regulating cellular functions, phenotypes, morphology, and differentiation potential.

In the regulation of cell behavior by light, photoisomerization groups can be introduced into hydrogels as light-switchable cross-linking points to flexibly control the stiffness of the matrix, thereby influencing cell morphology. For example, azobenzene, a molecule that undergoes light-induced trans–cis isomerization, was functionalized into the monomers of a biocompatible polyacrylamide-based hydrogel and coated with fibronectin to promote cell attachment. In the control group, cells exhibited a spread morphology. The distribution of the actin cytoskeleton could be observed using a green fluorescent signal produced by phalloidin staining. Upon UV light irradiation, the cross-linking density decreased, causing the hydrogel to soften. Consequently, cells in the group irradiated with UV light

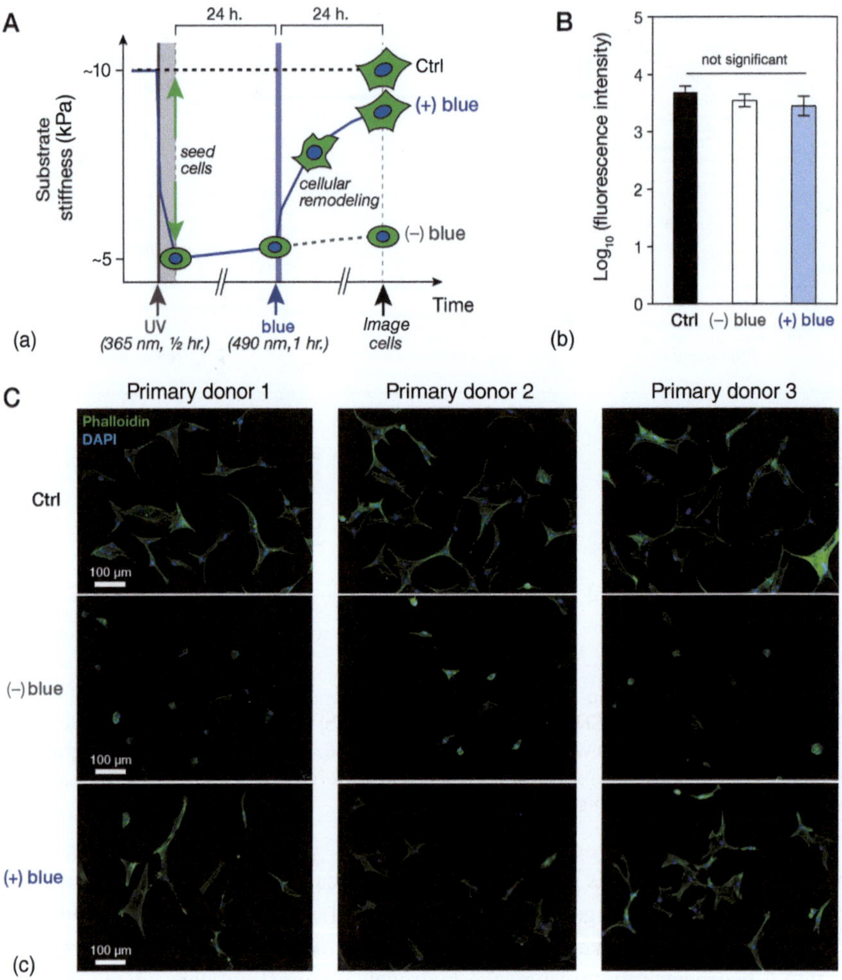

Figure 7.53 (a) Changes in hydrogel substrate stiffness under UV and blue light irradiation. (b) Cell viability indicated by fluorescence intensity in control, (−) blue, and (+) blue hydrogel groups. (c) Fluorescence images of mesenchymal stem cells from three human donors cultured on (−) blue and (+) blue hydrogels. *Source:* Lee et al. [81] / with permission of American Chemical Society.

(designated as the [−] blue group) exhibited a smaller volume. Subsequently, in the group irradiated with 490 nm blue light (designated as the [+]blue group), the cells recovered their spread morphology (Figure 7.53). Therefore, this hydrogel enables researchers to more precisely mimic in vivo-like microenvironmental changes experienced by cells, providing an important tool and methodology for cell biology research and tissue engineering applications [81].

Since cell-matrix interactions are dynamic processes, real-time detection of these interactions is crucial when studying hydrogels mimicking extracellular matrices. For instance, the fluorescent conjugated oligomer $OFBTCO_2Na$ is capable of forming hydrogels with poly(*N*-acryloyl glycinamide) (PNAGA) through hydrogen bonding

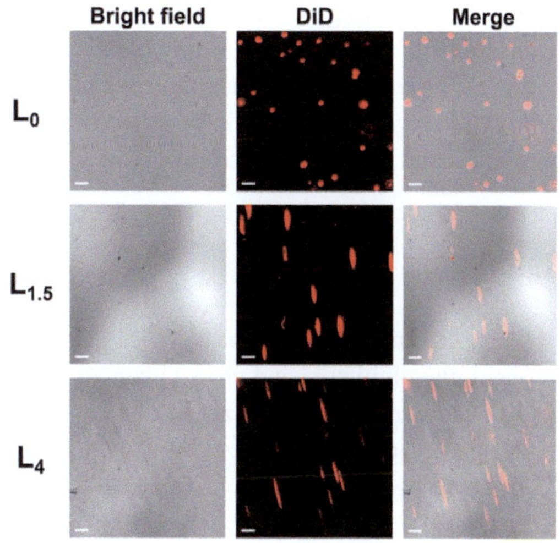

Figure 7.54 Fluorescent imaging of DiD-stained NIH-3T3 cells cultured for 7 days in L_0, $L_{1.5}$, and L_4 hydrogels. *Source:* Reproduced with permission from Zhang et al. [42] / American Chemical Society.

and hydrophobic interactions. These hydrogels not only exhibit high tensile strength and fluorescence properties but also serve as three-dimensional microporous scaffolds for encapsulating fibroblasts, thereby regulating cell growth.

As shown in Figure 7.54, DiD-stained cells exhibited significant morphological changes in red fluorescence signals after 7 days of mechanical stretching. Specifically, in L_0 hydrogel, the cell aspect ratio remained consistent with pre-long-term culture levels; in $L_{1.5}$ hydrogel, the aspect ratio reached 3.2; and in L_4 hydrogel, it further increased to 8.0. This indicates that cells can sense tensile forces within the scaffold and grow along the stretching direction. This hydrogel system not only provides a novel method for studying cell-matrix interactions but also serves as a controllable cell culture platform [42].

Furthermore, by leveraging the switchable host–guest interactions between cyclodextrin (β-CD) and azobenzene, optically tunable RGD pattern surfaces can be designed to induce macrophage phenotypic transformation. The method involves preparing norbornene-modified PEG and PEG-SH hydrogels via thioene click reactions, with SH-RGD attached at specific sites and SH-β-CD at other positions. Through host–guest recognition and azobenzene-modified RGD assembly, β-CD releases the assembled azobenzene-RGD upon UV irradiation, ultimately forming RGD stripe patterns.

Experimental results demonstrated that in the un-patterned U group, macrophages maintained a relatively low elongation factor (approximately 1.3) before and after UV irradiation, indicating no significant impact of surface treatment or UV exposure on cellular morphology elongation. In contrast, the patterned P group exhibited immediate elongation of cells, with the elongation factor reaching up to 4 after 20 h of culture, and no observed effect of UV irradiation on elongation rate. For the dynamic-responsive D group, BMDM showed an elongation coefficient of approximately 1.3 before UV exposure without RGD patterns. Post-irradiation, RGD patterns emerged, leading to gradual elongation over time and reaching peak elongation (around 4) at 48 h. This hydrogel system not only induces elongation in hydrogel-cultured macrophages but also promotes their polarization from pro-inflammatory to antiinflammatory phenotypes. This light-regulated hydrogel demonstrates significant potential for preventing immune-related diseases common in regenerative medicine, providing a novel research platform for immune regulation and cell behavior control [82].

References

1 Peng, R., Luo, Y., Cui, Q. et al. (2022). Covalent organic frameworks as efficient photoinitiators and cross-linkers to fabricate highly stretchable hydrogels. *ACS Applied Materials & Interfaces* 14 (43): 49254–49263.

2 Yang, N., Gong, F., Zhang, H. et al. (2024). Imidazole-stabilized gold nanoclusters as photocatalytic initiator and crosslinker to fabricate functional hydrogel. *Colloids and Surfaces A: Physicochemical and Engineering Aspects* 703: 135408.

3 Luo, Y., Peng, R., Zhang, H. et al. (2022). Graphitic carbon nitride colloid as one photoinitiator for two-step polymerization. *Colloids and Surfaces A: Physicochemical and Engineering Aspects* 650: 129615.

4 Strokov, K. and Galstyan, A. (2020). Chitosan-silicon phthalocyanine conjugate as effective photo-functional hydrogel for tracking and killing of bacteria. *European Journal of Organic Chemistry* 2020 (47): 7327–7332.

5 Wang, X., Guo, J., Zhang, Q. et al. (2020). Gelatin sponge functionalized with gold/silver clusters for antibacterial application. *Nanotechnology* 31 (13): 134004.

6 Zhang, Z., He, C., and Chen, X. (2018). Hydrogels based on pH-responsive reversible carbon-nitrogen double-bond linkages for biomedical applications. *Materials Chemistry Frontiers* 2 (10): 1765–1778.

7 Rong, L., Liu, Y., Fan, Y. et al. (2023). Injectable nano-composite hydrogels based on hyaluronic acid-chitosan derivatives for simultaneous photothermal-chemo therapy of cancer with anti-inflammatory capacity. *Carbohydrate Polymers* 310: 120721.

8 Dong, A., Xiao, W., Yuan, W., and Zuo, K. (2023). Self-healable and injectable nanocomposite hydrogel loading Iron-doped carbon dots for synergistic antibacterial peptide-photothermal-photodynamic antibacterial therapy. *ACS Applied Polymer Materials* 5 (11): 9564–9573.

9 Cho, S., Hwang, S.Y., Oh, D.X., and Park, J. (2021). Recent progress in self-healing polymers and hydrogels based on reversible dynamic B-O bonds: boronic/boronate esters, borax, and benzoxaborole. *Journal of Materials Chemistry A* 9 (26): 14630–14655.

10 Pandit, A.H., Nisar, S., Imtiyaz, K. et al. (2021). Injectable, self-healing, and biocompatible *N,O*-carboxymethyl chitosan/multialdehyde guar gum hydrogels for sustained anticancer drug delivery. *Biomacromolecules* 22 (9): 3731–3745.

11 Prasad, K., Izawa, H., Kaneko, Y., and Kadokawa, J.-i. (2009). Preparation of temperature-induced shapeable film material from guar gum-based gel with an ionic liquid. *Journal of Materials Chemistry* 19 (24): 4088–4090.

12 Yang, W., Zhang, Q., Zhou, J. et al. (2024). Self-healing guar gum-based nanocomposite hydrogel promotes infected wound healing through photothermal antibacterial therapy. *Biomacromolecules* 25 (6): 3432–3448.

13 Chen, S., Tang, F., Tang, L., and Li, L. (2017). Synthesis of Cu-nanoparticle hydrogel with self-healing and photothermal properties. *ACS Applied Materials & Interfaces* 9 (24): 20895–20903.

14 Bi, H., Zhang, X., Wang, Q. et al. (2023). Dynamic reversible disulfide bonds hydrogel of thiolated galactoglucomannan/cellulose nanofibril with self-healing property for protein release. *Industrial Crops and Products* 206: 117615.

15 van Kuringen, H.P.C. and Schenning, A.P.H.J. (2015). Hydrogen bonding in supramolecular nanoporous materials. In: *Hydrogen Bonded Supramolecular Materials* (ed. Z.-T. Li and L.-Z. Wu), 43–67. Berlin, Heidelberg: Springer Berlin Heidelberg.

16 Brunsveld, L., Folmer, B.J.B., Meijer, E.W., and Sijbesma, R.P. (2001). Supramolecular polymers. *Chemical Reviews* 101 (12): 4071–4098.

17 Ji, X., Shi, B., Wang, H. et al. (2015). Supramolecular construction of multifluorescent gels: interfacial assembly of discrete fluorescent gels through multiple hydrogen bonding. *Advanced Materials* 27 (48): 8062–8066.

18 Mai, W., Yu, Q., Han, C. et al. (2020). Self-healing materials for energy-storage devices. *Advanced Functional Materials* 30 (24): 1909912.

19 Yan, Q., Zhou, M., and Fu, H. (2020). Study on mussel-inspired tough TA/PANI@CNCs nanocomposite hydrogels with superior self-healing and self-adhesive properties for strain sensors. *Composites Part B-Engineering* 201: 108356.

20 Liu, C., Chen, Y., and Chen, J. (2010). Synthesis and characteristics of pH-sensitive semi-interpenetrating polymer network hydrogels based on konjac glucomannan and poly(aspartic acid) for in vitro drug delivery. *Carbohydrate Polymers* 79 (3): 500–506.

21 Liu, Y., Xi, Y., Zhao, J. et al. (2019). Preparation of therapeutic-laden konjac hydrogel for tumor combination therapy. *Chemical Engineering Journal* 375: 122048.

22 Gong, L., Liao, G., Luan, H. et al. (2020). Oil solubilization in sodium dodecylbenzenesulfonate micelles: new insights into surfactant enhanced oil recovery. *Journal of Colloid and Interface Science* 569: 219–228.

23 Gao, J., Tang, F., Li, Y. et al. (2025). Promoting the photoresponse rate with micelles in hydrogels. *Advanced Optical Materials* 13 (13): 2403352.

24 Zhang, X.N., Du, C., Du, M. et al. (2020). Kinetic insights into glassy hydrogels with hydrogen bond complexes as the cross-links. *Materials Today Physics* 15: 100230.

25 Wang, J., Zhao, X., Tang, F. et al. (2021). Synthesis of copper nanoparticles with controllable crystallinity and their photothermal property. *Colloids and Surfaces A: Physicochemical and Engineering Aspects* 626: 126970.

26 Hu, R., Yuan, H., Wang, B. et al. (2014). DNA hydrogel by multicomponent assembly for encapsulation and killing of cells. *ACS Applied Materials & Interfaces* 6 (15): 11823–11828.

27 Qin, H., Zhang, T., Li, H.-N. et al. (2017). Dynamic Au-thiolate interaction induced rapid self-healing nanocomposite hydrogels with remarkable mechanical behaviors. *Chem* 3 (4): 691–705.

28 Can, V., Kochovski, Z., Reiter, V. et al. (2016). Nanostructural evolution and self-healing mechanism of micellar hydrogels. *Macromolecules* 49 (6): 2281–2287.

29 Gao, Y., Gu, S., Duan, L. et al. (2019). Robust and anti-fatigue hydrophobic association hydrogels assisted by titanium dioxide for photocatalytic activity. *Soft Matter* 15 (19): 3897–3905.

30 Zhang, Y., Wu, M., Chen, J. et al. (2018). Tough, high stretched, self-healing C-dots/hydrophobically associated composited hydrogels and their use for a fluorescence sensing platform. *ChemistrySelect* 3 (21): 5756–5765.

31 Mantooth, S.M., Munoz-Robles, B.G., and Webber, M.J. (2019). Dynamic hydrogels from host-guest supramolecular interactions. *Macromolecular Bioscience* 19 (1): 1800281.

32 Jiang, X., Zeng, F., Yang, X. et al. (2022). Injectable self-healing cellulose hydrogel based on host-guest interactions and acylhydrazone bonds for sustained cancer therapy. *Acta Biomaterialia* 141: 102–113.

33 Liu, C., Guo, X., Ruan, C. et al. (2019). An injectable thermosensitive photothermal-network hydrogel for near-infrared-triggered drug delivery and synergistic photothermal-chemotherapy. *Acta Biomaterialia* 96: 281–294.

34 Kenawy, E.-R., Kamoun, E.A., Eldin, M.S.M., and El-Meligy, M.A. (2014). Physically crosslinked poly(vinyl alcohol)-hydroxyethyl starch blend hydrogel membranes: synthesis and characterization for biomedical applications. *Arabian Journal of Chemistry* 7 (3): 372–380.

35 Holloway, J.L., Lowman, A.M., and Palmese, G.R. (2013). The role of crystallization and phase separation in the formation of physically cross-linked PVA hydrogels. *Soft Matter* 9 (3): 826–833.

36 Adelnia, H., Ensandoost, R., Moonshi, S.S. et al. (2022). Freeze/thawed polyvinyl alcohol hydrogels: present, past and future. *European Polymer Journal* 164: 110974.

37 Shang, J., Le, X., Zhang, J. et al. (2019). Trends in polymeric shape memory hydrogels and hydrogel actuators. *Polymer Chemistry* 10 (9): 1036–1055.

38 Yang, L., Wang, Z., Fei, G., and Xia, H. (2017). Polydopamine particles reinforced poly(vinyl alcohol) hydrogel with NIR light triggered shape memory and self-healing capability. *Macromolecular Rapid Communications* 38 (23): 1700421.

39 Tang, F., Wang, X., Yao, C. et al. (2016). An emission-tunable fluorescent organic molecule for specific cellular imaging. *RSC Advances* 6 (81): 77745–77751.

40 Wang, J., Tang, F., Wang, Y. et al. (2020). Tunable single-molecule white-light emission in stimuli-responsive hydrogel. *Advanced Optical Materials* 8 (6): 1901571.

41 Chen, N., Yong, W., Xiong, T., and Fu, G. (2024). Fast photochromic and fluorescent switchable organohydrogels based on photoinduced electron transfer for display and storage. *Journal of Photochemistry and Photobiology A: Chemistry* 447: 115262.

42 Zhang, Q., Wang, X., Cong, Y. et al. (2022). Conjugated polymer-functionalized stretchable supramolecular hydrogels to monitor and control cellular behavior. *ACS Applied Materials & Interfaces* 14 (10): 12674–12683.

43 Wu, L., Gao, C., Li, Z., and Chen, G. (2017). Tunable photoluminescence from layered rare-earth hydroxide/polymer nanocomposite hydrogels by a cascaded energy transfer effect. *Journal of Materials Chemistry C* 5: 5207–5213.

44 Yan, X., Wang, T., Li, H. et al. (2022). Flexible aggregation-induced emission-active hydrogel for on-site monitoring of pesticide degradation. *ACS Nano* 16 (11): 18421–18429.

45 Wu, S., Yang, Y., Cheng, Y. et al. (2023). Fluorogenic detection of mercury ion in aqueous environment using hydrogel-based AIE sensing films. *Aggregate* 4 (3): e287.

46 Pourreza, N. and Ghomi, M. (2017). A novel metal enhanced fluorescence bio probe for insulin sensing based on poly vinyl alcohol-borax hydrogel functionalized by Ag dots. *Sensors and Actuators B-Chemical* 251: 609–616.

47 Thomson, L., Ginesi, R.E., Osborne, D.D. et al. (2023). Photothermal perylene bisimide hydrogels. *Chemistry - A European Journal* 29 (37): e202300663.

48 Yuan, Q., Huang, L.-Z., Wang, P.-L. et al. (2022). Cellulose nanofiber/molybdenum disulfide aerogels for ultrahigh photothermal effect. *Journal of Colloid and Interface Science* 624: 70–78.

49 Zhang, Q., Wang, X., Kang, Y. et al. (2022). Conjugated molecule-assisted supramolecular hydrogel with enhanced antibacterial and antibiofouling properties. *ACS Applied Bio Materials* 5 (6): 3107–3114.

50 Meng, Z., Zhou, X., Xu, J. et al. (2019). Light-triggered in situ gelation to enable robust photodynamic-immunotherapy by repeated stimulations. *Advanced Materials* 31 (24): e1900927.

51 He, Q.-T., Qian, P., Yang, X.-Y. et al. (2024). Rational design of bacteria-targeted and photo-responsive MOF gel with antibacterial and anti-inflammatory function for infected wound healing. *Chemical Engineering Journal* 493: 152760.

52 Wang, C., Dong, W., Li, P. et al. (2020). Reversible ion-conducting switch by Azobenzene molecule with light-controlled sol-gel transitions of the PNIPAm ion gel. *ACS Applied Materials & Interfaces* 12 (37): 42202–42209.

53 Wichterle, O. and Lím, D. (1960). Hydrophilic gels for biological use. *Nature* 185 (4706): 117–118.

54 Zhou, B., Rayner, A.W., Gregg, E.W. et al. (2024). Worldwide trends in diabetes prevalence and treatment from 1990 to 2022: a pooled analysis of 1108 population-representative studies with 141 million participants. *Lancet* 404 (10467): 2077–2093.

55 Kim, Y., Namgung, H., and Lee, T.S. (2016). Synthesis of a glucose oxidase-conjugated, polyacrylamide-based, fluorescent hydrogel for a reusable, ratiometric glucose sensor. *Polymer Chemistry* 7 (43): 6655–6661.

56 Veal, E.A., Day, A.M., and Morgan, B.A. (2007). Hydrogen peroxide sensing and signaling. *Molecular Cell* 26 (1): 1–14.

57 Lin, M.T. and Beal, M.F. (2006). Mitochondrial dysfunction and oxidative stress in neurodegenerative diseases. *Nature* 443 (7113): 787–795.

58 Jiang, X., Wang, H., Yuan, R., and Chai, Y. (2018). Functional three-dimensional porous conductive polymer hydrogels for sensitive electrochemiluminescence in situ detection of H_2O_2 released from live cells. *Analytical Chemistry* 90 (14): 8462–8469.

59 Jandl, B., Sedghiniya, S., Carstens, A., and Astakhova, K. (2019). Peptide-fluorophore hydrogel as a signal boosting approach in rapid detection of cancer DNA. *ACS Omega* 4 (9): 13889–13895.

60 Ehtesabi, H., Roshani, S., Bagheri, Z., and Yaghoubi-Avini, M. (2019). Carbon dots-sodium alginate hydrogel: a novel tetracycline fluorescent sensor and adsorber. *Journal of Environmental Chemical Engineering* 7 (5): 103419.

61 Stencel-Baerenwald, J.E., Reiss, K., Reiter, D.M. et al. (2014). The sweet spot: defining virus-sialic acid interactions. *Nature Reviews Microbiology* 12 (11): 739–749.

62 Randriantsilefisoa, R., Nie, C., Parshad, B. et al. (2020). Double trouble for viruses: a hydrogel nanocomposite catches the influenza virus while shrinking and changing color. *Chemical Communications* 56 (24): 3547–3550.

63 Aich, S., Delbaere, L.T.J., and Chen, R.D. (2001). Continuous spectrophotometric assay for β-glucuronidase. *BioTechniques* 30 (4): 846–850.

64 Ebrahimi, M.-M.S., Voss, Y., and Schoenherr, H. (2015). Rapid detection of *Escherichia coli* via enzymatically triggered reactions in self-reporting chitosan hydrogels. *ACS Applied Materials & Interfaces* 7 (36): 20190–20199.

65 Zhu, S., Wang, X., Cong, Y. et al. (2021). Free radical polymerization of gold nanoclusters and hydrogels for cell capture and light-controlled release. *ACS Applied Materials & Interfaces* 13 (16): 19360–19368.

66 Bermudez-Jimenez, C., Nino-Martinez, N., Patino-Marin, N. et al. (2020). Effective control of biofilms by photothermal therapy using a gold nanorod hydrogel. *Journal of Biomedical Materials Research Part B: Applied Biomaterials* 108 (2): 333–342.

67 Li, Y., Yan, Y., Wang, J. et al. (2022). Preparation of silver nanoparticles decorated mesoporous silica nanorods with photothermal antibacterial property. *Colloids and Surfaces A: Physicochemical and Engineering Aspects* 648: 129242.

68 Wang, H., Zhou, S., Guo, L. et al. (2020). Intelligent hybrid hydrogels for rapid *in situ* detection and photothermal therapy of bacterial infection. *ACS Applied Materials & Interfaces* 12 (35): 39685–39694.

69 Zhang, Y., Zhu, C., Zhang, Z. et al. (2021). Oxidation triggered formation of polydopamine-modified carboxymethyl cellulose hydrogel for anti-recurrence of tumor. *Colloids and Surfaces B: Biointerfaces* 207: 112025.

70 Ran, P., Xia, T., Zheng, H. et al. (2023). Light-triggered theranostic hydrogels for real-time imaging and on-demand photodynamic therapy of skin abscesses. *Acta Biomaterialia* 155: 292–303.

71 Bodugoz-Senturk, H., Macias, C.E., Kung, J.H., and Muratoglu, O.K. (2009). Poly(vinyl alcohol)-acrylamide hydrogels as load-bearing cartilage substitute. *Biomaterials* 30 (4): 589–596.

72 Lovrien, R. (1967). The photoviscosity effect. *Proceedings of the National Academy of Sciences of the United States of America* 57 (2): 236–242.

73 Zhao, W., Li, Y., Zhang, X. et al. (2020). Photo-responsive supramolecular hyaluronic acid hydrogels for accelerated wound healing. *Journal of Controlled Release* 323: 24–35.

74 Wu, K., Wu, X., Zhang, Y. et al. (2022). Semiconvertible hyaluronic hydrogel enabled red-light-responsive reversible mechanics, adhesion, and self-healing. *Biomacromolecules* 23 (3): 1030–1040.

75 Zhao, D., Tang, Q., Zhou, Q. et al. (2018). A photo-degradable injectable self-healing hydrogel based on star poly(ethylene glycol)-*b*-polypeptide as a potential pharmaceuticals delivery carrier. *Soft Matter* 14 (36): 7420–7428.

76 Li, B.-Y., Lin, T.-Y., Lai, Y.-J. et al. (2025). Engineering multiresponsive alginate/PNIPAM/carbon nanotube nanocomposite hydrogels as on-demand drug delivery platforms. *Small* 21 (12): e2407420.

77 Moorcroft, S.C.T., Roach, L., Jayne, D.G. et al. (2020). Nanoparticle-loaded hydrogel for the light-activated release and photothermal enhancement of antimicrobial peptides. *ACS Applied Materials & Interfaces* 12 (22): 24544–24554.

78 Wang, X., Wang, C., Wang, X. et al. (2017). A polydopamine nanoparticle-knotted poly(ethylene glycol) hydrogel for on-demand drug delivery and chemo-photothermal therapy. *Chemistry of Materials* 29 (3): 1370–1376.

79 DAC, W. and Yamada, K.M. (2020). The extracellular matrix in development. *Development* 147 (10): dev175596.

80 Mirbagheri, M., Adibnia, V., Hughes, B.R. et al. (2019). Advanced cell culture platforms: a growing quest for emulating natural tissues. *Materials Horizons* 6 (1): 45–71.

81 Lee, I.N., Dobre, O., Richards, D. et al. (2018). Photoresponsive hydrogels with photoswitchable mechanical properties allow time-resolved analysis of cellular responses to matrix stiffening. *ACS Applied Materials & Interfaces* 10 (9): 7765–7776.

82 Luo, Y., Zheng, X., Yuan, P. et al. (2021). Light-induced dynamic RGD pattern for sequential modulation of macrophage phenotypes. *Bioactive Materials* 6 (11): 4065–4072.

8

Future Development of Photofunctional Polymer Composites

Photofunctional polymer composites have shown great potential for biomedical applications in recent years. This type of material combines the precise spatiotemporal regulation ability of light with the actual demands of biomedicine, bringing new breakthroughs to fields such as disease diagnosis, treatment, and bioengineering. With the continuous advancement of technology, the development trend of photofunctional polymer composites in biological applications has shown multidimensional innovation and expansion.

8.1 Structural Optimization of Photofunctional Polymer Composites

8.1.1 Selection of Components for Intelligent Light Response

The intelligence of photofunctional polymer composites is one of the core directions for future development. At present, the demand for treatment methods in the biomedical field is increasing day by day, requiring that the treatment process not only precisely target the lesion site but also minimize damage to normal tissues to the greatest extent. Photofunctional polymer composites, with their unique light response characteristics, can precisely meet this demand. By designing and synthesizing photoresponsive components, such as photochromic molecules, photothermal conversion materials, and photosensitizers, these materials can activate or regulate their functions under light of specific wavelengths.

Photothermal conversion materials are a typical example. Under the irradiation of near-infrared light, they can efficiently convert light energy into thermal energy and then be used for precise tumor ablation treatment. Near-infrared light has excellent tissue penetration and can penetrate deep into the human body to reach tumor sites. When the photothermal conversion material accumulates in the tumor tissue, it only needs to be irradiated with near-infrared light of an appropriate wavelength and intensity, and the material can rapidly heat up, causing tumor cells to undergo coagulative necrosis due to high temperature and achieving the effect of tumor ablation. Compared with traditional surgical resection, this method has less trauma,

Photofunctional Polymer Composites for Bioapplications, First Edition. Xiaoyu Wang and Lidong Li.

faster recovery, and can precisely act on the tumor without damaging the surrounding normal tissues [1].

Photochromic materials also show great application potential. This type of material can change its chemical or physical properties when stimulated by light. For instance, certain polymers containing photochromic groups will undergo structural changes under light of specific wavelengths, thereby controlling the release of drugs. Researchers can preset the parameters of light stimulation according to the treatment needs to achieve precise spatiotemporal release of drugs. For instance, when treating local infections, photochromic materials carrying antibacterial drugs are made into dressings and applied to the infected area. Through intermittent light exposure, the drugs are released in the infected area at the predetermined time and dosage, maintaining an effective local drug concentration and enhancing the antibacterial effect. Meanwhile, reduce the possible side effects of systemic medication [2].

In addition to the above-mentioned materials, photosensitizers are important in the intelligent and precise development of photofunctional polymer composites. Photosensitizers can generate reactive oxygen species (ROS) under light conditions. These ROS have strong oxidizing properties and can damage key structures such as the cell membrane, mitochondria, and DNA of tumor cells, thereby achieving photodynamic therapy (PDT) [3]. By combining photosensitizers with targeted molecules to form composites, photosensitizers can specifically aggregate around tumor cells. During phototherapy, only tumor cells are attacked by high concentrations of ROS, while normal cells are protected from damage due to the low concentration of photosensitizers. This precise strike method has broad application prospects in the field of cancer treatment.

In terms of optimizing the composition structure, high-throughput screening technology should be combined to optimize the energy band structure of materials, develop modular functional units that can respond to different wavelengths (UV-Vis-NIR), and construct a programmable material library with efficient optical functions (Figure 8.1). At the same time, we can use the AI algorithms for in-depth analytical response of light components of the optical properties [4], matching the best illumination parameter automatically, so as to realize the intelligent material design and maximize the performance of ascension. These intelligent and precise photofunctional polymer composites are expected to lead a brand-new therapeutic revolution in the future biomedical field and safeguard human health.

8.1.2 In Situ Assembly Strategy

Self-assembly is a universal phenomenon in living organisms and is essential to control and regulate cellular functions. Although a series of photofunctional composite systems have been constructed based on self-assembly strategies, they have contributed significantly to the field of drug delivery. However, in complex biological environment systems, the nonspecific interactions between artificially assembled systems and biological macromolecules can make the performance of their functions unpredictable, such as the formation of a corona [5], which limits their functions.

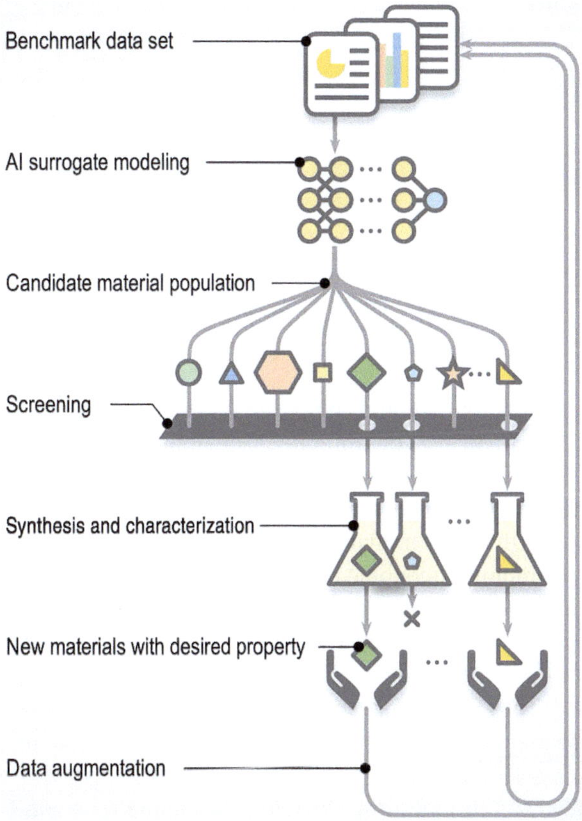

Figure 8.1 Data-driven artificial intelligence for creating novel materials with specific characteristics. *Source:* Kim [4] / with permission of American Chemical Society.

The in situ self-assembly of molecules within and outside cells provides new methods for regulating cell behavior, including inhibiting the proliferation of cancer cells. Many natural intracellular self-assembly behaviors have certain regulatory effects on cell behavior. For instance, misfolded proteins assemble into structures different from those of normally folded proteins, leading to the loss of protein activity and affecting cell viability [6]. Therefore, the introduction of the in situ cell assembly strategy can expand the biological functions of photofunctional polymer materials.

Peptide amphiphiles (PA) can be rationally designed to combine with peptides, nucleic acids, and polysaccharides, thereby forming assemblies of different structures. The most common ones are self-assembled into nanofibers. The universality of their design principles can be used to guide the construction of various nanomaterials, and their self-assembly patterns can be regulated by pH, divalent ion induction, and concentration [7]. Nanotechnology offers a variety of effective carriers for cancer treatment, with unique advantages including the ability to improve pharmacokinetics. PA represents a promising nanotechnology platform in

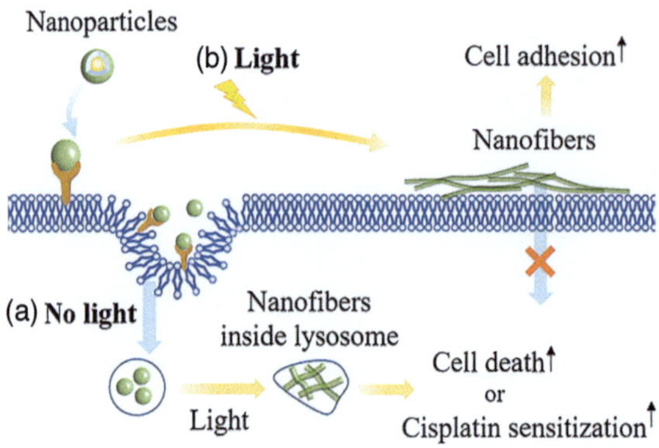

Figure 8.2 Light-induced self-assembly based on H-bond interactions and variations in molecular conformation. *Source:* Sun et al. [8] / with permission of American Chemical Society.

cancer treatment. To construct a photofunctional composite system by integrating the photoresponsive unit (3-methylene-2-(quinolin-8-yl)isoindolin-1-one) with the targeting peptide Arg-Gly-Asp (RGD) and the provided hydrogen-bonding peptide (Lys-Leu-Val-Phe-Phe, KLVFF) (Figure 8.2) [8]. By taking advantage of the conformational changes of MQIO units in living cells triggered by light, which cause molecular sliding, the hydrophilic/lipophilic balance is disrupted, leading to the decomposition of nanoparticles. Then, through the hydrogen bond interaction of KLVFF peptides, faster recombination is induced to form nanofibers. The morphological transformation from nonfunctionalized nanoparticles to functionalized nanofibers has been completed. The rapid self-assembly of phototriggered peptides can achieve temporal and spatial control, thereby inhibiting cell migration or cell death.

Lysosomes are closely related to the multidrug resistance of tumors. The construction of biomolecular aggregates through in situ self-assembly of peptide-based materials within lysosomes and is used to overcome drug resistance in tumor treatment [9]. Nitrobenzofurazan is environmentally sensitive. Attaching fluorescent molecules to the ε-amine of pH-responsive hexapeptide (LTP) can be used to image the molecular assembly in living systems. LTP was combined with sugar to form a peptide-based material. When this peptide-based material was taken up by tumor cells and entered the lysosome, LTP was affected by the low pH in the lysosome to trigger a phase transition, forming a hydrogel. This led to the expansion of the lysosome in tumor cells and an increase in permeability, resulting in the death of tumor cells. In addition, lysosomal in vivo assembly can significantly increase the killing efficiency of multidrug-resistant (MDR) cells and overcome their drug resistance.

The nucleus is one of the most important structures in a cell, controlling the genetic information and life activities of the cell. A pair of phosphopeptides rich in L- and D-leucine (1Lp and 1Dp), labeled with NBD dye molecules, reveals the assembly of intracellular polypeptides. Under the catalysis of alkaline phosphatase (ALP), leucine-rich 1Lp and 1Dp are dephosphorylated to form nanoribbon in the nucleus, selectively killing human osteosarcoma cells (SJSA-1) [10]. When phosphopeptides are dephosphorylated, peptide-based materials first transform into micelles and then into nanoribbons. Peptidyl assemblies first aggregate on the cell membrane, then, via endocytosis, enter the tumor cell and finally accumulate in the nucleolus of the cell nucleus. Proteomic analysis indicates that this pair of phosphopeptides, after forming nanoribbons and accumulating in the cell nucleus, can interact with histones within the cell nucleus, thus rapidly killing osteosarcoma cells without being toxic to normal cells. This work clarifies a new nuclear targeting mechanism, opening up new avenues for selectively killing osteosarcoma cells and reducing cell drug resistance.

The Golgi apparatus (GA) is an important organelle in eukaryotic cells, participating in posttranslational modifications of proteins, and is closely related to cell drug resistance. Enabling peptide-based materials to undergo in situ self-assembly and function in Golgi bodies is an important challenge. A fluorescent peptide thioester composed of aminoethyl thioester and labeled with nitrobenzofurazan at the terminal [11] can not only serve as a substrate for thioesterase but also target the cells' GA. GA-related thioesterases can hydrolyze peptide thioesters within cells, forming dimers and higher-order structures of thiols in GA. When GA is saturated, peptithioesters are enriched in the endoplasmic reticulum. Their accumulation in the endoplasmic reticulum and GA can disrupt protein transport, thereby causing cell death through multiple pathways. This work demonstrates a molecular platform based on peptithioester response to redox activity, which can target GA to disrupt protein transport and thereby kill tumor cells.

In addition, self-assembly has also made outstanding contributions in immunotherapy. Immunotherapy is one of the most promising emerging tumor treatment techniques [12]. By combining self-assembly with immunotherapy and taking advantage of the controllable deformation and release characteristics of self-assembly, the immune effect can be effectively enhanced. Hepatocellular carcinoma (HCC) has a poor prognosis and a high mortality rate. Immune checkpoint blockade therapy often fails in the treatment of HCC, which has also become a bottleneck for expanding HCC therapy. Functional high-density lipoprotein spheres (VG-sHDLs) were prepared by coassembling lipids and ApoA-1-mimetic peptides with the DiD dye, loading the esterase-reactive prodrug gemcitabine and vadimezan (Figure 8.3) [13]. DiD can label cells and monitor drug release. VG-sHDLs have the ability to cause HCC cells to generate damage-related cytokines and molecular patterns. In addition, VG-sHDLs can kill macrophages (M2) but have no obvious toxicity to macrophages (M1). In animal experiments, after intravenous injection, sHDL disassembled to release the drug. VG-sHDLs increased tumor-associated M1 and proinflammatory factors, but reduced tumor-associated M2 and the antiinflammatory cytokine interleukin-10. sHDL can recruit a large number of T cells and enhance their activity, thereby eradicating HCC and generating long-term tumor-specific immune memory to prevent HCC recurrence. Self-assembled nanomedicine sHDL lays the foundation for emerging immunomodulators in the immunotherapy of HCC. Cell-mediated drug delivery has received a lot of interest in cancer immunotherapy, as it can enhance the efficacy and specificity of treatment.

The assembly of DNA sequences is precisely controllable. Antitumor vaccines can be fabricated using DNA assembly nanomachines [14]. Two types of adjuvants and antigen peptides are coassembled with DNA nanostructures to form nanotubes.

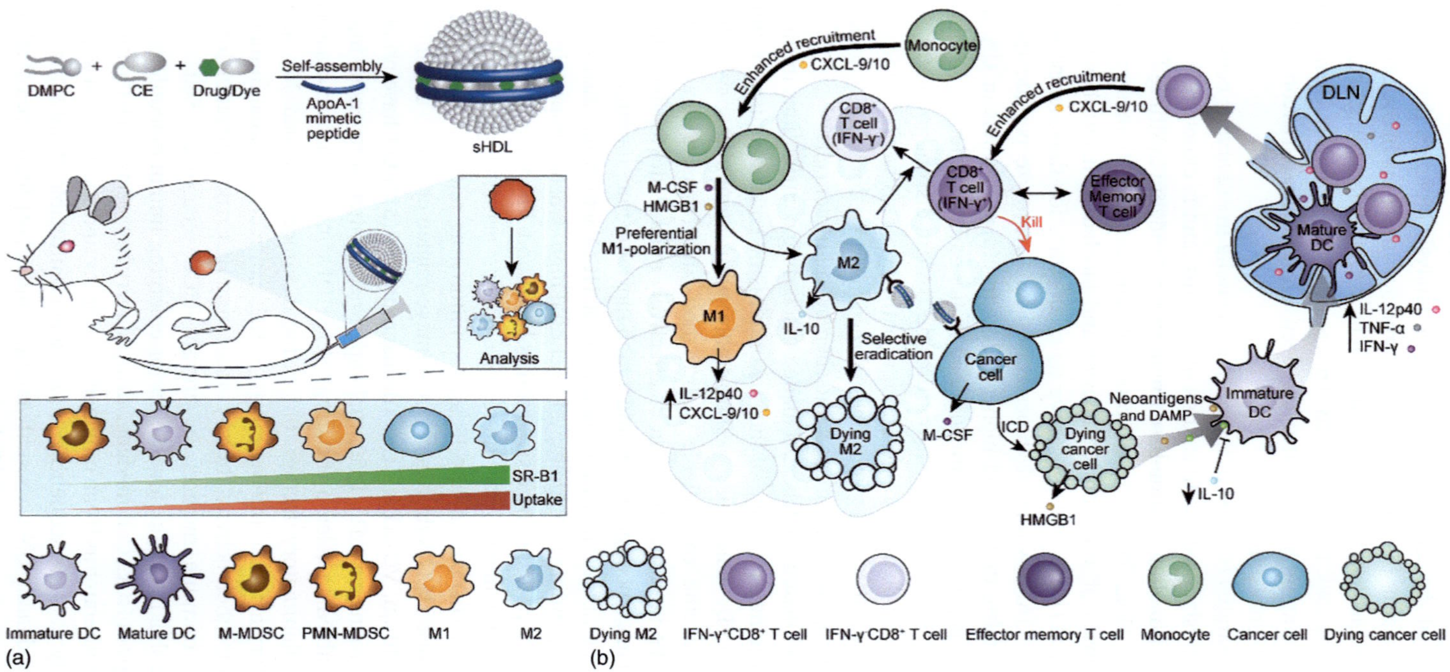

Figure 8.3 (a) Preparation and (b) mechanisms of immunotherapy of sHDL. *Source:* Wang et al. [13] / with permission of Elsevier.

The nanotubes are activated by the tumor subcellular microenvironment and carry multiple adjuvants and tumor antigens. This triggers specific T-cell immune responses and tumor cell death. The outer layer of the nanorobot is a low pH response "DNA lock strand". When this nanorobot is internalized into the lysosome of the cell, the low pH triggers its disassembly to expose tumor antigens and adjuvants, thereby activating the immune response of the cell. The signal response of DNA-labeled fluorescent dyes can be monitored to observe the conformational changes of "DNA lock," confirming the opening of DNA nanodevices and exposing adjuvants and tumor antigens. The DNA nanovaccine can activate an effective and specific T-cell response in mice, causing the tumors in tumor-bearing mice to stop growing or even regress. In addition, this assembled vaccine can stimulate a long-term T-cell response in mice and effectively protect them from secondary invasion by cancer cells. This DNA-assembled vaccine offers the possibility for future clinical trials with effective drug delivery, controllable drug release, and good therapeutic effects, laying the foundation for safe and effective combination therapy and the production of personalized cancer vaccines.

8.1.3 In Situ Aggregation Strategy

The modification of the cell membrane surface, as an important strategy for regulating cellular physiological metabolism, can protect cells from complex environments, enhance the efficiency of biosynthesis, and endow cells with nonnatural physiological functions [15]. Effective modification of the cell membrane surface can be achieved by introducing metal/semiconductor nanoparticles, high-molecular polymers, and metal complexes [16]. However, the introduction of complex functional materials from the outside, due to the uncontrollability of their introduction efficiency, makes it impossible to precisely regulate cell behavior. Therefore, the in situ synthesis of polymers on the cell membrane through extracellular electron transfer (EET) has broad application prospects in the field of regulating the physiological metabolic functions of cells.

Photoinduced electron/energy transfer (PET)-reversible addition–fragmentation chain transfer polymerization (RAFT) can achieve in situ controllable free radical polymerization on the surface of bacteria and cells [17]. First, monomers, chain transfer reagents, and photoinitiators were screened out, and the reaction conditions that can catalyze uniform molecular weight high-molecular polymers were optimized. Dibenzocyclooctyl was covalently modified on the bacterial surface through amide condensation, and then the chain transfer reagent containing azide groups was anchored to the bacterial surface through a copper-free click reaction. For the modification of cell surface polymers, the chain transfer containing long alkyl chain side chains was anchored to the cell surface through molecular fluidity. Furthermore, through light exposure, radicals are transferred along the chain to achieve the synthesis of functional polymers on the membrane. Compared with the introduction of functional polymers from the outside and uncontrollable free radical polymerization, in situ photocatalytic RAFT polymerization on the cell membrane surface can obtain functional polymers with more uniform and controllable molecular weights, while maintaining high cell viability. Due to the modification of the monomer side chain with azide groups and biotin, the functional groups of alkyne or streptavidin can be added to react with them, endowing the bacteria with other new functions.

Nonenzymatic catalytic reactions occurring on the cell surface can regulate cell functions through the photophysical properties and biological activities of the reaction products. The polymerization strategy catalyzed by biological palladium can be used to synthesize conjugated polymers in situ on the surface of living cells. As shown in Figure 8.4, based on the extracellular electron transport capacity of bacteria, the reducing property of *E. coli* was utilized to reduce Pd^{2+} to Pd, and then in situ catalyzed the Sonogashira coupling polymerization to generate poly(phenylene ethynylene) (PPE) with positive charges modified by side chains. The in situ synthesized PPE has a relatively good light absorption capacity and emits blue fluorescence on the surface of *E. coli* [18]. PPE has antibacterial properties against *E. coli* and can monitor the polymerization process with fluorescent signals. This in situ polymerization method is universal. When sodium ascorbate was reduced to Pd and anchored to the protein nucleus of microalgae, polymerization was induced. Through testing and analysis, it was found that PPE increased the

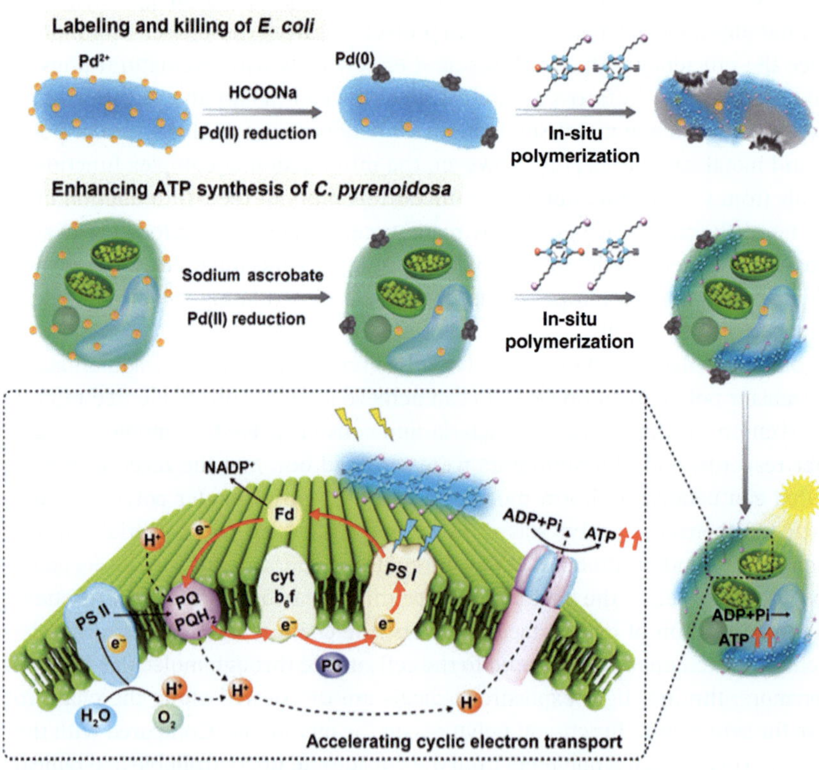

Figure 8.4 Cell surface in situ synthesis of PPE with different effects. *Source:* Qi et al. [18] / with permission of John Wiley & Sons.

light absorption of microalgae and simultaneously enhanced the transmission efficiency of the intermediate circulation electron flow, thereby promoting the light reaction rate of microalgae, increasing the production of the light reaction product ATP, and ultimately improving the photosynthetic efficiency of microalgae. This polymerization method has opened up a new path for regulating cell functions.

8.2 The Functional Improvement of Photofunctional Polymer Composites

8.2.1 Multifunctional Integration

Photofunctional polymer composites will not be confined to a single function but will develop in the direction of multifunctional integration. This trend is an inevitable outcome of the growing demand for comprehensive treatment and diagnostic solutions in the biomedical field. The multifunctional integrated photofunctional polymer composites can achieve the synergistic effect of multiple therapeutic mechanisms in a single treatment process. At the same time, combined with diagnostic functions, they provide a comprehensive solution for the treatment and monitoring of diseases.

Photodynamic therapy is widely recognized as a promising novel treatment method due to its noninvasiveness and high spatial selectivity. However, due to the limited penetration depth and transmission efficiency of light, the demand for external light sources hinders the application of PDT in deep tissue treatment [19]. To address such challenges, a PDT system based on in situ chemiluminescence was developed [20]. PDT of deep tissues is achieved by using the triple chemiluminescence system of luminol–hydrogen peroxide–hemoglobin. In addition to serving as a catalyst for the luminol system's chemiluminescence, hemoglobin also transports molecular oxygen. The chemiluminescence of luminol resonates induce the energy transfer to conjugated polymer MEH-PPV nanoparticle (Figure 8.5), promoting MEH-PPV to sensitize oxygen and generate ROS, which subsequently kill tumor cells. This novel system utilizes chemiluminescence instead of traditional light sources and uses hemoglobin to supply oxygen, addressing the difficulty of oxygen deficiency in hypoxic tumors and providing an effective strategy for PDT. This research offers evidence of concept for exploring an intelligent and multifunctional nanoplatform for efficient PDT.

Biological metabolism is the foundation of cellular activities, and the metabolic process involves a large number of chemical reactions. The redox balance within cells is crucial for their homeostasis and physiological functions. The out-of-control redox reactions within cells can induce many diseases, and the use of optical means to regulate these reaction processes can also be employed for the photocatalytic treatment of diseases. Generally speaking, the redox state within cells is related to nicotinamide adenine dinucleotide (NAD^+) and its reduced form, NADH, within the cells. Regulation of this process can either enhance cell function or cause cell apoptosis [21]. Introducing photocatalytic reactions into cells to affect NAD^+ is a

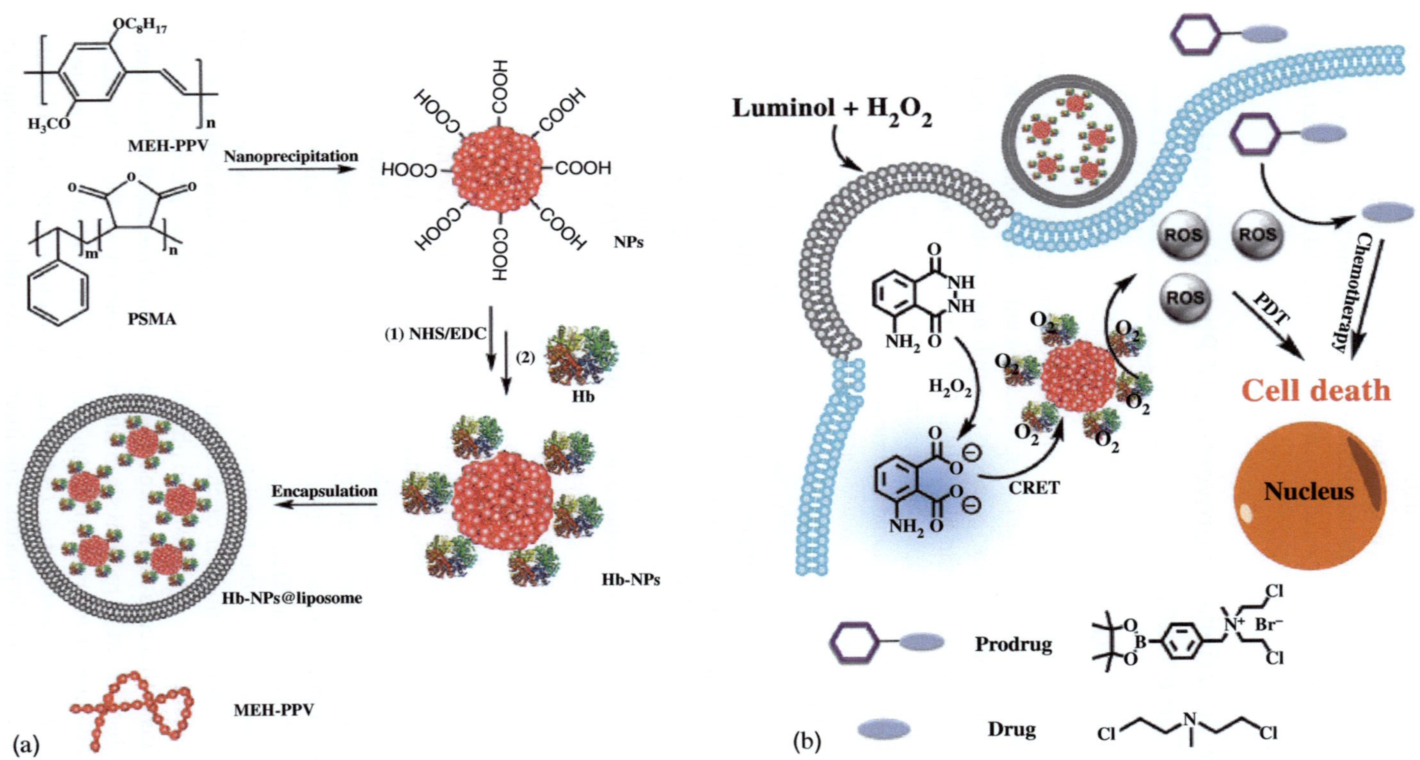

Figure 8.5 (a) The preparation and (b) anticancer action of photofunctional nanocomposites. *Source:* Jiang et al. [20] / with permission of John Wiley & Sons.

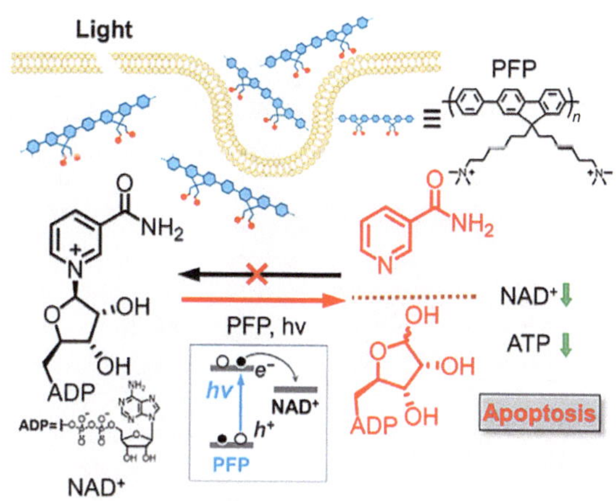

Figure 8.6 The regulation of the redox process by PFP in the cell. *Source:* Xia et al. [23] / Chinese Chemical Society. This figure has been published in CCS Chemistry [2023]; [Unexpected Photocatalytic Degeneration of NAD+ for Inducing Apoptosis of Hypoxia Cancer Cells] is available online at [10.31635/ccschem.022.202202463].

feasible approach for treating or killing cells [22]. Normal cells tend to maintain intracellular equilibrium, and the mutual conversion of NAD^+/NADH caused by exogenous factors will gradually return to normal levels. Selectively converting NAD^+ into other substances is a more direct regulatory approach. Cationic conjugated polymer poly(fluorene-co-phenylene) (PFP) is used as a photocatalyst to split NAD^+ into nicotinamide and ADP-ribose under light conditions, consuming NAD^+ within cancer cells and thereby affecting the dynamic balance of NAD^+/NADH within cells to achieve tumor treatment [23] (Figure 8.6). During this process, PFP is excited by light, and photoelectrons are transferred to NAD^+, subsequently generating ADP-ribosyl radicals and nicotinamide. ADP-ribosyl radicals are then quenched to form ADP-ribose. Constant NAD^+ consumption disrupts the coenzyme cycle's equilibrium, causing the mitochondrial membrane potential within cancer cells to decline, damaging the mitochondrial and mitochondrial electron respiratory chains, and ultimately promoting cell apoptosis. This process does not rely on oxygen and can achieve tumor killing under hypoxic conditions. By introducing photocatalytic reactions based on conjugated polymers into cells, it has achieved intervention in cellular physiological processes, providing a new strategy for the treatment of hypoxic tumors.

CO is an endogenous gas molecule that has a wide range of influences on apoptosis. Carbon monoxide has a pro-apoptotic effect on cancer cells and can exhibit an anti-apoptotic effect on normal cells [24]. To transport carbon monoxide to tissues and organs, there is currently a series of metal-carbonyl complex-based release molecules (CORMs) [25]. The application of traditional CORMs is limited due to their limited CO load, poor tumor targeting, and non-negligible toxicity. The multifunctional integrated photofunctional composite systems can effectively solve this problem. Carbon dot-modified C_3N_4 nanoparticles were compounded with Ag_3PO_4 nanoparticles (AgCCN), and then coupled with histidine-rich peptides to prepare HisAgCCN [26]. Figure 8.7 CO_2 can not only form carbamates by interacting with amino or imidazole groups in peptides, but also combine with protonated nitrogen atoms of imidazole groups through hydrogen

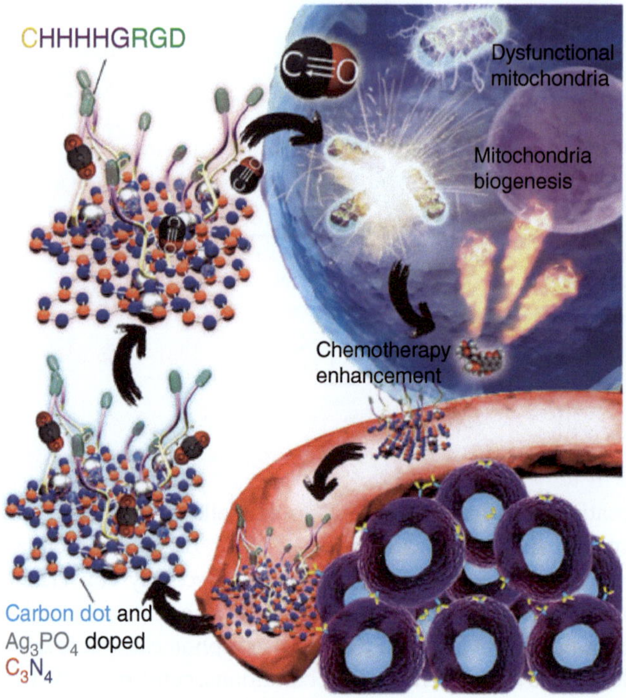

Figure 8.7 The design and action of HisAgCCN convert CO_2 into CO for killing tumors. *Source:* Zheng et al. [26] / with permission of John Wiley & Sons.

bonds. This process promotes the enrichment and capture of CO_2 and enhances the efficiency of AgCCN photocatalytic conversion of CO_2 to CO. The peptide's side chain RGD may target tumors by interacting with the highly expressed $\alpha_v\beta_3$ integrin on tumor vascular endothelial cells. While shielding healthy cells from the effects of chemotherapy, the CO generated by HisAgCCN may dramatically increase the susceptibility of tumor cells to DOX. Since CO may both prevent tissue damage in normal cells and enhance the formation of ROS in cancer cells with high oxidative stress, proteomics and metabolomics investigations have revealed that the duality of HisAgCCN is caused by CO's influence on mitochondria. Thus, using HisAgCCN as a photocatalyst to convert endogenous CO_2 in tumors into CO can significantly enhance the chemotherapy effect on prostate cancer cells, while the toxicity to normal cells is almost negligible.

In addition, combining photodynamic and photothermal treatment composites can achieve the synergistic effect of multiple therapeutic mechanisms under a single light exposure. Photothermal therapy directly destroys tumor cells by converting light energy into heat energy. PDT, on the other hand, utilizes photosensitizers to generate ROS, inducing apoptosis of tumor cells. When these two therapeutic mechanisms are integrated into the same material, they can function simultaneously under light conditions, significantly enhancing the killing effect on tumors. Additionally, this synergistic therapy technique lowers the dose needed for a single treatment, potentially reducing side effects [27].

Multifunctional integrated photofunctional polymer composites are still in the process of continuous innovation and development. In addition to the combination of photothermal and PDT, researchers are also exploring how to integrate other treatment methods, such as drug delivery and immunotherapy, with photofunctional polymer composites. For instance, composites capable of releasing immunomodulators under light control should be boosted to enhance the body's immune response to tumors, while combining photothermal or PDT to directly attack tumor cells. This multipronged treatment strategy is expected to further increase the success rate of tumor treatment.

Near-infrared light penetrates tissue more deeply and is less biologically toxic. Photofunctional composite nanoparticles were prepared by coprecipitating conjugated polymers based on pyrrolidone and thiophene with amphiphilic liposomes. 4T1 cells transfected with IFN-γ plasmids were coincubated with nanoparticles to construct an optogenetic platform. Photothermal polymers convert light energy into thermal energy, activating the heat-responsive promoter HSP70 and inducing the expression of downstream genes to cause cells to secrete IFN-γ (Figure 8.8) [28]. IFN-γ is a proinflammatory factor that can bind to tumor-associated macrophages (M2 subtype) and promote the polarization of M2 subtype macrophages to M1 subtype macrophages (the ratio of M1/M2 increased by 200% compared with the control group). Repolarized macrophages will secrete a series of proinflammatory

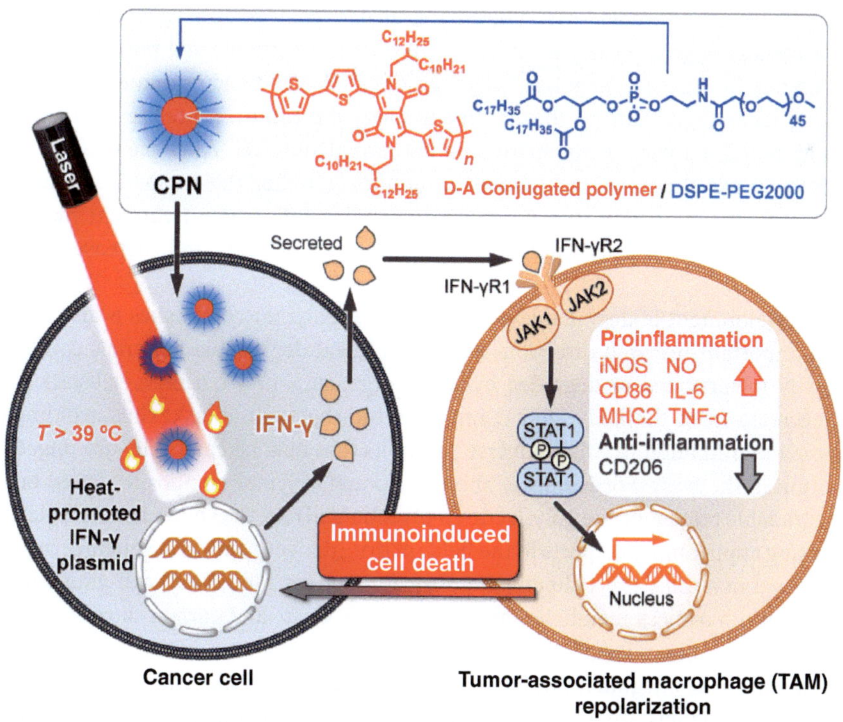

Figure 8.8 The immunotherapy mechanisms of photothermal conjugated polymer nanoparticles. *Source:* Fu et al. [28] / with permission of John Wiley & Sons.

cytokines such as TNF-α and IL-6. Alter the microenvironment around the tumor. This system also enhances the presentation efficiency of tumor-associated antigens and improves adaptive immune responses by upregulating the expression level of major histocompatibility complex class II proteins (MHC II). The activated M1 subtype macrophages can also repel cancer cells through secreted killing factors with phagocytosis. This strategy remotely activates the immune response with light, providing a path for noninvasive and precise cancer treatment.

In addition, photofunctional polymer composites can also be combined with imaging technology to achieve real-time monitoring and feedback of the treatment process. For instance, by introducing fluorescent probes or magnetic nanoparticles into the material, fluorescence imaging or magnetic resonance imaging can be carried out simultaneously with phototherapy, thus achieving the integration of diagnosis and treatment. Fluorescence imaging can provide high-resolution, real-time images, helping doctors precisely locate tumors and monitor treatment effects. Photoacoustic imaging, on the other hand, can provide deeper tissue information and assist in assessing the impact of treatment on surrounding tissues. This model that combines diagnosis and treatment not only enhances the accuracy of treatment but also provides doctors with immediate feedback, enabling them to make wiser decisions during the treatment process [29].

In practical applications, multifunctional integrated photofunctional polymer composites have also demonstrated great potential (Figure 8.9) [30]. Take tumor treatment as an example [31]. This material can achieve full coverage of the entire process from the early diagnosis of tumors to treatment and posttreatment monitoring. In the early diagnosis stage, tumors are located and evaluated through imaging technology. During the treatment stage, the tumor is precisely targeted through multiple mechanisms such as photothermal and PDT. After the treatment, continue to use the imaging function to monitor whether the tumor recurs and adjust the treatment plan in a timely manner. This full-course treatment model is expected to significantly improve the prognosis of patients and increase their survival rate.

The development of degradable photofunctional polymer composites is also one of the important research directions at present. Ideal degradable materials can be naturally metabolized or degraded by living organisms into harmless substances and excreted from the body after completing the therapeutic task, thus avoiding chronic inflammation or other adverse reactions that may be caused by prolonged retention in the body. For instance, photofunctional polymer composites based on biodegradable polymers like poly(lactic-*co*-glycolic acid) (PLGA) have demonstrated promising application prospects in the fields of drug delivery and tissue engineering. After these materials exert therapeutic functions under light, they will gradually degrade into harmless metabolic products such as water and carbon dioxide, and will not impose a long-term burden on the organism [32].

In addition to degradability, enhancing material stability and optimizing degradation products are also key factors in ensuring biocompatibility and safety. Photofunctional polymer composites need to maintain structural and functional stability in complex physiological environments to prevent premature failure

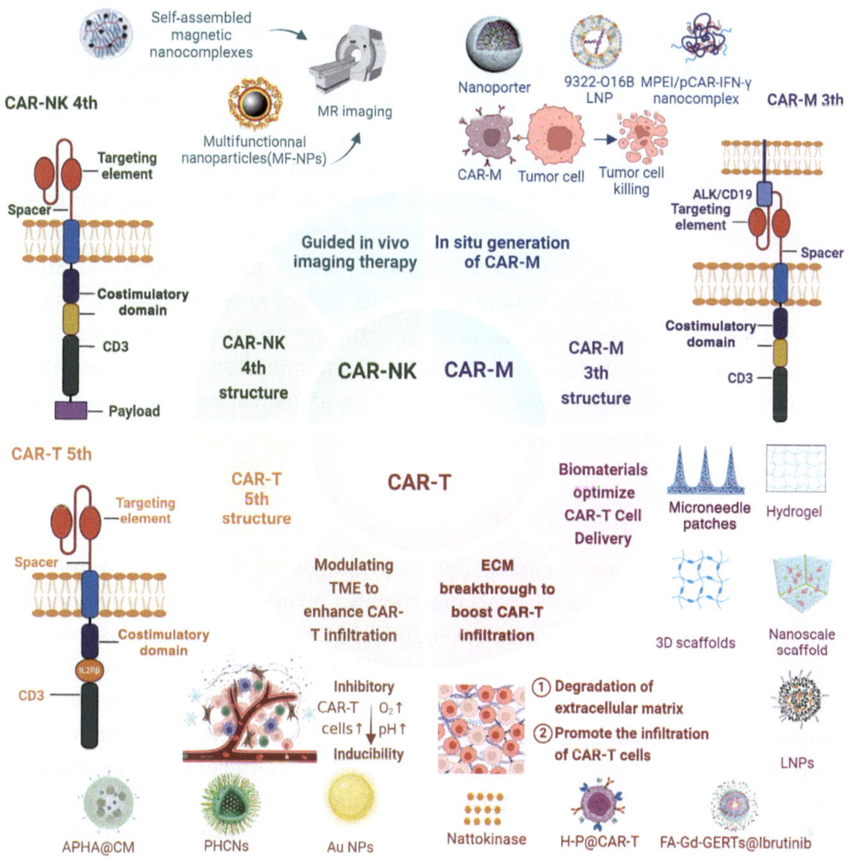

Figure 8.9 Multifunctional integrated biomaterials for CAR immune cell therapy. *Source:* Song et al. [30] / with permission of American Chemical Society.

or the release of harmful substances. At the same time, in-depth research on the degradation mechanism and metabolic pathways of materials to ensure that degradation products are easily processed and excreted by organisms is crucial for reducing long-term toxicity.

To comprehensively assess the biocompatibility and safety of photofunctional polymer composites, rigorous preclinical research and a standardized safety evaluation system are indispensable. This includes multiple steps, such as in vitro cytotoxicity tests, in vivo animal experiments, and long-term studies on biological distribution and metabolism. Through the evaluation of these systems, reliable safety data support can be provided for the clinical transformation of materials, ensuring their reliability and safety in practical applications.

However, multifunctional integrated photofunctional polymer composites also face some challenges in the process of moving toward clinical application. First of all, how to ensure that multiple functions work together in the material without interfering with each other is a key issue. Secondly, the biocompatibility and safety of the materials need to be strictly evaluated to guarantee their minimal toxicity to healthy tissues and long-term stability in the body. In addition, the preparation process of multifunctional materials is relatively complex, requiring precise regulation of the distribution and interaction of multiple components at the nanoscale, which poses higher demands on the field of materials science.

Overall, multifunctional integration is an important direction for the future development of photofunctional polymer composites. By integrating multiple therapeutic and diagnostic functions, these materials are expected to bring revolutionary progress to the biomedical field and provide more effective and comprehensive solutions for the treatment of complex diseases. As materials science, nanotechnology, and biomedical engineering continue to advance, it is believed that these challenges will be gradually solved, and multifunctional integrated photofunctional polymer composites will surely play an increasingly important role in clinical applications.

8.2.2 Interdisciplinary Integration

The development of photofunctional polymer composites will rely on interdisciplinary integration, a process that incorporates many disciplines, including materials science, biomedicine, chemical engineering, photophysics, and nanotechnology. Through in-depth collaboration among different disciplines, more efficient, safe, and multifunctional photofunctional polymer composites can be developed, bringing innovative solutions to the biomedical field. Among them, the combination of photofunctional polymer composites and precision manufacturing technology can achieve the construction of micro-nano structures, providing the possibility for customized optical functions and promoting the important role of photofunctional materials in more fields.

In tissue engineering, by integrating materials science and biomedical technology, biological scaffolds with photoresponsiveness can be developed [33]. By regulating the physical and chemical properties of the scaffold through light exposure, it promotes cell growth, differentiation, and tissue regeneration. In the field of bioimaging, the integration of photophysical technology and chemical engineering allows for deep tissue light signal penetration and high-resolution imaging in photofunctional polymer composites, making them an effective tool for disease detection and monitoring.

Chemical engineering contributes much to the preparation process and large-scale production of photofunctional polymer composites. Chemical engineers are dedicated to optimizing the synthetic routes and processing techniques of materials to enhance production efficiency, reduce costs, and ensure the consistency of product quality. For instance, by developing efficient nanofabrication techniques and large-scale synthesis methods, the industrial production of photofunctional polymer composites can be achieved, laying the foundation for their clinical application and commercial promotion [34].

Combining the high-speed and multifunctional development trend of electronic technology, photofunctional polymer composites have demonstrated significant potential in the development of miniaturized intelligent sensing systems. Among them, biooptoelectronic devices achieve precise control of light and electrical signals by integrating light sources (such as LEDs, OLEDs), photosensitive materials (such as photosensitizers, light-responsive molecules), and electronic components, and apply them to the monitoring and intervention of cells, tissues, and even individual behaviors [35]. Such devices not only possess high spatiotemporal resolution and good compatibility with biological systems but also integrate the advantages of rapid response, high precision, and nonintrusive nature of optoelectronic systems, demonstrating broad application prospects in fields such as optogenetics [36] and phototherapy [37].

The concept of bioelectronic devices has a long history. Its prototype can be traced back to the ancient Egyptian period, when surgical tools had already demonstrated initial attempts by humans to use tools to intervene in health [38]. After thousands of years of evolution, the concept and form of bioelectronic devices have gradually become clear in modern times. Thanks to the breakthroughs in early biomedical theories and the advancement of basic medical technologies, some prototypes of bioelectronic devices with practical application value, such as early cardiac stimulators, began to emerge [39]. After entering the 20th century, given the speed at which contemporary technology is developing and the profound integration of several fields, bioelectronic devices have witnessed explosive progress and have become an important milestone in the intersection of medicine and engineering.

By integrating micro-inorganic light-emitting diodes (μ-ILED), flexible biosensing systems, control modules, data analysis modules, etc., biooptoelectronic devices are fabricated to improve organ dysfunction (Figure 8.10) [40]. First, inhibitory opsins are expressed in the sensory afferent nerves of the bladder through viral vectors, and then the device is implanted. The flexible strain gauge wound around the bladder can accurately measure the filling and emptying of the bladder, and has no significant impact on the physiological function of the bladder. μ-ILED directly introduces light into the bladder and uses light to stimulate opsins, reducing the excitability of neurons. The wireless acquisition module can supply power to each part of the system. The software on handheld devices can collect, store, and analyze data in real time, and provide an automatic control interface for open-loop and closed-loop devices, as well as for regulating lighting parameters, etc. The experimental results show that in both normal and induced bladder dysfunction rats, the closed-loop system can automatically identify abnormal urination patterns and regulate bladder function through light stimulation [41]. For instance, in rats induced with bladder dysfunction, light stimulation significantly reduced the frequency of urination. This bioelectronic device that combines optogenetic technology with a wireless closed-loop system not only enhances the specificity and effectiveness of treatment but also reduces potential harm to patients. In addition, the design concept and core technology of this device can be extended to the research of other organs and diseases, providing new ideas and methods for the development of bioelectronic medicine.

Health detection is one of the most important applications of optoelectronic bioelectronic devices. Patterned gold electrodes and conductive polymer

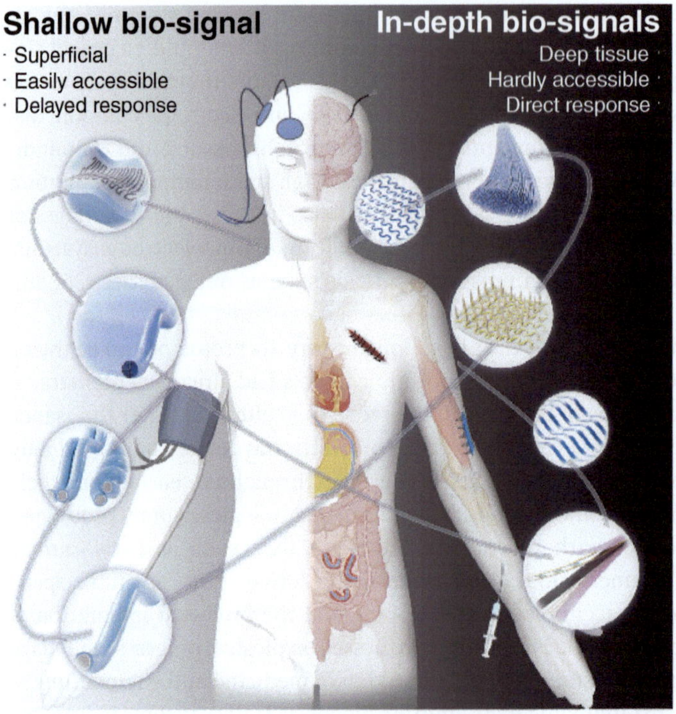

Figure 8.10 Different microinterventional sensors for monitoring biosignals. *Source:* Liu et al. [40] / with permission of American Chemical Society.

PEDOT:PSS were fabricated on ultrathin poly(*p*-xylene) substrates to prepare organic electrochemical transistors. Then, it may be combined with near-infrared inorganic micro-light-emitting diodes (μLED) to create bioelectronic devices that can wirelessly monitor sweat biomarkers such as glucose, lactic acid, and pH level [42]. As shown in Figure 8.11, the OECT's drain is linked in series with the μLED. The integrated device is powered by a battery. The OECT's source is linked to the positive terminal, while the μLED's n-contact electrode is connected

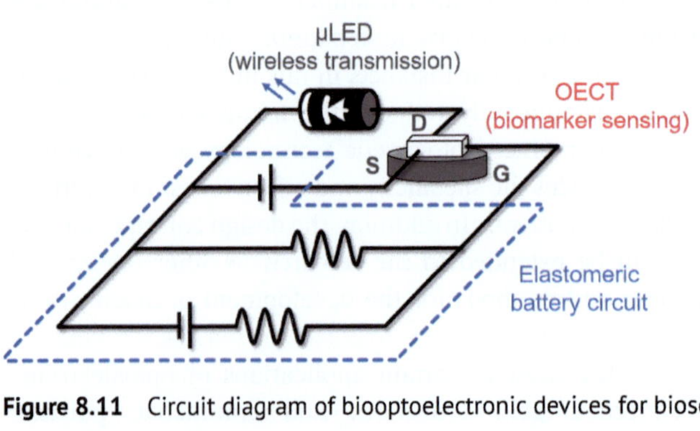

Figure 8.11 Circuit diagram of biooptoelectronic devices for biosensors.

to the negative terminal. The necessary gate voltage for the OECT is supplied by two resistors and an additional battery. Consequently, when the analyte drops onto the OECT sensor, the channel current can directly adjust the irradiance of the μLED. As a result, the μLED generates strong light and the OECT channel current stays high. For the pH value, the irradiance of the μLED gradually increases as the pH value decreases. By combining the device with a flexible battery circuit, a wearable detection patch can be fabricated and used for near-infrared image analysis. This wearable biomarker monitoring technology can also be extended to other biomedical application fields and has broad application prospects. By further optimizing the device design and integrating more sensors, multifunctional wearable devices capable of simultaneously monitoring multiple biomarkers can be developed, which holds significant value for remote medical monitoring and health management.

Photofunctional polymer composites, combined with 3D printing and microfluidic processing technology, have shown broad development prospects in the field of bioelectronic devices. In 3D printing, such materials can achieve high-precision and complex structure manufacturing, and are particularly suitable for the preparation of customized medical devices and bionic tissues. In the field of microfluidic processing, the light response characteristics of photofunctional polymer composites can be used to precisely control fluid flow, enhancing the performance and efficiency of microfluidic chips. The excellent photoelectric properties and biocompatibility of photofunctional polymer composites provide important support for the development of new sensors, wearable devices, and neural interfaces. In the future, with the advancement of materials science and the innovation of manufacturing technology, photofunctional polymer composites will drive more breakthrough applications in these fields, facilitating the integrated development of medical health, bioengineering, and electronic technology.

Originating in the 1980s, 3D printing is a cutting-edge manufacturing technique that creates three-dimensional items by layering materials [43]. The core principle is to decompose the digital three-dimensional model into thin-layer slices and use a printer to precisely deposit and solidify materials (such as plastics, metals, ceramics, and even biological materials), ultimately forming complex structures [44]. 3D bioprinting originated from traditional 3D printing technology, but focuses on using living cells, growth factors, and biocompatible matrices (such as hydrogels) as "inks" to imitate the microstructure and function of natural tissues [45]. 3D bioprinting typically employs methods such as inkjet and extrusion, enabling high-precision cell positioning and complex three-dimensional tissue construction. The medical industry has made extensive use of 3D (biological) printing technological advances in recent years. Its flexibility and innovative potential are driving revolutionary changes in manufacturing [46].

For instance, by compounding *N*-isopropylacrylamide (NIPAM) with carbon nanotubes, an intelligent hydrogel was developed, and was combined with a one-step 4D printing strategy (Figure 8.12) [47]. In addition to greatly increasing the composite's light absorption and thermal conductivity, uniform doping of single-walled carbon nanotubes also improves the cross-linked network's mechanical characteristics.

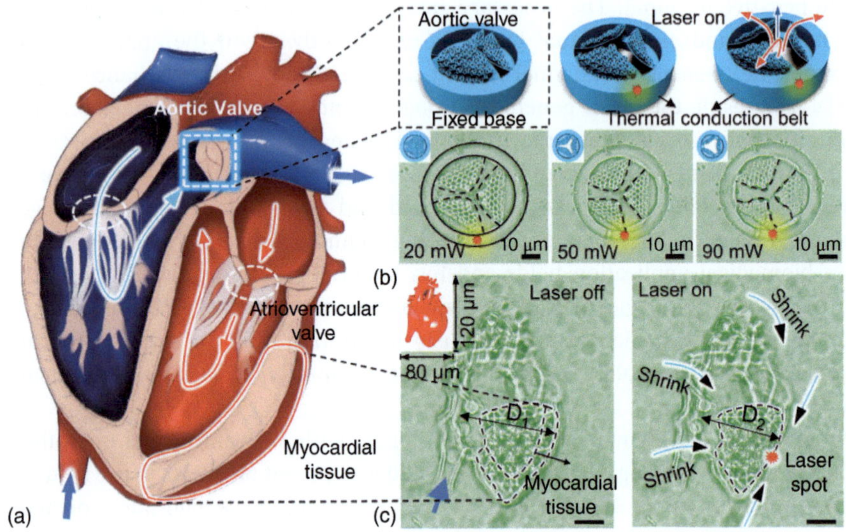

Figure 8.12 (a) The photoresponse micrographs of (b) aortic valve and (c) the heart. *Source:* Deng et al. [47] / with permission of John Wiley & Sons.

In addition, designing and assembling asymmetric metamaterial structural units is the key to achieving large-scale and precise 3D shape transformation.

The left ventricle of the tiny printed heart is surrounded by a composite structure that mimics the properties of cardiac tissue. This structure comprises a 4 μm-thick solid static layer and a dynamic layer composed of buckyball cells. The whole myocardial tissue exhibited a programmed response to light stimulation, contracting swiftly. This 4D printing method has opened up new paths for the construction of multifunctional intelligent micro-machines in fields such as bionics, drug delivery, and integrated microsystems.

Peripheral nerve interface (PNI) is a widely used neurointervention technique in clinical practice. It can induce specific responses in the body by stimulating nerves and can also be used to record compound action potentials to read sensory or motor signals. Current clinical applications mainly target larger nerves in the body, while intervention for small peripheral nerves with diameters ranging from a few to several 100 μm remains challenging at present. This is mainly due to the huge difficulties in design and implantation faced in developing small electrodes. Based on stereolithography 3D printing technology, an electrode for PNIs can be fabricated. Its structure consists of two layers of functional materials: the inner layer is a nonswelling flexible light-curable polyurethane that provides mechanical support; the outer layer is a highly swelling sodium acrylate-based super-absorbent hydrogel, serving as a responsive base. The electrode surface is gold-plated to enhance conductivity and ensure the effective transmission of neural signals. This electrode has excellent tensile and bending properties, making it convenient for surgical implantation. When the electrode comes into contact with water, the hydrogel rapidly swells and generates stress, driving the inner flexible polyurethane to undergo directional bending and folding. Eventually, it automatically forms a clamp structure, tightly wrapping the peripheral nerves and achieving precise stimulation of nerve signals. This design achieves structural self-assembly through differences in material swelling, providing a high-precision and low-damage solution for minimally invasive neural interfaces [48].

In addition, microfluidic bioelectronic devices ingeniously combine the core advantages of microfluidic technology and can achieve precise control of fluids at the micrometer scale [49]. This characteristic enables microfluidic bioelectronic devices to demonstrate great potential in a variety of application scenarios, especially in biomarker detection [50], drug screening [51], and cell culture [52], where significant progress has been made and wide application has been achieved.

For instance, a microfluidic chip can isolate monocytes from host blood and encapsulate them in sodium alginate gel droplets for the rapid production of monocyte vaccines (Figure 8.13) [53]. This chip consists of two areas: a microcolumn array for cell separation and a T-shaped structure for cell encapsulation. Through deterministic lateral displacement technology, the chip can efficiently separate monocytes. After optimizing the experimental parameters, the purity of monocyte separation exceeds 90%. The survival rate of monocytes after cross-linking in sodium alginate gel for 2 min and encapsulation can reach 85%, and it still remains at 95% after 3 days of culture. Combined with immunofluorescence, the cytoskeleton of cells can be observed at a higher resolution, and the effect of microfluidics on cells can be detected in real time. Using microfluidic chips, monocytes and varying concentrations of OVA-FITC (FITC-labeled ovalbumin) were enclosed in gel droplets. As time goes by, the fluorescence of monocytes increases, which means that antigens are gradually absorbed by monocytes. In addition to improving monocytes' ability to absorb antigens, the gel encapsulation encourages the spleen to produce more antigen-specific $CD8^+$ T lymphocytes. In practical applications, the monocyte vaccines produced have also shown good effects in preventing tumor occurrence, significantly prolonging the survival time of tumor-bearing mice.

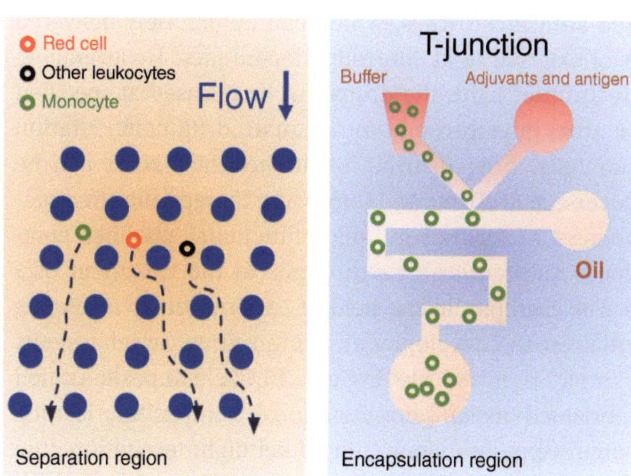

Figure 8.13 The construction and operation mechanism of the microfluidic device.

This research offers new ideas for the preparation and generation of customized cancer vaccines.

In conclusion, the development of photofunctional polymer composites cannot be separated from the integration of multiple disciplines. Through the collaboration of multiple disciplines such as materials science, biomedicine, chemical engineering, photophysics, and nanotechnology, more efficient, safe, and multifunctional photofunctional polymer composites can be continuously developed, promoting their wide application in the biomedical field. In the future, with the continuous deepening of interdisciplinary research, photofunctional polymer composites are expected to achieve more breakthroughs in the diagnosis, treatment, and monitoring of diseases, and make greater contributions to the cause of human health.

8.3 Expansion of Biological Applications

8.3.1 Personalized Medicine

The advancement of photofunctional polymer composites will be intricately associated with the progress in personalized medicine. This integration is an inexorable trend propelled by the escalating demand for precision and specificity in contemporary medical therapies. Personalized medicine aims to customize the most efficacious therapeutic approaches according to the individual patients' distinct physiological and pathological characteristics, as well as their specific disease states. Owing to their distinctive light-responsive properties, photofunctional polymer composites provide substantial technical assistance for achieving this goal.

Photofunctional polymer composites can be precisely tailored to fulfill the specific requirements of individual patients via accurate light stimulation. In the field of tumor treatment, medical practitioners can devise customized photofunctional polymer composites in accordance with the tumor's dimensions, location, and type, as well as the patient's distinct physiological traits. These materials possess the capacity to carry targeted anticancer drugs, which can be precisely delivered to the tumor site by means of external light-controlled techniques. By leveraging the spatial and temporal precision of light, the drugs can be released at specified locations within the tumor area, guaranteeing an adequate drug concentration in the affected tissue. Meanwhile, drug distribution in healthy tissues can be minimized, thus reducing adverse side effects and improving therapeutic efficacy.

With the continuous progress of technology, photofunctional polymer composites are expected to achieve substantial breakthroughs in the treatment of a wider spectrum of diseases. For example, in the field of cardiovascular medicine, these materials show potential for the development of innovative vascular stents (Figure 8.14) [54]. For example, a novel Yb/Er/CeO_2/PLGA composite-coated zinc-based scaffold was fabricated using photofunctional composites, leading to multiple performance improvements. Upon ultraviolet light excitation, the Yb/Er/CeO_2 layer generates an upconversion luminescence (UCL) effect, which facilitates the generation of ROS. This photochemical process notably enhances

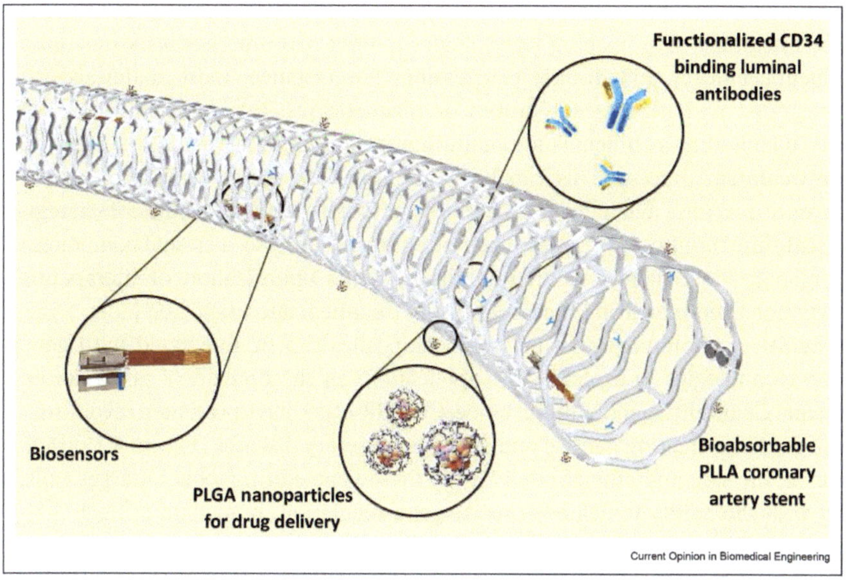

Figure 8.14 A diagram of a biosensor-based bioabsorbable PLLA. *Source:* Tetali et al. [54] / with permission of Elsevier.

the deposition efficiency of the PLGA layer and effectively enhances the corrosion resistance of the zinc scaffold. Furthermore, the mechanical adherence between the layers is strengthened by the addition of the $Yb/Er/CeO_2$ coating. More crucially, the coating makes it possible to control the kinetics of Zn^{2+} ion release, which improves corrosion resistance and greatly extends the scaffold's service life. This research presents an innovative combination of photofunctional composites with biodegradable metallic scaffolds, providing new technical perspectives and experimental bases for the development of next-generation biodegradable cardiac scaffolds, with promising prospects for clinical applications [55]. Furthermore, photofunctional polymer composites have been investigated in the research of neurodegenerative diseases for their potential applications in nerve protection and repair [56]. Regulating the activity and regenerative potential of nerve cells via light stimulation provides innovative perspectives and methodological strategies for the treatment of neurodegenerative diseases, including Alzheimer's disease and Parkinson's disease.

In the field of infectious disease management, photofunctional polymer composites have demonstrated remarkable antibacterial characteristics. Research findings suggest that photothermal and photodynamic effects are capable of effectively eradicating drug-resistant bacteria and other intractable pathogens. For instance, when these materials are fabricated into wound dressings or medical implants, they can induce localized hyperthermia or generate ROS, thereby effectively eliminating pathogens at the site of infection. At the same time, this approach helps reduce reliance on systemic antibiotics and mitigates the risk of developing antimicrobial resistance [57].

Throughout the course of the treatment, photofunctional polymer composites can be integrated with imaging technologies to offer real-time feedback regarding therapeutic efficacy and disease progression. For instance, these materials can be integrated with fluorescent probes or magnetic resonance imaging contrast agents, which allows clinicians to obtain high-resolution imaging data during the entire treatment process. This capability facilitates the timely evaluation of the therapeutic response and allows for necessary adjustments to the treatment strategy. This real-time monitoring and feedback mechanism serves as a crucial component of personalized medicine, facilitating the dynamic optimization of therapeutic interventions and increasing the likelihood of treatment success [58, 59].

In addition, photofunctional polymer composites can be integrated with gene editing technologies to expand their applications in the domain of personalized medicine. Gene editing tools, for instance, CRISPR-Cas9, have presented revolutionary methods for treating genetic and specific refractory diseases (Figure 8.15) [60]. However, the safe and efficient delivery of these gene editing tools to target cells, along with the ability to achieve precise gene regulation, remains a critical challenge. Photofunctional polymer composites present promising solutions to address this issue. For instance, upconversion nanomaterials coated with ultraviolet-sensitive conjugated polyelectrolytes have been employed to fabricate nanocarriers (UCNP-UVP-P), which exhibit a high density of positive charges. Upon near-infrared light irradiation, the polymer layer undergoes degradation, leading to a surface charge transition from positive to negative, accompanied by the release of plasmids. These plasmids exhibit efficient transfection of tumor cells in a time-controllable fashion (with a transfection efficiency of approximately

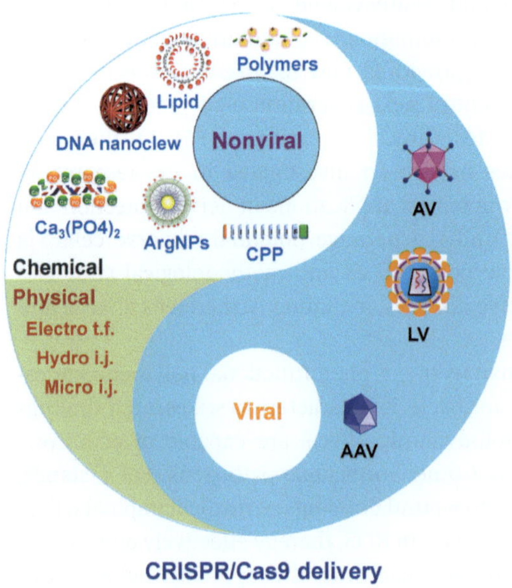

Figure 8.15 CRISPR/Cas9 delivery techniques. *Source:* Wang et al. [60] / with permission of American Chemical Society.

63% ± 4%). Moreover, functional CRISPR/Cas9 proteins can be effectively expressed within a defined near-infrared irradiation area. This light-responsive gene delivery system allows for the temporal and spatial regulation of gene editing tool release, thus promoting precise gene regulation and laying the foundation for innovative approaches in personalized gene therapy [61].

In conclusion, the integration of photofunctional polymer composites with personalized medicine represents a promising developmental direction in the field of biomedicine. Through the utilization of precise light stimulation and its integration with advanced technologies like gene editing, these materials possess the potential to provide more personalized and precise therapeutic approaches for patients. Future research should focus on further optimizing material design and performance, addressing existing challenges, and promoting the widespread application of photofunctional polymer composites in personalized medicine, ultimately contributing to improved patient outcomes and quality of life.

However, several technical and clinical challenges must be addressed to realize the broad application of photofunctional polymer composites in personalized medicine. Firstly, the individualized design and fabrication of these materials demand highly precise technological assistance, including accurate analysis of patients' distinctive physiological characteristics and finely tuned regulation of material properties. Secondly, ensuring the long-term stability and biocompatibility of these materials within the body, especially under extended light exposure during personalized treatment, still represents a crucial domain for further exploration. Moreover, the clinical translation of these materials requires the establishment of standardized manufacturing and quality control protocols to ensure their reliability and consistency among diverse patient populations.

In order to promote the clinical translation of photofunctional polymer composites, international cooperation and multidisciplinary collaboration are essential. Research teams from various countries and regions can share resources, technologies, and expertise to expedite the development and validation processes of these materials. Concurrently, close cooperation with clinicians, patient organizations, and regulatory authorities can contribute to the establishment of appropriate clinical trial protocols and regulatory approval procedures, thus promoting the translation of photofunctional polymer composites from laboratory environments to clinical applications.

In summary, the clinical translation and application of photofunctional polymer composites represent an inevitable trend in their developmental trajectory. Through the establishment of their safety and efficacy via clinical trials, and in combination with continuous technological progress and optimization, these materials are promising for attaining breakthroughs in the treatment of a wider spectrum of diseases, presenting new hope and enhanced medical solutions to patients. Looking ahead, as research deepens and technological maturity is achieved, it is anticipated that photofunctional polymer composites will be applied in the biomedical industry, bringing with them a new age of creative illness treatment.

8.3.2 Artificial Biological Photosynthesis

Optical functional polymer composites can integrate with complex biological systems to construct photosynthetic biohybrid systems. These synthetic materials possess customizable structures, enabling a broader light absorption range and enhanced light-harvesting capabilities [62]. In the selection of biological components, both enzymes and microorganisms serve as feasible alternatives. Enzymes possess merits such as high catalytic efficiency and specificity; nevertheless, they also exhibit limitations, including exacting environmental requirements and relatively low stability [63]. Conversely, microorganisms, facilitated by advancements in gene-editing technology, can regulate metabolic pathways, thus enabling the synthesis of more intricate compounds. Furthermore, microorganisms have the capacity for self-replication, an attribute that enzyme-based systems find difficult to emulate. By integrating the advantages of synthetic materials with those of biological systems, it is possible to overcome the limitations of individual components and provide innovative solutions for fields such as energy conversion, environmental remediation, and biomanufacturing.

The employment of photosynthetic biohybrid systems for solar-to-chemical energy conversion constitutes an efficient and environmentally sustainable methodology for energy generation. Among the diverse products of this process, hydrogen emerges as a crucial clean energy resource, presenting substantial economic and environmental advantages. A photosynthetic biohybrid system based on CdS and *E. coli*—referred to as CdS@*E. coli*—can be constructed by inducing the deposition of CdS quantum dots on the surface of *E. coli*. These nanoparticles possess a band structure conducive to photosynthetic activity. Exposure to light causes the CdS to produce photogenerated electrons, which boost the formation of pyruvate, inhibit the fermentation of lactic acid, increase the concentration of formate, promote hydrogenase activity, and raise NADH levels [64]. These improvements in metabolic activity empower the hybrid system to produce an approximately 30% augmentation in hydrogen yield when compared with numerous photoheterotrophic bacteria. The performance is notably better than that of traditional systems.

Furthermore, because of their great stability, reusability, low toxicity, and material engineering-tunable redox potential, carbon-based materials are thought to be excellent candidate materials. Specifically, iodine-doped hydrothermally carbonized carbon (I-HTCC) exhibits excellent charge transfer efficiency, favorable biocompatibility, and cost-effectiveness, making it well-suited for biological integration. Through a simple "attachment" approach, *E. coli* was integrated with I-HTCC to successfully construct a biohybrid system aimed at enhancing hydrogen production. Since the transfer of photoexcited electrons from I-HTCC to bacteria represents the rate-limiting step in hydrogen generation, photogenerated electrons are delivered via both direct injection and the NADH/NAD^+-mediated pathway. Hydrogen production is further regulated via the formate and NADH pathways. As a result, the hydrogen production efficiency of the hybrid system increased by 9.11% (Figure 8.16) [65]. The hybrid systems have exhibited significant advantages

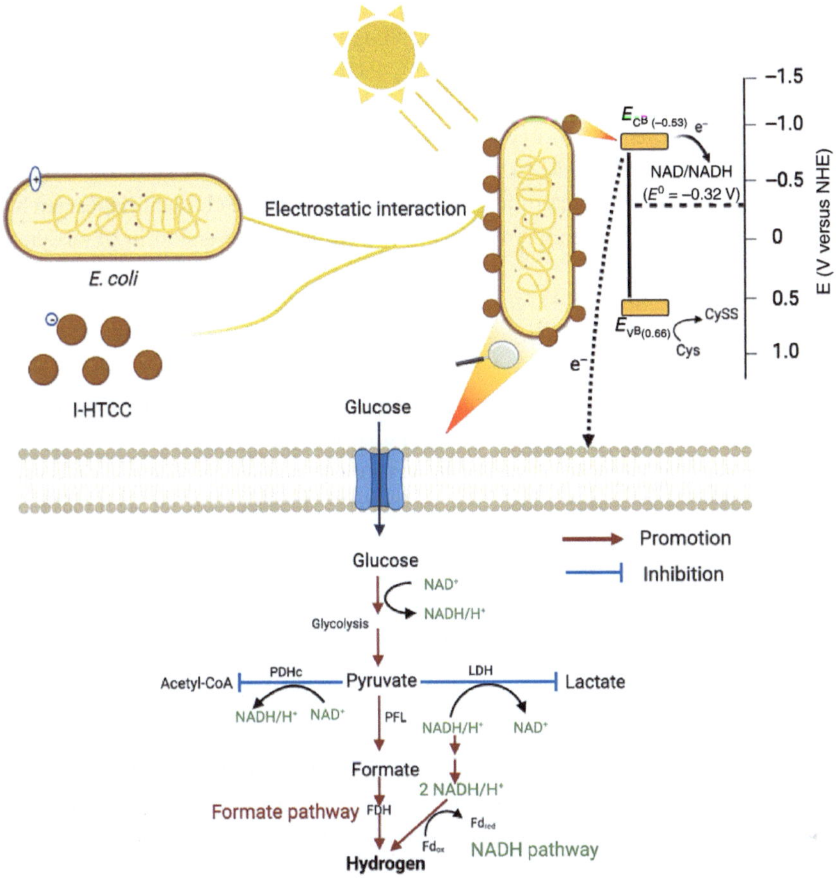

Figure 8.16 The action mechanism of the I-HTCC@*E. coli* biohybrid system for the photo-enhancing H_2 production. *Source:* Xiao et al. [65] / with permission of John Wiley & Sons.

in the transformation of chemical energy from solar energy, presenting a novel technical approach for the development of clean energy resources.

The significance of CO_2 in the context of climate change has spurred efforts to capture and convert carbon dioxide through innovative technological approaches. Notable progress has been made in the development of photosynthetic biohybrid systems, which integrate light-absorbing materials with microorganisms to transform CO_2 into valuable chemical products, such as acetic acid and methane. To facilitate the transformation of CO_2 into intricate organic molecules within these systems, several crucial factors need to be guaranteed: efficient electron transfer from the catalyst to CO_2, effective C—C bond formation, and high product selectivity. Photosynthetic biohybrid systems integrate highly efficient light-harvesting components with whole-cell biocatalysts, attaining CO_2 reduction efficiencies that approximate or exceed those of natural photosynthesis.

Certain bacteria, such as *Moorella thermoacetica* and *Sporomusa ovata*, are capable of converting CO_2 into acetic acid via the Wood–Ljungdahl pathway (WLP). The WLP is more energy efficient than other CO_2 fixation routes. Both wild-type and genetically modified microbes can also convert their intermediate and end products into valuable chemicals [66]. Moreover, some acetogenic bacteria possess the ability to perform EET, facilitating electron exchange between microbial cells and external solid-phase electron donors or acceptors [67]. Given these characteristics, such microorganisms represent promising biocatalysts for CO_2 conversion in photosynthetic biohybrid systems.

The photosensitization of *M. thermoacetica* with CdS NPs enables the construction of a highly efficient, photo-driven biohybrid system [68]. In this system, CdS nanoparticles are deposited on the surface of bacteria. Upon absorption of light energy, the nanoparticles generate photoelectrons that are subsequently converted into reducing equivalents [H], which are then directed into the WLP to reduce CO_2 into acetic acid. Experimental results demonstrate that under a 4-day light–dark cycle, the system can continuously and selectively produce acetic acid with a maximum quantum yield of 2.44%. This efficiency is comparable to the annual average photosynthetic efficiency observed in plants and algae, which ranges from 0.2% to 1.6%. Furthermore, the system possesses the distinctive advantage of self-replication, offering a novel approach for solar-driven CO_2 fixation and chemical energy conversion.

The differentiation in the photosynthetic mechanism between *M. thermoacetica* and CdS nanoparticles is delineated by the function of cysteine (Cys), which serves as a sacrificial reagent and undergoes oxidation to yield cystine (CySS) in a disulfide configuration. Nevertheless, numerous metal chalcogenides demonstrate restricted oxidative stability, which impedes their efficacy within such systems. To address this limitation, researchers have employed a biomimetic structure based on a tandem "Z-scheme" configuration, in which photoreduction and photooxidation are facilitated by two distinct light-absorbing components and photocatalysts [69]. As depicted in Figure 8.17, *M. thermoacetica* constructs a reduction module through the self-deposition of CdS nanoparticles (CdS NPs). In this system, photoexcited

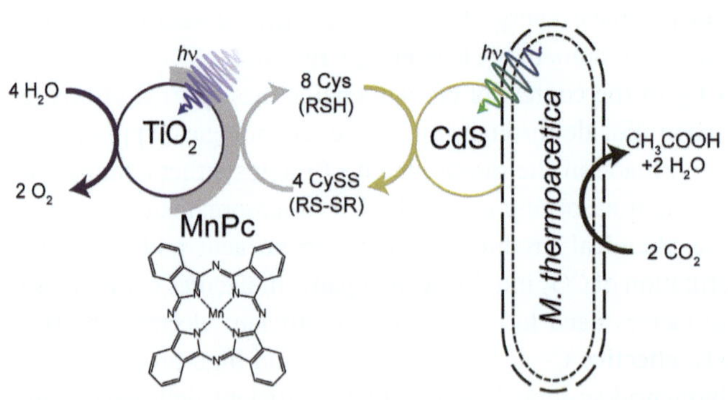

Figure 8.17 The working mechanism of *M. thermoacetica*-CdS/TiO_2-MnPc biohybrid system. *Source:* Sakimoto et al. [69] / with permission of American Chemical Society.

electrons are employed to reduce CO_2 to acetic acid, while cysteine is oxidized to cystine. A TiO_2-supported Mn(II) phthalocyanine (MnPc) catalyst serves as the oxidation module. Upon illumination, this module reduces cystine back to cysteine while simultaneously oxidizing water to produce O_2, thereby establishing a regenerative Cys/CySS redox cycle. The integrated photosynthetic biohybrid system, *M. thermoacetica*-CdS/TiO_2-MnPc, continuously generates acetic acid (equivalent to 8 e^-) and O_2 (4 e^-) over a period of 3.5 days. The molar ratio of the resulting products closely aligns with the theoretical value of 2:1, highlighting the potential of this biomimetic approach for solar-to-chemical energy conversion.

S. ovata is an electroactive bacterium with significant application potential in the field of bioelectrosynthesis. It is capable of efficiently converting CO_2 into multicarbon organic compounds. Its optimal growth temperature is 30 °C, a condition close to ambient temperature, which confers considerable advantages for large-scale microbial applications [70]. Based on this characteristic, researchers have developed a biohybrid system centered on bacteria that enables CO_2 reduction and light energy conversion by directly harnessing solar energy (Figure 8.18) [71]. The foundation of this system lies in the precise integration

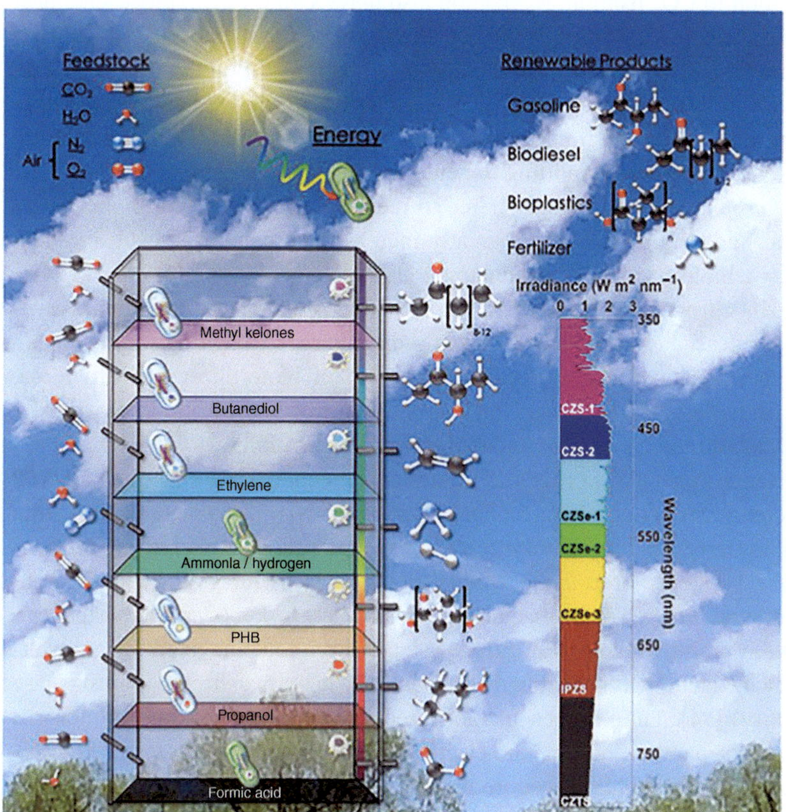

Figure 8.18 A "solar-powered factory" by combining different bacteria and quantum dots. *Source:* Ding et al. [71] / with permission of American Chemical Society.

of core-shell quantum dots with intracellular bacterial enzymes. Seven distinct types of quantum dots, characterized by excitation spectra encompassing the range from ultraviolet to near-infrared wavelengths, are selectively conjugated to specific enzyme sites within bacteria. This process results in the construction of a synthetic photosynthetic system: The zinc-rich shell of quantum dots promotes effective interaction with bacterial bioenzymes, enabling the light-driven transformation of CO_2, water, and nitrogen into renewable fuels and chemicals. This catalytic process exhibits a high conversion efficiency, yielding products such as isopropanol, 2,3-butanediol, formic acid, ammonia, ethylene, and the biodegradable biopolymer polyhydroxybutyrate. As a result, the suspension of nonphotosynthetic bacteria within a buffered solution and their subsequent integration into a "solar-powered factory" facilitates the direct conversion of renewable feedstocks into biofuels and specialty chemicals. These hybrid systems not only permit the production of high-value chemicals through the utilization of sunlight but also enable dynamic modulation of cellular functions in response to external stimuli, such as light, sound, and magnetic fields, thereby underscoring the promising potential of multimodal controllable nanobiotechnology.

In addition to CO_2 fixation, N_2 fixation also plays a vital role in biogeochemical cycles, exerting a significant influence on global ecosystems and agricultural economies [72]. A biohybrid system constructed by integrating the water-splitting catalyst CoP|Co–P with the autotrophic bacterium *Xanthobacter autotrophicus* has been developed to synthesize NH_3 from N_2 and H_2O [73]. In this system, CoP|Co–P catalyzes water splitting to generate H_2, which is subsequently utilized by *X. autotrophicus* as an energy source to reduce N_2 and CO_2 into biomass and ammonia. Radish growth may be greatly increased by using this hybrid system directly as a biofertilizer. By introducing a glutamate synthase inhibitor, the NH_3 produced by nitrogenase (Nzase) in *X. autotrophicus* is redirected from biomass synthesis and released extracellularly. This approach harnesses light energy to enable the gradual, sustained, and selective generation of biofertilizer and ammonia, offering a novel strategy for advancing the natural nitrogen cycle and improving the efficiency of ammonia synthesis and biofertilizer application.

Moreover, nitrogen-fixing bacteria can utilize light energy as a substitute for ATP to promote the reduction of atmospheric nitrogen to NH_3. This method shows the potential to transform nonphotosynthetic nitrogen-fixing bacteria into a photosynthetic hybrid system. For example, when cationic PFPs are integrated with the nonphotosynthetic nitrogen-fixing bacterium *Azotobacter chroococcum*, the system can substantially improve biological nitrogen fixation without the need for an external sacrificial agent, mainly through enhancing the activity of bacterial nitrogenase (Figure 8.19) [74]. The nitrogen fixation process is promoted when PFP absorbs light, charge separation occurs, and electrons are then transmitted to *A. chroococcum*'s redox proteins. This procedure results in an obvious increase in nitrogenase activity. The yields of L-amino acids and NH_4^+-N rise. Meanwhile, the expression levels of *nif*D, *nif*k, and other nitrogen fixation-related genes are significantly upregulated. Due to its stronger light-harvesting capability, more efficient charge transfer,

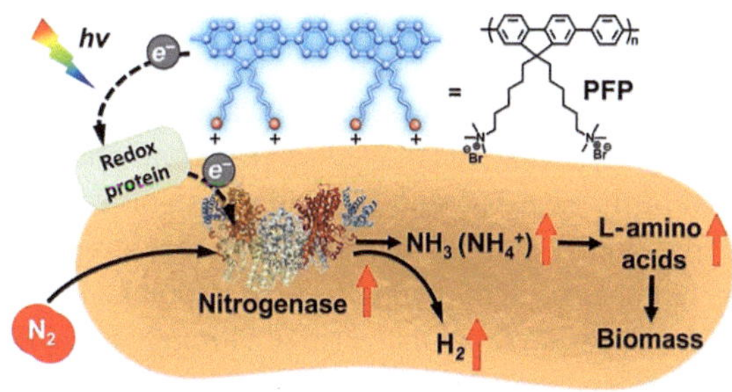

Figure 8.19 The work mechanism of PFP/*A. chroococcum* hybrid system for nitrogen fixation. *Source:* Zeng et al. [74] / with permission of John Wiley & Sons.

and enhanced metabolic activity, the photosynthetic organism hybrid system shows great potential for light-driven biological nitrogen fixation.

Inorganic photosynthetic biohybrid systems combine the superior light-harvesting capabilities of inorganic components with the distinct metabolic routes of living cells. As a result, they present substantial potential as sustainable and efficient platforms for biochemical synthesis. Microorganisms are widely employed in biomanufacturing owing to their rapid growth rates and the capability for genetically programmable, multistep catalytic processes, which facilitate the transformation of renewable carbon sources into higher-value chemical products. The integration of heterotrophic microorganisms with light-harvesting materials has been shown to enhance the efficiency of high-value chemical production [68].

Yeast *Saccharomyces cerevisiae* is widely employed in industrial applications, as it allows for the scalable production of target metabolites through well-developed genetic engineering tools [75]. In yeast, the pentose phosphate pathway serves as the primary source of NADPH. This pathway involves the oxidation of hexose with the concomitant release of two equivalents of CO_2, thereby reducing the theoretical carbon yield. Consequently, decoupling NADPH production from central carbon metabolism may offer a strategy to maximize carbon flux toward the synthesis of target metabolites. A promising methodology integrates highly efficient indium phosphide (InP) light-harvesting nanoparticles with genetically engineered *S. cerevisiae* to facilitate light-driven biosynthesis of valuable chemicals like shikimic acid. Deletion of the *ZWF1* gene, which encodes *glucose-6-phosphate* dehydrogenase, obstructs the pentose phosphate pathway and reduces endogenous NADPH production, thus forcing the cells to rely on photogenerated electrons for NADPH regeneration. InP nanoparticles function as photosensitizers due to their broad light absorption spectrum and oxygen tolerance. Upon light exposure, the electrons generated by InP nanoparticles are transferred into cells via cell wall-mediated electron transport or through soluble mediators, ultimately reducing $NADP^+$ to NADPH. This photogenerated NADPH fuels the final step of the shikimic acid biosynthetic pathway (conversion of *3-dehydroshikimate to shikimate*), thereby significantly enhancing product yield. The system enables intracellular NADPH regeneration via semiconductor-derived photoelectrons, effectively uncoupling biosynthesis from cofactor regeneration and thereby improving both carbon and energy utilization efficiency [76].

The photocatalytic system is an advanced technology that efficiently converts solar energy into other forms of energy, such as chemical energy. Against the backdrop of increasingly severe global energy crises and environmental challenges, photocatalytic systems that emulate natural photosynthesis to achieve solar energy conversion have become particularly significant. Beyond its applications in the energy sector, the photocatalytic system also exhibits substantial potential in the biomedical field, owing to its photobiological orthogonality, biocompatibility, and high spatiotemporal precision. Through in-depth investigation of photocatalytic reactions and their integration with various biological processes, the full utilization of solar energy can be achieved, thereby realizing its multifunctional value across multiple domains.

Optical functional polymer composites have emerged as highly promising candidates within the realm of photocatalysis, owing to their adjustable energy band gaps, facile chemical modification capabilities, and controllable solubility. These materials manifest outstanding photoelectrochemical performance and favorable biocompatibility, which have notably expanded their applications in biomedical and biological systems. Consequently, the development of a photocatalytic system founded on optical functional polymers and its integration into biomanufacturing and biological systems present novel and innovative strategies for the advancement of related disciplines. Such an artificial biosynthetic system not only promotes technological innovation but also lays the foundation for sustainable development, exhibiting considerable scientific value and promising application prospects.

References

1. Cui, X., Ruan, Q., Zhu, X. et al. (2023). Photothermal nanomaterials: a powerful light-to-heat converter. *Chemical Reviews* 123 (11): 6891–6952.
2. Karcher, J. and Pianowski, Z.L. (2018). Photocontrol of drug release from supramolecular hydrogels with green light. *Chemistry – A European Journal* 24 (45): 11605–11610.
3. Peng, N., Du, Y., Yu, G. et al. (2022). Light-activated reactive oxygen species-responsive nanocarriers for enhanced photodynamic immunotherapy of Cancer. *Langmuir* 38 (43): 13139–13149.
4. Kim, C. (2018). Accelerating photofunctional molecule discovery with artificial intelligence. *ACS Central Science* 4 (9): 1089–1091.
5. Monopoli, M.P., Aberg, C., Salvati, A., and Dawson, K.A. (2012). Biomolecular coronas provide the biological identity of nanosized materials. *Nature Nanotechnology* 7 (12): 779–786.
6. Svensson, M., Håkansson, A., Mossberg, A.K. et al. (2000). Conversion of α-lactalbumin to a protein inducing apoptosis. *Proceedings of the National Academy of Sciences of the United States of America* 97 (8): 4221–4226.
7. Hartgerink, J.D., Beniash, E., and Stupp, S.I. (2002). Peptide-amphiphile nanofibers: a versatile scaffold for the preparation of self-assembling materials. *Proceedings of the National Academy of Sciences of the United States of America* 99 (8): 5133–5138.

8 Sun, S., Liang, H.-W., Wang, H., and Zou, Q. (2022). Light-triggered self-assembly of peptide nanoparticles into nanofibers in living cells through molecular conformation changes and H-bond interactions. *ACS Nano* 16 (11): 18978–18989.

9 Wang, J., Hu, L., Zhang, H. et al. (2022). Intracellular condensates of oligopeptide for targeting lysosome and addressing multiple drug resistance of cancer. *Advanced Materials* 34 (1): e2104704.

10 Liu, S., Zhang, Q., He, H. et al. (2022). Intranuclear nanoribbons for selective killing of osteosarcoma cells. *Angewandte Chemie International Edition* 61 (44): e202210568.

11 Tan, W., Zhang, Q., Quinones-Frias, M.C. et al. (2022). Enzyme-responsive peptide thioesters for targeting Golgi apparatus. *Journal of the American Chemical Society* 144 (15): 6709–6713.

12 Rosenberg, S.A. (2005). Cancer immunotherapy comes of age. *Nature Clinical Practice Oncology* 2 (3): 115–115.

13 Wang, J., Zheng, C., Zhai, Y. et al. (2021). High-density lipoprotein modulates tumor-associated macrophage for chemoimmunotherapy of hepatocellular carcinoma. *Nano Today* 37: 101064.

14 Liu, S., Jiang, Q., Zhao, X. et al. (2021). A DNA nanodevice-based vaccine for cancer immunotherapy. *Nature Materials* 20 (3): 421–430.

15 Futaki, S. and Nakase, I. (2017). Cell-surface interactions on arginine-rich cell-penetrating peptides allow for multiplex modes of internalization. *Accounts of Chemical Research* 50 (10): 2449–2456.

16 Peschke, T., Rabe, K.S., and Niemeyer, C.M. (2017). Orthogonal surface tags for whole-cell biocatalysis. *Angewandte Chemie International Edition* 56 (8): 2183–2186.

17 Niu, J., Lunn, D.J., Pusuluri, A. et al. (2017). Engineering live cell surfaces with functional polymers via cytocompatible controlled radical polymerization. *Nature Chemistry* 9 (6): 537–545.

18 Qi, R., Zhao, H., Zhou, X. et al. (2021). In situ synthesis of photoactive polymers on a living cell surface via bio-palladium catalysis for modulating biological functions. *Angewandte Chemie International Edition* 60 (11): 5759–5765.

19 Wu, W., Shao, X., Zhao, J., and Wu, M. (2017). Controllable photodynamic therapy implemented by regulating singlet oxygen efficiency. *Advanced Science* 4 (7): 1700113.

20 Jiang, L., Bai, H., Liu, L. et al. (2019). Luminescent, oxygen-supplying, hemoglobin-linked conjugated polymer nanoparticles for photodynamic therapy. *Angewandte Chemie International Edition* 58 (31): 10660–10665.

21 Zhao, Y., Wang, A., Zou, Y. et al. (2016). *in vivo* monitoring of cellular energy metabolism using SoNar, a highly responsive sensor for NAD^+/NADH redox state. *Nature Protocols* 11 (8): 1345–1359.

22 Candish, L., Collins, K.D., Cook, G.C. et al. (2022). Photocatalysis in the life science industry. *Chemical Reviews* 122 (2): 2907–2980.

23 Xia, S., Bai, H., Zhang, E. et al. (2023). Unexpected photocatalytic degeneration of NAD^+ for inducing apoptosis of hypoxia cancer cells. *CCS Chemistry* 5 (10): 2324–2333.

24 Wegiel, B., Gallo, D., Csizmadia, E. et al. (2013). Carbon monoxide expedites metabolic exhaustion to inhibit tumor growth. *Cancer Research* 73 (23): 7009–7021.

25 He, Q., Kiesewetter, D.O., Qu, Y. et al. (2015). NIR-responsive on-demand release of CO from metal carbonyl-caged graphene oxide nanomedicine. *Advanced Materials* 27 (42): 6741–6746.

26 Zheng, D.-W., Li, B., Li, C.-X. et al. (2017). Photocatalyzing CO_2 to CO for enhanced cancer therapy. *Advanced Materials* 29 (44): 1703822.

27 Zhang, Y., Zhu, X., Zhang, J. et al. (2021). Synergistic upconversion photodynamic and photothermal therapy under cold near-infrared excitation. *Journal of Colloid and Interface Science* 600: 513–529.

28 Fu, X., Huang, Y., Zhao, H. et al. (2021). Near-infrared-light remote-controlled activation of cancer immunotherapy using Photothermal conjugated polymer nanoparticles. *Advanced Materials* 33 (34): e2102570.

29 Cao, Y., Chen, Z., and Ran, H. (2022). *in vivo* photoacoustic image-guided tumor photothermal therapy and real-time temperature monitoring using a core-shell polypyrrole@CuS nanohybrid. *Nanoscale* 14 (33): 12069–12076.

30 Song, Y., Wang, Y., Man, J. et al. (2025). Chimeric antigen receptor cells solid tumor immunotherapy assisted by biomaterials tools. *ACS Applied Materials & Interfaces* 17 (7): 10246–10264.

31 Shi, B., Li, D., Yao, W. et al. (2022). Multifunctional theranostic nanoparticles for multi-modal imaging-guided CAR-T immunotherapy and chemo-photothermal combinational therapy of non-Hodgkin's lymphoma. *Biomaterials Science* 10 (10): 2577–2589.

32 Rocha, C.V., Goncalves, V., da Silva, M.C. et al. (2022). PLGA-based composites for various biomedical applications. *International Journal of Molecular Sciences* 23 (4): 2034.

33 Luo, R., Xiang, X., Jiao, Q. et al. (2024). Photoresponsive hydrogels for tissue engineering. *ACS Biomaterials Science & Engineering* 10 (6): 3612–3630.

34 Paliwal, R., Babu, R.J., and Palakurthi, S. (2014). Nanomedicine scale-up technologies: feasibilities and challenges. *AAPS PharmSciTech* 15 (6): 1527–1534.

35 Lee, G.-H., Jeon, C., Mok, J.W. et al. (2022). Smart wireless near-infrared light emitting contact lens for the treatment of diabetic retinopathy. *Advanced Science* 9 (9): e2103254.

36 Won, S.M., Cai, L., Gutruf, P., and Rogers, J.A. (2023). Wireless and battery-free technologies for neuroengineering. *Nature Biomedical Engineering* 7 (4): 405–423.

37 Mineroff, J., Maghfour, J., Ozog, D.M. et al. (2024). Photobiomodulation CME part II: clinical applications in dermatology. *Journal of the American Academy of Dermatology* 91 (5): 805–815.

38 Barr, J. (2014). Vascular medicine and surgery in ancient Egypt. *Journal of Vascular Surgery* 60 (1): 260–263.

39 Greatbatch, W. and Holmes, C.F. (1991). History of implantable devices. *IEEE Engineering in Medicine and Biology Magazine* 10 (3): 38–41.

40 Liu, Y., Fan, P., Pan, Y., and Ping, J. (2025; https://doi.org/10.1021/acs.chemrev.5c00115). Flexible microinterventional sensors for advanced biosignal monitoring. *Chemical Reviews* 125: 8246–8318.

41 Mickle, A.D., Won, S.M., Noh, K.N. et al. (2019). A wireless closed-loop system for optogenetic peripheral neuromodulation. *Nature* 565 (7739): 361–365.

42 Kim, K.Y., Kang, J., Song, S. et al. (2024). An ultrathin organic-inorganic integrated device for optical biomarker monitoring. *Nature Electronics* 7 (10): 914–923.

43 Ligon, S.C., Liska, R., Stampfl, J. et al. (2017). Polymers for 3D printing and customized additive manufacturing. *Chemical Reviews* 117 (15): 10212–10290.

44 Lu, B., Li, D., and Tian, X. (2015). Development trends in additive manufacturing and 3D printing. *Engineering* 1 (1): 085–089.

45 Atala, A.G. (2020). Introduction: 3D printing for biomaterials. *Chemical Reviews* 120 (19): 10545–10546.

46 Lee, N. (2016). *The Lance*t technology: 3D printing for instruments, models, and organs? *Lancet* 388 (10052): 1368–1368.

47 Deng, C., Liu, Y., Fan, X. et al. (2023). Femtosecond laser 4D printing of light-driven intelligent micromachines. *Advanced Functional Materials* 33 (11): 2211473.

48 Hiendlmeier, L., Zurita, F., Vogel, J. et al. (2023). 4D-printed soft and stretchable self-folding cuff electrodes for small-nerve interfacing. *Advanced Materials* 35 (12): e2210206.

49 Chen, G., Zheng, J., Liu, L., and Xu, L. (2019). Application of microfluidics in wearable devices. *Small Methods* 3 (12): 1900688.

50 Yamaguchi, H. and Miyazaki, M. (2023). Enzyme-immobilized microfluidic devices for biomolecule detection. *Trac-Trends in Analytical Chemistry* 159: 116908.

51 Yang, Z., Zhou, Z., Si, T. et al. (2023). High throughput confined migration microfluidic device for drug screening. *Small* 19 (16): e2207194.

52 Zheng, C., Chen Gui, E., Pang, Y., and Huang, Y. (2012). An integrated microfluidic device for long-term culture of isolated single mammalian cells. *Science China: Chemistry* 55 (4): 502–507.

53 Tian, Y., Xu, C., Feng, J. et al. (2021). Personalized gel-droplet monocyte vaccines for cancer immunotherapy. *Lab on a Chip* 21 (22): 4414–4426.

54 Tetali, S.S.V., Fricker, A.T.R., van Domburg, Y.A., and Roy, I. (2023). Intelligent biomaterials for cardiovascular applications. *Current Opinion in Biomedical Engineering* 28: 100474.

55 Li, Y., Li, Y., and Qu, Y. (2025). Optimization of Zn biodegradable cardiac stent performance through near-infrared light-controlled release dual responsive cerium oxide loaded PLGA coating growth. *Applied Surface Science* 695: 162931.

56 Lee, H.P. and Gaharwar, A.K. (2020). Light-responsive inorganic biomaterials for biomedical applications. *Advanced Science* 7 (17): 2000863.

57 Lv, J., Qiu, Y., Pan, L. et al. (2024). Photothermal/photodynamic antibacterial hydrogel embedded with copper carbon dots and Au nanoparticles. *Nano TransMed* 3: 100034.

58 Jo, S.D., Ku, S.H., Won, Y.Y. et al. (2016). Targeted nanotheranostics for future personalized medicine: recent progress in cancer therapy. *Theranostics* 6 (9): 1362–1377.

59 Duan, Q.-J., Zhao, Z.-Y., Zhang, Y.-J. et al. (2023). Activatable fluorescent probes for real-time imaging-guided tumor therapy. *Advanced Drug Delivery Reviews* 196: 114793.

60 Wang, H.-X., Li, M., Lee, C.M. et al. (2017). CRISPR/Cas9-based genome editing for disease modeling and therapy: challenges and opportunities for nonviral delivery. *Chemical Reviews* 117 (15): 9874–9906.

61 Wu, Y., Zheng, J., Zeng, Q. et al. (2020). Light-responsive charge-reversal nanovector for high-efficiency in vivo CRISPR/Cas9 gene editing with controllable location and time. *Nano Research* 13 (9): 2399–2406.

62 Smith, P.T., Nichols, E.M., Cao, Z., and Chang, C.J. (2020). Hybrid catalysts for artificial photosynthesis: merging approaches from molecular, materials, and biological catalysis. *Accounts of Chemical Research* 53 (3): 575–587.

63 Farid, R.S., Moser, C.C., and Leslie, D.P. (1993). Electron transfer in proteins. *Current Opinion in Structural Biology* 3 (2): 225–233.

64 Wang, B., Zeng, C., Chu, K.H. et al. (2017). Enhanced biological hydrogen production from *Escherichia coli* with surface precipitated cadmium sulfide nanoparticles. *Advanced Energy Materials* 7 (20): 1700611.

65 Xiao, K., Tsang, T.H., Sun, D. et al. (2021). Interfacing iodine-doped hydrothermally carbonized carbon with *Escherichia coli* through an "add-on" mode for enhanced light-driven hydrogen production. *Advanced Energy Materials* 11 (21): 2100291.

66 Liu, C., Gallagher, J.J., Sakimoto, K.K. et al. (2015). Nanowire–bacteria hybrids for unassisted solar carbon dioxide fixation to value-added chemicals. *Nano Letters* 15 (5): 3634–3639.

67 Kato, S. (2015). Biotechnological aspects of microbial extracellular Electron transfer. *Microbes and Environments* 30 (2): 133–139.

68 Sakimoto, K.K., Wong, A.B., and Yang, P. (2016). Self-photosensitization of non-photosynthetic bacteria for solar-to-chemical production. *Science* 351 (6268): 74–77.

69 Sakimoto, K.K., Zhang, S.J., and Yang, P. (2016). Cysteine–cystine photoregeneration for oxygenic photosynthesis of acetic acid from CO_2 by a tandem inorganic–biological hybrid system. *Nano Letters* 16 (9): 5883–5887.

70 He, Y., Wang, S., Han, X. et al. (2022). Photosynthesis of acetate by *Sporomusa ovata*–CdS biohybrid system. *ACS Applied Materials & Interfaces* 14 (20): 23364–23374.

71 Ding, Y., Bertram, J.R., Eckert, C. et al. (2019). Nanorg microbial factories: light-driven renewable biochemical synthesis using quantum dot-bacteria nanobiohybrids. *Journal of the American Chemical Society* 141 (26): 10272–10282.

72 Gruber, N. and Galloway, J.N. (2008). An earth-system perspective of the global nitrogen cycle. *Nature* 451 (7176): 293–296.

73 Liu, C., Sakimoto, K.K., Colon, B.C. et al. (2017). Ambient nitrogen reduction cycle using a hybrid inorganic-biological system. *Proceedings of the National Academy of Sciences of the United States of America* 114 (25): 6450–6455.

74 Zeng, Y., Bai, H., Yu, W. et al. (2023). Increased nitrogenase activity in solar-driven biohybrids containing non-photosynthetic bacteria and conducting polymers. *Angewandte Chemie International Edition* 62 (30): e202303877.

75 Galanie, S., Thodey, K., Trenchard, I.J. et al. (2015). Complete biosynthesis of opioids in yeast. *Science* 349 (6252): 1095–1100.

76 Guo, J., Suástegui, M., Sakimoto, K.K. et al. (2018). Light-driven fine chemical production in yeast biohybrids. *Science* 362 (6416): 813–816.

Index

a

AAm, *N,N′*-methylenebis(acrylamide) (bisAA) 358
ABEI-Ag@PAni-PA composite 497, 498
acetic acid 75, 76, 266, 396, 397, 443, 551–553
acid-sensitive micelles 23
acrylamide (AM) 374
  molecules 454
  monomers 351, 481–482
2-acrylamido-2-methylpropanesulfonic acid (AMPS) 384
3-(acrylamido)phenylboronic acid (3-APBA) 374
adenosine triphosphate (ATP) 287, 326
Ag/$(PEI/PAA)_2$/anti-EpCAM film 436
aggregation-caused quenching (ACQ) 23, 94, 294
aggregation-induced emission (AIE)
  effect 147–148, 294–297
  micelles 23
  photosensitizer hydrogel 62
aggregation states of conjugated polymers
  dynamic aggregation 120–121
  influencing factors and regulatory strategies 122–123
  static aggregation 119–120
aging rate of photofunctional polymer materials 7
AgNPs-nanogels 382–384
alcohol dehydrogenase (ADH) 337
alkaline phosphatase (ALP) 274, 528
α- and β-cyclodextrins 221
α-cyclodextrin (α-CDs) 475
alpha-fetoprotein (AFP) 380
amine-terminated peptide nucleic acid 411
ammonium persulfate (APS) 466
amphiphilic conjugated polymer poly(phenylene)ethynylene (PPE) 409
amphiphilic molecules 190, 221, 222, 332, 409
amphiphilic polymers 225
anchoring effect 305, 306
9, 10-anthracenediyl-bis (methylene)dimalonic acid (ADPA) 491
antiadhesion effects 322
antibacterial effects 379, 459, 506, 526
antiepithelial cell adhesion molecule (anti-EpCAM) 436
antitumor effect 320, 333
antitumor vaccines 529
aromatic heterocycles 117
artificial photosynthesis system 25, 337, 443
atom transfer radical polymerization 411
atomic layer deposition (ALD) technology 349
Au(I)–thiolate species 148

Photofunctional Polymer Composites for Bioapplications, First Edition. Xiaoyu Wang and Lidong Li.

AuNC@histidine fluorescence emission 431
azobenzenes 25, 40, 44, 190, 191, 193, 422, 423, 493, 494, 496, 509, 510, 515, 517

b

BAsCNF films 433
1,3,5-benzenetriazole carboxylate (H_3BTT) 492
β-cyclodextrin 190, 221, 471, 472, 510, 517
BIMEG 287
biohybrid PFP/thylakoid electrode films 443–444
biological functionalization of CPNs 13, 14
biological metabolism 497, 533
biomacromolecules 257
biosensor-based bioabsorbable PLLA 547
2-biphenyl-4,6-bis(12-phenylindolo [2, 3-a]carbazole-11-yl)-1,3,5-triazine (PIC-TRZ) 97
4,4′-bipyridine (BPy) 296
N,*N*′-bis-acryloyl cystamine (BAC) 255, 469
bis-imidazole groups 288, 414
bisphenol A (BPA)
  concentrations 433
  quenching efficiency 432
bispillar[5]arene (BP5A) 302
block copolymers 23, 171, 189–191, 215, 227, 244
blue photochromic fibers (BPLF) 363
borate ester bonds 454, 461, 462, 485, 503, 512
boric acid 461
boron-based COFs 177
BPOTZ molecules 294
5-tert-butoxycarbonyl-5-methyl-1-pyrroline-*N*-oxide (BMPO) 301

c

calcium carbonate ($CaCO_3$) nanoparticles 274, 275
capillary-templated polymerization 355–357, 374
capsules 24–26
carbodiimide-mediated coupling reaction 457
carbon dots 154, 155, 459, 471–473, 499, 500, 535
carbon materials
  carbon dots 155
  carbon nanotubes (CNTs) 154–155
  fullerenes 156
  graphene 155–156
carbon monoxide (CO) 78, 535
carbon nanotubes (CNTs) 67, 154–155, 371, 512, 543
carbon nitride
  photocatalysis 159–160
  photoluminescence 160–161
  photothermal effect 161
carbon nitride nanosheets (CNNS) 239
carbon quantum dots (CQDs) 239, 347
carboxymethyl cellulose (CEC) 472, 505
cationic polymers 60, 115, 231
CDpP-B crystals 106
CdS/Pt-loaded RC film 446
CdS/RC composite film 444
cell culture 41, 314, 514–518, 545
cell imaging 14, 115, 123, 309–317
cell-mediated drug delivery 529
cellulose 403
  microstructure of 404
  molecular structure of 404
cellulose nanofibers (CNF) 373, 433, 490
chemical crosslinking methods
  dynamic covalent bond 457–464
  permanent covalent bond 454–457
chemical/physical synthesis methods 253
chemiluminescence 93, 104–106, 228, 301, 302, 497, 533
chitosan 396

hydrogel 44
molecular structure of 396
chitosan-tannic acid complex (CT) films 440
circularly polarized luminescence (CPL) signal regulation 416
CO_2-responsive block copolymer micelle 23
COF-enzyme composites 229
conduction band 40, 51, 70, 72, 151, 152, 290, 367, 447, 454, 479, 489, 492
conformational switching effect 118
conformation of conjugated polymers 116–119
conjugated polymer nanoparticles (CPNs) 13, 225, 226, 228, 272, 284, 299, 323, 537
conjugated polymer PPV-ST 333
conjugated polymers (CPs) 109, 212
aggregation states of 119–123
conformation of 116–119
luminescence sensing mechanism of
exciplexes and excimers 126
Förster resonance energy transfer (FRET) 124–125
metal-enhanced fluorescence (MEF) 126–127
photoinduced electron transfer (PET) 125
main-chain structure 109–113
side chain structure 113–116
copper nanoclusters 143–144, 415, 416
copper nanoparticles 127–129, 135–136, 138, 466, 468
coprecipitation approach 282
coprecipitation method 212–218, 235–236, 264, 265, 267, 282, 283, 318
core-shell structure 16, 27, 71, 122, 253–261, 273, 276–279, 347, 361, 368
covalent bonding 113, 156, 164, 194, 223, 285, 290, 407, 457, 493
covalent modification method 188, 228–229, 238–240
covalent organic framework (COF) 176, 212, 229
nanodots 62
optical properties of
heterojunction construction 184
photosensitive groups 183–184
coupling of functional building units 181–182
element doping 182–183
structure of
boron-based linkage 177–178
hydrazone linkage 179–181
imine linkage 178–179
triazine linkage 178
CPL-active film 417
CRISPR-Cas9 548, 549
critical micelle concentrations (CMC) 22, 225, 272
crystallization and cross-linking 476–477
crystal violet lactone (CVL) 363
CT-I/N composite film 441
CT-I/N patches 440
CT patch 440
Cu NC/polymer composite film 416
PL enhancement 416
PL spectra and decay profiles 416
Cu_2S nanoparticles 370
cyclic peptide RGD-modified nanoparticles 309
cyclodextrins 191, 220, 221, 224, 296, 458
cyclodextrin-drug complexes 225
cyclopenta-dithiophene (CDT) 292

d

DASA 333
data-driven artificial intelligence 527
degradable photofunctional polymer composite development 538
degree of polymerization 116, 117
dendrimers 190–194
N,N′-diaryl-dihydrodibenzo[a, c]phenazines (DPAC) 96

1,4-diazabicyclo[2.2.2]octane (DABCO) 365
dibenzo[a,c]phenazine-3,6-dinitrile 102
2′,7′-dichlorofluorescein (DCFH) 299, 300
2′,7′-dichlorofluorescein diacetate (DCFH-DA) 378
DiD-stained cells 481, 517
diethyl phthalate 301
9,10-dihydro-9,10-diboranthracene (DBA) 102
dihydroethidium 319
5,5-dimethyl-1-pyrroline-*N*-oxide (DMPO) 301
3,6-dimethyl-substituted diketopyrrolopyrrole (DPP) 318
dimethyl sulfoxide (DMSO) 354, 485
9,9-dioctylfluorene-co-benzothiadiazole (PFBT) 218
dip coating 406, 407
direct polymerization 212, 227–228
dispersion medium 18
disulfide bonds 255, 402, 454, 462–465, 469
dithienobenzothiadiazole (DBT) 287
dithienylethene (DTE) 283
DMCNPs 213
DNA detection 57–59, 498, 499
DNA/H/CCP fibers 355
DNA monolayers 225
DNA nanovaccine 531
DNA sequence detection 434
donor–acceptor–donor model 110
donor–acceptor–donor (D–A–D) molecular topology 113
DOTAP 21, 225, 271, 272
doxorubicin (DOX) 29, 224, 381, 382, 424, 512
DPFL-TPD binary nanoparticle films 421, 422
drug delivery 8, 14, 17, 20, 24, 25, 28, 29, 36, 40, 44, 47, 48, 51–54, 64, 115, 121, 129, 190, 222, 223, 238, 253, 274, 276, 292, 332–334, 382, 387, 395, 443, 453, 496, 507–509, 526, 529, 531, 537, 538, 544
dual-emission fluorescent nanoparticle system 288
dual-emissive film 430
dual-responsive intelligent surface for cancer cell capture and release 68
dye-coated silica 16
dynamic aggregation 120–121
dynamic covalent bond 457–464, 512

e

E. coli biohybrid system 550, 551
EDC/NHS mediated amination reaction 457
electrochemiluminescence (ECL) sensor 497
electroluminescence (EL) 93, 100–104, 427
electron-deficient property of benzothiadiazole 111
electron transfer mechanism 289–294, 419, 420, 432, 477–481
electrospinning method 347, 349
electrostatic adsorption 17, 233, 254, 280, 298
electrostatic interactions 14, 24, 30, 45, 49, 71, 185, 187, 218, 219, 221, 225, 231–233, 237, 243, 257, 258, 262, 271, 272, 274, 284, 288, 289, 292, 326, 330, 335, 383, 398, 408, 411, 422, 500, 504, 506
El-Sayed rule 98, 99
emission intensity 137, 168, 283, 287, 289, 292, 299, 319, 413, 431, 434, 481, 483, 485, 500
emulsion droplets 226, 267–270, 304, 305
emulsion microstructure 267
emulsion polymerization process 269
emulsions
- continuous phase 18
- dispersed phase 18
- NIR/visible light-induced phase inversion 19

oil/water 18
stability 18
water/oil 18
endocytosis mechanism 311
energy crisis 1, 2, 11, 72
energy/electron transfer effect 359–362, 416–422
energy transfer mechanism 281–289, 298, 416, 417, 419, 422, 481–483
environmental pollution 1, 11, 50, 70, 74
enzyme-based systems 550
enzyme-responsive self-assembly 223
ethylenediaminetetraacetic acid (EDTA) 274
excimers
exciplexes and 95
pyrene formation of 96
excited-state absorption (ESA) 291
external environmental factors 117, 278
extracellular matrix (ECM) 31, 323, 457, 514

f

fibrous-Au-GG film 425
film-mesolayer-nanoparticle 233
FITR spectroscopic analysis 466
flash nanoprecipitation (FNP) 215, 216
fluorene (P1) 110, 287, 299, 326, 420, 506
fluorescein isothiocyanate (FITC) 281, 433
fluorescence 37
amplification effect 298, 485
π-conjugated systems 93
restricted intramolecular rotation (RIR) mechanism 95
restricted intramolecular vibration (RIV) mechanism 95–97
fluorescence bioimaging 54
fluorescence detection 56–60, 126
fluorescence emission intensity 15, 146, 294, 432, 500
fluorescence enhancement effect 17, 94, 127, 412–416, 483–487
fluorescence imaging 2, 8, 14, 17, 22, 54–57, 61, 63, 311, 314–316, 333, 453, 538
fluorescence–phosphorescence emission system 427
fluorescence quenching effect 255, 276, 433
fluorescence resonance energy transfer (FRET) 16, 218, 231, 239, 410, 415, 481
fluorescence spectra, of PFDBT-BIMEG
on Ag/(PLL/PAA)/PLL-g-PEG film 414
on (PLL/PAA)/PLL-g-PEG film 414
fluorescent composite materials 12
fluorescent films 36–38, 43
fluorescent peptide thioester 529
fluorescent polymer nanoparticles 293, 332
fluorescent signals 54, 63, 222, 332, 515, 532
fluorinated ethylene propylene (FEP) 372
folded chain conformation 117, 119
Förster resonance energy transfer (FRET) 124–125, 281
free radical mechanism 60
freezing process 476
functional CRISPR/Cas9 proteins 549
functionalized particle emulsifiers 20

g

g-CN polymer hydrogel 240
GA/OA-g-CS polymer films 397
Gel/GSE/TiO_2 composite film 400
gelatin 400
molecular structure of 402
gelatin-based hydrogels 44
glucose oxidase (GOX) 229, 497
glutathione (GSH) 147, 231, 232, 262, 300, 313, 415, 457, 483, 507
glutathione (GSH)-modified gold nanoclusters 457

3-glycidoxypropyltrimethoxysilane 430
gold nanoclusters (AuNCs) 3, 5, 139–142, 146, 147, 231, 234, 239, 241, 267, 286, 290, 295, 305, 307, 312, 314, 316, 328, 430, 439, 455, 457, 483, 487, 502, 507
gold nanoflowers (AuNFs) 279, 280
gold nanoparticles (AuNPs) 129–133, 238, 241
  gold nanorods 129–130
  gold nanostars 132–133
  hollow gold nanoparticles 130–132
gold nanorods 43, 68, 129–130, 132, 133, 149, 233, 303, 304, 381, 424, 425, 487, 502, 513
gold nanostars 132–133
Golgi apparatus (GA) 529
graphitic carbon nitride (g-C_3N_4) 242, 456
green fluorescent protein (GFP) 187
GSH-AuNC 231, 232, 234, 439
guar gum (GG) 461, 503

h

H-aggregation 69, 119–121
H dimer nanospheres 70
heavy atom effect 61, 98, 298, 427
Heeger's research 115
hemagglutinin (HA) 500
N-heterocyclic aromatic hydrocarbon skeleton of carbazole 111
high-throughput screening technology 526
highest occupied molecular orbital (HOMO) 111, 290
hippocampal (HPC) 388, 389
HisAgCCN 535, 536
hollow gold nanoparticles 130–132
hollow structured nanocomposites 259
host–guest interactions 190, 220, 234, 297, 302, 303, 422–424, 458, 464, 472–476, 493, 510, 517
hyaluronic acid (HA) 63, 241, 262, 294, 458, 496, 502, 510
hybrid structure, nanocomposites 261–267
hybridized local and charge transfer (HLCT) 101, 103
hydrazone-linked COFs 179
hydrochloric acid 466, 471
hydrogel 43–45
  bond cross-linking 465–468
  preparation via polymerization 356
  scaffold 65
hydrogen bonding 77, 126, 181, 211, 219, 221, 233, 237, 266, 296, 318, 354, 383, 406, 407, 415, 427, 466, 475, 481, 507, 513, 516, 528
hydrogen peroxide (H_2O_2) 60, 150, 234, 301, 318, 491, 533
hydrophobic alkyl side chains 116
hydrophobic and electrostatic interactions 271, 284
hydrophobic effects 234, 262, 272, 296
hydrophobic interaction 45, 57, 62, 67, 69, 123, 212, 213, 219, 221, 225, 226, 262, 264, 266, 288, 313, 326, 328, 333, 398, 466, 470–472, 481, 506, 510, 517
hydrophobic monomers 470, 471
2-hydroxyethyl methacrylate (HEMA) 466
hydroxypropyl chitosan (HPCS) 458
hypochlorite ion (ClO^-) 329

i

IFN-γ cytokine 323
IFN-γ protein 324
immunotherapy 323–325, 529, 530, 537
indium phosphide (InP) light-harvesting nanoparticles 555
indocyanine green (ICG) 237, 270, 306, 440
infrared light-responsive drug release system 66
infrared thermal imaging 324, 488–490, 503
inorganic nonmetallic materials
  carbon materials 154–156

carbon nitride 156–161
perovskites 166–168
quantum dots (QDs) 151–154
titanium dioxide (TiO_2) 168–171
transition metal disulfides 161–166
inorganic/organic composites 4
inorganic photochromic functional films 39
inorganic photofunctional materials 2, 3, 230, 315
inorganic photosynthetic biohybrid systems 555
in situ assembly strategy 526–531
in situ polymerization method 532
in situ reaction method 240–243
interfacial active upconversion nano-phosphates 19
interfacial polyelectrolyte complexation (IPC) 30, 354
intersystem crossing (ISC) 61, 69, 93, 298, 365, 415, 490
intracellular NADPH regeneration 555
intracellular PFT photocatalytic CO production 78
intramolecular charge transfer (ICT) 42, 103, 188, 292, 293, 297, 478
ion cross-linking 398
ion-dipole interactions 296
ionogel 493, 494
iridium (III) complexes 298, 299, 331
iron-doped carbon dots 459
isothermal titration calorimetry (ITC) 313
ITO/PEDOT:PSS/conjugated polymer: [Ir(2-PhQ)$_2$acac]/SPPO13/Al device 427
ITO/ZnO/P3HT:PCBM/MoO_3 multilayer 444

j

Jablonski energy level diagram 37, 93, 418
J-aggregation 119–121, 318

k

konjac glucomannan (KGM) 365, 465

l

Langmuir–Blodgett (LB) film technology 409
layer-by-layer (LbL) self-assembly technique 407
films 424, 432
polyelectrolyte multilayer films 17
technology 24, 408, 432
light-controlled drug release 12, 171, 453, 507–515
light-controlled release
applications of 65–67
principle of 64–65
light-driven biotransformation 74–76
light-guiding properties 371–373
light-harvesting proteins 184–186
light-induced self-assembly 190, 528
light irradiation 9, 39, 47, 53, 66, 69, 71, 104, 149, 155, 156, 190, 193, 220, 239, 242, 257, 283, 284, 287, 300, 314, 321, 334, 363, 365, 366, 368, 370, 379, 423, 427, 437, 439, 440, 444, 455, 456, 462, 488, 490–494, 503, 506–511, 513, 515, 516, 548
light-responsive capsules 25
light-responsive materials 53, 60, 64, 194
light switch release mechanism 65
lignocellulose-based film (PR20) 404, 405
linear discriminant analysis (LDA) 57, 331
lipid nanoparticles (LNPs) 270
lipopolysaccharide (LPS) 329
liposomes 12, 20–22, 73, 78, 223–225, 270–272, 284–286, 312, 313, 337, 513, 537
liquid-phase exfoliation method 62
living radical polymerization 17
localized surface plasmon resonance (LSP) 137, 233

local surface plasmon resonance (LSPR) effect 7, 128
lowest unoccupied molecular orbital (LUMO) 111, 290, 419
luminescence sensing mechanism of conjugated polymers
  and excimers 126
  and exciplexes 126
  Förster resonance energy transfer (FRET) 124–125
  metal-enhanced fluorescence (MEF) 126–127
  photoinduced electron transfer (PET) 125
luminescent fiber fabrication 359
luminol 105, 228, 497, 533
lysosomes 222, 329, 334, 528, 531

m

macrocyclic confinement effect 297
macrocyclic molecules 234
MAPAH fibers 354
Maxwell's equations 134
mechanoluminescence (ML) 93, 106–108
11-mercaptoundecanoic acid (MUA) 262, 316
mesoporous silica nanorods (MSN) 502
metal-carbonyl complex-based release molecules (CORMs) 535
metal coordination effect 468–470
metal-enhanced fluorescence (MEF) effects 14–16, 124, 126–127, 136–139, 276–281, 320, 412
metal-organic frameworks (MOFs) 492
  optical properties of 172–176
  structure of 172
N,N′-methylenebis(acrylamide) (MBA) 356, 358, 384
methyl methacrylate (MMA) 227, 242
6-methyl-1,3,5-triazo-2,4-diamine (MTDA) 100
micelles
  photofunctional polymer composites 22–24
  photofunctional polymer nanocomposites 272–274
microemulsion method 13, 144, 225–227, 237, 238
microfibrillated cellulose (MFC) 405
microfluidic method 357–358
micropinocytosis 311
mixed matrix membranes (MMMs) 229, 243
molecular assembly methods 320
molybdenum disulfide (MoS_2) 161, 164, 245, 490
monodisperse silica nanoparticles 260
monolayer liposomes 20
montmorillonite (MMT) 269–270, 306
M. thermoacetica-CdS/TiO_2-MnPc biohybrid system 552
multichamber structure capsules 25
multifunctional integrated photofunctional polymer composites
  biocompatibility and safety of 539
  CAR immune cell therapy 539
  challenges 539
  degradable materials 538
  imaging technology 538
  immunomodulators 537
  immunotherapy mechanisms of 537
  photodynamic therapy 533
  tumor treatment 538
multifunctional micelles 24
multi-inlet vortex mixer (MIVM) 215, 216
multilayer composite film 407
multistimuli-responsive white-light hydrogel 47

n

Na-alg/GO nanocomposite films 398
N-acetyl-L-cysteine (NAC) 329
NADH oxidation 77, 337, 533, 550
nanoclusters 305

nanocomposites (Pdots) 322
  conjugated polymer nanoparticles (CNPs) 13
  conjugated polymers 12–13
nanomedicines 12, 25, 225, 529
nanotechnology 11, 53, 59, 60, 69, 127, 189, 253, 257, 346, 527, 540, 546
naphthalenediimide (NDI, P6) 112
natural polymer fibers 26, 27
near-infrared (NIR) absorption 36, 41–43, 69, 462
near-infrared light 7, 19, 25, 41–43, 48, 67, 147, 244, 258, 306, 324, 372, 525, 537
n-isopropylacrylamide (NIPAM) 456
nitrobenzofurazan 528, 529
nitrogen-fixation 554, 555
noble metal nanoclusters
  aggregation-induced emission (AIE) effect 147–148
  copper nanoclusters 143–144
  gold nanoclusters 139–142
  optical properties of 144–145
  photoluminescence 146
  photosensitive effect of 149–1150
  photothermal effect of 148–149
  silver nanoclusters 142–143
  two-photon absorption (TPA) 146–147
noble metal nanoparticles
  copper nanoparticles 135–136
  electronic structures of 128
  gold nanoparticles (AuNPs)
    gold nanorods 129–130
    gold nanostars 132–133
    hollow gold nanoparticles 130–132
  optical properties of
    metal-enhanced fluorescence effect (MEF) 136–138
    photothermal therapy (PTT) 138–139
  silver nanoparticles 133–135
nonenzymatic catalytic reactions 532
nonionic water-soluble conjugated polymers 115
nonsolvent induced phase separation (NIPS) technique 229, 352

o
OF-Si-doped silica nanoparticles 295, 296
oligoethylene glycol (EG) 287
one-pot method 457
optical detections 325–332
  and sensing 496–502
optical display 10, 383–386
optical fiber communication 2
optical fiber materials 26
optical function realization mechanism 12
optical functional polymer composites 550, 556
optical functional polymer films 427
optical sensing technology 374
optical therapies 317–325, 377–383, 437–443
optogenetics 30, 31, 35, 324, 374, 386–390, 537, 541
organic-inorganic hybrid solar cells 41
organic light-emitting diode (OLED) devices 2, 7, 93, 100, 427
organic materials, doping of 43
organic molecular-based fluorescence platforms 59
organic molecule-based fluorescence platforms 59
organic optoelectronic composites 75
organic photochromic film materials 39
organic photochromic functional films 39
organic photoelectric devices 4
organic photofunctional materials 3–5, 211, 212, 238
organic semiconductor-nonphotosynthetic bacterial hybrid system 76

ortho-nitrobenzyl 510, 512
oxidized dextran (OD) 459
oxidized hyaluronic acid (OHA) 458

p

PAA-C6-Azo conjugated polymer 422
PAA/PDDA multilayer film 408
palladium nanoparticles (Pd-NPs) 350, 351
PAMBI-2 copolymer films 427
- emission decay profiles of 428
- fluorescence and phosphorescence emissions of 428

PAMI-2 copolymer films 427
- emission decay profiles of 428
- fluorescence and phosphorescence emissions of 428

PAMMI-2 copolymer films 427
- emission decay profiles 428
- fluorescence and phosphorescence emissions 428

PAN-CQDs nanofibers 347, 348
PDA/AuNCs film 407, 408
Pdot-Qdot nanocomposites 236
$(PEI/PAA)_n$/anti-prostate-specific antigen (anti-PSA) multilayer films 413
permanent covalent bond 454–457
perovskites 166–168
PFBT-F127-SiO_2 NPs 55
PFBT-NH-PIMA dots 214
PFDBT-BIMEG, positively charged conjugated polymer 436
PFP/*A. Chroococcum* hybrid system for nitrogen fixation 554
PFP/PDI heterojunction 75
PFVCN
- absorption and emission spectra of 413
- fluorescence signal 435

phase transition mechanism 47, 48
phenylboronic acid (PBA) 266, 299, 309, 310, 320, 512
1, 4-phenylenediisothiocyanate 411
phenyl-modified carbon nitride (PhCN) 301, 310
phorbol 12-myristate 13-acetate (PMA) 497
phosphocreatine 337
phospholipids 20, 21, 116, 223, 225, 226, 237, 270–272, 317
phosphorescence 37
- ISC efficiency 98–99
- T_1 nonradiation 99–100

phosphorescence emission mechanism 427
phosphorescence intensity 300, 320
photoacid doping 50
photoacoustic effect 317
photoacoustic imaging 8, 41, 217, 307, 308, 315, 317, 538
photoacoustic signal intensity 308
photoactive nanocomponents 509
photobiological conversion technology 74, 75
photocatalytic CO_2 reduction 71–72, 78
photocatalytic degradation 2, 157, 169
photocatalytic effect 70–71, 170, 365–370
photocatalytic-Fenton reactions 368
photocatalytic H_2 generation mechanism, CdS/Pt-loaded RC film 446
photocatalytic therapy 76–78
photochromic effect 363–364
photochromic functional films
- applications 38
- color-changing mechanism 39

photochromic materials 39, 40, 43, 526
photochromic microcapsules (PM) 385
photodynamic sterilization 63, 64, 346, 507
photodynamic therapy (PDT)
- application of 61–64
- principle of 60–61

photoelectric conversion thin films 40–41

photofunctional composites 4–6, 11, 12, 17, 18, 43, 44, 46, 52, 227, 238, 240, 243, 244, 371, 381, 396, 404, 410, 412, 434, 435, 437, 440, 453, 526, 528, 535, 537, 546, 547
photofunctional conjugated polymers 217
photofunctional fibers 26
- natural material-based fibers 26–30
- synthetic polymer fibers 31–35

photofunctional hydrogel
- advantages of 44
- fluorescent hydrogel 46–47
- light-controlled release hydrogel 51–52
- nanoscale 44
- photochromic hydrogel 47–48
- photoelectric conversion hydrogel 50–51
- photothermal hydrogel 48–50
- physical properties 46
- structure and composition 45

photofunctional materials
- in biomedical fields 2
- and communication technology 2
- in information storage 2
- inorganic and organic 2, 3
- optical fiber communication 2
- photocatalytic degradation technology 2

photofunctional molecules 23, 264, 483, 485
photofunctional organic polymer composite system
- coprecipitation 212–218
- covalent modification 228–229
- direct polymerization 227–228
- microemulsion method 225–227
- physical mixing method 229–230
- self-assembly method 218–225

photofunctional organic/inorganic polymer composite systems
- coprecipitation method 235–236
- covalent modification 238–240
- in situ reaction method 240–243
- microemulsion 236–238
- physical mixing method 243–245
- self-assembly method 231–235

photofunctional polymer composite films
- biomedical field
  - advantages 429
  - energy storage 443–446
  - food packaging 446–448
  - optical detection 430–437
  - optical therapy 437–443
- functions
  - energy/electron transfer effect 416–422
  - fluorescence enhancement effect 412–416
  - photoisomerization effect 422–424
  - photothermal effect 424–426
  - self-supported films 396–406
  - substrate-supported films 406–412

photofunctional polymer composites
- advantages of 8
- artificial biological photosynthesis 550–556
- bioapplications
  - biological detection and imaging 54–60
  - light-controlled release 64–67
  - photocatalytic effect 70–78
  - photodynamic therapy (PDT) 60–64
  - photothermal therapy (PPT) 67–70
- biomedical application of 8–9
- in bone tissue regeneration application 9
- capsules 24–26
- clinical translation and application 549
- emulsion 18
- fibers 26–35
- gene editing technologies 548
- hydrogel composites 10
- in situ aggregation strategy 531–533
- in situ assembly strategy 526–531
- intelligent drug delivery systems 8
- intelligent light response 525–526
- interdisciplinary integration 540–546

photofunctional polymer composites (*contd.*)
interface engineering 7
liposomes 20–22
micelles 22–24
multifunctional integration 533–540
multifunctional synergies of 5, 6
nanocomposites
conjugated polymer nanoparticles (CPNs) 13
conjugated polymers 12–18
organic photovoltaic cells 6
performance enhancement 5
personalized medicine 546–549
in photothermal therapy (PPT) 9
size of 10
in soft tissue repair 9
structure and morphology of 9–11
systematic classification and analysis of 12
technical and clinical challenges 549
thin films composites 10
photofunctional polymer fiber composites
bioapplications
optical detection 374–377
optical display 383–386
optical therapy 377–383
optogenetics 386–390
functions
energy/electron transfer effect 359–362
light-guiding properties 371–373
photocatalytic effect 365–370
photochromic effect 363–364
photosensitization effect 364–365
photothermal conversion effect 370–371
structures of
capillary-templated polymerization 355–357
electrospinning 346–351
microfluidic method 357–358
stretching process 354–355
3D printing 351–352
wet spinning process 352–353
photofunctional polymer films
fluorescent films 36–38
near-infrared (NIR) absorption films 41–43
performance 35
photochromic films 38–40
photoelectric conversion thin films 40–41
photofunctional polymer materials
light absorption range 7
photocatalytic application of 7
photofunctional protein
green fluorescent protein (GFP) 187
light-harvesting proteins 184–186
polymers
block copolymers 189–191
dendrimers 191–194
photofunctional small molecules 214, 217
photogenerated electron-hole pairs 70, 159, 176, 181, 370, 446
photoinduced electron/energy transfer (PET)-reversible addition–fragmentation chain transfer polymerization (RAFT) 531
photoinduced electron transfer (PET) 124, 125, 290, 361, 419, 420, 480
photoinitiators 351, 352, 356, 371, 454, 455, 463, 471, 509, 531
photoisomerization effect 422–424, 493–495
photoisomerization groups 509, 515
photoluminescence
fluorescence
π-conjugated systems 93
restricted intramolecular rotation (RIR) mechanism 95
restricted intramolecular vibration (RIV) mechanism 95–97
phosphorescence
ISC efficiency 98–99
T_1 nonradiation 99–100
photopolymerization release mechanism 64

photoresponsive nanomaterials 9
photosensitization effect 128, 150, 298–303, 364–365, 419, 490–493
photosensitizer protoporphyrin IX (PpIX) 365
photosensitizers (PS) 21, 526
photoswitchable molecules 283
photosynthesis 334–337, 443, 550–556
photosynthetic biohybrid systems 550, 551, 553, 555
phototherapy 31, 33, 36, 68, 345, 346, 502–514, 526, 538, 541
photothermal agent 21, 67, 68, 138, 270, 306, 440, 459, 487, 502, 513
photothermal conversion effect 161, 370–371, 487–490
photothermal conversion efficiency 43, 68, 69, 129, 130, 132, 133, 138, 149, 161, 217, 236, 304–306, 322, 425, 488, 489, 503
photothermal conversion materials 11, 323, 453, 502, 525
photothermal effect 268, 303, 308, 424
photothermal–electrical cascade conversion 51
photothermal graphene oxide (GO) 380
photothermal therapeutic effects 9, 322
photothermal therapy (PTT) 9, 67–70, 138, 307, 322, 381, 453, 536
physical coating 406
physical crosslinking method
  crystallization and cross-linking 476–477
  host–guest interaction 472–476
  hydrogen bond cross-linking 465–468
  hydrophobic interaction 470–472
  metal coordination effect 468–470
physical mixing method 229–230, 243–245
Pickering emulsion 18–20, 237, 238, 267–270, 306, 310
π-conjugated system 42
plant seed germination 332
plasmon oscillation 137
plasmon resonance effect 276, 279, 380, 383
polyacrylamide (PAM) 281, 454, 455, 466, 468, 471, 487, 497, 515
poly(acrylic acid) (PAA) 237, 279, 304, 351, 412, 417, 434
polyacrylonitrile (PAN) 347, 359, 363
polyaniline (PANI) 59, 71, 239, 242, 306, 456, 488
polycaprolactone (PCL) 352, 371, 377, 380, 382
poly(dimethylsiloxane) (PDMS) composite films 31, 426, 439
poly(9, 9-dioctylfluorene) (PFO) 122
polydopamine (PDA) 233, 371, 407, 441, 458, 466, 505, 513
poly(dopamine) nanoparticles (PDAPs) 233, 407, 458, 476, 487, 513
polyelectrolyte multilayer capsules 274–276
polyelectrolyte multilayer film 17, 417
polyelectrolytes (PE) 17, 22, 30, 232, 233, 274, 289, 354, 408, 412, 417, 422, 432, 548
polyethylene glycol (PEG) 115, 149, 190, 212, 221, 223, 227, 238, 307, 356, 474, 496, 498, 500, 513
poly(ethylene glycol) diacrylate (PEGDA) 371, 374
poly(ethylene glycol) prepolymer (PEGDA) 352
polyfluorene 110, 114, 254, 330, 430
polyfluorene-based polymer (PFV) 254
polyhedral oligomeric silsesquioxane (POSS) 283, 382
poly(3-hexylthiophene) (P3HT) films 443
polylactic acid-glycolic acid (PLGA) 234
poly(L-glutamic acid) 221
poly(L-lactic acid) (PLLA) 31, 387
polymer dot formation 212, 213
polymer micelles 22, 23, 272
polymer molecular brushes 66
polymer optical fibers 26

poly[2-methoxy-5-((2-ethylhexyl)oxy)-phenylenevinylene] (MEH-PPV) 420
polymethyl acrylate (PMA) 354
polymethyl methacrylate (PMMA) 26, 228, 386, 416
poly(*N*-acryloyl glycinamide) (PNAGA) 516
poly(*para*-phenyleneethynylene) (PPE) 57
poly(phenylene ethynylene) (PPE) 409, 532
poly(*p*-phenylene vinyl) (PPV) 122, 220, 272
polysaccharides 29, 43, 45, 262, 309, 396, 398, 400, 457, 461, 465, 492, 527
poly(SP-*co*-NiPAAm) polymer brushes 67
poly(styrene-*co*-maleic anhydride) (PSMA) polymers 212, 213, 257, 265
polystyrene (PS) microspheres 269
polystyrene polyethylene glycol (PS-PEG) 212
polyvinyl alcohol (PVA) 31, 43, 45, 49, 237, 365, 415, 466, 476, 479, 485
poly(9-vinylcarbazole) (PVK) 265
polyvinylidene fluoride (PVDF) 386
porphyrin 62, 173, 176, 192, 193, 298
potassium persulfate (KPS) 469
PPE-TNT 420
precious metal composite nanoparticles 13
prodigiosin 403
protein detection 58, 59
  mechanism, of photofunctional composite film 434
proteomic analysis 529
protonation-sensitive fluorescent 478
PSVBA/PhCN nanoparticles 310, 311
Py-Bpin crystal 107
PyPE-TNT systems 420

q
quantum dots (QDs) 7, 8, 10, 17, 25, 36, 37, 52, 54, 57, 147, 151–155, 161, 194, 215, 218, 230, 231, 236, 239, 244, 245, 309, 330, 347, 359, 395, 454, 550, 553, 554
quantum yield (QY) 7, 55, 57, 74, 98, 102, 104, 109–111, 118, 119, 123, 146, 153, 156, 160, 162, 166, 187, 188, 215, 225, 230, 231, 257, 278, 286, 294–296, 327, 416, 439, 552
quaternary ammonium salt (QAS) 115, 291, 299, 320, 413, 417, 418, 443
quenching efficiency of BPA 432
quinoxoline 112

r
radiative decay rate 278, 294, 413
ratiometric fluorescent sensor 497
reactive oxygen species (ROS) 2, 105, 214, 298, 329, 364, 397, 456, 526
red photochromic fibers (RPLF) 363
redox balance 533
reduced graphene oxide (rGO) 244, 245
refractive index 1, 26, 32, 33, 168, 372, 374, 462
regenerated silk protein (RSF) 373
reprecipitation method 420
restricted intramolecular vibration (RIV) mechanism 95–97
restriction of intramolecular motion (RIM) 94
reversible addition–fragmentation chain transfer polymerization (RAFT) 17, 531
rGO@IOL composite films 440
room-temperature phosphorescence (RTP) 98, 101–103, 426
ROS-activatable photosensitizer composite materials 63
ROS production mechanism of photosensitizers 61
ROS-responsive polymer nanocomposites 234

s

sandwich configuration 101
Schaap's dioxetanes 105, 106
Schiff base reaction 457, 458
SD/BSA nanoparticles 443
SD/curcumin nanoparticles 443
SEGP composite film 403, 404
self-assembled monolayer (SAM) 410, 430
self-assembly 526
  of DNA and PFBT 411
  method 218–225, 231–235
self-polymerization 177, 407
self-quenching effect 63, 265, 361
self-supported films 395–406
SH-aPDT film 438, 439
sialic acid (SA) 266, 267, 269, 309, 310, 500
side chains of water-soluble conjugated polymers 115
silanization modification 411
silk fibroin 402
  film 403
  α-helix and β-sheet of 403
silver nanoclusters 142–143
silver nanoparticles 17, 133–135, 224, 231, 255, 256, 263, 275–278, 280, 281, 382, 409, 415, 417, 432, 436, 485, 492, 502
silylation reaction 410
single-step enzymatic reaction 433
singlet oxygen mechanism 61
SiO_2-dye composite nanoparticles 15
small organic molecules
  chemiluminescence 104–106
  electroluminescence (EL) 100–104
  mechanoluminescence (ML) 106–108
  photoluminescence
    fluorescence 93–97
    phosphorescence 97–100
sodium alginate (SA) 43, 49, 352, 368, 398, 400, 432, 434, 483, 499, 545
sodium hypophosphite (NaH_2PO_2) 136, 466
sodium polyacrylate (PAAS) 354
sodium salicylate (SA) 481
sodium tetraborate 462
solar-powered factory 553, 554
sol-gel method 254, 255
SP-8Y/RSF film 422
SP-COOH/RSF composite film 422, 423
spin coating 231, 243, 245, 406, 407, 427, 443
spin–orbit coupling (SOC) effect 97, 98, 102, 111, 427
spiropyran (SP) 19, 39, 47, 65, 66, 190, 286, 314, 363, 364, 385, 422, 466, 494–496
static aggregation 119–121
stretching process 354–355
substrate-supported films 395, 396, 406–412
sugar bridge interaction 320, 321
3-sulfopropyl methacrylate potassium salt (SPMA) 239
supramolecular composite photosensitizers 77
surface plasmon polarization wave (SPP) 233
surface plasmon resonance (SPR) 42, 127, 151, 230, 276, 383, 485, 487, 500
surface SPR effect 492
Suzuki coupling 110, 177
synergistic effect 5, 6, 111, 113, 217, 221, 253, 270, 276, 296, 370, 444, 471, 503, 533, 536
synthetic polymer hydrogels 44, 45

t

tannic acid (TA) 255, 368, 398, 440, 462
TEMPO-oxidized cellulose nanofibrils (T-CNF) 463
tetrabutylammonium hexafluorophosphate (TBAH) 361
tetracycline 159, 499, 500
tetraethyl orthosilicate (TEOS) 295

10,10′,11,11′-tetrahydro-5,5′-bidibenzo[a, d][7]annulenylidene (THBDBA) 95
tetraphenylethylene (TPE)-modified polymers 23
2,2,6,6-tetramethyl-4-piperidone hydrochloride (TEMP) 301
three-dimensional carbon nanosheets (3D-CNS) 236
3D printing technology 351, 352, 543, 544
thermal-activated delayed fluorescence (TADF) 97, 101
thermal response mechanism 64
thermal therapy 322
thin-film fluorescence sensing 37
thiolation reaction 286
thiol–metal interaction 469
titanium dioxide (TiO_2) 168–171
  crystal structure and morphology 170
  light response range 170–171
  nanoparticle size 170
tmf-BDP nanoparticles 69
TP-CS nanocomposite films 398
traditional materials
  in disease diagnosis and treatment 1
  limitations 1, 2
traditional phototherapy 345
transition metal disulfides (TMDs) 161–166
bis(2,4,6-trichlorophenyl) oxalate (TCPO) 301
4,4′,4-trimethyltriphenylamine (TTA) 359
triphenylphosphonium (TPP)-terminated peptide 307
triplet–triplet absorption (TTA) 291
triplet–triplet annihilation (TTA) 101, 103
tumor ablation treatment 525
twisted intramolecular charge transfer (TICT) 188, 293
two-photon absorption 55, 128, 146–147, 316
two-photon excitation (TPE) 147, 316
two-photon fluorescence imaging (TPFI) 55, 315, 316

u
ultra-flexible photothermal eutectogels 49
ultraviolet (UV) radiation 286, 462
unzipped multiwalled carbon nanotubes (uCNT) 371
upconversion nanoparticles (UCNPs) 8, 371, 375, 454
UV light irradiation 190, 193, 242, 287, 423, 462, 488, 494, 515

v
verteporfin (VP) 437
vibration-induced emission (VIE) 96, 332
4-vinylbenzeneboronic acid (VBA) 269
vinyl-modified spiropyran (SPMA) 466

w
water-soluble galactoglucomannan (GGM) 462
water-soluble polymer nanoparticles 226
wet chemical methods 135
wet spinning process 352–353
white-light-emitting nanofibers (WLENFs) 361
white organic light-emitting diodes (WOLEDs) 427
$WO_3/MoO_3/g\text{-}C_3N_4$ heterojunction 75

wool/acrylic (W/A) blended fibers 352
Wood–Ljungdahl pathway (WLP) 75, 336, 443, 552

y

Yb/Er/CeO_2/PLGA composite-coated zinc-based scaffold 546

z

zwitterionic polymer brushes, on ITO substrate 412
zwitterionic polymer synthesis 411